Yulei Yingji Shengwuxue

鱼类应激生物学

[美] 卡尔·B.施瑞克 [西] 路易斯·托特
[加] 安东尼·P.法雷尔 [加] 科林·J.布劳纳 编著

林浩然 张 勇 卢丹琪 李水生 译

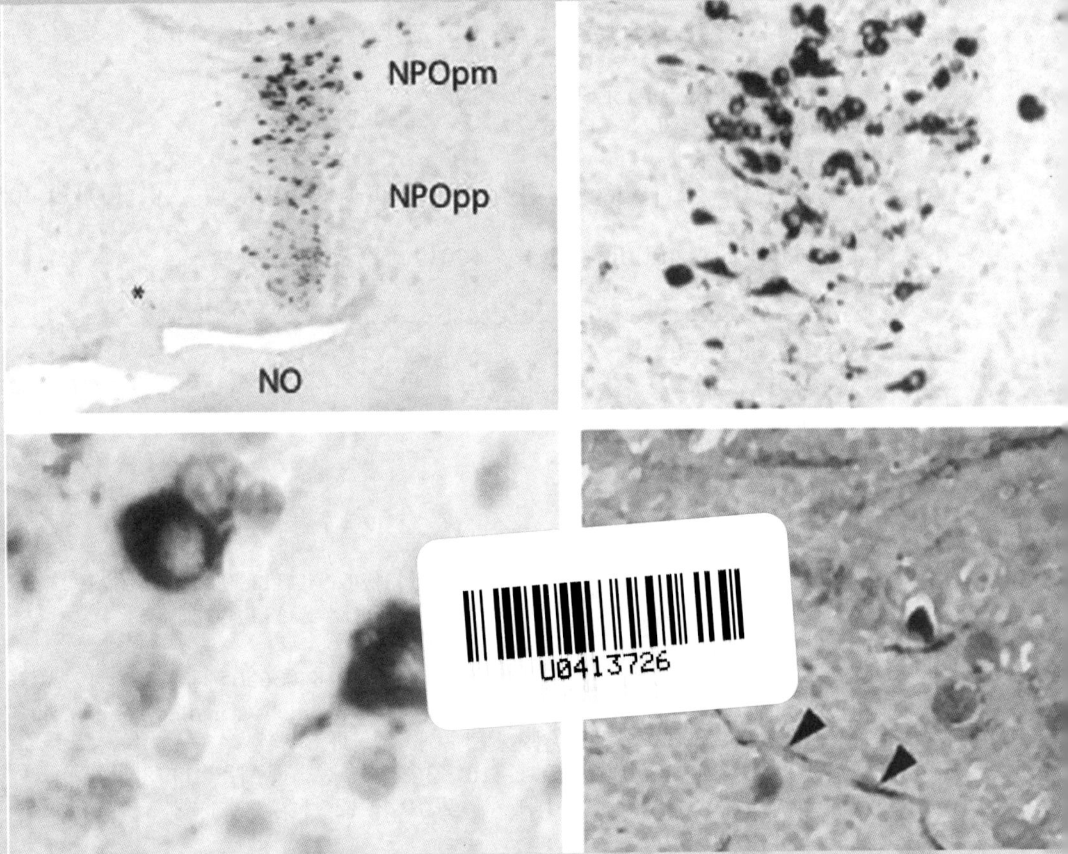

U0413726

中山大學出版社
SUN YAT-SEN UNIVERSITY PRESS
·广州·

Biology of Stress in Fish
Carl B. Schreck, Lluis Tort, Anthony P. Farrell and Colin J. Brauner
ISBN 978-0-12-802728-8

《鱼类应激生物学》（林浩然　张勇　卢丹琪　李水生　译）
ISBN 978-7-306-06706-7

注　意

本书涉及领域的知识和实践标准在不断变化。新的研究和经验拓展我们的理解，因此，须对研究方法、专业实践或医疗方法做出调整。从业者和研究人员必须始终依靠自身经验和知识来评估和使用本书中提到的所有信息、方法、化合物或本书中描述的实验。在使用这些信息或方法时，他们应注意自身和他人的安全，包括注意他们负有专业责任的当事人的安全。在法律允许的最大范围内，爱思唯尔、译文的原文作者、原文编辑及原文内容提供者均不对因产品责任、疏忽或其他人身或财产伤害及/或损失承担责任，亦不对由于使用或操作文中提到的方法、产品、说明或思想而导致的人身或财产伤害及/或损失承担责任。

图书在版编目（CIP）数据

鱼类应激生物学/［美］卡尔·B. 施瑞克（Carl B. Schreck）等编著；林浩然，等译. —广州：中山大学出版社，2019.12
书名原文：Biology of Stress in Fish
ISBN 978-7-306-06706-7

Ⅰ.①鱼…　Ⅱ.①卡…②林…　Ⅲ.①鱼类—生理应激　Ⅳ.①Q959.405

中国版本图书馆 CIP 数据核字（2019）第 206115 号

出 版 人：王天琪
策划编辑：廖丽玲
责任编辑：廖丽玲
封面设计：林绵华
责任校对：黄浩佳
责任技编：何雅涛
出版发行：中山大学出版社
电　　话：编辑部 020-84110283，84111996，84111997，84110771
　　　　　发行部 020-84111998，84111981，84111160
地　　址：广州市新港西路 135 号
邮　　编：510275　　　　传　真：020-84036565
网　　址：http://www.zsup.com.cn　　E-mail:zdcbs@mail.sysu.edu.cn
印 刷 者：广州市友盛彩印有限公司
规　　格：787mm×1092mm　1/16　36.5 印张　6 彩插　1178 千字
版次印次：2019 年 12 月第 1 版　　2019 年 12 月第 1 次印刷
定　　价：160.00 元

如发现本书因印装质量影响阅读，请与出版社发行部联系调换

内 容 简 介

本书为美国科学出版社 2016 年出版的“鱼类生理学”系列专著中的第 35 卷《鱼类应激生物学》(*Biology of Stress in Fish*)。它全面收集和总结了 30 多年来在鱼类应激生物学方面发表的科学著作和研究成果，共 13 章，内容包括：一、对应激的综合性论述——应激的概念、应激反应的生理特性、影响应激反应的因素；二、引起应激反应变动的生物性因素和非生物性因素；三、应激反应怎样产生以及如何进行调节；四、应激对鱼类生理和鱼体功能与行为表现的影响；五、鱼类应激的评估、应激管理、动物福利以及鱼类作为模式生物的应激；等等。

本书内容充实、系统全面、概念新颖、论述清晰，是一部学术水平很高的专著。

本书可供鱼类学、鱼类生理学、鱼类养殖生物学、鱼类内分泌学、鱼病学（鱼类病害学)、鱼类免疫学、鱼类分子生物学、鱼类遗传育种学等相关学科与研究领域的科学技术工作者和高等院校有关专业的师生学习参考，亦可作为专业的水产养殖工作者、观赏鱼类爱好者和关心动物健康和福利的人们的一本有用的参考书。

前　言

应激是求生存的一种固有的特性。鱼类对各种不同类型应激物（应激源）如何进行反应对于个体的存活和种群的繁殖能力都是至关重要的。对应激物（应激源）的反应亦是一种关键的变异性，它超越所有对鱼类实验生物学的研究并会影响到实际经营管理的几乎各个方面。实质上，应激能够影响所有的生理过程，因而亦会影响生物有机体的行为表现。因此，对于概括整个学科组成的研究学者来说，重要的是能够启动对鱼类或渔业生物学的研究，以便阐明应激以及它们对个体和种群的意义。出版这部《鱼类应激生物学》著作的目的就是提供这样的阐述。

本书力图满足各方面读者的需求。书中的各个章节都提供了最新的研究信息和思路，而在 1981 年由 A. D. 皮克林编辑出版的最后一本关于鱼类应激的综合性专著中所列举的对本领域极为必要的论述至今已经过去几十年了。《鱼类应激生物学》对应激生物学，包括本领域大多数最新的研究进展都做了综合的阐明。本书对于专业的鱼类生理学者以及需要了解应激及其相关内容的一般渔业生物学工作者来说是一部有用的参考书。专业的水产养殖工作者、观赏鱼类爱好者以及关心动物健康和福利的人们，亦会对本书感兴趣。本书开头是对应激的综合性论述——应激的概念、应激反应的生理特性、影响应激反应的因素；然后阐述影响应激反应变动的生物性因素和非生物性因素；接着讨论应激反应是怎样产生和调节的；随后各章论述应激对鱼类生理和鱼体功能与行为表现的影响；最后以有关鱼类应激的评估、应激管理、动物福利和鱼类作为模式生物的应激等应用性的章节而结束。为提供对有关鱼类应激知识状况的真实理解，本书清晰地标明我们对本领域理解的界线。当大胆设想（venturing）超出所理解的范围时，即使是有力支持的假设也会更多地基于主观推断、未经测试的模式和理论。关键的未知因素都被清楚地标明并将成为未来研究的课题。

《鱼类应激生物学》是一本创新性的著作，它至少起始于 6 年前，当时作为鱼类应激研究的专家，我们充分认识到在这方面必须有一本最新的图书。其结果就是出版了这部由 30 位专家撰写的有关他们各自研究领域的最新理论成果的共 13 章的专门著作。本书最后的结构和内容融合了许多人的讨论与评述，并且汇集了两个正式研讨会的学术成果，一个是在美国科瓦尔利斯的俄勒冈州立大学，另一个是在英国爱丁堡的赫里阿特-瓦特大学。本书编者对乐于承担各章编写的作者们和 25 位外界评论员对各章提出的积极、富有建设性和意义深刻的评论表示深切的谢意。最后，我们还要感谢爱思唯尔（Elsevier）出版社的各位成员，特别是帕特·冈扎列兹和克里斯蒂·戈麦斯所做出的不懈努力。

C. B. 施瑞克

L. 托特

本书作者一览表

N. Aluru（N. 阿路鲁），美国，伍兹霍尔海洋研究所

S. J. Cooke（S. J. 库克），加拿大，卡尔列敦大学

M. R. Donaldson（M. R. 唐纳森），加拿大，卡尔列敦大学

E. Faught（E. 福德），加拿大，卡尔加里大学

G. Flick（G. 弗利克），荷兰，拉德堡德大学

M. Gorissen（M. 戈里森），荷兰，拉德堡德大学

E. Höglund（E. 霍格伦特），挪威，挪威水研究所

P. -P. Huang（P. -P. 黄），中国台湾，台北，“中央研究院”

K. M. M. Jones（K. M. M. 琼斯），加拿大，布里通角大学

M. L. Kent（M. L. 肯特），美国，俄勒冈州立大学

C. Lawrence（C. 劳伦斯），美国，波士顿儿童医院

D. L. G. Noakes（D. L. G. 诺阿克斯），美国，俄勒冈州立大学

C. M. O' Connor（C. M. 奥康纳），加拿大，麦克马斯特大学

Ø. Øverli（Ø. 奥维里），挪威，挪威生命科学大学

N. W. Pankhurst（N. W. 潘克赫斯特），澳大利亚，格里菲斯大学

J. V. Planas（J. V. 普拉纳斯），西班牙，巴塞罗那大学

K. J. Rodnick（K. J. 罗德尼克），美国，爱达荷州立大学

B. Sadoul（B. 萨道尔），加拿大，卡尔加里大学

C. B. Schreck（C. B. 施瑞克），美国，俄勒冈州立大学

L. U. Sneddon（L. U. 斯纳顿），英国，利物浦大学

N. M. Sopinka（N. M. 索平卡），加拿大，卡尔列敦大学

S. Spagnoli（S. 斯帕诺里），美国，俄勒冈州立大学

C. D. Suski（C. D. 苏斯基），美国，伊利诺斯大学

Y. Takei（Y. 竹井祥郎），日本，东京大学

J. S. Thomson（J. S. 汤姆森），英国，利物浦大学

L. Tort（L. 托特），西班牙，巴塞罗那大学

M. M. Vijayan（M. M. 维扎延），加拿大，卡尔加里大学

S. Winberg（S. 温伯格），瑞典，厄萨拉大学

D. C. C. Wolfenden（D. C. C. 沃尔芬登），英国，埃利斯梅尔港蓝行星水族馆

T. Yada（T. 雅达），日本，国立水产研究所

目　　录

第 1 章　鱼类应激的概念

和所有脊椎动物一样，鱼类对威胁状态的一般性生理反应都可列为应激（stress）。它们在感受应激物（应激源）后几乎立刻就会启动应激反应。适度的应激状态可以产生有利的或者积极的效应（即正常应激，或良性应激，eustress）；而较为严重的状态引起适应性反应，但亦能产生不良的适应性或者负面结果（即痛苦，或不良应激，distress）。应激反应由两组激素系统启动和调控，引起糖皮质激素（主要是皮质醇）的产生和引起儿茶酚胺（如肾上腺素和去甲肾上腺素以及它们的前体多巴胺）的产生。和这些调节系统一起的次级应激反应因子能影响必需的资源分布，如将能量和氧气输送到维持生命所必需的身体部位，并协调水盐平衡和免疫系统。如果鱼类受到应激物的冲击而能抵御死亡，它们就能够恢复到类似或多少类似于稳态的标准。长期重复的应激或者长时间处于应激状态会对身体其他必要的生命机能（生长、发育、抗病力、行为、生殖等）负面影响，形成不良的适应性，而其大部分是由于和应激反应相联系的能量耗费（异常状态的负担）导致的。

鱼类对一种应激物的反应为何存在明显的变动，这是由于不同的分类单元以及在不同种类与品种内存在遗传差异。在应激反应内的变动是由鱼类的环境历史、目前的环境状况以及鱼类目前的生理状态所引起的。当前，鱼类生理学研究已经进展到让我们可以容易识别鱼类在哪里、鱼类在何时处于应激状态，但是，我们通常还不能够识别鱼类何时处在非应激状态，因为应激临床征兆的缺失并不经常与鱼类处于非应激状态相一致。另外，我们必须留意到有关应激临床征兆出现假阴性的可能性。还有，我们还不能够使用临床数据去精细或者准确地推断一个应激物的严重程度。

1.1　导　言

鱼的一生都充满着战胜、妥善应对和从威胁中复原的挑战。对脊椎动物健康的威胁通常会引起一系列生理活动过程，这些生理活动有助于生命有机体对应激物做出反应，从而使其能够妥善应付应激物或者能从对应激物的反应中恢复过来。Selye（1950）把动物对这些挑战的生理反应称为应激（stress）。他提到，相似的反应不管应激物的特性如何，都归类为普遍性适应综合征（general adaptation syndrome，GAS）。GAS 由一个对应激物产生所有各种反应的激素级联组成。他进一步提出，应激的完全丧失就是死亡。Selye（1976）提出的普遍性论点对鱼类是合适的，但对于其他脊椎动物，则需要许多防止误解的说明。显然，应激的生理反应对于全部生理系统都具有不可缺少的重要意义。所以，了解鱼类应激生物学将为鱼类生物学以及鱼类研究、经营管理和养殖产业提供重要的基础知识。

本书详细介绍了鱼类应激生理学的研究现状。它在 Pickering（1981）的《应激和鱼类》基础上扩展而形成了这部创新性的著作。要提供应激的整体综述，就不要聚焦于任何特别的应激物或者环境的耐受限度。我们认识到，许多环境因子（如温度、pH、混浊度、毒物、病原体、捕食者、人的手工操作），如果鱼类遭遇它们时达到或者超过鱼类正常的耐受能力，就会导致应激和功能异常。

本章的目的是对鱼类相关的应激概念进行概述。我们采用普通的术语分析讨论鱼类应激，而其他各章将提供目前所了解的有关鱼类应激的各个课题所必要的详细资料。从基础和应用前景两方面来了解我们知识的局限性亦很重要，本书的任务亦要提供这方面的知识。

1.1.1 什么是应激？

“应激”一词出人意料地难以界定。它是生物有机体的一种生理反应。但并不是环境的变异性引起这种生理反应的，我们将引起这种生理反应的事物称之为应激物或应激源。

1.1.1.1 应激的定义

“应激”一词来源于 Selye（1950，1973）提出的生理定义：应激是身体对于任何固有需求的非特异性反应。不过，我们在全书中所采用的定义是：“当生物有机体面临损害时试图抵御死亡或者重建稳态标准所发生的一连串生理活动”（Schreck，2000）。

我们这样阐述定义部分是由于 Selye 对涉及 GAS 的“应激”一词不适当的选用，这会对这个名词的定义产生一些混乱。传统上，物理学名词“应激”是指一个物体处在一种力量的影响中（GAS 的应激物）。“应变”（strain）对于 GAS 会是更好的选用，它是指一个物体由于一种力量而产生的畸变（distortion）（即对应激物的反应）。为了说明清楚，我们列举在文献中出现的许多关于“应激”一词的不同定义如下：

（1）由环境的或者其他因素导致超出适当性反应正常范围所产生的状态（Brett，1958）。

（2）当动物试图建立或者保持稳态时所发生的全部生理反应的总和（Wedemeyer，Mcleay，1981）。

（3）对长期生存可能受到损害而引起一个或多个生理变动的交替（Bayne，1985）。

（4）应激是由一个应激因子或应激物引起的从一个正常静止的状态或者稳态到偏离常规的状态（Barton，Iwama，1991）。

（5）当生物有机体面临超出正常范围的挑战并且试图恢复稳态时所发生的一系列生物学活动（Barton，1997）。

（6）受到威胁的稳态由一系列适应性反应而力图重建的状态（Chrousos，1998）。

（7）生物有机体旨在重新获得稳态的反应（Chrousos，2009）。

（8）应激是生物有机体对环境的需求超出其天然调控能力的状态（Koolhaas 等，

2011）。

贯穿所有定义的主题是：应激是对应激物的生理反应。有些定义把应激反应限定为神经内分泌引起的一连串反应。不论如何定义，这种连串反应倾向于非特异性，不管应激物的特征、类型和激烈程度如何，它们在性质上是相似的。但有各种理由说明应激反应在定量的量度方面有很大差别，Winberg 及其同事们（2016；见本书第 2 章）对此做了详细的讨论和综述。Levine（1985）对应激的定义做了一般性评论，而 McEwen 和 Lasley（2002）对于由什么组成应激进行了非常清晰的、较为现代的述评。

应激物可以是非常短促的（急性），例如，给网捕捉或者逃避捕食者；也可以是持久的（慢性），例如，在拥挤的水池中或处在社会等级的底层状态。所谓的“急性”和“慢性”取决于上下文，所以，并不容易界定。在考虑是急性应激还是慢性应激引起的作用时，不要以应激物的持续时间为依据，而应该以“应激物对动物的生理作用后果的持续时间为准则”（Boonstra，2013）。应激物的激烈程度亦是变化不定的。

1.1.1.2　生理学的应激反应

普遍性适应综合征（GAS）的概念（Selye，1950）就包含着应激是一种普遍性反应的意思。我们将在下文叙述这种生理学的应激反应，而其他作者在本书的其他章节中也做了进一步的阐述。一般来说，GAS 概念对于分析认识应激反应是很适合的，我们强调的反应并不全是现实性的。许多年前，Schreck（1981）就给出了一些关于鱼类的事例。例如，鱼类能被一些毒物如镉或者麻醉剂杀死而并不引起一种类-GAS 反应。不同的应激物和应激物的不同激烈程度亦能引起 GAS 成分明显不同的反应动态（Winberg 等，2016；见本书第 2 章）（Barton，2002）。

扩展到鱼类经历的日常正常范围以外的一些活动，能使我们感受到鱼类处于某种危险的情况下，而这就会导致应激反应。检测到一个真实的或者可察觉到的应激物对启动应激反应是必要的（Schreck，1981）。如果鱼并不能察觉到对生命有威胁的真实应激物，就不会启动传统的应激反应（Schreck，1981）。所以，正如 Ellis 等（2012）所指出的，应激反应的精神性（psychogenic）方面对于动物的健康特别重要。

应激反应有三个主要的阶段：警觉、抵抗和补偿或衰竭（死亡）（Selye，1950；Scheack，2000）。应激反应的特性（幅度和持续时间）取决于应激物的激烈程度和持续时间。在所有的情况下，警觉阶段基本上由参与逃逸、搏斗或竞争的系统上调组成。在抵抗阶段，鱼类或者完全克服应激物，从而重新恢复正常的稳态；或者充分地克服应激物，从而接近恢复正常（补偿）；或者开始跌入导致死亡的轨道。正如我们将在下文进一步讨论的，非常低水平的应激（正常应激）实质上是适应性的，而较高水平的应激（痛苦，亦是很难准确界定的名词）（Holden，2000）具有不良适应的和适应的成分。我们在这里是把应激和痛苦（distress）的含义分开来论述的。

基本上，对一个应激物感受的初级反应是引起神经内分泌的一连串反应，包括皮质类固醇激素（皮质醇和相关化合物）和儿茶酚胺（主要是肾上腺素和去甲肾上腺素）的合成和分泌。应激的内分泌学已经由 Wendelaar Bonga（1977）、Sumpter

（1997）和最近的 Pankhurst（2011）进行了阐述。对应激的主要皮质类固醇反应在七鳃鳗中是 11-脱氧皮质醇（Close 等，2010；Roberts 等，2014），在板鳃鱼类中是 1α-羟基皮质酮（Idler，Truscott，1966；Idler 等，1967a），而在软骨硬鳞类和硬骨鱼类中是皮质醇和其他相关的类固醇（Idler，Sangalang，1970；Webb 等，2007）。在鱼类中，随应激而产生的亦有大量血液循环浓度的可的松，它是一种皮质醇的代谢物（Patino 等，1987）。可以认为转化为可的松是起着皮质醇下调为一种没有活性代谢物的作用，但对这方面的研究很少。显然，鱼类对应激的反应亦能够产生许多其他的皮质类固醇，它们通常是低浓度的，而我们对它们的功能或者潜在作用还几乎一无所知。这些激素的主要作用是对参与逃逸、搏斗或竞争的系统给予可供利用的能量。图 1.1 描绘了组成这些应激反应的系列过程。Gorisson 和 Filk（2016；见本书第 3 章）提供了内分泌应激轴和应激反应调控的当代评论。

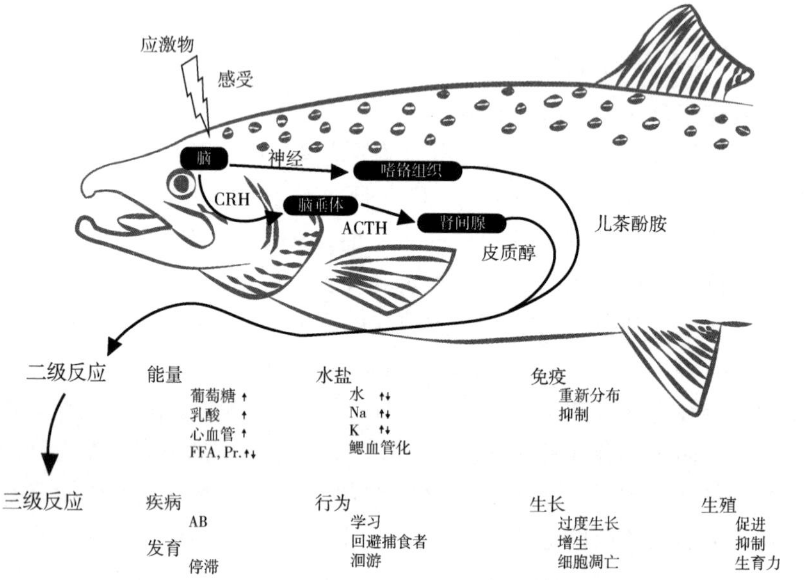

图 1.1　鱼类在痛苦（distress）期间的初级（在图中的鱼体内描绘的）、次级和三级反应

注：CRH，促肾上腺皮质激素释放激素；ACTH，肾上腺皮质激素；FFA，游离脂肪酸；Pr.，蛋白质；AB，抗体。

次级的反应包括心血管的和呼吸的反应（Rodnick，Planas，2016；见本书第 7 章），作为应激反应的结果，它们增加氧气以及能量底物的分布并释放到血液循环中。其他伴随的次级反应包括水盐的功能异常，由于肾上腺素影响鳃的血流型式和鳃的通透性，使得水分能够根据鱼所处的水环境盐度，顺着渗透梯度水分流入体内或者流出体外。所以，皮质醇在这方面合乎逻辑的作用是恢复渗透平衡（Takei，Hwang，2016；见本书第 6 章）。对皮质醇的另一种反应，或者其他的调控因子，是免疫抑制（immunosuppression）。通过在接触应激物时提供所必需的能量，这些激素对生物有机体基本上起着一种积极的作用。许多第三级反应，在整个动物的层次上是不良适应的。例如，健康和疾病抵抗力（Yada，Tort，2016；见本书第 10 章）、生殖能力（Pankhurst，2016；见本书第 8 章）、生长能力（Sadoul，Vijayan，2016；见本书第 5 章）、学习能力以及其他行为能力，如逃避捕食者的能力（Noakes，Jones，2016；见本书第 9 章）都受到损害。

1.1.1.3　应激和稳态

应激反应显然是鱼类克服艰难挑战的一种必要的作用机理，如果可能的话，鱼类能从应激反应中恢复稳态。不过，同样重要的是要知道在正常的非应激的情况下，血液循环中低水平的应激激素对于保持日常的生命机体能，生长（Schreck，1992）、免疫系统（Schreck，1992；Yada，Nakanishi，2002；Dhabhar，2008；Verburg-van Kemenade 等，2009；Tort，2011；Dhabhar，McEwen，2001）、发育（Schreck，1981）和学习（Sorensen 等，2013；Martins 等，2013）都是不可缺少的。也就是说，在激烈程度低的应激时，皮质醇和儿茶酚胺水平对日常的生命机能是起正面作用的，而在浓度升得较高和持续时间较长时就变为负面作用（Schreck，1992）。换句话说，参与应激反应的每一个成分都是构成稳态的一部分。

“稳态”（Cannon，1926，1932）是另一个难以界定的名词和不容易描述的概念。这个名词衍生自希腊语的词根“hemeo”，意思是“相似”或“类似”，用作词头，和 stasis，意为“稳定”或“停顿”，用以表示尽管生命有机体的外部环境不断变化，但仍然能够保持生命有机体内部环境的相对恒定。这种静态平衡的状态是由反馈系统维持的（Tortora，Grabowski，2000；Guyton，Hall，2011）。所以，稳态是一个动态的过程，难以辨别出一个稳定的分区，这也就增加了定义它的难度。Selye（1973）认识到在和稳态相关的概念内描述应激的难度。他提出“异型稳态”（heterostasis）一词，表示由外源因子刺激适应性的（非遗传性或补偿性的）作用机理而建立的一种新的稳态。通过发育和保持防御性组织的反应可以产生这种状态，而在此之前它们是暂停活动的。和应激造成的起伏变动相比较，鱼类内部环境所经历的变动是很小的。所以，我们可以得出结论：鱼类内部环境的成分超出稳态的范围时，鱼类就处在应激状态。这符合我们的定义。然而，在应激期间鱼类内部环境所经历的变动是相当大的；而如果我们认为鱼类只是处在稳态期间，排除应激和其他挑战迫使鱼类内部环境变动超出正常范围的话，鱼类在正常时期的内部环境亦会出现值得注意的变动。

当鱼类的系统经历正常状况时，应激的结果导致稳态的超负荷，正常的生理学作用机理不能够使鱼类的内部环境处于正常的范围之内。综合应激、异型稳态（见 1.1.1.4 节）和稳态等概念，Romero 等（2009）提出“反应的范围模型”（Reactive scope model）概念，用于说明在没有应激和应激之间的生理学界限。

从应激中恢复会使内部环境回复到稳态范围之内。如果应激物是急性的且不是过度伤害性的，鱼类就能够完全复原。然而，面对一个持久的慢性的应激物或者一个非常激烈的急性应激物时，鱼类就不可能恢复到完全相同的稳态特征，这种状态称之为补偿（Schreck，2000）。鱼类暂时补偿状态的一个很好的例子是在新的外界环境中驯化，例如对热驯化的补偿反应。然而，有两个重要的差别：驯化时的补偿反应是典型性有益的和可逆的。所以，在一个急性应激物之后，应激的许多临床症状将恢复到正常，而其他的生理系统可能长时间受到影响。例如，在短时间（几秒钟到几分钟）的手工操作应激物之后，鱼的皮质醇水平可能在几小时到一天时间内就回复到正常水平，而免疫系统至少要一个星期才能做到这样（Maule 等，1989）。因而，恢复过程是动态的，不能用单峰轨道（unimodal trajectory）来描述。当系统回复到非应激状态时，可能会超过原来的界限，随后会进行校正。在先前引用的事例中，发生应激后，紧接着获得性免疫系统立即受到抑制，第二天出现明显增强，并超过应激前状况，而在一星期内回复到应激前的水平（Maule 等，1989）。此外，尽管初级和次级应激因子恢复，但是鱼类可能不会在遭遇激烈的急性或慢性的应激物之后恢复过来。确实，在应激的临床指标（初级和次级应激反应的成分）之间是完全缺乏协调的，这表明恢复和延迟的死亡在一周或两周之后开始发生（如 Davis 等，2001；Davis，Schreck，2005）。

在这方面最早期的一些研究发现，强迫的活动导致所有受测试的鱼类血液循环中乳酸浓度显著提高，而提高的浓度是和鱼类在不同环境条件下（如温度）的应对能力相联系的（Black，1955；Black 等，1960）。此外，鱼类在肌肉活动后可能出现的延迟死亡，在活动停止大约 30 分钟之后变得明显，这可能是由于酸-碱平衡受到了破坏（Black，1958）。当传统的应激反应 GAS 指标不能表明鱼类超出稳态的范围时，临近死亡的最好预测因子实际上不是传统上认为属于 GAS 一部分的某些测定，而是和鱼类的反射动作相关（Davis，Ottmar，2006；Davis，2007，2010）。慢性应激物的应激反应因子如皮质醇的浓度之间亦可能缺乏协调，而不管鱼类是否处在应激状态。和稳态与异型稳态相联系的作用机理显然是造成初级的和三级的应激反应因子之间缺乏协调的原因。例如，Patino 等（1986）发现皮质醇水平在非应激的个体或者由于拥挤而处于慢性中等强度应激的个体中都是相似的，但后者对抵抗其他应激物如低氧水平、盐度提高或疾病侵入的能力是明显降低的。在慢性应激鱼类的例子中血浆皮质醇没有升高（称为假阴性），可以解释为它们已经提高了它们的皮质醇清除率，使得血液循环中的浓度下降到一个可以忍受的水平。Spagnoli 及其同事（2016；见本书第 13 章）详细地描述了应激的指标以及对数据的分析说明。

非特异性的应激反应由中枢神经系统释放激素开始，通过高度特异性受体或者细胞起作用。所以，靶器官和细胞类型的特异性以及上调或者下调的效应都是在细胞水平上决定的。例如，受体的特性能够决定一个细胞的反应速率；而膜受体介导的效应要比影响位于细胞核的基因快得多。一旦动作开始，应激反应就由许多作用过程调节，使鱼类试图恢复一个稳态的平衡：

（1）首要的是负反馈作用机理影响到所有参与的内分泌因子。

（2）它们自身受体的激素上调或下调，改变细胞反应的幅度。

（3）激素（或者它们的终产物）对代谢以及它们自身清除的影响。例如，诱导每种激素特异性的肝脏酶类，能使激素减少活性而变成一种能够排泄的类型，以防止组织长久地接触到激素的极高浓度，不然，任其持续下去，最终会导致死亡。

（4）激素-结合的血浆载体蛋白是轻度特异性的，能降低激素在靶组织和受体结合的有效性。确实，这些载体蛋白对于血管中自由溶解（即未结合）的激素（如皮质醇）具有很强的结合能力，但和哺乳动物类似的蛋白质相比，这种类固醇的结合亲和力相当低（Caldwell 等，1991）。高的结合能力意味着高的类固醇比例，例如皮质醇，它和这种载体蛋白结合后就不能和它们的特异性受体结合，其结果就会降低这种激素的效应。然而，低的结合亲和力暗示受体或许能够把激素从载体蛋白中脱离出来，从而降低载体蛋白的调节效率。虽然我们知道它们存在于板鳃鱼类（Idler 等，1967b；Idler，Freeman，1968）和硬骨鱼类（Freeman，Idler，1966；Idler，Freeman，1968）中，但涉及鱼类应激的激素载体蛋白的研究资料还很少。所以，我们还不了解这种载体蛋白在鱼类下调应激激素作用的调节机理中的重要性是怎样的。对于经历应激的雌鱼，这种激素-特异性结合蛋白质亦可能对发育中的卵起保护作用。结合蛋白质有助于保持血液循环中激素水平升高，例如在母体血液循环中疏水性的皮质醇，会显著减少它们渗入亲脂性卵子的数量（Schreck，2001）。

还要注意到的是，所有参与的调节作用机理，对每个应激反应因子在开始时和作用期间的反应型式可能互相之间都是不同的。Gorissen 和 Filk（2016；见本书第 3 章）对这个领域进行了一个较为完整的评述。

1.1.1.4　应激和能量学

应激物的生理学反应的最终结果是对整个身体的一种效应，所有生理学过程在某种程度上都是冲突的，因为必须提供能量去克服应激物。搏斗或者逃逸（Cannon，1926，1932）需要额外的能量来应对急性神经的和肌肉紧急的反应，而慢性接触应激物使其他的生命过程得不到能量则必须进行调整。这些代价可以看成是异型稳态的，是一尾鱼从应激物中恢复过来或者对付应激物所付出的代价（Schreck，2010）。Sterling 和 Eyer（1988）提出异型稳态范例为“通过变化达到稳定的能力”。其他的观点已在关于内温动物的研究中详尽阐述（如 Mc Ewen，1998；Mc Ewen，Wingfield，2003；Ashley，Wingfield，2012），而 Schreck（2010）针对鱼类讨论了这个概念。和能量学有关的还有 Sadoul 和 Vijayan（2016；见本书第 5 章）对代谢的讨论与 Rodnick 和 Planas（2016，本书第 7 章）对生理学影响以及对游泳和心血管反应的讨论。

对一个应激物的抵抗、恢复和应对（试图恢复稳态）的能量花费在能量学上是代价昂贵的，可以采用卡路里（热量）或者另一种代谢通货进行量化（例如，毫克/升耗氧量/小时/公斤鱼体重）。异型稳态的负荷有两种类型，1 型和 2 型（Mc Ewen，Wingfield，2003）。1 型是比较急性的，和紧急型的反应联系较为密切，例如逃避捕食者；而 2 型是比较接近应对型的反应，例如在养殖环境中持久的拥挤，以及较典型的慢性应激物（Schreck，2010）。此外，对一个应激物的顺应能从其他的生命过程中转移能量。在数学上，一个异型稳态负荷（$E_{异型稳态}$）的能量代价可以表示为：

$$E_{躯体生长} + E_{配子生长/成熟} + E_{活动性} = (E_{贮存} + E_{食物}) - (E_{STD} + E_{SDA} + E_{废物} + E_{异型稳态})$$

其中，$E_{躯体生长}$和 $E_{配子生长/成熟}$表示用于生长两种类型的能量，即身体和生殖（性腺、次级和三级性征）。

$E_{活动性}$包括用于所有各种活动需要的能量（如摄食、游泳以及和生殖有关的活动）。

（$E_{贮存}+E_{食物}$）表示可供鱼类使用的可转移型能量。它可以用于正面的机能，如两种类型的生长、活动和标准代谢（standard metabolism，STD）。STD 在这里是指瞬间的标准代谢，它亦可以认为是瞬间的标准代谢率（standard metabolicrate，SMR）。我们采用标记的 STD 以避免由于使用 SMR 而可能引起的混乱，因为 SMR 含有时间的成分。

E_{STD}表示和标准代谢相关的瞬时能量耗费（保持鱼类生存，如呼吸、血液循环和渗透压调节等所需要的最小量消费）。

E_{SDA}表示由于在摄食和食物加工过程中的能量耗费而未能利用的能量。代谢的低效能是这其中的一部分，可称之为熵（entropy），是指鱼类未能利用的部分，以热量的形式消失在周围环境中。

$E_{废物}$表示在排泄废物中的能量。

$E_{异型稳态}$表示对一个应激物抵抗或者恢复过来所需要的总能量。它包括直接用于抵抗和恢复过程的能量以及由于特异性动态活动所失去的能量。

有关应激能量代价的一般性文献资料非常少。大多数涉及应激的论文都和感染与免疫系统的能量需求有关（Sheldon，Verhulst，1996；Lochmiller，Deerenbery，2000；Sandland，Minchella，2003；Martin 等，2003，2008；Demas 等，2012）。只有少数论文研究鱼类的应激能量学，但我们知道那是代谢付出的高昂代价（Barton，Schreck，1987）。例如，短暂手工操作的应激物，其代价在整个代谢活动范围的 12%～30%之间（Davis，Schreck，1997）。亦有少量论文从能量学角度分析讨论环境的应激物（Glencross，Bermudes，2011；Ogoshia 等，2012）。

鱼类进行任何一种生命活动，如生长、发育、生殖、防御疾病以及一般的生命活动，最终都决定于它们的遗传特性。然而，鱼类实现其生命活动的能力，即它所表达的表现型，是由它们所处的外界环境决定的（距离每一种生命活动的最适状态有多近或者多远），而这种环境是鱼类在其生命的较早期曾经所处的，例如营养的历史（Schreck，1981；Schreck，Li，1991）或者行为的经历（Schjolden 等，2005）。正常应激能增加而痛苦能减小这种进行生命活动能力的范围（图 1.2）。和应激相联系的这种能量代价的概念化使我们联想到在急性应激的状态期间，$E_{异型稳态}$可能没有足够大的作用去干扰大范围的正常生命活动，如生长、生殖、洄游等。确实，在急性应激期间，鱼类几乎完全依赖于贮存形式的能量，在某种程度上是因为摄食行为受到破坏（如减少摄食、消化、同化作用）以及流向一定器官的血液被转向用于搏斗或者逃逸。此外，糖酵解产生腺苷三磷酸要比氧化磷酸化迅速得多。急性应激导致的死亡很可能是由于心血管系统衰竭，并在鱼类受到应激物影响的期间或者之后不久发生（延迟死亡）。如果死亡是在受到应激物影响之后的一周内或者更长的时间内发生，这很可能是鱼类感染的复制病原体（在宿主体内能够完成它们整个生命周期的病原体）引起的次生性疾病所造成的，因为皮质醇的免疫抑制作用会使鱼体易受病原体侵袭。不过，在急性应激物作用大约一个星期后开始引起死亡的原因还不能够确定（见第 1.1.1.3 节）。

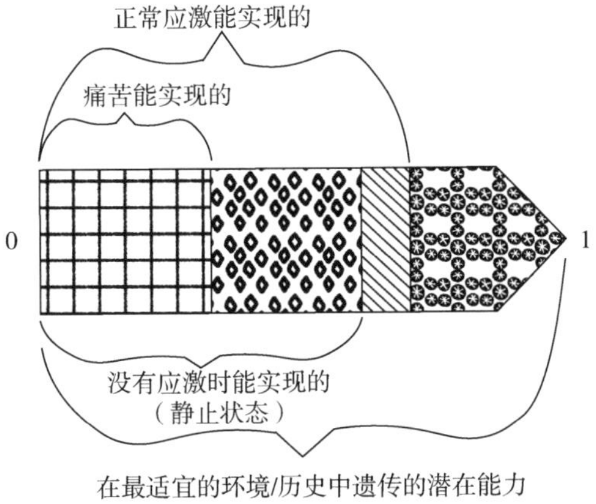

图 1.2　概念式地描绘鱼类实行其生命活动的能力

注：概念式地描绘鱼类实行其生命活动的能力，用一个载体表示由死亡，或者能力为0，到鱼类能够实现其遗传特性并且能够在各方面都是最适宜的环境中生存而得到最大的能力，其值为 1 的范围。这个载体能够描绘任何一种实行生命活动的性能。表示最大的实行生命活动的能力是从没有能力的（0）到有可能获得最大的能力（1）的范围，这取决于生命有机体处于一个完美的环境内。但是，设想环境从来不会是完美的，在没有应激（实际没有应激）的情况下，实际的实行能力要小于最大的潜在能力。能实现的能力可以看作是实际进行生命活动的能力范围。小量的应激（真正应激）能增加实际的实行生命活动能力的量度，而较严重的应激（痛苦）则减少其量度。所以，设想环境和一个应激物的量度能随着时间而变化，能实现的实行生命活动的能力亦是这样。在没有任何应激物的情况下，鱼类能实现的能力受到鱼类所处的环境和生命史（以及外因遗传学）的限定。应激进一步增强或减弱能实现的实行生命活动的能力。

在总的能量耗费方面，我们认为慢性应激会比急性应激经历期间造成明显更大的异型稳态负荷，尽管这些急性应激十分严重，但还不是伤害性的，这是因为慢性应激长时间需要能量。在严重的慢性应激情况下，鱼类不能完全补偿或者恢复，能量耗费会相当大，以至于贮存的能量代谢和食物获取都不能满足生长、发育和个体发育过程的需求。它们会受到负面的冲击或者是完全崩溃。在异型稳态负荷达到最高峰时，鱼类没有能量进行一般性活动，并且很可能由于其他因素的挑战而导致死亡，例如，由于免疫性能遭受破坏，它们不能承受随机遭遇的如一个捕食者或者一种感染物所造成的威胁。在这种情况下，复制的和非复制的寄生物（它们需要另一个宿主以完成其生命周期）都能导致鱼类死亡。这些范例为 Schreck（2010）的综述所证实。

1.1.2　应激反应的动态和对行为表现的效应

初级、次级和三级的应激反应因子对应激物的反应取决于应激物的激烈程度和持续时间。它们亦可以依鱼类的遗传特性、鱼类所处的环境、它们先前的经历以及它们遭遇应激物时的个体发育时期而定（Schreck，1981；Schreck，Li，1991）。图 1.3 描绘了对不同类型、激烈程度和接触应激物的初级、次级、三级应激反应因子的概念式反应。由于和应激反应的动态相联系的固有变异以及实验动物、研究设计和操作过程的差别，鱼类应激的研究根据应激反应动态的相互关系，在许多方面是不同的。在研究工作之间的这种差别是鱼类的不同起始状态（遗传特性、发育阶段、历史、一般的健康状况，等等）、参与的应激物性质、应激作用的持续时间以及取样的周期性等的函数。所以，只采用一个简单的取样时间来解释结果是特别有争议的。关于参与应激反应的短暂动态的描述，从这些混乱的比较中亦能够做出准确的判断。虽然我们能够描述应激反应的状态，但我们还不能以精确的时间或者量度来进一步解析它们。

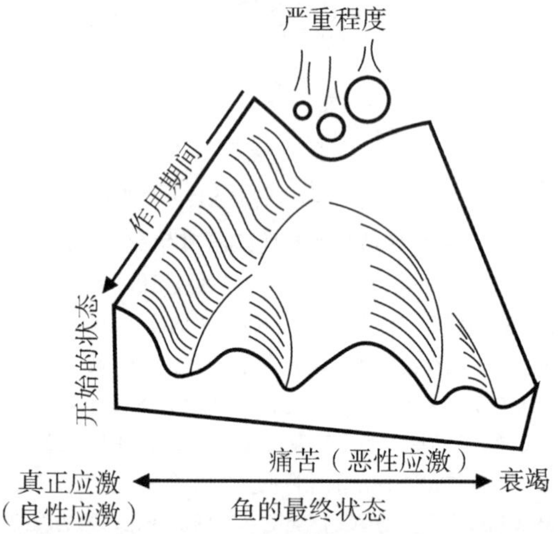

图 1.3　采用一个景观透视图，概念化地表示鱼类处在不同的起始生理状态和不同激烈程度的应激物接触并且经过不同时间历程的最终状态（对应激反应来说是健康的）是怎样的

注：图中设置不同质量（表现为形状大小）的球（表示鱼类不同生理状态）向下跌落到地面不均匀的斜坡景观中，球最终落在景观的不同底部。一个球落到景观底部有多远表示应激物的持续时间有多长，较短的持续时间表示较靠近底部（下坡）。景观的形状受到鱼起止状态的影响：起始的状态比较好，落入斜坡底部的最后状态也就比较好。如果球依轨道最终落入远处左侧（靠近双箭头直线的左侧），表示鱼经历真正的应激。一个温和而持续时间短的应激物能产生表示鱼的状态和面对应激物之前有些相似的景观（朝向左侧某处但还远没有达到沿着双箭头的真正应激）。增加应激物严重程度、应激的持续时间以及降低鱼类起始状态的质量会使景观下降轨道朝向鱼类不同程度消瘦的最后状态以及极端的应激衰竭（朝向双箭头直线的右端）。双向的箭头表示从最左边的最优真正应激到最右边的极端恶劣的健康与身体状况的连续范围。这个概念化的描绘和 Waddington（1977）的外遗传景观有些相似。

应激反应的动态有着很强的遗传成分（Fevolden 等，2002；Øverli 等，2005），在不同的种群之间差别很大（Ellis 等，2012）。所以，遗传力必然很高。虽然不是相当大，但在一个种群内的个体，它们应激反应的差别是明显的（Ellis 等，2012）。鱼类对应激物的反应亦存在着雌雄二态，范围从基因激活（Momoda 等，2007）到 GAS 因子反应的量度。所以，应激反应因子的静止值（resting value）、反应的量度、反应的暂时动态都可以依基因型（genotype）而明显变化。这些差别反应在不同的种群和同一种群内的不同个体对不同的应激源的耐受程度上；有些种类的耐性范围很小，而有些种类的耐性范围很广。此外，不同的种群和一个种群内的不同个体对不同应激物的反应并不一致。我们推测在一个分类单元内，鱼类的不同个体发育阶段之间存在着变异性，但这方面的数据非常少。Winberg 等（2016；见本书第2章）和 Faught 及同事（2016；见本书第4章）对应激反应的遗传学变异方面做了详细综述。

鱼类经受一个应激物时所处的周围环境能够影响生理学反应、耐性水平以及它们的速率。鱼类生活在一个变化的环境中，所以，潜在的应激（Schulte，2014）和环境变化能够不同程度地影响到应激反应，因为它们是通过不同的作用机理运行的。相关事例包括：①温度通过 Q_{10} 现象影响反应速率，从而影响应激反应成分的动态；②氧气浓度能影响代谢能力，因而影响异型稳态负荷的量度和持续时间；③社会环境能影响鱼类的等级结构，从而对一个共同应激物产生不同的反应；④消化状态（饱食对空胃）能影响应激耐性；⑤和健康个体相比较，病原体感染能影响应激反应和耐性。本书的第2、5、7、10、12章都以某种方式涉及环境的影响，包括鱼类福利。

鱼类先前的经历，特别是它们曾经栖息的环境，能够显著影响它们在应激期间的反应。要深刻了解这方面，一个简便的途径是考察在饥饿与饱食之间的差别。还有，先前经历的应激，能够有效地减弱对应激的反应和/或耐性，或者能够在体质上和精神上使鱼类变得更强，以便较好地应对另一个应激物。第2、7、9章进一步论述了鱼类状态的重要性。

一个新形成的领域是鱼类双亲的甚或祖父母的经历能够通过外遗传的作用而影响其对应激物的反应。有大量关于哺乳类和应激相关的外遗传的文献资料（Fish 等，2004），包括环境的应激物（Feil，Fraga，2012）。有一些应激的外遗传效应（epigenetic effect）是由皮质类固醇介导的（Lee 等，2010）。尽管有关鱼类外遗传的文献资料很少，但鱼类应激亦会有意义深远的外遗传后果。例如，Mommer 和 Bell（2014）以及 Mc Ghee 和 Bell（2014）发现鱼类具有对子代的应激母体效应，这和哺乳类一样，而目前已经证明后者的这种效应是通过 DNA 和组蛋白甲基化过程实现的。

鱼类生命周期的不同发育阶段对应激物具有不同的反应和耐性。明显的是，当鱼类在它们的生命周期中进行某种个体发育阶段的转换时，对应激物是更为敏感的（Barton 等，1985）。因此，当鱼类不处于这种个体发育的转换时期时，它们对应激反应的量度比较大，对应激物的耐性亦比较低。鱼类处在个体发育转换期时，对应激物的抵抗能力比较弱，这包括眼色素形成前的胚胎、孵化、开始经口摄食、变态（或者二龄鲑），以及生殖活动接近结束时（图1.4）（Feist，Schreck，2002）。对应激物的反应能力出现在鱼类发育的很早时期。鱼类胚胎大约在眼色素明显形成时开始产生皮质醇。在这个时期提高胚胎的拥挤程度会损害皮质醇的分泌和反馈作用（Ghaedi 等，2013，2014）。

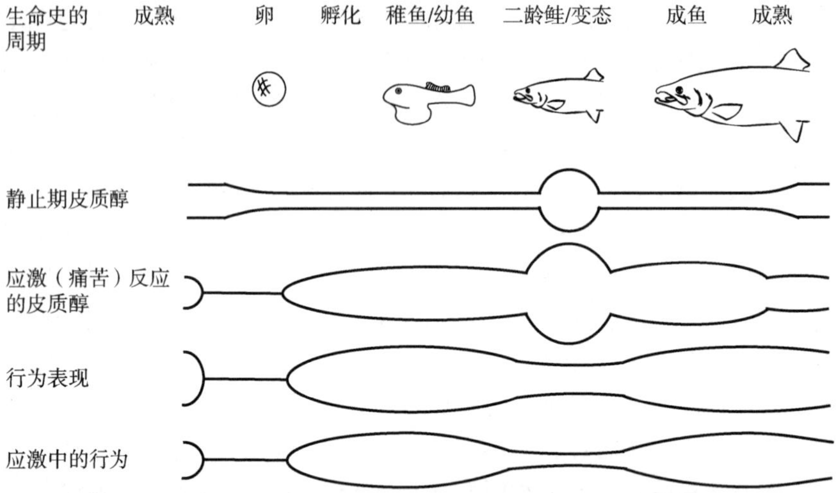

图 1.4　描绘鱼类整个生命史静止状态和痛苦状态血液循环中皮质醇水平以及静止状态（行为表现）与应激状态（应激中的行为表现）执行生命活动能力的概念式模型

注：在线条之间的空间表示血浆皮质醇浓度或者执行生命活动能力的量度。这个概念化的表示类似于 Hutchinson（1958）n-双向多维生态位概念中的范围。

应激能通过降低生育力（产卵数量）或者降低产出卵的质量，对鱼类成年雌鱼的生殖能力（reproductive fitness）产生负面影响（Schreck 等，2001）。我们把对应激的生殖反应概念化地表示为一系列的交替换位（tradeoff）。在应激情况下，雌鱼必须把可获得的能量用于从应激物中恢复过来或者忍受应激物的影响，和投入发育成熟的卵中以便在排卵前达到适宜的质量，在这两者之间进行分配。最后的交替换位是在雌鱼存活的可能性和净生存期间的生殖成效（后代的数量和质量）之间进行。然而，一尾受到应激反应的雌鱼可以不把能量投入到成熟卵和产卵中，使得产卵延缓，而在卵发育过程中，由于缺少能量投入，虽然有尽可能多的卵成熟，但大都营养不良，导致子代的数量较多而体质较差；或者，雌鱼的部分卵变成闭锁卵泡，把最大的生殖能量投入到较少量的卵中，使成熟卵有较高的营养含量并得到优质的子代。假设能量的有效性能对这些不同的交替换位是一种共同通货，对于处在这些轨道之一的雌鱼来说，营养应激显然是极为重要的。在卵发育成熟期间对付应激物的所有策略例子都可以在鱼类中找到。和应激反应相联系的母体转移的生物活性物质如皮质醇能在排卵前的卵子中出现。对于应激如何影响雄鱼的应激反应及其结果，我们了解得很少。第 2、8、12 章对个体发育的不同作用和应激反应做了较为详尽的综述。

1.1.2.1　生理学系统的反应

对一个应激物的感受启动-串联基本的直接的化学介导的反应（初级应激反应），

进而导致性质上多少相似的次级和三级应激反应。神经元信号在不到1秒钟内激发嗜铬组织分泌和产生儿茶酚胺以调整次级和三级反应。其他的神经肽，如多巴胺和γ-氨基丁酸（GABA）亦起作用。儿茶酚胺释放到血流中的速度很快（几秒之内），这使得难以在鱼类中对此进行研究，部分原因亦是由于在没有进一步应激的情况下（需要做好血管导管手术），很难从游动的鱼中取得血样。Perry 和 Bernier（1999）对鱼类儿茶酚胺应激反应做了综述。其他激素的释放稍为缓慢，持续的时间较长（几分钟或者持续几小时），并且是由脑的下丘脑细胞产生的促肾上腺皮质激素释放激素（简称 CRH；亦称为促肾上腺皮质激素释放因子，CRF）启动的。CRH 输送到前脑垂体的肾上腺皮质激素（ACTH）产生细胞，由它们特异性地产生和分泌 ACTH 进入血液循环中。ACTH 激发头肾的肾间腺细胞，使它们合成与分泌皮质醇进入血液循环中。CRH 和 ACTH 都能直接影响次级应激反应，ACTH 能直接在中枢作用于脑（Clemens 等，2001），而 CRH 和 ACTH 还能在外周作用于免疫系统（Schreck，Maule，2001）。实际上，和应激相关的内分泌系统和免疫系统的化学介体之间有许多互相交叉的联系（关于心理神经免疫学，见 Ader，2007）。

应激物的作用开始后需要多长时间才能观察到血液循环中皮质醇明显升高，这在不同的鱼类之间存在很大差别，而所研究过的大多数鱼类需要几分钟。Winberg 等（2016；见本书第2章）对观察到的各种鱼类不同应激因子动态的变化做了综述，而 Gorissen 和 Flik（2016；见本书第3章）概括介绍了参与这些作用的内分泌学内容。

应激反应的主要作用是使能量能够用于参加搏斗、逃逸或应对过程。贮存的能量是可代谢的，从肝脏释放到血液循环中成为可供组织利用的类型。例如，儿茶酚胺和皮质醇可促使肝脏糖原分解，使葡萄糖准备供给骨骼肌利用。此外，皮质醇能下调糖原异生系统，使葡萄糖可供组织利用。血液中的游离脂肪酸和蛋白质浓度亦可能受到影响，如同 Sadoul 和 Vijayan（2016；见本书第5章）所说明的。儿茶酚胺激发心脏与呼吸反应（充实鳃瓣；增加心输出量、血压和鳃换气作用），以便给组织提供更多的氧气；Rodnick 和 Planas（2016；见本书第7章）对这方面的应激反应做了综述。然而，有功能的鳃表面积增加会使水分和氧气流动增加，这称之为渗透-呼吸调和作用（osmorespiratory compromise）。于是，在淡水中的应激鱼类摄入水分以降低其渗透梯度，而在海水中的应激鱼类将丧失水分；前者因应激而增加体重，后者则丢失体重（Stevens，1972）。在渗透-呼吸调和作用的过程中，血液离子流亦明显变化，使离子浓度发生的任何变化都必须得到恢复。在和应激物作用期间及之后，不同的电解质浓度可能没有出现类似的反应型式，这似乎可以解释为和每种电解质相关的不同的细胞作用过程（Stewart 等，2016）。皮质醇的一个重要作用是激活鱼类参与将离子运输进入和送出鳃部、消化道和肾脏的离子泵。在应激使得水盐作用失去平衡之后，皮质醇的主要作用之一是使渗透和离子的平衡得到恢复。Takei 和 Hwang（2016；见本书第

6章）综述了应激对鱼类水盐平衡的影响。可测量的水盐干扰时间进程的变化相当大。

应激影响免疫系统。一个免疫反应的起始阶段是激活一定的非特异性反应，增强诸如细胞凋亡的过程。但是，甚至非常急性的应激物都能抑制高度特异性的免疫系统获得性分支。皮质醇能抑制淋巴细胞合成抗体的能力和诱导白细胞产生和转移的过程。这种抑制作用导致的不可避免的结果是鱼类受到应激物作用之后对感染物和疾病更加敏感，直到鱼类的免疫状态经过波动之后能够回复到应激前的免疫性能。

应激通过改变生物学过程的正常时间亦能影响这些过程的进展。Schreck 等（2001）和 Schreck（2010）论述了和生殖活动相关的这种现象。当鱼处在应激状态时，很可能有其他的“生物钟”受到干扰（Sanchez 等，2009）。

对应激物还有其他的一些生理反应是普遍性的，但并不完全符合 GAS 范围。例如，热休克蛋白（HSPs）是细胞应激反应的一部分。这些蛋白质亦称之为应激蛋白质，根据它们的分子量可以划分为几个蛋白质家族。每个家族具有一些不同的功能。这些蛋白质基本上是起保护性作用的。它们的主要功能之一是起伴侣蛋白（chaperone protein）的作用，帮助其他蛋白质防止应激物使之变形而保护其三级结构。这些蛋白质在一系列应激物，包括热、低氧、氧化应激物和病原体的反应中由细胞产生，并出现在许多硬骨鱼类中（Iwama 等，1998；Basua 等，2002；Kayhan，Duman，2010；Robert 等，2010；Currie，2011；Le Blanc 等，2012；Templeman 等，2014；Stitt 等，2014）。它们在板鳃鱼类中亦有类似的功能（Renshawa 等，2012）。

尽管我们知道在整个生物有机体水平上的许多生命过程，如生长、抗病、生殖、发育等都会不同程度地受到应激损害，但针对这些损害的型式特性或者损害的持续时间却还没有很好的研究例子。当然，在严重的应激情况下，鱼类将不能够恢复，其结果就是死亡。

1.1.2.2 急性的、慢性的和多次的应激物

应激生理反应各个成分的量度和持续时间在很大程度上取决于一个应激物的持续时间和激烈程度，这样，对一个急性的激烈应激物的反应量度就会和一个中等适度而持续的应激物一样大。虽然如此，警告阶段反应的动态和量度倾向于较为定型化（stereotypical）而不会顾及应激物的量度和持续时间，这和抵抗与补偿/恢复阶段不同（Schreck，2000）。对急性应激物的生理反应和对后续的鱼类进行必需的生命机能（如生长、发育、生殖、抵抗病原体）能力的影响在接触急性应激物（图 1.5）和接触慢性应激物（图 1.6）时是不同的。

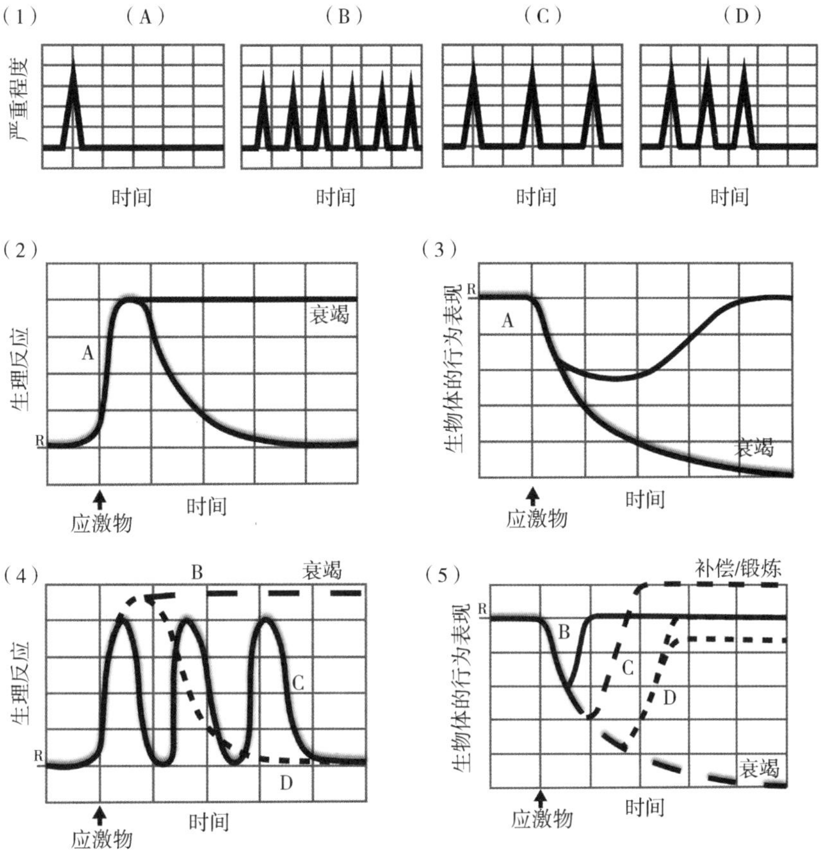

图 1.5　接触急性应激物的生理反应和行为表现型式

注：(1) 一个应激物的四个时间型式：(A) 一个单独的应激物；(B) 多个有顺序的应激物，每个之间的间隔时间相当短；(C) 有顺序的应激物，每个之间的间隔时间相当长；(D) 有顺序的急性应激物，每个之间的间隔时间较短。初级和次级应激因子的暂时动态 (2) 和行为表现或者生理反应 (Rx) (即执行生命机理的能力)。(3) 接触应激型式 A 时，应激物的作用是适度的，其结果是恢复到静止状态，或者应激物的作用是激烈的，导致衰竭 (即死亡)。接触应激型式 B、C、D 时的初级和次级应激因子的暂时动态 (4) 和行为表现 (5)。R 表示静止状态。单个的或有顺序的应激型 (A、B、C、D) 是在应激物箭头处开始的。应激物的效应范围从没有长期的行为表现结果到适度的导致补偿/变得结实强健 (即正常应激)，或者到严重程度，导致衰竭 (即死亡)。

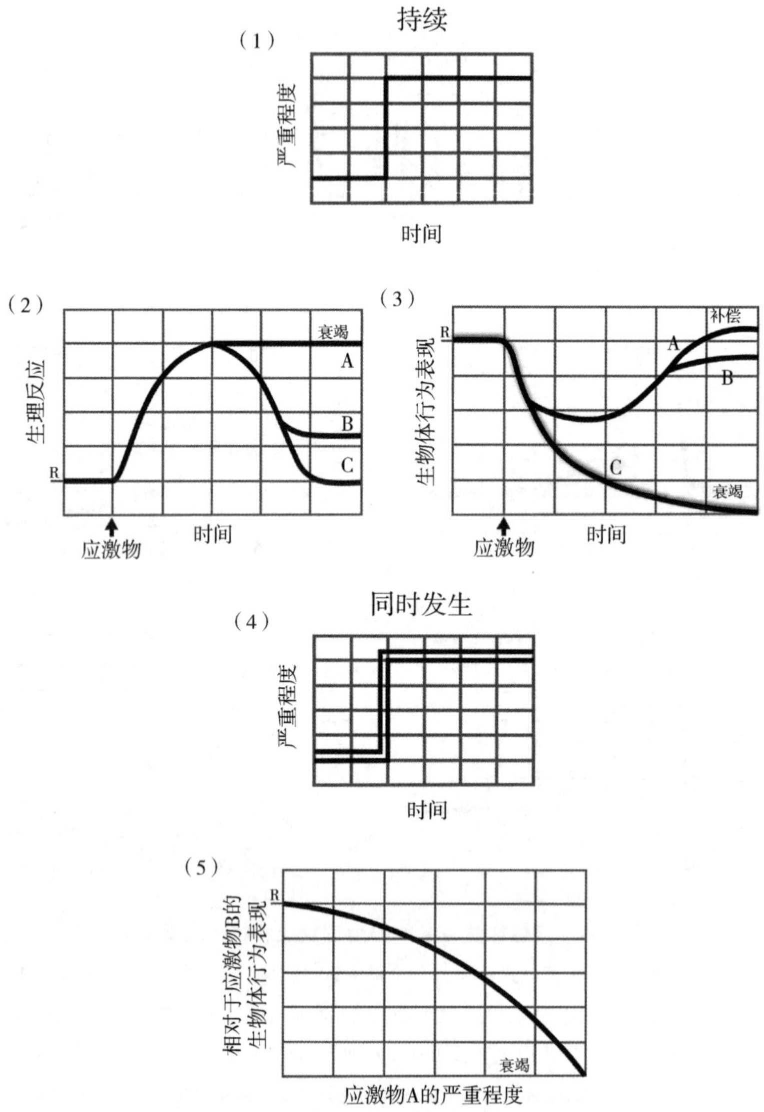

图 1.6　接触慢性应激物的生理反应和行为表现型式

注：（1）不同激烈程度的慢性连续应激物的时序型式。受到一个连续的应激物影响，应激物箭头为起始时间，（2）一定时间初级和次级应激因子的时序动态和（3）行为表现或生理反应（Rx）（执行生命活动的能力）。一个中等适度的应激物引起补偿作用和真正应激（A），激烈的应激物导致痛苦（B），一个极端激烈的应激物导致衰竭和死亡（C）。R 表示静止状态。应激物亦可能并发性和连续性发生（4）。应激物 A 增强的激烈程度对同时接触应激物 B 的鱼类行为活动和耐受能力的影响（5）。

关于慢性应激物，Boonstra（2013）依据他关于下丘脑-脑垂体-肾上腺（HPA）轴［在鱼类中是下丘脑-脑垂体-肾间腺（HPI）轴］功能的观点，坚持认为慢性应激物在自然界对野生动物并不会引起病理学作用。我们对此看法稍为不同，认为其他的应对作用机理可以保持生命，甚至一个有功能的 HPI 轴面对一个普通的应激物亦能这样（Patino 等，1986）。这并不是说在慢性应激情况下的应激反应不是适应的，它是的。然而，和生理学抵抗性与应对作用机理相联系的生物能量代价会使得执行其他生命功能的能力（如抵抗病原体的能力）变得较弱（见第 1.1.1.4 节）。我们怎样界定应激亦影响到我们怎样解析慢性应激物的影响。问题真正在于慢性应激是否损害动物的健康。它并不需要降低到零［正如 Boonstra（2013）所认为的］，实际上甚至不用降低很多，就会对主要的进化结果产生影响。曾经有学者认为在自然界，急性应激物要比慢性应激物普遍得多，至少对陆生动物是这样的（Boonstra，2013）。但是对鱼类可能不是这样，特别是在淡水环境中，鱼类可能经历持续的温度升高、水位降低或者受到污染物干扰，甚至在海水环境中会经历持续的寒冷或高温、海水酸化和低氧，这就迫使一些鱼类处在它们的耐性范围之外。和慢性应激相关的社会性状态的作用及其影响是重要的［见 Sapolsky（2005）关于灵长类有意义的剖析］。总之，有关鱼类在自然界的生理学资料非常少，而我们的想法绝大多数都是通过在实验室或养殖场的环境中进行的研究得来的。

鱼类经常受到一些有顺序的相同的急性应激物影响，如标记-重捕实验。我们知道这种情况会产生累积的效应。然而，目前我们对应激反应因子如何短暂反应以及多次性应激物的量度是怎样的都了解得很少。我们知道急性应激物经过数小时相隔，至少三次接触经历，会引起应激反应成分产生近似累加的效应，但我们不知道这种动向还能够继续多少次。我们亦不知道在一次经历而不是多次经历中，感受两个或更多个应激物之前在应激物之间作用的最短时间是多长。同样，我们不知道在有顺序的应激物之间，在前一个应激物的影响不再明显存在之前需要经过多长时间（Schreck，2000）（图 1.5）。

当两个应激物同时出现时（如鱼拥挤在低氧的环境中），一个应激物的存在能降低对另一个应激物的耐性（Schreck，2000）（图 1.6）。然而，鱼类在应对中逐渐下调应激反应以及对慢性应激物进行缓慢的补偿，并不会将它误作为明显的衰竭，或者可归因于缺乏对较多应激物产生生理反应的能力。在慢性应激期间，对其他遇到的应激物生理反应能力的减弱（Huntingford 等，2006；Pankhurst，2011；Zuberi 等，2014）由 Sneddon 等进行论述（Sneddon 等，2016；见在本书第 12 章）。对应激进行补偿期间，鱼类特意地下调应激反应，而因为应激物的超载和一个应激反应未能增强时，减弱的应激反应亦会出现。

我们还需要注意鉴别鱼类对应激物的适应情况。在先前经历过其他中等适度的有顺序的（或者每日的）应激物一段时间之后，应激反应的量度和持续时间能变小。这很可能是由于在较早接触先前的应激物期间产生了身体调节作用（physical conditioning）。对于先前曾经历过中等适度应激物的鱼类，一种心理（或精神）的成分可能起着增强行为表现的作用。对重复应激物的适应或驯化（habituation）是很好的例子。例如，在短暂而适度的每天应激经历后，给鱼投喂食物（一种奖赏）进行正面的调节以改善行为表现，要比随激烈的挑战后没有奖赏的情况好一些。经过积极调节的鱼对随后的应激活动能表现较低的生理反应，对其他的挑战如较低的溶解氧、升高的盐度和疾病等表现出较强的耐性（Schreck 等，1995）。Schreck（2010）把这种情况称之为“应激强化”（stress hardening）。Winderberg 等（2016；见本书第 2 章）讨论了鱼类对手工操作应激状态的反应是怎样变化的。

在概念化方面，我们把生理学应激因子和动物对应激状态的行为表现两者的动态都看成是一种多维家族的曲线，不应简单地描绘与展示。由一个应激物引起的反应曲线的特性受到多种因素包括鱼类的内在状态，如鱼类的遗传、发育阶段、鱼的先前经历（曾经过的环境和应激物经历）等的影响。它们亦受到外在因素，如应激物的激烈程度和持续时间、鱼的发育阶段以及鱼类现时所处环境的影响。这些因素共同影响鱼类对应激物生理反应的各个方面。它们亦影响鱼类执行生命机能的能力，亦就是说，它们影响到鱼类怎样良好地执行其他的生命机能（Schreck，Li，1991）。图 1.3 以一种简化的形式概念化地说明鱼类对急性和慢性应激物的反应型式和持续时间。

1.1.3 GAS 的当代看法：真正应激对痛苦

我们必须重新说明真正应激（eustress）和痛苦（distress）的概念，以便恰当地认识到鱼类对不同应激物的反应或者持续时间不是单峰的。双峰的反应是普通的，例如一个应激物低的激烈程度产生一种效应，而一个较高水平的激烈程度就产生相反的效应。对这种刺激作用（hormesis）的概念，Schreck（2010）做了较为充分的论述。可以认为，一个适应性反应是在一次稳态的初时被破坏之后能够诱导补偿的生物学过程。鱼类的免疫系统如何应对应激的不同激烈程度是这种现象的一个经典例子。低水平的应激使皮质醇水平略微升高，通过上调白细胞的重新部署、先天性免疫、效应细胞功能和细胞介导的免疫力而对免疫能力产生积极的效应。这种净效应能增强鱼类对感染物和肿瘤的抵抗力。应激的这种积极作用可称为真正应激（eustress）（Dhabhar，2008；Dhabhar，McEwen，2001；Schreck，2010）。

当应激物激烈程度加深，皮质醇水平升得较高并超过真正应激状态时，免疫系统摇摆不定，积极的作用开始消失，进而出现完全负面影响。这种情况是有回弹的，接着应激物的激烈程度进一步加深，前面提到的所有免疫因子都受到抑制。这种净效应就是抑制对感染物和肿瘤的抵抗力。这种负面效应的应激状态可称为痛苦

(distress)。

1.1.4 感觉系统和感受

如前所述，一个传统的应激反应只是启动一个对恐吓的感受。大多数涉及鱼类对应激物感受的文献资料都依随着应激物是通过激活传统的感觉器官（视觉、听觉、触觉、嗅觉以及鱼类的味觉）而感受的范例。这些激活的神经内分泌反应是在鱼的下丘脑内。Schreck（1981）以一种人为的（anthropogenic）观点提出在鱼类生命中的这些感受能引起惊吓、不舒服和痛苦而导致一次应激反应。然而，其他的感觉系统在应激感受以及随后的应激反应中亦明显起着作用。例如，处在和环境相关的酸化水平中（Barton 等，1985）、低氧、（O'Connor 等，2011）、碳酸过多（Schreck，1972；Sandblom 等，2013）、渗透应激（osmostress）（Takei，Hwang，2016；见本书第6章）以及病原体/寄生虫（Suzumoto 等，1977）都能使皮质醇水平升高。这些可以通过神经通道而不是传统的感觉器来影响应激反应系统。应激物能影响黏膜表面，如鳃、胃肠或者皮肤，能在这些部位产生反应，即首先发展为局部反应，随后引起内分泌的或者全身的反应。应激源如不均衡的或者不充足的饮食会影响消化道，并可能变成慢性应激物（Montero 等，2001）。同样，亚急性水平的污染物能局部影响鳃部。很明显，病原体作用于鳃、皮肤或消化道的侵入门户，能产生局部反应，包括细胞因子的表达和产生以及随后激活 HPI 轴。在一个病原体或者一个寄生虫侵入之后，很难准确地测定 HPI 轴被激活的时间，因为这决定于病原体的毒性和感染的强度。一旦 HPI 轴被激活，它们的信使成分就会通过它们的特异性受体在靶组织中发挥它们的作用。Gorissen 和 Flik（2016；见本书第 3 章）以及 Yada 和 Tort（2016；见本书第 10 章）在文中论述了应激激素和它们受体的作用。

1.1.5 应激反应的适应和非适应方面

生理学的适应是作用机理的总和，使生物有机体能够克服（调节或者恢复）由应激物造成的挑战。这些作用机理包含补偿的反应和/或适应（驯化）。它们包含所有的生理学反应作用机理，包括能量、免疫反应、学习和应对策略。

非适应性是指没有足够的资源去克服应激物和/或不可能控制的反应而造成的结果。事实上，鱼类和其他脊椎动物一样具有多种能力和作用机理去对付应激物的刺激和克服应激物造成的后果。如前所述，应激反应取决于应激物的类型、强度以及最后的反复结果。然而，两种因素，不可控制的和不可预知的，都非常难以确定应对的可能性，因为在这种情况下变异稳态的负荷将会增加（Korte 等，2007；Schreck，2010）。

养殖中的鱼类受到许多局限，特别是运动的限制和固定的环境限制。这意味着养殖鱼类在学习特征和锻炼结实方面不同于野生鱼类。

在监禁情况下对适当的鱼类进行应激管理，非常必要的是了解其主要的行为反应、应付能力以及鱼类反应的可预测性。实际上，近年在鱼类福利领域所做的大部分工作都是研究这些行为的特征；这些行为能决定鱼类应付应激物的能力以及福利的范围。模式动物能够提供一个更为完整的框架（Spagnoli 等，2016；见本书第 13 章）。

1.1.6 关键的未知因素

当代的研究途径已经为鱼类应激生物学提供了重要的新思路，并且显著拓展了如何描述鱼类对应激物反应的范例（表 1.1）。然而，还有许多未知的因素需要我们去研究解决，以便对鱼类应激有更为全面深入的理解。下面列出这些关键的未知因素和问题。

表 1.1 关于鱼类应激特征的传统的和当代的研究方法和思路

主题	传统的	当代的
行为反应	种类特异性的	种类特异性的，应对策略
生物学的序位体系	有机体的	有机体的，细胞的，分子的
内分泌应激反应	层次的（如下丘脑、脑垂体、肾间腺）	非层次的（多重方向）
学习和适应	立体型的	学习和适应
生理学调节	神经系统	神经系统
	内分泌系统	内分泌系统
		免疫系统
调节	稳态	稳态，异型稳态
反应动态	单峰的	单峰的，双峰的
反应变异	生理的	生理的
	代谢的	代谢的
		能量/资源分配
感觉	中枢的（神经系统）	多方面的（中枢和外周）
信号	激素	激素，细胞因子
应激效应	有机体的	有机体的，生理学的
	生理学的	痛苦，心理学的
应激物效应	痛苦	痛苦，真正应激
应激物性状	应激的特征	应激物特征
	应激物强度	应激物的强度
	应激物持续时间	应激物的持续时间
		应激物调控
		应激物可预知性
变异	种间的	种间的，个体的

我们预测鱼处于应激状态的能力相对较强。例如，应激激素的水平处在静止状态的浓度之上，可以有理由推测鱼处于应激状态中。同样，HSPs 浓度升高亦可以很明确地推测鱼所处的应激状态接近细胞损伤的程度，需要进行处理。但是，要预知鱼已经从接触急性或慢性应激物的应激状态中恢复过来，依据还是很少的。换句话说，应激激素或其他的应激因子在血液循环中的浓度未必能够表示鱼类已经完全恢复，它们只是表明鱼类或者已经恢复，或者已经能够降低这些激素的浓度。

关于应激反应因子的量度超过静止状态水平的推论还需要许多信息。例如，反应量变的轻微升高，如果它们导致一个真正应激状态，就可能是适应性的；而短期或长期略高的反应量度可能是不良适应。我们不知道这个转折点在哪里。还有，我们不知道是否较高的应激反应量度就表示鱼类要比处于刚好在真正应激之上的较低反应量度时受到更强的应激作用；同样，我们不知道对一系列激烈程度相同的或不同的应激物产生的反应量度是相同的还是不同的。另外的困难之处是，我们不知道通过不同的感觉系统感知应激物引起的是相似的还是不同的应激反应。需要信息来说明这个问题：对一个应激物的反应是否在一个等级的或者一个阈值的调控系统中进行？例如，是否连续升高的应激激素浓度使得应激反应水平不断增加？或者，对一个应激反应是否需要有最低量的激素浓度，而较高的激素浓度并没有附加的效应，一直到达可能存在的某个调节点（Set point）？

关于野生鱼类如何应对应激物的信息非常少。验证应激从监禁中的鱼类扩展到它们野外的同伙时需要了解些什么是重要的。确定如何研究野生鱼类应激的原始状态亦非常有用，这可以避免由于应激物和鱼发生的联系以及样品收集过程中可能带来的误差。

对鱼类进行的大量研究所提供的数据有助于建立应激因子和它们功能的正常而非应激范围的标准。然而，实际上是很有问题的，因为有许多因子影响到静止状态和反应水平以及它们和种内与种间差异偶联的相互作用，这基本上和普通的准则是相违背的。因而，涉及 GAS 的数据分析就不应简单从事。此外，数据混乱的解释就会出现这样的可能性：有些 GAS 因子更多的是通过阈值而不是通过连续的调控起作用，而在这方面没有确切的数据去建立综合的能够应用的阈值。底线就是如果能够建立一个有效的“食谱（cookbook）”，使应激测定能够得到解析，那将会是十分有用的。

C. B. 施瑞克　L. 托特　著
林浩然　译、校

参考文献

Ader, R. (2007). Psychoneuroimmunology, vols. 1 and 2. Burlington: Academic Press.

Ashley, N. T. and Wingfield, J. C. (2012). Sickness behavior in vertebrates: allostasis, life-history modulation, and hormonal regulation. In Ecoimmunology (eds. G. Demas and R. Nelson), pp. 45-91. New York, NY: Oxford University

Press.

Barton, B. A. (1997). Stress in finfish: past, present, and future-a historical perspective. In Fish Stress and Health in Aquaculture (eds. G. K. Iwama, A. D. Pickering, J. P. Sumpter and C. B. Schreck), pp. 1-33. Cambridge: Cambridge University Press.

Barton, B. A. (2002). Stress in fishes: a diversity of responses with particular reference changes in circulating corticosteroids. Integr. Comp. Biol. 42, 517-525.

Barton, B. A. and Iwama, G. (1991). Physiological changes in fish from stress in aquaculture with emphasis on the response and effects of corticosteroids. Ann. Rev. Fish Dis. 1, 3-26.

Barton, B. A. and Schreck, C. B. (1987). Metabolic cost of acute physical stress in juvenile steelhead. Trans. Am. Fish. Soc. 116, 257-263.

Barton, B. A., Schreck, C. B., Ewing, R. D., Hemmingsen, A. R. and Patino, R. (1985). Changes in plasma cortisol during stress and smoltification in coho salmon, Oncorhynchus kisutch. Gen. Comp. Endocrinol. 59, 468-471.

Basua, N., Todghama, A. E., Ackermana, P. A., Bibeaub, M. R., Nakanoa, K., Schulteb, P. M., et al. (2002). Heat shock protein genes and their functional significance in fish. Gene. 295, 173-183.

Bayne, B. L. (1985). Responses to environmental stress: tolerance, resistance and adaptation. In Marine Biology of Polar Regions and Effects of Stress on Marine Organisms (eds. J. S. Gray and M. E. Christiansen), pp. 331-349. New Jersey: Wiley.

Black, E. C. (1955). Blood levels of haemoglobin and lactic acid in some freshwater fishes following exercise. J. Fish. Res. Board Can. 12, 917-929.

Black, E. C. (1958). Hyperactivity as a lethal factor in fish. J. Fish. Res. Board Can. 15, 573-586. Black, E. C., Robertson, A. C., Hanslip, A. R. and Chiu, W. -G. (1960). Alterations in glycogen, glucose and lactate in rainbow and Kamloops trout, Salmo gairdneri following muscular activity. J. Fish. Res. Board Can. 17, 487-500.

Boonstra, R. (2013). Reality as the leading cause of stress: rethinking the impact of chronic stress in nature. Funct. Ecol. 27, 11-23.

Brett, J. R. (1958). Implications and assessment of environmental stress. In The Investigation of Fish-Power Problems (ed. P. A. Larkin), pp. 69-83. Vancouver: H. R. MacMillan Lectures in Fisheries, University of British Columbia.

Caldwell, C. A., Kattesh, H. G. and Strange, R. J. (1991). Distribution of cortisol among its free and protein-bound fractions in rainbow trout (Oncorhynchus mykiss): evidence of control by sexual maturation. Comp. Biochem. Physiol. 99A, 593-595.

Cannon, W. B. (1926). Physiological regulation of normal states: some tentative postulates concerning biological homeostatics. In A Charles Richet: ses amis, ses collègues, ses élèves (ed. A. Pettit), p. 91. Paris: Les éditions Médicales.

Cannon, W. B. (1932). The Wisdom of the Body. New York, NY: W. W. Norton & Company, Inc.

Chrousos, G. P. (1998). Stressors, stress, and neuroendocrine integration of the adaptive response. Ann. N. Y. Acad. Sci. 851, 311-335.

Chrousos, G. P. (2009). Stress and disorders of the stress system. Nat. Rev. Endocrinol. 5, 374.

Clements, S., Schreck, C. B., Larsen, D. A. and Dickhoff, W. W. (2001). Central administration of corticotrophin-releasing hormone stimulates locomotor activity in juvenile chinook salmon (Oncorhynchus tshawytscha). Gen. Comp. Endocrinol. 125, 319-327.

Close, D. A., Yun, S. -S., McCormick, S. D., Wildbill, A. J. and Li, W. (2010). 11-Deoxycortisol is a corticosteroid hormone in the lamprey. Proc. Natl. Acad. Sci. U. S. A. 107, 13942-13947.

Currie, S. (2011). Temperature. Heat shock proteins and temperature. In Encyclopedia of Fish Physiology, vol. 1 (eds. A. P. Farrell, E. D. Stevens, J. J. Cech, Jr. and J. G. Richards), pp. 1732-1737. San Diego: Academic Press.

Davis, L. E. and Schreck, C. B. (1997). The energetic response to handling stress in juvenile coho salmon. Trans. Am. Fish. Soc. 126, 248-258.

Davis, M. W. (2007). Simulated fishing experiments for predicting delayed mortality rates using reflex impairment in restrained fish. ICES J. Mar. Sci. 64, 1535-1542.

Davis, M. W. (2010). Fish stress and mortality can be predicted using reflex impairment. Fish Fisher 11, 1-11.

Davis, M. W. and Ottmar, M. L. (2006). Wounding and reflex impairment may be predictors for mortality in discarded or escaped fish. Fish. Res. 82, 1-6.

Davis, M. W. and Schreck, C. B. (2005). Responses by Pacific halibut to air exposure: lack of correspondence among plasma constituents and mortality. Trans. Am. Fish. Soc. 134, 991-998.

Davis, M. W., Olla, B. L. and Schreck, C. B. (2001). Stress induced by hooking, net towing, elevated sea water temperature and air in sablefish: lack of concordance between mortality and physiological measures of stress. J. Fish Biol. 58, 1-15.

Demas, G., Greive, T. J., Chester, E. and French, S. S. (2012). The energetics of immunity. In Ecoimmunology (eds. G. Demas and R. Nelson), pp. 259-296. New York, NY: Oxford University Press.

Dhabhar, F. S. (2008). Enhancing versus suppressive effects of stress on immune function: implications for immunoprotection versus immunopathology. Allergy Asthma

Clin. Immunol. 4, 2-11.

Dhabhar, F. S. and McEwen, B. S. (2001). Bidirectional effects of stress and glucocorticoid hormones on immune function: possible explanations for paradoxical observations. In Psychoneuroimmunology, vol. 1 (eds. R. Ader, D. L. Felten and N. Cohen), p. 301. San Diego: Academic Press.

Ellis, T., Yildiz, H. Y., López-Olmeda, J., Spedicato, M. T., Tort, L., Øverli, Ø., et al. (2012). Cortisol and finfish welfare. Fish Physiol. Biochem. 38, 163-188.

Faught, E., Aluru, N. and Vijayan, M. M. (2016). The Molecular Stress Response. In Fish Physiology-Biology of Stress in Fish, Vol. 35 (eds. C. B. Schreck, L. Tort, A. P. Farrell and C. J. Brauner), San Diego, CA: Academic Press.

Feil, R. and Fraga, M. F. (2012). Epigenetics and the environment: emerging patterns and implications. Nat. Rev. Genet. 13, 97-109.

Feist, G. and Schreck, C. B. (2002). Ontogeny of the stress response in chinook salmon, Oncorhynchus tshawytscha. Fish Physiol. Biochem. 2, 31-40.

Fevolden, S. E., Røed, K. H. and Gjerde, B. (2002). Genetic components of post-stress cortisol and lysozyme activity in Atlantic salmon; correlations to disease resistance. Fish Shellfish Immunol. 4, 507-519.

Fish, E. W., Shahrokh, D., Bagot, R., Caldji, C., Bredy, T., Szyf, M., et al. (2004). Epigenetic programming of stress responses through variations in maternal care. Ann. N. Y. Acad. Sci. 1036, 167-180.

Freeman, H. C. and Idler, D. R. (1966). Transcortin binding in Atlantic salmon (Salmo salar) plasma. Gen. Comp. Endocrinol. 7, 37-43.

Ghaedi, G., Yavari, V., Falahatkar, B., Nikbakht, G., Sheibani, M. T. and Salati, A. P. (2013). Whole egg and alevin changes of cortisol and interrenal tissue differences in rainbow trout Oncorhynchus mykiss exposed to different stocking densities during early development. Zoolog. Sci. 30, 1102-1109.

Ghaedi, G., Falahatkar, B., Yavari, V., Sheibani, M. T. and Broujeni, G. N. (2014). The onset of stress response in rainbow trout Oncorhynchus mykiss embryos subjected to density and handling. Fish Physiol. Biochem. 41, 485-493.

Glencross, B. D. and Bermudes, M. (2011). Adapting bioenergetic factorial modelling to understand the implications of heat stress on barramundi (Lates calcarifer) growth, feed utilisation and optimal protein and energy requirements-potential strategies for dealing with climate change? Aquacult. Nutr. 18, 411-422.

Gorissen, M. and Flik, G. (2016). Endocrinology of the Stress Response in Fish. In Fish Physiology-Biology of Stress in Fish, Vol. 35 (eds. C. B. Schreck, L. Tort, A. P. Farrell and C. J. Brauner), San Diego, CA: Academic Press.

Guyton, A. C. and Hall, J. E. (2011). Guyton and Hall Textbook of Medical Physiology. Philadelphia: Saunders Elsevier.

Holden, C. (2000). Laboratory animals. Researchers pained by effort to define distress precisely. Science 290, 1474-1475.

Huntingford, F. A., Adams, C., Braithwaite, V. A., Kadri, S., Pottinger, T. G., Sandoe, P., et al. (2006). Current issues in fish welfare. J. Fish Biol. 68, 332-372.

Idler, D. R. and Freeman, H. C. (1968). Binding of testosterone, 1α-hydroxycorticosterone and cortisol by plasma proteins of fish. Gen. Comp. Endocrinol. 11, 366-372.

Idler, D. R. and Sangalang, G. B. (1970). Steroids of a chondrostean: in-vitro steroidogenesis in yellow bodies isolated from kidneys and along the posterior cardinal veins of the American Atlantic sturgeon. Acipenser oxyrhynchus Mitchell. J. Endocrinol. 48, 627-637.

Idler, D. R. and Truscott, B. (1966). 1α-hydroxycorticosterone from cartilaginous fish: a new adrenal steroid in blood. J. Fish. Res. Bd. Can. 23, 615-619.

Idler, D. R., Freeman, H. C. and Truscott, B. (1967a). A preliminary communication on the biological activity of 1α-hydroxycorticosterone isolated from cartilaginous fish. J. Fish. Res. Bd. Can. 24, 205-206.

Idler, D. R., Freeman, H. C. and Truscott, B. (1967b). Biological activity and protein-binding of 1α-hydroxycorticosterone: an interrenal steroid in elasmobranch fish. Gen. Comp. Endocrinol. 9, 207-213.

Iwama, G. K., Thomas, P. T., Forsyth, R. and Vijayan, M. M. (1998). Heat shock proteins (chaperones) in fish and shellfish and their potential role in relation to fish health: a review. Rev. Fish Biol. Fish 8, 35-56.

Kayhan, F. E. and Duman, B. S. (2010). Heat shock protein genes in fish. Turkish J. Fish. Aquat. Sci. 10, 287-293.

Koolhaas, J. M., Bartolomucci, A., Buwalda, B., de Boer, S. F., Flügge, G., Korte, S. M., et al. (2011). Stress revisited: a critical evaluation of the stress concept. Neurosci. Biobehav. Rev. 35, 1291-1301.

Korte, S. M., Olivier, B. and Koolhaas, J. M. (2007). A new animal welfare concept based on allostasis. Physiol. Behav. 92, 422-428.

LeBlanc, S., Hoglund, E., Gilmour, K. M. and Currie, S. (2012). Hormonal modulation of the heat shock response: insights from fish with divergent cortisol stress responses. Am. J. Physiol. Regul. Integr. Comp. Physiol. 302, 184-192.

Lee, R. S., Tamashiro, K. L. K., Yang, X., Purcell, R. H., Harvey, A., Willour, V. L., et al. (2010). Chronic corticosterone exposure increases expression and decreases deoxyribonucleic acid methylation of Fkbp5 in mice. Endocrinology 151,

4332-4343.
Levine, S. (1985). A definition of stress? In Animal Stress (ed. G. P. Moberg), pp. 51-69. Bethesda: American Physiological Society.
Lochmiller, R. L. and Deerenberg, C. (2000). Trade-offs in evolutionary immunology: just what is the cost of immunity? Oikos 88, 87-98.
Martin, L. B., Scheuerlein, A. and Wikelski, M. (2003). Immune activity elevates energy expenditure if house sparrows: a link between direct and indirect costs? Proc. R. Soc. B 270, 153-158.
Martin, L. B., Weil, Z. M. and Nelson, R. J. (2008). Seasonal changes in vertebrate immune activity: mediation by physiological trade-offs. Proc. R. Soc. B 363, 321-339.
Martins, C. I. M., Galhardo, L., Noble, C., Damsgard, B., Spedicato, M. T., Zupa, W., et al. (2013). Behavioural indicators of welfare in farmed fish. Fish Physiol. Biochem. 38, 17-41.
Maule, A. G., Tripp, R. A., Kaattari, S. L. and Schreck, C. B. (1989). Stress alters immune function and disease resistance in chinook salmon (Oncorhynchus tshawytscha). J. Endocrinol. 120, 135-142.
McEwen, B. S. (1998). Protective and damaging effects of stress mediators. N. Engl. J. Med. 338, 171-179.
McEwen, B. S. and Lasley, E. N. (2002). The End of Stress As We Know It. Washington, D. C.: Joseph Henry Press.
McEwen, B. S. and Wingfield, J. C. (2003). The concept of allostasis in biology and biomedicine. Horm. Behav. 43, 2-15.
McGhee, K. E. and Bell, A. M. (2014). Paternal care in a fish: epigenetics and fitness enhancing effects on offspring anxiety. Proc. R. Soc. B 281, 20141146.
Mommer, B. C. and Bell, A. M. (2014). Maternal experience with predation risk influences genome-wide embryonic gene expression in threespined sticklebacks (Gasterosteus aculeatus). PLoS One 9 (6), e98564. http://dx.doi.org/10.1371/journal.pone.0098564.
Momoda, T. S., Schwindt, A. R., Feist, G. W., Gerwick, L., Bayne, C. J. and Schreck, C. B. (2007). Gene expression in the liver of rainbow trout, Oncorhynchus mykiss, during the stress response. Comp. Biochem. Physiol. Part D Genomics Proteomics 2, 303-315.
Montero, D., Tort, L., Robaina, L., Vergara, J. M. and Izquierdo, M. S. (2001). Low vitamin E in diet reduces stress resistance of gilthead sea bream (Sparus aurata) juveniles. Fish Shellfish Immunol. 11, 473-490.
Noakes, D. L. G. and Jones, K. M. M. (2016). Cognition, Learning, and Behavior. In

Fish Physiology-Biology of Stress in Fish, Vol. 35 (eds. C. B. Schreck, L. Tort, A. P. Farrell and C. J. Brauner), San Diego, CA: Academic Press.

O'Connor, E. A., Pottinger, T. G. and Sneddon, L. U. (2011). The effects of acute and chronic hypoxia on cortisol, glucose and lactate concentrations in different populations of threespined stickleback. Fish Physiol. Biochem. 37, 461-469.

Ogoshia, M., Katoa, K., Takahashia, H., Ikeuchid, T., Abea, T. and Sakamotoa, T. (2012). Growth, energetics and the cortisol-hepatic glucocorticoid receptor axis of medaka (Oryzias latipes) in various salinities. Gen. Comp. Endocrinol. 178, 175-179.

Øverli, O., Winberg, S. and Pottinger, T. G. (2005). Behavioral and neuroendocrine correlates of selection for stress responsiveness in rainbow trout—a review. Integr. Comp. Biol. 45, 463-474.

Pankhurst, N. W. (2011). The endocrinology of stress in fish: an environmental perspective. Gen. Comp. Endocrinol. 170, 265-275.

Pankhurst, N. W. (2016). Reproduction and Development. In Fish Physiology-Biology of Stress in Fish, Vol. 35 (eds. C. B. Schreck, L. Tort, A. P. Farrell and C. J. Brauner), San Diego, CA: Academic Press.

Patino, R., Schreck, C. B., Banks, J. L. and Zaugg, W. S. (1986). Effects of rearing conditions on the developmental physiology of smolting of coho salmon. Trans. Am. Fish. Soc. 115, 828-837.

Patino, R., Redding, J. M. and Schreck, C. B. (1987). Interrenal secretion of corticosteroids and plasma cortisol and cortisone concentrations after acute stress and during seawater acclimation in juvenile coho salmon (Oncorhynchus kisutch). Gen. Comp. Endocrinol. 68, 431-439.

Perry, S. F. and Bernier, N. J. (1999). The acute humoral and adrenergic stress response in fish: facts and fiction. Aquaculture 177, 285-295.

Pickering, A. D. (ed.). (1981). Stress and Fish. London: Academic Press.

Renshawa, G. M. C., Kuteka, A. K., Grantb, G. D. and Anoopkumar-Dukieb, S. (2012). Forecasting elasmobranch survival following exposure to severe stressors. Comp. Biochem. Physiol. Part A Mol. Integr. Physiol. 162, 101-112.

Roberts, B. W., Didier, W., Rai, S., Johnson, N. S., Libants, S., Yun, S.-S., et al. (2014). Regulation of a putative corticosteroid, 17, 21-dihydroxypregn-4-ene, 3, 20-one, in sea lamprey, Petromyzon marinus. Gen. Comp. Endocrinol. 196, 17-25.

Roberts, R. J., Agius, C., Saliba, C., Bossier, P. and Sung, Y. Y. (2010). Heat shock proteins (chaperones) in fish and shellfish and their potential role in relation to fish health: a review. J. Fish Dis. 33, 789-801.

Rodnick, K. J. and Planas, J. V. (2016). The Stress and Stress Mitigation Effects of Exercise: Cardiovascular, Metabolic, and Skeletal Muscle Adjustments. In Fish Physiology-Biology of Stress in Fish, Vol. 35 (eds. C. B. Schreck, L. Tort, A. P. Farrell and C. J. Brauner), San Diego, CA: Academic Press.

Romero, L. M., Dickens, M. J. and Cyr, N. E. (2009). The reactive scope model—a new model integrating homeostasis, allostasis, and stress. Horm. Behav. 55, 375-389.

Sadoul, B. and Vijayan, M. M. (2016). Stress and Growth. In Fish Physiology-Biology of Stress in Fish, Vol. 35 (eds. C. B. Schreck, L. Tort, A. P. Farrell and C. J. Brauner), San Diego, CA: Academic Press.

Sánchez, J. A., López-Olmeda, J. F., Blanco-Vives, B. and Sánchez-Vázquez, F. J. (2009). Effects of feeding schedule on locomotor activity rhythms and stress response in sea bream. Physiol. Behav. 98, 125-129.

Sandblom, E., Seth, H. and Sundh, H. (2013). Stress responses in Arctic char (Salvelinus alpinus L.) during hyperoxic carbon dioxide immobilization relevant to aquaculture. Aquaculture 414, 254-259.

Sandland, G. J. and Minchella, D. J. (2003). Cost of immune defense: an enigma wrapped in an environmental cloak? Trends Parasitol. 19, 571-574.

Sapolsky, R. M. (2005). The influence of social hierarchy on primate health. Science 308, 648-652.

Schjolden, J., Backström, T., Pulman, K. G. T., Pottinger, T. G. and Winberg, S. (2005). Divergence in behavioural responses to stress in two strains of rainbow trout (Oncorhynchus mykiss) with contrasting stress responsiveness. Horm. Behav. 48, 537-544.

Schreck, C. B. (1972). Steroid assays and their usefulness in fisheries research. Proc. Annu. Conf. Southeast Assoc. Fish Wildl. Agencies 26, 649-652.

Schreck, C. B. (1981). Stress and compensation in teleostean fishes: response to social and physical factors. In Stress and Fish (ed. A. D. Pickering), pp. 295-321. London: Academic Press.

Schreck, C. B. (1992). Glucocorticoids: metabolism, growth, and development. In The Endocrinology of Growth, Development and Metabolism in Vertebrates (eds. M. P. Schreibman, C. G. Scanes and P. K. T. Pang), pp. 367-392. New York, NY: Academic Press.

Schreck, C. B. (1996). Immunomodulation: endogenous factors. In Fish Physiology-The Fish Immune System: Organism, Pathogen, and Environment, Vol. 15 (eds. G. Iwama and T. Nakanishi), pp. 311-337. New York, NY: Academic Press.

Schreck, C. B. (2000). Accumulation and long-term effects of stress in fish. In The

Biology of Animal Stress: Basic Principles and Implications for Animal Welfare (eds. G. P. Moberg and J. A. Mench), pp. 147-158. Wallingford: CAB International.

Schreck, C. B. (2010). Stress and reproduction: the roles of allostasis and hormesis. Gen. Comp. Endocrinol. 165, 549-556.

Schreck, C. B. and Li, H. W. (1991). Performance capacity of fish: stress and water quality. In Aquaculture and Water Quality (eds. D. E. Brune and J. R. Tomasso), pp. 21-29. Baton Rouge: World Aquaculture Society, Advances in World Aquaculture 3.

Schreck, C. B. and Maule, A. G. (2001). Are the endocrine and immune systems really the same thing? In International Symposium on Comparative Endorinology (eds. H. J. T. Goos, R. K. Rostogi, H. Vaudry and R. Pierantoni), pp. 351-357. Naples: Monduzzi Editore.

Schreck, C. B., Jonsson, L., Feist, G. and Reno, P. (1995). Conditioning improves performance of juvenile chinook salmon, Oncorhynchus tshawytscha, to transportation stress. Aquaculture 135, 99-110.

Schreck, C. B., Contreras-Sanchez, W. and Fitzpatrick, M. S. (2001). Effects of stress on fish reproduction, gamete quality, and progeny. Aquaculture 197, 3-24.

Schulte, P. M. (2014). What is environmental stress? Insights from fish living in a variable environment. J. Exp. Biol. 217, 23-34.

Selye, H. (1950). Stress and the general adaptation syndrome. Br. Med. J. 1, 1383-1392.

Selye, H. (1973). Homeostasis and heterostasis. Perspect. Biol. Med. 441-445.

Selye, H. (1976). Stress in Health and Disease. Boston: Butterworth.

Sheldon, B. C. and Verhulst, S. (1996). Ecological immunology: costly parasite defenses and trade-offs in evolutionary ecology. Trends Ecol. Evol. 11, 317-321.

Sneddon, L. U., Wolfenden, D. C. C. and Thomson, J. S. (2016). Stress Management and Welfare. In Fish Physiology-Biology of Stress in Fish, Vol. 35 (eds. C. B. Schreck, L. Tort, A. P. Farrell and C. J. Brauner), San Diego, CA: Academic Press.

Sorensen, C., Johansen, I. B. and Øverli, Ø. (2013). Neural plasticity and stress coping in teleost fishes. Gen. Comp. Endocrinol. 181, 25-34.

Spagnoli, S., Lawrence, C. and Kent, M. L. (2016). Stress in Fish as Model Organisms. In Fish Physiology-Biology of Stress in Fish, Vol. 35 (eds. C. B. Schreck, L. Tort, A. P. Farrell and C. J. Brauner), San Diego, CA: Academic Press.

Sterling, P. and Eyer, J. (1988). Allostasis: a new paradigm to explain arousal pathology. In Handbook of Life Stress, Cognition and Health (eds. S. Fisher and J. Reason), pp. 629-649. New York, NY: John Wiley & Sons.

Stevens, E. D. (1972). Change in body weight caused by handling and exercise. J. Fish. Res. Board Can. 29, 202-203.

Stewart, H. A., Noakes, D. L. G., Cogliati, K. M., Peterson, J. T., Iverson, M. H. and Schreck, C. B. (2016). Salinity effects on plasma ion levels, cortisol, and osmolality in Chinook Salmon following lethal sampling. Comp. Biochem. Physiol. A Mol. Integr. Physiol. 192, 38-43.

Stitt, B. C., Burness, G., Burgomaster, K. A., Currie, S., McDermid, J. L. and Wilson, C. C. (2014). Intraspecific ariation in thermal tolerance and acclimation capacity in brook trout (Salvelinus fontinalis): physiological implications for climate change. Physiol. Biochem. Zool. 87, 15-29.

Sumpter, J. P. (1997). The endocrinology of stress. In Fish Stress and Health in Aquaculture (eds. G. K. Iwama, A. D. Pickering, J. P. Sumpter and C. B. Schreck), pp. 95-118. Cambridge: Cambridge University Press.

Suzumoto, B. K., Schreck, C. B. and McIntyre, J. D. (1977). Relative resistances of three transferrin genotypes of coho salmon (Oncorhynchus kisutch) and their hematological responses to bacterial kidney disease. J. Fish. Res. Board Can. 34, 1-8.

Takei, Y. and Hwang, P. -P. (2016). Homeostatic Responses to Osmotic Stress. In Fish Physiology-Biology of Stress in Fish, Vol. 35 (eds. C. B. Schreck, L. Tort, A. P. Farrell and C. J. Brauner), San Diego, CA: Academic Press.

Templeman, N. M., LeBlanc, S., Perry, S. F. and Currie, S. (2014). Linking physiological and cellular responses to thermal stress: β-adrenergic blockade reduces the heat shock response in fish. J. Comp. Physiol. B 184, 719-728.

Tort, L. (2011). Stress and immune modulation in fish. Dev. Comp. Immunol. 35, 1366-1375.

Tortora, G. J. and Grabowski, R. R. (2000). Principles of Anatomy and Physiology. New York, NY: John Wiley & Sons.

Verburg-van Kemenade, B. M. L., Stolte, E. H., Metz, J. R. and Chadzinska, M. (2009). Neuroendocrine-immune interactions in teleost fish. In Fish Physiology-Fish Neuroendocrinology, Vol. 28 (eds. N. J. Bernier, G. van der Kraak, A. P. Farrell and C. J. Brauner), pp. 313-364. London: Academic Press.

Waddington, C. H. (1977). Tools for Thought: How to Understand and Apply the Latest Scientific Techniques of Problem Solving. New York, NY: Basic Books.

Webb, M. A. H., Allert, J. A., Kappenman, K. M., Marcos, J., Feist, G. W., Schreck, C. B., et al. (2007). Identification of plasma glucocorticoids in pallid sturgeon in response to stress. Gen. Comp. Endocrinol. 154, 98-104.

Wedemeyer, G. A. and McLeay, D. J. (1981). Methods for determining the tolerance of fishes to environmental stressors. In Stress and Fish (ed. A. D. Pickering), pp. 295-321. London: Academic Press.

Wendelaar Bonga, S. E. (1977). The stress response in fish. Physiol. Rev. 77, 591-625.

Winberg, S., Höglund, E. and Øverli, Ø. (2016). Variation in the Neuroendocrine Stress Response. In Fish Physiology-Biology of Stress in Fish, Vol. 35 (eds. C. B. Schreck, L. Tort, A. P. Farrell and C. J. Brauner), San Diego, CA: Academic Press.

Yada, T. and Nakanishi, T. (2002). Interaction between endocrine and immune systems in fish. Int. Rev. Cytol. 220, 35-92. Yada, T. and Tort, L. (2016). Stress and Disease Resistance: Immune System and Immunoendocrine Interactions. In Fish Physiology-Biology of Stress in Fish, Vol. 35 (eds. C. B. Schreck, L. Tort, A. P. Farrell and C. J. Brauner), San Diego, CA: Academic Press.

Zuberi, A., Brown, C. and Ali, S. (2014). Effect of confinement on water-borne and whole body cortisol in wild and captive-reared rainbowfish (Melanotaenia duboulayi). Int. J. Agric. Biol. 16, 183-188.

第 2 章　神经内分泌应激反应的变动

在种间和种内，各个个体应对它们的应激反应和环境变动在程度上和时间上都是不同的。神经内分泌和生理学作用机理介导这种灵活性，并且大体上在整个脊椎动物亚门中是保守的，在这里特别就硬骨鱼类的适应性变动进行综述。遗传的和环境的因子影响，例如社会环境和营养物的有效性能够用来揭示基础的邻近作用机理。在应激反应性、行为和生命史特征之间的联系亦可以阐明对自然选择的认识和怎样保持个体的变动。

2.1　导　言

行为的、自主的和神经内分泌的应激反应全都表现出明显的种间和种内的差异（Barton，Iwama，1991；Barton，2002；Cockrem，2013）。种类的差别似乎部分地和生境与生活方式的不同有关（Wendelaar Bonga，1997；Mommsen，1999；Barton，2002；Pankhurst，2011）。然而，本章将着重于同一种类的不同个体鱼之间应激反应的差别。

这种差别可能和质量（如不同的个体表现出不同的行为和神经内分泌的型式）以及量度（如不同个体受到同一应激物作用后血浆皮质醇和儿茶酚胺不同程度地升高）有关。此外，一个动物的应激反应通常都和它的健康、营养以及一般的生理状态有关。再者，先前的经历、年龄、性别等因素对个体应激反应型式的形成亦是重要的。神经元的和神经内分泌系统是高度可塑性的，而应激反应的经历可以影响到对应激物的后续反应，这个过程称为适应/驯化（habituation）和致敏/敏化（sensitization）。通常，这些效应在生命周期的早期或者一定的生活时期，如溯河洄游鲑鳟鱼类进入二龄鲑时期，是最为明显的（Schreck，Tort，2016；见本书第 1 章）。

交感神经-嗜铬组织（SC）轴和下丘脑-脑垂体-肾间腺（HPI）轴的激活通常用来作为应激的指标（Wendelarr Bonga，1997；Mommsen，1999；Barton，2002）。然而，重要的是要注意到这些系统亦能为其他各种刺激如性行为和社会性取胜引起的反应所激活，正常来说，它们不能列入应激的行列中。糖皮质类固醇以及交感神经紧张和血液循环的儿茶酚胺对于能量转移和代谢调控都是重要的。所以，需要能量的行为，甚至不是由厌恶刺激所诱导，亦会激活这些系统。目前我们已经非常清楚，并不是一个厌恶刺激的物理性质导致应激，而是这个刺激能够预知和调控的程度在起作用。对于哺乳类动物，可以识别的认知功能对引起应激反应的作用状态是重要的（Paul 等，2005；Mendl 等，2010）。最近的研究表明，这也适用于硬骨鱼类和其他非哺乳类的脊椎动物（Oliveira，2013）。

此外，应激反应的型式是和行为特征相联系的，形成所谓的应对方式（coping style）（Koolhaas 等，1999）。鱼类对不同的应激物反应而显示大不相同的应激反应方式以及不同的行为型式，包括攻击水平、新奇事物恐怖症（neophobia）和一般运动能力以及建立行为常规的趋势（Øverli 等，2007）。应激的应对方式部分是由可遗传的因素调控的，但个别的应对方式亦受到先前经历的影响。此外，应对方式可能亦和生活史有关，例如开始内源性摄食的时间、性成熟的年龄、洄游的年龄等（Mommsen，1999；Wendelaar Bonga，1997）。

在本章，我们综述鱼类应激反应的个体发育和早先应激对后续应激反应型式影响的研究结果。我们特别着重于鱼类早前生活状况对后续应激反应的影响以及通过什么样的作用机理使这种影响发挥作用。我们亦论述硬骨鱼类致敏/敏化和适应/驯化作用机理的研究以及鱼类趋异的应激应对策略的研究。此外，对营养因子、社会状态和性成熟对应激影响的现有见识做了综述。

2.2　硬骨鱼类应激反应的个体发育

硬骨鱼类的不同种类之间从孵化到发育成熟期的变化很大。海洋的浮游性鱼类，如鳕科、鲽科和鲷科的鱼类产出很小的卵，孵出的幼鱼具有未发育成熟的器官。这些鱼类的 HPI 轴直到孵出后数天才出现（Tanaka 等，1995）。相反，鲑科鱼类产出大的卵，孵出的幼鱼已经处于发育较好的阶段并且具有功能的 HPI 轴（Barry 等，1995a）。

相关研究人员对斑马鱼（*Danio rerio*）HPI 轴的个体发育已经做了最为详尽的研究。采用整体原位杂交，Chandrasekar 等（2007）检测到斑马鱼受精后（hpf）24 小时视前区的皮质类固醇释放激素（CRH）阳性细胞。采用反转录多聚酶链式反应（RT-PCR），他获得一些证据表明受精后 1～5 天 CRH 的表达上调。然而，还不清楚这种现象是否反映了在这期间由于和应激物接触而使敏感性增强。

在受精后 24 小时，在正在发育的斑马鱼脑垂体中能检测到肾上腺皮质类固醇激素（ACTH）（Herzog 等，2003；Liu 等，2003）。ACTH 是由前体阿黑皮素原（POMC）衍生而来，POMC 亦产生其他的生物活性肽，如 α-黑色素细胞刺激激素（α-MSH）（Wendelaar Bonga，1997）。在受精后 48～64 小时能观察到 ACTH 远侧部和神经中间叶（α-MSH）的组织形成（Liu 等，2003）。然而，尽管 HPI 轴在孵化时（受精后 48 小时）已经有功能，但并不对应激物起反应而增加皮质醇的产生和释放。实际上，编码参与皮质醇生物合成的酶的基因以及肾间腺的黑皮质素受体 2（MC2R）基因都已经在孵化时表达，这时内源的皮质醇开始升高，但 HPI 轴没有显示出对应激物产生反应，直到孵化后 2 天，才表现为提高皮质醇的释放（Alsop，Vijayan，2009）。在胚胎开始发育期间，皮质醇浓度下降，但在孵化前开始增加（Alsop，Vijayan，2008）。基础的皮质醇水平开始增加时很可能就是生物合成内源皮质醇的开始。在发育早期观察到的皮质醇可能是来源于母体的。

HPI 轴开始对应激反应产生的时间存在着很大的种间差异。有些鱼类，如鲤鱼（*Cyprinus carpio*），HPI 轴在孵化时就立即对应激产生反应（Stouthart，1998；Flik 等，2002）；而其他鱼类，如黄鲈（*Perca flavescens*）、虹鳟（*Oncorhynchus mykiss*）、大鳞大麻哈鱼（*Oncorhynchurs tshawytscha*）（Feist，Schreck，2002）以及莫桑比克罗非鱼（*Oreochromis mossambicus*），应激并不会引起皮质醇分泌增加，一直到孵化后几天或几星期（Barry 等，1995a；Jentoft 等，2002；Pepels，Balm，2004）。在对虹鳟的研究中发现，受精后四星期基础的皮质醇水平升高，但是，直到孵化后两星期才观察到由应激诱导的皮质醇浓度升高（Barry 等，1995a）。还有，在孵化前虹鳟肾间腺细胞（或肾上腺同系物）已经能够对 ACTH 的刺激产生反应（Barry 等，1995b）。因而，可能下丘脑因子是有限的，或者，对应激感知所必需的感觉输入未能到达下丘脑（Alsop，Vijayan，2009）。

至少在斑马鱼胚胎发育时期的皮质醇信号主要是通过盐皮质素受体（MRs）介导的（Alsop，Vijayan，2009）。在斑马鱼整个发育期间，糖皮质类固醇受体（GR）的转录体减少而 MR mRNA 增加（Alsop，Vijayan，2008）。皮质醇以高亲和力和 MR 受体结合，而在一些种类，其他的 MR 配体，例如哺乳类的醛固酮，MR 通常和11β-羟基类固醇脱氧酶（11β-HSD2）共表达；而这种酶的作用是将皮质醇转化为皮质酮，消除了和 MR 的结合。然而，在斑马鱼胚胎中，11β-HSD2 表达很低（Alsop，Vijayan，2008）。值得注意的是，尽管大多数鱼类对应激的反应都不是皮质醇分泌升高，直到孵化后的某个时间，但早期受到的应激可以影响到生命中稍后显现的应激反应（Auperin，Geslin，2008）。例如，虹鳟在早期发育阶段受到短暂的应激物刺激（1 分钟在 0℃ 水中，1 分钟在水外），5 月龄的幼鱼表现为应激诱导的血浆皮质醇水平降低。

在个体发育早期经历的应激物对成年期神经内分泌应激反应的影响已经在哺乳类动物中进行广泛的研究。例如，出生前的应激反应影响脑的单胺能神经传递（Hayashi 等，1998；Ishiwata 等，2005）和下丘脑-脑垂体-肾上腺（HPA）轴的活性（见 Glover 等，2010 综述），而不良的亲代抚育会引起连贯的行为和神经内分泌的异常。此外，在断乳后与青春期之间处于社会隔离状态已经证明对鼠类神经内分泌应激反应有持久的影响，同时亦影响脑的结构和神经传递（Fone，Porkess，2008；Lukkes 等，2009）。这些影响当中有些还和脑的 5-羟色胺(5-HT)系统活性的变化有关。

鱼类从刚产出卵受精后到筑巢抚幼的亲本抚育水平有很大变化（Smith，Wootton，1995）。然而，至今有关应激物在鱼类个体发育早期影响到神经内分泌应激反应的研究成果，据我们所知，只局限于产出它们的子代而没有表现亲本抚育行为的种类。在 Auperin 和 Geslin（2008）的研究中，虹鳟胚胎（眼形成胚胎期）和刚孵出的幼苗受到冷刺激（0℃）并结合在空气中暴露的处理后，对孵化后 5 个月成为幼鱼时进行试验，它们对应激物的反应表现为皮质醇水平下降。HPI 轴反应性的这种减弱在用 ACTH 处理后并不变得明显，这表明早期应激对脑垂体上游的机制发挥了作用。

母体的应激可以影响子代的行为和应激反应，有证据表明母体的皮质醇是超越世代转移相关的环境信息作用机理的一部分。例如，雌性三刺鱼（*Gasterosteus aculeatus*）血浆和卵中皮质醇水平增加与较高的捕食压力（Predator pressure）相联系。而且，母体这种由捕食者诱导的皮质醇浓度升高和子代较明显的回避捕食者行为相关联（Giesing 等，2011）。同样，在大西洋鲑鱼（*Salmo salar*）中，母体的皮质醇处理能诱导子代对新奇事物产生较为明显的行为抑制作用（Eriksen 等，2006；Espmark 等，2008）。此外，卵经过皮质醇处理能对虹鳟幼鱼的学习（Sloman，2010）以及对应激的皮质醇反应（Auperin，Geslin，2008）产生影响。这表明亲本受到的应激能产生后成的（外遗传的）效应，并进而改变子代的应激应对能力。有关参与亲本皮质醇这种功能影响的作用机理信息才发表不久。参与基因转录的外遗传沉默因子，包括多个 DNA-甲基转移酶和两个组蛋白，能在受到应激作用的三刺鱼母体的胚胎中差异表达（Mommer，Bell，2014）。

除了前面讲述的激素功能之外，亲本的膳食可以影响子代的应激反应。例如，膳食中的脂肪酸组成能影响 HPI 轴的反应性（见第 2.6.2 节的讨论）和肌肉的脂肪酸组成。此外，肌肉的脂肪酸组成已证明能在卵母细胞中得到反应（Garrido 等，2007），表明亲本膳食能够影响子代 HPI 轴早期的个体发育。

2.3　应激的神经元底物和应激反应的变动

鱼类和其他脊椎动物一样，自主的、神经内分泌的和行为的应激反应是紧密协调一致的，但是，最近的研究结果已经清楚地显示神经作用机理的参与，而我们对它在这整个协调性中的作用还了解得不多。SC 轴和 HPI 轴的神经内分泌调控已经被广泛评述（Gorissen，Flik，2016；见本书第 3 章）（Wendelaar Bonga，1997；Mommsen，1999），而在本节中我们将着重阐述这些作用机理是怎样和应激反应中的个体差异相联系的。

血浆皮质醇水平的持久升高已被证明能抑制 HPI 轴的反应性（Øverli 等，1999a，b；Jeffrey 等，2014）。所以，对急性应激物反应的慢性应激的鱼和没有受到应激的对照鱼相比，具有较低的血浆皮质醇浓度（Barton 等，2005）。此外，在慢性应激的鱼中，应激诱导皮质醇反应的时间进程会显得比较慢些。慢性应激对于 HPI 轴功能的影响，对于社会性从属地位的鱼是明显的。例如，Øverli 等（1999a，b）报道社会性从属地位的北极红点鲑（*Salvelinus alpinus*）基础的血浆皮质醇水平升高，但与对一次急性拉网的应激反应和处于社会性优势地位的北极红点鲑相比，表现出较小而缓慢的皮质醇反应。Jeffrey 等（2014）在虹鳟的相关研究中得到同样的结果。然而，还不清楚 HPI 轴是处在什么样的水平以及这些差别是通过什么样的作用机理产生的。

HPI 轴是由一系列肽类作用于该轴的不同水平而调控的。皮质醇是在 ACTH 的刺激性调控下由硬骨鱼类的肾间腺细胞分泌出来，而 ACTH 是由脑垂体远侧部的促皮质激素细胞分泌的。来自下丘脑视前区（POA）的促肾上腺皮质激素释放因子（CRF）刺激 ACTH 的分泌活动。皮质醇的主要促分泌素 ACTH 由 POMC 产生；这种前激素原

亦产生 α-MSH 和 β-内啡呔，它们和 ACTH 一起作用以调控皮质醇的分泌。此外，几种其他的激素和神经内分泌因子，如硬骨鱼紧张肽 I（UI）、黑色素浓集激素（MCH）、精氨酸加压催产素（AVT）、硬骨鱼类催产素（IST）和神经肽 Y（NPY），对 HPI 轴的调控都起着重要作用。UI、AVT、IST 和 NPY 都曾被报道刺激脑垂体 ACTH 的释放，而 MCH 则显示出相反的作用（Wendelaar Bonga，1997）。UI 肽，由下丘脑的神经元以及由硬骨鱼的尾神经内分泌器官尾垂体产生，对脑垂体释放 ACTH 具有很强的刺激作用（Gorissen，Flik，2016；见本书第 3 章）。

在肾间腺组织的水平对 ACTH 的敏感性是不同的。例如，保持在高密度蓄养状态中的美洲红点鲑（*Salmo fontinalis*），其肾间腺组织对 ACTH 的敏感性要低于处在较低密度蓄养鱼的肾间腺组织，表明了慢性应激对 ACTH-诱导皮质醇分泌活动的影响（Vijayan，Leatherland，1990）。同样，禁闭应激使虹鳟的肾间腺组织对 ACTH 的敏感性降低（Balm，Pottinger，1995）。Sloman 等（2002）证明，当肾间腺细胞在原位受到 ACTH 控制时，从社会性从属地位的虹鳟中得到的细胞要比从优势地位虹鳟中得到的细胞表现更低的皮质醇分泌速率。此外，虹鳟选择性产生低水平应激后血浆皮质醇（低反应性，LR）的肾间腺对 ACTH 的敏感性要较低于产生高水平应激后血浆皮质醇（高反应性，HR）的肾间腺（Pottinger，Carrick，2001a，b）。一个合理的解释是，肾间腺对 ACTH 敏感性降低是由于黑皮质素Ⅱ（MCRⅡ，ACTH 受体）表达的下调和/或这些受体的脱敏作用。此外，血浆皮质醇水平升高可以加强 GR 的表达，而这就会增强皮质醇对肾间腺组织的超短负反馈作用。

慢性应激亦能影响脑垂体 ACTH 的合成与分泌。Winberg 和 Lepage（1998）曾报道虹鳟的社会性从属地位能引起脑垂体 POMC mRNA 表达的上调。然而，POMC mRNA 增加得最明显的部位是神经间叶，POMC 在这里加工为 α-MSH。

性类固醇激素亦能影响肾间腺对 ACTH 的敏感性，这提供了一个由性成熟引起的性别差异与调整应激反应的作用机理。在对虹鳟的研究中，Young 等（1996）曾报道 11-酮基睾酮（11-KT）抑制肾间腺 ACTH 分泌。

在下丘脑水平，社会等级影响 CRF 表达，通过这种作用亦能够介导社会性因素对应激反应的影响。但是，社会性因素诱导 CRF 表达发生变化的时间进程是复杂的。Doyon 等（2003）曾报道处于从属地位的虹鳟在 72 小时后视前区的 CRF 表达升高。Bernier 等（2008）发现处于从属地位的虹鳟 CRF 在 POA 的表达升高是暂时的，因为它们在社会性相互作用的 8 小时后升高，而 24 小时后就没有升高。此外，Jeffrey 等（2012）曾报道社会性相互作用 5 天后，在端脑或 POA、CRF 或 CRF 结合蛋白的表达，在处于优势地位的和处于从属地位的虹鳟之间并没有差别。

2.4　趋异的应激应付方式、动物的个性和行为综合征

尽管所有其他的因素如年龄和先前的经历都保持恒定，但在鱼类任何一个群体内的个体，它们对应激刺激的行为型式和生理反应都是有差别的。这种类型的差异已经越来越受到动物研究人员和人体生物医学研究人员的重视（Koolhaas 等，1999，2010；Korte 等，2005；Øverli 等，2007；Carere 等，2010；Coppens 等，2010；Cockrem，2013）。尽管在整个脊椎动物亚门很大程度上都是保守的作用机理和型式，但是，对鱼类的评论性研究是本节的重点。作为导言，我们简要地讲述一下名词。动物应对环境干扰的个体差异曾采用许多名词来描述，如行为综合征（Sih 等，2004a，b）、气质（temperament）（Francis，1996；Clark，Boinski，1995；Reale 等，2007）、动物品格（Gosling，2001）或应对方式（Koolhaas 等，1999，2010）。所有这些名词最终都是用来描述在特征（行为的、生理的，或二者）和健康之间的一贯联系。这种联系广泛分布在动物界，并曾经在无脊椎动物（Reichert，Hedrick，1993；Sinn 等，2006；Wilson，Krause，2012；Kralj-Fišer，Schuett，2014）、哺乳类（Réale 等，2000，2009；Boon 等，2007）、鸟类（Dingemanse 等，2004；Both 等，2005；David 等，2011）、爬行类（Stapley，Keogh，2005）和鱼类（如 Smith，Blumstein，2010；Wilson 等，2010）等的相关研究中进行描述。

Andrew Sih 和其合作者创造名词“行为综合征”，用来描述一组相联系的行为，反映超越多种状态的个体之间的连贯性（Sih 等，2004a，b）。如果在行为之间相同的联系出现在不同的生命史时期，行为综合征是稳定的（Bell，Stamps，2004）。重要的是，所有的行为都以神经的和神经内分泌的过程为先导，所以，行为的连贯性变化在某种程度上是衍生于神经内分泌活性的变化。Koolhass 等（1999）提出名词“应激应对方式”，用来描述对挑战性的行为和生理反应的互相联系的组合。这些作者把应激应对方式界定为“一组紧密联系的行为和生理的应激反应，它们在时间方面是连贯的，并表现出一定的个体群特征”。通常，不同的应对方式可以列为反应前的（proactive）和反应中的，每个都由一组不同行为与生理特征的反应所识别。反应前的应激应对方式的行为特征主要是相当于 Cannon（1929）原先描述的战斗-追逐反应，也就是高水平的活动和攻击、较大的冒险和大胆鲁莽，以及表示主动预防或操纵应激刺激的行为。处于反应前的个体比较倾向于常规-基础而灵活性低的行为，较多地依赖于先前的经历而不是当时实际的周围环境状况（Benus 等，1990；Bolhuis 等，2004；Ruiz-Gomez 等，2011）。反应中的应对特征是回避、不活动、低水平的攻击、胆怯受惊、较高的行为灵活性，动物能较好地感知周围环境的变化并产生相应的反应。这些行为特征的差别是和生理上的差别紧密联系的，反应前的个体，典型性的表现较高的交感神经反应性和儿茶酚胺释放；而反应中的个体表现较高的应激后血浆皮质醇或皮质酮水平（Koolhaas 等，1999；Øverli 等，2005）。

2.4.1 对比的应激应对方式的保守生理学

趋异的应对方式的个体变异指标在鱼类和哺乳类之间是相当保守的（Øverli 等，2005，2007；Silva 等，2010；Martins 等，2011a，b；Castanheira 等，2013；Sorensen 等，2013；Tudorache 等，2013）。趋异的皮质类固醇分泌活动是反应前的和反应中的应对方式的主要特征之一，反应前的个体比反应中的个体呈现更低的应激后血浆浓度。硬骨鱼类的生理应激反应和哺乳类以及其他陆栖脊椎动物的相似（Wendelaar Bonga，1999），表明它们在进化上是保守的。

在本节中，我们饶有兴趣地分析虹鳟在选择趋异的应激后皮质醇反应性（HR 和 LR 虹鳟）中的行为差别，例如，降低食欲和增强运动，这和在其他非哺乳类脊椎动物相关研究中报道的皮质醇的作用一致（Cash，Halberton，1999；Gregory，Wood，1999；Øverli 等，2002a）。选择育种证明皮质醇应激反应量度的遗传力能够产生鳟鱼的 HR 和 LR 品系（Pottinger，Carrick，1999）。在鲑鳟鱼模式系统中发现行为的、生理的和神经生物学的反应型式通常都和在哺乳类反应前的和反应中所发现的应对方式相一致（表 2.1），但对特异的环境因素和经历的敏感性，如营养状态，已经得到证实（Ruiz-Gomez 等，2008）。生理的-行为的相互的邻近依赖性特征亦值得关注，一些研究表明野生鲑鳟鱼类的 HPI 轴反应性和行为型式之间并没有联系（Brelin 等，2008；Thornqvist 等，2015）。然而，许多研究报道鳟鱼 HR 和 LR 品系有不同的生理与行为型式，它们反映了对反应中的和反应前选择的应激应对方式（Schjolden 等，2005，2006a），类似于在其他动物类群中所鉴别的。Pottinger 和 Carrick（2001a，b）开始检测和幼鱼为社会优势地位而进行双重斗争结果相关的虹鳟 HR 和 LR 品系之间的行为差别。虹鳟幼鱼和其他占优势地位的溪流栖息的鲑鳟鱼类一样，是高度区域性动物。当引入到一个地区时，它们就会采取对抗的行为，直到建立一个优势等级（dominance hierarchy）（Jonsson 等，1998；Winberg，Lepage，1998；Øverli 等，1999a，b）。Pottinger 和 Carrick（2001a，b）报道 LR 鱼在大多数的 HR-LR 配对中占有优势社会地位，表明了竞争能力和应激反应性之间的联系。和 HR 品系的鱼相比较，LR 品系的鱼是典型的冒险家，转移到一个新的环境中能较迅速地重新恢复摄食（Øverli 等，2002b；Ruiz-Gomez 等，2008）。Øverli 等（2002b）还报道了一个行为特点：当两个品系的鱼增强它们对地盘入侵者反应的活动性时，HR 品系鳟鱼的活动能力要明显强于 LR 品系鳟鱼。

表 2.1 鲑鳟鱼类模式种反应前和反应中的表现型生理应激反应的和行为的型式

性状	反应前	反应中	模式种类	参考文献
皮质醇基础	×	×	HR-LR 虹鳟	Pottinger，Carrick（1999）
皮质醇应激反应	低	高	HR-LR 虹鳟	Pottinger，Carrick（1999）
	低	高	虹鳟养殖品系	Øverli 等（2004）
	低	高	大西洋鲑养殖品系	Kittilsen 等（2001）

续表 2.1

性状	反应前	反应中	模式种类	参考文献
	低	高	北极红点鲑养殖品系	Backstrom 等（2014）
交感神经活性	高	低	HR-LR 虹鳟	Schjolden 等（2006a，b）
	高	低	虹鳟养殖品系	Van Rasij 等（1996）
	高	低	褐鳟野生型	Brelin 等（2005）
社会优势	优势者	从属者	HR-LR 虹鳟	Pottinger 和 Carrick（2001a，b）
	优势者	从属者	虹鳟养殖品系	Øverli 等（2004）
	从属者	优势者	HR-LR 虹鳟	Ruiz-Gomez 等（2008）
运动，基础	×	×	HR-LR 虹鳟	Øverli 等（2002a，b）
运动，急性应激	低	高	HR-LR 虹鳟	Øverli 等（2002a，b）
	低	高	大西洋鲑养殖品系	Kittilsen 等（2009）
在新环境中摄食	快	慢	HR-LR 虹鳟	Øverli 等（2002a，b）
	快	慢	虹鳟养殖品系	Øverli 等（2004）
	快	慢	大西洋鲑养殖品系	Kittilsen 等（2009）
	慢	快	HR-LR 虹鳟	Ruiz-Gomez 等（2008）
挣扎，网约束	高	低	北极红点鲑养殖品系	Magnhagen 等（2015）
常规形成	高	低	HR-LR 虹鳟	Ruiz-Gomez 等（2011）
条件反应的保存	高	低	HR-LR 虹鳟	Moreira 等（2004）

注：性状表示在研究之间明显的一致性，以及表示其他的模式系统（如，应用于欧洲鲈鱼的新建立的约束试验）。但要注意，在一项研究中（Ruiz-Gomez，2008）邻近依赖性效应是明显的，应激和饥饿后的补偿性生长是和对应激敏感的虹鳟 HR 品系的反应前行为相联系的。生活史性状、能量代谢，以及其他一些行为型式，如在新对象试验中的冒险和侵袭的潜在因素，表现出和应对方式不协调的和大量未能澄清的联系。

HR-LR 品系以皮质醇对应激反应的连贯性趋异进行选育，亦导致其他生理特征的趋异。和应激反应性的选择相关的是，在 HR 和 LR 虹鳟的成年雌鱼中开始研究行为和脑的单胺活性（Øverli 等，2001）。脑的单胺神经递质多巴胺（DA）、去甲肾上腺素（NE）和 5-羟色胺（5-HF）都深度参与硬骨鱼类和哺乳类行为的与生理的应激反应的调控和整合（Blanchard 等，1993；Stanford，Salmon，1993；Winberg，Nilsson，1993；Winberg 等，1997，2001；Øverli 等，1998，1999a，b；Höglund 等，2001，2002a，b，2005；Clements 等，2003；Larson 等，2003；Lepage 等，2005；Carpenter 等，2007；Lillesaar，2011；Medeiros，McDonald，2013；Chabbi，Ganesh，2015）。显然，鳟鱼的 HR 品系在应激的反应中增加 5-HT（脑干）、DA（脑干）和 NE（端脑，视顶盖）的浓度，而 LR 品系的鳟鱼并不这样（Øverli 等，2001）。HR 品系鳟鱼在禁闭的应激之后，脑干和视顶盖的单胺代谢物浓度升高，但 LR 品系鳟鱼没有这样。单胺和它们代谢物浓度同时升高表明这些神经递质的合成和代谢在 HR 品系鳟鱼经受应激后都有升高。然而，在下丘脑观察到一个不同的型式，即 LR 品系鳟鱼中表现出较高的羟色胺代谢物 5-羟基吲哚乙酸（5-HIAA）水平和去甲肾上腺素代谢物3-甲氧基-4-羟基苯乙醇水平。两个品系之间的端脑亦有所不同，在LR品系鱼中观察到5-HIAA/5-HT较高的基线比率，而在应激的LR和HR品系鳟鱼之间没有差别。神经递质代谢的这些差别似乎反映了这两种品系鱼功能递质释放的真实差异，因为在LR和HR品系的鳟鱼之间脑单胺氧化酶（MAO）活性没有明显区别（Schjolden等，2006b）。这些研究结果清楚地说明对虹鳟应激反应的选育亦和脑单胺能系统功能的变化相联系。

在虹鳟HR-LR品系的模式中，神经肽类如CRF似乎亦参与决定行为的型式（Backstrom等，2011）。在大鳞大麻哈鱼幼鱼脑腔内注射CRF能诱导超强的活动，这是一种依赖于5-HTT激活而并发的效应（Clements等，2003）。对蝾螈（*Taricha granulosa*）给予CRF能使背中下丘脑的DA浓度增加（Lowry等，2001）。应激诱导DA浓度增加并在HR品系虹鳟的脑干和视顶盖中周转，是依据应激后皮质醇水平进行选育的HR和LR品系之间主要的神经化学差别之一（Øverli等，2001）。DA合成与释放增加可能是糖皮质激素浓度急性升高的一种效应（Barrot等，2001）。所以，还不清楚HR-LR多巴胺能系统的差别是激素动态的一种起因还是一种结果。

基因决定脑羟色胺系统的变异是和动物品格、气质以及一种情绪失调发展倾向的变异相联系的。MAO的多态性和羟色胺转运蛋白（5-HTT）基因或者启动区（Lesch等，1996；Caspi等，2003；Fernandez等，2003）是在人和其他哺乳类动物中最显著的例子，而这些基因的变体亦已经在鱼类中报道（Elipot等，2014）。5-HTT的表达由糖皮质激素调节，但5-HTT“短”型对这些激素的反应是减弱的，表明在5-HTT多态性和动物品格性状之间的功能转接，在某种程度上可能取决于和糖皮质激素的相互作用。据我们所知，这种影响脑单胺能信号系统的潜在作用机理还未曾在鱼类相关的应激应对方式的模式中进行研究。

除了膜受体介导行为的迅速作用之外（Moore，Orchinik，1994），在脊椎动物中有两种受体类型介导皮质类固醇激素的作用，即MR受体和GR受体。关于哺乳类动物，有大量文献资料论述皮质类固醇受体在脑中的作用。糖皮质类固醇和MRs结合主要定位在边缘的脑结构（如海马），其结合亲和力比广泛分布在脑中的GRs高10倍（De Kloet，2004）。MRs和GRs在边缘脑结构的高度表达反映了它们参与记忆、学习和一般警戒的调控作用。起初，一个反应前的模式通过MRs介导而参与维持紧张的HPA轴的反应性和神经元的兴奋性。和MR的激活不同，GR广泛的激活而介导神经元兴奋性水平的降低。在这方面，皮质类固醇对神经元的兴奋性起着双相的作用。

在硬骨鱼类中，皮质醇受体对糖皮质类固醇的亲和力不同，而皮质醇的作用是通过GR1、GR2和MR介导（Colombe等，2000；Bury，2003）。在虹鳟中，MRs对皮质醇要比GRs有更高的亲和力（Bury，Sturm，2007）。还有，GR两个不同的种内同源物（paralog）对糖皮质类固醇的敏感性亦有所不同（Stolte等，2006）。由于硬骨鱼类主要缺乏常规的MR配体醛固酮，皮质醇似乎是鱼类最重要的MR配体（Wendelaar Bonga，1997；Stolte等，2008），包括虹鳟（Colombe等，2000）。虹鳟的

MR 以较低于 GRs 的皮质醇水平被激活，表明虹鳟 MR 在脑内具有和哺乳类 MR 相似的作用（Sturm 等，2005）。设想参与应对方式的 MRs 和 GRs 的表达不同，Johansen 等（2011）定量分析虹鳟应激的和非应激的 LR 和 HR 品系在不同脑区 GR1、GR2 和 MR 的 mRNA 表达。在所有分析测定的脑区，LR 品系鱼的 MR 表达都要比 HR 品系鱼的高。所以，改变 GR-MR 的平衡是和应对方式的表现型的表达相联系，包括 LR 品系鱼降低兴奋性和增强记忆力的保存。然而，受体的转录型式受到皮质醇以时间的和器官特异性方式所产生作用的调节（Teles 等，2013），因而，还不能简单地确定哪些性状是起因，哪些性状是激素水平改变后的结果。

2.4.2　应激、神经塑性和应对方式

越来越清楚的是，应激反应中的个体差异受到脑功能的影响。由于物理的挑战决定应激反应的量度，可预知性与调控以及心理过程是和动物如何确定一个应激状态相联系的，在它们所预期的结果中，各种因素都同样重要。对刺激后果的期待是强有力的应激反应的认知调节剂，如果我们对趋异的认知能力进行选育，应激应对方式的差别就会呈现（Giorgi 等，2003；Aguilar 等，2004；Steimer，Driscoll，2005）。相反的方面亦是一样，就是选育能产生具有分歧认知能力的应激反应动物。采用 HR-LR 鳟鱼品系，Moreira 等（2004）证明 LR 鳟鱼品系比 HR 鳟鱼品系保持较长时间的条件反应。这个研究结果引出的问题是：应激后低的皮质醇产量是否和通常较好的学习与记忆性能相联系？但稍后的试验表明情况并非如此。代替的是，相对于常规状态，选育的品系似乎包含对环境变化反应的行为灵活性的一般性差别（Ruiz-Gomez 等，2011）。稍后的研究表明，两个不同的试验操作，离开先前认知的位置重新设置食物的场所和引进一个新的目标物，HR-LR 品系的鱼会产生不同的反应。在认知原来食物位置的比率方面没有差别，但是，当食物位置重新设置后，在改变它们寻找食物的行为方面，反应前的 LR 品系鱼要比反应中的 HR 品系鱼慢得多。相反，LR 品系鱼主要是不理会那个破坏 HR 品系鱼摄食的新目标物（Ruiz-Gomez 等，2011）。所以，在两种品系之间的主要差别似乎不在于对完全新环境特征的学习能力，而在于影响动物诱导反应的环境中较为敏感的变化程度，这就偏离已经认知的常规状况。这个结论和对猪、啮齿类和鸟类的研究结果一致，表明在学习能力方面并没有不同，差别在于反应前和反应中的个体之间的常规能力（Benus，1991；Bolhuis 等，2004）。

现在还不清楚在认知和反映血液循环糖皮质激素急性作用的应激应对方式之间，以及反映个体发育或脑功能内在差别和不同激素水平慢性接触的组织与结构作用之间的联系程度如何。换句话说，一个变化的认知功能是不是一个变化的应激反应的起因或者结果？在应激应对、动物品格、行为的灵活性与多样性和脑功能之间的复杂相互关系是 2007 年《脑、行为和进化》杂志特刊的主题。该刊物列举了一些可遗传的和明显一致的性状关联（Schjolden，Winberg，2007；Veenema，Neumann，2007），以及一些急速而可更换的反应（Burmeister，2007）。此外，在动物的生命史中对社会挑战敏感性的变化（Wommack，Deville，2007）以及在等级制度形成的早期过程对于

神经生物学的影响可能要比在日常社会相互作用中发生的事态更为重要（Sorensen 等，2007）。最后，生理学中先存的差异可以决定社会等级（Korzan，Summers，2007），因而，并不需要简单地去理清动物生命中特殊活动的起因和结果。

Sorensen 等（2011）发现在社会性应激的虹鳟脑中，增殖能力的下降是由皮质醇介导的，这是一种类似于皮质醇在硬骨鱼类前脑抑制细胞增殖的作用，如同在哺乳类的海马脑区一样。这些研究结果成为在鳟鱼的 HR 和 LR 品系之间研究在脑细胞增殖速率和神经发生以及神经塑性的其他方面是否存在差异的根据，因为在关键脑区的速率或者结构总量的趋异能够很好地说明在认知能力、行为和应激反应方面的明显差别，而不必考虑引起这种神经元塑性趋异的作用机理。

2.4.3 个性（individuality）的遗传基础

一些种类的选育研究表明，对应激的生理的和行为的反应通常都和一个相关的可遗传的型式相联系（van Oers 等，2005；Øverli 等，2005）。各种性状（包括品格的性状）的遗传力随着条件的变化而变化（Charmantier，Garant，2005；Dingemanse 等，2009）。换句话说，经历的变化通过遗传决定行为的型式，但基因亦决定一个个体可能要经历什么样的状态。进一步的复杂情况是，基因如何影响敏感的个体面对环境的变化（Ruiz-Gomez 等，2011）。

行为的基本遗传基础不仅在选育和克隆的研究中得到了确切证明，而且事实上在鱼类种群内部和种群之间都出现了行为的变化。例如，网纹花鳉（*Poecilia retiuclata*）的种群，和捕食性 Charachid 科和丽鱼科鱼类的分布区是重叠的。这种花鳉显示出广泛的从分布区不重叠的花鳉种群到这些捕食者的不同行为型式（Seghers，1974），较为胆怯并且表现较强的成群趋势。这可以解释为两个生境中捕食压力的差别所致，这种假设为三刺鱼种群的研究所支持，它证明了行为中可遗传的、适应性抗捕食者的变异（Huntingford 等，1994；Bell 等，2010）。

遗传的、外遗传的（后生的）和环境的因子相互作用而形成个体的神经内分泌和行为的方式，从而授予对付应激、疾病与环境变化的易变而脆弱的能力。选择性的行为综合征和应激应对方式是怎样进化并为自然选择所保持，还是悬而未决的问题。本文将陈述在应激反应中个体的变化反映在两种硬骨鱼类，即虹鳟和大西洋鲑能看得见的外表特征中（Kitlilsen 等，2009）。鲑鱼和鳟鱼的肤色不同，从接近纯白色到稠密的斑点，其黑色斑点由产生真黑色素的载色素细胞形成。在虹鳟中，对 HR-LR 品系规格中分歧的 HPI 轴反应性进行的选育曾引起皮层色素型式的变化，低皮质醇-反应的鱼，皮层都有较多的斑点。这个性状还不只局限于和选育的方式相联系，因为大西洋鲑养殖种群中，具有较多斑点的个体对应激表现为减弱的生理和行为反应（Kittilsen 等，2009）（图 2.1）。这些结果证明可遗传的行为-生理的以及和形态性状的联系对于可选择的应对方式是特异性的；而在应激后皮质醇的产生和真黑色素的色素形成之间的联系，实际上是后来出现在其他种类的报道中的（Almasi 等，2010）。这类研究可以阐明不同应对方式与行为综合征的进化，因为在不同环境中出现的表现型以及它们对选择压力的反应能够精确而容易地记载下来。

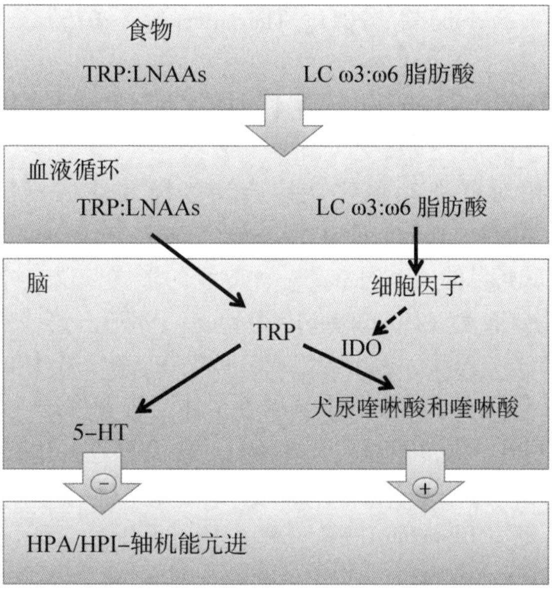

图 2.1　假设的在可供食用的脂肪酸、氨基酸和 HPA/HPI 轴活性之间相互作用的通道

注：色氨酸（TRP）和食物中其他大的中性氨基酸（LNAAs）之间的比率影响这些氨基酸的血浆浓度，它们转而影响 TRP 在脑部的可利用性。在长链 ω3 和 ω6 脂肪酸（LCω3：ω6）之间的食物比率影响这些脂肪酸的血浆浓度，进而影响细胞因子活性。这就可能对吲哚胺双加氧酶（IDO）的活性产生影响，而这种酶能把 TRP 转化为犬尿喹啉酸（kynurenic acid）和喹啉酸（quinolinicacid）。5-羟色胺（5-HT）曾表明能够阻抑由犬尿喹啉酸和喹啉酸引起的 HPA/HPI 轴机能亢进。虚线表示在鱼类中没有表示的通道。

2.4.4　应激的应对和生活史

对鲑鳟鱼类的研究已经认为应激的应对方式和生活史性状相联系，例如从鱼巢中刚孵化出的鱼苗，生长和进入二龄的稚鱼（Metcalfe，Thorpe，1992；Metcalfe 等，1995；Einum，Fleming，2000）。在这一科的鱼类中，雌鱼典型地把产出的卵埋压在河流底部砂石做成的巢内。孵化后，幼苗停留在砂石巢内，以卵黄为营养，直到它们出来开始保护领地并进行外源性摄食。这个过程的计时在同一个产卵巢内的个体当中可以有几个星期的变化（Mason，Chapman，1965；Brannas，1988），并且和行为性状的组成相关。早出来的鱼表现反应前的性状，如较为有侵占性行为、大胆冒失以及处于社会性优势地位（Metcalfe，Thorpe，1992；Metcatfe 等，1995）。相反，迟些出来的鱼通常胆怯，较少出现侵占行为，处于社会性从属地位（Metcalfe，Thorpe，1992；Metcatfe 等，1995），表现出一种反应应激应对方式的性状特征（Koolhass 等，1999）。这种从鱼巢中出来的计时和应激应对方式之间的关系为 Aberg-Andersson 等（2013）的研究所支持，他们证明虹鳟的 LR 品系鱼通常都要比 HR 品系鱼较早地从人造产卵巢中出来。但是，虽然最近的研究证明在早些和迟些从鱼巢中出来的鱼之间，在勇敢大胆（Vaz-Serrano 等，2011；Thornqvist 等，2015）和前脑基因表达（Thornqvist 等，2015）方面存在差别，但是从巢中出来的时间迟或早似乎并不和 HPI 轴的反应性相关（Vaz-Serrano 等，2011；Thornqvist 等，2015）。

2.5 对抗的相互作用：应激和攻击

对抗的相互作用对群居动物建立社会性等级和获取资源起着重要的作用（Bernstein，Gordon，1974；Huntingford，Turner，1987；Francis，1988）。在这些群居的动物中，优势者/从属者的关系通常是一种等级的形式，攻击冲突的结果是决定个体在等级中排序的主要因素（Huntingford，Turner，1987）。这些侵占性相互作用由一些行为，如炫耀、袭击、捏咬、追击等组成。鱼类在一个二分体的争夺中，典型的对抗行为必将产生一个优势者个体和一个从属者个体，后者将进一步抑制侵占性行为，并且逃避优势者（Øverli 等，1999a，b；Larson 等，2006）。社会性战败的行为效果包括食欲抑制（Meerlo 等，1997；Øverli 等，1998；Kramer 等，1999；Montero 等，2009）、减少侵占活动（Blanchard 等，1995；Hoglund 等，2001）和降低生殖行为（D' Amato，1988；Perret，1992）。

应激反应开始时（特别是 HPI 轴），两类个体都被激活，结果是在社会相互作用的起始阶段，优势者和从属者个体血液循环中皮质醇水平都升高，但在社会性等级建立数小时内优势者的皮质醇水平回复到底线（Ejike，Schreck，1980；Øverli 等，1999a，b）。对应激反应的血浆皮质醇水平升高是一种适应性反应，使生物有机体为环境变化和/或生理变化做好准备，因为皮质醇有助于维持稳态。从属者个体在社会性等级建立后，血浆皮质醇水平升高数天甚至数周（Winberg，Lapage，1998；Øverli 等，1999a，b；Sloman 等，2001）。皮质醇水平慢性的升高是和病理学相联系的，因而对动物有伤害。在血浆皮质醇水平和 ACTH 之间有一种负相关关系，即皮质醇水平升高导致 ACTH 浓度降低。这种情况适合于从属者个体和皮质醇喂食的鳟鱼（Jeffrey 等，2012），并且和另一项研究投喂皮质醇的鱼 ACTH 浓度低于对照组鱼的研究结果一致（Balm，Pottinger，1995）。已经证明皮质醇对社会性竞争中的行为表现有负的影响（Øverli 等，2002b）。以对应激高的或者低的皮质醇反应进行繁育的虹鳟，其中高血浆皮质醇水平反应的个体比较倾向于变成社会性从属者（Pottinger，Carrick，2001a，b）。

鱼类除血浆皮质醇水平外，在优势者和从属者之间，一些行为的和生理的特征有所不同。典型的优势者个体表现较明显的侵占行为（Larson 等，2006；Pavlidis 等，2011；Dahlbom 等，2012）。然而，从属者个体表现非常冷淡的行为（Pavlidis 等，2011）和逃避的行为（Larson 等，2006）、较高的 5-HIAA/5-HT 水平（Winberg，Lapage，1998；Sorensen 等，2011；Dahlbom 等，2012）、较高的 POMCA 和 POMCB 水平（ACTH 前体）（Winberg，Lapage，1998）、降低细胞增殖（Sorensen 等，2011）和减少食物摄取（Sorensen 等，2011）。

社会性相互作用是影响行为和生理的因子中研究得最好的一个。然而，还难以辨别生理的和行为的差别在优势者个体和从属者个体之间是造成社会等级的起因还是结果。这是一个复杂的课题，有一些因素在起作用，例如，在动物的生命史中对社会性挑战的敏感性变化（Wommack，Deville，2007）和在社会等级形成早期的经历对于社会性相互作用期间发生的事态在神经生物学方面的影响是重要的（Sorensen 等，2007）。此外，早先存在的生理学差别可能决定在社会中的等级（Korzan，Summers，2007），因而，并不需要简单地解析动物生命中特殊情况的起因和结果。在野生的和禁闭的情况下，一系列因素如年龄、个体大小、性别、亲缘关系、第二性征、先前的优势经历或者栖息在特殊的领地中等，都会影响到其在社会中的等级（Abbott 等，1985；Beacham，1988；Beaugrand，Cotnoir，1996；Sprague 1998；Cote，2000；Renison 等，2002）。一些研究表明，在可预知的和稳定的环境中，行为和生理的差别可以预先把动物归于一定的社会等级中（Morgan 等，2000；McCarthy，2001；Plusquellec 等，2001；Pottinger，Carrick，2001a，b）。在大多数情况下，优势者（取胜者）个体的应激反应是下调的。在一个稳定的等级中，解除应激的是优势的个体。然而，当环境发生变化时，如由稳定到不稳定时，优势者的个体就要比从属者个体处在较高的应激反应水平中。

在一场优势的搏斗中，一系列因素能影响社会等级的位置。较强的先天侵占能力（Holtby 等，1993；Adams，Huntingford，1996；Adams 等，1998；Cutts 等，1999b）、较高的摄食动机（Johnsson 等，1996）和较大的体型（Holtby 等，1993；Rhodes，Quinn，1998；Cutts 等，1999a）都和成为社会性优势者有关。在其他一些研究中，没有观察到体形和成为优势者趋势有联系（Huntingford 等，1990；Adams，Huntingford，1996；Yamamoto 等，1998）。在对丽体鱼（*Cichlasoma dimerus*）的研究中发现，体型决定雄鱼的社会地位，但对雌鱼没有作用（Alonso 等，2012）。早前的社会经历是成为优势者或者从属者（胜者/败者效应）的社会性相互作用结果的重要标志（Abbott，Dill，1985；Dugatkin，1997；Rhodes，Quinn，1998；Hsu，Wolf，1999；Johnsson 等，1999；Oliveira 等，2009；Oliveira 等，2011）。这可能是在早先的社会性相互作用中产生的神经化学或内分泌变化所造成的（Winberg，Nilsson，1993；Øverli 等，1999a，b；Hoglund 等，2001；Winberg 等，2001）。一些研究报道经过神经化学或内分泌学的调剂后产生可能成为优势者地位的变化。已经表明用生长激素处理能增强侵占行为和摄食动机，并进而能够形成社会等级的优势（Johnsson，Bjornsson，1994；Jonsson 等，1998，2003）；而使用 DA 的前体 L-DOPA 处理以提高脑的多巴胺活性能使北极红点鲑处于社会等级的优势位置（Winberg，Nilsson，1992）。遗传的低应激反应个体能预先安置处于社会等级的优势位置（Ruiz-Gomez 等，2008），不过并非经常是这样。在一项研究中，两个不同应激反应品系（HR 和 LR）的虹鳟频繁活动和挨饿，HR 品系个体超过 LR 品系个体成为优势者，而在正常情况下 LR 品系个体是优势者（Ruiz-Gomez 等，2008）。

在鱼类（Abbott，Dill，1985；Franck，Ribowski，1993；Huntingford 等，1993；

Winberg，Nilsson，1993；Winberg 等，1993；Nakano，1994，1995a，b；Gomez-Laplaza，Morgan，2003；Desjardins 等，2012）和其他脊椎动物（Raab 等，1986；Blanchard 等，1993；Albonetti，Farabollini，1994；Meerlo 等，1997；Engh 等，2005；Hsw 等，2006；Korzan，Summers，2007）中，从属者个体的行为受到普遍性抑制。这是由于抑制侵占和/或生殖行为、减少摄食、低的自发性运动性能和探索能力而形成的特征。从属者行为的抑制可以看作是一种被动的应对策略，以避免和优势者进行高代价的相互作用（Leshner，1980；Benus 等，1991）。

侵占行为可以是应激的和解除应激的。应激能诱导侵占行为（短暂的）或者抑制它（持久的）。良好的应激处理能力能预先把一个个体安置于侵占和优势者地位。在这方面，有一些设计杂乱的试验研究侵占对应激的影响（由于胜者和败者个体在最初的处境中有不同的应激反应）。不过，良好的应激处理能力在饥饿和重新安置的情况下，亦能预先使一个个体成为败者。

2.6　营养因素影响应激反应

食物的分量和组成能影响应激反应，并且成为应激反应型式当中产生种内变化的附加因素。特别是食物的氨基酸（AA）和脂肪酸（FA）组成是很重要的，因为 AA 和 FA 能影响神经递质合成、膜的组成和神经元的兴奋性。

2.6.1　氨基酸

一系列研究证明了食物操作中的 AA 色氨酸（TRP）和酪氨酸（TYR）对行为的和神经内分泌的影响（Fernstrom，1983；Fernstrom，Fernstrom，2007）。TRP 和 TYR 分别是 5-HT 和儿茶酚胺（多巴胺、去甲肾上腺素、肾上腺素）的前体，而 TRP 和 TYR 对应激反应和行为的影响是由这些单胺能的神经递质介导的。TRP 和 TYR 能通过相同的载体穿越血-脑屏障而和其他大的中性氨基酸（LNAA），如苯丙氨酸、亮氨酸、异亮氨酸、缬氨酸和蛋氨酸等竞争。因此，除食物中 TRP 和 TYR 的含量之外，这些其他的 LNAA 在食物中的含量亦会影响到神经内分泌的活动过程，包括对应激的行为和内分泌反应的整合（Markus，2008）。

5-HT 生物合成的限速步骤是由色氨酸羟化酶（TPH）催化的，这种酶有两种不同的类型，TPH1 和 TPH2（Lillesaar，2011）。TPH1 的 Km 值要低于 TPH2 的。事实上，TPH2 对 TRP 的 Km 值处在脑的 TRP 游离浓度的范围之内，使 5-HT 合成的速率明显受到 TRP 可用性的影响（McKinney 等，2001，2005）。在硬骨鱼的脑内，5-HT 细胞体位于后脑的缝核（raphe nucleus）以及一些间脑区。值得注意的是，TPH1 和 TPH2 呈现不同的表达，TPH2 在缝核表达，而 TPH1 主要在间脑和外周的 5-HT 细胞（如沿着味束）中表达（Lillesaar，2011）。因而，缝核 5-HT 合成速率严格地受到 TRP 可用性的限制，而在其他 5-HT 神经元群中的 5-HT 合成则不是这样的。然而，缝核 5-HT 神经元出现明显不同的突起型式，突起延伸到前脑区并向下到达延脑和脊髓。血浆 TRP 浓度以及脑的 TRP 可用性都受到食物的 TRP 含量、应激和免疫反应等因素的影响，并且还可以设想血浆 TRP 水平的变化可以作为调节中枢释放 5-HT 的一种外周信号（Russo 等，2009）。

和其他的 LNAA 不同，TRP 输送到血浆后和白蛋白结合，至少在哺乳类动物中是这样的（Fernstrom，1983；Fernstrom，Fernstrom，2007）。因而，食物中的碳水化合物亦影响到脑 TRP 的可用性，因为胰岛素的分泌使 AA 摄入到肌肉中。然而，由于 TRP 和白蛋白结合，血浆 TRP 浓度不会受到胰岛素的影响。替代的是，胰岛素分泌使脑对 TRP 的摄取增加，这是因为和 TRP 竞争相同载体的其他 LNAA 在血浆中的水平降低所致。然而，虹鳟的白蛋白并不存在和 TRP 的结合位点（Fuller，Roush，1973），所以，还不清楚其他硬骨鱼类的 TRP 是怎样输送到血浆中的。

针对许多鱼类的研究已经证明食物中补充 TRP 对脑 5-HT 代谢的影响（Winberg 等，2001；Lepage 等，2002；Hoglund 等，2005，2007；Basic 等，2013b；Martins 等，2013）。此外，这些研究表明 TRP 补充物起着行为的和生理的效应，提示 5-HT 信号的变化。例如，食物的 TRP 补充物能抑制侵占行为并有抗焦虑的作用，减轻应激引起的厌食作用（Winberg 等，2001；Hseu 等，2003；Hoglund 等，2005，2007；Wolkers 等，2012，2014）。还有，已经有研究证明，富含 TRP 的食物能以剂量依存的方式降低虹鳟（Lepage 等，2002）、大西洋鳕鱼（*Gadus morhua*）（Basic 等，2012）和大西洋鲑鱼（Basic 等，2013b）幼鱼应激后的血浆皮质醇水平。此外，Lepage 等（2003）还证明对 HPI 轴反应的抑制作用是出现在用富含 TRP 食物处理的 7 天后，而不是在处理 3 天和 28 天后。这说明 TRP 处理后是在一个颇为狭小的时间范围内才显示 TRP 降低应激反应的作用。目前还不清楚使用补充 TRP 的食物处理后能否对鱼类的行为和应激反应有某种持久的影响，因为在大多数情况下，所进行的 TRP 补充物产生效应的试验都是一直给鱼投喂这种食物（Lepage 等，2002，2003；Holglund 等，2005，2007）。不过，最近的研究证明，在结束富含 TRP 食物的处理期之后，食物中 TRP 的作用能持续达 10 天（Basic 等，2013a，b）。

参与介导 TRP 降低应激效应的作用机理目前尚未完全了解。值得重视的是，给虹鳟以选择性的羟色胺重摄取抑制剂（SSRI）citalopram 进行处理，能减少侵占行为和应激后血浆中的皮质醇，这和采用富含 TRP 食物处理产生的结果一样（Lepage，2005）。在哺乳类的压抑模型中，SSR1 的抗抑郁效应，例如压抑 HPA 轴的机能亢进，曾经归因于细胞外 5-HT 水平的增加和 5-HT 受体表达的变化。然而，通过由脑衍生的神经营养性因子（BDNF）和神经生长因子的刺激效应以促进神经发生（neurogenesis），5-HT 亦能够以神经塑性（neuroplasticity）作用机理而起作用（Dong-Ryulu 等，1999；van Donkelaar 等，2009）。5-HT 对神经塑性的作用最主要表现在海马区，一个对 HPA 轴活性具有抑制性作用的脑区（Mahar 等，2014）。

除 5-HT 合成之外，TRP 亦是其他能和神经内分泌应激反应发生相互作用的代谢通道的限速因素。在哺乳类动物中，犬尿喹啉通道通过吲哚胺双加氧酶（IDO）和色氨酸双加氧酶（TDO）将 TRP 转化为其他的生物活性物质，如犬尿喹啉酸和喹啉酸

(Maes 等，2009；Le Floc'h 等，2011)。TDO 和 IDO 酶分别为糖皮质类固醇和促炎细胞因子所激活。犬尿喹啉通道将 TRP 从 5-HT 的产生中分流出来（Russo 等，2009）。此外，犬尿喹啉酸和喹啉酸是神经活性物质，作用于兴奋性 AA 受体（AMPA、NMDA 和钾盐镁矾谷氨酸受体）（Zadori，2009），并可能对单胺能的信号起作用（Okuno 等，2011）。因此，应激和炎症过程能和食物 TRP 补充物的影响互相起作用，将可供利用的 TRP 分流到犬尿喹啉通道。然而，对哺乳类动物已进行了广泛的研究，而有关鱼类犬尿喹啉通道以及它如何和 HPI 轴互相作用的资料还很少。

儿茶酚胺生物合成的限速酶，酪氨酸羟化酶和它在体的底物酪氨酸处于充分饱和的状态。这表明食物中的 TYR 对儿茶酚胺的产生和释放没有影响，尽管 TYR 和 TRP 一样，都是必需氨基酸。但是，有一些研究报道施用 TYR 后，会引起行为的和内分泌的效应（Ferntrom，Fernstrom，2007）。有关食物 TYR 对鱼类生理作用的研究相当少（Li 等，2009）。在最近的研究中，Costas 等（2012）报道含有高水平 AA，包括 TYR 和 TRP 的食物，能使脑的 DA 和 5-HT 浓度增加。这个发现使作者们认为富含 TYR 的食物能刺激硬骨鱼类 DA 的产生。

2.6.2 脂肪酸

鱼类和其他脊椎动物一样，它们正常的生长和发育需要 ω-3FAs、二十二碳六烯酸（DHA）、十二碳五烯酸（EPA）、ω-6FA 和花生四烯酸（ARA）（Bell 等，1986）。一般来说，这三种长链多不饱和脂肪酸的生理功能在脊椎动物谱系中是相似的，它们都具有维持细胞膜结构与功能完整性的普遍作用，并具有成为类二十烷酸前体的特异性功能。类二十烷酸前体是一群旁分泌激素，包括前列腺素、血栓烷、白细胞三烯，以及各种羟基和过氧羟基脂肪酸（Wainwright，2002）。它们参与一系列的生理功能，而身体的几乎所有组织都能产生它们。它们通常是在应激状态的反应中在细胞的和身体整体的水平中合成的。

与哺乳类一样，鱼类类二十烷酸主要的前体是花生四烯酸（ARA）。以 EPA 作为前体的类二十烷酸，其生物活性通常都要低于由 ARA 合成的类二十烷酸。结果是，类二十烷酸活性受到组织中有效 FA 含量的影响。与此同时，有明显的证据表明，类二十烷酸的活性和产生能为食物中 ω-6 和 ω-3FAs 相对比例的变化所修饰（Tocher，2003）。然而，重要的是必须注意到许多脊椎动物，包括大多数淡水鱼类，能够从亚麻酸（LNA）产生 DHA 和 EPA，以及从亚油酸（LA）产生 ARA。在这些种类中，LNA 和 LA 是食物中的必需 FA，而这些 FA 在食物中的比例是组织中 DHA：EPA：ARA 最后比例的主要决定因子。然而，在海水鱼类中，LNA 和 LA 的转化率非常低，DHA、EPA 和 ARA 被认为是食物中的必需脂肪酸（Tocher，2003）。

食物中 FA 组成能在应激轴的不同水平影响神经内分泌的应激反应。对鼠类的研究已经证明 ARA 在环加氧酶（COX）的作用下转化为前列腺素、类二十烷酸，并在脑垂体和肾上腺水平刺激 HPA 轴的活性（Mulla，Buckingham，1999）。鱼类类似作用机理的指证已经在针对欧洲舌齿鲈（*Dicentrarchus labrax*）的研究中得到证实。给这种鱼类投喂的食物以 ω-3 欠缺的植物油置换一些富含 ω-3 的鱼油，结果出现较高的应激后血浆皮质醇水平（Montero 等，2003）。此外，对鲈鱼幼鱼投喂补充 ARA 的食物，在受到每天重复的应激刺激后出现较高的血浆皮质醇水平（Koven 等，2003），而灌注研究表明，EPA 和 ARA 能够刺激这种鱼类 HPI 轴的肾间腺产生皮质醇。此外，COX 抑制剂能降低这种刺激作用，表明 COX-衍生的类二十烷酸介导 EPA 和 ARA 刺激皮质醇产生的作用（Ganga 等，2006）。

在早期发育阶段，食物中的 DHA 尤为重要。一系列的研究已经证明它能促进成活、应激抵抗力（Lund，Steenfeldt，2011；Lund 等，2014）、幼鱼正常的脑发育、神经元迁移和神经生理功能（Benitez-Santana 等，2012）。但是，对于食物 DHA 和 HPI 轴活性之间联系的相关作用机理目前还不清楚。低水平的 ω-3FAs 和心理压抑有关（Hibbeln，1998，2002），包括伴随着的血浆皮质醇水平慢性升高。值得注意的是，富含鱼油的高水平 DHA 和 EPA 食物能够影响羟色胺能的传递（Vancassel 等，2008），刺激参与营养过程的基因表达（Wu 等，2004），这和抗抑郁的状态相似。对 ω-3FAs 的抗抑郁作用机理目前还没有完全了解清楚。可能的情况是，含有 ω-3∶ω-6 高比例的食物使激活 IDO 的类二十烷酸减少。这就转而从 TRP 的代谢物犬尿喹啉酸和喹啉酸分流出 Kyreni 通道，有助于 5-HT 产生（图 2. 1）。由于喹啉酸具有神经退行性变化的作用（Maes 等，2009），而 5-HT 能刺激神经发生，可以推想，这种 AA 和 FA 的相互作用能引起脑的结构过程，并进而产生对 HPA 轴机能亢进的压抑作用。至于 AA 和 FA 之间的这种关系以及 HPI 轴的活性是否存在于鱼类中还不清楚。但在所进行的试验中，给鱼投喂高比例植物油而含有低水平 ω-3FAs 的食物（图 2. 1），可以观察到提供富含 TRP 的食物是增强低应激耐性的一种途径。

在早期个体发育阶段，食物的营养组成特别重要，因为它能产生对神经内分泌应激反应的长期影响。对鼠类的研究表明，母乳的 FA 组成影响到成鼠的 HPA 轴功能（D' Asti 等，2010）。最近的研究证明，食物的 FAs 亦对鱼类产生长期的作用（Lund 等，2014）。在这些研究中，在梭鲈（*Sander lucioperca*）孵化后 7 ～ 27 天投喂不同 FA 组成的食物，在 85 天食物试验结束时，投喂食物中 DHA 含量低的梭鲈，在对新的试验处理反应中，其运动能力降低。然而，虽然这项研究表明食物的 FA 组成影响行为的应激反应，但还没有研究报道它对 HPI 轴的影响。

2. 7 未来的研究方向

鱼类神经内分泌应激反应是和遗传因素、当时的环境以及先前的经历相联系的。趋异的应激反应通常都和形成不同的应激应对方式的特异性行为型式有关。我们对于遗传因素怎样和环境互相作用以调控这些应激应对方式的发展，以及应对方式怎样和生活史的变化相联系，目前了解得非常有限。对抗的相互作用具有强烈的行为影响（即传统的胜者-败者效应）。正如前面所论述的，社会性应激以及其他的挑战已经表明至少能够改变和特异性应激应对方式相联系的行为性状。这些影响很可能是由外遗传的（后生的）作用介导的，至今还没有对其作用机理进行深入的探讨。业已清楚的是，应激应对和生活史的性状，如性成熟时间、洄游等有联系。但是，我们对于生态影响及其参与的作用机理亦了解不多。目前已经表明食物组成能够影响鱼类应激反应的型式和行为。饲料添加剂对于在水产养殖中培育具有胁迫抗性而没有侵占性的鱼种是一项可取的途径。另一个培育胁迫抗性养殖鱼种的策略是选择育种。能显示反应前的应激应对方式的鱼类显然最适合于养育在禁闭的水产养殖环境中。不过，反应前的鱼类在水产养殖生产中有一个明显的缺点——它们是侵占性的，还不清楚通过有目的的选育能否把侵占性从反应前表现型的其他性状中分离出来。所以，基础的和应用的研究领域要关注的是在何种程度和在何种情况下，通过人工的或自然的选育能够将相互关联的性状分离出来。

S. 温伯格　E. 霍格伦特　Ø. 奥维里　著

林浩然　译、校

参考文献

Abbott, J. C. and Dill, L. M. (1985). Patterns of aggressive attack in juvenile steelhead trout (Salmogairdneri). Can. J. Fish. Aquat. Sci. 42, 1702-1706.

Abbott, J. C., Dunbrack, R. L. and Orr, C. D. (1985). The interaction of size and experience in dominance relationships of juvenilesteelhead trout (*Salmogairdneri*). Behaviour92, 241-253.

Åberg-Andersson, M., Wahid Khanb, U., Øverli, Ø., Gjøen, H. M. and Höglund, E. (2013). Coupling between stress coping style and time of emergence from spawning nests in salmonid fishes: evidence from selected rainbow trout strains (Oncorhynchus mykiss). Physiol. Behav. 116, 30-34.

Adams, C. E. andHuntingford, F. A. (1996). What is a successful fish? Determinants of competitive success in Arctic char (Salvelinus alpinus) in different social contexts. Can. J. Fish. Aquat. Sci. 53, 2446-2450.

Adams, C. E., Huntingford, F. A., Turnbull, J. F. and Beattie, C. (1998). Alternative competitive strategies and the cost of food acquisition in juvenile Atlantic salmon (Salmo salar). Aquaculture 167, 17-26.

Aguilar, R., Gil, L., Fernández-Teruel, A. and Tobeña, A. (2004). Genetically-based behavioral traits influence the effects of shuttle box avoidance overtraining and extinction upon intertrial responding: a study with the Roman rat strains. Behav. Processes 66, 63-72.

Albonetti, M. E. and Farabollini, F. (1994). Social stress by repeated defeat: effects on social behaviour and emotionality. Behav. Brain. Res. 62, 187-193.

Almasi, B., Jenni, L., Jenni-Eiermann, S. and Roulin, A. (2010). Regulation of stress response is heritable and functionally linked to melanin-based coloration. J.

Evol. Biol. 23, 987-996.

Alonso, F., Honji, R. M., Moreira, R. G. and Pandolfi, M. (2012). Dominance hierarchies and social status ascent opportunity: anticipatory behavioral and physiological adjustments in a Neotropical cichlid fish. Physiol. Behav. 106, 612-618.

Alsop, D. andVijayan, M. M. (2008). Development of the corticosteroid stress axis and receptor expression in zebrafish. Am. J. Physiol. Regul. Integr. Comp. Physiol. 294, R2021-R2021.

Alsop, D. andVijayan, M. M. (2009). The zebrafish stress axis: molecular fallout from the teleost-specific genome duplication event. Gen. Comp. Endocrinol. 161, 62-66.

Auperin, B. and Geslin, M. (2008). Plasma cortisol response to stress in juvenile rainbow trout is influenced by their life history during early development and by egg cortisol content. Gen. Comp. Endocrinol. 158, 234-239.

Backström, T., Schjolden, J., Øverli, Ø., Thörnqvist, P.-O. and Winberg, S. (2011). Stress effects on AVT and CRF systems in two strains of rainbow trout (Oncorhynchus mykiss) divergent in stress responsiveness. Horm. Behav. 59, 180-186.

Backström, T., Brännäs, E., Nilsson, J. and Magnhagen, C. (2014). Behaviour, physiology and carotenoid pigmentation in Arctic charr (Salvelinus alpinus). J. Fish. Biol. 84, 1-9.

Balm, P. H. M. andPottinger, T. G. (1995). Corticotrope and melanotrope POMC-derived peptides in relation to interrenal function during stress in raibow trout (Oncorhynchus mykiss). Gen. Comp. Endocrinol. 98, 279-288.

Barrot, M., Abrous, D. N., Marinelli, M., Rougé-Pont, F., Le Moal, M. and Piazza, P. V. (2001). Influence of glucocorticoids on dopaminergic transmission in the rat dorsolateral striatum. Eur. J. Neurosci. 13, 812-818.

Barry, T. P., Malison, J. A., Held, J. A. and Parrish, J. J. (1995a). Ontogeny of the cortisol stress response in larval rainbow trout. Gen. Comp. Endocrinol. 97, 57-65.

Barry, T. P., Ochiai, M. and Malison, J. A. (1995b). In-vitro effects of acth on interrenal corticosteroidogenesis during early larval development in rainbow-trout. Gen. Comp. Endocrinol. 99, 382-387.

Barton, B. A. (2002). Stress infishes: a diversity of responses with particular reference to changes in circulating corticosteroids. Integr. Comp. Biol. 42, 517-525.

Barton, B. A. andIwama, G. K. (1991). Physiological changes in fish from stress in aquaculture with emphasis on the response and effects of corticosteroids. Annu. Rev.

Fish Dis. 1, 3-26.

Barton, B. A., Ribas, L., Acerete, L. and Tort, L. (2005). Effects of chronic confinement on physiological responses of juvenile gilthead sea bream, Sparus aurata L., to acute handling. Aquacult. Res. 36, 172-179.

Basic, D., Winberg, S., Schjolden, J., Krogdahl, A. and Höglund, E. (2012). Context-dependent responses to novelty in rainbow trout (Oncorhynchus mykiss), selected for high and low post-stress cortisol responsiveness. Physiol. Behav. 105, 1175-1181.

Basic, D., Krogdahl, A., Schjolden, J., Winberg, S., Vindas, M. A., Hillestad, M., et al. (2013a). Short-and long-term effects of dietary i-tryptophansupplementation on the neuroendocrine stress response in seawater-reared Atlantic salmon (Salmo salar). Aquaculture 388, 8-13.

Basic, D., Schjolden, J., Krogdahl, A., von Krogh, K., Hillestad, M., Winberg, S., et al. (2013b). Changes in regional brain monoaminergic activity and temporary down-regulation in stress response from dietary supplementation with L-tryptophan in Atlantic cod (Gadus morhua). Br. J. Nutr. 109, 2166-2174.

Beacham, J. L. (1988). The relative importance of body size and aggressive experience as determinants of dominance in pumpkinseed sunfish, Lepomis gibbosus. Anim. Behav. 36, 621-623.

Beaugrand, J. P. and Cotnoir, P. A. (1996). The role of individual differences in the formation of triadic dominance orders of male green swordtail fish (Xiphophorus helleri). Behav. Processes 38, 287-296.

Bell, A. M. and Stamps, J. A. (2004). Development of behavioural differences between individuals and populations of sticklebacks, Gasterosteus aculeatus. Anim. Behav. 68, 1339-1348.

Bell, A. M., Henderson, L. and Huntingford, F. A. (2010). Behavioral and respiratory responses to stressors in multiple populations of three-spined sticklebacks that differ in predation pressure. J. Comp. Physiol. B. 180, 211-220.

Bell, M. V., Henderson, R. J. and Sargent, J. R. (1986). The role of polyunsaturated fatty-acids in fish. Comp. Biochem. Physiol. B 83, 711-719.

BenÍtez-Santana, T., Juárez-Carrillo, E., Beatriz Betancor, M., Torrecillas, S., José Caballero, M. and Soledad Izquierdo, M. (2012). Increased mauthner cell activity and escaping behaviour in seabream fed long-chain PUFA. Br. J. Nutr. 107, 295-301.

Benus, R., Den Daas, S., Koolhaas, J. and van Oortmerssen, G. (1990). Routine formation and flexibility in social and non-social behaviour of aggressive and non-aggressive male mice. Behaviour 112, 176-193.

Benus, R. F., Bohus, B., Koolhaas, J. M. and van Oortmerssen, G. A. (1991). Heritable variation for aggression as a reflection of individual coping strategies. Experientia 47, 1008-1019.

Bernier, N. J., Alderman, S. L. and Bristow, E. N. (2008). Heads or tails? Stressor-specific expression of corticotropin-eleasing factor and urotensin I in the preoptic area and caudal neurosecretory system of rainbow trout. J. Endocrinol. 196, 637-648.

Bernstein, I. S. and Gordon, T. P. (1974). The function of aggression in primate societies: uncontrolled aggression may threaten human survival, but aggression may be vital to the establishment and regulation of primate societies and sociality. Am. Sci. 62, 304-311.

Blanchard, D. C., Sakai, R. R., McEwen, B., Weiss, S. M. and Blanchard, R. J. (1993). Subordination stress: behavioral, brain, and neuroendocrine correlates. Behav. Brain Res. 58, 113-121.

Blanchard, D. C., Spencer, R. L., Weiss, S. M., Blanchard, R. J., McEwen, B. and Sakai, R. R. (1995). Visible burrow system as a model of chronic social stress: behavioral and neuroendocrine correlates. Psychoneuroendocrinology 20, 117-134.

Bolhuis, J. E., Schouten, W. G., De Leeuw, J. A., Schrama, J. W. and Wiegant, V. M. (2004). Individual coping characteristics, rearing conditions and behavioural flexibility in pigs. Behav. Brain Res. 152, 351-360.

Boon, A. K., Réale, D. and Boutin, S. (2007). The interaction between personality, offspring fitness and food abundance in North American red squirrels. Ecol. Lett. 10, 1094-1104.

Both, C., Dingemanse, N. J., Drent, P. J. and Tinbergen, J. M. (2005). Pairs of extreme avian personalities have highest reproductive success. J. Anim. Ecol. 74, 667-674.

Brännäs, E. (1988). Emergence of Baltic salmon, Salmo salar L., in relation to temperature—a laboratory study. J. Fish. Biol. 33, 589-600.

Brelin, D., Petersson, E. and Winberg, S. (2005). Divergent stress coping styles in juvenile brown trout (Salmo trutta). Ann. N. Y. Acad. Sci. 1040, 239-245.

Brelin, D., Petersson, E., Dannewitz, J., Dahl, J. and Winberg, S. (2008). Frequency distribution of coping strategies in four populations of brown trout (Salmo trutta). Horm. Behav. 53, 546-556.

Burmeister, S. S. (2007). Genomic responses to behavioral interactions in an African cichlidfish: mechanisms and evolutionary implications. Brain Behav. Evol. 70, 247-256.

Bury, N. R., Sturm, A., Le Rouzic, P., Lethimonier, C., Ducouret, B., Guiguen,

Y. , et al. (2003). Evidence for two distinct functional glucocorticoid receptors in teleost fish. J. Mol. Endocrinol. 31, 141-156.

Bury, N. R. and Sturm, A. (2007). Evolution of the corticosteroid receptor signalling pathway infish. Gen. Comp. Endocrinol. 153, 47-56.

Cannon, W. B. (1929). Bodily changes in pain, fear, hunger, and rage. New York: Appleton.

Carere, C. , Caramaschi, D. and Fawcett, T. W. (2010). Covariation between personalities and individual differences in coping with stress: converging evidence and hypotheses. Curr. Zool. 56, 728-740.

Carpenter, R. E. , Watt, M. J. , Forster, G. L. , Øverli, Ø. , Bockholt, C. , Renner, K. J. , et al. (2007). Corticotropin releasing factor induces anxiogenic locomotion in trout and alters serotonergic and dopaminergic activity. Horm. Behav. 52, 600-611.

Cash, W. B. andHolberton, R. L. (1999). Effects of exogenous corticosterone on locomotor activity in the red-eared slider turtle, Trachemys scripta elegans. J. Exp. Zool. 284, 637-644.

Caspi, A. , Sugden, K. , Moffitt, T. E. , Taylor, A. , Craig, I. W. , Harrington, H. , et al. (2003). Influence of life stress on depression: moderation by a polymorphism in the 5-HTT gene. Science 301, 386-389.

Castanheira, M. F. , Herrera, M. , Costas, B. , Conceic-ão, L. E. and Martins, C. I. (2013). Linking cortisol responsiveness and aggressive behaviour in gilthead seabream Sparus aurata: indication of divergent coping styles. Appl. Anim. Behav. Sci. 143, 75-81.

Chabbi, A. and Ganesh, C. (2015). Evidence for the involvement of dopamine in stress-induced suppression of reproduction in the cichlid Fish Oreochromis mossambicus. J. Neuroendocrinol. 27, 343-356.

Chandrasekar, G. , Lauter, G. and Hauptmann, G. (2007). Distribution of corticotropinreleasing hormone in the developing zebrafish brain. J. Comp. Neurol. 505, 337-351.

Charmantier, A. and Garant, D. (2005). Environmental quality and evolutionary potential: lessons from wild populations. Proc. R. Soc. Lond. B Biol. Sci. 272, 1415-1425.

Clarke, A. S. andBoinski, S. (1995). Temperament in nonhuman primates. Am. J. Primatol. 37, 103-125.

Clements, S. , Moore, F. L. and Schreck, C. B. (2003). Evidence that acute serotonergic activation potentiates the locomotor-stimulating effects of corticotropin-releasing hormone in juvenile Chinook salmon (Oncorhynchus tshawytscha). Horm. Behav. 43, 214-221.

Cockrem, J. F. (2013). Individual variation in glucocorticoid stress responses in animals. Gen. Comp. Endocrinol. 181, 45-58.

Colombe, L., Fostier, A., Bury, N., Pakdel, F. and Guiguen, Y. (2000). A mineralocorticoidlike receptor in the rainbow trout, Oncorhynchus mykiss: cloning and characterization of its steroid binding domain. Steroids 65, 319-328.

Coppens, C. M., De Boer, S. F. and Koolhaas, J. M. (2010). Coping styles and behavioural flexibility: towards underlying mechanisms. Phil. Trans. R. Soc. Lond. B Biol. Sci. 365, 4021-4028.

Costas, B., Aragao, C., Soengas, J. L., Miguez, J. M., Rema, P., Dias, J., et al. (2012). Effects of dietary amino acids and repeated handling on stress response andbrain monoaminergic neurotransmitters in Senegalese sole (Solea senegalensis) juveniles. Comp. Biochem. Physiol. A Mol. Integr. Physiol. 161, 18-26.

Cote, S. D. (2000). Dominance hierarchies in female mountain goats: stability, aggressiveness and determinants of rank. Behaviour 137, 1541-1566.

Cutts, C. J., Metcalfe, N. B. and Taylor, A. C. (1999a). Competitive asymmetries in territorial juvenile Atlantic salmon, Salmo salar. Oikos 86, 479-486.

Cutts, C. J., Brembs, B., Metcalfe, N. B. and Taylor, A. C. (1999b). Prior residence, territory quality and life-history strategies in juvenile Atlantic salmon (Salmo salar L.). J. Fish. Biol. 55, 784-794.

Dahlbom, S. J., Backström, T., Lundstedt-Enkel, K. and Winberg, S. (2012). Aggression and monoamines: effects of sex and social rank in zebrafish (Danio rerio). Behav. Brain. Res. 228, 333-338.

D'Amato, F. R. (1988). Effects of male social status on reproductive success and on behavior in mice (Mus musculus). J. Comp. Psychol. 102, 146-151.

D'Asti, E., Long, H., Tremblay-Mercier, J., Grajzer, M., Cunnane, S. C., DiMarzo, V., etal. (2010). Maternal dietary fat determines metabolic profile and the magnitude of endocannabinoid inhibition of the stress response in neonatal rat offspring. Endocrinology 151, 1685-1694.

David, M., Auclair, Y. and Cézilly, F. (2011). Personality predicts social dominance in female zebra finches, Taeniopygia guttata, in a feeding context. Anim. Behav. 81, 219-224.

DeKloet, E. R. (2004). Hormones and the stressed brain. Ann. N. Y. Acad. Sci. 1018, 1-15.

Desjardins, J. K., Hofmann, H. A. and Fernald, R. D. (2012). Social context influences aggressive and courtship behavior in a cichlid fish. PLoS One 7, e32781.

Dingemanse, N. J., Both, C., Drent, P. J. and Tinbergen, J. M. (2004). Fitness consequences of avian personalities in a fluctuating environment. Proc. R. Soc. Lond.

Ser. B Biol. Sci. 271, 847-852.

Dingemanse, N. J. , Van Der Plas, F. , Wright, J. , Réale, D. , Schrama, M. , Roff, D. A. , et al. (2009). Individual experience and evolutionary history of predation affect expression of heritable variation in fish personality and morphology. Proc. R. Soc. Lond. B Biol. Sci. 282. http://dx. doi. org/10. 1098/rspb. 2008. 1555.

Dong-Ryulu, L. , Semba, R. , Kondo, H. , Goto, S. and Nakano, K. (1999). Decrease in the levels of NGF and BDNF in brains of mice fed a tryptophan-deficient diet. Biosci. Biotechnol. Biochem. 63, 337-340.

Doyon, C. , Gilmour, K. M. , Trudeau, V. L. and Moon, T. W. (2003). Corticotropin-releasing factor and neuropeptide Y mRNA levels are elevated in the preoptic area of socially subordinate rainbow trout. Gen. Comp. Endocrinol. 133, 260-271.

Dugatkin, L. A. (1997). Winner and loser effects and the structure of dominance hierarchies. Behav. Ecol. 8, 583-587.

Einum, S. and Fleming, I. A. (2000). Selection against late emergence and small offspring in Atlantic salmon (Salmo salar). Evolution 54, 628-639.

Ejike, C. and Schreck, C. B. (1980). Stress and social hierarchy rank in coho salmon. Trans. Am. Fish. Soc. 109, 423-426.

Elipot, Y. , Hinaux, H. , Callebert, J. , Launay, J. -M. , Blin, M. and Rétaux, S. (2014). A mutation in the enzyme monoamine oxidase explains part of the Astyanax cavefish behavioural syndrome. Nat. Commun. 5, 3647.

Engh, A. L. , Siebert, E. R. , Greenberg, D. A. and Holekamp, K. E. (2005). Patterns of alliance formation and post conflict aggression indicate spotted hyaenas recognize third-party relationships. Anim. Behav. 69, 209-217.

Eriksen, M. S. , Bakken, M. , Espmark, A. , Braastad, B. O. and Salte, R. (2006). Prespawning stress in farmed Atlantic salmon Salmo salar: maternal cortisol exposure and hyperthermia during embryonic development affect offspring survival, growth and incidence of malformations. J. Fish. Biol. 69, 114-129.

Espmark, A. M. , Eriksen, M. S. , Salte, R. , Braastad, B. O. and Bakken, M. (2008). A note on pre-spawning maternal cortisol exposure in farmed Atlantic salmon and its impact on the behaviour of offspring in response to a novel environment. Appl. Anim. Behav. Sci. 110, 404-409.

Feist, G. andSchreck, C. B. (2002). Ontogeny of the stress response in Chinook salmon, Oncorhynchus tshawytscha. Fish Physiol. Biochem. 25, 31-40.

Fernandez, F. , Sarre, S. , Launay, J. M. , Aguerre, S. , Guyonnet-Dupérat, V. , Moisan, M. P. , et al. (2003). Rat strain differences in peripheral and central serotonin transporter protein expression and function. Eur. J. Neurosci. 17, 494-506.

Fernstrom, J. D. (1983). Role of precursor availability in control of monoamine biosynthesis in brain. Physiol. Rev. 63, 484-546.

Fernstrom, J. D. (1990). Aromatic-amino-acid and monoamine syntesis in the central-nervoussystem—influence of the diet. J. Nutr. Biochem. 1, 508-517.

Fernstrom, J. D. and Fernstrom, M. H. (2007). Tyrosine, phenylalanine, and catecholamine synthesis and function in the brain. J. Nutr. 137, 1539-1547.

Flik, G., Stouthart, X. J. H. X., Spanings, F. A. T., Lock, R. A. C., Fenwick, J. C. and Wendelaar Bonga, S. E. (2002). Stress response to waterborne Cu during early life stages of carp, Cyprinus carpio. Aquat. Toxicol. 56, 167-176.

Fone, K. C. F. and Porkess, M. V. (2008). Behavioural and neurochemical effects of post-weaning social isolation in rodents—relevance to developmental neuropsychiatric disorders. Neurosci. Biobehav. Rev. 32, 1087-1102.

Francis, R. C. (1988). On the relationship between aggression and social dominance. Ethology 78, 223-237.

Francis, R. C. (1990). Temperament inafish: a longitudinal studyof the development of individual differences in aggression and social rank in the Midas cichlid. Ethology 86, 311-325.

Franck, D. andRibowski, A. (1993). Dominance hierarchies of male green swordtails (Xiphophorus helleri) in nature. J. Fish. Biol. 43, 497-499.

Fuller, R. W. and Roush, B. W. (1973). Binding of tryptophan to plasma-proteins in several species. Comp. Biochem. Physiol. B. 46, 273-276.

Ganga, R., Tort, L., Acerete, L., Montero, D. and Izquierdo, M. S. (2006). Modulation of ACTH-induced cortisol release by polyunsaturated fatty acids in interrenal cells from gilthead seabream, Sparus aurata. J. Endocrinol. 190, 39-45.

Garrido, S., Rosa, R., Ben-Hamadou, R., Cunha, M. E., Chicharo, M. A. and van der Lingen, C. D. (2007). Effect of maternal fat reserves on the fatty acid composition of sardine (Sardina pilchardus) oocytes. Comp. Biochem. Physiol. B 148, 398-409.

Giesing, E. R., Suski, C. D., Warner, R. E. and Bell, A. M. (2011). Female sticklebacks transfer information via eggs: effects of maternal experience with predators on offspring. Proc. R. Soc. B 278, 1753-1759.

Giorgi, O., Lecca, D., Piras, G., Driscoll, P. and Corda, M. (2003). Dissociation between mesocortical dopamine release and fear-related behaviours in two psychogenetically selected lines of rats that differ in coping strategies to aversive conditions. Eur. J. Neurosci. 17, 2716-2726.

Glover, V., O'Connor, T. G. and O'Donnell, K. (2010). Prenatal stress and the programming of the HPA axis. Neurosci. Biobehav. Rev. 35, 17-22.

Gómez-Laplaza, L. M. and Morgan, E. (2003). The influence of social rank in the angelfish, Pterophyllum scalare, on locomotor and feeding activities in a novel environment. Lab. Anim. 37, 108-120.

Gorissen, M. and Flik, G. (2016). Endocrinology of the stress response in fish. In fish physiology-biology of stress in fish, Vol. 35 (eds. C. B. Schreck, L. Tort, A. P. Farrell and C. J. Brauner), San Diego, CA: Academic Press.

Gosling, S. D. (2001). From mice to men: what can we learn about personality from animal research? Psychol. Bull. 127, 45-86.

Gregory, T. R. and Wood, C. M. (1999). The effects of chronic plasma cortisol elevation on the feeding behaviour, growth, competitive ability, and swimming performance of juvenile rainbow trout. Physiol. Biochem. Zool. 72, 286-295.

Hayashi, A., Nagaoka, M., Yamada, K., Ichitani, Y., Miake, Y. and Okado, N. (1998). Maternal stress induces synaptic loss and developmental disabilities of offspring. Int. J. Dev. Neurosci. 16, 209-216.

Herzog, W., Zeng, X., Lele, Z., Sonntag, C., Ting, J.-W., Chang, C.-Y., et al. (2003). Adenohypophysis formation in the zebrafish and its dependence on sonic hedgehog. Dev. Biol. 254, 36-49.

Hibbeln, J. R. (1998). Fish consumption and major depression. Lancet 351, 1213.

Hibbeln, J. R. (2002). Seafoodconsumption, the DHA content of mothers'milk and prevalencerates of postpartum depression: a cross-national, ecological analysis. J. Affect. Disord. 69, 15-29.

Höglund, E., Kolm, N. and Winberg, S. (2001). Stress-induced changes in brain serotonergic activity, plasma cortisol and aggressive behavior in Arctic charr (Salvelinus alpinus) is counteracted by L-DOPA. Physiol. Behav. 74, 381-389.

Höglund, E., Balm, P. H. and Winberg, S. (2002a). Behavioural and neuroendocrine effects of environmental background colour and social interaction in Arctic charr (Salvelinus alpinus). J. Exp. Biol. 205, 2535-2543.

Höglund, E., Balm, P. H. and Winberg, S. (2002b). Stimulatory and inhibitory effects of 5-HT 1A receptors on adrenocorticotropic hormone and cortisol secretion in a teleost fish, the Arctic charr (Salvelinus alpinus). Neurosci. Lett. 324, 193-196.

Höglund, E., Bakke, M. J., Øverli, Ø., Winberg, S. and Nilsson, G. E. (2005). Suppression of aggressive behaviour in juvenile Atlantic cod (Gadus morhua) byi-tryptophan supplementation. Aquaculture 249, 525-531.

Höglund, E., Sørensen, C., Bakke, M. J., Nilsson, G. E. and Øverli, Ø. (2007). Attenuation of stress-induced anorexia in brown trout (Salmo trutta) by pre-treatment with dietary i-tryptophan. Br. J. Nutr. 97, 786-789.

Holtby, L. B., Swain, D. P. and Allan, G. M. (1993). Mirror-elicited agonistic behavior and body morphology as predictors of dominance status in juvenile coho salmon (Oncorhynchus kisutch). Can. J. Fish. Aquat. Sci. 50, 676-684.

Hseu, J. R., Lu, F. I., Su, H. M., Wang, L. S., Tsai, C. L. and Hwang, P. P. (2003). Effect of exogenous tryptophan on cannibalism, survival and growth in juvenile grouper, Epinephelus coioides. Aquaculture 218, 251-263.

Hsu, Y. and Wolf, L. L. (1999). The winner and loser effect: integrating multiple experiences. Anim. Behav. 57, 903-910.

Hsu, Y., Earley, R. L. and Wolf, L. L. (2006). Modulation of aggressive behaviour by fighting experience: mechanisms and contest outcomes. Biol. Rev. Camb. Philos. Soc. 81, 33-74.

Huntingford, F. A. and Turner, A. K. (1987). Animal Conflict. London: Chapman & Hall/CRC.

Huntingford, F. A., Metcalfe, N. B., Thorpe, J. E., Graham, W. D. and Adams, C. E. (1990). Social dominance and body size in Atlantic salmon parr, Salmo salar L. J. Fish. Biol. 36, 877-881.

Huntingford, F. A., Metcalfe, N. B. and Thorpe, J. E. (1993). Social status and feeding in Atlantic salmon Salmo salar parr: The effect of visual exposure to a dominant. Ethology 94, 201-206.

Huntingford, F. A., Wright, P. andTierney, J. (1994). Adaptive Variation in Antipredator Behaviour in Threespine Stickleback. The Evolutionary Biology of the Threespine Stickleback. Oxford: Oxford University Press.

Ishiwata, H., Shiga, T. and Okado, N. (2005). Selective serotonin reuptake inhibitor treatment of early postnatal mice reverses their prenatal stress-induced brain dysfunction. Neuroscience 133, 893-901.

Jeffrey, J. D., Esbaugh, A. J., Vijayan, M. M. and Gilmour, K. M. (2012). Modulation of hypothalamic-pituitary-interrenal axis function by social status in rainbow trout. Gen. Comp. Endocrinol. 176, 201-210.

Jeffrey, J. D., Gollock, M. J. and Gilmour, K. M. (2014). Social stress modulates the cortisol response to an acute stressor in rainbow trout (Oncorhynchus mykiss). Gen. Comp. Endocrinol. 196, 8-16.

Jentoft, S., Held, J. A., Malison, J. A. and Barry, T. P. (2002). Ontogeny of the cortisol stress response in yellow perch (Perca flavescens). Fish Physiol. Biochem. 26, 371-378.

Johansen, I. B., Sandvik, G. K., Nilsson, G. E., Bakken, M., Øverli, Ø., et al.

(2011). Cortisol receptor expression differs in the brains of rainbow trout selected for divergent cortisol responses. Comp. Biochem. Physiol. Part D Genomics Proteomics 6, 126-132.

Johnsson, J. I. and Björnsson, B. T. (1994). Growth hormone increases growth rate, appetite and dominance in juvenile rainbow trout, Oncorhynchus mykiss. Anim. Behav. 48, 177-186.

Johnsson, J. I., Jonsson, E. and Björnsson, B. T. (1996). Dominance, nutritional state, and growth hormone levels in rainbow trout (Oncorhynchus mykiss). Horm. Behav. 30, 13-21.

Johnsson, J. I., Nöbbelin, F. and Bohlin, T. (1999). Territorial competition among wild brown trout fry: effects of ownership and body size. J. Fish. Biol. 54, 469-472.

Jönsson, E., Johnsson, J. I. and Björnsson, B. T. (1998). Growth hormone increases aggressive behavior in juvenile rainbow trout. Horm. Behav. 33, 9-15.

Jönsson, E., Johansson, V., Björnsson, B. T. and Winberg, S. (2003). Central nervous system actions of growth hormone on brain monoamine levels and behavior of juvenile rainbow trout. Horm. Behav. 43, 367-374.

Kittilsen, S., Schjolden, J., Beitnes-Johansen, I., Shaw, J., Pottinger, T. G., Sørensen, C., et al. (2009). Melanin-based skin spots reflect stress responsiveness in salmonid fish. Horm. Behav. 56, 292-298.

Koolhaas, J. M., Korte, S. M., De Boer, S. F., Van Der Vegt, B. J., Van Reenen, C. G., Hopster, H., et al. (1999). Coping styles in animals: current status in behavior and stressphysiology. Neurosci. Biobehav. Rev. 7, 925-935.

Koolhaas, J., De Boer, S., Coppens, C. and Buwalda, B. (2010). Neuroendocrinology of coping styles: towards understanding the biology of individual variation. Front Neuroendocrinol. 31, 307-321.

Korte, S. M., Koolhaas, J. M., Wingfield, J. C. and Mcewen, B. S. (2005). The Darwinian concept of stress: benefits of allostasis and costs of allostatic load and the trade-offs in health and disease. Neurosci. Biobehav. Rev. 29, 3-38.

Korzan, W. J. and Summers, C. H. (2007). Behavioral diversity and neurochemical plasticity: selection of stress coping strategies that define social status. Brain Behav. Evol. 70, 257-266.

Koven, W., van Anholt, R., Lutzky, S., Ben Atia, I., Nixon, O., Ron, B., et al. (2003). The effect of dietary arachidonic acid on growth, survival, and cortisol levels in different-age gilthead seabream larvae (Sparus auratus) exposed to handling or daily salinity change. Aquaculture 228, 307-320.

Kralj-Fišer, S. and Schuett, W. (2014). Studying personality variation in invertebrates: why bother? Anim. Behav. 91, 41-52.

Kramer, M., Hiemke, C. and Fuchs, E. (1999). Chronic psychosocial stress and antidepressant treatment in tree shrews: time-dependent behavioral and endocrine effects. Neurosci. Biobehav. Rev. 23, 937-947.

Larson, E., Norris, D. and Summers, C. (2003). Monoaminergic changes associated with socially induced sex reversal in the saddleback wrasse. Neuroscience 119, 251-263.

Larson, E. T., O'Malley, D. M. andMelloni, R. H. (2006). Aggression and vasotocinare associated with dominant-subordinate relationships in zebrafish. Behav. Brain Res. 167, 94-102.

Le Floc'h, N., Otten, W. and Merlot, E. (2011). Tryptophan metabolism, from nutrition to potential therapeutic applications. Amino. Acids 41, 1195-1205.

Lepage, O., Tottmar, O. and Winberg, S. (2002). Elevated dietary intake of i-tryptophan counteracts the stress-induced elevation of plasma cortisol in rainbow trout (Oncorhynchus mykiss). J. Exp. Biol. 205, 3679-3687.

Lepage, O., Vilchez, I. M., Pottinger, T. G. and Winberg, S. (2003). Time-course of the effect of dietary i-tryptophan on plasma cortisol levels in rainbow trout Oncorhynchus mykiss. J. Exp. Biol. 206, 3589-3599.

Lepage, O., Larson, E. T., Mayer, I. and Winberg, S. (2005). Serotonin, but not melatonin, plays a role in shaping domninant-subordinate relationships and aggression in rainbow trout. Horm. Behav. 48, 233-242.

Lesch, K.-P., Bengel, D., Heils, A., Sabol, S. Z., Greenberg, B. D., Petri, S., et al. (1996). Association of anxiety-related traits with a polymorphism in the serotonin transporter gene regulatory region. Science 274, 1527-1531.

Leshner, A. I. (1980). The interaction of experience and neuroendocrine factors in determining behavioural adaptions to aggression. Prog. Brain Res. 53, 427-438.

Li, P., Mai, K. S., Trushenski, J. and Wu, G. Y. (2009). New developments in fish amino acid nutrition: towards functional and environmentally oriented aquafeeds. Amino. Acids 37, 43-53.

Lillesaar, C. (2011). The serotonergic system in fish. J. Chem. Neuroanat. 41, 294-308.

Liu, N.-A., Huang, H., Yang, Z., Herzog, W., Hammerschmidt, M., Lin, S., et al. (2003). Pituitary corticotroph ontogeny and regulation in transgenic zebrafish. Mol. Endocrinol. 17, 959-966.

Lowry, C. A., Burke, K. A., Renner, K. J., Moore, F. L. and Orchinik, M. (2001). Rapid changes in monoamine levels following administration of corticotropin-releasing factor or corticosterone are localized in the dorsomedial hypothalamus. Horm. Behav. 39, 195-205.

Lukkes, J., Vuong, S., Scholl, J., Oliver, H. and Forster, G. (2009). Corticotropin-releasing factor receptor antagonism within the dorsal raphe nucleus reduces social anxiety-like behavior after early-life social isolation. J. Neurosci. 29, 9955-9960.

Lund, I. andSteenfeldt, S. J. (2011). The effects of dietary long-chain essential fatty acids on growth and stress tolerance in pikeperch larvae (Sander lucioperca L.). Aquacult. Nutr. 17, 191-199.

Lund, I., Höglund, E., Ebbesson, L. O. E. and Skov, P. V. (2014). Dietary LC-PUFA deficiency early in ontogeny induces behavioural changes in pike perch (Sander lucioperca) larvae and fry. Aquaculture 432, 453-461.

Maes, M., Yirmiya, R., Noraberg, J., Brene, S., Hibbeln, J., Perini, G., et al. (2009). The inflammatory & neurodegenerative (I&ND) hypothesis of depression: leads for future research and new drug developments in depression. Metab. Brain Dis. 24, 27-53.

Magnhagen, C., Backström, T., Øverli, Ø., Winberg, S., Nilsson, J., Vindas, M., et al. (2015). Behavioural responses in a net restraint test predict interrenal reactivity in Arctic charr (Salvelinus alpinus). J. Fish. Biol. 87, 88-99.

Mahar, I., Bambico, F. R., Mechawar, N. and Nobrega, J. N. (2014). Stress, serotonin, and hippocampal neurogenesis in relation to depression and antidepressant effects. Neurosci. Biobehav. Rev. 38, 173-192.

Markus, C. R. (2008). Dietary amino acids and brain serotonin function; implications forstressrelated affective changes. Neuromol. Med. 10, 247-258.

Martins, C. I., Castanheira, M. F., Engrola, S., Costas, B. and Conceição, L. E. (2011a). Individual differences in metabolism predict coping styles in fish. Appl. Anim. Behav. Sci. 130, 135-143.

Martins, C. I., Silva, P. I., Conceição, L. E., Costas, B., Höglund, E., Øverli, Ø., et al. (2011b). Linking fearfulness and coping styles in fish. PLoS One 6, e28084.

Martins, C. I. M., Silva, P. I. M., Costas, B., Larsen, B. K., Santos, G. A., Conceicao, L. E. C., et al. (2013). The effect of tryptophan supplemented diets on brain serotonergic activity and plasma cortisol under undisturbed and stressed conditions in grouped-housed Nile tilapia Oreochromis niloticus. Aquaculture 400, 129-134.

Mason, J. C. and Chapman, D. W. (1965). Significance of early emergence, environmental rearing capacity and behavioral ecology of juvenile coho salmon in stream channels. J. Fish. Res. Fish. Res. Board Can. 22, 173-190.

McCarthy, I. D. (2001). Competitive ability is related to metabolic asymmetry in juvenile rainbow trout. J. Fish. Biol. 59, 1002-1014.

McKinney, J., Teigen, K., Froystein, N. A., Salaun, C., Knappskog, P. M., Haavik, J., et al. (2001). Conformation of the substrate and pterin cofactor bound to human tryptophan hydroxylase. ImportantroleofPhe313insubstratespecificity. Biochemistry40, 15591-15601.

McKinney, J., Knappskog, P. M. and Haavik, J. (2005). Different properties of the central and peripheral forms of human tryptophan hydroxylase. J. Neurochem. 92, 311-320.

Medeiros, L. R. and McDonald, M. D. (2013). Cortisol-mediated downregulation of the serotonin 1A receptor subtype in the Gulf toadfish, Opsanus beta. Comp. Biochem. Physiol. A Mol. Integr. Physiol. 164, 612-621.

Meerlo, P., Overkamp, G. J. F. and Koolhaas, J. M. (1997). Behavioural and physiological consequences of a single social defeat in Roman high-and low-avoidance rats. Psychoneuroendocrinology 22, 155-168.

Mendl, M., Burman, O. H. and Paul, E. S. (2010). An integrative and functional framework for the study of animal emotion and mood. Proc. R. Soc. B 277, 2895-2904.

Metcalfe, N. B. and Thorpe, J. E. (1992). Early predictors of life-history events—link between 1st feeding date, dominance and seaward migration in Atlantic salmon, Salmo salar L. J. Fish. Biol. 41, 93-99.

Metcalfe, N. B., Taylor, A. C. and Thorpe, J. E. (1995). Metabolic-rate, social-status andlifehistory strategies in Atlantic salmon. Anim. Behav. 49, 431-436.

Mommer, B. C. and Bell, A. M. (2014). Maternal experience with predation risk influences genome-wide embryonic gene expression in threespined sticklebacks (Gasterosteus aculeatus). PLoS One 9, e98564. http://dx. doi. org/10. 1371/journal. pone. 0098564.

Mommsen, R. P. (1999). Cortisol in teleosts: dynamics, mechanisms of action, and metabolic regulation. Rev. Fish Biol. Fish. 9, 211-268.

Montero, D., Kalinowski, T., Obach, A., Robaina, L., Tort, L., Caballero, M. J., et al. (2003). Vegetable lipid sources for gilthead seabream (Sparus aurata): effects on fish health. Aquaculture 225, 353-370.

Montero, D., Lalumera, G., Izquierdo, M. S., Caballero, M. J., Saroglia, M. and Tort, L. (2009). Establishment of dominance relationships in gilthead sea bream

(Sparus aurata) juveniles during feeding: effects on feeding behaviour, feed utilization, and fish health. J. Fish. Biol. 74, 1-16.

Moore, F. L. andOrchinik, M. (1994). Membrane receptors for corticosterone: a mechanism for rapid behavioral responses in an amphibian. Horm. Behav. 28, 512-519.

Moreira, P., Pulman, K. and Pottinger, T. (2004). Extinction of a conditioned response in rainbow trout selected for high or low responsiveness to stress. Horm. Behav. 46, 450-457.

Morgan, D., Grant, K. A., Prioleau, O. A., Nader, S. H., Kaplan, J. R. and Nader, M. A. (2000). Predictors of social status in cynomolgus monkeys (Macaca fascicularis) after group formation. Am. J. Primatol. 52, 115-131.

Mulla, A. and Buckingham, J. C. (1999). Regulation of the hypothalamo-pituitary-adrenal axis by cytokines. BestPract. Res. Clin. Endocrinol. Metab. 13, 503-521.

Nakano, S. (1994). Variation in agonistic encounters in a dominance hierarchy of freely interacting red-spottedmasu salmon (Oncorhynchus masouishikawai). Ecol. Freshw. Fish 3, 153-158.

Nakano, S. (1995a). Individual differences in resource use, growth and emigration under the influence of a dominance hierarchy in fluvial red-spotted masu salmon in a natural habitat. J. Anim. Ecol. 64, 75-84.

Nakano, S. (1995b). Competitive interactions for foraging microhabitats in a size-structuredinterspecific dominance hierarchy of two sympatric stream salmonids in a natural habitat. Can. J. Zool. 73, 1845-1854.

Okuno, A., Fukuwatari, T. and Shibata, K. (2011). High tryptophan diet reduces extracellular dopamine release via kynurenic acid production in rat striatum. J. Neurochem. 118, 796-805.

Oliveira, R. F. (2009). Social behavior in context: hormonal modulation of behavioral plasticity and social competence. Integr. Comp. Biol. 49, 423-440.

Oliveira, R. F. (2013). Mind thefish: zebrafish as a model in cognitive social neuroscience. Front. Neural Circuits. 7, 131. http://dx.doi.org/10.3389/fncir.2013.00131.

Oliveira, R. F., Silva, J. F. and Simões, J. M. (2011). Fighting zebrafish: characterization of aggressive behavior and winner-loser effects. Zebrafish 8, 73-81.

Øverli, Ø., Winberg, S., Damsård, B. and Jobling, M. (1998). Food intake and spontaneous swimming activity in Arctic char (Salvelinus alpinus): role of brain serotonergic activity and social interactions. Can. J. Zool. 76, 1366-1370.

Øverli, Ø., Olsen, R. E., Løvik, F. and Ringø, E. (1999a). Dominance hierarchies in Arctic charr, Salvelinus alpinus L. : differential cortisol profiles of dominant and subordinate individuals after handling stress. Aquacult. Res. 30, 259-264.

Øverli, Ø. , Harris, C. A. and Winberg, S. (1999b). Short-term effects of fights for social dominance and the establishment of dominant-subordinate relationships on brain monoamines and cortisol in rainbow trout. Brain Behav. Evol. 54, 263-275.

Øverli, Ø. , Pottinger, T. G. , Carrick, T. R. , Øverli, E. and Winberg, S. (2001). Brain monoaminergic activity in rainbow trout selected for high and low stress responsiveness. Brain Behav. Evol. 57, 214-224.

Øverli, Ø. , Kotzian, S. and Winberg, S. (2002a). Effects of cortisol on aggression and locomotor activity in rainbow trout. Horm. Behav. 42, 53-61.

Øverli, Ø. , Pottinger, T. G. , Carrick, T. R. , Øverli, E. and Winberg, S. (2002b). Differences in behaviour between rainbow trout selected for high-and low-stress responsiveness. J. Exp. Biol. 205, 391-395.

Øverli, Ø. , Korzan, W. J. , Höglund, E. , Winberg, S. , Bollig, H. , Watt, M. , et al. (2004). Stress coping style predicts aggression and social dominance in rainbow trout. Horm. Behav. 45, 235-241.

Øverli, Ø. , Winberg, S. and Pottinger, T. G. (2005). Behavioral and neuroendocrine correlates of selection for stress responsiveness in rainbow trout—a review. Integr. Comp. Biol. 45, 463-474.

Øverli, Ø. , Sørensen, C. , Pulman, K. G. , Pottinger, T. G. , Korzan, W. , Summers, C. H. , et al. (2007). Evolutionary background for stress-coping styles: relationships between physiological, behavioral, and cognitive traits in non-mammalian vertebrates. Neurosci. Biobehav. Rev. 31, 396-412.

Pankhurst, N. W. (2011). The endocrinology of stress infish: an environmental perspective. Gen. Comp. Endocrinol. 170, 265-275.

Paul, E. S. , Harding, E. J. and Mendl, M. (2005). Measuring emotional processes in animals: the utility of a cognitive approach. Neurosci. Biobehav. Rev. 29, 469-491.

Pavlidis, M. , Sundvik, M. , Chen, C. Y. and Panula, P. (2011). Adaptive changes in zebrafish brain in dominant-subordinate behavioral context. Behav. Brain Res. 225, 529-537.

Pepels, P. P. L. M. and Balm, P. H. M. (2004). Ontogeny of corticotropin-releasing factor and of hypothalamic-pituitary-interrenal axis responsiveness to stress in tilapia (Oreochromis mossambicus; Teleostei). Gen. Comp. Endocrinol. 139, 251-265.

Perret, M. (1992). Environmental and social determinants of sexual function in the male lesser mouse lemur (Microcebus murinus). Folia Primatol. 59, 1-25.

Plusquellec, P. , Bouissou, M. F. and Le Pape, G. (2001). Early predictors of dominance ability in heifers (Bos taurus L.) of the Herens breed. Behaviour 138, 1009-1031.

Pottinger, T. and Carrick, T. (1999). Modification of the plasma cortisol response to stress in rainbow trout by selective breeding. Gen. Comp. Endocrinol. 116, 122-132.

Pottinger, T. G. and Carrick, T. R. (2001a). ACTH does not mediate divergent stress responsiveness in rainbow trout. Comp. Biochem. Physiol. A: Mol. Integr. Physiol. 129, 399-404.

Pottinger, T. G. and Carrick, T. R. (2001b). Stress responsiveness affects dominant—subordinate relationships in rainbow trout. Horm. Behav. 40, 419-427.

Raab, A., Dantzer, R., Michaud, B., Mormede, P., Taghzouti, K., Simon, H., et al. (1986). Behavioral, physiological and immunological consequences of social status and aggression in chronically coexisting resident intruder dyads of male rats. Physiol. Behav. 36, 223-228.

Réale, D., Gallant, B. Y., Leblanc, M. and Festa-Bianchet, M. (2000). Consistency of temperament in bighorn ewes and correlates with behaviour and life history. Anim. Behav. 60, 589-597.

Réale, D., Reader, S. M., Sol, D., Mcdougall, P. T. and Dingemanse, N. J. (2007). Integrating animal temperament within ecology and evolution. Biol. Rev. 82, 291-318.

Réale, D., Martin, J., Coltman, D., Poissant, J. and Festa-Bianchet, M. (2009). Male personality, life-history strategies and reproductive success in a promiscuous mammal. J. Evol. Biol. 22, 1599-1607.

Renison, D., Boersma, D. and Martella, M. B. (2002). Winning and losing: causes for variability in outcome of fights in male Magellanic penguins (Spheniscus magellanicus). Behav. Ecol. 13, 462-466.

Rhodes, J. S. and Quinn, T. P. (1998). Factors affecting the outcome of territorial contests between hatchery and naturally reared coho salmon parr in the laboratory. J. Fish. Biol. 53, 1220-1230.

Riechert, S. E. and Hedrick, A. V. (1993). A test for correlations among fitness-linked behavioural traits in the spider Agelenopsis aperta (Araneae, Agelenidae). Anim. Behav. 46, 669-675.

Ruiz-Gomez, M. D., Kittilsen, S., Hoglund, E., Huntingford, F. A., Sørensen, C., Pottinger, T. G., et al. (2008). Behavioral plasticity in rainbow trout (Oncorhynchus mykiss) with divergent coping styles: when doves become hawks. Horm. Behav. 54, 534-538.

Ruiz-Gomez, M. D., Huntingford, F. A., Øverli, Ø., Thörnqvist, P.-O. and

Höglund, E. (2011). Response to environmental change in rainbow trout selected for divergent stress coping styles. Physiol. Behav. 102, 317-322.

Russo, S., Kema, I. P., Bosker, F., Haavik, J. and Korf, J. (2009). Tryptophan as an evolutionarily conserved signal to brain serotonin: molecular evidence and psychiatric implications. World J. Biol. Psychiatry 10, 258-268.

Schjolden, J. and Winberg, S. (2007). Genetically determined variation in stress responsiveness in rainbow trout: behavior and neurobiology. Brain Behav. Evol. 70, 227-238.

Schjolden, J., Backström, T., Pulman, K. G., Pottinger, T. G. and Winberg, S. (2005). Divergence in behavioural responses to stress in two strains of rainbow trout (Oncorhynchus mykiss) with contrasting stress responsiveness. Horm. Behav. 48, 537-544.

Schjolden, J., Pulman, K., Pottinger, T., Metcalfe, N. and Winberg, S. (2006a). Divergence in locomotor activity between two strains of rainbow trout (Oncorhynchus mykiss) with contrasting stress responsiveness. J. Fish. Biol. 68, 920-924.

Schjolden, J., Pulman, K. G., Pottinger, T. G., Tottmar, O. and Winberg, S. (2006b). Serotonergic characteristics of rainbow trout divergent in stress responsiveness. Physiol. Behav. 87, 938-947.

Schreck, C. B. and Tort, L. (2016). The Concept of Stress in Fish. In Fish Physiology-Biology of Stress in Fish, Vol. 35 (eds. C. B. Schreck, L. Tort, A. P. Farrell and C. J. Brauner), San Diego, CA: Academic Press.

Seghers, B. H. (1974). Schooling behavior in the guppy (Poecilia reticulata): an evolutionary response to predation. Evolution 28, 486-489.

Sih, A., Bell, A. and Johnson, J. C. (2004a). Behavioral syndromes: an ecological and evolutionary overview. Trends Ecol. Evol. 19, 372-378.

Sih, A., Bell, A. M., Johnson, J. C. and Ziemba, R. E. (2004b). Behavioral syndromes: an integrative overview. Q. Rev. Biol. 79, 241-277.

Silva, P. I. M., Martins, C. I., Engrola, S., Marino, G., Øverli, Ø. and Conceição, L. E. (2010). Individual differences in cortisol levels and behaviour of Senegalese sole (Solea senegalensis) juveniles: evidence for coping styles. Appl. Anim. Behav. Sci. 124, 75-81.

Sinn, D., Apiolaza, L. and Moltschaniwskyj, N. (2006). Heritability and fitness-related consequences of squid personality traits. J. Evol. Biol. 19, 1437-1447.

Sloman, K. A. (2010). Exposure of ova to cortisol pre-fertilisation affects subsequent behaviour and physiology of brown trout. Horm. Behav. 58, 433-439.

Sloman, K. A., Metcalfe, N. B., Taylor, A. C. and Gilmour, K. M. (2001). Plasma cortisol concentrations before and after social stress in rainbow trout and brown trout. Physiol. Biochem. Zool. 74, 383-389.

Sloman, K. A., Montpetit, C. J. and Gilmour, K. M. (2002). Modulation of catecholamine release and cortisol secretion by social interactions in the rainbow trout, Oncorhynchus mykiss. Gen. Comp. Endocrinol. 127, 136-146.

Smith, C. and Wootton, R. J. (1995). The costs of parental care in teleost fishes. Rev. Fish Biol. Fish. 5, 7-22.

Smith, B. R. and Blumstein, D. T. (2010). Behavioral types as predictors of survival in Trinidadian guppies (Poecilia reticulata). Behav. Ecol. 21, 919-926. http://dx.doi.org/10.1093/beheco/arq084.

Sørensen, C., Øverli, Ø., Summers, C. H. and Nilsson, G. E. (2007). Social regulation of neurogenesis in teleost. Brain Behav. Evol. 70, 239-246.

Sørensen, C., Bohlin, L. C., Øverli, Ø. and Nilsson, G. E. (2011). Cortisol reduces cell proliferation in the telencephalon of rainbow trout (Oncorhynchus mykiss). Physiol. Behav. 102, 518-523.

Sørensen, C., Johansen, I. B. and Øverli, Ø. (2013). Neural plasticity and stress coping in teleost fishes. Gen. Comp. Endocrinol. 181, 25-34.

Sprague, D. S. (1998). Age, dominance rank, natal status, and tenure among male macaques. Am. J. Phys. Anthropol. 105, 511-521.

Stanford, S. and Salmon, P. E. (1993). Stress: From Synapse to Syndrome. London: Academic Press.

Stapley, J. and Keogh, J. S. (2005). Behavioral syndromes influence mating systems: floater pairs of a lizard have heavier offspring. Behav. Ecol. 16, 514-520.

Steimer, T. and Driscoll, P. (2005). Inter-individual vs line/strain differences in psychogenetically selected Roman High- (RHA) and Low- (RLA) Avoidance rats: neuroendocrine and behavioural aspects. Neurosci. Biobehav. Rev. 29, 99-112.

Stolte, E. H., van Kemenade, B. M. L. V., Savelkoul, H. F. J. and Flik, G. (2006). Evolution of glucocorticoid receptors with different glucocorticoid sensitivity. J. Endocrinol. 190, 17-28.

Stolte, E. H., Nabuurs, S. B., Bury, N. R., Sturm, A., Flik, G., Savelkoul, H. F. J., et al. (2008). Stress and innate immunity in carp: corticosteroid receptors and pro-inflammatory cytokines. Mol. Immunol. 46, 70-79.

Stouthart, A. J. H. X., Lucassen, E. C. H. E. T., van Strien, F. J. C., Balm, P. H. M., Lock, R. A. C. and Wendelaar Bonga, S. E. (1998). Stress responsiveness of the pituitaryinterrenal axis during early life stages of common carp (Cyprinus carpio). J. Endocinol. 157, 127-137.

Sturm, A. , Bury, N. , Dengreville, L. , Fagart, J. , Flouriot, G. , Rafestin-Oblin, M. , et al. (2005). 11-deoxycorticosterone is a potent agonist of the rainbow trout (Oncorhynchus mykiss) mineralocorticoid receptor. Endocrinology 146, 47-55.

Tanaka, M. , Tanangonan, J. B. , Tagawa, M. , de Jesus, E. G. , Nishida, H. , Isaka, M. , et al. (1995). Development of the pituitary, thyroid and interrenal glands and applications of endocrinology to the improved rearing of marine fish larvae. Aquaculture 135, 111-126.

Teles, M. , Tridico, R. , Callol, A. , Fierro-Castro, C. and Tort, L. (2013). Differential expression of the corticosteroid receptors GR1, GR2 and MR in rainbow trout organs with slow release cortisol implants. Comp. Biochem. Physiol. A Mol. Integr. Physiol. 164, 506-511.

Thörnqvist, P. -O. , Höglund, E. and Winberg, S. (2015). Natural selection constrains personality and brain gene expression differences in Atlantic salmon (Salmo salar). J. Exp. Biol. 218, 1077-1083.

Tocher, D. R. (2003). Metabolism and functions of lipids and fatty acids in teleostfish. Rev. Fish. Sci. 11, 107-184.

Tudorache, C. , Schaaf, M. J. and Slabbekoorn, H. (2013). Covariation between behaviour and physiology indicators of copingstyle in zebrafish (*Danio rerio*). J. Endocrinol. 219, 251-258.

Vancassel, S. , Leman, S. , Hanonick, L. , Denis, S. , Roger, J. , Nollet, M. , et al. (2008). n-3 Polyunsaturated fatty acid supplementation reverses stress-induced modifications on brain monoamine levels in mice. J. Lipid. Res. 49, 340-348.

vanDonkelaar, E. L. , van den Hove, D. L. A. , Blokland, A. , Steinbusch, H. W. M. and Prickaerts, J. (2009). Stress-mediated decreases in brain-derived neurotrophic factor as potential confounding factor for acute tryptophan depletion-induced neurochemical effects. Eur. Neuropsychopharmacol. 19, 812-821.

vanOers, K. , De Jong, G. , Van Noordwijk, A. J. , Kempenaers, B. and Drent, P. J. (2005). Contribution of genetics to the study of animal personalities: a review of case studies. Behaviour 142, 1185-1206.

vanRaaij, M. T. , Pit, D. S. , Balm, P. H. , Steffens, A. B. and Van Den Thillart, G. E. (1996). Behavioral strategy and the physiological stress response in rainbow trout exposed to severe hypoxia. Horm. Behav. 30, 85-92.

Vaz-Serrano, J. , Ruiz-Gomez, M. L. , Gjøen, H. M. , Skov, P. V. , Huntingford, F. A. , Øverli, Ø. , et al. (2011). Consistent boldness behaviour in early emerging fry of domesticated Atlantic salmon (Salmo salar): decoupling of behavioural and physiological traits of the proactive stress coping style. Physiol. Behav. 103, 359-364.

Veenema, A. H. and Neumann, I. D. (2007). Neurobiological mechanisms of aggression and stress coping: a comparative study in mouse and rat selection lines. Brain Behav. Evol. 70, 274-285.

Vijayan, M. M. and Leatherland, J. F. (1990). High stocking density affects cortisol secretion and tissue distribution in brook charr, Salvelinus fontinalis. J. Endocrinol. 124, 311-318.

Wainwright, P. E. (2002). Dietary essential fatty acids and brain function: a developmental perspective on mechanisms. Proc. Nutr. Soc. 61, 61-69.

Wendelaar Bonga, S. E. (1997). The stress response in fish. Physiol. Rev. 77, 591-624.

Wilson, A. D. and Krause, J. (2012). Metamorphosis and animal personality: a neglected opportunity. Trends Ecol. Evol. 27, 529-531.

Wilson, A. D., Godin, J. G. J. and Ward, A. J. (2010). Boldness and reproductive fitness correlates in the eastern mosquitofish, Gambusia holbrooki. Ethology 116, 96-104.

Winberg, S. and Lepage, O. (1998). Elevation of brain 5-HT activity, POMC expression, and plasma cortisol in socially subordinate rainbow trout. Am. J. Physiol. 274, 645-654.

Winberg, S. and Nilsson, G. E. (1992). Induction of social dominance by L-dopa treatment in Arctic charr. Neuroreport 3, 243-246.

Winberg, S. and Nilsson, G. E. (1993). Roles of brain monoamine neurotransmitters in agonistic behavior and stress reactions, with particular reference to fish. Comp. Biochem. Physiol. C Pharmacol. Toxicol. Endocrinol. 106, 597-614.

Winberg, S., Nilsson, G. E., Spruijt, B. M. and Höglund, U. (1993). Spontaneous locomotor activity in Arctic charr measured by a computerized imaging technique—role of brain serotonergic activity. J. Exp. Biol. 179, 213-232.

Winberg, S., Nilsson, A., Hylland, P., Söderstöm, V. and Nilsson, G. E. (1997). Serotonin as a regulator of hypothalamic-pituitary-interrenal activity in teleost fish. Neurosci. Lett. 230, 113-116.

Winberg, S., Øverli, Ø. and Lepage, O. (2001). Suppression of aggression in rainbow trout (Oncorhynchus mykiss) by dietary L-tryptophan. J. Exp. Biol. 204, 3867-3876.

Wolkers, C. P. B., Serra, M., Hoshiba, M. A. and Urbinati, E. C. (2012). Dietary i-tryptophan alters aggression in juvenile matrinxa Brycon amazonicus. Fish Physiol. Biochem. 38, 819-827.

Wolkers, C. P. B., Serra, M., Szawka, R. E. and Urbinati, E. C. (2014). The time course of aggressive behaviour in juvenile matrinxa Brycon amazonicus fed with dietary i-tryptophan supplementation. J. Fish. Biol. 84, 45-57.

Wommack, J. C. and Deville, Y. (2007). Stress, aggression, and puberty: neuroendocrine correlates of the development of agonistic behavior in golden hamsters. Brain Behav. Evol. 70, 267-273.

Wu, A., Ying, Z. and Gomez-Pinilla, F. (2004). Dietary omega-3 fatty acids normalize BDNF levels, reduce oxidative damage, and counteract learning disability after traumatic brain injury in rats. J. Neurotrauma 21, 1457-1467.

Yamamoto, T., Ueda, H. and Higashi, S. (1998). Correlation among dominance status, metabolic rate and otolith size in masu salmon. J. Fish. Biol. 52, 281-290.

Young, G., Thorarensen, H. and Davie, P. S. (1996). 11-ketotestosterone suppresses interrenal activity in rainbow trout (Oncorhynchus mykiss). Gen. Comp. Endocrinol. 103, 301-307.

Zadori, D., Klivenyi, P., Vamos, E., Fulop, F., Toldi, J. and Vecsei, L. (2009). Kynurenines in chronic neurodegenerative disorders: future therapeutic strategies. J. Neural. Transm. 116, 1403-1409.

第3章　鱼类应激反应的内分泌学
——一种适应性-生理学的观点

任何生物有机体应对环境的挑战，恰当的处理应激状态对其成活都是至关重要的。现存的鱼类是地球上最早的脊椎动物，它们能够很好地适应环境的变迁，因而得到广阔的有时是快速的传播发展。祖先的基因组扩展（二次或三次全基因组复制周期）和稳定的水环境使鱼类得以成功地进化并最终发展成为四足类动物。精密的内分泌组织结构提供了一个下丘脑式的化学媒介体，综合各种信号以便合理利用能量，并在遇到应激情况时进行搏斗或者逃避。我们论述鱼类前脑和（非详尽无遗的）下丘脑所展现的最新见解，主要都是在对斑马鱼的研究中获得的。促肾上腺皮质激素释放因子、肾上腺皮质激素（ACTH）、α-黑色素细胞激素、肾上腺素和皮质醇，对这些下丘脑-脑垂体-肾间腺（HPI）的关键化学介导物都做了综述并且展现在应激反应的异型稳态调节的文本中。我们把本章献给 S. E. Wendelaar Bonga，我们的朋友和导师，他带领我们进入应激的理念之中，并教导我们如何去研究它。

3.1　导　言

认识到鱼类的巨大多样性以及在生理的应激反应中复杂多样的变化结果是非常重要的。我们所知道的哺乳类应激反应生理学在很大程度上都是从模式动物，如小鼠和大鼠获得的，它们是通过近亲交配以消除动物之间的变异；而鱼类应激的研究是以分类学上差异很大的种类以及在一个种类和一个种群内差异很大的个体为基础进行的。我们可能会忘记我们所了解的关于人体应激生理学的许多方面都是和单个种类即人（*Homo sapiens*）的生理学相联系的。

现存的硬骨鱼类①，是最主要的栖息在世界海洋中的脊椎动物。现存鱼类的巨大发展，估计达到35000种，通常都归因于在早期脊椎动物进化发展过程中至少两次全基因组复制（whole genome duplication，WGDs）历程（Dehal，Boore，2005）。WGDs产生一个丰富的遗传库，编码新的蛋白质，以形成精密而不断增长的对水环境中可供

---

①　现存硬骨鱼类，*Osteichthys*，是用鳃呼吸的脊椎动物，具有硬骨胳，硬骨鳍条（鳞质鳍条，lepidotrichia），衍生成对的肺或气鳔，包括叶（肉）鳍鱼类（*Sarcopterygii*）（脊椎动物鳍的结构是肱骨，相当于胸鳍，而股骨相当于其腹鳍）和辐鳍鱼类（*Actinopterygii*）（脊椎动物具单个背鳍，菱形鳞片，内骨胳和肌肉并不延伸到鳍内，有长而柔性的鳍条形成的鳍）。叶（肉）鳍鱼类由两种矛尾鱼（*coelacanths*）和三种肺鱼（澳大利亚肺鱼目，*Ceratodontiformes*）组成。至今已知有35000种辐鳍鱼类，约占已知脊椎动物种类的70%（见柏克莱网址：http：//evolution. berkeley. edu/evolibrary/article/fishtree-01）。

利用而丰富的小生境的适应性。适应于小生境需要多种多样的行为进化（从游离的生活方式到成群结队和精巧的亲代抚育），这取决于它们的生物复杂性。的确，我们发现鱼类适应于地球上的每一种小生境，包含水生的以及水生之外的。值得注意的是，在最早的鱼类出现在大约5亿年前的奥陶纪（Ordovicium）之后，至少发生过五次动物区系性灭绝（mass extinction），这是全世界范围的灾难，而明显的是，鱼类存活下来了（Long，1995）。有记载的是，所有牙形石（conodont）和一些谱系的鱼类如甲胄鱼类（ostracoderms）、棘鱼类（acanthodians）（棘鲨）和盾皮鱼类（placoderms）（如 *Dunkleosteus*）都已灭绝。我们可以推测，相对稳定的海水环境提供了一个缓冲的介体，以对付威胁到生存的激烈环境变化。海水环境通过其缓冲的作用、缓慢的变化，为叶（肉）鳍鱼类和辐鳍鱼类的适应赢得了时间。鱼类强大的进化能力亦就是由于它们能够产生可遗传的表现型变异和丰富的生理多样化，鲈形目鱼类（10033种现存的鱼类）非凡的传播发展（距今约1亿1千万年）就是很好的说明。而在近期（距今约100万年），非洲马拉维湖的丽鱼科鱼类已经从一个祖先繁衍发展超过500种鱼类（Albertson等，1999）。

早先的和现今的脊椎动物能够获得进化的成功，至关重要的一种适应性是能够应对以任何应激形式（真正应激或痛苦）出现的动态性变化。对由持续变化的环境产生的刺激必须妥善处理并且学会应付重复出现的状况。在一个异型稳态的范围内（Schreck，Tort，2016；见本书第1章），我们认为当一个应激物负面影响一个生命有机体时就会出现痛苦的状态。其结果是，它们不能够发动适宜而有效的在正常情况下由肾上腺素和皮质醇介导的内分泌反应，以及迅速转移糖（肾上腺素）和调整能量分配以满足新的需求（皮质醇），从而未能恢复到应激前的状态。接着，对应激物的成功反应，肾上腺素和皮质醇的大量分泌及时而恰当地制止应激物并重新恢复正常生理状态，因而是有助于生存的。我们必须强调的是，要能恰当地应对应激，就需要学习和记忆，能够辨别危险的和安全的状况，确定厌恶的和非厌恶的刺激，并且意识到来自世界各处的感知。鱼类具有这些性能吗？

1997年，我们的同事Wendelaar Bonga发表了一篇经典式的论文《鱼类应激反应》（Wendelaar Bonga，1997）。在本章中，我们着重于鱼类内分泌研究的发展，而在那篇创新性的论文中，由斑马鱼研究提供的近期进展占了很大的篇幅。Wendelaar Bonga的主要结论是，儿茶酚胺能的信使和类固醇的应激通道以及它们的功能（刺激氧的摄取和输送，转移能量并从生长与生殖中将能量重新分配，免疫抑制），在所有脊椎动物中，从鱼类到人类，大都是相似的。明显的差别是涉及应激干扰的全身的水盐平衡；它们因为肾上腺素扰乱上皮（皮肤和鳃）对水分与离子的可渗透性和皮质醇的作用（通过特异性受体特征而作用于它们的靶目标，如糖皮质类固醇和盐皮质类固醇），并且和其他的激素如催乳激素和生长激素（GH）相互作用而受到影响。重要的是，还要考虑到海水鱼类（和海水中的广盐性鱼类）必须饮水以保持水盐平衡。但是，在应激时，淡水鱼类亦会饮水（Takei，2002；Huising等，2003；Nobata等，2013）；从渗透压调节的观点来看，这是一个并不恰当的行为，由于明显的渗透作用会使水流入淡水鱼类体内（即它们并不需要饮水）。接着，鱼类栖息的水中离子组成，包括其中潜在的污染物，不仅影响广阔的鳃表面，亦经常影响到精细密致的肠道。显然，水中的污染物可以起到毒物和诱发内分泌反应的应激物作用。水温、PH、离子组成和钙含量都可以缓和或者加剧水盐的扰乱，而所有这些在研究应激物的强度和性质时都必须考虑到。水环境和生活方式对一个生命有机体应激反应的复杂性来说，无疑起着重要作用。

自 1997 年起，不断变化发展的资源目录把已经估计的鱼类种类数量从大约 20000 种增加到 35000 种，特别是由于增长的注释基因组数量和现代分子生物学的仪器设备，这将会对增长的应激适应性进行更为深刻的解析。经常想到“陆地/离开水生境”亦是重要的，这可以导致鱼类和陆栖脊椎动物在应激反应方面相比较时出现差别。

3.1.1 鱼的前脑

要懂得脊椎动物的应激反应，重要的是要认识和掌握在反应中复杂行为的作用。正如前面所述，在一个适宜的应激反应中，记忆、学习、辨别和预知对于应对一个动态的环境都是重要的，需要脑的结构主宰这些行为。从进化的观点看，就鱼类的成功而论，脊椎动物脑的基本结构图型符合这些性能就必须发育到相当大的复杂程度，因而寻求一个统一实体以进行这些复杂的行为是必要的。外行的鱼类生理学者可能会对现代生物学发现鱼类简单的脑而感到惊奇，而它所表现的关键作用是在斑马鱼的研究中取得的，斑马鱼现在是研究发育生物学和生理学最好的模式脊椎动物（Kalueff 等，2014；Stewart 等，2014）。

早在 19 世纪末（Gage，1893；Studnicka，1896），就研究人员就已经认识到鱼类前脑的发育和较高等的脊椎动物有所不同，它是由神经管的最吻端部分外翻（向外弯曲）而形成一对结实的叶状体，而由神经组织围绕的侧脑室完全缺失。这种结构保留在所有的辐鳍鱼类中，包括硬骨鱼类超目（各种硬骨鱼类）、多鳍鱼类（多鳍鱼和芦鳗）、软骨硬鳞类（鲟鱼和匙吻鲟）、鳞骨鱼类（雀鳝）和弓鳍鱼类（弓鳍鱼）。外翻是辐鳍鱼类前脑基本的形态发生过程（Nieuwenhuys 等，1998；Nieuwenhuys，2009）。直到 1960 年，流行的推测都认为鱼类（实际上是所有爬行类前的脊椎动物）前脑的发育和组织结构是嗅觉输入的线路占有优势（Herrick，1933）。然而，我们现在从一项连续流动的研究中认识到硬骨鱼类端脑的发育要复杂得多，并且是高度变化的。从视觉、听觉、电感器、嗅觉和体感器官系统的输入通过前小球（preglomerular）核和丘球核而转导到前脑的亚区（Ito，Yamamoto，2009）。连贯性形成的图像表明，所有脊椎动物前脑同源物的趋同进化，不仅整合必要的生理化学输入（从外界环境）和内在环境的信号（如能量状态，氧需要量，内部环境），而且支配着应对生物群落多变的信息而产生反应的复杂行为。这些信息可能是有利的，保留回报的期望；或者是不利的，意味着危险带来的损害、焦虑、害怕以及持久的慢性应激。的确，已有的证据就是鱼类（辐鳍鱼类和多鳍鱼类）的前脑包含着和哺乳类的海马、扁桃体、梨形皮层和异皮层（isocortex）同源的脑区（Portavella 等，2002，2004；Northcutt，2006；Mueller 等，2011；Mueller，2012）。我们认为，从进化的观点看，应对正面和负面不断增强的环境信号，应付真正的应激和痛苦，在充满竞争的脊椎动物世界中已经能够完美地成活下来。最早期的脊椎动物，对付应激相关的内分泌系统亦已发展到相当完善的程度。我们强调应激生理学的生态-生理学观点的重要性。这就是要考虑到任何生物有机体的生态学，包括从水中过渡到陆地，或者在复杂性不断增长的脊椎动物社会群落中，包括从鱼类到人类。

3.1.2　应激

在任何有关应激的论文中，应激的定义都会再次被提到，因为这个名词运用于生命科学的许多学科中。“应激”一词原先来自物理学，用以描述钢条对弯曲力的抗性，或者换个说法，钢条对影响它的力量产生的恢复力。心理学科使用这个名词描述对一个生物系统的压力，并且包括在生命有机体压力中的人的情绪性成分。这当然就引起了复杂而又混淆不清的局面。Schreck 和 Tort（2016；见本书第 1 章）把应激定义为：“当生命有机体面临损害时力图抵抗死亡或重建正常稳态所发生的一连串生理过程。”在本章中，针对化学介导物，我们把狭义的应激定义为：由一个因子（应激物）诱导的可能是有利的和不利的一种内分泌反应的状态。当应激反应导致及时地回复到应激前的状态并恢复稳态的调节，我们把这种情况划分为真正应激或良性应激（的确，不是所有的应激都是不好的）。当应激物变为慢性，皮质醇并不能恢复到应激前的状态，如果内分泌反应不能及时地回复到应激前的状态，痛苦或非良性应激就出现，稳定的调节持续被破坏，随之就进入病理状态了。真正应激和痛苦的概念对异型稳态的概念来说是重要的，在本章的结尾，我们还要进一步详细论述。

重要的是，我们把应激反应的肾上腺素能作用组合看作是恰当的应激反应，正如肾上腺素立即释放能量以进行搏斗或者逃逸（Selye，1950，1973）。肾上腺素作用增强之后，反应的皮质固醇成分（皮质醇）实际上对适应有着密切关系：皮质醇适当地起作用，搏斗或者逃避后随之较多的持续能量分配并重建应激前的状态，包括重新设置稳态调控的调节点（Korte 等，2007；Koolhaas 等，2011）。通过它在恢复通透性调控的干扰中与肾上腺素的拮抗作用以及肾上腺素大量分泌所引起的细胞凋亡，可以充分了解皮质醇的作用（皮质醇是催乳激素和 GH 作用的许可性因子）。最后，适当的能量平衡就是应激调节所要求达到的。我们用两个例子来简单阐明这点：渗透压调节和摄食的能量代价。

鱼类受到水生境影响，它们具有非角质化的皮肤，经常受到明显的耗费能量的应激。大多数鱼类保持大约 300 mOsmol/kg 的血浆重量摩尔渗透压浓度，这是指在淡水中（重量摩尔渗透压浓度大约是 20 mOsmol/kg）；而在海水中（重量摩尔渗透压浓度大约是 1000 mOsmol/kg），渗透梯度分别为 280 mOsmol/kg 和 700 mOsmol/kg。生活在淡水或海水中的鱼类需要保持它们的渗透平衡，能量消耗的过程取决于主动的离子转运（由腺苷三磷酸 ATP-消费酶类介导和次级的主动交换作用机理）；显然，半咸水

(重量摩尔渗透压浓度大约是 300 mOsmol/kg) 使渗透挑战减少到最小，而适应在半咸水中生活的鱼类可以预期在对付应激方面会有较大的能量预算。确实，在等渗性的介质中，渗透压调节的代价是最低的（Febry，Lutz，1987)。亚洲河口红树林地带高效率的有鳍鱼类产量（高生育力和良好的保护条件）进一步证明了这种见解，那里是处于等渗性的（Primavera，2005)。半咸水的地区历来都是适宜于开展水产养殖的。此外，广盐性鱼类通常都会有一段时间生活在河口生境中，特别是由幼鱼过渡到成鱼的脆弱阶段和快速生长时期。这些重要的生活时期有利于将渗透调节的能量消耗减低到最少并且能够优先把能量安排投入到生长和对付应激当中。特别是，鱼类能量预算中渗透调节的后果是我们在应激生理学的论述中所涉及的。关于应激和渗透调节的详细阐述见于 Takei 和 Hwang（2016；见本书第 6 章)。

摄食的能量可采用名词“特种动力作用”（specific dynamic action，SDA）来表示，它是一餐的摄入、消化、吸收和同化作用的能量消耗总和。SDA 的特征是餐后迅速增加代谢速率达到峰值，而后缓慢地回复到餐前水平。由于消化作用而增加的代谢速率在鱼类可以高达 136%（Fitzgibbon 等，2007)。获得能量需要付出代价，一点都不少，因为消化作用是高能量耗费的。此外，摄食的细致精确调控，包含了过多的内分泌反应，表现为内分泌应激轴各种成分的相互作用和双向交流沟通。在本章，我们将注重前面曾经论述过的和鱼类特别有关方面的内分泌应激轴的各种成分。

3.2 应激和脑：神经-内分泌下丘脑

早期鱼类在海洋的两个主要带，即远洋带（pelagic zone）和海底带（benthic zone）扩展。在海底带，底栖生活的叶（肉）鳍鱼类（成为后来所有四足类动物的基础）繁荣发展；而在远洋带，辐鳍鱼类迅速繁衍传播而获取巨大的发展。我们发现鱼类大体上有两种不同的下丘脑-脑垂体解剖构造，一种具有门脉系统（即正中隆起，如矛尾鱼和肺鱼类)，而另一种没有（所有现存的硬骨鱼类)。现代的硬骨鱼类失去了早期某些辐鳍鱼类和叶（肉）鳍鱼类（和所有较高等的脊椎动物，即叶鳍鱼类的后裔四足类动物）具有的较为原始的鲨型脑垂体门脉系统。

这种解剖学的差别就使得具有下丘脑门脉系统的鱼类，其下丘脑能前馈（feedforward）控制脑垂体的排出物而起着一个真正内分泌器官的作用；而那些没有门脉系统鱼类的下丘脑，其信号是神经递质特性的。这种情况影响到脑垂体作为靶标的反应性，因为通常在突触间隙的受体，其结合的亲和力都比较低，而传统的受体对内分泌信号具有高的亲和力。体液信号在血液循环中受到高度的稀释而需要受体的适应。血管床容量决定激素稀释的程度，这就是激素释放后在血液循环中的浓度，因而受体的亲和力随之要调整；而在一种内分泌型的状态下，可以预期其亲和力是高的。在一个神经递质型的状态下，激素释放于突触间隙而后才达到其目标，可以设想在突触间隙中较高的激素浓度是决定受体质量的因素。此外，在后一种状态中，是轴突-

靶细胞的解剖学结构决定着调控的作用；而在内分泌型的状态中，是靶细胞的受体型式决定着调控的作用。

在辐鳍鱼类的硬骨鱼类（见前述），视前区神经元发出含有促肾上腺皮质激素释放因子（CRF）的轴突直接到达脑垂体ACTH细胞［以及促黑激素（MSH）细胞，见稍后一节］。所以，现代硬骨鱼类从下丘脑发出两个分开的神经内分泌通道，即CRF通道到达脑垂体远侧部，而另一个通道携带精氨酸加压素（AVP），硬骨鱼类催产素（IST）和CRF到达脑垂体神经部和MSH细胞。后一种通道对MSH细胞进行直接的神经元调控（类似于在ACTH细胞的情况），以及对外周目标的神经内分泌调控，而使脑垂体中间部起着神经血器官的作用。除了通过脑垂体神经部释放下丘脑的CRF之外，鱼类还有独特的非脑的CRF［硬骨鱼紧张肽Ⅰ（UI）］表达部位，即尾神经内分泌系统（CNSS）（Lu等，2004；Craig等，2005；Bernier等，2008）。这种神经内分泌器官（CNSS）位于脊髓的尾端，它释放CRF和UI，其实质的作用是控制硬骨鱼类的脑垂体和调节应激反应（Bernier等，2008）。此外，头肾的嗜铬组织产生和释放CRF和CRF-结合蛋白（CRFBP）（Huising等，2007；见第3.4.3节）。通过原位杂交研究确切说明的是CRF受体在头肾的细胞定位，以便证实是否存在（超-）短环反馈作用机理。一系列外周器官表达CRF_1受体，包括鳃、（头）肾脏、脾脏和心脏（Huising等，2007），而这些组织就是体液的CRF和相关肽的靶标。

除CRF之外，CRF-相关肽对外周的调控可能起着很大的作用。值得一提的是终身一次产卵的太平洋马苏大麻哈鱼（*Oncorhynchus masou*），它们在溯河上游洄游期间达到完全的性成熟，产卵一次然后迅速衰老而死亡。这个过程的特征是显著的高皮质醇血症（hypercortisolinemia），它可以认为是造成鱼死亡的原因。产卵大麻哈鱼的皮质醇水平比起幼鱼和未产卵鱼的非应激水平（典型是<10ng/ml）要增长5～7倍（Westring等，2008）。Westring和他的同事表明是UI而不是CRF对这种高皮质醇血症起着更为直接的作用，因为UI的表达水平一直升高并和皮质醇分泌的增长紧密联系，而CRF的表达峰值出现两次，和皮质醇的分泌活动联系不明显。换句话说，在成熟的马苏大麻哈鱼中，调控皮质醇产生的作用可能是有差别的，即UI调控基本组成的，而CRF调控应激的。这种对皮质醇产生的差别调控状况可以联想到莫桑比克罗非鱼处在酸化水中慢性应激时所观察到的对头肾的独立调控状况（Lamers等，1994）。

外部环境由许多传感器持续地进行着监测。在记录到对内在环境的外来威吓（如出现捕食者）、水质的变化或者和水产养殖相关的应激物时，对这种潜在威胁的感知就会激活下丘脑，而在下丘脑中最重要的就是应激轴的视前核（NPO）和外侧结节核（NLT）。由这些核产生下丘脑调节脑垂体的重要信号（Cerda-Reverter，Canosa，2009）。

在对斑马鱼幼鱼的研究中（Fernandes等，2013），表明旁系同源的orthopedia转录因子（orthopedia transcription factors，Otps；斑马鱼otpa和otpb基因的产物）和其他的一起，特别是间脑的下丘脑神经元亚型产生CRF和精氨酸加压催产素（AVT）

以及腹间脑的多巴胺（DA）产生细胞。这个作用过程并不局限于斑马鱼，在所有脊椎动物，包括人类都是保守的。在所有的脊椎动物中，CRF/AVT和DA神经元的Otp-依存性证实了一个共同的基本结构图型的见解，它们都处在下丘脑功能解剖的精细水平上。我们在斑马鱼的下丘脑确实发现小细胞和大细胞的细胞群。前者突出到脑垂体远侧部，在那里释放CRF、生长抑素和促甲状腺素释放激素（TRH），因而在内分泌应激轴中起关键作用。后者产生IST（在斑马鱼中称为催产素OXT，因为其基因为催产素直向同源物）和AVT（在斑马鱼中称为加压素AVP，因为其基因为加压素直向同源物）并释放到主要位于脑垂体中间部的神经血区。要注意的是，这些细胞经常共同表达下丘脑调控脑垂体的信号（主要是CRF加上AVT、缩胆囊肽加上脑啡肽-B；Herget，Ryu，2015），而这可能会妨碍我们对单个信号的了解。由下丘脑的核内紧密混杂的细胞亚型产生一系列下丘脑调控脑垂体的信号是极为普遍的，而我们要了解这些多因素的信号混合物只有期待通过分析这些细胞的同源异型基因框（homebox gene）的产物-依赖性、多基因的原位杂交、它们的蛋白质产物，以及在差别信号释放基础上的受体型式，以鉴别各种单个细胞的特性才能实现。现在对斑马鱼引人注目的研究成果已经迅速出现在文献资料中，它们对脊椎动物比较内分泌学的贡献是无可估量的（Machluf 等，2011；Fernandes 等，2013；Herget 等，2014；Herget，Ryu，2015）。

3.2.1 基本轴的相互作用

下丘脑整合对应激处理、生长和生殖基本过程的调控作用。四个三联的（下丘脑、脑垂体、外周内分泌腺体）神经内分泌轴参加这些调控作用，它们都存在于所有的脊椎动物中。我们要强调的是应激轴和其他非应激轴是交叉作用的。这个见解有助于阐明一些复杂的现象，如生长的突然迸发（季节的或昼夜的）、代谢的抑制和性成熟（雄性化和雌性化）。生命意味着能量消耗，所以，对付应激和生长的激烈能量消耗过程在保持存活的优先调控内容中占据着最重要的位置。

下丘脑调控生长的关键激素是生长激素释放激素、生长素释放肽和生长抑素，而这些激素决定脑垂体GH以及最后肝脏的胰岛素样生长因子的产生。对于应激的处理，下丘脑产生的主要激素是CRF、AVT、黑色素浓集激素（MCH）和DA，它们决定脑垂体ACTH以及肾间腺（肾上腺）皮质醇的产生；皮质醇的作用对于将能量分配到急需能量的器官是必不可少的；如同前文所陈述的，环境对生命有机体发起挑战时，需要重新分配能量以恢复到应激前的状态。和应激轴平行的第二个轴是甲状腺轴，下丘脑产生的TRH、CRF和DA促使脑垂体分泌促甲状腺激素，它调控甲状腺产生四碘甲腺原氨酸（T_4）和三碘甲腺原氨酸（T_3）。甲状腺激素是基础代谢和变态相关过程的必需调节剂，并经常在能量消耗活动中起着允许作用。在生殖轴中，促性腺激素-释放激素、亲吻激素（kisspeptin）和促性腺激素-抑制激素从下丘脑分泌产生，作用于脑垂体，促使脑垂体产生促黄体激素和卵泡-刺激激素，最终调控精巢产生睾酮和卵巢产生雄二醇。

在论述应激/能量的有效性时，另一个必须注意的概念是鱼类和许多动物的生命方式一样是变温动物（外温动物），定温运物（内温运物）在生长早期消耗的能量要比变温动物高三倍左右。尽管生产子代的能量消耗（估计每天每公斤卵质量、孵化物或幼仔为250 kJ）在变温动物和定温动物是相近的，但鱼类可能消耗35%的可代谢能量，而定温动物只是2%～6%（Wieser，1985）。接着，在能量消耗过程的调节方面包括对付应激，鱼类和我们研究中所了解的哺乳类之间会有很大差别，重要的是在致力于对付应激时需要考虑到生命史阶段和生殖状态。在涉及变温动物方面另一个需要考虑的是鱼类可能出现行为和情绪的发热（Boltanna 等，2013；Rey 等，2015）。例如斑马鱼表现热的偏爱选择，在接触病毒感染或应激时，它们较多时间停留在较高温度中（+2℃～4℃）。这个反应是真正意义的发热，它激活了和哺乳类相同的免疫-生理的与代谢的通道。这个反应是即时的并延续达8小时。我们应该考虑到进行应激相关试验的结果要少于8小时的时限，这就可以防止动物充分利用它们的适应性能，而它们可能只是在最适度的范围以内起作用。

对于鱼类的应激轴，我们现在将CRF和AVT看成是主要的正面的前馈信号，而把DA和MCH看成是负面的前馈信号。实际的情况则要复杂得多，如在脑垂体的吻端远侧部（rPD），它包含ACTH和催乳激素细胞，接受除CRF、AVT、DA和MCH之外的许多其他的ACTH和MCH释放-调节因子，包括IST、生长激素释放激素、脑垂体腺苷酸环化酶激活多肽（PACAP）、生长抑素和甘丙肽。Cerda-Reverter和Canosa（2009），Bernier等（2009）对这些下丘脑调控脑垂体的因子进行了详尽的综述。

3.2.2　CRF系统

鱼类和哺乳类一样，CRF（在本章中，我们依然采用国际药理学联盟制定的常规命名法；Hauger 等，2003）是调控应激轴的主要下丘脑因子（Vale 等，1981；Wendelaar Bonga，1997；Filk 等，2006）。CRF从下丘脑的视前区释放，促使脑垂体远侧部释放ACTH（见第3.3.1节），进而从头肾释放皮质醇（见第3.4.2节）。CRF是41个氨基酸组成的多肽，由一个160个氨基酸的前体衍生而来（Vale 等，1981）。在整个脊椎动物当中是明显保守的（Huising 等，2004；Huising，2006）。除CRF外，CRF家族还包括UI（Pearson 等，1980），哺乳类的尿皮质素-1（Ucn1）（Vanghon 等，1995）、尿皮质素-2和尿皮质素-3（Pearson 等，1980；Coulouarn 等，1998；Lovejoy，Balment，1999；Lewis 等，2001）。我们认为Boorse和Denver（Boorse 等，2005）对脊椎动物CRF做了一个出色的系统发生总论。大多数脊椎动物都有两个CRF受体，即CRF_1受体和CRF_2受体（第三个CRF受体只在对鲶鱼的研究中描述过；Arai 等，2001），它们对CRF家族不同成员的亲和力明显不同。然而CRF_1受体对CRF和UI/Ucn1的亲和力相等（Arai 等，2001；Coste 等，2002；Manuel 等，2014c），而CRF_2受体对UI和尿皮质素要比对CRF有更高的亲和力（Wei 等，1998；

Hsu, Hsueh, 2001; Manuel 等, 2014c)。

CRF 生物利用率的关键调节物是 CRFBP。CRFBP 是一个由 322 个氨基酸组成的蛋白质，其结构和 CRF 受体没有联系（Potter 等，1991），它亦存在于无脊椎动物当中（Huising, Filk, 2005）。无脊椎动物并不分泌产生 CRF，这表明 CRFBP 是作为伴侣蛋白以及在无脊椎动物和脊椎动物的功能中起着不同的作用。实际上，昆虫的 CRFBP 和调控水分平衡的利尿激素（DH）是蛋白伴侣。陆生环境及其潜在的干旱对入侵陆地昆虫的水分平衡是一种严重的应激源，DH-CRFBP 的相互协同作用可以对调控这类应激源发挥作用。我们把在脊椎动物内分泌应激轴中的作用看作是 CRFBP 和 CRF 作为伴侣的趋同新功能。值得注意的是，DH 和 CRF 以及相关蛋白质在寒武纪（即 7 亿～9.93 亿年前）有一个共同的祖先（Huising, Filk, 2005），这和 CRFBP 起源于非常早的时期是一致的。$CRFBP_S$和 CRF 与 UI/Ucn1 有不同的结合部位（Huising 等，2008），并且它们和这些配体结合的亲和力要较高于 CRF 受体的结合亲和力（Sutton 等，1995；Manual 等，2014c）。这些结合亲和力的型式表明 CRFBP 和它的配体之间的平衡是朝向结合蛋白-配体复合物移动，而 CRFBP 对于决定 CRF（及其相关物）游离而具有生物活性的部分起重要作用。有关受体结合部位的状况/微环境以及这三种蛋白质类群当中的相互作用还了解得很少。在应激的生理学研究中评价蛋白质的基因表达，通常都可以看到 CRFBP 的表达能很好地说明预期的内分泌应激物活性（Geven 等，2006）：在应激状态中，CRFBP 表达受抑制，在基础的非应激状态中，CRFBP 的表达上调。

3.2.3 CRF 系统的个体发育

在鱼的卵中证明 CRF mRNA 的存在似乎是不可思议的事。Chandrasekar 和他的同事（Chandrasekar 等，2007）通过原位杂交在受精后 24 小时的斑马鱼显示 CRF mRNA；而采用较为敏感的半定量 RT-PCR 技术甚至在未受精卵中证明 crf，其受体 crf1 和 crf2 以及 crfbp 的 mRNA 的存在（Alderman, Bernier, 2009）。然而，引起的问题是，如此明显的母体沉积的信号对胚胎的作用何在？虽然我们并不知道这种信号在卵中（即卵黄或胚盘）的分布情况，但我们清楚地知道受精和第一次卵裂后所有的细胞和卵黄都可以通过细胞质敞开接触，而小的分子能在动物极的卵黄与细胞之间移动（Kimmel, Law, 1985；Kimmel 等，1995）。实际上，一些母体沉积的 RNAs 受精后可能在动物极的胎盘中混合在一起（如 Theusch 等，2006；Kosaka 等，2007）。显然，母体的 CRF 信号在早期发育中已经起作用，而在胚胎的应激生理中并不需要：尽管 CRF_1受体亦存在于未受精卵中，但这个信号在受精后 6 ～ 12 小时之间消失（Alderman, Bernier, 2009）。这可以解释为一些鱼类在孵化期间由无应答性到发生应激（以皮质醇的产生而测定出来）的基本作用机理，包括虹鳟（*Oncorhynchus mykiss*）（Barry 等，1995a，b）、大鳞大麻哈鱼（*Oncorhynchus tshawytscha*）（Feist,

Schreck，2002）、斑马鱼（Alsop，Vijayan，2009）和欧洲鲈鱼（*Dicentrarchus labrax*）（Tsalafouta 等，2014），但并不出现在所有鱼类，如鲤鱼（*Cyprinus carpio*）（Stouthart 等，1998）和莫桑比克罗非鱼（*Oreochromis mossambicus*）（Pepels 等，2004）在孵化前就没有应激反应的表现。要注意的是应激反应（即皮质醇产生）并不需要 CRF-依赖性（Fuzzen 等，2011；Tsalafouta 等，2014）。大多数鱼类的生存策略可能在它们早期生命阶段的应激反应都是进行二叉分枝式（dichotomy）。

3.2.4　对脑垂体的调控

采用特异性的免疫组织化学方法（Huising 等，2004；图 3.1）进行检测，在脑垂体 ACTH 细胞邻近发现 CRF 和 CRFBP 这些蛋白质，它们还大量存在于脑垂体中间部。在应激轴，CRF 和 CRFBP 同时发生（不需要在较高的端脑区和视叶；Alderman，Bernier，2007）。目前，我们还缺少一种操作或技术来观察和定量测定 CRF 及其在体的结合蛋白，我们只能够推测这些分子在应激生理学中的生理意义，尽管分子分析测定（在应激范例中基因上调和下调）提供了这些信号参与应激反应的明显证据（Bernier 等，1999；Huising 等，2004；Bernier，Craig，2005；Madaro 等，2015，2016）；然而，这些蛋白质作用的真实数据在相关的文献资料中是没有的。mRNA 和蛋白质分子半寿期的差别进一步增加了我们研究的复杂性（Maier 等，2009；Schwanhausser 等，2011）。不过，在对一些鱼类采用（基因重组的）蛋白质/肽对脑垂体组织进行灌注的研究中证明了这些作用，但在所有的研究实例中，对于高度复杂的由多种因子调控的应激类固醇激素产生的实际情况来说，都是过于简单化了。能说明这种复杂性的一项研究是鲤鱼脑垂体释放 ACTH 和 MSH。和 DA 支配调控在体 ACTH 细胞的作用一样（Metz 等，2005；见第 3.3 节），在灌流系统中的脑垂体外植体（即从下丘脑多巴胺能的调控中移取出来）表明能增加 ACTH 的释放。在这样一种装置中我们不能证明 CRF-刺激 ACTH 释放，除非将脑垂体组织安置在 DA 的紧张性状态影响之中（Metz 等，2005）。显然，多巴胺能的环-AMP 第二信使通道和 CRF（亦是 C-AMP 通道）的互相依存决定着 ACTH 的胞吐作用，而这看来是在体内发生的。

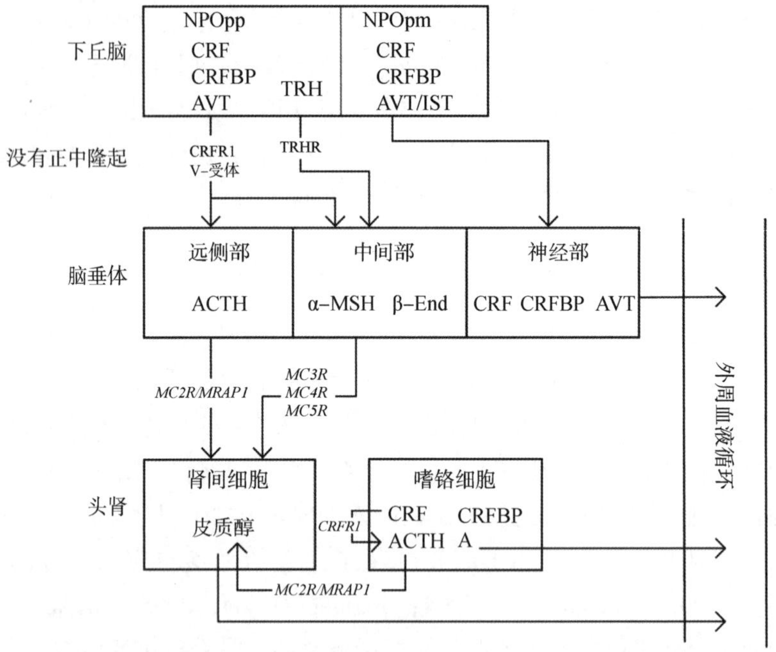

图 3.1 辐鳍鱼类/硬骨鱼类下丘脑-脑垂体-肾间腺（HPI）轴及其内分泌应激反应的化学介体的图解

注：在下丘脑神经内分泌细胞的腹方、视前核（NPOpp）小细胞群发出含有促肾上腺皮质激素释放因子（CRF）的轴突到脑垂体肾上腺皮质激素（ACTH）细胞（没有正中隆起）；接着，转运主要的前馈作用（feedforward）到脑垂体以启动内分泌应激反应；而一些尚未能明确界定的（位于 NPO 的边缘）成群细胞突起，并行一起将 CRF 结合蛋白（CRFBP）输送到 ACTH 细胞。CRF 和 CRFBP 之间的平衡决定 CRF 的生物活性，亦就是激活脑垂体 CRF-受体（$CRFR_1$）。NPOpp 细胞和精氨酸加压素（AVT）可能还有硬骨鱼类催产素（IST）共表达。脑垂体中间部的促黑色素细胞刺激激素（α-MSH）产生细胞受到下丘脑的 CRF/CRFBP 和促甲状腺素释放激素（TRH，通过其受体 TR）的调控。ACTH 信号通过 MC2R-MRAP1-复合物（2MC2Rs 和 4MRAP1s）而处在其靶标中，这些受体对 ACTH 是高度特异性的（和其他 α-MSH-序列没有或者很少混杂）。应激到来时释放的 α-MSH 能刺激肾间腺皮质醇的组成性释放，并起着（应激相关的）厌食性信号作用（设想环状的 α-MSH 分子能容易地通过血脑屏障）。在 NPO（NPOpm）的背侧大细胞群中的神经内分泌细胞伸入到神经血的脑垂体中间部，将信号送到身体血液循环中。本图解由图 3.2 的免疫化学研究结果加以说明。

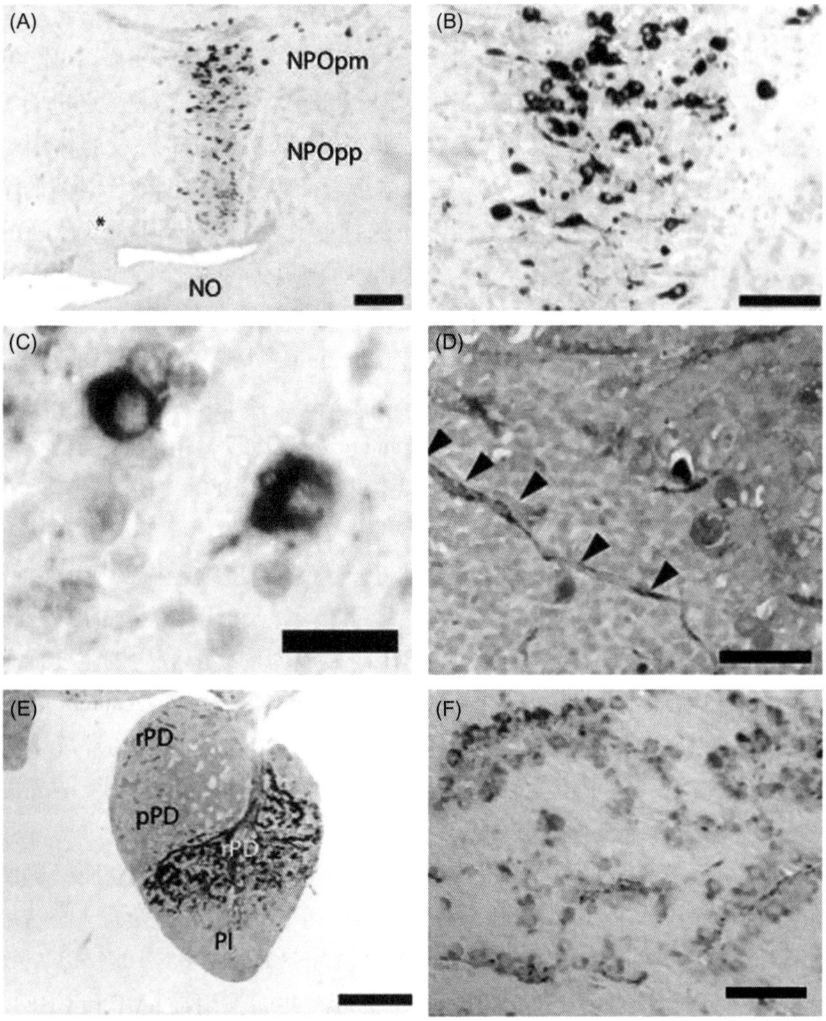

图 3.2　鲤鱼（*Cyprinus carpio*）下丘脑-脑垂体复合体的免疫组织化学研究

注：（A）视前核用兔抗羊促肾上腺素皮质激素释放因子（CRF）血清染色的矢切面，表示背侧大细胞群（NPOpm）；（B）视前核腹侧小细胞群（NPOpp）的放大；NPO 位于视神经（NO）背侧室空间上方。在（A）中的星号表示 CRF-结合蛋白（CRFBP）表达细胞的位置，它们表示于（C）。在（D）中有一束含有 CRFBP 的轴突（箭头所指）伸向远侧部［PD，见（E）］；CRFBP 采用兔抗人 CRFBP 抗血清（#5144；1∶1000；由已故 Wylie Vale 博士赠送）。（E）表示鲤鱼脑垂体用抗-CRF 血清（和 A 相同）染色后的矢切面。注意深度着色的神经部（PN）伸入到中间部（PI），以及染色的神经纤维出现在吻端远侧部（rPD），但没有出现在近端远侧部（PPD）。（F）蓝色 CRF 轴突伸入到远侧部的促肾上腺皮质激素细胞（褐色，ACTH；ACTH 抗体是抗鲤鱼 Cys-ACTH［10-23］；Metz 等，2004）。初级抗体用羊抗兔 IgG-生物素（1∶200，Bio-rad）以 Vectastain ABC 扩增试剂盒进行扩增。图（A-E）引自 Huising M O，Metz J R，van Schooten C，Taverne-Thiele A J，Hermsen T，Verburgvan Kemanade B M L，等（2004）。鲤鱼（*Cyprinus carpio*）CRH、CRH-BP 和 CRH-R1 的结构特征以及这些蛋白质在急性应激反应中的作用。J. Mol. Endocrinol. 32，627 – 648，经许可引用。（见书后彩图）

在几种鱼类的相关研究中已确认 CRF 的 ACTH 释放功能，包括金鱼（Fryer 等，1984）、鳟鱼（Baker 等，1996）、金头鲷（Rotllant 等，2000）、莫桑比克罗非鱼（van Enckevort 等，2000）和鲤鱼（Metz 等，2005）。然而，对鲤鱼进行试验时，Metz 和他的同事使用离体的灌流系统未能证明 CRF 刺激 ACTH-释放的作用。在他们的试验中，基础的 ACTH 分泌由游离的脑垂体缓慢而稳定地超过预定时间的增加。他们的结论是 ACTH 的在体释放必定是在紧张的抑制作用（离体时因下丘脑的联系被切断而失去这种作用）调控之下，这为非应激动物提供一种保持血液循环中 ACTH 低水平的作用机理。在视前区的多巴胺能神经元伸入到脑垂体（Kah 等，1986），通过 DA-受体（D2 型）抑制 ACTH 分泌（Selbie 等，1989；Civelli 等，1993；Stefaneanu 等，2001）。因此，当 DA 以能够激活 D-2 受体的浓度共同施用时，才可以证明 CRF-刺激离体的脑垂体释放 ACTH（Lamers 等，1997）。所以，至少对于鲤鱼，要诱发导致 ACTH 分泌的细胞内通道，就需要激活 CRF_1 和 D-2 受体（Metz 等，2005）。

莫桑比克罗非鱼处在酸化（pH 4.0）水中 5 天引起持续轻微的应激（或者是实际上的真正应激），结果对脑垂体分泌物产生明显不同和出人意料的调控作用。其次，刺激性的 DA-受体（D-1 型）表达，TRH 明显成为皮质醇产生的主要而适度的调节剂，脑垂体中间部产生的双乙酰化-MSH 成为脑垂体的介体（Lamers 等，1997）。这种状态会使得 CRF/ACTH/皮质醇通道在一个为多种应激物挑战的环境中，免除一种较快速/急性的应激反应。应激轴的这种双重组织似乎是受到水环境的支配，并为鱼类生理适应策略提供了良好的条件。

Baker 和同事们（Baker 等，1996）证明鳟鱼 CRF 和 AVT 能独立而又协调地作用在离体刺激 ACTH 分泌。Person 和合作者（Person 等，1996）证实了这些结果，其研究亦包括 IST 是另一个刺激因子。CRF 的作用最强（ED_{50}：8×10^{-7}M），AVT 次之（ED_{50}：2×10^{-10}M），接着是 IST（ED_{50}：10^{-7}M）。共同施用（AVT 和 CRF）并不影响脑垂体产生 MSH；然而，和 ACTH 共同施用时，AVT 能增强皮质醇的释放，这和它们从脑垂体中间部 MSH 与生长乳素相联系的（神经血的）神经部释放出来后的主要体液作用是一致的（Jerez-Cepa 等，2016）。

CRF-和 UI-表达神经元可以募集起来，从而在不同应激物的反应中以相似的方式起作用（Bernier 等，2008）。实际上，下丘脑的 UI-表达神经元将神经分布到鱼类的脑垂体（Lederis 等，1994），但还不清楚是否从这些细胞释放 UI 直接调控 ACTH 的分泌（Bernier，Craig，2005）。在后一项研究中，Bernier 和 Craig 发现低氧能刺激虹鳟下丘脑视前区 CRF 和 UI 的基因表达，但不影响 UI 和 ACTH 之间的联系。UI 在鳟鱼脑腔内注射后所起的厌食性信号的效能是 CRF 的 5 倍多（Bernier，2006；Ortega 等，2013）。已经清楚的是，应激的鱼类会失去食欲与停止摄食。这种在脊椎动物脑内应激和摄食双重线路的调控网络不仅强调双向交流沟通的重要性，亦说明相互独立的能量消耗（应激）和供给（摄食）生理学所需要的时间。对于调控和应激相关的摄食抑制的神经回路（Volkoff 等，2005；Bernier，2006；Gorissen 等，2006）目前还很少了解，CRF 受体在下丘脑、较高的脑中心以及脑干的功能特征还有待深入阐明。

3.2.5　CRF 和行为

CRF 除了对应激轴的调节起着已经清楚阐明的作用之外，它（似乎是独立的）还影响鱼类的行为。CRF 能减弱竞争能力，CRF 和 UI 能引起焦虑（Backstrom 等，2011）和虹鳟通氧过度（Le Mevel 等，2009）。此外，处于社会从属地位使鳟鱼下丘脑视前区 CRF mRNA 水平增加（Doyon 等，2005），表明行为本身亦能影响下丘脑的 CRF 活性。有关 CRF 和行为的详细论述，可参阅 Noakes 和 Jones（2016；见本书第 9 章）。

3.3　应激和脑垂体

脑垂体通常被认为是主要的内分泌腺体，是脑和外周内分泌调节系统的相交处。来自较高的脑中枢（包括端脑的皮质中心）的信息到达脑垂体后，由它发出化学信号通过血液输送到外周的内分泌组织，这些化学信号携带综合的信使以调控生长、应激处理、能量代谢、生殖和免疫。在鱼类的早期发育中，gh（最早和最丰富的）和阿黑皮素原（POMC）基因（时间和丰度是其次的）表达（Pogoda，Hammerschmidt，2009）和它们的生理调控作用一致：起初是生长，其次是功能的应激轴以应对变动而有危险的环境。

信号的特异性最终是由外周靶目标的受体类型决定的。必须考虑到多种激素在受体通道之间的相互作用，而要了解这些相互作用需要精准确定靶目标的受体类型。在斑马鱼和青鳉（*Oryzias latipes*；Temminck，Schlegel），它们是基因组已经完全注释的模式种类，我们现在应该转回头来研究它们的各种激素，并且着重于受体的类型，以阐明靶目标的作用过程以及在第二信使通道的互相交叉作用和双向的交流沟通。

对付应激是一个需要能量的过程，其中两种激素起关键性作用：肾上腺素和皮质醇。肾上腺素在几秒到几分钟内，调动能量如葡萄糖以准备搏斗或者逃逸；而皮质醇实际上是适应性的激素，在体内安排能量流，在几小时到几天的时间内通过能量在体内的重新分配而回复到应激前的状态（Korte 等，2007；Koolhaas 等，2011）。依据这个看法，皮质醇的产生必须调控在这样的时间框架内，并且需要既保持基本的维持状态，取决于异型稳态的负荷而进行不同水平的组成性释放，亦应满足较为急性的需求，即所谓的峰涌（peak surge）。脑垂体含有两个细胞群体表达阿黑皮素原（POMC），它们在应激和能量家室中起重要作用；在肾上腺皮质激素细胞中，POMC 加工为 ACTH 和 β-促脂解素（LPH），而在促黑色素激素细胞中，POMC 加工为 α-MSH、中间叶促皮质样肽（CLIP）和 β-内啡肽。ACTH 和 α-MSH 是出自高度多效和多因子黑皮质素系统的两种信号，包括多种配体和多种明确界定的受体（Metz 等，2006；Takahashi 等，2006）。值得注意的是，α-MSH 和 β-内啡肽的乙酰化决定它们的生物活性；乙酰化的 α-MSH（Lamers 等，1992）比非乙酰化的肽有更强的效价，

但乙酰化使 β-内啡肽的作用减弱（Wilkinson，2006）。所以，翻译后乙酰化的机制对脑垂体中间部的排出物起着重要作用。稍后将论述 ACTH 和乙酰化 α-MSH 在鱼类应激反应中的作用。在斑马鱼两个 POMC 细胞群体的发育期间，发现它们源自迁入脑垂体的垂体基板（hypophysial placode）：先是 ACTH 细胞，其次是 α-MSH 细胞（Herzog 等，2003）。我们可能想知道在这种分配基础上的 POMC 分子究竟提供哪些信息。激素原转变酶（aka 枯草蛋白酶，枯草杆菌产生丝氨酸蛋白酶）最终确定 POMC 加工的程度和加工的产物（Zhou 等，1993；Tanaka，2003）。LPH 与 CLIP 的信息和应激生理学的联系很少，这些信号将不在本章论述。最近对 β-内啡肽的了解比较多些，阿片样物质受体的所有组成都存在于脑和外周器官中（Chadzinska 等，2009）。β-内啡肽在哺乳类急性和慢性应激期间升高，并且在应激反应结束时起重要作用（Drolet 等，2001）。这种作用在鱼类应激生理学中尚未确定。在对鲤鱼的试验中曾观察到不同的多少是截短的 β-内啡肽起着突触前的抑制性调节作用（van den Burg 等，2001），看来这好像是脊椎动物的一种特性。

3.3.1 促肾上腺皮质激素（ACTH）

到达前脑垂体后，CRF-CRFBP 复合体必须离解，使游离的 CRF 能够激活其相关的 CRF_1 受体并在促肾上腺皮质激素细胞中表达（Wynn 等，1983；Chen 等，1993；Huising 等，2004），从而刺激 POMC-衍生肽，如 ACTH、β-内啡肽和其他（Benjannet 等，1991；Castro，Morrison，1997）的合成（Bruhn 等，1984）与释放（River 等，1982）。血液循环 ACTH 的靶标是头肾的肾间腺细胞，它们合成与分泌糖皮质激素皮质醇（Filk 等，2006）。这就是内分泌的串联反应过程，我们称之为内分泌应激轴。

除了已经明确阐明的调控皮质醇产生与释放之外，ACTH 还有什么作用呢（Dores 等，2014）？要回答这个问题，我们就要先问问：我们在哪里能找到 ACTH 受体，即所谓的黑皮质素受体 2（MC2R）？在鱼类中，MC2R 实际上几乎全部都存在于头肾内（Metz 等，2005，2006），可能在头肾的肾间腺细胞内，但还没有确定。头肾组织包含多种不同的细胞类型，包括免疫系统和应激轴互相作用的细胞系（Verburg-van Kemenade，Schreck，2007）。在红鳍东方鲀（*Takifugu rubripes*），MC2R 出现在端脑和下丘脑组织中（Klovins，2003），而在其他鱼类，则和 MC4R 与 MC5R 一起（Metz 等，2006）。

由于缺乏合适的抗体/抗血清以检测血浆的激素水平，对鱼类 ACTH 功能的研究很少。此外，POMC-转录丰度分析结果的转达只限于在翻译后发生信号多样性的有关资料中记载。在 mRNA 和蛋白质水平将 ACTH 从 α-MSH（以及 POMCs）区分出来的仪器设备是必要的。幸运的是，免疫组织化学和原位杂交已经证明 ACTH 和 α-MSH 的生产细胞在解剖和结构方面是分开的，这意味着适当的分离这些细胞群是可行的，但目前还很少应用。在 3.4.4 节，我们将详尽介绍瘦素（瘦蛋白）对应激轴的作用，并且证明这个能量成分的信号通过直接影响 ACTH 和皮质醇的产生与分泌而减弱应激轴的活性（Gorissen 等，2012）。

3.3.2　α-MSH

在慢性应激期间，血浆的 α-MSH 水平升高（Lamers 等，1994；Arends 等，2000；Metz 等，2005）。研究表明，α-MSH 对处于慢性应激的莫桑比克鱼（Lamers 等，1997）和处于海水中的条斑星鲽（*Verasper moseri*）（Kobayashi 等，2011）起着促皮质素作用；海水高的重量摩尔渗透压浓度（1000 mOsmol/kg）给鱼类造成不可回避的渗透应激负荷。但是，在鲤鱼的相关试验中，α-MSH 的任何与应激相关的作用都未能得到证明，不管是单独的 α-MSH，或者是 α-MSH 连同共同释放的 POMC-衍生肽，特别是截短的 β-内啡肽（Metz 等，2005；van den Burg 等，2005）。这表明 α-MSH 对皮质醇释放的特异性作用不仅是种类特异性的，而且是生活-阶段依赖性的（如淡水对海水的生活阶段）。可以用几种解释来说明 α-MSH 缺少促皮质素作用。α-MSH 是一个环形多肽，而 ACTH（它包含其前面 13 个位置的 α-MSH 序列）是一个线形多肽，因此，受体-结合部分是不同的。α-MSH 和 ACTH 都具有结合模体 HFRW，其中 W 对于配体受体互相作用是最重要的氨基酸（Liang 等，2015）；而推想 ACTH 还具有一个附加而独特的 RKPRP 结合模体（Dores 等，2014）。另外，黑皮质素受体（MCRs）能展现其明显相似的多样性和特异性，因此，这就是为何使得它和配体混杂的结合？最重要的是，ACTH 需要共表达与共同调节的蛋白质黑皮质素 2 受体附属蛋白质 1（MRAP1），以便能够和其相关的 MC2R 结合（Agulleiro 等，2010；Liang 等，2015；Dores 等，2016）（Faught 等，2016；见本书第 4 章）。这似乎排除了α-MSH 对 ACTH 受体通道的作用。然而，在虹鳟（Haitina 等，2004）、鲤鱼（Metz 等，2005）和条纹星鲽（Kobayashi 等，2011）头肾组织中发现潜在的其他 α-MSH-敏感的 MCRs（MC1R、MC3R、MC4R 和 MC5R）亦可能存在于莫桑比克罗非鱼中，从而说明 α-MSH 的促皮质素作用。

3.4　应激和头肾

头肾组织存在于鱼类的谱系中（Hofmann 等，2010），包含嗜铬细胞、肾间腺组织、红细胞生成的和白细胞生成的细胞系、免疫系统的次级淋巴部分。在所有的脊椎动物中，脾脏监察着血液循环中可能潜伏的病原体，亦是血液红细胞成熟和凋亡后被清除的场所。胸腺细胞来自头肾，在胸腺中成熟；而产生抗体的 B 细胞在头肾产生与成熟。前面描述的是成鱼的状态，而在早期（斑马鱼）的发育阶段，在到达成鱼阶段前曾发生两次免疫细胞祖先迁移的高潮（Chen，Zon，2009）。

近年的许多研究致力于下丘脑和脑垂体内分泌学。但近来对鱼类肾间腺的专门研究不多。肾间腺组织通常都被描述为应激轴串联过程中的决定性部位和皮质醇产生的位置，而有些研究是在类固醇-生成细胞和脑垂体-肾间腺相互作用的早期发育阶段进行的（Hsu 等，2004；Liu，2001；To 等，2007）。常见的研究主要是描述不同种鱼类肾间腺解剖结构的差异，但并没有发现有意义的内分泌新见解。要深入了解应激轴的激活，必须详细阐明 HPI 轴各个成分（特别是肾上腺素和皮质醇-产生细胞之间的交流沟通）的交流沟通状况。关于嗜铬细胞对皮质醇-产生细胞的影响，或者反过来，皮质醇产生细胞对嗜铬细胞的影响，我们知道些什么呢?

3.4.1 儿茶酚胺生产细胞

硬骨鱼类和哺乳类肾上腺髓质的同种组织是嗜铬细胞，它们和头肾的肾间腺组织混合在一起。嗜铬细胞来源于神经嵴细胞，和皮质醇-产生细胞一起排列在后主静脉周围（在各种硬骨鱼类当中有所不同)。鲤科鱼类，如鲤鱼和斑马鱼，肾间腺组织形成颇为结实的独立器官，而大多数其他鱼类，肾间腺组织松散地排列在后主静脉周围，在前肾的肾小管之间，位于鳃室之后的体腔背侧。嗜铬组织的得名来源于其采用传统的组织化学染色技术证明它们能和铬酸盐与次-铬酸盐结合的特性。

3.4.2 类固醇-生成细胞

硬骨鱼类和哺乳类肾上腺皮质的同种组织是沿着头肾内的后主静脉分布的类固醇-生成细胞，头肾是硬骨鱼类特有的器官，来源于前肾（pronephros)。皮质类固醇生成细胞的主要产物是皮质醇，这和人类一样。然而，鱼类缺乏醛固酮合成酶以产生一种特异性的盐皮质类固醇。皮质醇的功能既是一种糖皮质类固醇信号，亦是一种盐皮质类固醇信号；是靶细胞的受体构成决定由皮质醇受体复合物（F-GR 或 F-MR）解读基因组的某一个区段。此外，受体/转录因子通过它们和基因的皮质醇-应答元件相互作用决定应答基因是被激活还是被抑制。皮质醇作用的主要靶组织包括鳃、肝脏、肠上皮和脑。

所以，我们在鱼类头肾看到靠得很近的产生儿茶酚胺和皮质醇的细胞，以及免疫细胞的细胞因子和相关信号的所有组成成分，它们能够进行局部的和外周的相互作用与协调，并且表现互相交叉作用。我们可以参阅 Yada 和 Tort（2016；见本书第 10 章）关于和免疫系统相互作用的详细论述。

3.4.3 在头肾内的交流沟通

关于鱼类儿茶酚胺贮存与释放的一篇典范性综述（Reid 等，1998）谈到肾上腺素能反应的输入支进行生物合成与胞吐释放的过程。他们的结论是皮质醇只是在较长的期限（注射含有皮质醇的椰子油沉积物后 3 ～7 天；Reid 等，1998）才会影响肾上腺素的贮存与释放。皮质醇能敏化肾上腺素释放，但不能刺激儿茶酚胺在原位的分泌活动。对虹鳟的研究（Reid 等，1996）证明皮质醇并不影响苯基乙醇胺-N-甲基转移酶；这种酶的作用是把去甲肾上腺素甲基化为肾上腺素（和在哺乳类研究的结果不同，Betito 等，1992)。这种情况是否在所有鱼类都一样，还需要对 Blaschko 通道(Blaschko，1939）基因编码酶类中的皮质醇应答元件做进一步的分析研究。

鲤鱼 CRF 和 CRFBP 存在于嗜铬细胞的一个亚群中，和它们的神经嵴来源相吻合(Huising 等，2007)，但 CRF 和 CRFBP 在这些细胞中的作用还不清楚。在离体用 8-

溴-AMP 刺激后，CRF 能从头肾组织中释放出来。从鲤鱼的 CRF 系统可以联想到哺乳类肾上腺内的 CRF 系统，它参与调控成束的糖皮质激素释放和肾上腺血流控制。在皮质细胞缺少 CRF_1 受体就会联想到必须鉴别一种髓质的成分以阐明 CRF 的这种间接的作用。实际上，ACTH 是和 CRF 一起产生与释放的（Suda 等，1984；Ehrhart-Bornstein 等，2000）。我们认为，在鱼类中，CRF 可能起着一种自分泌的前馈作用，促使 ACTH 产生和分泌；然后 ACTH 和肾间腺细胞的 MC2R 结合而产生和分泌皮质醇。ACTH 直接而剂量依存地刺激鳟鱼释放肾上腺素（Reid 等，1996），这种情况可能出现在所有鱼类中，并且和这两种类型细胞紧密的解剖组织结构相符合。这个系统或许能够对中枢启动的应激反应结果进行局部的精细调整，它是一种明显存在于鱼类和哺乳类共同祖先（～4. 5 亿年前）的性状（Huising 等，2007）。

3. 4. 4　应激和能量

所有的应激反应都使短暂的高血糖达到高峰以便给生命有机体/鱼类提供燃料进行搏斗或者逃逸。下丘脑对嗜铬细胞直接的神经作用，促使糖原分解的肾上腺素以 1 秒到 1 分钟的速度释放。然后，皮质醇起作用以维持高血糖状态数分钟到数小时，对能量进行适应性的重新分布和重新配置。实际上，肾上腺素是真正的应激激素，皮质醇是适应性激素（Korte 等，2007；Koolhaas 等，2010，2011）。

通过对虹鳟的一系列研究，Soengas 实验室（西班牙的 Vigo）证明葡萄糖能调节 ACTH-诱导皮质醇的释放（Conde-Sieira 等，2013）。高的葡萄糖水平能增强离体的皮质醇释放，而松胞菌素-B（一种葡萄糖转运的抑制剂）减弱其释放。这对鱼类糖血状态和非胰腺激素释放之间的联系提供了基本的证据。接着，在急性应激状况下，高血糖和皮质醇的合成和释放密切相关。在这种情况下，肾上腺素不仅使葡萄糖分离出来，还作为一种高血糖的信号，起着 ACTH-介导皮质醇释放的必要条件的作用。对头肾碎片的离体研究证明了一种葡萄糖传感蛋白（特异性的葡萄激酶/己糖激酶Ⅳ），并为头肾肾间腺细胞的免疫组织化学研究所确认。这些研究结果证实了关于肾间腺的葡萄糖敏感性必定是皮质醇生成细胞一种固有特性的见解。应激轴和能量代谢是明显牵连在一起的，可以用一种调控食物/能量摄入的重要因子瘦素（瘦蛋白）的作用进来一步说明。瘦素通过直接抑制脑垂体组成性的和 CRF-调节的 ACTH 释放以及肾间腺皮质醇的释放而抑制应激轴的活性（Gorissen 等，2012）。提高鱼类（鲤科、鲑科）瘦素水平的一种方式是让它们处于应激的低氧状态中（Bernier 等，2012；MacDonald 等，2014）。取决于低氧的程度，能量消耗或多或少降低。在这种情况下发生大的应激反应可能会产生相反的结果（counterproductive），因为应激和能量代谢是紧密联系的。有学者曾经提到瘦素能把能量成分的信息转达到中枢脑区并且降低应激反应。瘦素以这种方式在协调真正应激与痛苦之间的精细平衡中发挥重要作用（Gorissen，Filk，2014）。

3.5 综述和展望

异型稳态（McEwen，Wingfield，2003）定义为“保持或重建稳态的主动过程”（Romero 等，2009）。接着，我们可以把异型稳态看作是生物有机体在这个主动过程中产生化学介体的能力，也就是和应激相关的激素（肾上腺素和皮质醇）、细胞因子、副交感神经活性（Romero 等，2009），以便适应新的状态或者一个环境的挑战。个体在感知（即鉴别）方面有很大差异，而异型稳态能使个体对变化环境的生理反应得到恰当的定制（customization）。从正常环境状态的偏离产生一种称为异型稳态的状态，在这种状态下，动物的调节能力发生变化（Koob，LeMoal，2001；Korte 等，2007；McEwen，Wingfield，2010）。这种情况的特征是增加异型稳态的主要化学介体，并且如同前面曾界定的，只能保持一个相当短的时期。尽管应激反应最初是一种适应性反应，持续受到应激物的影响就会导致一个称为异型稳态负荷的状态，疾病和病理发生，结果是身体衰竭（McEwen，Wingfield，2003，2010；Korte 等，2007；Schreck，2010）。根据先前的经历，预知一个现行状态结果的能力，对于异型稳态的原理是十分重要的（McEwen，Wingfield，2003），这使得生命有机体能够重新安排行为的和生理调节点，以便较好地应对挑战或者变化的环境状况。业已清楚的是，异型稳态的概念对于我们了解在应激反应性、应对方式、动物特征以及基本的化学介体等方面的个体差异起了重要的作用。

Manuel 和他的合作者（Manuel 等，2014a，b；Gorissen 等，2015；Manuel 等，2016）进行的一系列研究深入阐述了异型稳态模型的功能，他们采用不同年龄、不同品系（AB 和 Tupfel 长鳍），先前经过富集和不可预知的慢性应激（接受变数）处理的斑马鱼进行抑制性回避畏惧的实验范例（操作变数：休克次数和休克强度）。动物对付应激的能力是由被试验者的特征和环境特征之间的相互作用决定的。当环境状况适宜时（如发育期间营养丰富的环境），鱼类的应对能力增强；而在不良的环境状况下（慢性应激、贫瘠状态等）就会削弱应对能力（Manuel，2015；图 3.3）。这种多维的应激对付异型稳态的观点加深了我们对一般脊椎动物特别是鱼类复原能力的理解。比起在教科书中我们研究哺乳类与人类应激轴所观察到的自上而下的传统调控作用，鱼类作为脊椎动物最早期的代表，实际上所包含的内容要丰富得多。有一系列的因子参与，包括 HPI 轴在肾间腺组织中局部产生的成分（CRF 和 ACTH）、肾间腺葡萄糖传感蛋白的确证以及能量-感知的作用机制和葡萄糖的利用。这些都说明应激生理学的复杂性、多效性，甚至丰余性。复杂的调控作用必定是早期脊椎动物创建的，在这个基础上，鱼类和所有四足动物得以成功地进化与发展。

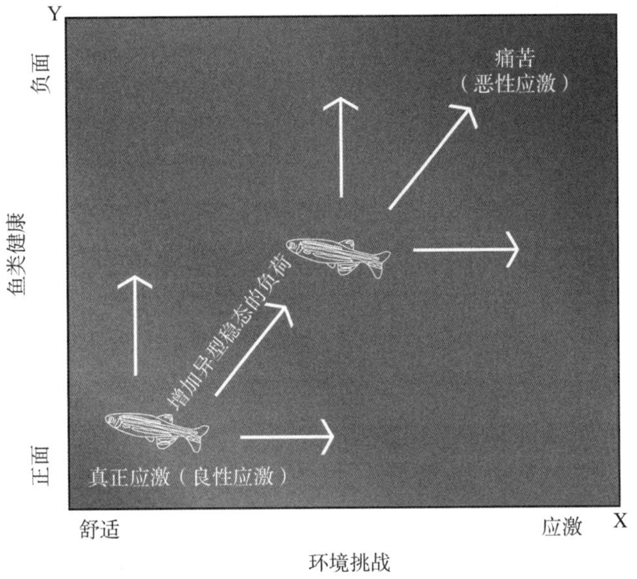

图 3.3　由鱼类适合度（健康）（Y 轴）和环境挑战（X 轴）所表示的异型应激景观的图解

注：绿色反映真正应激的情况，红色表示痛苦。从真正应激到痛苦的过渡是多因素的和二维的。很明显，一个应激的环境对不健康的鱼要比非常健康的鱼有更大的影响。健康的鱼能够应对较多的环境挑战。在实际生活中，环境和鱼类特征形成连续统一体并一起决定异型稳态的负荷。对鱼类增加异型稳态负荷（如频繁的分选和运输的应激操作过程）能很快进入痛苦状态。应激意味着能量的耗费和重新安置，因此，感知它们所处的环境对于产生恰当的应激反应是很重要的：肾上腺素和皮质醇介导新的需求并且能够重新设定生理的调节点。（见书后彩图）

M. 戈里森　G. 弗利克　著

林浩然　译、校

参考文献

Agulleiro, M. J., Roy, S., Sanchez, E., Puchol, S., Gallo-Payet, N. and Cerda-Reverter, J. M. (2010). Role of melanocortin receptor accessory proteins in the function of zebrafish melanocortin receptor type 2. Mol. Cell. Endocrinol. 320, 145-152.

Albertson, R. C., Markert, J. A., Danley, P. D. and Kocher, T. D. (1999). Phylogeny of a rapidly evolving clade: the cichlidfishes of Lake Malawi, East Africa. Proc. Natl. Acad. Sci. U. S. A. 96, 5107-5110.

Alderman, S. L. and Bernier, N. J. (2007). Localization of corticotropin-releasing factor, urotensin I, and CRF-binding protein gene expression in the brain of the zebrafish, Danio rerio. J. Comp. Neurol. 502, 783-793. http://dx.doi.org/10.1002/cne.21332.

Alderman, S. L. and Bernier, N. J. (2009). Ontogeny of the corticotropin-releasing factor system in zebrafish. Gen. Comp. Endocrinol. 164, 61-69. http://dx.doi.org/10.1016/j.ygcen.2009.04.007.

Alsop, D. and Vijayan, M. M. (2009). Molecular programming of the corticosteroid stress axis during zebrafish development. Comp. Biochem. Physiol. Mol. Integr. Physiol. 153, 49-54. http://dx.doi.org/10.1016/j.cbpa.2008.12.008.

Arai, M., Assil, I. Q. and Abou-Samra, A. B. (2001). Characterization of three corticotropinreleasing factor receptors in catfish: a novel third receptor is predominantly expressed in pituitary and urophysis. Endocrinology 142, 446-454. http://dx.doi.org/10.1210/ endo.142.1.7879.

Arends, R. J., Rotllant, J., Metz, J. R., Mancera, J. M., Wendelaar Bonga, S. E. and Flik, G. (2000). alpha-MSH acetylation in the pituitary gland of the sea bream (Sparus aurata L.) in response to different backgrounds, confinement and air exposure. J. Endocrinol. 166, 427-435.

Backstrom, T., Pettersson, A., Johansson, V. and Winberg, S. (2011). CRF and urotensin I effects on aggression and anxiety-like behavior in rainbow trout. J. Exp. Biol. 214, 907-914. http://dx.doi.org/10.1242/jeb.045070.

Baker, B. I., Bird, D. J. and Buckingham, J. C. (1996). In the trout, CRH and AVT synergize to stimulate ACTH release. Regul. Pept. 67, 207-210.

Barry, T. P., Malison, J. A., Held, J. A. and Parrish, J. J. (1995a). Ontogeny of the cortisol stress response in larval rainbow trout. Gen. Comp. Endocrinol. 97, 57-65. http://dx.doi.org/ 10.1006/gcen.1995.1006.

Barry, T. P., Ochiai, M. and Malison, J. A. (1995b). In vitro effects of ACTH on interrenal corticosteroidogenesis during early larval development in rainbow trout. Gen. Comp. Endocrinol. 99, 382-387. http://dx.doi.org/10.1006/gcen.1995.1122.

Benjannet, S., Rondeau, N., Day, R., Chre 'tien, M. and Seidah, N. G. (1991). PC1 and PC2 are proprotein convertases capable of cleaving proopiomelanocortin at distinct pairs of basic residues. Proc. Natl. Acad. Sci. U. S. A. 88, 3564-3568.

Bernier, N. J. (2006). The corticotropin-releasing factor system as a mediator of the appetitesuppressing effects of stress infish. Gen. Comp. Endocrinol. 146, 45-55. http://dx.doi.org/ 10.1016/j.ygcen.2005.11.016.

Bernier, N. J. and Craig, P. M. (2005). CRF-related peptides contribute to stress response and regulation of appetite in hypoxic rainbow trout. Am. J. Physiol. Regul. Integr. Comp. Physiol. 289, R982-R990. http://dx.doi.org/10.1152/ajpregu.00668.2004.

Bernier, N. J., Lin, X. W. and Peter, R. E. (1999). Differential expression of corticotropinreleasing factor (CRF) and urotensin I precursor genes, and evidence of

CRF gene expression regulated by cortisol in goldfish brain. Gen. Comp. Endocrinol. 116, 461-477. http: //dx. doi. org/10. 1006/gcen. 1999. 7386.

Bernier, N. J. , Alderman, S. L. and Bristow, E. N. (2008). Heads or tails? Stressor-specific expression of corticotropin-releasing factor and urotensin I in the preoptic area and caudal neurosecretory system of rainbow trout. J. Endocrinol. 196, 637-648. http: //dx. doi. org/ 10. 1677/JOE-07-0568.

Bernier, N. J. , Flik, G. and Klaren, P. H. M. (2009). Regulation and Contribution of The Corticotropic, Melanotropic and Thyrotropic Axes to The Stress Response in Fishes. In Fish Physiology-Fish Neuroendocrinology, Vol. 28 (eds. N. J. Bernier, G. Van der Kraak, A. P. Farrell and C. J. Brauner), pp. 235-311. London: Academic Press.

Bernier, N. J. , Gorissen, M. and Flik, G. (2012). Differential effects of chronic hypoxia and feed restriction on the expression of leptin and its receptor, food intake regulation and the endocrine stress response in common carp. J. Exp. Biol. 215, 2273-2282. http: //dx. doi. org/10. 1242/jeb. 066183.

Betito, K. , Diorio, J. , Meaney, M. J. and Boksa, P. (1992). Adrenal phenyle thanolamine Nmethyltrans feraseinduction in relationtog lucocorticoidrece ptordynamics: evidenceth atacute exposure to high cortisol levels is sufficient to induce the enzyme. J. Neurochem. 58, 1853-1862.

Blaschko, H. (1939). The specific action of L-Dopa decarboxylase. J. Physiol. (Lond.) 96, 50P-51P.

Boltaña, S. , Rey, S. , Roher, N. , Vargas, R. , Huerta, M. , Huntingford, F. A. , et al. (2013). Behavioural fever is a synergic signal amplifying the innate immune response. Proc. Biol. Sci. 280, 20131381. http: //dx. doi. org/10. 1098/rspb. 2013. 1381.

Boorse, G. C. , Crespi, E. J. , Dautzenberg, F. M. and Denver, R. J. (2005). Urocortins of the South African clawed frog, Xenopus laevis: conservation of structure and function in tetrapod evolution. Endocrinology 146, 4851-4860. http: //dx. doi. org/10. 1210/en. 20050497.

Bruhn, T. O. , Plotsky, P. M. and Vale, W. W. (1984). Effect of paraventricular lesions on corticotropin-releasing factor (CRF) -like immunoreactivity in the stalk-median eminence: studies on the adrenocorticotropin response to ether stress and exogenous CRF. Endocrinology 114, 57-62. http: //dx. doi. org/10. 1210/endo-114-1-57.

Castro, M. G. and Morrison, E. (1997). Post-translational processing of proopiomelanocortin in the pituitary and in the brain. Crit. Rev. Neurobiol. 11, 35-57.

Cerdá-Reverter, J. M. and Canosa, L. F. (2009). Neuroendocrine systems of the fish brain. In fish physiology-fish neuroendocrinology, Vol. 28 (eds. N. J. Bernier, G. Van der Kraak, A. P. Farrell and C. J. Brauner), pp. 3-74. London: Academic Press.

Chadzinska, M., Hermsen, T., Savelkoul, H. F. J. and Kemenade, B. M. L. V. -V. (2009). Cloning of opioid receptors in common carp (Cyprinus carpio L.) and their involvement in regulation of stress and immune response. Brain Behav. Immun. 23, 257-266. http://dx.doi.org/10.1016/j.bbi.2008.10.003.

Chandrasekar, G., Lauter, G. and Hauptmann, G. (2007). Distribution of corticotropinreleasing hormone in the developing zebrafish brain. J. Comp. Neurol. 505, 337-351./http://dx.doi.org/10.1002/cne.21496S.

Chen, A. T. and Zon, L. I. (2009). Zebrafish blood stem cells. J. Cell. Biochem. 108, 35-42. http://dx.doi.org/10.1002/jcb.22251.

Chen, R., Lewis, K. A., Perrin, M. H. and Vale, W. W. (1993). Expression cloning of a human corticotropin-releasing-factor receptor. Proc. Natl. Acad. Sci. U. S. A. 90, 8967-8971.

Civelli, O., Bunzow, J. R. and Grandy, D. K. (1993). Molecular diversity of the dopamine receptors. Annu. Rev. Pharmacol. Toxicol. 33, 281-307. http://dx.doi.org/10.1146/annurev.pa.33.040193.001433.

Conde-Sieira, M., Alvarez, R., López-Patiño, M. A., MÍguez, J. M., Flik, G. and Soengas, J. L. (2013). ACTH-stimulated cortisol release from head kidney of rainbow trout is modulated by glucose concentration. J. Exp. Biol. 216, 554-567. http://dx.doi.org/10.1242/jeb.076505.

Coste, S. C., Quintos, R. F. and Stenzel-Poore, M. P. (2002). Corticotropin-releasing hormonerelated peptides and receptors: emergent regulators of cardiovascular adaptations to stress. Trends Cardiovasc. Med. 12, 176-182.

Coulouarn, Y., Lihrmann, I., Jegou, S., Anouar, Y., Tostivint, H., Beauvillain, J. C., et al. (1998). Cloning of the cDNA encoding the urotensin II precursor in frog and human reveals intense expression of the urotensin II gene in motoneurons of the spinal cord. Proc. Natl. Acad. Sci. U. S. A. 95, 15803-15808.

Craig, P. M., Al-Timimi, H. and Bernier, N. J. (2005). Differential increase in forebrain and caudal neurosecretory system corticotropin-releasing factor and urotensin I gene expression associated with seawater transfer in rainbow trout. Endocrinology 146, 3851-3860. http://dx.doi.org/10.1210/en.2005-0004.

Dehal, P. and Boore, J. L. (2005). Two rounds of whole genome duplication in the ancestral vertebrate. PLoS Biol. 3, 1700-1708. http://dx.doi.org/10.1371/journal.pbio.0030314.

Dores, R. M. , Londraville, R. L. , Prokop, J. , Davis, P. , Dewey, N. and Lesinski, N. (2014). Molecular evolution of GPCRS: melanocortin/melanocortin receptors. J. Mol. Endocrinol. 52, T29-T42. /http: //dx. doi. org/10. 1530/JME-14-0050.

Doyon, C. , Trudeau, V. L. and Moon, T. W. (2005). Stress elevates corticotropin-releasing factor (CRF) and CRF-binding protein mRNA levels in rainbow trout (Oncorhynchus mykiss). J. Endocrinol. 186, 123-130. http: //dx. doi. org/10. 1677/joe. 1. 06142.

Dores, R. M. , Liang, L. , Hollmann, R. E. , Sandhu, N. and Vijayan, M. M. (2016). Identifying the activation motif in the N-terminal of rainbow trout and zebrafish melanocortin-2 receptor accessory protein 1 (MRAP1) orthologs. General and Comparative Endocrinology In press. http: //dx. doi. org/10. 1016/j. ygcen. 2015. 12. 031.

Drolet, G. , Dumont, E. C. , Gosselin, I. , Kinkead, R. , Laforest, S. and Trottier, J. F. (2001). Role of endogenous opioid system in the regulation of the stress response. Prog. Neuropsychopharmacol. Biol. Psychiatry 25, 729-741.

Ehrhart-Bornstein, M. , Haidan, A. , Alesci, S. and Bornstein, S. R. (2000). Neurotransmitters and neuropeptides in the differential regulation of steroidogenesis in adrenocorticalchromaffin co-cultures. Endocr. Res. 26, 833-842.

Faught, E. , Aluru, N. and Vijayan, M. M. (2016). The Molecular Stress Response. In Fish Physiology-Biology of Stress in Fish, Vol. 35 (eds. C. B. Schreck, L. Tort, A. P. Farrell and C. J. Brauner), San Diego, CA: Academic Press.

Febry, R. and Lutz, P. (1987). Energy partitioning infish: the activity-related cost of osmoregulation in a euryhaline cichlid. J. Exp. Biol. 128, 63-85.

Feist, G. and Schreck, C. B. (2002). Ontogeny of the stress response in chinook salmon, Oncorhynchus tshawytscha. Fish Physiol. Biochem. 25, 31-40.

Fernandes, A. M. , Beddows, E. , Filippi, A. and Driever, W. (2013). Orthopedia transcription factor otpa and otpb paralogous genes function during dopaminergic and neuroendocrine cell specification in larval zebrafish. PLoS One 8, e75002. http: // dx. doi. org/10. 1371/ journal. pone. 0075002.

Fitzgibbon, Q. P. , Seymour, R. S. and Ellis, D. (2007). The energetic consequence of specific dynamic action in southern bluefin tuna Thunnus maccoyii. J. Exp. Biol. 210 (Pt 2), 290-298.

Flik, G. , Klaren, P. H. M. , Van den Burg, E. H. , Metz, J. R. and Huising, M. O. (2006). CRF and stress infish. Gen. Comp. Endocrinol. 146, 36-44. http: //dx. doi. org/10. 1016/j. ygcen. 2005. 11. 005.

Fryer, J. , Lederis, K. and Rivier, J. (1984). Cortisol inhibits the ACTH-releasing activity of urotensin-I, CRF and sauvagine observed with superfused goldfish pituitary-cells. Peptides 5, 925-930.

Fuzzen, M. L. M., Alderman, S. L., Bristow, E. N. and Bernier, N. J. (2011). Ontogeny of the corticotropin-releasing factor system in rainbow trout and differential effects of hypoxia on the endocrine and cellular stress responses during development. Gen. Comp. Endocrinol. 170, 604-612. http://dx.doi.org/10.1016/j.ygcen.2010.11.022.

Gage, S. P. (1893). The Brain of Diemyctilus Viridescens From Larval to Adult Life and Comparison With the Brain of Amia and Petromyzon. Ithaca: The Wilder Quarter Century Book.

Geven, E. J. W., Verkaar, F., Flik, G. and Klaren, P. H. M. (2006). Experimental hyperthyroidism and central mediators of stress axis and thyroid axisactivity in common carp (Cyprinus carpio L.). J. Mol. Endocrinol. 37, 443-452.

Gorissen, M. and Flik, G. (2014). Leptin in teleosteanfish, towards the origins of leptin physiology. J. Chem. Neuroanat. 61-62, 200-206. http://dx.doi.org/10.1016/j.jchemneu.2014.06.005.

Gorissen, M., Marnix, H. A. G., Flik, G. and Huising, M. O. (2006). Peptides and proteins regulating food intake: a comparative view. Anim. Biol. 56, 447-473.

Gorissen, M., Bernier, N. J., Manuel, R., de Gelder, S., Metz, J. R., Huising, M. O., et al. (2012). Recombinant human leptin attenuates stress axis activity in common carp (Cyprinus carpio L.). Gen. Comp. Endocrinol. 178, 75-81. http://dx.doi.org/10.1016/j.ygcen.2012.04.004.

Gorissen, M., Manuel, R., Pelgrim, T. N. M., Mes, W., de Wolf, M. J. S., Zethof, J., et al. (2015). Differences in inhibitory avoidance, cortisol and brain gene expression in TL and AB zebrafish. Genes Brain Behav. 14, 428-438. http://dx.doi.org/10.1111/gbb.12220.

Haitina, T., Klovins, J., Andersson, J., Frederiksson, R., Lagerström, M. C., Larhammar, D., et al. (2004). Cloning, tissue distribution, pharmacology and three-dimensional modelling of melanocortin receptors 4 and 5 in rainbow trout suggest close evolutionary relationship of these subtypes. Biochemical Journal 380, 475-486.

Hauger, R. L., Grigoriadis, D. E., Dallman, M. F., Plotsky, P. M., Vale, W. W. and Dautzenberg, F. M. (2003). International Union of Pharmacology. XXXVI. Current status of the nomenclature for receptors for corticotropin-releasing factor and their ligands. Pharmacol. Rev. 55, 21-26. http://dx.doi.org/10.1124/pr.55.1.3.

Herget, U. and Ryu, S. (2015). Coexpression analysis of nine neuropeptides in the neurosecretory preoptic area of larval zebrafish. Front. Neuroanat. 9, 2. http://dx.doi.org/10.3389/fnana.2015.00002.

Herget, U., Wolf, A., Wullimann, M. F. and Ryu, S. (2014). Molecular neuroanatomy and chemoarchitecture of the neurosecretory preoptic-hypothalamic area in zebrafish larvae. J. Comp. Neurol. 522, 1542-1564. http://dx. doi. org/10. 1002/cne. 23480.

Herrick, C. J. (1933). The functions of the olfactory parts of the cerebral cortex. Proc. Natl. Acad. Sci. U. S. A. 19, 7-14.

Herzog, W., Zeng, X., Lele, Z., Sonntag, C., Ting, J.-W., Chang, C.-Y., et al. (2003). Adenohypophysis formation in the zebrafish and its dependence on sonic hedgehog. Dev. Biol. 254, 36-49.

Hofmann, J., Greter, M., Pasquier Du, L. and Becher, B. (2010). B-cells need a proper house, whereas T-cells are happy in a cave: the dependence of lymphocytes on secondary lymphoid tissues during evolution. Trends Immunol. 31, 144-153. http://dx. doi. org/10. 1016/j. it. 2010. 01. 003.

Hsu, S. Y. and Hsueh, A. J. (2001). Human stresscopin and stresscopin-related peptide are selective ligands for the type 2 corticotropin-releasing hormone receptor. Nat. Med. 7, 605-611. http://dx. doi. org/10. 1038/87936.

Hsu, H. J., Lin, G. and Chung, B. C. (2004). Parallel early development of zebrafish interrenal glands and pronephros: differential control by wt1 and ff1b. Endocr. Res. 30, 803.

Huising, M. O. (2006). Communication in the Endocrine and Immune Systems, pp. 1-298. Nijmegen.

Huising, M. O. and Flik, G. (2005). The remarkable conservation of corticotropin-releasing hormone (CRH) -binding protein in the honeybee (Apis mellifera) dates the CRH system to a common ancestor of insects and vertebrates. Endocrinology 146, 2165-2170. http://dx. doi. org/10. 1210/en. 2004-1514.

Huising, M. O., Guichelaar, T., Hoek, C., Verburg-van Kemenade, B. M. L., Flik, G., Savelkoul, H., et al. (2003). Increased efficacy of immersion vaccination in fish with hyperosmotic pretreatment. Vaccine 21, 4178-4193. http://dx. doi. org/10. 1016/S0264410X (03) 00497-3.

Huising, M. O., Metz, J. R., van Schooten, C., Taverne-Thiele, A. J., Hermsen, T., Verburg-van Kemanade, B. M. L., et al. (2004). Structural characterisation of a cyprinid (Cyprinus carpio L.) CRH, CRH-BP and CRH-R1, and the role of these proteins in the acute stress response. J. Mol. Endocrinol. 32, 627-648.

Huising, M. O., van der Aa, L. M., Metz, J. R., de Fátima Mazon, A., Verburg-van Kemenade, B. M. L. and Flik, G. (2007). Corticotropin-releasing factor (CRF) and CRF-binding protein expression in and release from the head kidney of common carp: evolutionary conservation of the adrenal CRF system. J. Endocrinol.

193, 349-357. http://dx.doi.org/10.1677/JOE-07-0070.

Huising, M. O., Vaughan, J. M., Shah, S. H., Grillot, K. L., Donaldson, C. J., Rivier, J., et al. (2008). Residues of corticotropin releasing factor-binding protein (CRF-BP) that selectively abrogate binding to CRF but not to urocortin 1. J. Biol. Chem. 283, 8902-8912. http://dx.doi.org/10.1074/jbc.M709904200.

Ito, H. and Yamamoto, N. (2009). Non-laminar cerebral cortex in teleostfishes? Biol. Lett. 5, 117-121. http://dx.doi.org/10.1098/rsbl.2008.0397.

Jerez-Cepa, I., Mancera, J. M., Flik, G. and Gorissen, M. (2016). Vasotinergic and isotonergic co-regulation in stress response of common carp (Cyprinus carpio L.). Adv. Comparative Endocrinol. VII (in press).

Kah, O., Dubourg, P., Onteniente, B., Geffard, M. and Calas, A. (1986). The dopaminergic innervation of the goldfish pituitary. An immunocytochemical study at the electron-microscope level using antibodies against dopamine. Cell Tissue. Res. 244, 577-582.

Kalueff, A. V., Echevarria, D. J. and Stewart, A. M. (2014). Gaining translational momentum: more zebrafish models for neuroscience research. Progress Neuro-Psychopharmacol. Biol. Psychiatry 55, 1-6. http://dx.doi.org/10.1016/j.pnpbp.2014.01.022.

Kimmel, C. B. and Law, R. D. (1985). Cell lineage of zebrafish blastomeres. I. Cleavage pattern and cytoplasmic bridges between cells. Dev. Biol. 108, 78-85.

Kimmel, C. B., Ballard, W. W., Ullmann, B. and Schilling, T. F. (1995). Stages of embryonicdevelopment of the zebrafish. Dev. Dynam. 203, 253-310. http://dx.doi.org/10.1002/aja.1002030302.

Klovins, J. (2003). The melanocortin system in fugu: determination of POMC/AGRP/MCR gene repertoire and synteny, as well as pharmacology and anatomical distribution of the MCRs. Mol. Biol. Evol. 21, 563-579. http://dx.doi.org/10.1093/molbev/msh050.

Kobayashi, Y., Chiba, H., Yamanome, T., Schiöth, H. B. and Takahashi, A. (2011). Melanocortin receptor subtypes in interrenal cells and corticotropic activity of alphamelanocyte-stimulating hormones in barfin flounder, Verasper moseri. Gen. Comp. Endocrinol. 170, 558-568. http://dx.doi.org/10.1016/j.ygcen.2010.11.019.

Koob, G. F. and Le Moal, M. (2001). Drug addiction, dysregulation of reward, and allostasis. Neuropsychopharmacology 24, 97-129. http://dx.doi.org/10.1016/S0893-133X (00) 00195-0.

Koolhaas, J. M., de Boer, S. F., Coppens, C. M. and Buwalda, B. (2010). Neuroendocrinology of coping styles: towards understanding the biology of individual variation. Front Neuroendocrinol. 31, 307-321. http://dx. doi. org/10. 1016/j. yfrne. 2010. 04. 001.

Koolhaas, J. M., Bartolomucci, A., Buwalda, B., de Boer, S. F., Flügge, G., Korte, S. M., et al. (2011). Stress revisited: a critical evaluation of the stress concept. Neurosci. Biobehav. Rev. 35, 1291-1301. http://dx. doi. org/10. 1016/j. neubiorev. 2011. 02. 003.

Korte, S. M., Olivier, B. and Koolhaas, J. M. (2007). A new animal welfare concept based on allostasis. Physiol. Behav. 92, 422-428. http://dx. doi. org/10. 1016/j. physbeh. 2006. 10. 018.

Kosaka, K., Kawakami, K., Sakamoto, H. and Inoue, K. (2007). Spatiotemporal localization of germ plasm RNAs during zebrafish oogenesis. Mech. Dev. 124, 279-289. http://dx. doi. org/10. 1016/j. mod. 2007. 01. 003.

Lamers, A. E., Flik, G., Atsma, W. and Wendelaar Bonga, S. E. (1992). A role for di-acetyl alpha-melanocyte-stimulating hormone in the control of cortisol release in the teleost Oreochromis mossambicus. J. Endocrinol. 135, 285-292.

Lamers, A. E., Flik, G. and Wendelaar Bonga, S. E. (1994). A specific role for Trh in release of diacetyl alpha-MSH in tilapia stressed by acid water. Am. J. Physiol. Regul. Integr. Comp. Physiol. 267, R1302-R1308.

Lamers, A. E., Brugge Ter, P. J., Flik, G. and Wendelaar Bonga, S. E. (1997). Acid stress induces a D1-like dopamine receptor in pituitary MSH cells of Oreochromis mossambicus. Am. J. Physiol. 273, R387-R392.

Lederis, K., Fryer, J. N., Okawara, Y., Schonrock, C. and Richter, D. (1994). Corticotropinreleasing factors acting on thefish pituitary: experimental and molecular analysis. In Fish Physiology-Molecular Endocrinology of Fish, Vol. 13 (eds. N. M. Sherwood and C. L. Hew), pp. 67-100. San Diego, CA: Academic Press.

Le Mevel, J. C., Lancien, F., Mimassi, N. and Conlon, J. M. (2009). Central hyperventilatory action of the stress-related neurohormonal peptides, corticotropin-releasing factor and urotensin-I in the trout Oncorhynchus mykiss. Gen. Comp. Endocrinol. 164, 51-60. http:// dx. doi. org/10. 1016/j. ygcen. 2009. 03. 019.

Lewis, K., Li, C., Perrin, M. H., Blount, A., Kunitake, K., Donaldson, C., et al. (2001). Identification of urocortin III, an additional member of the corticotropin-releasing factor (CRF) family with high affinity for the CRF2 receptor. Proc. Natl. Acad. Sci. U. S. A. 98, 7570-7575. http://dx. doi. org/10. 1073/pnas. 121165198.

Liang, L., Schmid, K., Sandhu, N., Angleson, J. K., Vijayan, M. M. and Dores, R. M. (2015). Structure/function studies on the activation of the rainbow trout melanocortin 2 receptor. Gen. Comp. Endocrinol. 210, 145-151.

Liu, Y. -W. (2007). Interrenal organogenesis in the zebrafish model. Organogenesis 3, 44-48. Long, J. A. (1995). The Rise of Fishes. London: Johns Hopkins Press Ltd.

Lovejoy, D. A. and Balment, R. J. (1999). Evolution and physiology of the corticotropinreleasing factor (CRF) family of neuropeptides in vertebrates. Gen. Comp. Endocrinol. 115, 1-22. http://dx. doi. org/10. 1006/gcen. 1999. 7298.

Lu, W., Dow, L., Gumusgoz, S., Brierley, M. J., Warne, J. M., McCrohan, C. R., et al. (2004). Coexpression of corticotropin-releasing hormone and urotensin I precursor genes in the caudal neurosecretory system of the euryhalineflounder (Platichthys flesus): a possible shared role in peripheral regulation. Endocrinology 145, 5786-5797. http://dx. doi. org/ 10. 1210/en. 2004-0144.

MacDonald, L. E., Alderman, S. L., Kramer, S., Woo, P. T. K. and Bernier, N. J. (2014). Hypoxemia-induced leptin secretion: a mechanism for the control of food intake in diseasedfish. J. Endocrinol. 221, 441-455. http://dx. doi. org/10. 1530/ JOE-13-0615.

Machluf, Y., Gutnick, A. and Levkowitz, G. (2011). Development of the zebrafish hypothalamus. Ann. N. Y. Acad. Sci. 1220, 93-105. http://dx. doi. org/10. 1111/j. 17496632. 2010. 05945. x.

Madaro, A., Olsen, R. E., Kristiansen, T. S., Ebbesson, L. O. E., Nilsen, T. O., Flik, G., et al. (2015). Stress in Atlantic salmon: response to unpredictable chronic stress. J. Exp. Biol. 218, 2538-2550. http://dx. doi. org/10. 1242/ jeb. 120535.

Madaro, A., Olsen, R. E., Kristiansen, T. S., Ebbesson, L. O. E., Flik, G. and Gorissen, M. (2016). A comparative study of the response to repeated chasing stress in Atlantic salmon (Salmo salar L.) parr and post-smolts. Comp. Biochem. Physiol. A Mol. Integr. Physiol. 192, 7-16. http://dx. doi. org/10. 1016/j. cbpa. 2015. 11. 005.

Maier, T., Güell, M. and Serrano, L. (2009). Correlation of mRNA and protein in complex biological samples. FEBS Lett. 583, 3966-3973. http://dx. doi. org/10. 1016/ j. febslet. 2009. 10. 036.

Manuel, R. (2015). Biology of welfare infish (Ph. D. thesis). Nijmegen: Radboud University.

Manuel, R., Gorissen, M., Roca, C. P., Zethof, J., van de Vis, H., Flik, G., et al. (2014a). Inhibitory avoidance learning in zebrafish (Danio rerio): effects of shock intensity and unraveling differences in task performance. zebrafish 11, 341-352. http://dx. doi. org/ 10. 1089/zeb. 2013. 0970.

Manuel, R., Gorissen, M., Zethof, J., Ebbesson, L. O. E., van de Vis, H., Flik, G., et al. (2014b). Unpredictable chronic stress decreases inhibitory avoidance learning in Tuebingen long-fin zebrafish: stronger effects in the resting phase than in the active phase. J. Exp. Biol. 217, 3919-3928. http://dx.doi.org/10.1242/jeb.109736.

Manuel, R., Metz, J. R., Flik, G., Vale, W. W. and Huising, M. O. (2014c). Corticotropinreleasing factor-binding protein (CRF-BP) inhibits CRF-and urotensin-I-mediated activation of CRF receptor-1 and-2 in common carp. Gen. Comp. Endocrinol. 202, 69-75. http://dx.doi.org/10.1016/j.ygcen.2014.04.010.

Manuel, R., Gorissen, M. and van den Bos, R. (2016). Relevance of test-and subject-related factors on inhibitory avoidance (performance) of zebrafish for psychopharmacology studies. Current Psychopharmacology 5 (2), epub.

McEwen, B. S. and Wingfield, J. C. (2003). The concept of allostasis in biology and biomedicine. Horm. Behav. 43, 2-15.

McEwen, B. S. and Wingfield, J. C. (2010). What is in a name? Integrating homeostasis, allostasis and stress. Horm. Behav. 57, 105-111. http://dx.doi.org/10.1016/j.yhbeh.2009.09.011.

Metz, J. R., Huising, M. O., Meek, J., Taverne-Thiele, A. J., Wendelaar Bonga, S. E. and Flik, G. (2004). Localization, expression and control of adrenocorticotropic hormone in the nucleus preopticus and pituitary gland of common carp (Cyprinus carpio L.). The Journal of Endocrinology 182, 23-31.

Metz, J. R., Geven, E. J. W., van den Burg, E. H. and Flik, G. (2005). ACTH, alpha-MSH, and control of cortisol release: cloning, sequencing, and functional expression of the melanocortin-2 and melanocortin-5 receptor in Cyprinus carpio. Am. J. Physiol. Regul. Integr. Comp. Physiol. 289, R814-R826. http://dx.doi.org/10.1152/ajpregu.00826.2004.

Metz, J. R., Peters, J. J. M. and Flik, G. (2006). Molecular biology and physiology of the melanocortin system infish: a review. Gen. Comp. Endocrinol. 148, 150-162. http://dx.doi.org/10.1016/j.ygcen.2006.03.001.

Mueller, T. (2012). What is the thalamus in zebrafish? Front. Neurosci. 6, 1-14. http://dx.doi.org/10.3389/fnins.2012.00064.

Mueller, T., Dong, Z., Berberoglu, M. A. and Guo, S. (2011). The dorsal pallium in zebrafish, Danio rerio (Cyprinidae, Teleostei). Brain Res. 1381, 95-105. http://dx.doi.org/10.1016/j.brainres.2010.12.089.

Nieuwenhuys, R. (2009). The forebrain of actinopterygians revisited. Brain Behav. Evol. 73, 229-252. http://dx.doi.org/10.1159/000225622.

Nieuwenhuys, R., ten Donkelaar, H. J. and Nicholson, C. (1998). The Central Nervous System of Vertebrates. Berlin Heidelberg: Springer-Verlag.

Noakes, D. L. G. and Jones, K. M. M. (2016). Cognition, Learning, and Behavior. In

Fish Physiology-Biology of Stress in Fish, Vol. 35 (eds. C. B. Schreck, L. Tort, A. P. Farrell and C. J. Brauner), San Diego, CA: Academic Press.

Nobata, S., Ando, M. and Takei, Y. (2013). Hormonal control of drinking behavior in teleostfishes; insights from studies using eels. Gen. Comp. Endocrinol. 192, 214-221. http://dx.doi.org/10.1016/j.ygcen.2013.05.009.

Northcutt, R. G. (2006). Connections of the lateral and medial divisions of the goldfish telencephalic pallium. J. Comp. Neurol. 494, 903-943. http://dx.doi.org/10.1002/cne.20853.

Ortega, V. A., Lovejoy, D. A. and Bernier, N. J. (2013). Appetite-suppressing effects and interactions of centrally administered corticotropin-releasing factor, urotensin I and serotonin in rainbow trout (Oncorhynchus mykiss). Front. Neurosci. 7, 196. http://dx.doi.org/10.3389/fnins.2013.00196.

Pearson, D., Shively, J. E., Clark, B. R., Geschwind, I. I., Barkley, M., Nishioka, R. S., et al. (1980). Urotensin II: a somatostatin-like peptide in the caudal neurosecretory system offishes. Proc. Natl. Acad. Sci. U. S. A. 77, 5021-5024.

Pepels, P., van Helvoort, H., Wendelaar Bonga, S. E. and Balm, P. H. M. (2004). Corticotropinreleasing hormone in the teleost stress response: rapid appearance of the peptide in plasma of tilapia (Oreochromis mossambicus). J. Endocrinol. 180, 425-438. http://dx.doi.org/10.1677/joe.0.1800425.

Pierson, P. M., Guibbolini, M. E. and Lahlou, B. (1996). A V1-type receptor for mediating the neurohypophysial hormone-induced ACTH release in trout pituitary. J. Endocrinol. 149, 109-115.

Pogoda, H.-M. and Hammerschmidt, M. (2009). How to make a teleost adenohypophysis: molecular pathways of pituitary development in zebrafish. Mol. Cell. Endocrinol. 312, 2-13. http://dx.doi.org/10.1016/j.mce.2009.03.012.

Portavella, M., Vargas, J. P., Torres, B. and Salas, C. (2002). The effects of telencephalic pallial lesions on spatial, temporal, and emotional learning in goldfish. Brain Res. Bull. 57, 397-399.

Portavella, M., Torres, B. and Salas, C. (2004). Avoidance response in goldfish: emotional and temporal involvement of medial and lateral telencephalic pallium. J. Neurosci. 24, 2335-2342./http://dx.doi.org/10.1523/JNEUROSCI.4930-03.2004S.

Potter, E., Behan, D. P., Fischer, W. H., Linton, E. A., Lowry, P. J. and Vale, W. W. (1991). Cloning and characterization of the cDNAs for human and rat corticotropin releasing factor-bindingproteins. Nature 349, 423-426. http://dx.doi.org/10.1038/349423a0.

Primavera J. H. (2005). Mangroves and Aquaculture in Southeast Asia.

Reid, S. G., Vijayan, M. M. and Perry, S. F. (1996). Modulation of catecholamine storage and release by the pituitary interrenal axis in the rainbow trout, Oncorhynchus mykiss. J. Comp. Physiol. B 165, 665-676.

Reid, S. G., Bernier, N. J. and Perry, S. F. (1998). The adrenergic stress response infish: control of catecholamine storage and release. Comp. Biochem. Physiol. C Pharmacol. Toxicol. Endocrinol. 120, 1-27.

Rey, S., Huntingford, F. A., Boltaña, S., Vargas, R., Knowles, T. G. and MacKenzie, S. (2015). Fish can show emotional fever: stress-induced hyperthermia in zebrafish. Proc. R. Soc. B Biol. Sci. 282, 20152266-20152267. http://dx.doi.org/10.1098/rspb.2015.2266.

Rivier, C., Brownstein, M., Spiess, J., Rivier, J. and Vale, W. (1982). In vivo corticotropinreleasing factor-induced secretion of adrenocorticotropin, beta-endorphin, and corticosterone. Endocrinology 110, 272-278. http://dx.doi.org/10.1210/endo-110-1-272.

Romero, L. M., Dickens, M. J. and Cyr, N. E. (2009). The reactive scope model-a new model integrating homeostasis, allostasis, and stress. Horm. Behav. 55, 375-389. http://dx.doi.org/10.1016/j.yhbeh.2008.12.009.

Rotllant, J., Balm, P. H., Ruane, N. M., Pérez-Sánchez, J., Wendelaar Bonga, S. E. and Tort, L. (2000). Pituitary proopiomelanocortin-derived peptides and hypothalamus-pituitary-interrenal axis activity in gilthead sea bream (Sparus aurata) during prolonged crowding stress: differential regulation of adrenocorticotropin hormone and alpha-melanocyte-stimulating hormone release by corticotropin-releasing hormone and thyrotropin-releasing hormone. Gen. Comp. Endocrinol. 119, 152-163. http://dx.doi.org/10.1006/gcen.2000.7508.

Schreck, C. B. (2010). Stress andfish reproduction: the roles of allostasis and hormesis. Gen. Comp. Endocrinol. 165, 549-556. http://dx.doi.org/10.1016/j.ygcen.2009.07.004.

Schreck, C. B. and Tort, L. (2016). The Concept of Stress in Fish. In Fish Physiology-Biology of Stress in Fish, Vol. 35 (eds. C. B. Schreck, L. Tort, A. P. Farrell and C. J. Brauner), San Diego, CA: Academic Press.

Schwanha ̈usser, B., Busse, D., Li, N., Dittmar, G., Schuchhardt, J., Wolf, J., et al. (2011). Global quantification of mammalian gene expression control. Nature 473, 337-342. http://dx.doi.org/10.1038/nature10098.

Selbie, L. A., Hayes, G. and Shine, J. (1989). The major dopamine D2 receptor: molecular analysis of the human D2A subtype. DNA (Mary Ann Liebert, Inc.) 8, 683-689.

Selye, H. (1950). Stress and the general adaptation syndrome. Br. Med. J. 1,

1383-1392.

Selye, H. (1973). The evolution of the stress concept. Am. Sci. 61, 692-699.

Stefaneanu, L., Kovacs, K., Horvath, E., Buchfelder, M., Fahlbusch, R. and Lancranjan, L. (2001). Dopamine D2 receptor gene expression in human adenohypophysial adenomas. Endocrine 14, 329-336.

Stewart, A. M., Braubach, O., Spitsbergen, J., Gerlai, R. and Kalueff, A. V. (2014). Zebrafish models for translational neuroscience research: from tank to bedside. Trends Neurosci. 37, 264-278. http://dx.doi.org/10.1016/j.tins.2014.02.011.

Stouthart, A. J. H. X., Lucassen, E., van Strien, F., Balm, P., Lock, R. A. C. and Wendelaar Bonga, S. E. (1998). Stress responsiveness of the pituitary-interrenal axis during early life stages of common carp (Cyprinus carpio). J. Endocrinol. 157, 127-137.

Studnička, F. K. (1896). Beiträge zur Anatomie und Entwickelungsgeschichte des Vorderhirns der Cranioten. Sitzungsber K-Bohm Gesell-Sch Wissensch Mathem Naturw Kl 15, 1-32.

Suda, T., Tomori, N., Tozawa, F., Demura, H. and Shizume, K. (1984). Effects of corticotropin-releasing factor and other materials on adrenocorticotropin secretion from pituitary glands of patients with cushing's disease in vitro. J. Clin. Endocrinol. Metabol. 59, 840-845. http://dx.doi.org/10.1210/jcem-59-5-840.

Sutton, S. W., Behan, D. P., Lahrichi, S. L., Kaiser, R., Corrigan, A., Lowry, P., et al. (1995). Ligand requirements of the human corticotropin-releasing factor-binding protein. Endocrinology 136, 1097-1102. http://dx.doi.org/10.1210/endo.136.3.7867564.

Takahashi, A., Amano, M., Amiya, N., Yamanome, T., Yamamori, K. and Kawauchi, H. (2006). Expression of three proopiomelanocortin subtype genes and mass spectrometric identification of POMC-derived peptides in pars distalis and pars intermedia of barfin flounder pituitary. Gen. Comp. Endocrinol. 145, 280-286. http://dx.doi.org/10.1016/j.ygcen.2005.09.005.

Takei, Y. (2002). Hormonal control of drinking in eels: an evolutionary approach. Symp. Soc. Exp. Biol. 61-82.

Takei, Y. and Hwang, P.-P. (2016). Homeostatic Responses to Osmotic Stress. In Fish Physiology-Biology of Stress in Fish, Vol. 35 (eds. C. B. Schreck, L. Tort, A. P. Farrell and C. J. Brauner), San Diego, CA: Academic Press.

Tanaka, S. (2003). Comparative aspects of intracellular proteolytic processing of peptide hormone precursors: studies of proopiomelanocortin processing. Zool. Sci. 20, 1183-1198.

Theusch, E. V., Brown, K. J. and Pelegri, F. (2006). Separate pathways of RNA recruitment lead to the compartmentalization of the zebrafish germ plasm. Dev. Biol. 292, 129-141. http://dx.doi.org/10.1016/j.ydbio.2005.12.045.

To, T. T., Hahner, S., Nica, G., Rohr, K. B., Hammerschmidt, M., Winkler, C., et al. (2007). Pituitary-interrenal interaction in zebrafish interrenal organ development. Mol. Endocrinol. (Baltimore, Md.) 21, 472-485. http://dx.doi.org/10.1210/me.2006-0216.

Tsalafouta, A., Papandroulakis, N., Gorissen, M., Katharios, P., Flik, G. and Pavlidis, M. (2014). Ontogenesis of the HPI axis and molecular regulation of the cortisol stress response during early development in Dicentrarchus labrax. Sci. Rep. 4. http://dx.doi.org/10.1038/ srep05525.

Vale, W., Spiess, J., Rivier, C. and Rivier, J. (1981). Characterization of a 41-residue ovine hypothalamic peptide that stimulates secretion of corticotropin and beta-endorphin. Science (New York, N. Y.) 213, 1394-1397.

van den Burg, E. H., Metz, J. R., Arends, R. J., Devreese, B., Vandenberghe, I., Van Beeumen, J., et al. (2001). Identification of beta-endorphins in the pituitary gland and blood plasma of the common carp (Cyprinus carpio). J. Endocrinol. 169, 271-280.

van den Burg, E. H., Metz, J. R., Spanings, F. A. T., Wendelaar Bonga, S. E. and Flik, G. (2005). Plasma alpha-MSH and acetylated beta-endorphin levels following stress vary according to CRH sensitivity of the pituitary melanotropes in common carp, Cyprinus carpio. Gen. Comp. Endocrinol. 140, 210-221. http://dx.doi.org/10.1016/j.ygcen.2004.11.010.

van Enckevort, F. H., Pepels, P. P., Leunissen, J. A., Martens, G. J., Wendelaar Bonga, S. E. and Balm, P. H. (2000). Oreochromis mossambicus (tilapia) corticotropin-releasing hormone: cDNA sequence and bioactivity. J. Neuroendocrinol. 12, 177-186. http://dx.doi.org/10.1046/j.1365-2826.2000.00434.x.

Vaughan, J., Donaldson, C., Bittencourt, J., Perrin, M. H., Lewis, K., Sutton, S., et al. (1995). Urocortin, a mammalian neuropeptide related tofish urotensin I and to corticotropinreleasing factor. Nature 378, 287-292. /http://dx.doi.org/10.1038/378287a0.

Verburg-van Kemenade, B. M. L. and Schreck, C. B. (2007). Immune and endocrine interactions. Gen. Comp. Endocrinol. 152, 352. http://dx.doi.org/10.1016/j.ygcen.2007.05.025.

Volkoff, H., Canosa, L. F., Unniappan, S., Cerda-Reverter, J. M., Bernier, N. J., Kelly, S. P., et al. (2005). Neuropeptides and the control of food intake infish. Gen. Comp. Endocrinol. 142, 3-19. http://dx.doi.org/10.1016/j.ygcen.2004.

11.001.

Wei, E. T., Thomas, H. A., Christian, H. C., Buckingham, J. C. and Kishimoto, T. (1998). D-amino acid-substituted analogs of corticotropin-releasing hormone (CRH) and urocortin with selective agonist activity at CRH1 and CRH2beta receptors. Peptides 19, 1183-1190.

Wendelaar Bonga, S. E. (1997). The stress response infish. Physiol. Rev. 77, 591-625.

Westring, C. G., Ando, H., Kitahashi, T., Bhandari, R. K., Ueda, H., Urano, A., et al. (2008). Seasonal changes in CRF-I and urotensin I transcript levels in masu salmon: correlation with cortisol secretion during spawning. Gen. Comp. Endocrinol. 155, 126-140. http://dx.doi.org/10.1016/j.ygcen.2007.03.013.

Wieser, W. (1985). A new look at energy conversion in ectothermic and endothermic animals. Oecologia 66 (4), 506-510.

Wilkinson, C. W. (2006). Roles of acetylation and other post-translational modifications in melanocortin function and interactions with endorphins. Peptides 27, 453-471. http://dx.doi.org/10.1016/j.peptides.2005.05.029.

Wynn, P. C., Aguilera, G., Morell, J. and Catt, K. J. (1983). Properties and regulation of highaffinity pituitary receptors for corticotropin-releasing factor. Biochem. Biophys. Res. Commun. 110, 602-608.

Yada, T. and Tort, L. (2016). Stress and Disease Resistance: Immune System and Immunoendocrine Interactions. In Fish Physiology-Biology of Stress in Fish, Vol. 35 (eds. C. B. Schreck, L. Tort, A. P. Farrell and C. J. Brauner), San Diego, CA: Academic Press.

Zhou, A., Bloomquist, B. T. and Mains, R. E. (1993). The prohormoneconvertases PC1 and PC2 mediate distinct endoproteolytic cleavages in a strict temporal order during proopiomelanocortin biosynthetic processing. J. Biol. Chem. 268, 1763-1769.

第 4 章　分子应激反应

在本章，我们综合论述参与鱼类下丘脑-脑垂体-肾间腺（HPI）轴的调控以及由皮质类固醇信号调节的应激反应基因调控所观察到的基因转录多度（transcript abundance）的关键型式。血浆中的皮质醇水平，是硬骨鱼类主要的皮质类固醇，在受到应激物影响后几分钟内升高，而这种类固醇的作用主要是由包含核受体家族如糖皮质类固醇受体（GR）和盐皮质类固醇受体（MR）的基因组信号所介导。由于使用受体拮抗物和基因击倒技术的与日俱增，已经清楚证明在靶组织特异性的分子反应是和这些皮质类固醇受体的激活密切联系的。虽然对应激的分子调节取决于种类、发育阶段、应激物的类型和作用期限，但总而言之，皮质醇的作用是减少能量需要的过程，增加能量的转移和重新安排。最近的研究亦注意到母体的皮质醇和 GR 信号对鱼类的发育是不可缺少的，但其基本的作用机理目前还不是很了解。转录多度是皮质醇调节通道一个很好的指标，但这些数据本身还缺少生理方面的联系，除非是伴随着下游蛋白质和/或代谢物的变化。随着分子生物学领域的不断发展，新的研究途径包括下一代基因测序和基因编辑技术等，使得甚至在非模式的鱼类中都能够产生转基因动物，以阐明应激适应中不可缺少的分子作用机理。这些基本的理论知识对于在水产养殖中改善生命有机体的福利和保护生态系统的健康与生物多样性都有密切关系。

4.1　导　言

激素的应激反应已经在硬骨鱼类中进行了深入的研究，它是生理调节以重新建立稳态的关键介体（Wenderaar Bonga，1997；Mommsen 等，1999；Barton，2002；Vijayan 等，2010）。受到应激物影响的内分泌反应是血浆中主要的儿茶酚胺即肾上腺素和主要的糖皮质类固醇即皮质醇水平迅速升高（Reid 等，1998；Mommsen 等，1999）。从生理调控功能的观点看，应激诱导血浆中皮质醇和肾上腺素的反应是暂时分开的，肾上腺素在几秒钟内增加，并且迅速在血浆循环中清除掉，而皮质醇是在对应激挑战的反应中由几分钟到几小时内升高［图 4.1（A）；Vijayan 等，2010］。肾上腺素的作用包括迅速转移葡萄糖成为燃料以应对应激物诱导的能量需求（Fabbri，Moon，2015）。此外，靶组织对皮质醇刺激的反应是比较缓慢的，通常包括作为效应物的蛋白质合成以促进能量底物的转移和重新安排，包括补充鱼体内缺失的糖原贮存（Mommsen 等，1999；Vijayan 等，2010）。

肾上腺素信号通过 G-偶联蛋白（GDCRs）家族中的 β-肾上腺素受体调节靶标蛋白质的磷酸化/去磷酸化的细胞反应［Fabbri，Moon，2015；图 4.1（B）］。虽然肾上腺素影响转录的变化，包括参与代谢（Ings 等，2012）和免疫反应（Castillo 等，2009；Chadzinska

等，2012）的基因，这种激素的主要作用是独立于鱼类转录调节作用之外的（Fabbri 等，1998）。在哺乳动物的模型中，肾上腺素激活 cAMP 反应结合蛋白（CREB），它是在应激期间参与代谢调节的关键转录因子（Thiel 等，2005）。但是，CREB 在鱼类的模型由肾上腺素介导的应激信号中的作用还不清楚（Fabbri，Moon，2015）。

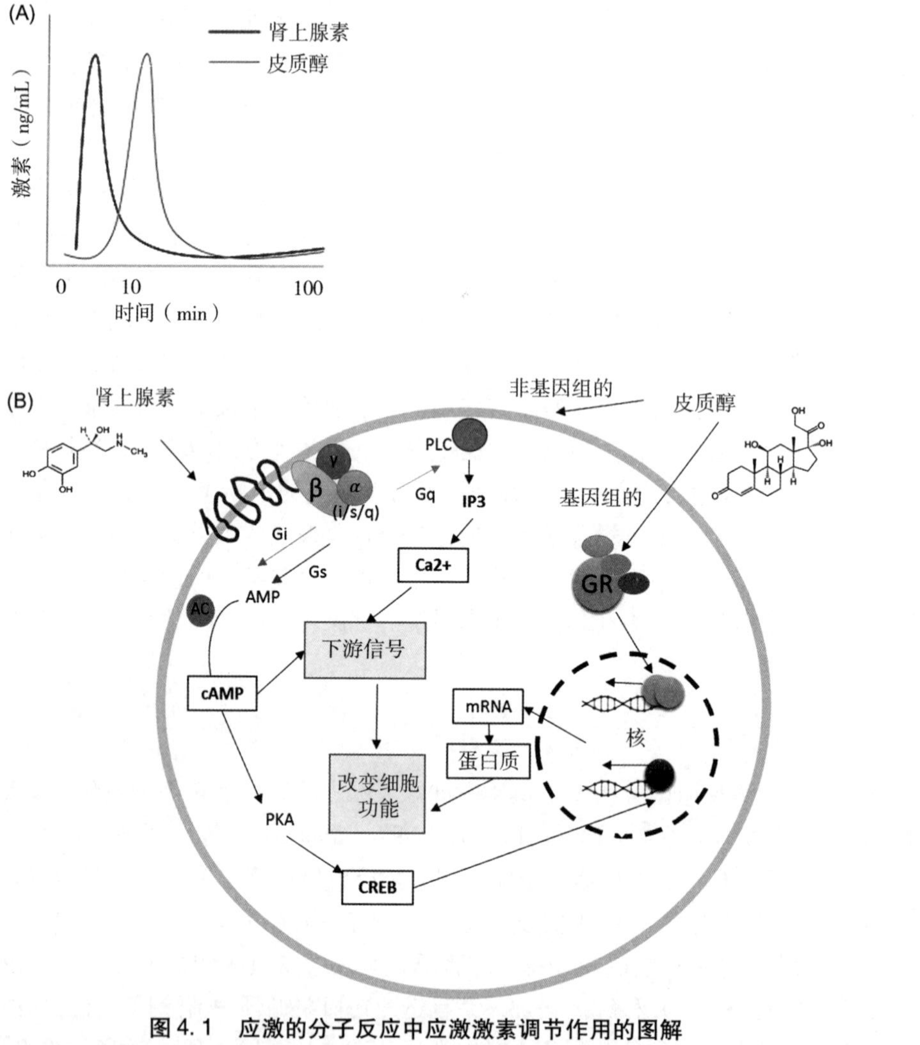

图 4.1　应激的分子反应中应激激素调节作用的图解

注：（A）应激物介导的血浆肾上腺素反应快（秒钟到分钟）于皮质醇反应（分钟到小时），前者引起迅速的细胞反应，而类固醇的反应主要是参与基因组信号。（B）通过 G-蛋白偶联的肾上腺系统受体（α，β），肾上腺素信号以组织特异性方式表达。G-蛋白偶联受体（GPCRs）形成次级的信号串联，激活（Gs）或者抑制（Gi）产生 cAMP 的腺苷环化酶。随后的 PKA 磷酸化激活转录因子 cAMP 反应元件结合蛋白（CREB），引起结合 CREB 反应元件（CREs）的转录调节作用。基因组的皮质醇信号是研究得最广泛的应激信号通道。然而在应激期间亦出现非基因组的信号，但它们的作用机理还不清楚。在基因组的皮质醇信号中，类固醇和它位于胞质的核受体结合，并和其他蛋白质以杂络物（heterocomplex）的形式存在着；由两个这样的配体结合受体形成二聚体，移位到细胞核，和糖皮质类固醇反应元件（GREs）结合而诱导靶基因的转录。

皮质醇的作用是重要的，但不是唯一的，还有基因组作用包括基因转录和蛋白质合成［图4.1（B）；Mommsen等，1999；Vijayan等，2005］。在鱼类中，皮质醇的基因组作用通过GRs和/或MRs的激活而介导（见第4.3节）。自2000年以来，作为应激和/或皮质醇刺激的分子标记物的靶基因表达已经在硬骨鱼类当中广泛使用，而在这当中基因组分析的定量工具包括定量实时聚合酶链反应（PCR）和微点阵（microarray）的出现，起着最重要的推动作用（Aluru，Vijayan，2009）。大多数研究都是检测稳态的转录水平，而在鱼类中，观察应激反应作用机理的基本资料还很缺乏（Aluru，Vijayan，2009）。尽管这样，介导应激物的分子反应加强了代谢的变化，包括对应激适应不可缺少的能量底物重新分配（Alura，Vijayan，2009）。然而，这亦可能发生于其他能量-需求的通道，包括生长、健康和生殖（Vijayan等，2010；Philip，Vijayan，2015）。由于血浆皮质醇升高是硬骨鱼类主要应激反应的特征，而至今对它们大部分作用的研究都涉及转录的调控，本章将着重于鱼类由这种应激类固醇介导的分子反应。此外，我们亦将集中注意力于基因操作、基因组和外基因组分析的最新进展，包括反向基因技术和下一代基因测序技术，它们是阐明激素作用方式和相关的应激-应对作用机理的重要手段。

4.2 下丘脑-脑垂体-肾间腺（HPI）轴的分子调控

硬骨鱼类的HPI轴和哺乳类的下丘脑-脑垂体-肾上腺（HPI）轴是同义的（Wendelaar Bonga，1997）。鱼类虽然缺少一个分离的肾上腺，但类固醇生成细胞围绕头肾区（肾间组织）内的后主静脉周围分布（Wendelaar Bonga，1997）。应激物诱导的HPI轴激活是高度保守的，并且使硬骨鱼类血液循环中基本的肾上腺皮质类固醇-皮质醇的释放达到最高点（Wendelaar Bonga，1997；Mommsen等，1999；Barton，2002）。简言之，应激物的输入刺激促肾上腺皮质激素释放因子（CRF）从下丘脑释放，这种肽类激素随之刺激促肾上腺皮质激素（ACTH）从前脑垂体的促肾上腺皮质激素生成细胞释放出来（Wendelaar Bonga，1997）。ACTH是由前体蛋白质阿黑皮素原（POMC）经过翻译后修饰而产生。POMC裂解后产生多个和应激相关的肽，包括ACTH、α-黑色素细胞刺激激素（α-MSH）和β-内啡肽（Takahashi，Kawauchi，2006；Dores，2013）。这些肽在血液循环中的水平曾用作鱼类急性应激反应的指标（Wendelaar Bonge，1997）。虽然有几种激素刺激鱼类头肾释放皮质醇，ACTH是皮质类固醇生成的主要促分泌素（Wendelaar Bonga，1997；Mommsen等，1999）。ACTH和肾间腺组织中类固醇生成细胞的黑皮质素受体2（MC2R）结合以激活信号串联而使皮质醇产生（Aluru，Vijayan，2008）。这些分子反应，包括参与HPI轴调控关键基因的转录变化，以及靶组织对皮质醇刺激的反应，都曾在硬骨鱼类中进行深入的研究（图4.2），并将在下节中进行概述。

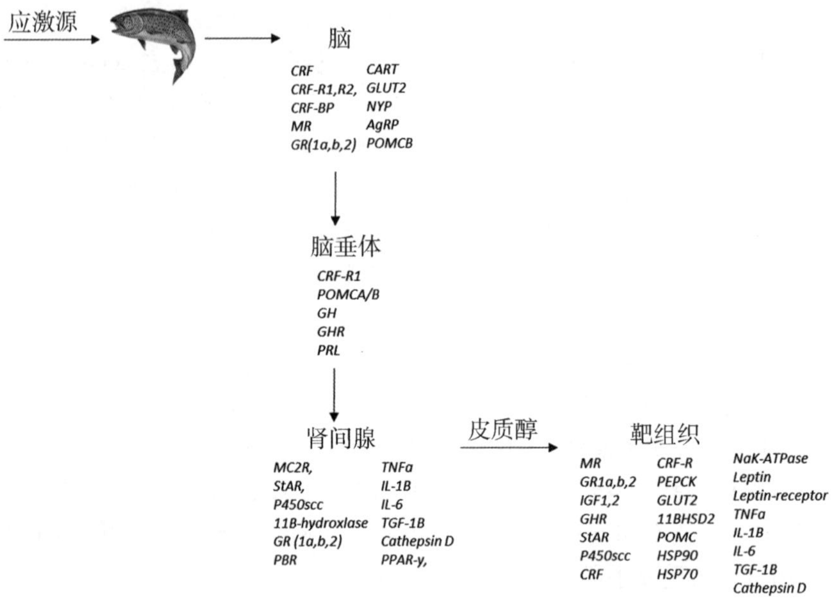

图 4.2　对 HPI 轴激活应答进行调节的应激-相关基因

注：CRF，促肾上腺皮质激素释放激素；CRF-R1、R2，CRF-受体；CRF-BP，CRF-结合蛋白；MR，盐皮质激素受体；GR（1a、b、2），糖皮质激素受体；CART，可卡因-安非他明-调节转录体；GLUT2，葡萄糖转运蛋白 2；NPY，神经肽 Y；AgRP，野灰蛋白相关蛋白；POMCA/B，阿黑皮质素原；GH，生长激素；GHR，生长激素受体；PRL，催乳素；MC2R，黑皮质素 2 受体；StAR，类固醇生成急性调节蛋白；$P450_{scc}$，P450 侧链裂解；PBR，外周型苯二氮杂受体；TNFa，肿瘤坏死因子 a；IL，白细胞介素；TGF，转化生长因子；PPAR，过氧化物增殖物激活受体。

4.2.1　下丘脑

在下丘脑中应激物介导的 CRF 上调是 HPI 轴激活的第一步（Filk 等，2006）。和哺乳类一样，这个神经肽不仅在硬骨鱼类的中枢神经系统广泛表达（Alderman，Bernier，2007），亦在外周组织，包括头肾表达（Huising 等，2007）。但是，下丘脑调控脑垂体 CRF 的主要部位是视前区（POA），它和哺乳类室旁核的下丘脑调控脑垂体的 CRF 神经元是类似的（Pepels 等，2002；Bernier 等，2008）。硬骨鱼类还有另外来源的 CRF 相关肽，即位于脊髓末端的尾神经内分泌系统（CNSS）（Bernier 等，2008）。在应激期间尾神经内分泌肽类的生理作用尚有待于确定，但由 CNSS 产生的硬骨鱼紧张肽 I（UI）在 HPI 轴的调控中是起作用的（Craig 等，2005；Bernier 等，2008）。UI 能直接刺激肾间腺产生皮质醇（Arnold-Reed，Balment，1994；Kelsall，Balmant，1998；Huising 等，2007），亦能够间接通过增加脑垂体分泌 ACTH 而促进肾间腺产生皮质醇（Fryer 等，1983）。这些研究结果证明 CRF-相关肽在生命有机体的应激反应中起着某种作用，而大多数研究都着重于和鱼类 HPI 轴功能相关的 CRF 转录的变化（Filk 等，2006）。

CRF 的转录水平受到应激的调节，但这种变化并不是一致的，可能反映了所使用的应激物模型的差别。例如，有些研究表明应激物-介导 POA 中 CRF mRNA 水平升高（Ando 等，1999；Huising 等，2004；Doyon 等，2005，2006；Craig 等，2005；Fuzzen 等，2010；Aruna 等，2012；Wunderink 等，2012）；但其他的一些研究则没有观察到 CRF 转录水平出现任何变化（Bernier 等，2012；Ghisleni 等，2012）。同样，提高血浆皮质醇水平，既能增加 CRF 的转录（Madison 等，2015），也可以没有影响（Doyon 等，2005），或者抑制 CRF 的转录水平（Bernier，Peter，2001；Bernier 等，2004；Yeh，2015）。但是，一些研究表明，在虹鳟（*Oncorhynchus mykiss*）中，GR 信号对于保持 CRF 转录的丰度和应激-介导的皮质醇产生是不可缺少的（Doyon 等，2006；Alderman，2012）。例如，使用 mifepristone 阻抑 GR 信号能降低鳟鱼（Alderman 等，2012）和斑马鱼（Yeh，2015）POA 中 CRF 转录的丰度，从而形成这样的设想，其他的调控作用机理可能是适当的，包括 MR 信号在恢复皮质醇稳态的负反馈调节中的作用（Alderman 等，2012；Alderman，Vijayan，2012）。一些研究亦表明，在应激期间 CRF 结合蛋白对 CRF 生物有效度（bioavailability）的调节起着某种作用（Filk 等，2006；Bernier，2006；Alderman，Bernier，2009；Bernier 等，2012；Wunderink 等，2012），但这还有待于通过实验去证实。

营养状态亦影响鱼类下丘脑的 CRF 转录水平。例如，CRF 和 UI 升高使食物摄取减弱，其中包括 CRF 受体的激活（Bernier，Peter，2001；Ortega 等，2013）。此外，CRF 结合蛋白能减轻对食物缺失的反应，表明 CRF 生物有效度增加能抑制摄食活动（Bernier 等，2012；Wunderink 等，2012）。在这些研究中，血浆皮质醇水平虽然并不和基因的变化相关，但可以合理地设想 CRF 升高反映了 HPI 轴激活的增强。正如血浆皮质醇水平升高能够调整编码摄食-相关肽的基因，包括 POMC、神经肽 Y 和可卡因-安非他明-调节转录体在下丘脑的丰度（Gesto 等，2014）以及肝脏的瘦素（Madison 等，2015），其结果说明了鱼类的食欲相关肽和应激轴的 CRF 之间功能的相互作用。最近我们的研究证明用 venlafaxine，一种强有力的选择性 5-羟色胺与去甲肾上腺素重摄取的抑制剂进行处理，能提高鳟鱼脑的 5-羟色胺、去甲肾上腺素和多巴胺的含量（Melnyk Lamont 等，2014）。这种处理使 POA 有较高的 CRF 转录丰度并且减少了食物摄取（Melnyk-Lamont 等，2014）。在金鱼（*Carassius auratus*）中，使用氟西汀（fluoxetine），一种选择性 5-羟色胺重摄取的抑制剂，亦观察到 POA 的 CRF 转录水平同样的增加（Mennigen，2010）。这些研究结果都证明单胺类和其他的神经肽类一起对鱼类应激反应中的 CRF 转录水平起着调节作用（Bernier 等，2012；Ortega 等，2013）。

4.2.2　脑垂体

前脑垂体是在应激物介导的 CRF 刺激反应中 ACTH 产生的主要来源（Filk 等，2006；Dores，2013）。在腺脑垂体，CRF 和一种 GPCR，即 CRF-受体（CRF-Rs）结合而刺激 ACTH 释放（Baker 等，1996）。鱼类的 CRF-Rs 只有两种同种型，即 CRF-

R1 和 CRF-R2，但它们在应激反应中独特的功能作用还不清楚（Dores，2013；Manuel 等，2014）。CRF 信号引起 POMC 基因的反式激活作用，POMC 基因编码的前体蛋白经过翻译后修饰而产生黑皮质素家族的激素，包括 ACTH 和 α、β、γ、δ-MSH（Dores，Baron，2011）。血浆中皮质醇和 ACTH 水平升高都能调节脑垂体中 POMC 的转录水平（Nakano 等，2013；Laiz-Carrion 等，2009；Aruna 等，2012；Madison 等，2015），表明是激素的负反馈调节作用（Mommsen 等，1999）。值得注意的是，对慢性应激物包括低氧和禁食的反应中，脑垂体 POMC 转录水平在应激物作用后数天仍然保持升高状态（Bernier 等，2012）。长时间禁食亦能连续地上调鳟鱼下丘脑 POMCmRNA 的丰度，证明这些蛋白质/肽类作为饱食信号对降低能量耗费起着某种作用（Jorgensen 等，2015）。所以，在应激反应期间，这种基因转录的变化可能在食欲和能量稳态的调控中起重要作用。

斑马鱼（*Danio rerio*）是一种研究下丘脑和脑垂体功能作用机理的良好模式鱼类。这是因为它的胚胎易于操作，能实行反向遗传研究。因此，我们最近的研究表明击倒斑马鱼 GR（通过吗啉寡核苷酸）而影响几种分子通道，包括一些参与应激轴功能的通道，表明皮质醇信号在调节 CRF 和 POMC 转录水平中起着关键作用（Nesan，Vijayan，2013a，b）。此外，受精卵皮质醇含量的操作处理，模拟皮质醇的母体转移，能够调节斑马鱼胚胎的 CRF 和 POMCA 的转录水平（Nesen，Vijayan，2016）。特别是，较低的受精卵皮质醇含量能够导致斑马鱼胚胎较高的 CRF 和 POMCA 转录水平，而随着皮质醇含量升高，两者的转录水平都比较低。在早期发育期间，这些关键的应激反应基因的变化是和斑马鱼幼鱼皮质醇对应激物反应的变化相一致的，表明这些转录体在斑马鱼孵化后应激轴功能的程序中起着作用（Nesan，Vijayan，2016）。

对 rx3 斑马鱼突变体（表 4.1）的研究亦证明脑垂体在应激调节中的重要性，这种突变体在视网膜同源框基因 3（由 chokh 基因编码；Dickmeis 等，2007）携带一个突变。它们的眼睛和视网膜发育受到严重损害，脑垂体缺少促肾上腺皮质激素细胞和促黑色素激素细胞（Dickmeis，2009）。这个突变体表现较低的皮质醇水平，这和它们缺少促肾上腺皮质激素细胞以及相关的没有 POMC 表达是一致的。使用这个突变体是为了着重说明 HPI 轴激活和皮质醇在外周昼夜节律调节中的重要作用（Dickmeis 等，2007；Dickmeis，2009）。特别是，皮质醇在产生细胞周期性节律的外周生物钟的调节中起着关键作用（Dickmeis 等，2007）。此外，肾上腺的昼夜生物钟可能调节类固醇生成细胞对 ACTH 刺激的敏感性，从而引起皮质醇释放的节律性（Dickmeis，2009）。斑马鱼幼鱼已经表现出皮质醇释放的节律性（Yeh，2015），rx3 突变体显然是研究早期发育期间 POMC 在应激轴中功能作用的最好模型。出人意料的是，一直没有人利用这个突变体来研究 HPI 轴激活和皮质醇的调节作用，部分原因可能是它短的寿命（3～4 星期）不宜于长时间的研究。总之，对于阐明 CRF 和 POMC 在 HPI 轴发育和功能调节中的作用机理这个难题，斑马鱼是一个极好的模式动物。

表 4.1　解析应激的分子反应的药理学与基因组学的技术

试剂/方法	方法/拮抗物	靶标	来源
药理学的：			
RU486 (Mifepristone)	皮质醇	GR	Medeiros 等（2014）；Philip 等（2012）；Alderman 等（2012）；Li 等（2012）；Tipsmark 等（2009）；Aluru，Vijayan（2007）
螺甾内酯	皮质醇	MR	Kulerich 等（2007）
	11-脱氧皮质酮		
18β-甘草亭酸	催化（皮质醇-可的松）	11βHSD2	Alderman 等（2012）
甲吡酮	催化（11-脱氧皮质醇-皮质醇）	11β-羟化酶	Rodela 等（2011）
基因组学的：			
吗啉	GR 吗啉	GR	Pikulkaew 等（2011）；Nesen，Vijayan（2012）；Chatzopoulou 等（2015）
TALENS	SR4G 转基因品系	GRE 报道基因品系	Krug 等，(2014)
锌指核酸酶	GR^{s35T}	GR 破裂	未发表
大型突变（乙基亚硝基脲）	（先前的 utouto）	（等位基因 357）	Muto 等（2005）；Ziv 等（2013）；Muto 等（2013）；Griffiths 等（2012）
大型突变（乙基亚硝基脲）	Chokh/rx3	前脑垂体的促肾上腺皮质激素细胞减少（chk 基因座的等位基因 s399）	Loosli 等（2003）；Dickmeis 等（2007）；Dickmeis 等（2011）
CRISPR/Cas9			未发表

4.2.3　头肾（肾间腺组织）

头肾是低等脊椎动物皮质类固醇的主要来源，它类似于哺乳类的肾上腺皮质（Wendelaar Bonga，1997；Mommsen 等，1999）。ACTH 和散布在头肾中的类固醇生成细胞的 MC2R 结合，它是一种 GPCR（Aluru，Vijayan，2008；Dores 等，2014）。尽管和黑皮质素家族其他蛋白质多个配体结合，但 MC2R 独特的只有一个配体（Aluru，

Vijayan，2008；Dores 等，2014）。最近的研究表明 MC2R 在鱼类中的功能是和两个黑皮质素受体辅助蛋白（MRAPs），即 MRAP1 和 MRAP2 的有效性相联系的（Cerda-Reverter 等，2013；Dores，Garcia，2015）。对 ACTH 刺激的反应使受体激活必须有 MRAP1 的存在，但不是 MRAP2（Liang 等，2013；Agulleiro 等，2013；Dores 等，2016）。MC2R 的 MPAP-非依赖性信号只曾在米氏叶吻银鲛（*Callorhynchus milli*）中报道，MC2R 基因在 CHO 细胞中表达（Reinick 等，2012）。总而言之，这些研究结果着重证明 ACTH 对 MC2R 激活的特异作用，以及 MC2R 和 MPAPs 在调节应激轴功能协同进化中的重要作用（Cerda-Reverter 等，2013；Dores，Garcia，2015）。所有这些鱼类 MC2R 激活的研究都是离体使用哺乳类细胞的报道试验（reporter assay）进行的（Liang 等，2015；Dores，Garcia 等，2015；Dores 等，2016），而对鱼类 MRAPs 在应激反应中调节作用的在体研究还非常少。

最近，我们证明在虹鳟的急性应激反应中头肾 MRAP1 而不是 MRAP2 的转录水平暂时的升高（N. Sandhu，L. Liang，B. Dores，M. M. Vijayan，在整理中）。虹鳟在应激中 MRAP1 转录水平紧随着血浆皮质醇水平暂时增加而升高，因而可以推想 MRAP1 的调节可能对急性应激反应中 MC2R 的激活与皮质醇的产生起重要作用（Sandhu 等，在整理中）。研究表明 MC2R 转录水平受到应激物的影响，包括对鱼类的污染物（Aluru，Vijayan，2008；Hontela，Vijayan，2009；Sandhu 等，2014；Jeffrey 等，2014）。这样就使得在体应激物介导皮质醇产生的减弱和离体 ACTH 刺激皮质醇的产生互相联系起来（Hontela，Vijayan，2009；Alure，Vijayan，2008；Sandhu，Vijayan，2011）。然而，导致 MC2R 信号受到损害和损坏皮质醇对污染物反应的作用机理还不清楚。如果 MC2R 信号是受到它们的辅助蛋白牢固的调节，那么 MC2R 激活的损害就可能有多个靶标参与，包括 MRAPs，但这还有待于证实。MRAP2 在 MC2R 信号中的作用一直不清楚。然而，最近的一项研究表明 MC4R 在斑马鱼体内和 MRAP2 共表达可能起着 ACTH 受体的作用（Agulleiro 等，2013），这说明有多种调节的作用机理调控鱼类皮质类固醇生成。

对 ACTH 信号反应的皮质醇产生虽然在体与离体都进行了深入的研究，但对参与肾间腺类固醇生成调节的信号通道还了解得很少（Hontela，Vijayan，2009）。cAMP 类似物和钙调节皮质醇的合成（Sandhu，Vijayan，2011；Alsop 等，2009；Lacroix，Hontela，2006；Hontela，Vijayan，2009），表明在皮质醇生物合成的调节中多种信号通道互相交叉。MC2R 的拮抗物还不清楚，因而还难以阐明参与下游信号配体后（postligand）结合的关键作用机理。有些哺乳类的研究报道称一个截短的 ACTH 分子能够起着部分抵消受体的功能（Bouw 等，2014），但这还没有在鱼类中进行试验。大多数研究都是使用 cAMP 类似物（8-溴-cAMP）或者腺苷酶激活剂毛喉素（Alsop 等，2009）去刺激 ACTH-诱导的皮质醇在肾间组织中的产生（Hontela，Vijayan，2009）。在鱼类中，ACTH 激活 MC2R 是和类固醇生成急性调节蛋白（StAR）的转录丰度相联系的，而这是一种对转运胆固醇由外到内穿过线粒体膜起作用的限速步骤；亦和 P450 侧链裂解（P450scc）相联系，它是在类固醇生成中催化胆固醇转变为孕烯醇酮的限速酶；亦和催化 11-脱氧皮质醇转变为皮质醇这一终端步骤的 11β-羟化酶（cyp11A1）相联系（Kusakabe 等，2009；Geislin，Auperin，2004；Hagen 等，2006；Aluru，Vijayan，2006，2008；Castillo 等，2008；Fuzzen 等，2010）。所有这些类固醇生成的基因通常都是保守的（Arukwe，2008；Dores，2013），而 StAR 在所预期的虹鳟和人类之间蛋白激酶 A 磷酸化位点中表现 100% 的同源性（Arekwe，2008）。

类固醇生成的基因表达型式并不是经常一致的，例如有些研究对在体的急性应激物反应或者离体的 ACTH 刺激中并没有观察到类固醇生成基因转录丰度的增加（Geislin，Auperin，2004；Hagen 等，2006）。目前我们对应激的这种基因转录反应不一致性的原因还不清楚，可能有几种因素的影响，包括类固醇的水平，它可能调节转录的丰度（Castillo 等，2008）。在硬骨鱼类，遭遇急性应激物大约 1～4 小时后，血浆中主要的皮质类固醇浓度都达到峰值，并在 24 小时内回复到基础浓度（Vijayan，2010）。在肾间腺组织水平存在一个对皮质醇产生的超短环调控作用（Bradford 等，1992），而这可能包含类固醇生成基因的调控，我们在虹鳟中观察到 GR 信号直接参与头肾 StAR 转录水平的调控，可以支持这个说法（Alderman 等，2012）。不过，还不清楚 GR 信号是如何调节 StAR 转录丰度的，而这个基因启动区所推定的糖皮质激素反应元件（GRE）还未曾报道。有一种可能性就是皮质醇通过快速的非基因组通道而起作用，包括 CREB 的调节作用，影响 StAR 转录（Manna 等，2003；Clark，Cochrum，2007）。支持这个论点的是，用皮质醇进行处理能迅速增强鳟鱼肝细胞 CREB 的磷酸化作用（L. Dindia，M. M. Vijayan，在整理中），但这是否包含在皮质醇刺激的反应中 StAR 转录的调控作用，还有待于确证。

类固醇生成基因，包括 StAR 和 P450scc 持续的阻抑，可以在肾上腺毒素阻抑鱼类皮质类固醇生成的反应中观察到（Hontela，Vijayan，2009）。这些基因对于阻抑化学应激物介导的鱼类皮质醇产生是关键的靶标（Aluru 等，2005；Aluru，Vijayan，2006；Gravel，Vijayan，2006；Sandhu，Vijayan，2011；Best 等，2014；Sandhu 等，2014）。目前我们对污染物影响类固醇生成关键步骤的作用方式还不了解。最近我们的研究表明，镉（Cd）在 MC2R 信号水平上破坏类固醇生成，而在这种金属镉处理的反应中，cAMP 类似物和毛喉素能够挽回 StAR 和 P450scc 的转录水平（Sandhu，Vijayan，2011）。因为污染物对上游靶标的影响可能会限制下游蛋白质的活性，我们不应低估对靶蛋白的直接影响，特别是化学应激物能直接影响 StAR 的转录（Clark，Cochrum，2007）。

采用污染物进行较长期的研究揭示皮质醇生物合成作用持续的激活，这可以由虹鳟较高的 StAR 和 P450scc 转录水平而得到证明（Ings 等，2011a；Best 等，2014；Sandhu 等，2014）。然而，这和升高的皮质醇水平无关，表明类固醇的产生和类固醇的生成能力并不协调配合。目前一个主要的限制条件是缺乏测定这些靶基因在鱼类蛋白质中表达的工具，由于转录的变化可能并不一定反映组织的作用能量。这是因为基因表达可能取决于影响 mRNA 稳定性与周转的因子（Roy，Jacobson，2013），而这是在鱼类中还没有进行研究的领域。关于类固醇生成能力和类固醇产生之间未能协调配合的另一个可能的原因是 11β-羟类固醇脱氧酶 2 型（11βHSD2）参加调节作用；在长期和污染物接触期间，它是催化皮质醇转化为没有生物活性的可的松的关键酶。实际上，用镉处理哺乳类的绒毛膜癌细胞，在培养介质中观察到较低的皮质醇浓度，而这和较高的 11βHSD2 丰度相关（Ronco 等，2010）。总而言之，应激物激活 HPI 轴，引起分子反应以支撑提高鱼类头肾组织皮质类固醇生成的能力。污染物损害皮质醇的产生，这显然是沿着皮质类固醇生成通道多个位点的靶标反应。在采用转录反应作为类固醇生成的分子标志时，应激物的类型和强度以及其作用的持续时间都会影响到它们的表达型式。长时间引起类固醇生成转录体持续性上调的作用机理以及皮质醇生物合成能力的后果如何，都有待于确证。

4.3 基因组的皮质醇信号

关于鱼类皮质类固醇受体和皮质醇信号已经发表了好几篇综述性论文（Vijayan 等，2005；Prunet 等，2006；Stolte 等，2006；Bury，Sturm，2007；Alsop，Vijayan，2009；Aluru，Vijayan，2009；Schaaf 等，2009；Nesan，Vijayan，2013a）。通常，类固醇的细胞作用包含基因组的和非基因组的信号组合［图 4.1（B）］。GR 和 MR 都包含在基因组的皮质醇信号内。目前对硬骨鱼类的作用机理和细胞的重要性还不清楚（Borski，2000；Dindia 等，2012，2013）。最近的研究指明快速的皮质醇介导细胞效应的多种作用方式，包括对快速激活信号通道的膜生物物理变化（Dindia 等，2012，2013）和膜受体介导的应激信号通道的激活（Borski，2000）。但是，至今尚未发现一个皮质类固醇的膜受体（Dindia 等，2015）。总之，所报道的皮质醇大多数的作用都是由基因组信号包括 GR 和 MR 激活所介导的。

4.3.1 糖皮质激素受体（GR）

GR 是硬骨鱼类糖皮质激素作用的主要受体，它最早在虹鳟中克隆和测序，并在各种组织中表达（Ducouert 等，1995）。随后，GR 的同种型在许多鱼类中克隆和测序，包括伯氏朴丽鱼（*Haplochromis burtoni*；Greenwood 等，2003）、舌齿鲈（*Dicentrarchus labrax*）、胖头鲅（*Pimephales promelas*；Filby 和 Tyler，2007）、金头鲷（*Sparus aurata*；Acerete 等，2007）、鲤鱼（*Cyprinus carpio*；Stolte 等，2008）和斑光蟾鱼（*Porichthys notatus*；Arterbery 等，2010）。这些同种型是在大约 3.5 亿年前由整个基因组复制过程中产生的（Alsop，Vijayan，2009）。硬骨鱼类有两个基因，GR1 和 GR2，编码 GR（Bury 等，2003；Vijayan 等，2005；Prunet 等，2006），例外的是斑马鱼，它在基因组内只有一个 GR（Alsop，Vijayan，2008，2009；Schaaf 等，2008）。在伯氏朴丽鱼（*Haplochromis burtoni*；Greenwood 等，2003）中已鉴别 GR2 的一个剪接变体，α 和 β，而在斑马鱼中有一个 GRB 剪接变体（Schaaf 等，2008；Alsop，Vijayan，2009）。GR1 和 GR2 之间的亲和力不同，GR2 对较低浓度的皮质醇和地塞米松较为敏感（Bury 等，2003）。虽然 GR 的拮抗物 mifepristone 都能抑制 GR1 和 GR2

的反式激活作用的活性，但GR1对这种拮抗物较为敏感（Bury等，2003）。受体亲和力和配体特异性的这些差别主要都是由离体的报道试验所确认（Bury等，2003），而在应激期间两种GR不同的功能作用在鱼类中还没有清晰阐明（Prunet等，2006）。最近一项使用GR击倒的研究方法首次证明在斑马鱼胚胎特异性GRα和GRβ激活的转录物组和代谢物组的变化（Chatzopoulou等，2015）。这项研究着重阐明GRβ在斑马鱼GRα信号负调控中起着关键性作用（Chatzopoulou等，2015）。

在哺乳类中，GR是和二聚体Hsp90、每个Hsp70的一分子、抑免蛋白以及P 23等结合成杂络物而存在于细胞质内（Pratt等，2006）。在没有类固醇的细胞中，受体将连续地穿梭来回于细胞质和细胞核之间（Madon，DeFranco，1993）；但是，类固醇和它们的受体结合将会引起快速的易位而和GREs结合。传统的非配体的GR存在于细胞质内，而MR出现在细胞质和细胞核内。皮质醇扩散经过质膜而在细胞质内和受体-蛋白伴侣复合物结合。然后蛋白伴侣和受体离解分开，配体-受体复合物易位到细胞核并和靶基因的GRE结合以引发分子反应（Pratl等，2006；图4.1）。有关鱼类GR杂络物的资料很少，而GRE、特异的15bp、不完全的回文的DNA模体（AGAACA nnnn TGTTCT）是高度保守的（Esbaugh，Walsh，2009）。GRE通过提供一个支架在正确位置结合GR而起到变构激活体（allosteric activator）的作用（Schoneveld等，2004）。GR亦能通过调节其他转录因子包括NF-kB和/或AP-1的反式激活特性而起到阻抑基因转录的作用（Schoneveld等，2004）。少数研究曾经鉴别糖皮质类固醇对靶基因调节作用的特性（Esbaugh，Walsh，2009），生物信息学分析鉴定鱼类上游GREs的一些基因（Aluru，Vijayan，2007；Philip，Vijayan，2015；Alderman等，2012；表4.2）。这样，随着基因组分析（第4.6节）新的仪器设备投入使用，在不久的未来将会提供良好的研究途径以鉴定鱼类在应激期间皮质醇对靶基因调节的作用机理特征。

表4.2 鱼类基因在启动区推定的糖皮质类固醇反应元件（GREs）

基因	种类	来源
11BHSD2	斑马鱼（*Danis rerio*）	Alderman等（2012）
乳酸脱氨酶2	底鳉（*Fundulus heteroclitus*）	Schalte等（2000）
α-淀粉酶	尖吻鲈（*Lates calcarifer*）	Ma等（2004）
生长激素2[a]	虹鳟（*Oncorhynchus mykiss*）	Yang等（1997）
催乳激素	虹鳟（*Oncorhynchus mykiss*）	Argenton等（1996）
细胞因子1、2的阻抑剂	虹鳟（*Oncorhynchus mykiss*）	Philip，Vijayan（2015）
金属硫蛋白1	虹鳟（*Oncorhynchus mykiss*）	Olsson等（1995）
PEPCK[b]	虹鳟（*Oncorhynchus mykiss*）	Aluru，Vijayan（2007）
谷氨酰胺合成酶1、2、4[b]		

续表 4.2

基因	种类	来源
精氨酸酶[b]		
谷氨酰胺脱羧酶 65[b]		
富含半胱氨酸蛋白[b]		
Hsp90[b]		
甲状腺素受体 a[b]	虹鳟（*Oncorhynchus mykiss*）	Aluru，Vijayan（2007）
金属硫蛋白[b]		
P4502M1[b]		
成纤维细胞生长因子[b]		
脂蛋白受体[b]		
芳香碳氢化合物受体-α + β[b]		
卵黄膜蛋白-B[b]		
糖皮质类固醇受体	伯氏朴丽鱼（*Haplochromis burtoni*）	Greenwood 等（2003）
谷氨酰胺合成酶	海湾豹蟾鱼（*Opsanus beta*）	Esbaugh，Walsh（2009）
细胞色素 P4501A（CYP1A）	川鲽（*Platichrhys flesus*）	Williams 等（2000）

注：[a]在启动区亦含有 cAMP 反应元件（CRE）。

[b]认为是 GR-依赖性的调节基因，为皮质醇和 RU486 的结合，清除皮质醇-介导的转录（没有进行启动区的生物信息学分析）。

GR 的转录丰度受到应激的不同程度的调节，许多研究表明在应激的反应中和/或用皮质醇处理后都上调 GR 的转录水平（Vijayan 等，2005，2010；Prunet 等，2006；Bury，Sturm，2007；Stolte 等，2008；Alderman 等，2012）。鱼类 GR 的两个种内同源基因（paralog）通常都表现出相似的表达型式，但亦曾经报道在对应激物的反应中，这两个种内同源基因以组织特异性的方式起着有差别的调节作用（Stolte 等，2008）。然而，和 GR 同种型差别表达相联系的功能作用还有待确定。在应激中 GR 转录的上调可能是皮质醇水平升高以及相关的 GR 转录作用激活的直接结果（Sathiyaa，Vijayan，2003）。研究结果表明，GR 转录水平虽然升高，但 R 蛋白质的表达在皮质醇刺激的反应中是下调的（Vijayan 等，2003；Sathiyaa，Vijayan，2003）。GR 蛋白质下调是由于皮质醇介导的蛋白体降解增加（Sathiyaa，Vijayan，2003）。值得注意的是，蛋白体分解的阻抑能消除皮质醇介导的 GR 蛋白质下调，并上调 GR 的转录水平。鳟鱼在体的肝脏和离体的肝细胞在皮质醇处理的反应中都可以观察到 GR 的转录水平和 GR 蛋白质表达之间不协调配合的情况，说明 GR 作为鱼类应激中一个重要的细胞适应性因子的自身调节作用（autoregulation）（Vijayan 等，2003；Sathiyaa，Vijayan，2003）。总之，GR 在靶组织中的表达可能受到一个负反馈作用机理的紧密调控，使这个受体在广泛的功能范围内起着重要作用，包括生长、代谢、免疫、应激以及渗透压和离子调节的行为表现（Vijayan等，2010）。但是，在作用机理的水平上，在应激反应和糖皮质类固醇的刺激中，GR在靶细胞内的调节作用以及相关的细胞反应，都有待于在鱼类模型中阐明。

4.3.2　盐皮质激素受体（MR）

硬骨鱼类缺乏醛固酮，它是哺乳类主要的MR信号的配体（Funder，1997）。MR虽然能以高亲和力与皮质醇结合，但MR在哺乳类作用的靶组织具有11βHSD2，它把皮质醇降解为可的松，由此而得到醛固酮-MR信号（Funder，1997）。如同在哺乳类中一样，11βHSD2在调节鱼类皮质醇的生物有效性中亦起着重要作用（Kusakabe等，2003）。这个基因在它的启动区含有GRE，并在过量皮质醇刺激的反应中上调（Alderman，Vijayan，2012；Li等，2012）。这个酶在斑马鱼早期发育中调节皮质醇水平（Alsop，Vijayan，2008），并在鳟鱼卵细胞生成期间控制母体皮质醇渗入到卵黄中（Li等，2012）起重要作用。

MR在鱼类组织中广泛表达，研究表明，11-脱氧皮质甾酮可能起着MR激动剂的作用（Milla等，2008；Stolte等，2008；Pippal等，2011）。然而，这个激素作用的生理意义在鱼类中还不清楚（Stolte等，2008）。MR已经在一些鱼类中克隆与测序，包括虹鳟（Colombe等，2000）、斑马鱼（Alsop，Vijayan，2008）、鲤鱼（Stolte等，2008）和朴丽鱼（*Haplochromis*）（Greenwood等，2003），但是，对MR的作用和它的生理效能，特别是离子调节作用方面，还没有阐明（McCormick等，2008；Cruz等，2013）。尽管经历基因组复制，大多数鱼类只有一个MR，表明从硬骨鱼类的基因组中丢失了一个额外的受体（Alsop，Vijayan，2008）。对于皮质醇通过GR和相关的靶基因调节的基因组作用已经进行研究，但对有关鱼类MR-介导的皮质醇信号还了解得不多（Takahashi，Sakamoto，2013）。大量研究数据都表明皮质醇对糖皮质激素和盐皮质激素起着双重的作用，其中有些作用可能是由MR信号介导的，特别是在组织中以及迫切需要保持渗透压平衡的情况下（Takahashi，Sakamoto，2013；Kiilerich等，2007；McCormick等，2008；Kelly，Chasiotis，2011）。最近的一项研究证明GR和MR至少是在离体的情况下能相互作用以调节转录的调控，这提示在鱼类应激期间这两种受体亦可能参与调节靶组织的皮质醇反应（Kiilerich等，2015）。在体的情况下看来似乎是可能的，这两类受体都能在鱼类许多组织中表达（Aruna等，2012）。

最近的研究亦提到在鱼类应激期间，皮质醇信号通过MR激活皮质醇的负反馈调节而发挥作用（Alderman等，2012；Alderman，Vijayan，2012），以及MR在早期发育（Alsop，Vijayan，2008；Nesan，Vijayan，2013a）和行为（Manuel等，2014，2015）中的作用。GR转录体由母体沉积在斑马鱼的胚胎内，它们的水平一直到孵化都不升高（Alsop，Vijayan，2008）；MR转录体沉积的量还很少，但在胚胎发育期间却逐渐增加。在胚胎发育期间GR和MR转录水平的不一致性可以推想到MRs可能在鱼类早期发育中起重要作用（Alsop，Vijayan，2008；Nesan，Vijayan，2013a），但这还有待于验证。总之，目前对MR信号的作用、介导反应的配体以及受影响的靶基因等都还没有了解清楚，但至今的研究结果似乎可以证明皮质醇是 MR 作用于鱼类的关键介体。

4.4 皮质醇的基因组效应

基因组的皮质醇信号在鱼类中具有广泛范围的作用（Aluru，Vijayan，2009）；本书在渗透压调节和酸碱平衡（Takei，Huang，2016；本书第 6 章）、生殖（Pankhurst，2016；本书第 8 章）和健康（Yada，Tort，2016；本书第 10 章）等方面进行论述。为了避免和其他各章重复，本章将着重介绍皮质醇介导的分子反应以及在早期发育和急性应激行为表现方面的基本内容。

4.4.1 应激轴的发育

最近的研究清楚地表明皮质醇在硬骨鱼类早期发育的基本分子调节中起着重要作用（Nesan，Vijayan，2013a）。主要的研究都在斑马鱼中进行，因为它们发育快速，胚胎透明，利用这种鱼类能进行正向和反向的遗传学研究（Nesan，Vijayan，2013a）。HPI 轴基因表达按年月次序的排列，包括 CRF（Chandrasekar 等，2007）、POMC（Herzog 等，2003，Liu 等，2003）、StAR、MC2R（To 等，2007）、11β-羟化酶、MR、GR 和 11β-HSD2（Alsop，Vijayan，2008）等，都曾在斑马鱼早期发育期间进行描述（Alsop，Vijayan，2009；Nesan，Vijayan，2013a）。虽然 HPI 轴功能的所有分子作用机理都出现在斑马鱼孵化时（受精后 48 小时）（Alsop，Vijayan，2009），但最早的应激反应只在受精后大约 72 小时才发生，表明 HPI 轴激活的延迟（Alderman，Bernier，2009；Alsop，Vijayan，2009；Nesan，Vijayan，2013a）。这些研究成果支持了有关在胚胎发育过程应激处于低反应期的论点，而这对模式动物适宜的胚胎发育过程是必不可少的（Nesan，Vijayan，2013a）。

GR 和皮质醇都是母体转移来的，而这个蛋白质和类固醇的从头合成分别开始于原肠胚形成（＞受精后 8 小时）和孵化后（Nesan，Vijayan，2013a）。实际上，母体提供的 GR 和皮质醇对于发育的进程是必要的（Nesan 等，2012；Nesan，Vijayan，2012，2013a，b；Pikulkaew 等，2011）。GR 信号的重要性是通过使用吗啉寡核苷酸将 GR 暂时击倒以及相关的遗传型与表现型发生变化的研究而确定的（Pikulkaew 等，2011；Nesan，Vijayan，2012，2013b）。由母体皮质醇刺激的 GR 信号可能影响早期的基因组变化，并且在发育基因转录调控的关键时期——中囊胚转换期（MBT）中发挥作用（Pikulkaew 等，2011；Nesan 等，2012）。此外，采用微点阵（microagray）研究基因组形式的变化，证明 GR 信号参与斑马鱼后 MBT 阶段一系列关键发育基因的调节作用（Nesan，Vijayan，2013b）。胚胎发育期间（受精后 36 小时）的转录物组分析阐明了由 GR 信号调节的几个通道，包括神经系统发育、细胞迁移、细胞对细胞的信号、心血管发育、器官形态以及骨骼/肌肉系统发育（Nesan，Vijayan，2013a，b）。虽然还缺少作用机理的研究以证明皮质醇信号直接调节这些通道，但已有一些研究结果证明在斑马鱼早期发育期间，母源皮质醇调节的骨骼形态发生蛋白（BMPs）可能是一种关键的形态发生要素（Nesan，Vijayan，2012）。BMPs 参与斑马鱼早期发育的许多方面，包括器官形成、前后内胚层模式（Tiso 等，2002；Poulain 等，2006）、中胚层模式（Neave 等，1997），以及内耳和侧线的形成（Mowbray 等，2001），这就说明母源皮质醇在发育过程中起着重要的作用（Nesan，Vijayan，2013a，b）。除了用吗啉暂时击倒的实验外，最近亦建立了稳定的转基因斑马鱼敲除品系（表 4.1）。特别是 GRs357 突变体，它是在结合 GR 区的 DNA 第二个锌指模体内以精氨酸替换半胱氨酸，能够清除突变体 GR 和靶基因 GREs 结合的能力（Ziv 等，2013）。这个突变体的 CRF 和皮质醇水平慢性升高，基本上酷似慢性应激的状态，表现出和压抑相似的行为表现型，包括减少探索行为和损害反复处于引起焦虑环境中的适应能力（Ziv 等，2013）。这说明皮质醇在脑的发育和行为的表现型中起重要的作用，然而，和遗传型与表现型相联系的分子作用机理还有待于阐明。

母体的皮质醇对发育的进程是必不可少的，但受精卵不正常的皮质醇含量，类似于一个母源的应激和过多的皮质醇转移，对子代是有害的（Nesan，Vijayan，2012，2016；Wilson 等，2013，2015）。在鱼类中，母源的应激和增加的皮质醇水平会使皮质醇过多地转移到胚胎中（Geising 等，2011；Stratholt 等，1997；McCormick，1999；Eriksen 等，2006；Kleppe 等；2013；Faught 等，2016）。最近的研究表明母源皮质醇对良好的心脏发育特别重要（Nesan，Vijayan，2012；Wilson 等，2013，2015）。例如，给一个细胞的胚胎微量注射皮质醇以提高基础的皮质醇水平，模拟母体的应激，会引起心脏畸形，而这和关键的心脏基因受到阻抑有关，包括 nkx2.5 心肌球蛋白轻链 1、心肌钙蛋白型 T2A 和钙转运 ATP 酶（Nesan，Vijayan，2012）。这些研究结果表明皮质醇在心脏发育中起的作用，但是，还不清楚在早期发育期间，过多的皮质醇是如何抑制这些基因的（Nesan，Vijayan，2012）。我们亦观察到受精卵中过多的皮质醇含量酷似母源的应激，使斑马鱼增强前脑的神经发生，而这就相当于增加了斑马鱼胚胎神经原性标志物神经 D（neuroD）和 orthopediab 的空间丰度（C. Best，D. Kurassch，M. M. Vijayan，整理中）。此外，这些鱼是抗焦虑的（anxiolytic），因此可以推想母源的皮质醇介导的神经发生可能在行为表现型的发育过程中起重要作用。这些研究结果着重说明在良好的发育过程中，母源皮质醇严密的调节作用是很重要的，因为应激和过多的皮质醇沉积会影响脑的功能，导致行为的和心脏的表现型畸形。

最近的研究亦关注到 20β-羟类固醇脱氨酶 2 型（20βHSD2）的作用，它在斑马鱼中将可的松分解为 20β-羟基可的松（Tokarz 等，2012，2013）。在应激反应/用皮质醇处理时，这个酶的转录是上调的，而击倒 20βHSD2 会引起性质截然不同的应激表现型，表明这个酶在皮质醇应激反应中起一定的作用（Tokarz 等，2013）。由于母源沉积的皮质醇严密调节作用对良好的发育十分重要，皮质醇的降解酶类，包括 11βHSD2 和 20βHSD2，在阻抑斑马鱼胚胎过多皮质醇的影响中所起的重要作用就不足为奇了（Tokarz 等，2012，2013）。此外，我们的研究亦表明过多的皮质醇，酷似母源的应激，在离体的卵巢滤泡中上调 11βHSD2 的转录丰度，这可能是斑马鱼胚胎皮质醇含量严密调节的作用机理（Faught 等，2016）。总之，这些研究结果说明受精卵皮质醇含量的严密调节作用对斑马鱼良好的发育进程，包括肌肉生长和发育都是必不可少的（Nesan，Vijayan，2012，2013a，b）。利用斑马鱼击倒 GR 的突变体可以深刻阐明皮质醇在调节发育基本的分子作用机理中所起的作用（Nesan，Vijayan，2013b；Ziv 等，2013；Chatzopoulou 等，2015）。

最近的研究深入阐明了斑马鱼胚胎 GR 激活后转录物组和代谢物组的反应（Chatzopoulou 等，2015）。采用击倒 GRα 和 GRβ 的研究方法，这项研究首次证明和这两个 GR 剪接变体的激活相联系的特异性分子与代谢的反应。作者们使用剪接的阻抑性吗啉和合成的糖皮质激素地塞米松进行处理，以检测 α 和 βGR 信号的差别调节作用。根据转录物组和代谢物组的数据，这两个不同的基因群既在基础的情况下亦在应激的情况下被激活（Chatzopoulou 等，2015）。在地塞米松处理组，葡萄糖的增加是和一些关键的糖异生基因相联系的，包括 pck1、pck2、pfkrb41 和 g6pca（Chatzopoulou 等，2015）。pck 基因编码烯醇丙酮酸磷酸羟激酶（PEPCK），并在肝细胞生成之前在多种细胞类型中表达（Jurczyk 等，2011）。pck 基因在启动区有 GREs（表 4.2），它们在发育期间通过皮质醇水平的变化而得到调节。于是，斑马鱼孵化后皮质醇含量的增加可以说明 GR 受体在利用内源性（卵黄）营养物向外源性摄食的过渡期间，在能量重新分配的方面起着关键性作用。皮质醇亦能增加和蛋白质分解代谢有关的基因（Wiseman 等，2007），包括增加参与丝氨酸蛋白质水解酶和蛋白酶体通道（ube2a/b myliqb，foxredz）的基因转录丰度（Chatzopoulou 等，2015），这样，两种 GR 受体之间的相互作用对能量代谢的调节就值得重视。有意思的是，斑马鱼 HPI 轴的激活和内源营养物资源（卵黄）的减少相一致，因而，我们可以推想斑马鱼孵化后皮质醇的升高对于能量底物的分配和外源摄食的开始可能起着一种刺激作用。

4.4.2 应激期间的分子调节

肝脏是激素作用的主要靶器官，使动物能够通过代谢作用对付应激。葡萄糖是基本的能量燃料，在面临应激物的损害而重建稳态的过程中是不可缺少的（Mommsen 等，1999；Enes 等，2009）；葡萄糖或者通过糖原分解迅速在肝脏产生，或者通过糖原异生作用的延缓激活而在肝脏产生（Vijayan 等，2010）。传统的逃逸或搏斗反应包括肾上腺素介导的糖原分解作用加强，从而使葡萄糖迅速输送到血液循环中（Mommsen 等，1999；Fabbri，Moon，2015）。应激引起的延缓葡萄糖反应通常是由糖皮质激素激活的 GR 介导的，从而使包括糖原异生作用的代谢基因上调，这不仅使应激介导的血液循环中葡萄糖水平增加，而且亦使得应激作用后衰竭的糖原贮存得到补充（Mommsen 等，1999；Leung，Woo，2010；Vijayan 等，2010）。在鱼类的应激反应和皮质醇刺激中，一系列代谢通道被激活，包括提高糖原异生作用，糖酵解以及脂肪酸和氨基酸的代谢作用（Mommsen 等，1999）。在肝脏的水平，皮质醇主要的代谢功能是围绕着葡萄糖的产生以及相关底物的重新分配而使这个代谢物葡萄糖能够得以产生。一些研究表明皮质醇能增加葡萄糖生产，而用 GR 的拮抗物 mifepristone 能抵消皮质醇的这种作用，这就证实了 GR 信号的这种直接效应（Aluru，Vijayan，2009；Philip 等，2012；Philip，Vijayan，2015；图 4.3）。虽然研究结果证明皮质醇在糖原转移以增加葡萄糖产量中发挥作用，但可以认为这更像是一个快速的非基因组反应而不是一个基因组的效能（Mommsen 等，1999；Vijayan 等，2010）。应激后糖原的减少主要是肾上腺素介导的（Mommsen 等，1999；Reid 等，1998；Fabbri 等，1998）；然而，我们不能低估皮质类固醇在应激期间引发肝细胞对肾上腺素功能中所起的作用（Reid 等，1992；Mommsen 等，1999）。其他的激素，主要是胰高血糖素，亦可能在应激期间促进糖原代谢和增强葡萄糖排出量方面起作用。但是，在急性应激期间调节糖原耗尽和应激后恢复期间糖原的补充中，调节葡萄糖的激素动态之间的相互作用或者信号的作用机理，还了解得很少（Mommsen 等，1999；Vijayan 等，2010）。

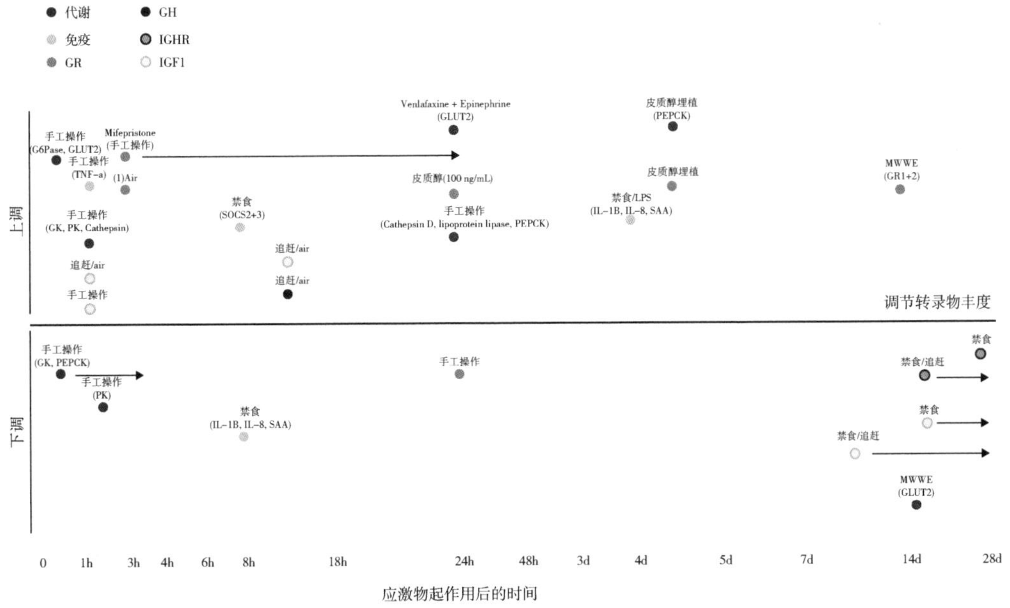

图 4.3　在应激物处理的反应中，肝脏时序基因的转录变化

注：用于引起基因转录发生变化的应激物在图上方表示为不同颜色的圆点。对特异性的种类，可参看文中。StAR，类固醇生成急性调节蛋白；P450scc，P450 侧链裂解；MC2R，黑皮质素 2 受体；GR，糖皮质激素受体；GLUT2，葡萄糖转运蛋白；GH，生长激素；GRH，生长激素受体；PRL，催乳激素；MC2R，黑皮质素 2 受体；IGF-1，胰岛素样生长因子 1；IGF-R，胰岛素样生长因子受体；PEPCK，烯醇丙酮酸磷酸羧激酶；G6Pase，葡萄糖-6-磷酸酶；GK，葡萄糖激酶；GS，谷氨酸合成酶。＊见 Wiseman 等，2007 的代谢基因目录。注意这不是一个应激在肝脏内所有基因变化的详尽无遗的目录，进一步详细说明见于文中。箭头表示在实验的进程中基因保持上调或下调的持续时间。（见书后彩图）

经常观察到的皮质醇分子反应是上调鱼类肝脏内的糖异生，PEPCK 激活关键限速步骤中的转录丰度和酶活性（Sathiyan，Vijayan，2003；Vijayan 等，2003；Aluru，Vijayan，2007；图 4.3）。研究已经表明氨基酸能优先用于鱼类肝脏内的糖异生作用，而甘油和乳酸盐在皮质醇刺激的反应中亦用于葡萄糖的产生（Mommsen 等，1999）。除 PEPCK 外，已糖激酶（在肝脏的 HK 和葡萄糖激酶 GK）和丙酮酸激酶（PK）都经常测定，虽然后两种酶在糖酵解中是限速步骤的，而且不同于 PEPCK，是葡萄糖利用的一种指征。HK 催化葡萄糖磷酸化为葡萄糖-6-磷酸盐，在糖酵解中是第一个限速步骤（Mommsen 等，1999），而 PK 催化磷酸烯醇丙酮酸转化为丙酮酸。在应激期间这些转录体的调节还很不清楚，但是，经常和应激物作用相关的营养状态变化，将会调节鱼类 HK 和 GK 的 mRNA 丰度和活性（Soengas 等，2006；Perez- Jimenez 等，2007；图 4. 3）。

在应激期间受到皮质醇刺激影响的其他值得注意的组织包括白肌，它提供氨基酸和乳酸盐以提高肝脏的葡萄糖异生作用（Milligan，1997，2003；Frolow，Milligan，2004）。研究结果已经证实 GR 信号参与这个反应，但是，在应激期间参与肌肉分解代谢和能量底物调整的转录体和蛋白质的调节作用，还一直没有在鱼类模型中进行机理研究。为了对应激期间的转录调节作用进行更加全面的阐明，最近的一些研究采用基因组测序技术，包括微点阵和在代谢的活组织中（包括肝脏和肌肉）进行的第二代测序技术（表 4. 3）。Wiseman 等（2007）采用靶标鳟鱼的微点阵证明参与蛋白质分解代谢的基因转录丰度增加，包括组织蛋白酶 D、精氨酸酶和谷氨酸合成酶。组织蛋白酶 D 是一种天冬氨酸蛋白酶，参与非特异性蛋白质的降解作用，是上调的并且可以说明基因上调和蛋白质减少之间的不一致性，因为在应激期间蛋白质稳定性可以非特异性地受到影响（Mommsen 等，1999；图 4. 3）。应激物作用后经常变化的基因可以大致分类为参与组织代谢的酶类和转运蛋白，包括葡萄糖、蛋白质和氨基酸的代谢，以及免疫反应的，包括主要组织相容性复合体-2、白细胞介素、急性期蛋白和肿瘤坏死因子（表 4. 3）。转录反应通常是变动的，在应激的反应中或者皮质醇刺激的情况下，有些基因转录体在短时间内表达，大多数都调整超过 24 小时，亦有些调整超过一个较长的时期（图 4. 3）。总之，在应激的适应或者不良适应中这些转录变动的作用还不清楚，而大多数研究对转录变化的稳定水平已经深入阐明。此外，在应激期间和/或由皮质醇刺激而通过转录的和翻译的调控基因变化的作用机理还很不清楚。

表 4. 3　大规模转录组分析（cDNA 微点阵或 RNA-测序）评估应激/皮质醇对基因变化的影响

应激物	种类	组织	方法	基因影响	来源
手工操作/拥挤	虹鳟（*Oncorhynchus mykiss*）	肝脏	RNA 测序 [Illumina Hiseq 2000]	总基因变化	Liu 等（2014）
LPS（禁食，盐度）	尖吻鲈（*Latescalcarifer*）	肠	RNAseg [Roche 454]	总基因变化（非特异性）	Xia 等 [2013]

续表 4.3

应激物	种类	组织	方法	基因影响	来源
铝	大西洋鲑 (*Salm salar*)	肌肉	RNAseg [Illumina 基因分析仪ⅡX]	miRNA 变化	Kure 等 [2013]
手工操作	刀鲚 (*Coilia nasus*)	肝脏	RNAseg [Illu mina Hiseg (2000)]	总基因变化	Du 等 (2014)
皮质醇埋植	金头鲷 (*Sparus aurata*)	肝脏	微点阵/qPCR	总基因变化 [GR，醇脱氢酶，MHC Ⅱ抗原-联系不变链，嗜血的，MHC 级Ⅱ抗原 β 键，T-细胞受体，CIDE3，Pitrilysin 金属蛋白酶 1，cdc48，葡萄糖磷酸变位酶，锌-结合 1]	Teles 等 (2013)
皮质醇	虹鳟 (*Oncorhynchus mykiss*)	肝细胞	微点阵/qPCR	总基因变化	Aluru，Vijayan (2007)
皮质醇	斑马鱼 (*Danio rerio*)	胚胎	微点阵/qPCR	总基因变化	Nesan，Vijayan (2013b)
皮质醇埋植	大西洋鳕 (*Gadus morhua*)	胚胎	微点阵/qPCR	总基因变化 [stxbp6，fbxw2，capn12，thbs4，syt12，cocolc，sel113，ipo7]	Kleppl 等 (2013)
手工操作/拥挤	虹鳟 (*Oncorhynchus mykiss*)	肝脏	微点阵/qPCR	总基因变化，在应激物作用后 1 小时和 24 小时测定	Wiseman 等 (2007)
手工操作/暴露空气中	虹鳟 (*Oncorhynchus mykiss*)	肝脏	微点阵/qPCR	总基因变化	Momoda 等 (2007)

续表 4.3

应激物	种类	组织	方法	基因影响	来源
	虹鳟（*Oncorhynchus mykiss*）	肝脏	微点阵/qPCR	触珠蛋白，补体因子H，CIHBP，B-凝血因子，TC8422，抗-胰蛋白酶，血浆铜蓝蛋白，血清白蛋白，TC8200，EST10729，血清浆黏蛋白，14-3-3	Cairns 等（2008）
手工操作	虹鳟（*Oncorhynchus mykiss*）	脑和肾脏	微点阵/qPCR	总基因变化	Krasnov 等（2005）
禁食/投喂	金头鲷（*Sparus aurata*）	心脏和骨骼肌	微点阵/qPCR	Makroinl. 丝氨酸蛋白抑制蛋白 hl 前体，δ-9-去饱和酶 1，骨骼肌钠通道，神经细胞附着分子，膜-型基质金属蛋白酶，膜联蛋白 2a，丙酮酸脱氢酶，同工酶 2，脂连蛋白受体蛋白 1	Calduch-Giner 等（2014）
LPS	虹鳟（*Oncorhynchus mykiss*）	单核细胞/巨噬细胞原代培养	微点阵/qPCR	HSp70，Jun-B，iNFkBm，MMP13	Mackenzie 等（2006a）
LPS	虹鳟（*Oncorhynchus mykiss*）	卵巢	微点阵/qPCR	总基因变化	Mackenzie 等（2006b）
Lepeophtheirus salmonis/皮质醇	大西洋鲑（*Salmo salar*）	皮肤	微点阵/qPCR	总基因变化	Krasnov 等（2012）
热应激	虹鳟（*Oncorhychus mykiss*）	RBCs	微点阵/qPCR	总基因变化	Lewis 等（2010）
热/盐度	虹鳟（*Oncorhynchus mykiss*）	鳃、脑、肝脏、脾脏、肾脏、肌肉	微点阵/qPCR	全基因变化［casp3，casp8，p53，nupr，Hsp70b，nakatpala，naka-tpalb，nakatpalc，nakat-pa3］	Sanchez 等（2011）

续表 4.3

应激物	种类	组织	方法	基因影响	来源
热	大西洋鳕（*Gadus morhua*）	脾脏	微点阵/qPCR	全基因变化［DHX58，STAT1，RF7，ISGL5，RSAD2，IKBa］	Hori 等（2012）
普萘洛尔	胖头鲹（*Pimephales promelas*）	脑	微点阵/qPCR	全基因变化	Lorenzi 等（2012）
Elyoxetine nalafaxin，carbamazepine	胖头鲹（*Pimephales promelas*）	整体	微点阵/qPCR	和人类疾病相关的基因（阿尔茨海默氏病，双极 schizophrenia，ADHD，帕金森氏病）	Thomas，Klaper（2012）
MWWE	虹鳟（*Oncorhynchus mykiss*）	肝脏	微点阵/qPCR	全基因变化	Ings 等（2011b）
莠去净 nonylpenol	虹鳟（*Oncorhynchus mykiss*）	肝脏	微点阵/qPCR	全基因变化	Shelley 等（2012）

注：LPS，脂多糖类；MWWE，城市废水流出物。

靶组织的分子反应，包括提高基因转录和翻译，可以有助于增加和应激相联系的能量需求。应激实际上增强鱼类的代谢率（Barton，Iwama，1991），而基因和蛋白质表达对鱼类代谢需求的作用还不清楚。和蛋白质合成相关的能量需求主要是有助于增加总的细胞代谢率（Iwama 等，2006；Vijayan 等，2010），但转录的机理对总能量收支所起的作用还不清楚。在鱼类中，基因的转录并不经常和相应的蛋白质合成密切联系，这一点很重要（Sathiyaa，Vijayan，2003；表 4.4）。一般都认为靶基因上调和蛋白质表达对于动物应对应激物是至关紧要的，应激物会在应激期间引起代谢需求的增加。然而，有些应激物降低代谢作用而成为一种适应性反应，包括低氧/缺氧和冷休克（Storey，1999）。在这种状态下，即使总的蛋白质合成能力缩小，而基因转录和蛋白质合成仍按照预定的目标增加，这是对付应激物所必需的。这说明应激期间在靶细胞水平有一种调控作用机理可以调节能量需求的通道。在应激适应期间支配这种能量底物重新分配的各种因素还没有深入阐明（Vijayan 等，2010）。

表 4.4 皮质醇-反应基因（G）和相应的蛋白质（P）变化

应激物	种类	组织	基因（G）	蛋白质（P）	G-P 相关?	来源
低氧/追赶/禁闭/皮质醇埋植	虹鳟（*Oncorhynchus mykiss*）	松果体	Aanat2	AANAT2 酶活性	Y	Lopez-Patino 等（2014）
手工操作应激	虹鳟（*Oncorhynchus mykiss*）	脑 肝脏	GR GR	GR GR	N N	Alderman 等（2012）
皮质醇埋植	虹鳟（*Oncorhynchus mykiss*）	肝脏	PEPCK GR HSP90	PEPCK 酶 GR HSP90	N N Y	Vijayan 等（2003）
手工操作应激	虹鳟（*Oncorhynchus mykiss*）	肝脏	GK PEPCK G6Pase GLUT2 PK	PEPCK 活性 PK	N Y	Lopez-Patino 等（2014）
MWWE	虹鳟（*Oncorhynchus mykiss*）	肝脏	GR1，GR2 HSP90	GR HSP90 [D100%]	Y/N N	Ings 等（2011b）
皮质醇/尿素	海湾豹蟾鱼（*Opsanusbeta*）	鳃	GS 普遍存在-GS [U 未熔组（没有 F）]	GS 活性 [U 集群的]	N	McDonald 等（2009）
皮质醇	虹鳟（*Oncorhynchus mykiss*）	肝细胞	GR [UF; DRU 组]	GR [DF; DRU 组]	N	Aluru，Vijayan（2007）
皮质醇-诱导禁食	莫桑比克罗非鱼（*Oreochromis Mossambicus*）	胃	生长素释放肽 [DF]	血浆生长素释放肽 [DF]	Y	Janzen 等（2012）
皮质醇注射	斑马鱼（*Danio rerio*）	胚胎	GR [NC]	GR [D48h]	N	Nesan，Vijayan（2012）
GR 击倒	斑马鱼 *Danio rerio*）	胚胎	GR [NC]	GR [DMo 处理]	N	Nesan，Vijayan（2012）
禁食（没有应激激素测定）	金头鲷（*Sparus aurata*）	肝脏	GK [U]	GK 活性 [U]	Y	Caseras 等（2000）

注：G-P 代表（Y = 正的，N = 负的）所研究的基因和蛋白质表达或酶活性之间的相关联系。

由于在应激期间能量需求会增加，分配到其他通道的能量，包括生长和免疫反应的，可能在转录水平中下调（Tort，2011；Reindl，Sheridan，2012）。在应激的反应中，生长激素受体和IGF-1转录水平都下调，这在鱼类在体和离体的肝细胞中都曾观察到（Small等，2006；Nakano等，2013）。虽然一些研究已经检测慢性应激和皮质醇处理对生长的影响（Bernier，Peter，2001；Bernier等，2004），但对它们的作用机理尚未阐明。已经清楚肌抑制素（myostatin）对脊椎动物的肌肉生长起负的调节作用。哺乳类肌抑制素的启动子中含有GRE，这说明皮质醇对生长起着直接的调节作用。然而，皮质醇的这种反应并没有在硬骨鱼类中保留下来（Biga等，2004；Galt等，2014）。用皮质醇处理能下调罗非鱼肌抑制素的表达（Radgers等，2003）。我们亦已经知道用皮质醇能够减弱肝脏内GH诱导的IGF-1 mRNA丰度，表明在应激和生长抑制之间存在着联系（Leung等，2008；Philip，Vijayan，2015）。然而，在银大麻哈鱼（*Oncorhynchus kisutch*）（Pierce等，2010）和罗非鱼（*Oreochromis mossambicus*）（Pierce等，2011）中，皮质醇实际上能刺激IGF-2的转录水平。最近的研究亦证明慢性皮质醇处理能增强肝脏igfbpl，暗示修饰的IGFs功能起着抑制生长的潜在作用（Madison等，2015）。IGF-1和IGF-2都是GH促进效能的主要介体，并对其他组织包括性腺和心脏起着下游的作用。例如：在斑马鱼再生期间，IGF-2b在心脏中是上调的（Huang等，2013），而在成年欧洲褐鳟（*Salmo trutta*）曾报道在心体指数和皮质醇水平之间是正相关的联系，表明皮质醇在心肌重建中起作用（Johansen等，2011）。在促生长轴和皮质醇介导对心肌发育的效能之间的相互作用还需做进一步研究，特别是因为在发育期间过多的皮质醇对斑马鱼心肌发育和功能有非常明显的影响（Nesan，Vijayan，2012；Wilson等，2013）。

最近对虹鳟肝细胞进行的研究可以说明应激期间生长和免疫轴之间抑制作用的基本机理。细胞因子信号的阻抑基因（SOCS）在它们的启动区含有GRE，能负调节生长激素，而细胞因子信号受到JAK-STAT通道的阻抑（Philip，Vijayan，2015；图4.3）。脂多糖类（LPS）的刺激能上调细胞因子基因的转录丰度，包括鳟鱼肝细胞的IL-6和IL-8；但是，用皮质醇处理只能阻抑由LPS-诱导的IL-6而不是IL-8的转录丰度（Philip等，2012；Philip，Vijayan，2015）。由于LPS-介导IL-6而不是IL-8表达以参与JAK-STAT通道，这些研究结果表明在细胞因子的阻抑中皮质醇诱导SOCS的上调起着作用（Philip，Vijayan，2015）。实际上，皮质醇处理使SOCS-1和SOCS-2转录丰度增加，而这为肝细胞内的mifepristone所抵消（Philip等，2012；图4.4）。皮质醇处理亦阻抑急性生长激素刺激的IGF-1转录丰度，而这和鳟鱼肝脏薄片中减弱JAK-STAT的激活是相关联的（Philip，Vijayan，2015）。因而，皮质醇对生长和免疫调节的变化看起来是和SOCS的上调以及JAK-STAT通道的阻抑相联系的（Philip，Vijayan，2015；图4.4）。综合这些研究结果可以说明，在鱼类应激期间，在肝脏水平存在着一种有效能的作用机理调节能量-需求的通道。我们认为，在鱼类应激期间，皮质醇作为细胞能量重新分配的关键介体对SOCS进行调节（Philip，Vijayan，2015）。总之，皮质醇的靶组织分子效能还只是刚开始被引起重视，而有迹象表明这种类固醇在应激物-介导的对生长和免疫功能的作用中起着关键作用。

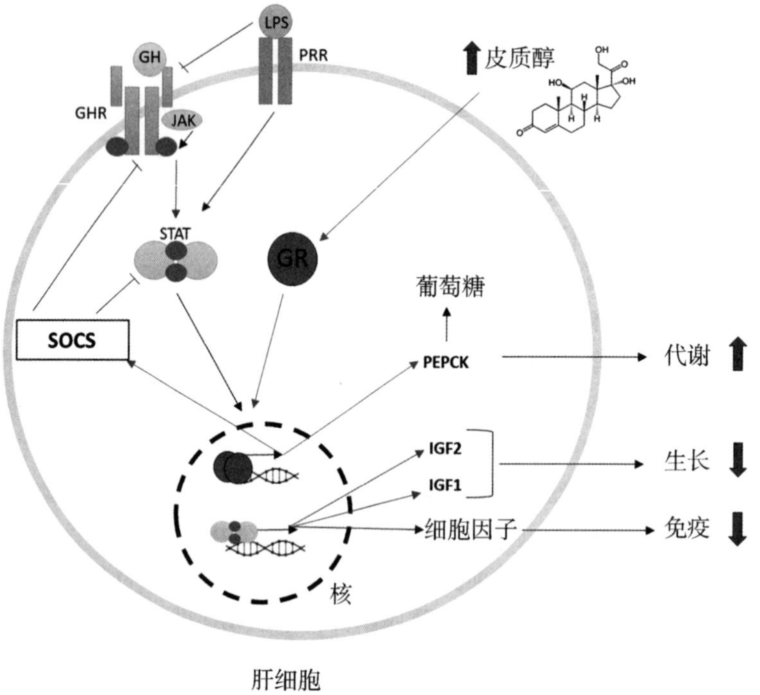

图 4.4 在硬骨鱼类的肝细胞内，应激、生长和免疫通道之间的细胞内相互作用

注：皮质醇扩散通过皮质膜而和糖皮质类固醇受体（GR）结合。一旦结合发生，它传导到核并通过和靶基因的糖皮质类固醇反应元件（GRE）结合而诱导基因转录，包括细胞因子信号的阻抑基因（SOCS）和烯醇丙酮酸磷酸羧激酶（PEPCK）。生长激素（GH）信号通过生长激素受体（GHR）激活 JAK-STAT 通道。GHR 与配体结合，JAK（詹纳斯激酶）磷酸化，信号转录蛋白与转录激活剂（STAT）和 JAK 的磷酸酪氨酸结合。JAK 磷酸化 STAT，它们二聚化并传导到核，启动靶基因转录，包括胰岛素样生长因子 1（IGF-1）和 2（IGF-2）。LPS 和模式识别受体结合，它亦激活 JAK-STAT 信号，并引起细胞因子的合成。绿色箭头表示皮质醇信号；红色箭头表示 GH 信号；黄色箭头表示 LPS 信号。（见书后彩图）

4.4.3 细胞的调节

关于鱼类的细胞应激反应，特别是热休克反应已经发表了一些综述（Iwama 等，1998；Ackerman 等，2000a，b；Basu 等，2002；Iwama，2004；Deane，Woo，2011）。在细胞水平，一些研究已经检测几个级别热休克蛋白（HSPs）的调节作用，它们通常被归类为热休克反应（Richter 等，2010）。热休克蛋白几乎可以在每一个细胞类型中表达，而且它们提供的细胞保护作用可以抵消在应激反应中产生的蛋白毒性（Iwama 等，1998；Deane，Woo，2011），因而使它们成为细胞应激反应中最合适的指标。HSPs 根据其功能可以大致分为 7 类，包括传统的分子蛋白伴侣，蛋白质水解系统的成分，RNA 和 DNA 修饰酶，代谢酶，转录因子，激酶以及转运、解毒和膜调节蛋白质（Richter 等，2010）。热休克反应在不同种类当中是保守的，但每类蛋白质的组成和上调速率有所不同（Richter 等，2010）。主要作用是分子伴侣的蛋白质类别按其分子量分为 5 个家族，包括 Hsp27、60、70、90 和 110（Kregel，2002；Richter 等，2010）。这些应激蛋白在一个非应激的细胞中的组成占总蛋白的 5%～10%，但在应激的时间内能诱导高达总蛋白的 15% 左右（Calderwood 等，2007；Pockley 等，2008）。通过激活热休克因子（HSFs）引起热休克反应的调节作用与 HSPs 的合成，而在应激期间 HSFs 和热休克基因的启动区结合，然后启动转录。有 4 种可能的热休克因子（HSFs），而 HSF1 是哺乳类和低等脊椎动物包括鱼类的主要转录因子（Deane，Woo，2011），在没有应激时，HSF1 以一种潜在的单体状态和 Hsp70 结合而存在于细胞质内（Kregel，2002）。在应激的刺激而影响蛋白质的作用时，Hsp70 或者泛素（ubiquitin）从 HSF1 移除，导致同源三聚体化（homotrimerization），随后转运到核而为几种激酶引起高磷酸化（Richter 等，2010）。三聚体和 Hsp70 基因启动区的热休克元件结合，诱导转录。HSF1 和热休克元件结合的调节是由热休克结合蛋白 1 介导的，而它亦受到 Hsp70 的负反馈调节，因而抑制合成作用（Shi 等，1998）。

HSPs，特别是 Hsp70 在鱼类细胞内的蛋白伴侣作用已经深入阐明（Iwama 等，1998；Deane，Woo，2011）。然而，主要的研究都是检测在不同的应激物作用下和不同的鱼类与细胞系统中这种蛋白质表达的时间格局（Vijayan 等，2005；Deane，Woo，2011）。意外的是，有个别研究确实关注到这些蛋白质对应激物反应提供细胞保护的作用机理。蛋白质毒性对 HSPs 的表达是一个重要的刺激（Iwama 等，1999；Basu 等，2002），而蛋白伴侣作用对于 HSPs 在鱼类恢复蛋白质稳态中所起的作用还不清楚。一些研究已经建立在生物有机体的应激反应和相应的细胞应激反应的调节之间的联系（Vijayan 等，2005；Iwama 等，2006）。皮质醇能够减轻虹鳟的鳃和肝脏对热休克的反应（Basu 等，2001；Boone，Vijayan，2002b）以及鲤鱼对铜处理的反应（De Boeck 等，2003）；然而，外源的皮质醇处理并不影响平鲷（*Sparus sarba*）离体的和克氏鳟（*Oncorhynchus clarkii*）在体的 Hsp70 表达的肝脏水平，或者 Hsp70mRNA 水平（Ackerman 等，2000a，b）。皮质醇亦能降低鳟鱼肝细胞原代培养物的 Hsp90mRNA 丰度（Sathiyaa 等，2003）。而 Hsp90 在鳟鱼体内肝脏的蛋白表达是升高的（Vijayan 等，2003），除皮质醇外，GH 和催乳激素亦能使平鲷肝脏的 Hsp70 转录丰度和蛋白质水平降低（Deane 等，1999）。

这些变异着重说明建立一个作用模式以研究生物有机体的应激反应如何影响细胞的应激反应是必要的（Deane 和 Woo，2011）。例如，还不清楚皮质醇如何调节细胞内的热休克反应，因为在 Hsp70 或 Hsp90 基因的启动子还未曾报道一个推定的 GRE。在虹鳟中，糖皮质类固醇介导的 Hsp70 表达的减弱有蛋白体参与（Boone，Vijayan，2002a）。还有，最近一项研究表明，糖皮质类固醇诱导的微小 RNA（miRNAs）调节作用降低了哺乳类一个细胞系统的 HSP70 蛋白质丰度（Kukreti 等，2013）。HSPs 亦能影响 GR 信号，因为 Hsp90 和 Hsp70 两者都是 GR 杂络物的重要成分，对于保持受体的配体-结合构象是不可缺少的（Collingwood 等，1999；Morishima 等，2000）。这些研究着重阐明由应激物诱导的皮质醇刺激所引起的细胞应激反应受到多种因素调控的可能性。我们对皮质醇-GR结合和信号作用模式的主要见解都是以哺乳动物的模式为基础（Vijayan 等，2005），而针对较低等脊椎动物，将生物有机体应激反应和细胞应激适应性连接起来进行作用机理的研究还是很欠缺的。

4.5 分子反应的意义

本综述虽然主要着重于应激的分子反应，但是在应激期间确定转录丰度是否转译为蛋白质表达，亦是同样重要的。蛋白质的表达是转录反应功能相关的最好指标。研究结果已经表明，在应激反应或者皮质醇刺激中，转录水平与蛋白质表达和/或酶活性之间是正面相符合的（表4.4），包括 Hsp70 和 Hsp90（Vijayan 等，2003）、PK（Lopez-Patino 等，2014）、生长素释放肽（Janzen 等，2012）和 GK（Caseras 等，2000）。虽然一些研究已经把基因转录丰度和它们相关的蛋白质表达/酶活性联系起来，但亦有一些例子在转录丰度和相应的蛋白质表达之间是不相匹配的（Sathiyaa 和 Vijayan，2003）。例如，鳟鱼肝脏在皮质醇处理后的反应中，GR 转录水平和蛋白质表达出现一种相反的关系（Vijayan 等，2003）。这种不相匹配的情况亦出现在用应激物进行处理的实验中。例如，对虹鳟进行手工操作的应激（Alderman 等，2012），或者用皮质醇处理（肝细胞）（Aluru，Vijayan，2007），在肝脏和脑内都观察到 GR 转录丰度增加；然而，蛋白质水平在这些研究中都明显下降。酶类亦表现出相关性的缺少。尽管虹鳟在手工操作的应激后 15 分钟 GK 和 PEPCK 的转录丰度快速下调（Lopez-Patino 等，2014），而在应激物作用后 240 ～ 480 分钟 PEPCK 的活性上调（Lopez-Patino 等，2014）。在蛋白质表达和 mRNA 水平之间缺乏相关性可能是由于 mRNA 周转的变化，正如在鳟鱼肝细胞中曾经提到的 GR 自身调节的作用机理那样（Sathiyaa，Vijayan，2003）。显然，GR 被皮质醇激活后通过蛋白酶体通道而增加受体降解，进而引发 GR 转录作用，导致 GR 转录水平和蛋白质表达的不相匹配（Sathiyaa，Vijayan，2003）。这种在靶细胞水平的反馈调节作用机理可能适应于延长靶组织的皮质醇反应性。

基因的转录水平通常都用来作为应激物处理的标志。它们越来越多地应用于毒理学中，特异性的基因转录水平信号受到特异性污染物包括多氯联苯（PCBs）（细胞色素 P4501A）和金属（金属硫蛋白）的影响，但这并不是作用效应的表现，除非还伴随着其他蛋白质和生物化学的标志物（Sarkar 等，2006）。测定 mRNA 转录丰度虽然提供了有关通道激活的精确指证，但在推断信息以归因于特异性效应时，必须对结果做仔细的分析说明。我们知道有好几类基因转录的调节剂，包括参与组蛋白修饰的辅激活物与蛋白质和甲基化因子（Sexton，Cavalli，2015；Venkatesh，Workman，2015），但有少数研究关注到这些蛋白质在模式鱼类的调节作用。此外，调节 mRNA 稳定性和周转的因素（Roy，Jacobson，2013）可能在应激反应中影响基因表达和蛋白质合成，但和应激与适应性的关系还了解得很少。然而，有关转录调控的一些有意义的途径已开始起作用，包括外因基因组（epigenome）修饰和通过非孟德尔方法的世代遗传（Zhang 等，2013）。实际上，除激素之外，包括皮质醇，已经阐明哺乳类 CpG 岛（基因组中富含 CpG 的单拷贝非甲基化基因座）甲基化型式的改变（Zhang 等，2013）。应激或者皮质醇处理对外遗传改变（表观改变）的影响还未曾在较低等脊椎动物中研究。至今，涉及鱼类分子反应的大量研究都是使用定量 PCR 或者大量的基因组分析技术，包括微点阵或第二代基因测序来定量测定基因转录丰度。使用基因敲除和基因击倒技术以确定应激相关蛋白质的功能作用还只局限于斑马鱼和青鳉（Pikulkaew 等，2011；Nesan，Vijayan，2012；Wilson 等，2013；Cruz 等，2013；Benato 等，2014；Chatzopoulou 等，2015；Ishikawa 等，2013），而且主要都是以早期发育为出发点来进行的。具有可行性的新技术，包括第二代测序技术和簇状正规间隙短回文重复（CRISPR）-Cas9 基因-编辑工具（Hwang 等，2013；Auer 等，2014；Ablain 等，2015），可以用来对非模式生物有机体以组织特异性的方式进行定向的基因敲除研究，从而可以对应激轴功能以及动物健康的相关方面进行作用机理的鉴别分析。

4.6　研究应激的分子反应的方法

直到最近，我们对非模式鱼类应激分子反应的大部分了解都是来自使用定量 PCR 或者蛋白质印迹法对个别基因或者蛋白质表达的定量分析。随着寡核苷酸和 cDNA 微点阵的出现，这种情况发生了明显改变，因为它们能够对基因表达的总体变化进行定量分析（Aluru，Vijayan，2009；Rise 等，2004；Villeneuve 等，2008；Whitehead 等，2011）。然而，这些方法还不能稳定而广泛地使用于非模式种类，因为它们缺少基因组的信息。这些局限性随着新测序技术的出现而得以解决。主要变化发生在过去的十年，第二代的或深层次的测序技术能够应用于非模式种类的基因组和转录组的测序。这些技术能够在公平公正的情况下对定量分析基因表达型式提供价格低廉而高效率的技术手段。使用这些技术方法已经测序几种鱼类的转录组（Kolmakov 等，2008；Dleksiak 等，2011；Salem 等，2015；Salem 等，2010；Wang 等，2014）。转录组的研究，如果设计得好，就能够测试某种假设，亦可以有助于发现鱼类对应激适应以及导致不良适应的基本通道。

4.6.1　使用定向诱变剂的机理研究

为了研究应激反应的机理基础，在哺乳类的系统中广泛使用基因击倒和基因敲除的方法。在鱼类中，以吗啉寡核苷酸为基础的基因击倒方法已经广泛使用。直到最近，基因敲除或者反向遗传学方法（失活一个基因然后评估其表现型的后果）还没有在鱼类的研究中得到广泛使用。这种情况随着能编程序的核酸酶（programable nucleases）出现而发生了变化，它能够低成本和高效率地进行定向诱变（Kim，Kim，2014）。这些定向的基因组-编辑工具包括锌指核酸酶、转录激活剂样效应器核酸酶和 RNA-引导的基因工程核酸酶（普遍称为 CRISPR-Cas9）。这些技术能够在除了斑马鱼这种生物医学模式动物之外的许多种鱼类中产生突变体（Aluru 等，2015a；Ansai 等，2012，2013，2014；Dong 等，2011；Edvardsen 等，2014；Li 等，2013，2014；Wang，Hong，2014；Yano 等，2014；表 4.1 中应激特异性的基因敲除）。迄今为止，许多学者使用这些技术方法在一些世代时间短的鱼类包括斑马鱼、罗非鱼、青鳉中已经获益良多，因为只需要很短的时间来建立潜在的建立者（founder）和杂合的突变体。随着高效率 CRISPR-Cas9 方法的建立，在注射的胚胎本身能够产生双等位的基因敲除，使这种状况亦有所改变（Jao 等，2013）。这种方法能够对任何有兴趣种类的基因功能进行研究。除了进行基因敲除，CRISPR-Cas9 方法还能够进行基因插入（gene insertion）、基因矫正（gene correction）以及在任何有兴趣区域的染色体重新排列（Kim 和 Kim，2014）。总之，定向突变能够应用于了解基因的功能，特别是在鱼类特异性谱系中复制的那些基因。定向基因编辑的出现，和高通量测序技术包括 RNAseq 一起，已经为对任何有兴趣的鱼类进行作用机理的研究打开了大门(表 4.1)。

4.6.2 应激反应的外遗传（表观遗传）调节

来自生物医学研究不断充实的证据表明，外遗传的作用机理在基因调节中起着非常重要的作用。在有些例子中，这些作用机理表现为对表现型特征的转世代（transgenerational）遗传起着一定的作用。在哺乳类和其他的种类中，对环境因素研究得比较深入，包括营养、行为、应激和化学因子，它们在发育的重要时期影响外因基因组（epigenome）（Turechi，Meaney，2014；Yan 等，2014；Yano 等，2014；Zannas，West，2015；Zheng 等，2014）。相比起来，对较低等脊椎动物在应激反应中外因基因组变化的了解还非常少。一个广泛受到测试的假设认为，在早期发育期间，环境的影响改变外因基因组并且导致在整个生命过程中出现复杂的表现型和疾病的易感性。掌握环境因素对外遗传作用机理的影响对于预测基因的表达和表现型的变化是必要的。在本节中，我们概述有关外遗传作用机理的研究进展并阐明潜在的环境因素对它们的影响。外遗传是研究初级 DNA 序列独立发生变动的基因功能中可遗传的变化（Laird，Jeanisch，1996）。外遗传修饰研究得最深入的是 DNA 甲基化、组蛋白修饰和非编码 RNAs。对外遗传作用机理紧密地调控早期胚胎发育过程已经有充分的论证（Reik 等，2003）。在哺乳类中，外遗传修饰是重要发育进程的关键调节剂，包括 X（染色体）-失活、基因组的印记和终末分化（Callinan，2006；Reik 等，2003）。

4.6.2.1 DNA 甲基化作用

DNA 甲基化是增加一个甲基族的共价修饰。DNA-甲基转移酶催化转移一个甲基族到胞嘧啶的 CpG 二核苷酸-5 位置的反应。DNA 甲基化被认为是直接干扰转录因子结合 DNA 识别位点的阻抑标志，或者是募集包藏组蛋白脱乙酰酶或组蛋白甲基转移酶的协阻抑物复合物的阻抑标志（Cedar，Bergman，2012；Fuks，2005）。在脊椎动物中，可遗传的甲基化主要发生在 CpG 二核苷酸，而 DNA 的甲基化能够改变调节区的功能状态。由于 DNA 甲基化在发育中的重要性，任何外源因子改变这些作用机理都可能影响发育的轨道和表现型。目前我们对 DNA 甲基化的大部分了解都是来自对哺乳类的研究，而低等脊椎动物的研究资料非常少。至今还很少有人研究应激物对鱼类 DNA 甲基化的影响（Aluru 等，2015b；Liu 等，2014；Pierron 等，2014a，b）。化学应激物（特别是毒物包括 PAHs、PCBs 和异源雌激素）已经被证明能改变一定的靶基因的 DNA 甲基化型式（Aluru 等，2015b；Corrales 等，2014；Fang 等，2013a，b）。为了确定其他的应激物调节甲基化型式的作用，有必要进行类似的研究。例如，必须确定环境因素，包括温度、pH 和溶解氧对 DNA 甲基化型式的影响。这些环境因素对脊椎动物生长和生殖产物的影响已经了解得很清楚。然而，对这些变化的机理基础还没有很好研究。最近的一些研究表明，遗传的和外遗传的作用机理参与决定这些复杂的表现型。此外，DNA 甲基化是一种和水产养殖密切相关的重要外遗传作用机理，因为有证据表明它对于稍后开始的表现型已具有发育的基础（Li 等，2010）。例如，在孵化和鱼苗培育的生产实践中，苗种的营养和环境中的化合物能够影响到成鱼的生长和抗病能力。哺乳类的研究已经证明各种环境因素能影响成鱼的行为和健康，不只是影响到现在所处的一代，还会影响到接着的下一代（Skinner，2014）。这些研究结果清楚地说明这些长期的或者转世代的作用具有外遗传的基础，特别是 DNA 甲基化。在鱼类进行的这类研究必将会清楚地显示早期生命培育的实践对生长、抗病力和生殖力的影响。对于哺乳类的 HPA 轴，已经非常清楚地证明脑内 GR 的启动子甲基化对应激物是敏感的和有反应的（Weaver 等，2004；McGo Wan 等，2009）。硬骨鱼类是否存在着类似的作用机理还有待于阐明。

4.6.2.2　非编码 RNAs

直到最近，非编码 RNAs（ncRNAs）被认为是在细胞内起着种属的功能（generic function），包括 mRNA 翻译（rRNAs 和 tRNAs）、剪接（小核 RNAs；snRNAs）和核糖体 RNAs（小核仁 RNAs；snoRNAs）的修饰（Mattick，2006）。在过去的十年，随着一系列 ncRNAs 在发育、生理和抗病中起着重要作用的发现而改变了这个观点（Berezikov，2011）。非编码 RNA 是一个非特异性名词，用来描述任何一种 RNA，它不编码功能的蛋白质，小于 200～300 核苷酸长度。按照大小和功能，它们进一步划分为 MiRNAs、短干扰 RNAs（siRNAs）、piwi-互相作用 RNAs（piRNAs）、双链 RNAs（dsRNAs）和长非编码 RNAs（IncRNAs）。由于 ncRNAs 能够调节几种外遗传现象以建立基因表达的长期影响，它们被认为在外遗传调节中起重要作用（Berezikov，2011）。有关非编码 RNA 在鱼类应激反应调节中所起作用的研究资料非常少。随着高通量测序技术的出现，对各种非编码 RNAs 进行分类引起了极大的关注。未来的研究应该着重于对水产养殖中起重要作用的和在不同环境状态中与环境密切联系的非编码 RNA 进行测序。一旦鉴别非编码 RNA 的重要类群，反向遗传学的技术方法就可以用来敲除特异性的 ncRNAs，进而鉴别它们的功能。对于应激的反应，鉴别 ncRNAs 表达的时序和组织特异性变化以及它们调节的作用机理是至关重要的。

4.7 结束语和未知的方面

许多广泛的应激生理反应都已经阐明，从丰富的基因转录数据中呈现的状况是皮质醇不仅在应激适应期间，而且亦在动物行为表现的其他方面包括生长和发育中，起着调节分子作用机理的关键作用。实际上，当鱼类受到应激物干扰时，皮质醇的作用普遍影响到动物行为表现的各个方面；然而，造成各种不同生理变化的作用模式和特异性信号通道仍然不是很清楚。大多数涉及应激生理反应的研究都是以血浆的皮质醇水平来说明。在所有实例中，这也许不是特别适用的，除非我们已经了解到激素的动态以及相关的皮质类固醇受体调节作用对靶组织的影响。掌握靶组织皮质类固醇受体的调节与激活，以及皮质醇降解酶类的调节，对于阐明和皮质醇水平升高相联系的生理反应是非常重要的。例如，如果靶组织的受体是下调的，或者，如果是相应增加皮质醇的降解酶类，那么，慢性升高的血浆皮质醇水平可能就不会反映对代谢的负作用。在鱼类中观察到升高皮质醇水平使 11βHSD2 和 20βHSD2 转录水平增加（Li 等，2012；Alderman，Vijayan，2012；Tokarz 等，2013）以及降低 GR 蛋白质表达（Vijayan 等，2003，2005）的情况可以支持上述说法。

虽然已经有丰富的转录数据，但还有许多方面尚未阐明。在应激期间，应激激素影响靶组织的特异性作用机理基本上是不清楚的。部分原因是鱼类的研究极少采用药理学的技术或者分子技术（基因敲除/吗啉代）来了解基本的作用机理。当前大部分研究通常都是证明应激激素的变化和相关的基因变化之间的联系。使用变异体品系吗啉代的基因击倒和救援的研究，或者较广泛地使用受体已知的拮抗物和其他信号分子将有助于避免做出推测性的结论，并且能够较为具体地界定应激激素在调节应激信号通道和靶组织反应中的作用。例如，最近一项研究（Chatzopoulou 等，2015）使用基因击倒和转录组与代谢组方法，在斑马鱼胚胎剪接变体-特异性反应对应激期间，鉴定这些受体激活的相关功能作用提供了深刻的见解。虽然这类研究需要使用模式生物有机体，但是在这些研究中鉴别的应激反应通道和候选基因亦能够在非模式天然的以及和水产养殖相关的种类中进行测试。使用突变体来回答有关作用机理的问题目前还局限于硬骨鱼类的发育模式种类（斑马鱼和青鳉）。然而，最近基因组编辑取得的进展，包括 CRISPR/Cas9，将会产生定向的基因敲除和基因敲入技术，并且将会对各种非模式动物，包括水产养殖相关的和/或生态敏感的种类，在应激期间更为深入地鉴别其作用机理提供高效的技术手段。

尽管本章着重阐述了皮质醇在应激的分子调节中起的关键作用，但还有许多未知的方面，包括：

（1）皮质醇如何引起一个快速的非基因组的反应？特别是，GR 信号有一个推定的膜受体吗？

（2）在应激适应中，GR 和 MR 信号的作用是什么？

（3）母源的应激会影响其下一个世代吗？它的作用机理如何？

（4）环境因子（pH、温度、氧）改变外遗传机理的作用是什么?

（5）外遗传的作用机理在发育期间受到应激长期影响所引起的作用如何?

（6）DNA 甲基化能调节和 HPI 轴相联系的基因表达吗?

（7）参与 HPI 轴相联系的基因调节作用中的候选微小 RNAs 和长非编码 RNAs 是哪些?

（8）急性或慢性应激能够改变在类固醇生成组织/细胞以及脑内的 DNA 甲基化型式和/或非编码 RNAs 吗?

E. 福德　N. 阿路鲁　M. N. 维扎延　著

林浩然　译、校

参考文献

Ablain, J., Durand, E. M., Yang, S., Zhou, Y. and Zon, L. I. (2015). A CRISPR/Cas9 vector system for tissue-specific gene disruption in zebrafish. Dev. Cell. 32, 756-764.

Acerete, L., Balasch, J. C., Castellana, B., Redruello, B., Roher, N., Canario, A. V., et al. (2007). Cloning of the glucocorticoid receptor (GR) in gilthead seabream (Sparus aurata). Differential expression of GR and immune genes in gilthead seabream after an immune challenge. Comp. Biochem. Physiol. B Biochem. Mol. Biol. 148, 32-43.

Ackerman, P. A., Forsyth, R. B., Mazur, C. F. and Iwama, G. K. (2000a). Stress hormones and the cellular stress response in salmonids. Fish. Physiol. Biochem. 23, 327-336.

Ackerman, P. A., Forsyth, R. B., Mazur, C. F. and Iwama, G. K. (2000b). Stress hormones and the cellular stress response in salmonids. Stress Int. J. Biol. Stress. 327-336.

Agulleiro, M. J., Sánchez, E., Leal, E., Cortés, R., Fernández-Durán, B., Guillot, R., et al. (2013). Molecular characterization and functional regulation of melanocortin 2 receptor (MC2R) in the sea bass. A putative role in the adaptation to stress. PLoS One 8, e65450.

Alderman, S. L. and Bernier, N. J. (2007). Localization of corticotropin-releasing factor, urotensin I and CRF-binding protein gene expresion in the brain of the zebrafish, Danio rerio. J. Comp. Neurol. 502, 783-793.

Alderman, S. L. and Bernier, N. J. (2009). Ontogeny of the corticotropin-releasing factor system in zebrafish. Gen. Comp. Endocrinol. 164, 61-69.

Alderman, S. L. and Vijayan, M. M. (2012). 11β-Hydroxysteroid dehydrogenase type 2 in zebrafish brain: a functional role in hypothalamus-pituitary-interrenal axis regulation. J. Endocrinol. 215, 393-402.

Alderman, S. L. , McGuire, A. , Bernier, N. J. and Vijayan, M. M. (2012). Central and peripheral glucocorticoid receptors are involved in the plasma cortisol response to an acute stressor in rainbow trout. Gen. Comp. Endocrinol. 176, 79-85.

Alsop, D. and Vijayan, M. M. (2008). Development of the corticosteroid stress axis and receptor expression in zebrafish. Am. J. Physiol. Regul. Integr. Comp. Physiol. 294, R711-R719.

Alsop, D. and Vijayan, M. M. (2009). Molecular programming of the corticosteroid stress axis during zebrafish development. Comp. Biochem. Physiol. A Mol. Integr. Physiol. 153, 49-54.

Alsop, D. , Ings, J. S. and Vijayan, M. M. (2009). Adrenocorticotropic hormone suppresses gonadotropin-stimulated estradiol release from zebrafish ovarian follicles. PLoS One 4, e6463.

Aluru, N. and Vijayan, M. M. (2006). Aryl hydrocarbon receptor activation impairs cortisol response to stress in rainbow trout by disrupting the rate-limiting steps in steroidogenesis. Endocrinology 147, 1895-1903.

Aluru, N. and Vijayan, M. M. (2007). Hepatic transcriptome response to glucocorticoid receptor activation in rainbow trout. Physiol. Genomics 31, 483-491.

Aluru, N. and Vijayan, M. M. (2008). Molecular characterization, tissue-specific expression, and regulation of melanocortin 2 receptor in rainbow trout. Endocrinology 149, 4577-4588.

Aluru, N. and Vijayan, M. M. (2009). Stress transcriptomics infish: a role for genomic cortisol signaling. Gen. Comp. Endocrinol. 164, 142-150.

Aluru, N. , Renaud, R. , Leatherland, J. F. and Vijayan, M. M. (2005). Ah receptor-mediated impairment of interrenal steroidogenesis involves StAR protein and P450scc gene attenuation in rainbow trout. Toxicol. Sci. 84, 260-269.

Aluru, N. , Karchner, S. I. , Franks, D. G. , Nacci, D. , Champlin, D. and Hahn, M. E. (2015a). Targeted mutagenesis of aryl hydrocarbon receptor 2a and 2b genes in Atlantic killifish (Fundulus heteroclitus). Aquat. Toxicol. 158, 192-201.

Aluru, N. , Kuo, E. , Helfrich, L. W. , Karchner, S. I. , Linney, E. A. , Pais, J. E. , et al. (2015b). Developmental exposure to 2, 3, 7, 8-tetrachlorodibenzo-p-dioxin alters DNA methyltransferase (dnmt) expression in zebrafish (Danio rerio). Toxicol. Appl. Pharmacol. 284, 142-151.

Ando, H. , Hasegawa, M. , Ando, J. and Urano, A. (1999). Expression of salmon corticotrophin releasing hormone precursor gene in the preoptic nucleus in stressed rainbow trout. Gen. Comp. Endocrinol. 113, 87-95.

Ansai, S. , Ochiai, H. , Kanie, Y. , Kamei, Y. , Gou, Y. , Kitano, T. , Yamamoto, T. and Kinoshita, M. (2012). Targeted disruption of exogenous EGFP gene in medaka using zinc-finger nucleases. Dev. Growth Differ. 54, 546-556.

Ansai, S., Sakuma, T., Yamamoto, T., Ariga, H., Uemura, N., Takahashi, R., et al. (2013). Efficient targeted mutagenesis in medaka using custom-designed transcription activatorlike effector nucleases. Genetics 193, 739-749.

Ansai, S., Inohaya, K., Yoshiura, Y., Schartl, M., Uemura, N., Takahashi, R., et al. (2014). Design, evaluation, and screening methods for efficient targeted mutagenesis with transcription activator-like effector nucleases in medaka. Dev. Growth Differ. 56, 98-107.

Argenton, F., Ramoz, N., Charlet, N., Bernardini, S., Colombo, L., Bortolussi, M., et al. (1996). Mechanisms of transcriptional activation of the promoter of the rainbow trout prolactin gene by GHF1/Pit1 and glucocorticoid the pituitary-restricted POU domain transcription factor GHF1/Pit1 is required for the expression of the growth hormone (GH). Biochem. Biophys. Res. Commun. 66, 57-66.

Arnold-Reed, D. E. and Balment, R. J. (1994). Peptide hormones influence in vitro interrenal secretion of cortisol in the trout, Oncorhynchus mykiss. Gen. Comp. Endocrinol. 96, 85-91.

Arterbery, A. S., Deitcher, D. L. and Bass, A. H. (2010). Corticosteroid recep-tor expression in ateleost fish that displays alternative male reproductive tactics. Gen. Comp. Endocrinol. 165, 83-90.

Arukwe, A. (2008). Steroidogenic acute regulatory (StAR) protein and cholesterol side-chain cleavage (P450scc) -regulated steroidogenesis as an organ-specific molecular and cellular target for endocrine disrupting chemicals in fish. Cell Biol. Toxicol. 24, 527-540.

Aruna, A., Nagarajan, G. and Chang, C. F. (2012). Involvement of corticotrophin-releasing hormone and corticosteroid receptors in the brain-pituitary-gill of tilapia during the course of seawater acclimation. J. Neuroendocrinol. 24, 818-830.

Auer, T. O., Duroure, K., De Cian, A., Concordet, J. and Del Bene, F. (2014). Highly efficient CRISPR/Cas9-mediated knock-in in zebrafish by homology-independent DNA repair. Genome Res. 24 (1), 142-153.

Baker, B. I., Bird, D. J. and Buckingham, J. C. (1996). In the trout, CRH and AVT synergize to stimulate ACTH release. Regul. Pept. 67, 207-210.

Barton, B. A. (2002). Stress infishes: a diversity of responses with particular reference to changes in circulating corticosteroids. Integr. Comp. Biol. 42, 517-525.

Barton, B. A. and Iwama, G. K. (1991). Physiological changes infish from stress in aquaculture with emphasis on the response and effects of corticosteroids. Annu. Rev. Fish Dis. 1, 3-26.

Basu, N., Nakano, T., Grau, E. G. and Iwama, G. K. (2001). The effects of cortisol on heat shock protein 70 levels in twofish species. Gen. Comp. Endocrinol. 124, 97-105.

Basu, N., Todgham, A. E., Ackerman, P. A., Bibeau, M. R., Nakano, K., Schulte, P. M., et al. (2002). Heat shock protein genes and their functional significance in fish. Gene 295, 173-183.

Benato, F., Colletti, E., Skobo, T., Moro, E., Colombo, L., Argenton, F., et al. (2014). A living biosensor model to dynamically trace glucocorticoid transcriptional activity during development and adult life in zebrafish. Mol. Cell. Endocrinol. 392 (1-2), 60-72.

Berezikov, E. (2011). Evolution of microRNA diversity and regulation in animals. Nat. Rev. Genet. 12, 846-860.

Bernier, N. J. (2006). The corticotropin-releasing factor system as a mediator of the appetitesuppressing effects of stress infish. Gen. Comp. Endocrinol. 146, 45-55.

Bernier, N. J. and Peter, R. E. (2001). The hypothalamic-pituitary-interrenal axis and the control of food intake in teleostfish. Comp. Biochem. Physiol. B Biochem. Mol. Biol. 129, 639-644.

Bernier, N. J., Bedard, N. and Peter, R. E. (2004). Effects of cortisol on food intake, growth, and forebrain neuropeptide Y and corticotropin-releasing factor gene expression in goldfish. Gen. Comp. Endocrinol. 135, 230-240.

Bernier, N. J., Alderman, S. L. and Bristow, E. N. (2008). Heads or tails? Stressor-specific expression of corticotropin-releasing factor and urotensin I in the preoptic area and caudal neurosecretory system of rainbow trout. J. Endocrinol. 196, 637-648.

Bernier, N. J., Gorissen, M. and Flik, G. (2012). Differential effects of chronic hypoxia and feed restriction on the expression of leptin and its receptor, food intake regulation and the endocrine stress response in common carp. J. Exp. Biol. 215, 2273-2282.

Best, C., Melnyk-Lamont, N., Gesto, M. and Vijayan, M. M. (2014). Environmental levels of the antidepressant venlafaxine impact the metabolic capacity of rainbow trout. Aquat. Toxicol. 155, 190-198.

Biga, P. R., Cain, K. D., Hardy, R. W., Schelling, G. T., Overturf, K., Roberts, S. B., et al. (2004). Growth hormone differentially regulates muscle myostatin1 and-2 and increases circulating cortisol in rainbow trout (Oncorhynchus mykiss). Gen. Comp. Endocrinol. 138, 32-41.

Boone, A. N. and Vijayan, M. M. (2002a). Glucocorticoid-mediated attenuation of the Hsp70 response in trout hepatocytes involves the proteasome. Am. J. Physiol. Regul. Integr. Comp. Physiol. 283, R680-R687.

Boone, A. N. and Vijayan, M. M. (2002b). Constitutive heat shock protein 70 expression in rainbow trout hepatocytes: effect of heat shock and heavy metal exposure. Comp. Biochem. Physiol. C Toxiocol. Pharmacol. 132 (2), 223-233.

Borski, R. J. (2000). Nongenomic membrane actions of glucocorticoids in vertebrates. Trends Endocrinol. Metab. 11, 427-436.

Bouw, E., Huisman, M., Neggers, S. J. C. M. M., Themmen, A. P. N., van der Lely, A. J. and Delhanty, P. J. D. (2014). Development of potent selective competitive-antagonists of the melanocortin type 2 receptor. Mol. Cell. Endocrinol. 394, 99-104.

Bradford, C. S., Fitzpatrick, M. S. and Schreck, C. B. (1992). Evidence for ultra-short-loop feedback in ACTH-induced interrenal steroidogenesis in coho salmon: acute selfsuppression of cortisol secretion in vitro. Gen. Comp. Endocrinol. 87, 292-299.

Bury, N. R. and Sturm, A. (2007). Evolution of the corticosteroid receptor signalling pathway infish. Gen. Comp. Endocrinol. 153, 47-56.

Bury, N. R., Sturm, A., Le, R. P., Lethimonier, C., Ducouret, B., Guiguen, Y., et al. (2003). Evidence for two distinct functional glucocorticoid receptors in teleostfish. J. Mol. Endocrinol. 31, 141-156.

Cairns, M. T., Johnson, M. C., Talbot, A. T., Pemmasani, J. K., McNeill, R. E., Houeix, B., et al. (2008). A cDNA microarray assessment of gene expression in the liver of rainbow trout (Oncorhynchus mykiss) in response to a handling and confinement stressor. Comp. Biochem. Physiol. Part D Genomics Proteomics 3, 51-66.

Calderwood, S. K., Mambula, S. S. and Gray, P. J. (2007). Extracellular heat shock proteins in cell signaling and immunity. Ann. N. Y. Acad. Sci. 39, 28-39.

Calduch-Giner, J. A., Echasseriau, Y., Crespo, D., Baron, D., Planas, J. V., Prunet, P., et al. (2014). Transcriptional assessment by microarray analysis and large-scale meta-analysis of the metabolic capacity of cardiac and skeletal muscle tissues to cope with reduced nutrient availability in gilthead sea bream (Sparus aurata L.). Mar. Biotechnol. 16, 423-435.

Callinan, P. A. (2006). The emerging science of epigenomics. Hum. Mol. Genet. 15, R95-R101. Cedar, H. and Bergman, Y. (2012). Programming of DNA methylation patterns. Annu. Rev. Biochem. 81, 97-117.

Caseras, A., Metón, I., Fernández, F. and Baanante, I. V. (2000). Glucokinase gene expression is nutritionally regulated in liver of gilthead sea bream (Sparus aurata). Biochim. Biophys. Acta Gene. Struct. Expr. 1493, 135-141.

Castillo, J., Castellana, B., Acerete, L., Planas, J. V., Goetz, F. W., Mackenzie, S., et al. (2008). Stress-induced regulation of steroidogenic acute regulatory protein expression in head kidney of Gilthead seabream (Sparus aurata). J. Endocrinol. 196, 313-322.

Castillo, J., Teles, M., Mackenzie, S. and Tort, L. (2009). Stress-related hormones modulate cytokine expression in the head kidney of gilthead seabream (Sparus aurata). Fish Shellfish Immunol. 27, 493-499.

Cerdá-Reverter, J. M., Agulleiro, M. J., Cortés, R., Sánchez, E., Guillot, R., Leal, E., et al. (2013). Involvement of melanocortin receptor accessory proteins (MRAPs) in the function of melanocortin receptors. Gen. Comp. Endocrinol. 188, 133-136.

Chadzinska, M., Tertil, E., Kepka, M., Hermsen, T., Scheer, M. and Lidy Verburg-van Kemenade, B. M. (2012). Adrenergic regulation of the innate immune response in common carp (Cyprinus carpio L.). Dev. Comp. Immunol. 36, 306-316.

Chandrasekar, G., Lauter, G. and Hauptmann, G. (2007). Distribution of corticotropinreleasing hormone in the developing zebrafish brain. J. Comp. Neurol. 505, 337-351.

Chatzopoulou, A., Roy, U., Meijer, A. H., Alia, A., Spaink, H. P. and Schaaf, M. J. M. (2015). Transcriptional and metabolic effects of glucocorticoid receptor a and b signaling in zebrafish. Endocrinology 156, 1757-1769.

Clark, B. J. and Cochrum, R. K. (2007). The steroidogenic acute regulatory protein as a target of endocrine disruption in male reproduction. Drug Metab. Rev. 39, 353-370.

Collingwood, T. N., Urnov, F. D. and Wolffe, A. P. (1999). Nuclear receptors: coactivators, corepressors and chromatin remodeling in the control of transcription. J. Mol. Endocrinol. 23, 255-275.

Colombe, L., Fostier, A., Bury, N., Pakdel, F. and Guiguen, Y. (2000). A mineralocorticoidlike receptor in the rainbow trout, Oncorhynchus mykiss: cloning and characterization of its steroid binding domain. Steroids 65, 319-328.

Corrales, J., Fang, X., Thornton, C., Mei, W., Barbazuk, W. B., Duke, M., et al. (2014). Effects on specific promoter DNA methylation in zebrafish embryos and larvae following benzo [a] pyrene exposure. Comp. Biochem. Physiol. Part C 163, 37-46.

Craig, P. M., Al-Timimi, H. and Bernier, N. J. (2005). Differential increase in forebrain and caudal neurosecretory systemcorticotropin-releasing factor and urotensin I gene expression associated with seawater transfer in rainbow trout. Endocrinology 146, 3851-3860.

Cruz, S. A., Lin, C.-H., Chao, P. L. and Hwang, P. P. (2013). Glucocorticoid receptor, but not mineralocorticoid receptor, mediates cortisol regulation of epidermal ionocyte development and ion transport in zebrafish (Danio rerio). PLoS One 8, e77997.

Deane, E. E. and Woo, N. Y. S. (2011). Advances and perspectives on the regulation and expression of piscine heatshock proteins. Rev. Fish Biol. Fish. 21, 153-185.

Deane, E. E., Kelly, S. P., Lo, C. K. M. and Woo, N. Y. S. (1999). Effects of GH, prolactin and cortisol on hepatic heat shock protein 70 expression in a marine teleost Sparus sarba. J. Endocrinol. 161, 413-421.

De Boeck, G., De Wachter, B., Vlaeminck, A. and Blust, R. (2003). Effect of cortisol treatment and/or sublethal copper exposure on copper uptake and heat shock protein levels in common carp, Cyprinus carpio. Environ. Toxicol. Chem. 22 (5), 1122-1126.

Dickmeis, T. (2009). Glucocorticoids and the circadian clock. J. Endocrinol. 200, 3-22.

Dickmeis, T. and Foulkes, N. S. (2011). Glucocorticoids and circadian clock control of cell proliferation: At the interface between three dynamic systems. Mol. Cell. Endocrinol. 331, 11-22.

Dickmeis, T., Lahiri, K., Nica, G., Vallone, D., Santoriello, C., Neumann, C. J., et al. (2007). Glucocorticoids play a key role in circadian cell cycle rhythms. PLoS Biol. 5, e78.

Dindia, L., Murray, J., Faught, E., Davis, T. L., Leonenko, Z. and Vijayan, M. M. (2012). Novel nongenomic signaling by glucocorticoid may involve changes to liver membrane order in rainbow trout. PLoS One 7, e46859.

Dindia, L., Faught, E., Leonenko, Z., Thomas, R. and Vijayan, M. M. (2013). Rapid cortisol signaling in response to acute stress involves changes in plasma membrane order in rainbow trout liver. Am. J. Physiol. Endocrinol. Metab. 304, E1157-E1166.

Dong, Z., Ge, J., Li, K., Xu, Z., Liang, D., Li, J., et al. (2011). Heritable targeted inactivation of myostatin gene in yellow catfish (Pelteobagrus fulvidraco) using engineered zinc finger nucleases. PLoS One 6, e28897.

Dores, R. M. (2013). Observations on the evolution of the melanocortin receptor gene family: distinctive features of the melanocortin-2 receptor. Front. Neurosci. 7, 1-12.

Dores, R. M. and Baron, A. J. (2011). Evolution of POMC: origin, phylogeny, posttranslational processing, and the melanocortins. Ann. N. Y. Acad. Sci. 1220, 34-48.

Dores, R. M. andGarcia, Y. (2015). Viewsontheco-evolutionofthemelanocortin-2receptor, MRAPs, and the hypothalamus/pituitary/adrenal-interrenal axis. Mol. Cell. Endocrinol. 408, 12-22.

Dores, R. M. , Londraville, R. L. , Prokop, J. , Davis, P. , Dewey, N. and Lesinski, N. (2014). Molecular evolution of GPCRs: melanocortin/melanocortin receptors. J. Mol. Endocrinol. 52, T29-T42.

Dores, R. M. , Liang, L. , Hollmann, R. E. , Sandhu, N. and Vijayan, M. M. (2016). Identifying the activation motif in the N-terminal of rainbow trout and zebrafish melanocortin-2 receptor accessory protein 1 (MRAP1) orthologs. Gen. Comp. Endocrinol. http: //dx. doi. org/10. 1016/j. ygcen. 2015. 12. 031.

Doyon, C. , Trudeau, V. L. and Moon, T. W. (2005). Stress elevates corticotropin-releasing factor (CRF) andCRF-binding protein mRNA levels in rainbow trout (Oncorhynchus mykiss). J. Endocrinol. 186, 123-130.

Doyon, C. , Leclair, J. , Trudeau, V. L. and Moon, T. W. (2006). Corticotropin-releasing factor and neuropeptide Y mRNA levels are modified by glucocorticoids in rainbow trout, Oncorhynchus mykiss. Gen. Comp. Endocrinol. 146, 126-135.

Du, F. , Xu, G. , Nie, Z. , Xu, P. and Gu, R. (2014). Transcriptome analysis gene expression in the liver of Coilia nasus during the stress response. BMC Genomics 15, 558.

Ducouret, B. , Tujague, M. , Ashraf, J. , Mouchel, N. , Servel, N. , Valotaire, Y. , et al. (1995). Cloning of a teleostfish glucocorticoid receptor shows that it contains a deoxyribonucleic acid-binding domain different from that of mammals. Endocrinology 136, 3774-3783.

Edvardsen, R. B. , Leininger, S. , Kleppe, L. , Skaftnesmo, K. O. and Wargelius, A. (2014). Targeted mutagenesis in Atlantic salmon (Salmo salar L.) using the CRISPR/Cas9 system induces complete knockout individuals in the F0 generation. PLoS One 9, e108622.

Enes, P. , Panserat, S. , Kaushik, S. and Oliva-Teles, A. (2009). Nutritional regulation of hepatic glucose metabolism infish. Fish Physiol. Biochem. 35, 519-539.

Eriksen, M. S. , Bakken, M. , Espmark, A. , Braastad, B. O. and Salte, R. (2006). Prespawning stress in farmed Atlantic salmon Salmo salar: maternal cortisol exposure and hyperthermia during embryonic development affect offspring survival, growth and incidence of malformations. J. Fish Biol. 69, 114-129.

Esbaugh, A. J. and Walsh, P. J. (2009). Identification of two glucocorticoid response elements in the promoter region of the ubiquitous isoform of glutamine synthetase in gulf toadfish, Opsanus beta. Am. J. Physiol. Regul. Integr. Comp. Physiol. 297, R1075-R1081.

Fabbri, E. and Moon, T. W. (2015). Adrenergic signaling in teleostfish liver, a challenging path. Comp. Biochem. Physiol. B Biochem. Mol. Biol. http: //dx. doi. org/10. 1016/j. cbpb. 2015. 10. 002. [Epub ahead of print].

Fabbri, E. , Capuzzo, A. and Moon, T. W. (1998). The role of circulating catecholamines in the regulation offish metabolism: an overview. Comp. Biochem. Physiol. Part C 120, 177-192.

Fang, X. , Corrales, J. , Thornton, C. , Scheffler, B. E. and Willett, K. L. (2013a). Global and gene specific DNA methylation changes during zebrafish development. Comp. Biochem. Physiol. Part B Biochem. Mol. Biol. 166, 99-108.

Fang, X. , Thornton, C. , Scheffler, B. E. and Willett, K. L. (2013b). Benzo [a] pyrene decreases global and gene specific DNA methylation during zebrafish development. Environ. Toxicol. Pharmacol. 36, 40-50.

Faught, E. , Best, C. and Vijayan, M. M. (2016). Maternal stress-associated cortisol stimulation may protect embryos from cortisol excess in zebrafish. R. Soc. Open Sci. 3, 160032.

Filby, A. L. and Tyler , C. R. (2007). Cloningandcharacterizationofc DNA sforhormonesand/or receptors of growth hormone, insulin-like growth factor I, thyroid hormone, and corticosteroid and the gender, tissue, and developmental specific expression of their mRNA transcripts in fathead minnow (Pimephales promelas). Gen. Comp. Endocrinol. 150, 151-163.

Flik, G. , Klaren, P. H. M. , Van den Burg, E. H. , Metz, J. R. and Huising, M. O. (2006). CRF and stress infish. Gen. Comp. Endocrinol. 146, 36-44.

Frolow, J. and Milligan, C. L. (2004). Hormonal regulation of glycogen metabolism in white muscle slices from rainbow trout (Oncorhynchus mykiss Walbaum). Am. J. Physiol. Integr. Comp. Physiol. 287, 1344-1353.

Fryer, J. , Leders, K. and Rivier, J. (1983). Urotensin I, a CRF-like neuropeptide stimulates ACTH release from the teleost pituitary. Endrocrinology 113, 2308-2310.

Fuks, F. (2005). DNA methylation and histone modifications: teaming up to silence genes. Curr. Opin. Genet. Dev. 15, 490-495.

Funder, J. W. (1997). Glucocorticoid and mineralocorticoid receptors: biology and clinical relevance. Annu. Rev. Med. 48, 231-240.

Fuzzen, M. L. M. , Van Der Kraak, G. and Bernier, N. J. (2010). Stirring up new ideas about the regulation of the hypothalamic-pituitary-interrenal axis in zebrafish (Danio rerio). Zebrafish 7, 349-358.

Galt, N. J. , Froehlich, J. M. , Remily, E. A. , Romero, S. R. and Biga, P. R. (2014). The effects of exogenous cortisol on myostatin transcription in rainbow trout, Oncorhynchus mykiss. Comp. Biochem. Physiol. A Mol. Integr. Physiol. 175, 57-63.

Geislin, M. and Auperin, B. (2004). Relationship between changes in mRNAs of the genes encoding steroidogenic acute regulatory protein and P450 cholesterol side chain cleavage in head kidney and plasma levels of cortisol in response to different kinds of acute stress in the rainbow trout (Oncorhynchus mykiss). Gen. Comp. Endocrinol. 135, 70-80.

Gesto, M., Soengas, J. L., RodrÍguez-Illamola, A. and MÍguez, J. M. (2014). Arginine vasotocin treatment induces a stress response and exerts a potent anorexigenic effect in rainbow trout, Oncorhynchus mykiss. J. Neuroendocrinol. 26, 89-99.

Ghisleni, G., Capiotti, K. M., Da Silva, R. S., Oses, J. P., Piato, ? . L., Soares, V., et al. (2012). The role of CRH in behavioral responses to acute restraint stress in zebrafish. Prog. Neuropsychopharmacol. Biol. Psychiatry 36, 176-182.

Geising, E. R., Suski, C. D., Warner, R. E. and Bell, A. M. (2011). Female sticklebacks transfer information via eggs: effects of maternal experience with predators on offspring. Proc. Biol. Sci. 278 (1712), 1753-1759.

Gravel, A. and Vijayan, M. M. (2006). Salicylate disrupts interrenal steroidogenesis and brain glucocorticoid receptor expression in rainbow trout. Toxicol. Sci. 93, 41-49.

Greenwood, A. K., Butler, P. C., White, R. B., DeMarco, U., Pearce, D. and Fernald, R. D. (2003). Multiple corticosteroid receptors in a teleostfish: distinct sequences, expression patterns, and transcriptional activities. Endocrinology 144, 4226-4236.

Griffiths, B. B., Schoonheim, P. J., Ziv, L., Voelker, L., Baier, H. and Gahtan, E. (2012). A zebrafish model of glucocorticoid resistance shows serotonergic modulation of the stress response. Front Behav. Neurosci. 6, 1-10.

Hagen, I. J., Kusakabe, M. and Young, G. (2006). Effects of ACTH and cAMP on steroidogenic acute regulatory protein and P450 11b-hydroxylase messenger RNAs in rainbow trout interrenal cells: relationship with in vitro cortisol production. Gen. Comp. Endocrinol. 145, 254-262.

Herzog, W., Zeng, X., Lele, Z., Sonntag, C., Ting, J. W., Chang, C. Y., et al. (2003). Adenohypophysis formation in the zebrafish and its dependence on sonic hedgehog. Dev. Biol. 254, 36-49.

Hontela, A. and Vijayan, M. M. (2009). Adrenocortical toxicology infishes. In: Adrenal Toxicology, Target Organ Toxicology Series (eds. P. W. Harvey, D. J. Everett and C. J. Springall), pp. 233-256. London: Informa Healthcare USA, Inc.

Hori, T. S., Gamperl, A., Booman, M., Nash, G. W. and Rise, M. L. (2012). A moderate increase in ambient temperature modulates the Atlantic cod (Gadus morhua) spleen transcriptome response to intraperitoneal viral mimic injection. BMC Genomics 13, 431.

Huang, Y., Harrison, M. R., Osorio, A., Kim, J., Baugh, A., Duan, C., et al.

(2013). Igf signaling is required for cardiomyocyte proliferation during zebrafish heart development and regeneration. PLoS One 8, e67266.

Huising, M. O., Metz, J. R., van Schooten, C., Taverne-Thiele, A. J., Hermsen, T., Verburg-van Kemenade, B. M. L., et al. (2004). Structural characterisation of a cyprinid (Cyprinus carpio L.) CRH, CRH-BP and CRH-R1, and the role of these proteins in the acute stress response. J. Mol. Endocrinol. 32, 627-648.

Huising, M. O., van der Aa, L. M., Metz, J. R., de Fátima Mazon, A., Kemenade, B. M. L. V. and Flik, G. (2007). Corticotropin-releasing factor (CRF) and CRF-binding protein expression in and release from the head kidney of common carp: evolutionary conservation of the adrenal CRF system. J. Endocrinol. 193, 349-357.

Hwang, W. Y., Fu, Y., Reyon, D., Maeder, M. L., Tsai, S. Q., Sander, J. D., et al. (2013). Efficient genome editing in zebrafish using a CRISPR-Cas system. Nat. Biotechnol. 31, 227-229.

Ings, J. S., Servos, M. R. and Vijayan, M. M. (2011a). Exposure to municipal wastewater effluent impacts stress performance in rainbow trout. Aquat. Toxicol. 103, 85-91.

Ings, J. S., Servos, M. R. and Vijayan, M. M. (2011b). Hepatic transcriptomics and protein expression in rainbow trout exposed to municipal wastewater effluent. Environ. Sci. Technol. 45, 2368-2376.

Ings, J. S., George, N., Peter, M. C. S., Servos, M. R. and Vijayan, M. M. (2012). Venlafaxine and atenolol disrupt epinephrine-stimulated glucose production in rainbow trout hepatocytes. Aquat. Toxicol. 106-107, 48-55.

Ishikawa, T., Okada, T., Ishikawa-Fujiwara, T., Todo, T., Kamei, Y., Shigenobu, S., et al. (2013). ATF6a/b-mediated adjustment of ER chaperone levels is essential for development of the notochord in medakafish-supplemental material. Mol. Biol. Cell 24, 1387-1395.

Iwama, G. K. (2004). Are hsps suitable for indicating stressed states infish? J. Exp. Biol. 207, 15-19.

Iwama, G. K., Thomas, P. T., Forsyth, R. B. and Vijayan, M. M. (1998). Heat shock protein expression infish. Rev. Fish Biol. Fish. 8, 35-56.

Iwama, G. K., Vijayan, M. M., Forsyth, R. B. and Ackerman, P. A. (1999). Heat shock proteins and physiological stress infish. Integr. Comp. Biol. 39, 901-909.

Iwama, G. K., Afonso, L. O. B. and Vijayan, M. M. (2006). Stress infishes. In The Physiology of Fishes (eds. D. H. Evans and J. B. Claiborne), pp. 319-342. Boca Raton, FL: CRC Press.

Janzen, W. J., Duncan, C. A. and Riley, L. G. (2012). Cortisol treatment reduces ghrelin signaling and food intake in tilapia, Oreochromis mossambicus. Domest.

Anim. Endocrinol. 43, 251-259.

Jao, L.-E., Wente, S. R. and Chen, W. (2013). Efficient multiplex biallelic zebrafish genome editing using a CRISPR nuclease system. Proc. Natl. Acad. Sci. 110, 13904-13909.

Jeffrey, J. D., Gollock, M. J. and Gilmour, K. M. (2014). Social stress modulates the cortisol response to an acute stressor in rainbow trout (Oncorhynchus mykiss). Gen. Comp. Endocrinol. 196, 8-16.

Johansen, I. B., Sandvik, G. K., Nilsson, G. E., Bakken, M. and Overli, O. (2011). Cortisol receptor expression differs in the brains of rainbow trout selected for divergent cortisol responses. Comp. Biochem. Physiol. Part D Genomics Proteomics 6, 126-132.

Jorgensen, E. H., Bernier, N. J., Maule, A. G. and Vijayan, M. M. (2015). Effect of long-term fasting and a subsequent meal on mRNA abundances of hypothalamic appetite regulators, central and peripheral leptin expression and plasma leptin levels in rainbow trout. Peptides http://dx. doi. org/10. 1016/j. peptides. 2015. 08. 010.

Jurczyk, A., Roy, N., Bajwa, R., Gut, P., Lipson, K., Yang, C., et al. (2011). Dynamic glucoregulation and mammalian-like responses to metabolic and developmental disruption in zebrafish. Gen. Comp. Endocrinol. 170, 334-345.

Kelly, S. P. and Chasiotis, H. (2011). Glucocorticoid and mineralocorticoid receptors regulate paracellular permeability in a primary cultured gill epithelium. J. Exp. Biol. 214, 2308-2318.

Kelsall, C. J. and Balment, R. J. (1998). Native urotensins influence cortisol secretion and plasma cortisol concentration in the euryhaline flounder, Platichthys flesus. Gen. Comp. Endocrinol. 112, 210-219.

Kiilerich, P., Kristiansen, K. and Madsen, S. S. (2007). Cortisol regulation of ion transporter mRNA in Atlantic salmon gill and the effect of salinity on the signaling pathway. J. Endocrinol. 194, 417-427.

Kiilerich, P., Triqueneaux, G., Christensen, N. M., Trayer, V., Terrien, X., Lombès, M., et al. (2015). Interaction between the trout mineralocorticoid and glucocorticoid receptors in vitro. J. Mol. Endocrinol. 55, 55-68.

Kim, H. and Kim, J. S. (2014). A guide to genome engineering with programmable nucleases. Nat. Rev. Genet. 15, 321-334.

Kleppe, L., Karlsen, O., Edvardsen, R. B., Norberg, B., Andersson, E., Taranger, G. L., et al. (2013). Cortisol treatment of prespawning female cod affects cytogenesis related factors in eggs and embryos. Gen. Comp. Endocrinol. 189, 84-95.

Kolmakov, N. N., Kube, M., Reinhardt, R. and Canario, A. V. (2008). Analysis of the goldfish Carassius auratus olfactory epithelium transcriptome reveals the presence of numerous nonolfactory GPCR and putative receptors for progestin pheromones. BMC Genomics 9, 429.

Krasnov, A., Koskinen, H., Pehkonen, P., Rexroad, C. E., Afanasyev, S. and Mölsä, H. (2005). Gene expression in the brain and kidney of rainbow trout in response to handling stress. BMC Genomics 6, 3.

Krasnov, A., Skugor, S., Todorcevic, M., Glover, K. A. and Nilsen, F. (2012). Gene expression in Atlantic salmon skin in response to infection with the parasitic copepod Lepeophtheirus salmonis, cortisol implant, and their combination. BMC Genomics 13, 130.

Kregel, K. C. (2002). Heat shock proteins: modifying factors in physiological stress responses and acquired thermotolerance. J. Appl. Physiol. 92, 2177-2186.

Krug, R. G., Poshusta, T. L., Skuster, K. J., Berg, M. R., Gardner, S. L. and Clark, K. J. (2014). A transgenic zebrafish model for monitoring glucocorticoid receptor activity. Genes Brain Behav. 478-487.

Kukreti, H., Amuthavalli, K., Harikumar, A., Sathiyamoorthy, S., Feng, P. Z., Anantharaj, R., et al. (2013). Muscle-specific MicroRNA1 (miR1) targets heat shock protein 70 (HSP70) during dexamethasone-mediated atrophy. J. Biol. Chem. 288, 6663-6678.

Kure, E. H., Sæbø, M., Stangeland, A. M., Hamfjord, J., Hytterød, S., Heggenes, J., et al. (2013). Molecular responses to toxicological stressors: profiling microRNAs in wild Atlantic salmon (Salmo salar) exposed to acidic aluminum-rich water. Aquat. Toxicol. 138-139, 98-104.

Kusakabe, M., Nakamura, I. and Young, G. (2003). 11β-Hydroxysteroid dehydrogenase complementary deoxyribonucleic acid in rainbow trout: cloning, sites of expression, and seasonal changes in gonads. Endocrinology 144, 2534-2545.

Kusakabe, M., Zuccarelli, M. D., Nakamura, I. and Young, G. (2009). Steroidogenic acute regulatory protein in white sturgeon (Acipenser transmontanus): cDNA cloning, sites of expression and transcript abundance in corticosteroidogenic tissue after an acute stressor. Gen. Comp. Endocrinol. 162, 233-240.

Lacroix, A. and Hontela, A. (2006). Role of calcium channels in cadmium-induced disruption of cortisol synthesis in rainbow trout (Oncorhynchus mykiss). Comp. Biochem. Physiol. C Toxicol. Pharmacol. 144, 141-147.

Laird, P. W. and Jaenisch, R. (1996). The role of DNA methylation in cancer genetics and epigenetics. Annu. Rev. Genet. 30, 441-464.

Laiz-Carrión, R., Fuentes, J., Redruello, B., Guzmán, J. M., MartÍn del RÍo, M. P., Power, D., et al. (2009). Expression of pituitary prolactin, growth hormone and somatolactin is modified in response to different stressors (salinity, crowding and food-deprivation) in gilthead sea bream Sparus auratus. Gen. Comp. Endocrinol. 162, 293-300.

Leung, L. Y. and Woo, N. Y. (2010). Effects of growth hormone, insulin-like growth factor 1, triiodothyronine, thyroxine, and the cortisol on gene expression of carbohydrate metabolic enzymes in sea bream hepatocytes. Comp. Biochem. Physiol. A Mol. Integr. Physiol. 157, 272-282.

Leung, L. Y., Kwong, A. K., Man, A. K. and Woo, N. Y. (2008). Direct actions of cortisol, thyroxine and growth hormone on IGF-1 mRNA expression in sea bream hepatocytes. Comp. Biochem. Physiol. A Mol. Integr. Physiol. 151, 705-710.

Lewis, J. M., Hori, T. S., Rise, M. L., Walsh, P. J. andCurrie, S. (2010). Transcriptome responses to heat stress in the nucleated red blood cells of the rainbow trout (Oncorhynchus mykiss). Physiol. Genomics 42, 361-373.

Li, C. C., Maloney, C. A., Cropley, J. E. and Suter, C. M. (2010). Epigenetic programming by maternal nutrition: shaping future generations. Epigenomics 2, 539-549.

Li, M., Christie, H. L. and Leatherland, J. F. (2012). The in vitro metabolism of cortisol by ovarian follicles of rainbow trout (Oncorhynchus mykiss): comparison with ovulated oocytes and pre-hatch embryos. Reproduction 144, 713-722.

Li, M. H., Yang, H. H., Li, M. R., Sun, Y. L., Jiang, X. L., Xie, Q. P., et al. (2013). Antagonistic roles of Dmrt1 and Foxl2 in sex differentiation via estrogen production in tilapia as demonstrated by TALENs. Endocrinology 154, 4814-4825.

Li, M., Yang, H., Zhao, J., Fang, L., Shi, H., Li, M., et al. (2014). Efficient and heritable gene targeting in tilapia by CRISPR/Cas9. Genetics 197, 591-599.

Liang, B., Wei, D.-L., Cheng, Y.-N., Yuan, H.-J., Lin, J., Cui, X.-Z., et al. (2013). Restraint stress impairs oocyte developmental potential: role of CRH-Induced apoptosis of ovarian cells. Biol. Reprod. 89, 1-12.

Liang, L., Schmid, K., Sandhu, N., Angleson, J. K., Vijayan, M. M. and Dores, R. M. (2015). Structure/function studies on the activation of the rainbow trout melanocortin-2 receptor. Gen. Comp. Endocrinol. 210, 145-151.

Liu, N. A., Huang, H., Yang, Z., Herzog, W., Hammerschmidt, M., Lin, S., et al. (2003). Pituitary corticotroph ontogeny and regulation in transgenic zebrafish. Mol. Endocrinol. 17, 959-966.

Liu, Y., Yuan, C., Chen, S., Zheng, Y., Zhang, Y., Gao, J., et al. (2014). Global and cyp19a1a gene specific DNA methylation in gonads of adult rare minnow Gobiocypris rarus under bisphenol A exposure. Aquat. Toxicol. 156, 10-16.

Loosli, F., Staub, W., Finger-Baier, K. C., Ober, E. A., Verkade, H., Wittbrodt,

J. , et al. (2003). Loss of eyes in zebrafish caused by mutation of chokh/rx3. EMBO Rep. 4, 894-899.

Lopez-Patino, M. , Gesto, M. , Conde-Siera, M. , Soengas, J. L. and Miguez, J. M. (2014). Stress inhibition of melatonin synthesis in the pineal organ of rainbow trout (Oncorhynchus mykiss) is mediated by cortisol. J. Exp. Biol. 217, 1407-1416.

Lorenzi, V. , Mehinto, A. C. , Denslow, N. D. and Schlenk, D. (2012). Effects of exposure to the b-blocker propranolol on the reproductive behavior and gene expression of the fathead minnow, Pimephales promelas. Aquat. Toxicol. 116-117, 8-15.

Ma, P. , Liu, Y. , Reddy, K. P. , Chan, W. K. and Lam, T. J. (2004). Characterization of the seabass pancreatic a-amylase gene and promoter. Gen. Comp. Endocrinol. 137, 78-88.

MacKenzie, S. , Iliev, D. , Liarte, C. , Koskinen, H. , Planas, J. V. , Goetz, F. W. , et al. (2006a). Transcriptionalanalysis of LPS-stimulatedactivation of trout (Oncorhynchusmykiss) monocyte/ macrophage cells in primary culture treated with cortisol. Mol. Immunol. 43, 1340-1348.

MacKenzie, S. , Montserrat, N. , Mas, M. , Acerete, L. , Tort, L. , Krasnov, A. , etal. (2006b). Bacterial lipopolysaccharide induces apoptosis in the trout ovary. Reprod. Biol. Endocrinol. 4, 46.

Madan, A. P. and DeFranco, D. B. (1993). Bidirectional transport of glucocorticoid receptors across the nuclear envelope. Proc. Natl. Acad. Sci. U. S. A. 90, 3588-3592.

Madison, B. M. , Tavakoli, S. , Kramer, S. and Bernier, N. J. (2015). Chronic cortisol and the regulation of food intake and the encordine growth axis in rainbow trout. J. Endocrinol. 226, 103-119.

Manna, P. R. , Eubank, D. W. , Lalli, E. , Sassone-Corsi, P. and Stocco, D. M. (2003). Transcriptional regulation of the mouse steroidogenic acute regulatory protein gene by the cAMP responseelement binding protein and steroidogenic factor 1. J. Mol. Endocrinol. 30, 381-397.

Manuel, R. , Gorissen, M. , Zethof, J. , Ebbesson, L. O. E. , van de Vis, H. , Flik, G. , et al. (2014). Unpredictable chronic stress decreases inhibitory avoidance learning in Tuebingen long-fin zebrafish: stronger effects in the resting phase than in the active phase. J. Exp. Biol. 217, 3919-3928.

Manuel, R. , Zethof, J. , Flik, G. and van den Bos, R. (2015). Providing a food reward reduces inhibitory avoidance learning in zebrafish. Behav. Process. 120, 69-72.

Mattick, J. S. (2006). Non-coding RNA. Hum. Mol. Genet. 15, R17-R29.

Mccormick, M. I. (1999). Experimental test of the effect of maternal hormones on larval quality of a coral reeffish. Oecologia 118, 412-422.

McCormick, S. D., Regish, A., O'Dea, M. F. and Shrimpton, J. M. (2008). Are we missing a mineralocorticoid in teleost fish? Effects of cortisol, deoxycorticosterone and aldosterone on osmoregulation, gill Na^+, K^+-ATPase activity and isoform mRNA levels in Atlantic salmon. Gen. Comp. Endocrinol. 157, 35-40.

McDonald, M. D., Vulesevic, B., Perry, S. F. and Walsh, P. J. (2009). Urea transporter and glutamine synthetase regulation and localization in gulf toadfish gill. J. Exp. Biol. 212, 704-712.

McGowan, P. O., Sasaki, A., D'Alessio, A. C., Dymov, S., Labonté, B., Szyf, M., et al. (2009). Epigenetic regulation of the glucocorticoid receptor in human brain associates with childhood abuse. Nat. Neurosci. 12, 342-348.

Medeiros, L. R., Cartolano, M. C. and McDonald, M. D. (2014). Crowding stress inhibits serotonin 1A receptor-mediated increases in corticotropin-releasing factor mRNA expression and adrenocorticotropin hormone secretion in the Gulf toadfish. J. Comp. Physiol. B Biochem. Syst. Environ. Physiol. 184, 259-271.

Melnyk-Lamont, N., Best, C., Gesto, M. and Vijayan, M. M. (2014). The antidepressant venlafaxine disrupts brain monoamine levels and neuroendocrine responses to stress in rainbow trout. Environ. Sci. Technol. 48, 13434-13442.

Mennigen, J. A., Lado, W. E., Zamora, J. M., Duarte-Guterman, P., Langlois, V. S., Metcalfe, C. D., et al. (2010). Waterborne fluoxetine disrupts the reproductive axis in sexually mature male goldfish, Carassius auratus. Aquat. Toxicol. 100 (4), 354-364.

Milla, S., Terrien, X., Sturm, A., Ibrahim, F., Giton, F., Fiet, J., et al. (2008). Plasma 11deoxycorticosterone (DOC) and mineralocorticoid receptor testicular expression during rainbow trout Oncorhynchus mykiss spermiation: implication with 17α, 20β-dihydroxyprogesterone on the milt fluidity? Reprod. Biol. Endocrinol. 6, 19.

Milligan, C. L. (1997). The role of cortisol in amino acid mobilization and metabolism following exhaustive exercise in rainbow trout (Oncorhynchus mykiss Walbaum). Fish Physiol. Biochem. 16, 119-128.

Milligan, C. L. (2003). A regulatory role for cortisol in muscle glycogen metabolism in rainbow trout Oncorhynchus mykiss Walbaum. J. Exp. Biol. 206, 3167-3173.

Mommsen, T. P., Vijayan, M. M. and Moon, T. W. (1999). Cortisol in teleosts: dynamics, mechanisms of action, and metabolic regulation. Rev. Fish Biol. Fish. 9, 211-268.

Momoda, T. S., Schwindt, A. R., Feist, G. W., Gerwick, L., Bayne, C. J. and

Schreck, C. B. (2007). Gene expression in the liver of rainbow trout, Oncorhynchus mykiss, during the stress response. Comp. Biochem. Physiol. Part D Genomics Proteomics 2, 303-315.

Morishima, Y., Murphy, P. J. M., Li, D.-P., Sanchez, E. R. and Pratt, W. B. (2000). Stepwise assembly of a glucocorticoid receptor hsp90 heterocomplex resolves two sequential ATP-dependent events involving first Hsp70 and then Hsp90 in opening of the steroid binding pocket. J. Biol. Chem. 275, 18054-18060.

Mowbray, C., Hammerschmidt, M. and Whitfield, T. T. (2001). Expression of BMP signalling pathway members in the developing zebrafish inner ear and lateral line. Mech. Dev. 108, 179-184.

Muto, A., Orger, M. B., Wehman, A. M., Smear, M. C., Kay, J. N., Page-McCaw, P. S., et al. (2005). Forward genetic analysis of visual behavior in zebrafish. PLoS Genet. 1, e66.

Muto, A., Taylor, M. R., Suzawa, M., Korenbrot, J. I. andBaier, H. (2013). Glucocorticoidreceptor activity regulates light adaptation in the zebrafish retina. Front Neural. Circuits 7, 145.

Nakano, T., Afonso, L. O. B., Beckman, B. R., Iwama, G. K. and Devlin, R. H. (2013). Acute physiological stress down-regulates mRNA expressions of growth-related genes in coho salmon. PLoS One 8, e71421.

Neave, B., Holder, N. and Patient, R. (1997). A graded response to BMP-4 spatially coordinates patterning of the mesoderm and ectoderm in the zebrafish. Mech. Dev. 62, 183-195.

Nesan, D. and Vijayan, M. M. (2012). Embryo exposure to elevated cortisol level leads to cardiac performance dysfunction in zebrafish. Mol. Cell. Endocrinol. 363, 85-91.

Nesan, D. and Vijayan, M. M. (2013a). Role of glucocorticoid in developmental programming: evidence from zebrafish. Gen. Comp. Endocrinol. 181, 35-44.

Nesan, D. and Vijayan, M. M. (2013b). The transcriptomics of glucocorticoid receptor signaling in developing zebrafish. PLoS One 8, e80726.

Nesan, D. and Vijayan, M. M. (2016). Maternal cortisol mediates hypothalamus-pituitaryinterrenal axis development in zebrafish. Sci. Rep. 6, 22582.

Nesan, D., Kamkar, M., Burrows, J., Scott, I. C., Marsden, M. and Vijayan, M. M. (2012). Glucocorticoid receptor signaling is essential for mesoderm formation and muscle development in zebrafish. Endocrinol. 153, 1288-1300.

Oleksiak, M. F., Karchner, S. I., Jenny, M. J., Franks, D. G., Welch, D. B. M. and Hahn, M. E. (2011). Transcriptomic assessment of resistance to effects of an aryl hydrocarbon receptor (AHR) agonist in embryos of Atlantic killifish (Fundulus heteroclitus) from a marine Superfund site. BMC Genomics 12, 263.

Olsson, P. E., Kling, P., Erkell, L. J. and Kille, P. (1995). Structural and functional analysis ofthe rainbow trout (Oncorhyncus mykiss) metallothionein-A gene. Eur. J. Biochem. 230, 344-349.

Ortega, V. A., Lovejoy, D. A. and Bernier, N. J. (2013). Appetite-suppressing effects and interactions of centrally administered corticotropin-releasing factor, urotensin I and serotonin in rainbow trout (Oncorhynchus mykiss). Front Neurosci. 7, 196.

Pankhurst, N. W. (2016). Reproduction and Development. In Fish Physiology-Biology of Stress in Fish, Vol. 35 (eds. C. B. Schreck, L. Tort, A. P. Farrell and C. J. Brauner), San Diego, CA: Academic Press.

Pepels, P. P. L. M., Meek, J., Wendelaar Bonga, S. E. and Balm, P. H. M. (2002). Distribution and quantification of the corticotropn-releasing hormone in the brain of the teleost fish Oreochromis mossambicus (tilapia). J. Comp. Neurol. 453, 247-268.

Pérez-Jiménez, A., Guedes, M. J., Morales, A. E. and Oliva-Teles, A. (2007). Metabolic responses to short starvation and refeeding in Dicentrarchus labrax. Effect of dietary composition. Aquaculture 265, 325-335.

Philip, A. M. and Vijayan, M. M. (2015). Stress-immune-growth interactions: cortisol modulates suppressors of cytokine signaling and JAK/STAT pathway in rainbow trout liver. PLoS One 10, e0129299.

Philip, A. M., Daniel Kim, S. and Vijayan, M. M. (2012). Cortisol modulates the expression of cytokines and suppressors of cytokine signaling (SOCS) in rainbow trout hepatocytes. Dev. Comp. Immunol. 38, 360-367.

Pierce, A. L., Dickey, J. T., Felli, L., Swanson, P. and Dickhoff, W. W. (2010). Metabolic hormones regulate basal and growth hormone-dependent igf 2 mRNA level in primary cultured coho salmon hepatocytes: effects of insulin, glucagon, dexamethasone, and triiodothyronine. J. Endocrinol. 204, 331-339.

Pierce, A. L., Breves, J. P., Moriyama, S., Hirano, T. and Grau, G. E. (2011). Differential regulation of Igf1 and Igf2 mRNA levels in tilapia hepatocytes: effects of insulin and cortisol on GH sensitivity. J. Endocrinol. 211, 187-200.

Pierron, F., Baillon, L., Sow, M., Gotreau, S. and Gonzalez, P. (2014a). Effectof low-dose cadmium exposure on DNA methylation in the endangered European eel. Environ. Sci. Technol. 48, 797-803.

Pierron, F., Bureau du Colombier, S., Moffett, A., Caron, A., Peluhet, L.,

Daffe, G. , et al. (2014b). Abnormal ovarian DNA methylation programming during gonad maturation in wild contaminatedfish. Environ. Sci. Technol. 48, 11688-11695.

Pikulkaew, S. , Benato, F. , Celeghin, A. , Zucal, C. , Skobo, T. , Colombo, L. , et al. (2011). The knockdown of maternal glucocorticoid receptor mRNA alters embryo development in zebrafish. Dev. Dyn. 240, 874-889.

Pippal, J. B. , Cheung, C. M. I. , Yao, Y. -Z. , Brennan, F. E. and Fuller, P. J. (2011). Characterization of the zebrafish (Danio rerio) mineralocorticoid receptor. Mol. Cell. Endocrinol. 332, 58-66.

Pockley, A. G. , Muthana, M. and Calderwood, S. K. (2008). The dual immunoregulatory roles of stress proteins. Trends Biochem. Sci. 33, 71-79.

Poulain, M. , Fürthauer, M. , Thisse, B. , Thisse, C. and Lepage, T. (2006). Zebrafish endoderm formation is regulated by combinatorial Nodal, FGF and BMP signalling. Development 133, 2189-2200.

Pratt, W. B. , Morishima, Y. , Murphy, M. and Harrell, M. (2006). Chaperoning of glucocorticoid receptors. Handb. Exp. Pharmacol. 111-138.

Prunet, P. , Sturm, A. and Milla, S. (2006). Multiple corticosteroid receptors infish: from old ideas to new concepts. Gen. Comp. Endocrinol. 147, 17-23.

Reid, S. D. , Moon, T. W. and Perry, S. F. (1992). Rainbow trout hepatocyte adrenoceptors, catecholamine responsiveness and effects of cortisol. Am. Physiol. Soc. 262, 794-799.

Reid, S. G. , Bernier, N. J. and Perry, S. F. (1998). The adrenergic stress response infish: control of catecholamine storage and release. Comp. Biochem. Physiol. Part C 120, 1-27.

Reik, W. , Santos, F. and Dean, W. (2003). Mammalian epigenomics: reprogramming the genome for development and therapy. Theriogenology 59, 21-32.

Reindl, K. M. and Sheridan, M. A. (2012). Peripheral regulation of the growth hormoneinsulin-like growth factor system infish and other vertebrates. Comp. Biochem. Physiol. A Mol. Integr. Physiol. 163, 231-245.

Reinick, C. L. , Liang, L. , Angleson, J. K. and Dores, R. M. (2012). Identification of an MRAPindependent melanocortin-2 receptor: functional expression of the cartilaginous fish, Callorhinchus milii, melanocortin-2 receptor in CHO cells. Endocrinology 153, 4757-4765.

Rise, M. L. , Jones, S. R. M. , Brown, G. D. , von Schalburg, K. R. , Davidson, W. S. and Koop, B. F. (2004). Microarray analyses identify molecular biomarkers of Atlantic salmon macrophage and hematopoietic kidney response to Piscirickettsia salmonis infection. Physiol. Genomics 20, 21-35.

Richter, K., Haslbeck, M. and Buchner, J. (2010). The heat shock response: life on the verge of death. Mol. Cell 40, 253-266.

Rodela, T. M., Esbaugh, A. J., McDonald, M. D., Gilmour, K. M. and Walsh, P. J. (2011). Evidence for transcriptional regulation of the urea transporter in the gill of the Gulf toadfish, Opsanus beta. Comp. Biochem. Physiol. B Biochem. Mol. Biol. 160, 72-80.

Rodgers, B. D., Weber, G. M., Kelley, K. M. and Levine, M. A. (2003). Prolonged fasting and cortisol reduce myostatin mRNAlevels in tilapia larvae; short-term fasting elevates. Am. J. Physiol. Regul. Integr. Comp. Physiol. 284, R1277-R1286.

Ronco, A. M., Llaguno, E., Epuñan, M. J. and Llanos, M. N. (2010). Effect of cadmium on cortisol production and 11b-hydroxysteroid dehydrogenase 2 expression by cultured human choriocarcinoma cells (JEG-3). Toxicol. Vitro 24, 1532-1537.

Roy, B. and Jacobson, A. (2013). The intimate relationships of mRNA decay and translation. Trends Genet 29, 691-699.

Salem, M., Xiao, C., Womack, J., Rexroad, C. E. and Yao, J. (2010). A microRNA repertoire for functional genome research in rainbow trout (Oncorhynchus mykiss). Mar. Biotechnol. 12, 410-429.

Salem, M., Paneru, B., Al-Tobasei, R., Abdouni, F., Thorgaard, G. H., Rexroad, C. E., et al. (2015). Transcriptome assembly, gene annotation and tissue gene expression atlas of the rainbow trout. PLoS One 10, 1-27.

Sánchez, C. C., Weber, G. M., Gao, G., Cleveland, B. M., Yao, J. and Rexroad, C. E. (2011). Generation of a reference transcriptome for evaluating rainbow trout responses to various stressors. BMC Genomics 12, 626.

Sandhu, N. and Vijayan, M. M. (2011). Cadmium-mediated disruption of cortisol biosynthesis involves suppression of corticosteroidogenic genes in rainbow trout. Aquat. Toxicol. 103, 92-100.

Sandhu, N., McGeer, J. C. and Vijayan, M. M. (2014). Exposure to environmental levels of waterborne cadmium impacts corticosteroidogenic and metabolic capacities, and compromises secondary stressor performance in rainbow trout. Aquat. Toxicol. 146, 20-27.

Sarkar, A., Ray, D., Shrivastava, A. N. and Sarker, S. (2006). Molecular biomarkers: their significance and application in marine pollution monitoring. Ecotoxicology 15, 333-340.

Sathiyaa, R. and Vijayan, M. M. (2003). Autoregulation of glucocorticoid receptor by cortisol in rainbow trout hepatocytes. Am. J. Physiol. Cell Physiol. 284, 1508-1515.

Sathiyaa, R. , Campbell, T. and Vijayan, M. M. (2001). Cortisol modulates HSP90 mRNA expression in primary cultures of trout hepatocytes. Comp. Biochem. Physiol. B Comp. Biochem. 129, 679-685.

Schaaf, M. J. M. , Champagne, D. , van Laanen, I. H. C. , van Wijk, D. C. W. A. , Meijer, A. H. and Meijer, O. C. (2008). Discovery of a functional glucocorticoid receptor B-isoform in zebrafish. Endocrinology 149, 1591-1599.

Schaaf, M. J. M. , Chatzopoulou, A. and Spaink, H. P. (2009). The zebrafish as a model system for glucocorticoid receptor research. Comp. Biochem. Physiol. Part A 153, 75-82.

Schoneveld, O. J. L. M. , Gaemers, I. C. and Lamers, W. H. (2004). Mechanisms of glucocorticoid signalling. Biochim. Biophys. Acta Gene Struct. Expr. 1680, 114-128.

Schulte, P. M. , Glemet, H. C. , Fiebig, A. A. and Powers, D. A. (2000). Adaptive variation in lactate dehydrogenase-B gene expression: role of a stress-responsive regulatory element. Proc. Natl. Acad. Sci. 97, 6597-6602.

Sexton, T. and Cavalli, G. (2015). The role of chromosome domains in shaping the functional genome. Cell 160, 1049-1059.

Shelley, L. K. , Ross, P. S. , Miller, K. M. , Kaukinen, K. H. and Kennedy, C. J. (2012). Toxicity of atrazine and nonylphenol in juvenile rainbow trout (Oncorhynchus mykiss): effects on general health, disease susceptibility and gene expression. Aquat. Toxicol. 124-125, 217-226.

Shi, Y. , Mosser, D. D. and Morimoto, R. I. (1998). Molecular chaperones as HSF1-specific transcriptional repressors. Genes Dev. 12, 654-666.

Skinner, M. K. (2014). Environmental stress and epigenetic transgenerational inheritance. BMC Med. 12, 153.

Small, B. C. , Murdock, C. A. , Waldbieser, G. C. and Peterson, B. C. (2006). Reduction in channel catfish hepatic growth hormone receptor expression in response to food deprivation and exogenous cortisol. Domest. Anim. Endocrinol. 31, 340-356.

Soengas, J. L. , Polakof, S. , Chen, X. , Sangiao-Alvarellos, S. and Moon, T. W. (2006). Glucokinase and hexokinase expression and activitis in rainbow trout tissues: changes with food depravation and refeeding. Am. J. Physiol. 291, R810-R821.

Stolte, E. H. , van Kemenade, B. M. L. V. , Savelkoul, H. F. J. and Flik, G. (2006). Evolution of glucocorticoid receptors with different glucocorticoid sensitivity. J. Endocrinol. 190, 17-28.

Stolte, E. H. , Nabuurs, S. B. , Bury, N. R. , Sturm, A. , Flik, G. , Savelkoul, H. F. J. , et al. (2008). Stress and innate immunity in carp: corticosteroid receptors and pro-inflammatory cytokines. Mol. Immunol. 46, 70-79.

Storey, K. B. (1997). Organic solutes in freezing tolerance. Comp. Biochem. Physiol. Part A Physiol. 117, 319-326.

Stratholt, M. L., Donaldson, E. M. and Liley, N. R. (1997). Stress induced female coho reflected in appear elevation of plasma cortisol in adult salmon (Oncorhynchus kisutch), is egg cortisol content, but does not to affect early development. Aquaculture 158, 141-153.

Takahashi, A. and Kawauchi, H. (2006). Evolution of melanocortin systems infish. Gen. Comp. Endocrinol. 148, 85-94.

Takahashi, H. and Sakamoto, T. (2013). The role of "mineralocorticoids" in teleost fish: relative importance of glucocorticoid signaling in the osmoregulation and "central" actions of mineralocorticoid receptor. Gen. Comp. Endocrinol. 181, 223-228.

Takei, Y., and Hwang, P.-P. (2016). Homeostatic Responses to Osmotic Stress. In Fish Physiology-Biology of Stress in Fish, Vol. 35 (eds. C. B. Schreck, L. Tort, A. P. Farrell, and C. J. Brauner), San Diego, CA: Academic Press.

Teles, M., Boltaña, S., Reyes-López, F., Santos, M. A., Mackenzie, S. and Tort, L. (2013). Effects of chronic cortisol administration on global expression of GR and the liver transcriptome in Sparus aurata. Mar. Biotechnol. (NY) 15, 104-114.

Terova, G., Gornati, R., Rimoldi, S., Bernadini, G. and Saroglia, M. (2005). Quantification of a glucocorticoid receptor in sea bass (Dicentrarchus labrax L.) reared at high stocking density. Gene 357, 144-151.

Thiel, G., Al Sarraj, J. and Stefano, L. (2005). cAMP response element binding protein (CREB) activates transcription via two distinct genetic elements of the human glucose-6phosphatase gene. BMC Mol. Biol. 6, 1-4.

Thomas, M. A. and Klaper, R. D. (2012). Psychoactive pharmaceuticals inducefish gene expression profiles associated with human idiopathic autism. PLoS One 7, e32917.

Tipsmark, C. K., Jørgensen, C., Brande-Lavridsen, N., Engelund, M., Olesen, J. H. and Madsen, S. S. (2009). Effects of cortisol, growth hormone and prolactin on gill claudin expression in Atlantic salmon. Gen. Comp. Endocrinol. 163, 270-277.

Tiso, N., Filippi, A., Pauls, S., Bortolussi, M. and Argenton, F. (2002). BMP signalling regulates anteroposterior endoderm patterning in zebrafish. Mech. Dev. 118, 29-37.

To, T. T., Hahner, S., Nica, G., Rohr, K. B., Hammerschmidt, M., Winkler, C., et al. (2007). Pituitary-interrenal interaction in zebrafish interrenal organ development. Mol. Endocrinol. 21, 472-485.

Tokarz, J., Mindnich, R., Norton, W., Möller, G., Hrabě de Angelis, M., et al. (2012). Discovery of a novel enzyme mediating glucocorticoid catabolism in fish: 20betahydroxysteroid dehydrogenase type 2. Mol. Cell. Endocrinol. 349, 202-213.

Tokarz, J., Norton, W., Möller, G., Hrabě de Angelis, M. and Adamski, J. (2013). Zebrafish 20b-hydroxysteroid dehydrogenase type 2 is important for glucocorticoid catabolism in stress response. PLoS One 8, e54851.

Tort, L. (2011). Stress and immune modulation infish. Dev. Comp. Immunol. 35, 1366-1375. Turecki, G. and Meaney, M. (2014). Effects of the social environment and stress on glucocorticoid receptor gene methylation: a systematic review. Biol. Psychiatry 1-10.

Venkatesh, S. and Workman, J. L. (2015). Histone exchange, chromatin structure and the regulation of transcription. Nat. Rev. Mol. Cell Biol. 16, 178-189.

Villeneuve, D. L., Knoebl, I., Larkin, P., Miracle, A. L., Carter, B. J., Denslow, N. D., et al. (2008). Altered gene expression in the brain and liver of female fathead minnows Pimephales promelas Rafinesque exposed to fadrozole. J. Fish Biol. 72, 2281-2340.

Vijayan, M. M., Raptis, S. and Sathiyaa, R. (2003). Cortisol treatment affects glucocorticoid receptor and glucocorticoid-responsive genes in the liver of rainbow trout. Gen. Comp. Endocrinol. 132, 256-263.

Vijayan, M. M., Prunet, P. and Boone, A. N. (2005). Xenobiotic impact on corticosteroid signalling. In Biochemical and Molecular Biology of Fishes, Vol 6. Environmental Toxicology (eds. T. W. Moon and T. P. Mommsen), pp. 365-394. Amsterdam: Elsevier.

Vijayan, M. M., Aluru, N. and Leatherland, J. F. (2010). Stress response and the role of cortisol. In Fish Diseases and Disorders. Vol 2: Non-Infectious Disorders (eds. J. F. Leatherland and P. Woo), pp. 182-201. Oxfordshire: CABI.

Wang, R.-L., Bencic, D. C., Garcia-Reyero, N., Perkins, E. J., Villeneuve, D. L., Ankley, G. T., et al. (2014). Natural Variation in Fish Transcriptomes: Comparative Analysis of the Fathead Minnow (Pimephales promelas) and Zebrafish (Danio rerio). PLoS One 9, e114178.

Wang, T. and Hong, Y. (2014). Direct gene disruption by TALENs in medaka embryos. Gene 543, 28-33.

Weaver, I. C. G., Cervoni, N., Champagne, F. A. D.', Alessio, A. C., Sharma, S., Seckl, J. R., et al. (2004). Epigenetic programming by maternal behavior. Nat. Neurosci. 7, 847-854.

Wendelaar Bonga, S. E. (1997). The stress response infish. Physiol. Rev. 77, 591-625.

Whitehead, A., Roach, J. L., Zhang, S. and Galvez, F. (2011). Genomic mechanisms of evolved physiological plasticity in killifish distributed along an environmental salinity gradient. Proc. Natl. Acad. Sci. 108, 6193-6198.

Williams, T. D., Lee, J. S., Sheader, D. L. and Chipman, J. K. (2000). The cytochrome P450 1A gene (CYP1A) from Europeanflounder (Platichthys flesus), analysis of regulatory regions and development of a dual luciferase reporter gene system. Mar. Environ. Res. 50, 1-6.

Wilson, K. S., Matrone, G., Livingstone, D. E. W., Al-Dujaili, E. A. S., Mullins, J. J., Tucker, C. S., et al. (2013). Physiological roles of glucocorticoids during early embryonic development of the zebrafish (Danio rerio). J. Physiol. 591, 6209-6220.

Wilson, K. S., Baily, J., Tucker, C. S., Matrone, G., Vass, S., Moran, C., et al. (2015). Early-life perturbations in glucocorticoid activity impacts on the structure, function and molecular composition of the adult zebrafish (Danio rerio) heart. Mol. Cell. Endocrinol. 414, 120-131.

Wiseman, S., Osachoff, H., Bassett, E., Malhotra, J., Bruno, J., VanAggelen, G., et al. (2007). Gene expression pattern in the liver during recovery from an acute stressor in rainbow trout. Comp. Biochem. Physiol. Part D Genomics Proteomics 2, 234-244.

Wunderink, Y. S., MartÍnez-RodrÍguez, G., Yúfera, M., MartÍn Montero, I., Flik, G., Mancera, J. M., et al. (2012). Food deprivation induces chronic stress and affects thyroid hormone metabolism in Senegalese sole (Solea senegalensis) post-larvae. Comp. Biochem. Physiol. Part A Mol. Integr. Physiol. 162, 317-322.

Xia, J. H., Liu, P., Liu, F., Lin, G., Sun, F., Tu, R., et al. (2013). Analysis of stress-responsive transcriptome in the intestine of Asian seabass (Lates calcarifer) using RNA-Seq. DNA Res. 2, 449-460.

Yada, T. and Tort, L. (2016). Stress and Disease Resistance: Immune System and Immunoendocrine Interactions. In Fish Physiology-Biology of Stress in Fish, Vol. 35 (eds. C. B. Schreck, L. Tort, A. P. Farrell and C. J. Brauner), San Diego, CA: Academic Press.

Yan, H., Simola, D. F., Bonasio, R., Liebig, J., Berger, S. L. and Reinberg, D. (2014). Eusocial insects as emerging models for behavioural epigenetics. Nat. Rev. Genet. 15, 677-688.

Yang, B.-Y., Chan, K.-M., Lin, C.-M. and Chen, T. T. (1997). Characterization of rainbow trout (Oncorhynchus mykiss) growth hormone 1 gene and the promoter region of growth hormone 2 hene. Arch. Biochem. Biophys. 340, 359-368.

Yano, A., Nicol, B., Jouanno, E. and Guiguen, Y. (2014). Heritable targeted inactivation of the rainbow trout (Oncorhynchus mykiss) master sex-determining gene using zinc-finger nucleases. Mar. Biotechnol. 16, 243-250.

Yeh, C.-M. (2015). The basal NPO crhfluctuation is sustained under compromised glucocorticoid signaling in diurnal zebrafish. Front Neurosci. doi: 10.3389/fnins.2015.00436.

Zannas, A. S. and West, A. E. (2015). Epigenetics and the regulation of stress vulnerability and reslience. Neuroscience 264, 157-170.

Zhang, T. Y., Labonté, B., Wen, X. L., Turecki, G. and Meaney, M. J. (2013). Epigenetic mechanisms for the early environmental regulation of hippocampal glucocorticoid receptor gene expression in rodents and humans. Neuropsychopharmacology 38, 111-123.

Zheng, J., Xiao, X., Zhang, Q. and Yu, M. (2014). DNA methylation: the pivotal interaction between early-life nutrition and glucose metabolism in later life. Br. J. Nutr. 112, 1850-1857.

Ziv, L., Muto, A., Schoonheim, P. J., Meijsing, S. H., Strasser, D., Ingraham, H. A., et al. (2013). An affective disorder in zebrafish with mutation of the glucocorticoid receptor. Mol. Psychiatry 18, 681-691.

第5章　应激与生长

在鱼类中，其肌肉重量占总体重的一半以上，因此，它们的器官系统的大小变化被认为对其生长至关重要。肌肉生长是一系列复杂过程之后的最终结果，动物首先从环境中吸收营养，再将这些营养适当分配以增加肌细胞数量和大小。应激会影响这些过程，包括能量的利用、吸收和分配，从而减缓肌肉生长。应激可以发生在野外环境，如躲避捕食者行为或者水质恶化；或者发生在饲养环境，如手工操作、拥挤环境、分类分级和标粗，这些都会对动物产生应激，使其重新分配生长能量以应对应激。在这里，我们会着重指出皮质醇（硬骨鱼类中的主要糖皮质激素）在调节过程中导致肌肉生长受到抑制的潜在作用。通过应激的介导，血浆皮质醇浓度会升高，这会影响能量的摄入、吸收和利用，包括蛋白质周转和保护性蛋白质表达的调节，所有这些都导致肌肉生长可用的能量减少。皮质醇也会调整肌肉生长调节因子，包括生长因子和转录因子，从而导致生长受到抑制。总体而言，关于应激影响生长的机制信息目前还很少，但现有文献表明，应激源（应激物）介导血液循环中皮质醇水平升高和调节鱼类肌肉蛋白质增加起着关键作用。

5.1　导　言

在许多种鱼类中，生长是一种高度遗传的特性。通过挑选一些生长速率更快的鱼类可以增加鱼类产量（Dunham，2011）。在水产养殖中，鱼类的快速生长是最重要的选育性状之一（Gjedrem，2005）。然而，采用所有鱼类个体都含有相同基因组的同基因鱼类家系而带来的弊端，也说明了环境和生活史对于鱼类生长轨迹的影响是很重要的（Dupont-Nivet 等，2009）。应激是影响生长最重要的生理因素之一，而一些生物类应激源和非生物类应激源，包括一些常规孵化操作，如手工操作和分类分级、水质不良和拥挤环境都能够抑制鱼类生长（McCormick 等，1998；DiBattista 等，2006）。对付应激时，由于动物激活了一系列复杂的能量消耗途径以恢复体内稳态并保持其功能完整性，因而改变了鱼类体内的能量状态（Barton，2002）。由于在一定的时间内动物可利用的生长能是有限的（Guderley，Pörtner，2010），应激的应对会浪费部分用于生长的能量底物，从而导致鱼类产量减少。动物面对所暴露的应激源时，能量需求的急剧增加为应激反应相关激素通道的激活所介导，包括下丘脑-垂体-肾间组织轴，导致皮质醇增加。反过来，为了恢复稳态，这些反应又会将能量底物动员和重新分配（Faught 等，2016；本书第4章）。因此，应激和生长之间的联系错综复杂。本章着重介绍我们目前对于硬骨鱼应激介导的生长抑制的研究，重点是皮质醇在肌肉中介导这些效应时所发挥的作用。我们将本章分为3个主要部分：第一部分注重使用模拟方法分析生长的资源分配；第二部分和第三部分分别描述应激和/或皮质醇影响能量分配和调节生长启动子的潜在分子机制，从而影响肌肉的生长。本章还确定了主要的知识空白和未来的挑战方向。

5.2 生长的一种概念组成

由于配子的生长已为Pankhurst论述（2016；见本书第8章），在这里，我们主要关注体细胞生长。鱼类的生长是指鱼体长度和重量的增加，这是一个复杂的过程，为觅食活动、营养同化、能量底物分配和利用等诸多因素所影响。对于鱼类，其肌肉重量超过体重的一半，因此，这种器官系统的大小变化对生长至关重要（Mommsen，Moon，2001）。与其他动物模式不同，鱼类的生长是无限制的，既包括了新肌纤维的形成（增生），也包括了原有肌纤维大小的增加（肥大）（Mommsen，2001）。虽然在大多数其他脊椎动物类群中，肌肉增生仅限于早期发育时期，但在硬骨鱼类中，成鱼50%以上的生长都是肌肉增生导致的（Mommsen，2001）。因此，从生长的角度来看，鱼体结构的改变或许能最好地反映在肌肉蛋白质的增加中。肌细胞中的大部分蛋白质合成发生在肌原纤维中并且与肌球蛋白重链（myosin heavy chain，MHC）蛋白质有关。因此，MHC丰度和支持结构物如骨骼和软骨已被认为是结构生长的标志物（Mommsen，2001）。关于鱼类生长的肌肉发育的作用机理、结构和生理学已有一些优秀的论述（Mommsen，2001；Mommsen，Moon，2001；Rescan，2005；Johnston等，2011）。在本章中，我们主要关注结构蛋白沉积时骨骼肌的生长，并且提出应激影响整体能量分配的见解。

5.2.1 生长的动态能量收支模型

为了预测一定投喂量下的鱼体体重，如今已有多种模型用于表述生长（Dumas等，2010；Seginer，2016）。其中，基于热力学定律的生物能模型可以定量评估动物生长期间不同过程中的能量分配，包括维持过程、生长过程、繁殖过程。这些预测模型能很好地符合许多种鱼类的生长曲线（Warren，Gerald，1967；Kitchell等，1977；Pauly，1981；Essington等，2001）。最近，动态能量收支模型（DEB）对于生长预测愈加受到欢迎（Kooijman，2000，2010；Nisbet等，2000），因为该模型涵盖了生物体的整个生命周期并考虑到环境变量的影响（Sousa等，2010）。这种模型已被广泛应用于包括硬骨鱼在内的多种脊椎动物物种（表5.1）。

表5.1 动态能量收支模型（DEB）适用的鱼类种类

品种	学名	参考文献
欧鲽	*Pleuronectes platessa*	van der Veer等（2001）
越洋公鱼	*Hypomesus transpacificus*	Fujiwara等（2005）

续表 5.1

品种	学名	参考文献
川鲽	*Platichthys flesus*	van der Veer 等（2001），Freitas 等（2010）
欧洲黄盖鲽	*Limanda limanda*	van der Veer 等（2001），Freitas 等（2010）
欧洲鳎	*Solea solea*	van der Veer 等（2001），Freitas 等（2010），Eichinger 等（2010）
欧洲无须鳕	*Merluccius merluccius*	Bodiguel 等（2009）
胖头鲹	*Pimephales promelas*	Jager 等（2009）
小眼长臀鰕虎鱼	*Pomatoschistus microps*	Freitas 等（2010）
小长臀虾虎鱼	*Pomatoschistus minutus*	Freitas 等（2010）
绵鳚	*Zoarces viviparous*	Freitas 等（2010）
短角床杜父鱼	*Myoxocephalus scorpius*	Freitas 等（2010）
狼鲈	*Dicentrarchus labrax*	Freitas 等（2010）
大西洋鳕鱼	*Gadus morhua*	Freitas 等（2010），Klok 等（2014）
大西洋鲱鱼	*Clupea harengus*	Freitas 等（2010）
黍鲱鱼	*Sprattus sprattus*	Freitas 等（2010）
欧洲鳀鱼	*Engraulis encrasicolus*	Pecquerie 等（2009），Freitas 等（2010），Pethybridge 等（2013）
毛鳞鱼	*Mallotus villosus*	Einarsson 等（2011）
斑马鱼	*Danio rerio*	Augustine 等（2011，2012）
东方金枪鱼	*Thunnus orientalis*	Jusup 等（2011）
细鳞大麻哈鱼	*Oncorhynchus gorbuscha*	Pecquerie 等（2011）
红大麻哈鱼	*Oncorhynchus nerka*	Pecquerie 等（2011）
银大麻哈鱼	*Oncorhynchus kisutch*	Pecquerie 等（2011）
大麻哈鱼	*Oncorhynchus keta*	Pecquerie 等（2011）
大鳞大麻哈鱼	*Oncorhynchus tshawytscha*	Pecquerie 等（2011）
沙重牙鲷	*Diplodus sargus*	Serpa 等（2013）
金头鲷	*Sparus aurata*	Serpa 等（2013）
林氏泥荫鱼	*Umbra limi*	Filgueira 等（2016）

DEB 描述了动物能量同化及其将能量分配给维持、生长和繁殖的速率，图 5.1 表示其能量通路。该模型假设能量供应由同化的营养素持续供应，且动物最大同化速率与其体表面积成正比（Kooijman，2010）。能量同化是食物供应、摄取（取决于动物的食欲和活动）和肠道营养吸收能力共同的结果。食物供应由外在因素决定，包括环境中的食物量以及因同类物种的密度和社会地位相关因素造成的直接的食物竞争。相反，摄食和营养吸收能力由内在因素决定，例如动物的食欲和活动，并受到神经和内分泌系统的调节。能量会被持续性分配到两个主要的区域，即身体结构和成熟度（稍后再做描述），其中每个区域还需要分配一部分固定的维持能量（图 5.1）。

结构区域包括组成动物身体的所有成分，如肌肉和骨骼（Kooijman，2010）；而成熟度区域是指表示动物复杂性的一些组成成分，例如，脑部、免疫系统和性腺。尽管后者仅占动物体重的一小部分，但它们却消耗了大部分的能量。例如，虽然人类大脑只占其体重的2%，但其消耗的能量占了能量消耗总量的20%（Mink等，1981）。金鱼（*Carassius auratus*）也是如此，其大脑占总体质量的1%，但消耗的能量却超过总能量预算的7%（Mink等，1981）。同样的，免疫系统的激活对于动物的质量所占比例很小，但其能量需求很高（Wolowczuk等，2008；Rauw，2012）。因此，成熟度所分配能量的增加对动物的良好发育是至关重要的，它会减少身体结构的资源分配额，从而影响动物整体的生长。

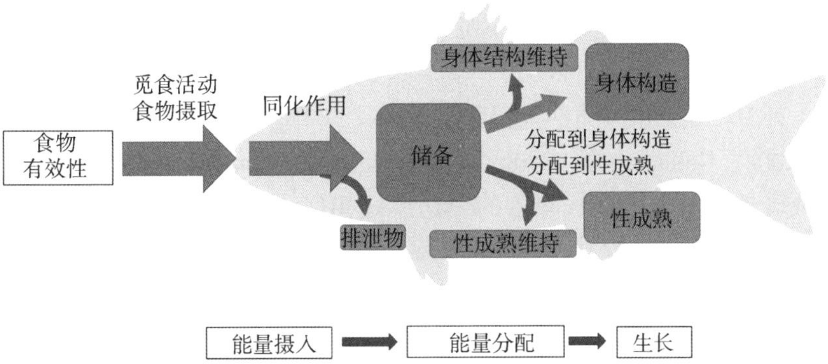

图5.1 基于动态能量收支理论的鱼类能量流动图

注：通过觅食、食物摄取和同化，从环境中获取能量并将其转移到储备区室。然后将能量分配到身体结构或成熟度，并将其分配给维护部分（maint.）。在所提出的模型中，生长主要与身体结构所分配的能量额有关，因此，其与蓝色能量路径（蓝色箭头和区室）正相关。相反，分配给橙色区室的能量增加时（通过橙色箭头）会对生长直接起反作用。该图基于Kooijman（2010）提出的动态能量收支模型。（见书后彩图）

DEB模型还假设把分配给身体结构和成熟度的一部分能量用作它们的维持能量（如渗透压调节、呼吸和蛋白质周转），并且重要的是，这部分能量往往是优先考虑的（Kooijman，2010）。这可以在限制投喂的鱼类中得到说明，这时鱼类的生长和性成熟都会严重受损（Bromley等，2000；Zhu等，2015），表明在低能量摄入时，几乎没有能量分配给身体结构和成熟度。不论是在常规的生长环境下，还是在变化（应激性）的环境中，DEB模型都通过可以充分了解的分配给身体结构、成熟度和维持的能量额，为动物生长的预测提供了一种可行的工具（Jager，Selck，2011；Augustine）。在本综述中，动物从环境中转移能量以形成肌肉结构的能力是鱼类生长最重要的一方面。当应激增加能量消耗时，DEB模型或许可以提供一个概念性的框架，以了解应激源如何影响能量分配，继而影响生长。

5.2.2 肌细胞生长

细胞生长要经历一系列复杂的过程，包括前体细胞的特化、增殖和分化，从而形成肌细胞，并且互相融合而形成多核细胞（Bentzinger 等，2012）。同样的，细胞肥大基本上是蛋白质沉积增加的结果，由蛋白质合成和蛋白质水解之间的平衡所调控。我们目前还只是开始了解鱼类肌肉增生和肥大的机制。

5.2.2.1 成肌细胞的分化和融合

与哺乳动物不同，鱼类增生性肌肉生长会发生在其整个生命周期中（Mommsen，2001）。肌肉增生包含新生肌纤维的形成，并受到生肌调节因子（MRFs）的刺激作用，包括肌源性分化蛋白（MYOD）、MYF5、肌细胞生成素和 MRF4。在哺乳动物中，MRF 基因 *MYOD* 和 *MYF5* 是干细胞特化为肌源性祖细胞所必需的（Weintraub 等，1991；Rudnicki 等，1993），并且 *MYOD/MYF5* 双敲除的小鼠会导致完全缺乏骨骼肌（Rudnicki 等，1993）。在骨骼肌生成期间，*MYF5* 被发现是最先表达的，*MYOD* 紧接其后。在之后的肌肉发育期间，肌细胞生成素和 MRF4 调控从成肌细胞到肌细胞的分化（Sumariwalla，Klein，2001），其后是细胞融合，这是肌细胞生成过程的最后一步。这 4 种 MRFs 已在鱼类中得到鉴定（Rescan 等，1995；Rescan，Gauvry，1996；Tan，Du，2002；Galloway 等，2006；García de la serrana 等，2014），并且它们的表达模式与哺乳动物肌肉生成过程中的表达模式相似（Vélez 等，2015）。肌细胞的融合是肌肉形成和发育过程中的基本步骤，也是一个高度调控的过程（Kim 等，2015）。最近的研究表明，nephrin、KIRREL3L、myomaker 和复杂的 JAMB / JAMC 蛋白在调节斑马鱼（*Danio rerio*）肌细胞融合中起着关键作用（Srinivas 等，2007；Sohn 等，2009；Powell，Wright，2011；Landemaine 等，2014）。然而，关于应激和/或皮质醇刺激对于影响鱼类肌细胞生长和分化中所发挥作用的相关信息还很少。

5.2.2.2 肌肉蛋白酶解

组织蛋白质不断合成和降解，两个过程之间的平衡决定了肌肉的生长（Jagoe 等，2002；Lai 等，2004）。鱼类肌细胞中的蛋白质水解似乎遵循在哺乳动物中观察到的一般机制，其中 4 种高度保守的系统发挥着最重要的作用，包括泛素-蛋白酶体系统（UPS）、自噬-溶酶体系统（ALS）、钙蛋白酶-钙蛋白酶抑制剂系统，以及凋亡蛋白酶系统（Salem 等，2006；Seiliez 等，2008a，2010；Salmerón 等，2013）。其中，UPS 和 ALS 最受关注，其占无血清鱼肌细胞总降解蛋白质的 55% 左右（Seiliez 等，2014）。研究表明，这 2 个系统受环境因素的调控，包括营养可用性和生长因子，这表明环境因素在肌肉生长和维持过程中发挥着关键作用（Cleveland，Weber，2010；Seiliez 等，2012a）。

UPS 对于受损或存活时间较短的蛋白质降解至关重要（Attaix 等，2005），其涉及 2 个主要步骤：（1）通过多泛素化标记底物；（2）通过蛋白酶体复合物降解。第一步使 UPS 成为选择性靶向蛋白，并通过涉及泛素连接酶的多步通道进行，包括 2 种在鱼组织中表达的蛋白质：ATROGIN-1 和肌肉 RING 指蛋白-1（MURF1）（Seiliez 等，2008b；Wang 等，2011）。然而，除了它们的表达之外，关于在鱼类中起作用的 UPS 的潜在分子机制，目前获得的信息有限。

与UPS不同，ALS是一种复杂的，绝大部分具有非特异性的系统，它可以降解细胞质的部分物质，并且在回收大型蛋白质和不是UPS目标的细胞器中起主要作用（Sandri，2010；Noda，Inagaki，2015）。首先，它产生双脂质膜并在细胞内底物周围延伸，形成自噬体囊泡，将底物与细胞质隔离。ALS所涉及的分子机制在酵母（*Saccharomyces cerevisiae*）中进行了深入研究，证明了自噬相关蛋白（ATGs）对于囊泡的形成所发挥的核心作用。自噬体的形成始于ATG1复合物的组装，从而产生噬菌体组装位点或前自噬体结构（Suzuki等，2001）。然后由酶Vps34，也称为III类磷脂酰肌醇3-激酶（PI3K）与ATG6（或者哺乳动物中的BECN1）相结合来驱动吞噬体的延长，形成酵母和哺乳动物自噬所需的复合物（Morris等，2015）。复合物ATG2-ATG18／WIPI被招募并且是自噬体成核所必需的（Velikkakath等，2012）。在自噬体扩张期间，ATG8通过ATG12／ATG5复合物与磷脂酰乙醇胺结合，使ATG8锚定在自噬体膜上并控制扩张（Xie等，2008）。哺乳动物表达了几种ATG8直系同源物，其中最重要的是LC3和GABARAP（Shpilka等，2011）。所有这些ATG在真核生物中都是高度保守的，并且最近已在硬骨鱼类中进行了鉴定（表5.2）。

表5.2 在鱼类中检测到的自噬相关蛋白（ATG）及其在酵母中的同源物

酵母蛋白	鱼类相关蛋白	品种	组织	参考文献
ATG8	LC3B	虹鳟	原代培养的成肌细胞和白肌	Seiliez等（2010）
	LC3B	斑马鱼	胚胎细胞	Yabu等（2012）
	LC3	斑马鱼	胚胎	He等（2009）
	GABARAPL1	虹鳟	原代培养的成肌细胞和白肌	Seiliez等（2010）
	GABARAP	斑马鱼	胚胎	He等（2009）
ATG12	ATG12L	虹鳟	原代培养的成肌细胞和白肌	Seiliez等（2010）
	ATG12	斑马鱼	胚胎	Hu等（2011）
ATG4	ATG4B	虹鳟	原代培养的成肌细胞和白肌	Seiliez等（2010）
ATG5	ATG5	斑马鱼	胚胎	Hu等（2011）
	ATG5	智利羽鼬鳚	原代培养的成肌细胞和白肌	Aedo等（2015）
ATG7	ATG7	斑马鱼	胚胎	Hu等（2011）
ATG6	POBECLIN1	牙鲆	鳃	Kong等（2011）
	BECLIN1	斑马鱼	胚胎	He等（2009）
	BECLIN1	鲤鱼	肾脏	Gao等（2014）
ATG1	ULK1A	斑马鱼	胚胎	He等（2009）
	ULK1B	斑马鱼	胚胎	He等（2009）
ATG9	ATG9A	斑马鱼	胚胎	He等（2009）
	ATG9A	斑马鱼	胚胎	He等（2009）
ATG16	ATG16L1	智利羽鼬鳚	原代培养的成肌细胞和白肌	Aedo等（2015）

虽然 ALS 在鱼类肌肉生长调节中的作用尚不清楚，但其中重要蛋白质的高度保守性表明调节机制可能与哺乳动物相似。一旦自噬体完成，它就会沿着微管向溶酶体转运，它们融合形成自噬溶酶体，导致囊泡内生物物质的酶促降解。尽管对其融合和降解所涉及的机制知之甚少，但 SNARE 蛋白、RAB7 和 HOPS 复合物（Gutierrez 等，2004；Fader 等，2009；McEwan 等，2015）、GABARAP，以及酶磷脂酰肌醇 4-激酶 α（PI4Kα）等被认为在哺乳动物细胞系统中起着重要作用（Wanget 等，2015）。总之，蛋白质降解途径的抑制可能是肌肉生长的一个关键方面，而应激介导激活蛋白质分解途径可能是生长抑制的一种机制，但这还有待于在鱼类中进行验证。

5.2.2.3 肌肉蛋白质动力学的调控

尽管在哺乳动物中描述了涉及蛋白质合成和蛋白质水解的分子调节机制，但在鱼类中尚不清楚。研究表明，营养和生长因子是鱼类生长的重要调节因子，并且这和蛋白质动力学的调节有关（图 5.2）。具体而言，转化生长因子-β（TGFs-beta）、胰岛素、胰岛素样生长因子（IGFs）、其结合蛋白（IGFBPs）和生长激素（GH）等都是调节鱼类肌肉生成的关键因素（Mommsen，Moon，2001；Reindl，Sheridan，2012）。

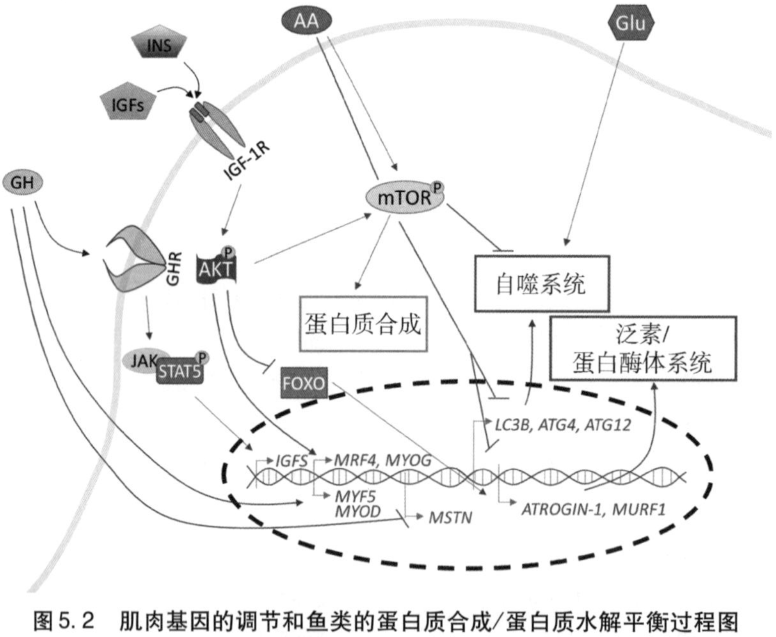

图 5.2 肌肉基因的调节和鱼类的蛋白质合成/蛋白质水解平衡过程图

注：氨基酸（AAs）下调自噬相关基因的表达并增加雷帕霉素机制性靶标（mTOR）的活性，从而抑制自噬激活。同时，胰岛素（INS）和胰岛素样生长因子（IGFs）可以结合 IGF-1 受体（IGF1R），磷酸化 AKT 导致 mTOR 活性增加。磷酸化的 AKT 也可能抑制 FOXO 导致泛素/蛋白酶体系统基因的下调。生长激素（GH）与 GHR 结合并激活 JAK / STAT5 通道后，会导致 IGF-1 表达的上调。调节基因是肌源性分化蛋白（*MYOD*）、胰岛素样生长因子（*IGFs*）、肌源性调节因子 4（*MRF*4）、肌细胞生成素（*MYOG*）、肌源性因子 5（*MYF*5）、肌肉生长抑制素（*MSTN*）、微管相关蛋白 *LC3B*、自噬相关基因 4 和 12（*ATG*4 和 *ATG*12）、*ATROGIN*-1 和肌肉 *RING*-指蛋白-1（*MURF*1）。

在哺乳动物中，肌肉生长抑制素是一种属于 TGF-β 超家族的蛋白质，是肌肉生长的强抑制剂，其基因突变会导致一些物种出现肥大（Kambadur 等，1997；McPherron 等，1997；Mosher 等，2007）。硬骨鱼类具有多达 4 种肌肉生长抑制素的同源物（Rescan 等，2001；Roberts，Goetz，2001；Rodgers 等，2007）。鱼类中肌肉生长抑制素的功能类似于哺乳动物，因为该蛋白质敲除后，会导致斑马鱼体重显著增加（Acosta 等，2005；Lee 等，2009）。此外，减少肌肉生长抑制素的蛋白表达不会增加斑马鱼和青鳉（*Oryzias latipes*）的体重，但会导致其肌肉增生（Xu 等，2003；Sawatari 等，2010）。类似地，一种肌肉生长抑制素的拮抗剂，促滤泡素抑制素的过表达，会增加转基因虹鳟（*Oncorhynchus mykiss*）的肌肉增生（Medeiros 等，2009），这都证明肌肉生长抑制素在肌肉生长调节中发挥着关键作用。肌肉生长抑制素抑制生长和细胞增殖（Seiliez 等，2012b）并和鱼类蛋白质降解的刺激（Seiliez 等，2013b）有关。如同在哺乳动物中，肌肉生长抑制素在鱼类中的作用方式是通过磷酸化的 SMAD3（Seiliez 等，2012b；Fuentes 等，2013a）和 SMAD2（Seiliez 等，2013b），和激活素型受体结合，激活细胞内信号串联 SMAD，导致抑制 AKT / mTOR 通路（Seiliez 等，2013b）。

AKT /mTOR 途径在调节哺乳动物的蛋白质合成/蛋白质水解平衡中起重要作用（Laplante，Sabatini，2009），鱼类也是这种情况，它对于肌肉生长必不可少（Seiliez 等，2008a，2012a）。实际上，经历磷酸化后，雷帕霉素（mTOR）蛋白的机制性靶标激活蛋白质合成并抑制自噬，导致肌肉组织中蛋白质的沉积增加（图 5. 2）。最近的研究表明，能量底物在调节鱼类肌肉细胞的 mTOR 活性中起作用。例如，在虹鳟鱼和鲷鱼（*Sparus aurata L.*）的培养肌细胞中，氨基酸（AAs）中度缺失会减弱 mTOR 活性（Seiliezet 等，2012a；Vélez 等，2014），导致 LC3B2 / LC3B1 比率（一种常用的自噬标志物）增加（Seiliez 等，2012a；Yabu 等，2012），而氨基酸的补充通过 mTOR 非依赖性途径会减少自噬相关基因的表达（Seiliez 等，2012a）。然而，与哺乳动物不同，葡萄糖未能对离体的鳟鱼和鲷鱼肌细胞中的 mTOR 及其下游途径产生任何影响（Belghit 等，2013；Vélez 等，2014）；但是，葡萄糖在这些细胞中显示出不依赖于 mTOR 的自噬激活（Belghit 等，2013）。这些结果表明血浆营养水平也可能是肌肉细胞中蛋白质平衡的关键调节因子（图 5. 2）。

肌肉蛋白质合成和蛋白质水解也受磷酸肌醇 3-激酶/ AKT（PI3K / AKT）信号通道的调节（Rommel 等，2001）。该通道抑制与蛋白质水解相关的基因表达，并增加与增生相关的 MRF 基因的表达。PI3K / AKT 通道的激活导致叉头框类 O（FOXO）转录因子的磷酸化和失活，降低泛素连接酶的表达（Mammucari 等，2008）。与哺乳动物一样，PI3K / AKT 通道也可通过磷酸化鱼类的 mTOR 而起作用（Fuentes 等，2013b）。例如，可以通过对虹鳟进行再投喂（Seiliez 等，2008b，2013a；Belghit 等，2013）或改变其饮食成分来调节 PI3K / AKT 通道（Seiliez 等，2011a）。通过 PI3K/ AKT 途径 FOXO 的磷酸化也观察到虹鳟 ATROGIN-1 和 MOUR1 的相关下调（Cleveland，Weber，2010；Seiliez 等，2010）。由于 IGF-1 和胰岛素是鱼类 PI3K /

AKT 途径的强激活剂（Castillo 等，2006；Montserrat 等，2007；Cleveland，Weber，2010；Seiliez 等，2010；Vélez 等，2014），这些激素的生长促进作用主要是由于蛋白质合成的增强以及对蛋白质水解的抑制（McCormick 等，1992；Wood 等，2005；Reindl，Sheridan，2012；图 5.2）。

除胰岛素和 IGF-1 外，GH 亦可通过上调 IGF-1 表达直接或间接地在肌肉生长中发挥作用。GH 已显示可增强鲷鱼培养的肌细胞中 MRF（MYOD2，MRF5）和 IGF-1 的表达（Fuentes 等，2013b）。GH 对 IGF-1 表达的上调是由于 GH 受体（GHR）激活，从而导致 JAK2-STAT5 信号通路激活的结果（Fuentes 等，2011）。最近的研究还发现，GH 处理对鳟鱼肌细胞中肌生长因子，MSTN1A 和 MSTN1B 的表达具有负面影响（Seiliez 等，2011b），表明可能还存在着其他控制因子调节肌肉生长的 GH 信号传导。应激以及 HPI 轴的相关激活导致血液循环中的皮质醇浓度升高，会干扰许多这方面的机制，从而影响生长。我们提出，应激对鱼类生长抑制（肌肉结构）的影响涉及 HPI 轴的两个方面作用：（1）减少生长能量的分配；（2）肌肉形成促进因子的下调（图 5.3）。

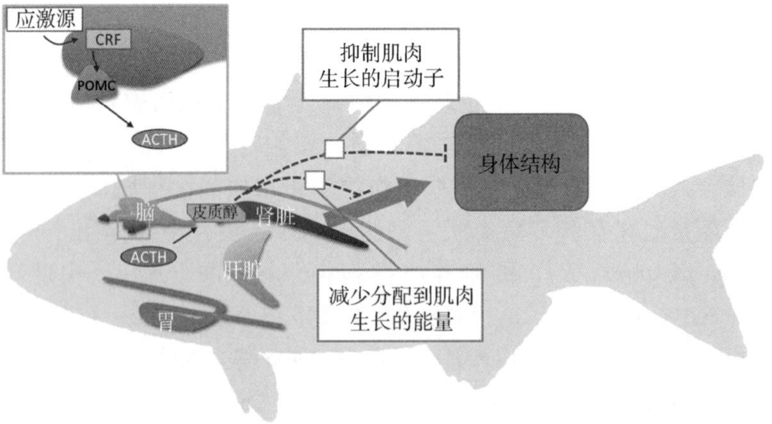

图 5.3 应激对生长影响的模型图

注：面临应激源时，下丘脑-垂体-肾间组织轴（HPI）被激活，导致下丘脑中促肾上腺皮质激素释放因子（CRF）的产生，它刺激垂体前叶合成阿黑皮素原（POMC）。被翻译的 POMC 修饰为具生物活性的促肾上腺皮质激素（ACTH），它刺激位于头肾的肾间组织释放皮质醇。皮质醇通过减少用于肌肉生长的能量分配和抑制肌肉生长的促进因子来影响肌肉结构。

5.3 应激对生长可利用能量的影响

Faught 及其同事报道了大量应激及其相关的 HPI 轴激活导致皮质醇分泌的研究（2016 年；本书第 4 章）。简而言之，应激源诱导 HPI 轴的活化，下丘脑分泌促肾上腺皮质激素释放因子（CRF），CRF 会通过垂体前叶进行翻译后修饰来刺激阿黑皮素原（POMC）的合成和促肾上腺皮质激素（ACTH）的分泌。ACTH 反过来会刺激主要位于硬骨鱼头肾区域的类固醇生成细胞中皮质醇的生物合成（Faught 等，2016；本书第 4

章)。本章中我们关注的是应激过程中，皮质醇在影响鱼类肌肉生长方面的作用。

肌肉结构的增长在很大程度上取决于能量底物到达最终的结构部分。这种分配取决于进入鱼类的总体营养量，以及其他身体功能所需的能量，包括 DEB 模型所描述的维持能、成熟能和繁殖能。动物暴露于应激源及其相关的血液循环皮质醇浓度升高可以通过各种机制改变这些能量流，包括减少食物摄取、限制肠道对食物的吸收以及增加维持过程的能量分配，这些都会导致结构部分所分配能量的减少（图 5.4）。

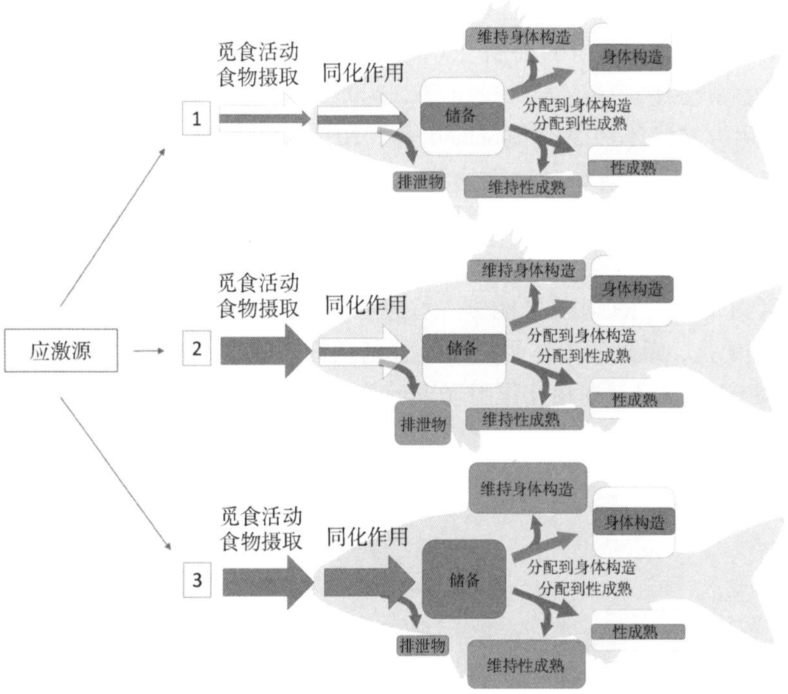

图 5.4　应激影响结构的总体资源分配的拟建模型图

注：应激源减少生长可用能量的三个主要影响方面如下：(1) 由于觅食活动减少而导致能量供应减少；狭窄箭头表示食物摄入和同化的减少，小蓝框表示总体能量储备的减少，导致成熟（较小橙框）和结构（较小蓝框）所分配能量的减少。(2) 由于食物同化减少，导致能量供应减少；食物摄入由正常蓝色箭头表示，但是由窄蓝色箭头表示的同化部分减少，导致粪便中的能量损失增加（较大橙框）；这也降低了总体能量储备（小蓝框），导致成熟度（较小橙框）和结构（较小蓝框）的能量分配减少。(3) 维护能的需求增加；食物摄入和同化由正常蓝色箭头表示，导致总体能量储备正常（蓝色框），但维持能（较大橙框）分配更多时也限制了成熟（较小橙框）和结构（较小蓝框）的能量分配。（见书后彩图）

5.3.1　食物摄取

应激对鱼类的一个基本影响就是食欲不振。例如，大西洋鲑（*Salmo salar*）暴露于应激源 17 天后，其食物消耗减少了 62%（McCormick 等，1998），虹鳟在缺氧时食物消耗减少了 50%（Bernier，Craig，2005）。应对应激导致的食物摄取量减少会降低鱼类储备的能量，有限的可用能量额会优先于其他功能而分配给维护，导致生长减少

[图 5.4（1）]。已有几项研究试图探讨摄食减少的机制问题，包括对神经肽和肠肽的作用，但其机制至今尚不清楚。

食物摄取由下丘脑的食欲中枢和厌食中枢调控，分别具有刺激和抑制食欲和摄食行为的作用（Volkoff 等，2005）。鱼的脑部含有许多关键分子，包括食欲激素神经肽 Y（NPY），以及 HPI 轴中抑制鱼类食欲的两种分子 CRF 和 POMC。脑室内注射 NPY 已证明可增加金鱼、斑点鮰（*Ictalurus punctatus*）和虹鳟的食物消耗（López-Patiño 等，1999；Silverstein，Plisetskaya，2000；Aldegunde，Mancebo，2006）。相反，注射 NPY 抑制剂可以减少食物摄入量（Aldegunde，Mancebo，2006）。对哺乳动物的研究表明，NPY 的产生受到瘦素的控制（Wang 等，1997）。在鱼类中，瘦素最近被克隆并显示主要在肝脏中产生（Kurokawa 等，2005；Huising 等，2006）。然而，瘦素的调节及其作用方式尚不清楚（Copeland 等，2011；Gorissen，Flik，2014）。实际上，中枢或腹腔注射同源瘦蛋白对动物的食物摄入具有强烈的食欲抑制作用（Murashita 等，2008；Li 等，2010；Gonget 等，2015）。瘦素对摄食量的影响可能涉及 NPY 的下调（图 5.5），以及鱼类中 POMC 的上调（Murashita 等，2008，2011；Li 等，2010）。另外，瘦素对 POMC 的刺激可能涉及鱼类中 PI3K / AKT 通道的激活（Gong 等，2015）。有趣的是，最近的离体研究还通过刺激肝细胞 IGF-1 和 GHRs 的基因表达证明了 GH-IGF 轴与瘦素存在相互作用（Won 等，2016）。

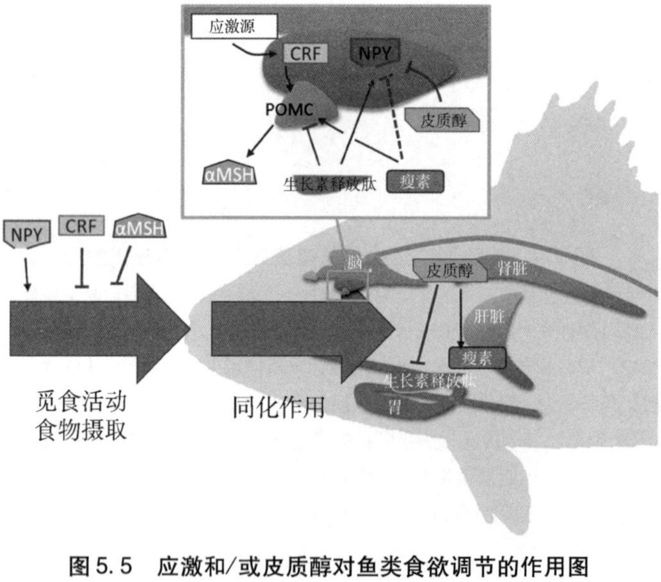

图 5.5 应激和/或皮质醇对鱼类食欲调节的作用图

注：神经肽 Y（NPY）是一种强烈的食欲激活剂。下丘脑的促肾上腺皮质激素释放因子（CRF）和 α-黑素细胞刺激激素（αMSH）（由来自垂体的 POMC 调控生成），都是应激源激活 HPI 轴而释放的食欲抑制分子。HPI 轴激活后产生的皮质醇下调生长素释放肽（一种抑制 POMC 同时促进 NPY 分泌的促食欲激素）的产生。皮质醇还可刺激肝脏中瘦素的产生。瘦素被认为是食欲抑制激素并且可以刺激 HPI 轴，可能还下调 NPY 的作用。

与哺乳动物一样，瘦素已被确定为鱼类应激轴的调节因子（Roubos 等，2012；Gorissen，Flik，2014）。最近的一项研究表明，慢性皮质醇给药会增加虹鳟肝脏瘦素的基因表达（Madison 等，2015）。在离体肝细胞的原代培养物中也观察到这种效应，并且通过添加 GR 抑制剂 RU486 可以阻断该作用，这也支持皮质醇介导 GR 信号传导的说法（Madison 等，2015）。此外，病原体感染引起的慢性应激会减少摄食量和上调下丘脑 POMC 转录水平，继而提高鱼体内血浆瘦素水平（MacDonald 等，2014）。

鱼类胃和肠道会合成生长素释放肽（Kaiya 等，2003），它是鱼类中属于强烈的促食欲激素（Unniappan 等，2004；Shepherdet 等，2007；Penney，Volkoff，2014；Tinoco 等，2014）。这种促食欲效应和在哺乳动物中的研究相一致，并且是通过刺激金鱼下丘脑（Miura 等，2006）和虹鳟（Velasco 等，2016）中的 NPY 表达（图 5.5）来介导。中枢注射生长素释放肽也会降低虹鳟 POMC 的基因表达（Velasco 等，2016；图 5.5）。然而，生长素释放肽对食物摄取和 NPY 表达的影响在鱼类中并不一致。研究表明，生长素释放肽还可以对食物摄入产生抑制作用（Jönsson 等，2010；Schroeter 等，2015），这被认为是由于虹鳟 CRF 表达增加而间接介导所致（Jönsson 等，2010；Jönsson，2013）。监禁应激会导致血浆皮质醇水平升高，这对幼年鲑鱼的血浆生长素释放肽水平没有影响或抑制作用，但对食物摄入往往具有抑制作用（Pankhurst 等，2008a，b）。最近对罗非鱼（*Oreochromis mossambicus*）进行皮质醇注射的研究显示，它会导致血浆生长素释放肽的水平降低，这和胃部生长素释放肽基因表达的下调有关（Janzen 等，2012）。总体而言，应激可以调节血浆生长素释放肽水平，这可能是由皮质醇介导的，但在鱼类中和停止进食相关的分子机制尚不清楚。

在鱼类中已鉴定出几种厌食性多肽（Volkoff 等，2005），包括 HPI 轴的一些关键激素。实际上，与假注射的对照组相比，注射 CRF 会大大减少鱼类的食物摄入量（Bernier，Peter，2001）。处于低氧条件下的鱼类注射 CRF 受体拮抗剂后，可以降低其食物摄入的抑制作用，这表明在应激条件下，CRF 在食物摄入调节过程中具有核心作用（Bernier，Craig，2005）。另外，POMC 蛋白裂解后产生的 α-MSH 会抑制鱼类的食欲（Cerdá-Reverter 等，2003）。因此，应激和 HPI 轴的激活将通过直接产生厌食性多肽或者间接通过皮质醇的作用来减少食物摄入量（图 5.5），而这将限制生长所需的能量［图 5.4（1）］。

5.3.2　肠道能量底物的吸收

一旦摄入食物后，胃肠道便开始消化和吸收营养素，为身体储备可用的能量。消化道结构和功能特征可通过环境变化和内部因素（包括鱼类的繁殖）而得到明显的改变（Habibi，Ince，1984；Pankhurst，Sorensen，1984；Madsen 等，2015）。除了减少食物摄入量外，还可以通过降低鱼类吸收营养的能力来抑制其消化功能［图 5.4（2）］。鱼类肠道结构和微生物群体的分析结果表明，15 分钟的急性应激可以暂时影响鲑鱼的肠道超微结构及其功能（Olsen 等，2002，2005）。即使摄食量正常，应激也会抑制尼罗罗非鱼（*Oreochromis niloticus*）胃酸的分泌，从而降低其消化效率

(Moriarty, 1973)。在盐度驯化期间，皮质醇会增加前肠的上皮周转（Takahashi 等，2006b），这有利于鱼类应对盐度环境。皮质醇水平也已证实可以改变肠液的转运（Veillette 等，1995）。类似地，鱼肠中糖皮质激素受体（GR）转录物的存在也证实了皮质醇的直接作用（Takahashi 等，2006a）。鳟鱼慢性皮质醇摄入对肠道结构的影响以及相关的消化效率降低也证实了该结论（Barton 等，1987）。这些结果都支持皮质醇应激轴的激活和鱼类应激期间营养吸收受损之间可能存在联系。但遗憾的是，现在仍然不清楚皮质醇影响鱼类肠道功能的分子机制。因此，应激可以影响鱼类消化系统的结构和功能，从而减少食物能量的同化［图 5.4（2）］。

5.3.3 维持能的需求

暴露于应激源而导致的稳态重建涉及能量需求过程的激活，包括离子调节、代谢底物可用性以及应激蛋白的运输和合成。所有这些活动都会转化为增加 DEB 模型中结构和成熟度所分配的维持能［图 5.4（3）］。这些能量消耗限制了结构的产生，但同时增加了对应激源进行应对和存活的机会。

应激情形下增加的维持能一般是很难进行测量的，但呼吸、通气和心血管活动的变化，甚至新陈代谢率的变化都是一些较好的观测指标。不同类型的应激源，包括社会性应激源，会增加鱼类的氧气消耗（Barton，Schreck，1987；Sloman 等，2000；Sadoul 等，2015a），这与心呼吸系统的亢奋相吻合（Laitinen，Valtonen，1994）。同样，外源性皮质醇给药也可导致代谢率的增加，这表明应激诱导的血浆皮质醇水平升高可能在提高动物的代谢率方面发挥作用（Morgan，Iwama，1996）。此外，还有研究表明应激和/或皮质醇刺激会增加鱼的代谢范围（Mommsen 等，1999；Aluru，Vijayan，2007，2009），这也支持了上述说法。皮质醇刺激的关键代谢反应之一是肝脏代谢能力的提高，包括 AA 分解代谢的增强和 C3 底物引导的氧化和糖合成（Mommsen 等，1999；Aluru，Vijayan，2007，2009），而 GR 拮抗剂米非司酮存在时这些反应停止（Aluru，Vijayan，2007）。这些代谢变化以及能量需求的相关增加对于动物对付应激是必不可少的。然而，稳态恢复所需能量的急剧增加可能是以降低生长能量的可用性为代价的［图 5.4（3）］。

在细胞水平上，从存活途径到细胞凋亡和取代，这一系列反应的激活都会增加其维持的成本。所有这些机制都是耗能的，因为它们涉及新蛋白质的合成、细胞成分的替代，或者组织水平上的细胞替代（Fulda 等，2010）。热休克蛋白（HSPs）是一种蛋白质家族的统称，其对于应激的和非应激的正常细胞功能十分重要（Schreck，Tort，2016；本书第 1 章），这在鱼类综述中得到了广泛的讨论（Iwama 等，1998，2004；Deane，Woo，2010）。HSP 在细胞中具有保护作用，包括螯合功能，维持蛋白质稳定以应对细胞应激，并且是应激类固醇信号通路的必要组分（Pratt，1993；Vijayan 等，2005；Deane，Woo，2010）。HSP 的合成，特别是处于应激期间，其能量耗费巨大但同时也是必不可少的，并且可能以其他能量的需求为代价而发生。例如，在鳟鱼肝细胞中，热休克诱导的 HSPs 家族 70kDa（千道尔顿）蛋白的表达上调和皮质醇介导的葡萄糖生成能力降低相对应（Boone 等，2002），在该能量需求中，每产生 1 摩尔葡萄糖便要至少消耗 10 分子 ATP。因此，应激源介导的蛋白质稳态变化（对保护至关重要）显得十分耗能，并且发生在其他能量的需求途径中，包括生长。

细胞蛋白质稳态也受蛋白质分解代谢增加的影响。应激和/或皮质醇刺激和参与分解代谢途径相关蛋白质的基因上调有关，例如组织蛋白酶 D、其他的组织蛋白酶和转氨酶（Mommsen 等，1999；Aluru，Vijayan，2007，2009）。面对营养性应激源时，鱼类可通过增加自噬和 UPS 对营养性应激源的反应来加速蛋白质的分解代谢（Seiliez 等，2008b；Yabu 等，2012；Belghit 等，2013，2014）。类似地，手工操作应激也可增加智利羽鼬鳚（*Genypterus chilensis*）（Aedo 等，2015）中自噬相关基因（*ATG*9，*ATG*5，*ATG*161，*ATG*4，*ATG*7 和 *LC*3*Bi*）的上调。相同物种的培养性肌细胞也显示出其自噬相关基因的上调和皮质醇有关，而这与更高水平的泛素化蛋白相关（Aedo 等，2015）。因此，为了应对皮质醇而上调蛋白质水解过程，并且导致肌细胞体积减小（Aedo 等，2015）。值得注意的是，这些效应在 RU486 预处理的细胞中受到抑制，这表明皮质醇在蛋白质水解调节中具有直接的基因组作用（Aedo 等，2015）。另外，由于应激源的强度和持续时间，会激活细胞死亡（包括细胞凋亡）途径。在鱼类中，这些效应主要发生在接触毒物的过程中（Krumschnabel 等，2005；Wu 等，2015；Morcillo 等，2016），但最近的转录组分析也显示由于手工操作应激而导致凋亡相关的基因表达增加（Du 等，2014）。这些研究都支持该观点，即应激诱导的皮质醇水平升高会增加肌细胞蛋白质降解和/或细胞死亡。总之，从个体水平到细胞水平，一系列维持机制被激活以应对应激源，都可能是由皮质醇信号介导的。由于这些通路的激活需要能量，因此，维持能量的重新分配将会减少分配给生长的能量［图 5.4（3）］。

5.4　应激对肌肉形成的启动子影响

令人惊讶的是，很少研究会涉及应激和/或皮质醇刺激对鱼类肌肉功能的影响。尽管大多数研究都是以肝脏作为应激适应和皮质醇作用的代谢目标，但其中研究肌肉的颇少，且其研究主要是描述性的（Milligan，1997；Mommsen 等，1999；Milligan 等，2000）。基于肝脏中的反应，研究已经提供皮质醇刺激对肌肉功能的可能影响，特别是应激期间肌肉作为皮质醇介导蛋白质水解的目标（Mommsen 等，1999；Aluru，Vijayan，2009）。综上所述，我们可以有把握地设想应激和/或皮质醇作用将以多种方式影响肌肉功能，包括蛋白质合成和蛋白质分解途径，以及通过降低肌原性基因的表达和减少 GH-IGF 轴的作用来影响肌肉的形成。了解潜在的分子机制对于确立应激和生长之间的因果关系至关重要。

5.4.1 应激对于肌细胞生长的影响

皮质醇作用的分子机制已在本书中详细描述（Faught 等，2016；本书第 4 章）。简单来说，皮质醇的基因组作用包括通过和糖皮质反应元件（GRE）的相互作用来结合 GR 和激活靶基因。在哺乳动物模型中，皮质醇降低了肌细胞生成蛋白基因 *MYOD* 和 *MYF5* 的表达，这表明应激类固醇激素具有下调肌细胞生长的作用（Pandurangan 等，2014）。在鱼类中，关于应激对 MRFs 影响的资料较少。最近，手工操作应激降低了智利羽鼬鳚中 MYOD1 的表达，而在体外与皮质醇一起孵育的肌细胞未显示 MYOD 转录水平上的任何变化（Aedo 等，2015）。*MSTN* 基因启动子区域的分析结果显示在几种鱼类中存在 GRE（Garikipati 等，2006；Funkenstein 等，2009；De Santis，Jerry，2011），表明皮质醇可能直接调节肌肉生长。为此，GR 的击倒降低了斑马鱼中的 *MSTN* 表达（Nesan 等，2012）。然而，应激源或皮质醇给药对 *MSTN* 水平的影响在不同研究或者不同物种中并不一致（表 5.3）。例如，尽管虹鳟 *MSTN* 基因缺乏 GRE（Garikipati 等，2006，2007），但最近的一项研究显示，皮质醇处理离体的鳟鱼成肌细胞后 *MSTN1B* 转录水平会增加（Galt 等，2014），表明皮质醇对 *MSTN* 表达的影响可能是 GR 不依赖性的。皮质醇刺激导致 *MSTN* 表达的过程是否涉及鱼类的盐皮质激素受体激活和/或非基因组信号通路仍有待观察（皮质醇信号的综述参见 Faught 等，2016；本书第 4 章）。

表 5.3　应激源或糖皮质激素处理对鱼类肌肉生长抑制素（*MSTN1* or *MSTN2*）表达的影响

［下调（－），上调（＋），或者无显著差异（NS）］

鱼类品种	应激源或糖皮质激素处理	组织	与对照组相比，肌肉生长抑制素表达水平		参考文献
			MSTN1	MSTN2	
斑马鱼	高密度养殖	肌肉	－		Vianello 等（2003）
罗非鱼	禁食 3 天	全身	－		
	禁食 6 天	全身	－		
	禁食 3 天（试验 2）	全身	＋		
	禁食 9 天（试验 2）	全身	－		Rodgers 等（2003）
	皮质醇注射 3 小时或 6 小时	全身	－		
	禁食 28 天	白肌	NS		
鲶鱼	禁食 1 天	白肌	－		
	禁食 14 天和 28 天	白肌	NS		
	低温处理 1 天	白肌	－		Weber，Bosworth（2005）
	低温处理 14 天和 28 天	白肌	＋		

续表5.3

鱼类品种	应激源或糖皮质激素处理	组织	与对照组相比，肌肉生长抑制素表达水平		参考文献
			MSTN1	MSTN2	
鲶鱼	地塞米松注射后12小时	白肌	−		
	地塞米松注射后24小时	白肌	NS		Weber等（2005）
虹鳟	饥饿30天	白肌	−	NS	Johansen，Overturf（2006）
斑马鱼	高密度暂养3天	肌肉	NS	NS	Helterline等（2007）
		大脑	NS	NS	
		脾脏	+	+	
尖吻鲈	禁食30天	白肌	+	NS	De Santis，Jerry（2011）
		肝脏	+	+	
		大脑	−	NS	
		鳃	−	+	
虹鳟	皮质醇注射12小时或24小时后	红肌	NS	+（2A）	Galt等（2014）
		白肌	NS	NS	
	皮质醇孵育24小时	培养的成肌细胞	+（1B）	NS	

总体而言，皮质醇对肌肉生长调节的影响似乎和几个因素有关，包括物种差异、发育阶段和动物的营养状况，这还需要在鱼类中得到很好的验证。皮质醇除了对肌肉功能的影响外，这种类固醇可能还会调节肌肉生长过程中涉及的营养素、生长因子和激素的水平，从而间接影响生长（Mommsen，2001）。

5.4.2 GH-IGF轴的应激调控

GH主要由脑垂体分泌，并由下丘脑的多肽、神经递质和激素所调控（Björnsson等，2002；Wong等，2006）。在包括鱼类在内的脊椎动物中，GH的分泌由一系列复杂的调控机制所控制，其中涉及多种GH释放因子，包括多巴胺、生长素释放肽、胆囊收缩素和促性腺激素释放激素，还有几种生长抑素（SSTs）和其他抑制剂，如5-羟色胺或去甲肾上腺素（Wong等，1993；Kaiya等，2003；Gahete等，2009；

Sheridan, Hagemeister, 2010)。在哺乳动物中, GH 分泌也被证明是在糖皮质激素的控制下 (Mazziotti, Giustina, 2013), 但还缺乏类似于鱼类机制的直接证据。急性应激可调节鱼体内 GH 的血液循环水平, 表明鱼类的应激轴和 GH 轴之间存在相互作用 (Deane, Woo, 2008)。监禁应激、温度的突然变化或细菌感染会降低鱼类的 GH 水平或 GH mRNA 水平 (Pickering 等, 1991; Auperin 等, 1997; Rotllant 等, 2000, 2001; Deane, Woo, 2006)。在鱼类中, 这些反应是否与 SST 分泌的增加、生长激素释放激素 (GHRH) 分泌的减少相关, 或者是一个不同的作用通路, 都仍有待研究。有趣的是, 社会中处于从属地位的伯氏朴丽鱼 (*Haplochromis burtoni*) 体内皮质醇水平较高 (Fox 等, 1997), 同时它们具有较大的含有生长抑素的神经元 (Hofmann, Fernald, 2000), 这种情况支持皮质醇对 SST 分泌具有正调节作用的说法 (图 5.6)。据我们所知, 应激对于鱼类 GHRH 影响的相关资料还知之甚少。因此, 虽然有证据表明应激会影响 GH-IGF 轴, 并且这可能涉及皮质醇的刺激作用, 但其中相关的机制有待于在鱼类中阐明。

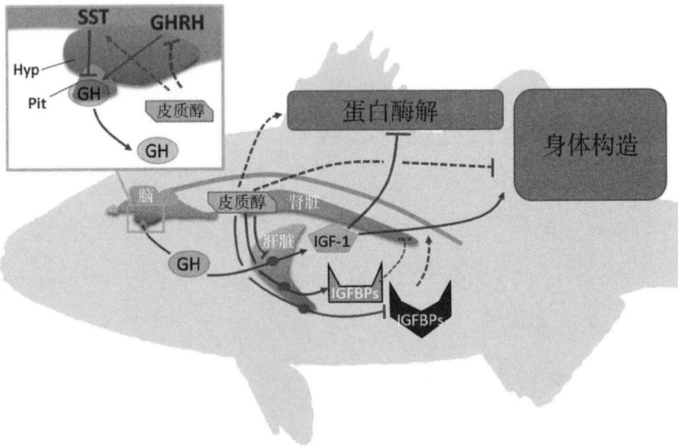

图 5.6 皮质醇对鱼类 GH-IGF 轴的影响

注: 生长抑素 (SST) 和生长激素释放激素 (GHRH) 分别抑制和上调垂体 (Pit) 中生长激素 (GH) 的表达。然后 GH 在循环系统中释放并主要在肝脏中激活胰岛素样生长因子 (IGF) 的产生。IGF 会抑制蛋白质水解并增强结构的产生。IGF 结合蛋白 (IGFBPs) 调节 IGFs 活性, 小 IGFBPs (橙色) 抑制 IGF 作用, 大 IGFBPs (黑色) 增加 IGF 活性。皮质醇通过抑制 GHRH 分泌和增加 SST 的产生从而与 GH-IGF 轴相互作用。皮质醇还抑制肝脏中的 IGF-1 表达并调节 IGFBP 表达 (抑制大 IGFBP 的分泌, 但促进小 IGFBP 的分泌)。实线表示已知的关系, 而虚线表示拟提出的关系。(见书后彩图)

GH 合成和储存后, 通过与 GH 受体相互作用而释放到血浆中, 从而在整个身体, 主要是在鱼类的肝脏和肌肉中发挥作用 (Jiao 等, 2006)。与 GH 受体结合后, GH 激活 IGF-1 的产生, 它主要由肝细胞和部分的肌细胞分泌, 这可以通过转录组丰度的增加以及该生长因子的血液循环水平来证明 (Reindl, Sheridan, 2012)。如第 5.2.2.3 节所述, 升高的血浆 IGF-1 水平可增加肌肉生长。另外, 一些研究表明, 糖皮质激素可以影响 IGF 对 GH 的敏感性以及 IGF-1 的产生。例如, 罗非鱼腹腔注射皮质醇会降低肝脏中 IGF-1 的表达及其血浆水平而不影响血浆的 GH 水平（Kajimura 等，2003）。同样，在鲑鱼（Pierce 等，2005）和鳟鱼（Philip，Vijayan，2015）的肝细胞中，地塞米松或皮质醇会抑制 GH 介导的 IGF-1 上调表达。IGF 对 GH 的敏感性降低可能与皮质醇刺激鱼类肝脏中 GH 受体表达的降低有关（Small 等，2006）。在体外，皮质醇会增加细胞因子信号转导抑制因子（SOCS），如 SOCS-1 和 SOCS-2 的表达（Philip 和 Vijayan，2015），这两种基因以其对 JAK / STAT 通道的抑制作用而闻名，该通道是导致 IGF-1 产生的主要 GH 信号通道（图 5.2）。在分离的肝细胞中，皮质醇直接降低 IGF-1 的表达，表明皮质醇可以在鱼类肝细胞中调节 IGF-1 的产生而不依赖于 GH（Leung 等，2008）。总之，这些结果着重说明皮质醇在应激过程中通过调节 GH-IGF 轴直接或间接影响鱼类肌肉生长的关键作用（图 5.6）。

IGF 对肌肉功能的活性取决于高亲和力的 IGFBPs。在哺乳动物中，现已发现六种不同大小的 IGFBPs，它们可调节 IGFs 的分布和稳定性以及和 IGF 受体的结合力（Rosenzweig，2004）。在不同的实验条件下，IGFBP 可增强或者抑制 IGFs 的作用，并且 IGFBP 的产生受糖皮质激素的调节（Conover 等，1993；Okazaki 等，1994；Gabbitas，Canalis，1996；Pereira 等，1999）。例如，IGFBP-1 高度磷酸化或过量时可抑制 IGFs 的作用（Cox 等，1994；Yu 等，1998），而在低浓度或去磷酸化时，IGFBP-1 增加 IGFs 活性（Yu 等，1998）。针对哺乳动物的研究还表明，注射皮质醇会增加体内 IGFBP-1 的产生（Conover 等，1993），并且增加 IGFBP-1 mRNA 在离体成骨样细胞中的表达（Okazaki 等，1994）。类似地，IGFBP-3 在 IGF 作用的调节中起重要作用，因为它将 IGFs 运输到靶组织并且将其半衰期从几分钟增加到 12 小时。在体外，地塞米松会降低大鼠 IGFBP-3 的表达（Villafuerte 等，1995）。总而言之，IGFBPs 可能在 IGF 的作用及其有效性中发挥关键作用（图 5.6），并且可能是应激期间皮质醇作用的靶标，但在硬骨鱼类中这方面的研究还很少。

在鱼类中，由于三轮的基因组复制和鲑鳟鱼类中的四轮基因组复制，研究分别报道了 9～11 种和 19 种不同的 IGFBPs（Daza 等，2011；Macqueen 等，2013）。罗非鱼注射皮质醇后，血浆中四种 IGFBPs 的水平迅速增加（2 小时内），而两种 IGFBPs 减少（Kajimura 等，2003）。该反应取决于蛋白质的大小，其中 24～32 kDa 的 IGFBPs 显示为增加，而≥40kDa 的 IGFBPs 显示为减少（Kajimura 等，2003）。类似地，监禁应激降低了相当于哺乳动物 IGFBP-3 的 IGFBP（33kDa）的血浆水平，而另外两个较小的 IGFBPs（24kDa 和 28kDa）会在阳光鲈鱼（金眼狼鲈 *Morone chrysops* 与条纹狼鲈 *Morone saxatilis* 的杂交种）中增加表达（Davis，Peterson，2006）。然而，在同一个研究中，血浆皮质醇水平的增加并不影响 IGFBPs 的水平。目前尚不清楚生长快速鱼类中较高水平的大分子 IGFBPs 和生长缓慢鱼类中较低水平的小分子 IGFBPs 是否与应激和/或皮质醇刺激有关（Shimizu 等，1999；Peterson，Small，2004）。我们得出的结论是，GH-IGF 轴受鱼类应激的调节，导致皮质醇影响 GH 分泌、IGF 对 GH 的敏感性以及 IGFBPs 的产生，而所有这些都将有助于减少生长的能量需求。

5.5 结束语和知识空白

在生长能量的分配期间，应激对众多步骤产生广泛而普遍的影响。在鱼类的各种组织水平下，HPI 轴活化与皮质醇水平的相应增加和生长相关的生理机制紧密相互作用。应激反应机制的激活是需要大量能量的，并且发生在包括生长在内的其他能量需求途径的成本中。生长在肌肉细胞中被描述为蛋白质增加的过程，这是结构蛋白质合成和结构蛋白质分解之间正平衡的结果。应激通过激活维持的过程（如蛋白质水解和应激应对机制），减少食物摄入和能量吸收来降低总体的可用能量，将这种平衡转向结构蛋白质分解而减少生长（图 5.7）。此外，应激还可以直接抑制肌细胞生成过程的分子机制和生长的激素调节，促进结构蛋白质合成转变为结构蛋白质分解代谢，从而影响生长（图 5.7）。

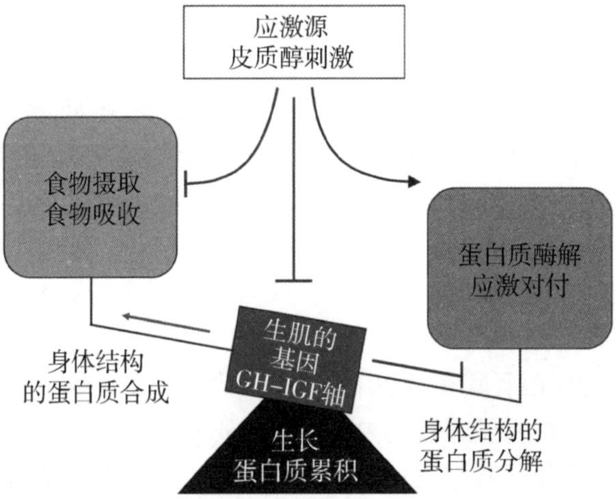

图 5.7 皮质醇影响肌肉生长的蛋白质平衡模型图

注：肌肉生长基本上是结构蛋白质增加的结果，是蛋白质合成和蛋白质分解途径之间的良好平衡。我们设想应激和皮质醇刺激会使平衡转向蛋白质分解方向，从而导致生长抑制。蛋白质合成的增加和分解的减少会导致蛋白质沉积的增加。当摄取食物、食物同化以及生长促进因子（包括 GH-IGF 轴）的刺激增加时，更多能量分配给肌肉生长，便会出现上述情况。另一方面，当平衡转向蛋白质分解时，如当发生应激时（皮质醇刺激），肌肉蛋白质水解增强，以及应激应对机制的激活，最终导致蛋白质沉积减少。皮质醇抑制机制会导致生长所分配的能量增加，而其促进机制会导致生长可利用的能量减少。

这些减少肌肉生长的机制被认为有助于动物应对应激源并在其中存活下来。在水产养殖中，生长是人们感兴趣的主要表型，所以应尽可能避免应激反应机制的激活。这可以通过严格的水质管理、限制或去除病原体以及减少物理应激因素（包括手工操作和标粗）来实现，但这可能是一个利润较低的选择（Bostock 等，2010）。在育种计划中包含一些稳定特征可能是一个更加可持续性发展的解决办法，以便选择一些抗逆性强的动物（Sadoul 等，2015a，b）。然而，要提供稳定性状的相关生物标记，还需要确定应激影响生长的潜在机制。

从本章可以清楚地看出，关于应激对肌肉生长的影响，我们的知识还存在很大不足。具体而言，我们还缺乏关于应激和/或皮质醇刺激如何导致肌肉蛋白质动态变化途径机制的见解。这里，我们确定了一些关键的知识空白：

（1）皮质醇浓度的含义和/或这种激素的作用机制是什么？促使身体平衡从良性应激（真正应激）、促进新陈代谢和生长，到非良性应激（痛苦）状态，这个过程是分解代谢和抑制生长吗？

（2）应激和/或皮质醇影响肌肉蛋白质合成和蛋白质水解的分子机制是什么？

（3）皮质醇对自噬和/或泛素/蛋白酶体系统有直接影响吗？

（4）皮质醇如何在鱼类的应激期间重新分配能量底物？SOCS 是生长能量分配的关键调节者吗？

（5）应激会影响肠道微生物群落，导致肠道功能和营养吸收的变化吗？

（6）皮质醇调节鱼类食欲的机制是什么？

（7）皮质醇在调节鱼类的肠道结构和肠道多肽过程中起作用吗？

B. 萨道尔　M. M. 维扎延　著

李水生　译

林浩然　校

参考文献

Acosta, J., Carpio, Y., Borroto, I., González, O. and Estrada, M. P. (2005). Myostatin gene silenced by RNAi show a zebrafish giant phenotype. J. Biotechnol. 119, 324-331.

Aedo, J. E., Maldonado, J., Aballai, V., Estrada, J. M., Bastias-Molina, M., Meneses, C., et al. (2015). mRNA-seq reveals skeletal muscle atrophy in response to handling stress in a marine teleost, the red cusk-eel (Genypterus chilensis). BMC Genomics 16, 1024.

Aldegunde, M. and Mancebo, M. (2006). Effects of neuropeptide Y on food intake and brain biogenic amines in the rainbow trout (Oncorhynchus mykiss). Peptides 27, 719-727.

Aluru, N. and Vijayan, M. M. (2007). Hepatic transcriptome response to glucocorticoid receptor activation in rainbow trout. Physiol. Genomics 31, 483-491.

Aluru, N. and Vijayan, M. M. (2009). Stress transcriptomics infish: a role for genomic cortisol signaling. Gen. Comp. Endocrinol. 164, 142-150.

Attaix, D., Ventadour, S., Codran, A., Béchet, D., Taillandier, D. and Combaret, L. (2005). The ubiquitin-proteasome system and skeletal muscle wasting. Essays. Biochem. 41, 173-186.

Augustine, S., Gagnaire, B., Floriani, M., Adam-Guillermin, C. and Kooijman, S.

A. L. M. (2011). Developmental energetics of zebrafish, Danio rerio. Comp. Biochem. Physiol. A Mol. Integr. Physiol. 159, 275-283.

Augustine, S., Gagnaire, B., Adam-Guillermin, C. and Kooijman, S. A. L. M. (2012). Effects of uranium on the metabolism of zebrafish, Danio rerio. Aquat. Toxicol. 118-119, 9-26.

Auperin, B., Baroiller, J. F., Ricordel, M. J., Fostier, A. and Prunet, P. (1997). Effect of confinement stress on circulating levels of growth hormone and two prolactins in freshwater-adapted tilapia (Oreochromis niloticus). Gen. Comp. Endocrinol. 108, 35-44.

Barton, B. A. (2002). Stress infishes: a diversity of responses with particular reference to changes in circulating corticosteroids. Integr. Comp. Biol. 42, 517-525.

Barton, B. A. and Schreck, C. B. (1987). Metabolic cost of acute physical stress in juvenile steelhead. Trans. Am. Fish. Soc. 116, 257-263.

Barton, B. A., Schreck, C. B. and Barton, L. D. (1987). Effects of chronic cortisol administration and daily acute stress on growth, physiological conditions, and stress responses in juvenile rainbow trout. Dis. Aquat. Organ. 2, 173-185.

Belghit, I., Panserat, S., Sadoul, B., Dias, K., Skiba-Cassy, S. and Seiliez, I. (2013). Macronutrient composition of the diet affects the feeding-mediated down regulation of autophagy in muscle of rainbow trout (O. mykiss). PLoS ONE. 8, e74308.

Belghit, I., Skiba-Cassy, S., Geurden, I., Dias, K., Surget, A., Kaushik, S., et al. (2014). Dietary methionine availability affects the main factors involved in muscle protein turnover in rainbow trout (Oncorhynchus mykiss). Br. J. Nutr. 112, 493-503.

Bentzinger, C. F., Wang, Y. X. and Rudnicki, M. A. (2012). Building muscle: molecular regulation of myogenesis. Cold Spring Harb. Perspect. Biol. 4, a008342.

Bernier, N. J. and Peter, R. E. (2001). Appetite-suppressing effects of urotensin I and corticotropin-releasing hormone in goldfish (Carassius auratus). Neuroendocrinology 73, 248-260.

Bernier, N. J. and Craig, P. M. (2005). CRF-related peptides contribute to stress response and regulation of appetite in hypoxic rainbow trout. Am. J. Physiol. -Regul. Integr. Comp. Physiol. 289, R982-R990.

Björnsson, B. T., Johansson, V., Benedet, S., Einarsdottir, I. E., Hildahl, J., Agustsson, T., et al. (2002). Growth hormone endocrinology of salmonids: regulatory mechanisms and mode of action. Fish Physiol. Biochem. 27, 227-242.

Bodiguel, X., Maury, O., Mellon-Duval, C., Roupsard, F., Le Guellec, A. -M. and Loizeau, V. (2009). A dynamic and mechanistic model of PCB bioaccumulation in the European hake (Merluccius merluccius). J. Sea Res. 62, 124-134.

Boone, A. N., Ducouret, B. and Vijayan, M. M. (2002). Glucocorticoid-induced glucose release is abolished in trout hepatocytes with elevated hsp70 content. J. Endocrinol. 172, R1-R5.

Bostock, J., McAndrew, B., Richards, R., Jauncey, K., Telfer, T., Lorenzen, K., et al. (2010). Aquaculture: global status and trends. Philos. Trans. R. Soc. Lond. B Biol. Sci. 365, 2897-2912.

Bromley, P. J., Ravier, C. and Witthames, P. R. (2000). The influence of feeding regime on sexual maturation, fecundity and atresia in first-time spawning turbot. J. Fish. Biol. 56, 264-278.

Castillo, J., Ammendrup-Johnsen, I., Codina, M., Navarro, I. and Gutiérrez, J. (2006). IGF-I and insulin receptor signal transduction in trout muscle cells. Am. J. Physiol. -Regul. Integr. Comp. Physiol. 290, R1683-R1690.

Cerdá-Reverter, J. M., Schiöth, H. B. and Peter, R. E. (2003). The central melanocortin system regulates food intake in goldfish. Regul. Pept. 115, 101-113.

Cleveland, B. M. and Weber, G. M. (2010). Effects of insulin-like growth factor-I, insulin, and leucine on protein turnover and ubiquitin ligase expression in rainbow trout primary myocytes. Am. J. Physiol. -Regul. Integr. Comp. Physiol. 298, R341-R350.

Conover, C. A., Divertie, G. D. and Lee, P. D. (1993). Cortisol increases plasma insulin-like growth factor binding protein-1 in humans. Acta Endocrinol. (Copenh). 128, 140-143.

Copeland, D. L., Duff, R. J., Liu, Q., Prokop, J. and Londraville, R. L. (2011). Leptin in teleost fishes: an argument for comparative study. Front. Physiol. 2, 26.

Cox, G. N., McDermott, M. J., Merkel, E., Stroh, C. A., Ko, S. C., Squires, C. H., et al. (1994). Recombinant human insulin-like growth factor (IGF) -binding protein-1 inhibits somatic growth stimulated by IGF-I and growth hormone in hypophysectomized rats. Endocrinology 135, 1913-1920.

Davis, K. B. and Peterson, B. C. (2006). The effect of temperature, stress, and cortisol on plasma IGF-I and IGFBPs in sunshine bass. Gen. Comp. Endocrinol. 149, 219-225.

Daza, D. O., Sundström, G., Bergqvist, C. A., Duan, C. and Larhammar, D. (2011). Evolution of the insulin-like growth factor binding protein (IGFBP) family. Endocrinology 152, 2278-2289.

Deane, E. E. and Woo, N. Y. S. (2006). Molecular cloning of growth hormone from silver sea bream: effects of abiotic and biotic stress on transcriptional and translational expression. Biochem. Biophys. Res. Commun. 342, 1077-1082.

Deane, E. E. and Woo, N. Y. S. (2008). Modulation offish growth hormone levels by salinity, temperature, pollutants and aquaculture related stress: a review. Rev. Fish Biol. Fish. 19, 97-120.

Deane, E. E. and Woo, N. Y. S. (2010). Advances and perspectives on the regulation and expression of piscine heat shock proteins. Rev. Fish Biol. Fish. 21, 153-185.

De Santis, C. and Jerry, D. R. (2011). Differential tissue-regulation of myostatin genes in the teleost fish lates calcarifer in response to fasting. Evidence for functional differentiation. Mol. Cell. Endocrinol. 335, 158-165.

DiBattista, J. D., Levesque, H. M., Moon, T. W. and Gilmour, K. M. (2006). Growth depression in socially subordinate rainbow trout Oncorhynchus mykiss: more than a fasting effect. Physiol. Biochem. Zool. 79, 675-687.

Du, F., Xu, G., Nie, Z., Xu, P. and Gu, R. (2014). Transcriptome analysis gene expression in the liver of Coilia nasus during the stress response. BMC Genomics 15, 558.

Dumas, A., France, J. and Bureau, D. (2010). Modelling growth and body composition infish nutrition: where have we been and where are we going? Aquacul. Res. 41, 161-181.

Dunham, R. A. (2011). Aquaculture and fisheries biotechnology: genetic approaches (second ed.). Wallingford, Oxfordshire, UK: Cabi.

Dupont-Nivet, M., Médale, F., Leonard, J., Le Guillou, S., Tiquet, F., Quillet, E., et al. (2009). Evidence of genotype-diet interactions in the response of rainbow trout (Oncorhynchus mykiss) clones to a diet with or without fishmeal at early growth. Aquaculture 295, 15-21.

Eichinger, M., Loizeau, V., Roupsard, F., Le Guellec, A. M. and Bacher, C. (2010). Modelling growth and bioaccumulation of polychlorinated biphenyls in common sole (Solea solea). J. Sea Res. 64, 373-385.

Einarsson, B., Birnir, B. and Sigurjsson, S. (2011). A dynamic energy budget (DEB) model for the energyusage and reproduction of the icelandic capelin (Mallotus villosus). J. Theor. Biol. 281, 1-8.

Essington, T. E., Kitchell, J. F. and Walters, C. J. (2001). The von Bertalanffy growth function, bioenergetics, and the consumption rates offish. Can. J. Fish. Aquat. Sci. 58, 2129-2138.

Fader, C. M., Sánchez, D. G., Mestre, M. B. and Colombo, M. I. (2009). TI-VAMP/VAMP7 and VAMP3/cellubrevin: two v-SNARE proteins involved in specific steps of the autophagy/multivesicular body pathways. Biochim. Biophys. Acta 1793, 1901-1916.

Faught, E., Aluru, N. and Vijayan, M. M. (2016). The Molecular Stress Response.

In Fish Physiology-Biology of Stress in Fish, Vol. 35 (eds. C. B. Schreck, L. Tort, A. P. Farrell and C. J. Brauner), San Diego, CA: Academic Press.

Filgueira, R., Chapman, J. M., Suski, C. D. andCooke, S. J. (2016). The influenceofwatershedland use cover on stream fish diversity and size-at-age of a generalistfish. Ecol. Indic. 60, 248-257.

Fox, H. E., White, S. A., Kao, M. H. F. and Fernald, R. D. (1997). Stress and dominance in a social fish. J. Neurosci. 17, 6463-6469.

Freitas, V., Cardoso, J. F. M. F., Lika, K., Peck, M. A., Campos, J., Kooijman, S. A. L. M., et al. (2010). Temperature tolerance and energetics: a dynamic energy budget-based comparison of North Atlantic marine species. Philos. Trans. R. Soc. Lond. B Biol. Sci. 365, 3553-3565.

Fuentes, E. N., Einarsdottir, I. E., Valdes, J. A., Alvarez, M., Molina, A. and Björnsson, B. T. (2011). Inherent growth hormone resistance in the skeletal muscle of the fine flounder is modulated by nutritional status and is characterized by high contents of truncated GHR, impairment in the JAK2/STAT5 signaling pathway, and low IGF-I expression. Endocrinology 153, 283-294.

Fuentes, E. N., Pino, K., Navarro, C., Delgado, I., Valdés, J. A. and Molina, A. (2013a). Transient inactivation of myostatin induces muscle hypertrophy and overcompensatory growth in zebrafish via inactivation of the SMAD signaling pathway. J. Biotechnol. 168, 295-302.

Fuentes, E. N., Safian, D., Einarsdottir, I. E., Valdés, J. A., Elorza, A. A., Molina, A., et al. (2013b). Nutritional status modulates plasma leptin, AMPK and TOR activation, and mitochondrial biogenesis: implications for cell metabolism and growth in skeletal muscle of the fine flounder. Gen. Comp. Endocrinol. 186, 172-180.

Fujiwara, M., Kendall, B. E., Nisbet, R. M. and Bennett, W. A. (2005). Analysis of size trajectory data using an energetic-based growth model. Ecology 86, 1441-1451.

Fulda, S., Gorman, A. M., Hori, O. and Samali, A. (2010). Cellular stress responses: cell survival and cell death. Int. J. Cell Biol. 2010, 214074.

Funkenstein, B., Balas, V., Rebhan, Y. and Pliatner, A. (2009). Characterization and functional analysis of the 5uflanking region of Sparus aurata myostatin-1 gene. Comp. Biochem. Physiol. A Mol. Integr. Physiol. 153, 55-62.

Gabbitas, B. and Canalis, E. (1996). Cortisol enhances the transcription of insulin-like growth factor-binding protein-6 in cultured osteoblasts. Endocrinology 137, 1687-1692.

Gahete, M. D., Durán-Prado, M., Luque, R. M., MartÍnez-Fuentes, A. J., Quintero, A., Gutiérrez-Pascual, E., et al. (2009). Understanding the multifactorial control of growth hormone release by somatotropes. Ann. N. Y. Acad. Sci. 1163, 137-153.

Galloway, T. F., Bardal, T., Kvam, S. N., Dahle, S. W., Nesse, G., Randøl, M., et al. (2006). Somite formation and expression of MyoD, myogenin and myosin in Atlantic halibut (Hippoglossus hippoglossus L.) embryos incubated at different temperatures: transient asymmetric expression of MyoD. J. Exp. Biol. 209, 2432-2441.

Galt, N. J., Froehlich, J. M., Remily, E. A., Romero, S. R. and Biga, P. R. (2014). The effects of exogenous cortisol on myostatin transcription in rainbow trout, Oncorhynchus mykiss. Comp. Biochem. Physiol. A Mol. Integr. Physiol. 175, 57-63.

Gao, D., Xu, Z., Kuang, X., Qiao, P., Liu, S., Zhang, L., et al. (2014). Molecular characterization and expression analysis of the autophagic gene Beclin 1 from the purse red common carp (Cyprinus carpio) exposed tocadmium. Comp. Biochem. Physiol. C Toxicol. Pharmacol. 160, 15-22.

GarcÍa de la serrana, D., Codina, M., Capilla, E., Jiménez-Amilburu, V., Navarro, I., Du, S.-J., et al. (2014). Characterisation and expression of myogenesis regulatory factors during in vitro myoblast development and in vivo fasting in the gilthead sea bream (Sparus aurata). Comp. Biochem. Physiol. Part A Mol. Integr. Physiol. 167, 90-99.

Garikipati, D. K., Gahr, S. A. and Rodgers, B. D. (2006). Identification, characterization, and quantitative expression analysis of rainbow trout myostatin-1a and myostatin-1b genes. J. Endocrinol. 190, 879-888.

Garikipati, D. K., Gahr, S. A., Roalson, E. H. and Rodgers, B. D. (2007). Characterization of rainbow trout myostatin-2 genes (rtMSTN-2a and-2b): genomic organization, differential expression, and pseudogenization. Endocrinology 148, 2106-2115.

Gjedrem, T. (ed.). (2005). Selection and breeding programs in aquaculture. p. 364. Dordrecht, The Netherlands: Springer.

Gong, N., Jönsson, E. and Bjornsson, B. T. (2015). Acute anorexigenic action of leptin in rainbow trout is mediated by the hypothalamic Pi3k pathway. J. Mol. Endocrinol. 15-0279.

Gorissen, M. and Flik, G. (2014). Leptin in teleostean fish, towards the origins of leptin physiology. J. Chem. Neuroanat. 61-62, 200-206.

Guderley, H. and Pörtner, H. O. (2010). Metabolic power budgeting and adaptive strategies in zoology: examples from scallops and fish. Can. J. Zool. 88, 753-763.

Gutierrez, M. G., Munafó, D. B., Berón, W. and Colombo, M. I. (2004). Rab7 is required for the normal progression of the autophagic pathway in mammalian cells. J.

Cell. Sci. 117, 2687-2697.

Habibi, H. R. and Ince, B. W. (1984). A study of androgen-stimulated l-leucine transport by the intestine of rainbow trout (Salmo gairdneri Richardson) in vitro. Comp. Biochem. Physiol. Part A Physiol. 79, 143-149.

He, C., Bartholomew, C. R., Zhou, W. and Klionsky, D. J. (2009). Assaying autophagic activity in transgenic GFP-Lc3 and GFP-Gabarap zebrafish embryos. Autophagy 5, 520-526.

Helterline, D. L. I., Garikipati, D., Stenkamp, D. L. and Rodgers, B. D. (2007). Embryonic and tissue-specific regulation of myostatin-1 and-2 gene expression in zebrafish. Gen. Comp. Endocrinol. 151, 90-97.

Hofmann, H. A. and Fernald, R. D. (2000). Social status controls somatostatin neuron size and growth. J. Neurosci. 20, 4740-4744.

Hu, Z., Zhang, J. and Zhang, Q. (2011). Expression pattern and functions of autophagy-related gene atg5 in zebrafish organogenesis. Autophagy 7, 1514-1527.

Huising, M. O., Geven, E. J. W., Kruiswijk, C. P., Nabuurs, S. B., Stolte, E. H., Spanings, F. A. T., et al. (2006). Increased leptin expression in common carp (Cyprinus carpio) after food intake but not after fasting or feeding to satiation. Endocrinology 147, 5786-5797.

Iwama, G. K., Thomas, P. T., Forsyth, R. B. and Vijayan, M. M. (1998). Heat shock protein expression infish. Rev. Fish Biol. Fish. 8, 35-56.

Iwama, G. K., Afonso, L. O. B., Todgham, A., Ackerman, P. and Nakano, K. (2004). Are hsps suitable for indicating stressed states infish? J. Exp. Biol. 207, 15-19.

Jager, T. and Selck, H. (2011). Interpreting toxicity data in a DEB framework: a case study for nonylphenol in the marine polychaete Capitella teleta. J. Sea Res. 66, 456-462.

Jager, T., Vandenbrouck, T., Baas, J., Coen, W. M. D. and Kooijman, S. A. L. M. (2009). A biology-based approach for mixture toxicity of multiple endpoints over the life cycle. Ecotoxicology 19, 351-361.

Jagoe, R. T., Lecker, S. H., Gomes, M. and Goldberg, A. L. (2002). Patterns of gene expression in atrophying skeletal muscles: response to food deprivation. FASEB J. 16, 1697-1712.

Janzen, W. J., Duncan, C. A. and Riley, L. G. (2012). Cortisol treatment reduces ghrelin signaling and food intake in tilapia, Oreochromis mossambicus. Domest. Anim. Endocrinol. 43, 251-259.

Jiao, B., Huang, X., Chan, C. B., Zhang, L., Wang, D. and Cheng, C. H. K. (2006). The coexistence of two growth hormone receptors in teleostfish and their differential signal transduction, tissue distribution and hormonal regulation of expression in seabream. J. Mol. Endocrinol. 36, 23-40.

Johansen, K. A. and Overturf, K. (2006). Alterations in expression of genes associated with muscle metabolism and growth during nutritional restrictionand refeeding in rainbow trout. Comp. Biochem. Physiol. Part B Biochem. Mol. Biol. 144, 119-127.

Johnston, I. A., Bower, N. I. and Macqueen, D. J. (2011). Growth and the regulation of myotomal muscle mass in teleostfish. J. Exp. Biol. 214, 1617-1628.

Jönsson, E. (2013). The role of ghrelin in energy balance regulation in fish. Gen. Comp. Endocrinol. 187, 79-85.

Jönsson, E., Kaiya, H. and Björnsson, B. T. (2010). Ghrelin decreases food intake in juvenile rainbow trout (Oncorhynchus mykiss) through the central anorexigenic corticotropinreleasing factor system. Gen. Comp. Endocrinol. 166, 39-46.

Jusup, M., Klanjscek, T., Matsuda, H. and Kooijman, S. A. L. M. (2011). A full lifecycle bioenergetic model for bluefin tuna. PLoS ONE 6, e21903.

Kaiya, H., Kojima, M., Hosoda, H., Moriyama, S., Takahashi, A., Kawauchi, H., et al. (2003). Peptide purification, complementary deoxyribonucleic acid (DNA) and genomic DNA cloning, and functional characterization of ghrelin in rainbow trout. Endocrinology 144, 5215-5226.

Kajimura, S., Hirano, T., Visitacion, N., Moriyama, S., Aida, K. and Grau, E. G. (2003). Dual mode of cortisol action on GH/IGF-I/IGF binding proteins in the tilapia, Oreochromis mossambicus. J. Endocrinol. 178, 91-99.

Kambadur, R., Sharma, M., Smith, T. P. L. and Bass, J. J. (1997). Mutations in myostatin (GDF8) in double-muscled Belgian blue and piedmontese cattle. Genome Res. 7, 910-915.

Kim, J. H., Jin, P., Duan, R. and Chen, E. H. (2015). Mechanisms of myoblast fusion during muscle development. Curr. Opin. Genet. Dev. 32, 162-170.

Kitchell, J. F., Stewart, D. J. and Weininger, D. (1977). Applications of a bioenergetics model to yellow perch (Percaflavescens) and walleye (Stizostedion vitreum vitreum). J. Fish. Res. Board Can. 34, 1922-1935.

Klok, C., Nordtug, T. and Tamis, J. E. (2014). Estimating the impact of petroleum substances on survival in early life stages of cod (Gadus morhua) using the dynamic energy budget theory. Mar. Environ. Res. 101, 60-68.

Kong, H. J., Moon, J. -Y., Nam, B. -H., Kim, Y. -O., Kim, W. -J., Lee, J. -H., et al. (2011). Molecular characterization of the autophagy-related gene beclin-1 from the olive flounder (Paralichthys olivaceus). Fish. Shellfish. Immunol. 31, 189-195.

Kooijman, S. A. L. M. (2000). Dynamic energy and mass budgets in biological systems. Cambridge, United Kingdom: Cambridge University Press.

Kooijman, S. A. L. M. (2010). Dynamic energy budget theory for metabolic organisation. Cambridge, United Kingdom: Cambridge University Press.

Krumschnabel, G., Manzl, C., Berger, C. and Hofer, B. (2005). Oxidative stress, mitochondrial permeability transition, and cell death in Cu-exposed trout hepatocytes. Toxicol. Appl. Pharmacol. 209, 62-73.

Kurokawa, T., Uji, S. and Suzuki, T. (2005). Identification of cDNA coding for a homologue to mammalian leptin from pufferfish, Takifugu rubripes. Peptides 26, 745-750.

Lai, K.-M. V., Gonzalez, M., Poueymirou, W. T., Kline, W. O., Na, E., Zlotchenko, E., et al. (2004). Conditional activation ofAKT in adult skeletal muscle induces rapid hypertrophy. Mol. Cell. Biol. 24, 9295-9304.

Laitinen, M. and Valtonen, T. (1994). Cardiovascular, ventilatory and total activity responses of brown trout to handling stress. J. Fish. Biol. 45, 933-942.

Landemaine, A., Rescan, P.-Y. and Gabillard, J.-C. (2014). Myomaker mediates fusion of fast myocytes in zebrafish embryos. Biochem. Biophys. Res. Commun. 451, 480-484.

Laplante, M. and Sabatini, D. M. (2009). mTOR signaling at a glance. J. Cell. Sci. 122, 3589-3594.

Lee, C.-Y., Hu, S.-Y., Gong, H.-Y., Chen, M. H.-C., Lu, J.-K. and Wu, J.-L. (2009). Suppression of myostatin with vector-based RNA interference causes a double-muscle effect in transgenic zebrafish. Biochem. Biophys. Res. Commun. 387, 766-771.

Leung, L. Y., Kwong, A. K. Y., Man, A. K. Y. and Woo, N. Y. S. (2008). Direct actions of cortisol, thyroxine and growth hormone on IGF-I mRNA expression in sea bream hepatocytes. Comp. Biochem. Physiol. A Mol. Integr. Physiol. 151, 705-710.

Li, G.-G., Liang, X.-F., Xie, Q., Li, G., Yu, Y. and Lai, K. (2010). Gene structure, recombinant expression and functional characterization of grass carp leptin. Gen. Comp. Endocrinol. 166, 117-127.

López-Patiño, M. A., Guijarro, A. I., Isorna, E., Delgado, M. J., Alonso-Bedate, M. and de Pedro, N. (1999). Neuropeptide Y has a stimulatory action on feeding behavior in goldfish (Carassius auratus). Eur. J. Pharmacol. 377, 147-153.

MacDonald, L. E., Alderman, S. L., Kramer, S., Woo, P. T. K. and Bernier, N. J. (2014). Hypoxemia-induced leptin secretion: a mechanism for the control of food intake in diseasedfish. J. Endocrinol. 221, 441-455.

Macqueen, D. J., Garcia de la Serrana, D. and Johnston, I. A. (2013). Evolution of ancient functions in the vertebrate insulin-like growth factor system uncovered by study of duplicated salmonidfish genomes. Mol. Biol. Evol. 30, 1060-1076.

Madison, B. N. , Tavakoli, S. , Kramer, S. and Bernier, N. J. (2015). Chronic cortisol and the regulation of food intake and the endocrine growth axis in rainbow trout. J. Endocrinol. 226, 103-119.

Madsen, S. S. , Weber, C. , Nielsen, A. M. , Mohiseni, M. , Bosssus, M. C. , Tipsmark, C. K. , et al. (2015). Sexual maturation and changes in water and salt transport components in the kidney and intestine of three-spined stickleback (Gasterosteus aculeatus L.). Comp. Biochem. Physiol. A Mol. Integr. Physiol. 188, 107-119.

Mammucari, C. , Schiaffino, S. and Sandri, M. (2008). Downstream of AKT: FoxO3 and mTOR in the regulation of autophagy in skeletal muscle. Autophagy 4, 524-526.

Mazziotti, G. and Giustina, A. (2013). Glucocorticoids and the regulation of growth hormone secretion. Nat. Rev. Endocrinol. 9, 265-276.

McCormick, S. D. , Kelley, K. M. , Young, G. , Nishioka, R. S. and Bern, H. A. (1992). Stimulation of coho salmon growth by insulin-like growth factor I. Gen. Comp. Endocrinol. 86, 398-406.

McCormick, S. D. , Shrimpton, J. M. , Carey, J. B. , O'Dea, M. F. , Sloan, K. E. , Moriyama, S. , et al. (1998). Repeated acute stress reduces growth rate of Atlantic salmon parr and alters plasma levels of growth hormone, insulin-like growth factor I and cortisol. Aquaculture 168, 221-235.

McEwan, D. G. , Popovic, D. , Gubas, A. , Terawaki, S. , Suzuki, H. , Stadel, D. , et al. (2015). PLEKHM1 regulates autophagosome-lysosome fusion through HOPS complex and LC3/ GABARAP proteins. Mol. Cell. 57, 39-54.

McPherron, A. C. , Lawler, A. M. and Lee, S. -J. (1997). Regulation of skeletal muscle mass in mice by a new TGF-p superfamily member. Nature 387, 83-90.

Medeiros, E. F. , Phelps, M. P. , Fuentes, F. D. and Bradley, T. M. (2009). Overexpression of follistatin in trout stimulates increased muscling. Am. J. Physiol. - Regul. Integr. Comp. Physiol. 297, R235-R242.

Milligan, C. L. (1997). The role of cortisol in amino acid mobilization and metabolism following exhaustive exercise in rainbow trout (Oncorhynchus mykiss Walbaum). Fish Physiol. Biochem. 16, 119-128.

Milligan, C. L. , Hooke, G. B. and Johnson, C. (2000). Sustained swimming at low velocity following a bout of exhaustive exercise enhances metabolic recovery in rainbow trout. J. Exp. Biol. 203, 921-926.

Mink, J. W. , Blumenschine, R. J. and Adams, D. B. (1981). Ratio of central nervous system to body metabolism in vertebrates: its constancy and functional basis. Am. J. Physiol. 241, R203-R212.

Miura, T., Maruyama, K., Shimakura, S.-I., Kaiya, H., Uchiyama, M., Kangawa, K., et al. (2006). Neuropeptide Y mediates ghrelin-induced feeding in the goldfish, Carassius auratus. Neurosci. Lett. 407, 279-283.

Mommsen, T. P. (2001). Paradigms of growth infish. Comp. Biochem. Physiol. B: Biochem. Mol. Biol. 129, 207-219.

Mommsen, T. P. and Moon, T. W. (2001). Hormonal Regulation of Muscle Growth. In Fish Physiology-Muscle Development and Growth, Vol. 18 (ed. I. E. Johnston), London: Academic Press.

Mommsen, T. P., Vijayan, M. M. and Moon, T. W. (1999). Cortisol in teleosts: dynamics, mechanisms of action, and metabolic regulation. Rev. Fish Biol. Fish. 9, 211-268.

Montserrat, N., Sánchez-Gurmaches, J., Garcia de la Serrana, D., Navarro, M. I. and Gutiérrez, J. (2007). IGF-I binding and receptor signal transduction in primary cell culture of muscle cells of gilthead sea bream: changes throughout in vitro development. Cell Tissue Res. 330, 503-513.

Morcillo, P., Esteban, M. Á. and Cuesta, A. (2016). Heavy metals produce toxicity, oxidative stress and apoptosis in the marine teleost fish SAF-1 cell line. Chemosphere 144, 225-233.

Morgan, J. D. and Iwama, G. K. (1996). Cortisol-induced changes in oxygen consumption and ionic regulation in coastal cutthroat trout (Oncorhynchus clarki clarki) parr. Fish Physiol. Biochem. 15, 385-394.

Moriarty, D. J. W. (1973). The physiology of digestion of blue-green algae in the cichlidfish, Tilapia nilotica. J. Zool. 171, 25-39.

Morris, D. H., Yip, C. K., Shi, Y., Chait, B. T. and Wang, Q. J. (2015). Beclin 1-vps34 complex architecture: understanding the nuts and bolts of therapeutic targets. Front. Biol. 10, 398-426.

Mosher, D. S., Quignon, P., Bustamante, C. D., Sutter, N. B., Mellersh, C. S., Parker, H. G., et al. (2007). A mutation in the myostatin gene increases muscle mass and enhances racing performance in heterozygote dogs. PLoS Genet. 3, e79.

Murashita, K., Uji, S., Yamamoto, T., Rønnestad, I. and Kurokawa, T. (2008). Production of recombinant leptin and its effects on food intake in rainbow trout (Oncorhynchus mykiss). Comp. Biochem. Physiol. Part B Biochem. Mol. Biol. 150, 377-384.

Murashita, K., Jordal, A.-E. O., Nilsen, T. O., Stefansson, S. O., Kurokawa, T.,

Björnsson, B. T. , et al. (2011). Leptin reduces Atlantic salmon growth through the central pro-opiomelanocortin pathway. Comp. Biochem. Physiol. A Mol. Integr. Physiol. 158, 79-86.

Nesan, D. , Kamkar, M. , Burrows, J. , Scott, I. C. , Marsden, M. and Vijayan, M. M. (2012). Glucocorticoid receptor signaling is essential for mesoderm formation and muscle development in zebrafish. Endocrinology 153, 1288-1300.

Nisbet, R. M. , Muller, E. B. , Lika, K. and Kooijman, S. A. L. M. (2000). From molecules to ecosystems through dynamic energy budget models. J. Anim. Ecol. 69, 913-926.

Noda, N. N. and Inagaki, F. (2015). Mechanisms of autophagy. Ann. Rev. Biophys. 44, 101-122.

Okazaki, R. , Riggs, B. L. and Conover, C. A. (1994). Glucocorticoid regulation of insulin-like growth factor-binding protein expression in normal human osteoblast-like cells. Endocrinology 134, 126-132.

Olsen, R. E. , Sundell, K. , Hansen, T. , Hemre, G. I. , Myklebust, R. , Mayhew, T. M. , et al. (2002). Acute stress alters the intestinal lining of Atlantic salmon, Salmo salar L. : an electron microscopical study. Fish Physiol. Biochem. 26, 211-221.

Olsen, R. E. , Sundell, K. , Mayhew, T. M. , Myklebust, R. and Ringo, E. (2005). Acute stress alters intestinal function of rainbow trout, Oncorhynchus mykiss (Walbaum). Aquaculture 250, 480-495.

Pandurangan, M. , Moorthy, H. , Sambandam, R. , Jeyaraman, V. , Irisappan, G. and Kothandam, R. (2014). Effects of stress hormone cortisol on the mRNA expression of myogenenin, MyoD, Myf5, PAX3 and PAX7. Cytotechnology 66, 839-844.

Pankhurst, N. W. (2016). Reproduction and Development. In Fish Physiology-Biology of Stress in Fish, Vol. 35 (eds. C. B. Schreck, L. Tort, A. P. Farrell and C. J. Brauner), San Diego, CA: Academic Press.

Pankhurst, N. W. and Sorensen, P. W. (1984). Degeneration of the alimentary tract in sexually maturing European Anguilla anguilla (L.) and American eels Anguilla rostrata (LeSueur). Can. J. Zool. 62, 1143-1149.

Pankhurst, N. W. , King, H. R. and Ludke, S. L. (2008a). Relationship between stress, feeding and plasma ghrelin levels in rainbow trout, Oncorhynchus mykiss. Marine Freshwater Behav. Physiol. 41, 53-64.

Pankhurst, N. W. , Ludke, S. L. , King, H. R. and Peter, R. E. (2008b). The relationship between acute stress, food intake, endocrine status and life history stage in juvenile farmed Atlantic salmon, Salmo salar. Aquaculture 275, 311-318.

Pauly, D. (1981). The relationship between gill surface area and growth performance infishes: a generalization of von Bertalanffy's theory of growth. Berichte des Deutschen Wissenschaftlichen Kommission fur Meeresforschung 28, 251-282.

Pecquerie, L., Petitgas, P. and Kooijman, S. A. L. M. (2009). Modelingfish growth and reproduction in the context of the dynamic energy budget theory to predict environmental impact on anchovy spawning duration. J. Sea Res. 62, 93-105.

Pecquerie, L., Johnson, L. R., Kooijman, S. A. L. M. and Nisbet, R. M. (2011). Analyzing variations in life-history traits of pacific salmon in the context of dynamic energy budget (DEB) theory. J. Sea Res. 66, 424-433.

Penney, C. C. and Volkoff, H. (2014). Peripheral injections of cholecystokinin, apelin, ghrelin and orexin in cavefish (Astyanax fasciatus mexicanus): effects on feeding and on the brain expression levels of tyrosine hydroxylase, mechanistic target of rapamycin and appetiterelated hormones. Gen. Comp. Endocrinol. 196, 34-40.

Pereira, R. C., Blanquaert, F. and Canalis, E. (1999). Cortisol enhances the expression of mac25/insulin-like growth factor-binding protein-related protein-1 in cultured osteoblasts. Endocrinology 140, 228-232.

Peterson, B. C. and Small, B. C. (2004). Effects of fasting on circulating IGF-binding proteins, glucose, and cortisol in channel catfish (Ictalurus punctatus). Domest. Anim. Endocrinol. 26, 231-240.

Pethybridge, H., Roos, D., Loizeau, V., Pecquerie, L. and Bacher, C. (2013). Responses of european anchovy vital rates and population growth to environmentalfluctuations: an individual-based modeling approach. Ecol. Model. 250, 370-383.

Philip, A. M. and Vijayan, M. M. (2015). Stress-immune-growth interactions: cortisol modulates suppressors of cytokine signaling and JAK/STAT pathway in rainbow trout liver. PLoS ONE 10, e0129299.

Pickering, A. D., Pottinger, T. G., Sumpter, J. P., Carragher, J. F. and Le Bail, P. Y. (1991). Effects of acute and chronic stress on the levels of circulating growth hormone in the rainbow trout, Oncorhynchus mykiss. Gen. Comp. Endocrinol. 83, 86-93.

Pierce, A. L., Fukada, H. and Dickhoff, W. W. (2005). Metabolic hormones modulate the effect of growth hormone (GH) on insulin-like growth factor-I (IGF-I) mRNA level in primary culture of salmon hepatocytes. J. Endocrinol. 184, 341-349.

Powell, G. T. and Wright, G. J. (2011). Jamb and Jamc are essential for vertebrate myocyte fusion. PLoS Biol. 9, e1001216.

Pratt, W. B. (1993). The role of heat shock proteins in regulating the function, folding, and trafficking of the glucocorticoid receptor. J. Biol. Chem. 268, 21455-21458.

Rauw, W. M. (2012). Immune response from a resource allocation perspective. Front. Genetics3, 267.

Reindl, K. M. and Sheridan, M. A. (2012). Peripheral regulation of the growth hormoneinsulin-like growth factor system infish and other vertebrates. Comp. Biochem. Physiol. A Mol. Integr. Physiol. 163, 231-245.

Rescan, P. Y. (2005). Muscle growth patterns and regulation duringfish ontogeny. Gen. Comp. Endocrinol. 142, 111-116.

Rescan, P. -Y. and Gauvry, L. (1996). Genome of the rainbow trout (Oncorhynchus mykiss) encodes two distinct muscle regulatory factors with homology to MyoD. Comp. Biochem. Physiol. B Biochem. Mol. Biol. 113, 711-715.

Rescan, P. -Y., Gauvry, L. and Paboeuf, G. (1995). A gene with homology to myogenin is expressed in developing myotomal musculature of the rainbow trout and in vitro during the conversion of myosatellite cells to myotubes. FEBS Lett. 362, 89-92.

Rescan, P. -Y., Jutel, I. and Rallière, C. (2001). Two myostatin genes are differentially expressed in myotomal muscles of the trout (Oncorhynchus mykiss). J. Exp. Biol. 204, 3523-3529.

Roberts, S. B. and Goetz, F. W. (2001). Differential skeletal muscle expression of myostatin across teleost species, and the isolation of multiple myostatin isoforms. FEBS Lett. 491, 212-216.

Rodgers, B. D., Weber, G. M., Kelley, K. M. and Levine, M. A. (2003). Prolonged fasting and cortisol reduce myostatin mRNA levels in tilapia larvae; short-term fasting elevates. Am. J. Physiol. -Regul. Integr. Comp. Physiol. 284, R1277-R1286.

Rodgers, B. D., Roalson, E. H., Weber, G. M., Roberts, S. B. and Goetz, F. W. (2007). A proposed nomenclature consensus for the myostatin gene family. Am. J. Physiol. -Endocrinol. Metabol. 292, E371-E372.

Rommel, C., Bodine, S. C., Clarke, B. A., Rossman, R., Nunez, L., Stitt, T. N., et al. (2001). Mediation of IGF-1-induced skeletal myotube hypertrophy by PI (3) K/AKT/mTOR and PI (3) K/AKT/GSK3 pathways. Nat. Cell. Biol. 3, 1009-1013.

Rosenzweig, S. A. (2004). What's new in the IGF-binding proteins? Growth Hormone IGF Res. 14, 329-336.

Rotllant, J., Balm, P. H. M., Wendelaar-Bonga, S. E., Pérez-Sánchez, J.

andTort, L. (2000). Adrop in ambient temperature results in a transient reduction of interrenal ACTH responsiveness in the gilthead sea bream (Sparus aurata, L.). Fish Physiol. Biochem. 23, 265-273.

Rotllant, J., Balm, P. H. M., Pérez-Sánchez, J., Wendelaar-Bonga, S. E. and Tort, L. (2001). Pituitary and interrenal function in gilthead sea bream (Sparus aurata L., Teleostei) after handling and confinement stress. Gen. Comp. Endocrinol. 121, 333-342.

Roubos, E. W., Dahmen, M., Kozicz, T. and Xu, L. (2012). Leptin and the hypothalamopituitary-adrenal stress axis. Gen. Comp. Endocrinol. 177, 28-36.

Rudnicki, M. A., Schnegelsberg, P. N. J., Stead, R. H., Braun, T., Arnold, H.-H. and Jaenisch, R. (1993). MyoD or Myf-5 is required for the formation of skeletal muscle. Cell 75, 1351-1359.

Sadoul, B., Leguen, I., Colson, V., Friggens, N. C. and Prunet, P. (2015a). A multivariate analysis using physiology and behavior to characterize robustness in two isogenic lines of rainbow trout exposed to a confinement stress. Physiol. Behav. 140, 139-147.

Sadoul, B., Martin, O., Prunet, P. and Friggens, N. C. (2015b). On the use of a simple physical system analogy to study robustness features in animal sciences. PLoS ONE 10, e0137333.

Salem, M., Kenney, P. B., Rexroad, C. E. and Yao, J. (2006). Molecular characterization of muscle atrophy and proteolysis associated with spawning in rainbow trout. Comp. Biochem. Physiol. Part D Genomics Proteomics 1, 227-237.

Salmerón, C., García de la serrana, D., Jiménez-Amilburu, V, Fontanillas, R., Navarro, I, Johnston, I. A., Gutiérrez, J. and Capilla, E. (2013). Characterisation and expression of calpain family members in relation to nutritional status, diet composition and flesh texture in gilthead sea bream (Sparus aurata). PLoS ONE 8, e75349.

Sandri, M. (2010). Autophagy in skeletal muscle. FEBS Lett. 584, 1411-1416.

Sawatari, E., Seki, R., Adachi, T., Hashimoto, H., Uji, S., Wakamatsu, Y., et al. (2010). Overexpression of the dominant-negative form of myostatin results in doubling of musclefiber number in transgenic medaka (Oryzias latipes). Comp. Biochem. Physiol. A Mol. Integr. Physiol. 155, 183-189.

Schreck, C. B. and Tort, L. (2016). The Concept of Stress in Fish. In Fish Physiology-Biology of Stress in Fish, Vol. 35 (eds. C. B. Schreck, L. Tort, A. P. Farrell and C. J. Brauner), San Diego, CA: Academic Press.

Schroeter, J. C., Fenn, C. M. and Small, B. C. (2015). Elucidating the roles of gut neuropeptides on channel catfish feed intake, glycemia, and hypothalamic NPY and POMC expression. Comp. Biochem. Physiol. A Mol. Integr. Physiol. 188, 168-174.

Seginer, I. (2016). Growth models of gilthead sea bream (Sparus aurata L.) for aquaculture: a review. Aquacul. Eng. 70, 15-32.

Seiliez, I, Gabillard, J.-C., Skiba-Cassy, S., Garcia-Serrana, D., Gutiérrez, J., Kaushik, S., Panserat, S. and Tesseraud, S. (2008a). An in vivo and in vitro assessment of TOR signaling cascade in rainbow trout (Oncorhynchus mykiss). Am. J. Physiol. Regul. Integr. Comp. Physiol. 295, R329-R335.

Seiliez, I, Panserat, S., Skiba-Cassy, S., Fricot, A., Vachot, C., Kaushik, S. and Tesseraud, S. (2008b). Feeding status regulates the polyubiquitination step of the ubiquitin-proteasomedependentproteolysisinrainbowtrout (Oncorhynchusmykiss) muscle. J. Nutr. 138, 487-491.

Seiliez, I., Gutierrez, J., Salmerón, C., Skiba-Cassy, S., Chauvin, C., Dias, K., et al. (2010). An in vivo and in vitro assessment of autophagy-related gene expression in muscle of rainbow trout (Oncorhynchus mykiss). Comp. Biochem. Physiol. B Biochem. Mol. Biol. 157, 258-266.

Seiliez, I., Panserat, S., Lansard, M., Polakof, S., Plagnes-Juan, E., Surget, A., et al. (2011a). Dietary carbohydrate-to-protein ratio affects TOR signaling and metabolism-related gene expression in the liver and muscle of rainbow trout after a single meal. Am. J. Physiol. -Regul. Integr. Comp. Physiol. 300, R733-R743.

Seiliez, I, Sabin, N. and Gabillard, J.-C. (2011b). FoxO1 is not a key transcription factor in the regulation of myostatin (mstn-1a and mstn-1b) gene expression in trout myotubes. Am. J. Physiol. -Regul. Integr. Comp. Physiol. 301, R97-R104.

Seiliez, I., Gabillard, J.-C., Riflade, M., Sadoul, B., Dias, K., Avérous, J., et al. (2012a). Amino acids downregulate the expression of several autophagy-related genes in rainbow trout myoblasts. Autophagy 8, 364-375.

Seiliez, I., Sabin, N. and Gabillard, J.-C. (2012b). Myostatin inhibits proliferation but not differentiation of trout myoblasts. Mol. Cell. Endocrinol. 351, 220-226.

Seiliez, I., Médale, F., Aguirre, P., Larquier, M., Lanneretonne, L., Alami-Durante, H., et al. (2013a). Postprandial regulation of growth-and metabolism-related factors in zebrafish. Zebrafish 10, 237-248.

Seiliez, I., Taty Taty, G. C., Bugeon, J., Dias, K., Sabin, N. and Gabillard, J.-C. (2013b). Myostatin induces atrophy of trout myotubes through inhibiting the TORC1 signaling and promoting ubiquitin-proteasome and autophagy-lysosome degradative pathways. Gen. Comp. Endocrinol. 186, 9-15.

Seiliez, I., Dias, K. and Cleveland, B. M. (2014). Contribution of the autophagy-lysosomal and ubiquitin-proteasomal proteolytic systems to total proteolysis in rainbow trout (Oncorhynchus mykiss) myotubes. Am. J. Physiol. -Regul. Integr. Comp. Physiol. 307, R1330-R1337.

Serpa, D., Ferreira, P. P., Ferreira, H., da Fonseca, L. C., Dinis, M. T. and Duarte, P. (2013). Modelling the growth of white seabream (Diplodus sargus) andgilthead seabream (Sparus aurata) in semi-intensive earth production ponds using the dynamic energy budget approach. J. Sea Res. 76, 135-145.

Shepherd, B. S., Johnson, J. K., Silverstein, J. T., Parhar, I. S., Vijayan, M. M., McGuire, A., et al. (2007). Endocrine and orexigenic actions of growth hormone secretagogues in rainbow trout (Oncorhynchus mykiss). Comp. Biochem. Physiol. A Mol. Integr. Physiol. 146, 390-399.

Sheridan, M. A. and Hagemeister, A. L. (2010). Somatostatin and somatostatin receptors infish growth. Gen. Comp. Endocrinol. 167, 360-365.

Shimizu, M., Swanson, P. and Dickhoff, W. W. (1999). Free and protein-bound insulin-like growth factor-I (IGF-I) and IGF-binding proteins in plasma of coho salmon, Oncorhynchus kisutch. Gen. Comp. Endocrinol. 115, 398-405.

Shpilka, T., Weidberg, H., Pietrokovski, S. and Elazar, Z. (2011). Atg8: an autophagy-related ubiquitin-like protein family. Genome. Biol. 12, 226.

Silverstein, J. T. and Plisetskaya, E. M. (2000). The effects of NPY and insulin on food intake regulation infish. Am. Zool. 40, 296-308.

Sloman, K. A., Motherwell, G., O'Connor, K. I. and Taylor, A. C. (2000). The effect of social stress on the Standard Metabolic Rate (SMR) of brown trout, Salmo trutta. Fish Physiol. Biochem. 23, 49-53.

Small, B. C., Murdock, C. A., Waldbieser, G. C. and Peterson, B. C. (2006). Reduction in channel catfish hepatic growth hormone receptor expression in response to food deprivation and exogenous cortisol. Domest. Anim. Endocrinol. 31, 340-356.

Sohn, R. L., Huang, P., Kawahara, G., Mitchell, M., Guyon, J., Kalluri, R., et al. (2009). A role for nephrin, a renal protein, in vertebrate skeletal muscle cell fusion. Proc. Natl. Acad. Sci. 106, 9274-9279.

Sousa, T., Domingos, T., Poggiale, J.-C. and Kooijman, S. A. L. M. (2010). Dynamic energy budgettheoryrestorescoherenceinbiology. Philos. Trans. R. Soc. BBiol. Sci. 365, 3413-3428.

Srinivas, B. P., Woo, J., Leong, W. Y. and Roy, S. (2007). A conserved molecular pathway mediates myoblast fusion in insects and vertebrates. Nat. Genet. 39, 781-786.

Sumariwalla, V. M. and Klein, W. H. (2001). Similar myogenic functions for myogenin and MRF4 but not MyoD in differentiated murine embryonic stem cells. Genesis 30, 239-249.

Suzuki, K., Kirisako, T., Kamada, Y., Mizushima, N., Noda, T. and Ohsumi, Y. (2001). The pre-autophagosomal structure organized by concerted functions of APG genes is essential for autophagosome formation. EMBO J. 20, 5971-5981.

Takahashi, H., Sakamoto, T., Hyodo, S., Shepherd, B. S., Kaneko, T. and Grau, E. G. (2006a). Expression of glucocorticoid receptor in the intestine of a euryhaline teleost, the Mozambique tilapia (Oreochromis mossambicus): effect of seawater exposure and cortisol treatment. Life Sci. 78, 2329-2335.

Takahashi, H., Takahashi, A. and Sakamoto, T. (2006b). In vivo effects of thyroid hormone, corticosteroids and prolactin on cell proliferation and apoptosis in the anterior intestine of the euryhaline mudskipper (Periophthalmus modestus). Life Sci. 79, 1873-1880.

Tan, X. and Du, S. (2002). Differential expression of two MyoD genes in fast and slow muscles of gilthead seabream (Sparus aurata). Dev. Genes. Evol. 212, 207-217.

Tinoco, A. B., Näslund, J., Delgado, M. J., de Pedro, N., Johnsson, J. I. and Jönsson, E. (2014). Ghrelin increases food intake, swimming activity and growth in juvenile brown trout (Salmo trutta). Physiol. Behav. 124, 15-22.

Unniappan, S., Canosa, L. F. and Peter, R. E. (2004). Orexigenic actions of ghrelin in goldfish: feeding-induced changes in brain and gut mRNA expression and serum levels, and responses to central and peripheral injections. Neuroendocrinology 79, 100-108.

van der Veer, H. W., Kooijman, S. A. L. M. and van der Meer, J. (2001). Intra-and interspecies comparison of energyflow in North Atlantic flatfish species by means of dynamic energy budgets. J. Sea Res. 45, 303-320.

Veillette, P. A., Sundell, K. and Specker, J. L. (1995). Cortisol mediates the increase in intestinalfluid absorption in Atlantic salmon during parr-smolt transformation. Gen. Comp. Endocrinol. 97, 250-258.

Velasco, C., Librán-Pérez, M., Otero-Rodiño, C., López-Patiño, M. A., Míguez, J. M., CerdaáReverter, J. M., et al. (2016). Ghrelin modulates hypothalamic fatty acid-sensing and control of food intake in rainbow trout. J. Endocrinol. 228, 25-37.

Vélez, E. J., Lutfi, E., Jiménez-Amilburu, V., Riera-Codina, M., Capilla, E., Navarro, I., et al. (2014). IGF-I and amino acids effects through TOR signaling on proliferation and differentiation of gilthead sea bream cultured myocytes. Gen. Comp. Endocrinol. 205, 296-304.

Vélez, E. J. , Lutfi, E. , Azizi, S. , Montserrat, N. , Riera-Codina, M. , Capilla, E. , et al. (2015). Contribution of in vitro myocytes studies to understanding fish muscle physiology. Comp. Biochem. Physiol. B Biochem. Mol. Biol. Available at: http: // www. sciencedirect. com/ science/article/pii/S1096495915002158.

Velikkakath, A. K. G. , Nishimura, T. , Oita, E. , Ishihara, N. and Mizushima, N. (2012). Mammalian Atg2 proteins are essential for autophagosome formation and important for regulation of size and distribution of lipid droplets. Mol. Biol. Cell. 23, 896-909.

Vianello, S. , Brazzoduro, L. , Valle, L. D. , Belvedere, P. and Colombo, L. (2003). Myostatin expression during development and chronic stress in zebrafish (Danio rerio). J. Endocrinol. 176, 47-59.

Vijayan, M. M. , Prunet, P. and Boone, A. N. (2005). Xenobiotic impact on corticosteroid signaling. In Biochemistry and Molecular Biology of Fishes. (eds. T. W. Moon and T. P. Mommsen), pp. 365-394. Environmental Toxicology. Amsterdam, The Netherlands: Elsevier.

Villafuerte, B. C. , Koop, B. L. , Pao, C. I. and Phillips, L. S. (1995). Glucocorticoid regulation of insulin-like growth factor-binding protein-3. Endocrinology 136, 1928-1933.

Volkoff, H. , Canosa, L. F. , Unniappan, S. , Cerdá-Reverter, J. M. , Bernier, N. J. , Kelly, S. P. , et al. (2005). Neuropeptidesandthecontroloffoodintakeinfish. Gen. Comp. Endocrinol. 142, 3-19.

Wang, Q. , Bing, C. , Al-Barazanji, K. , Mossakowaska, D. E. , Wang, X. M. , McBay, D. L. , et al. (1997). Interactions between leptin and hypothalamic neuropeptide Y neurons in the control of food intakeand energy homeostasis in the rat. Diabetes 46, 335-341.

Wang, J. , Salem, M. , Qi, N. , Kenney, P. B. , Rexroad, C. E. , III and Yao, J. (2011). Molecular characterization of the MuRF genes in rainbow trout: potential role in muscle degradation. Comp. Biochem. Physiol. B Biochem. Mol. Biol. 158, 208-215.

Wang, H. , Sun, H. -Q. , Zhu, X. , Zhang, L. , Albanesi, J. , Levine, B. , et al. (2015). GABARAPs regulate PI4P-dependent autophagosome: lysosome fusion. Proc. Natl. Acad. Sci. 112, 7015-7020.

Warren C. E. and Gerald, D. E. (1967). Laboratory studies on the feeding, bioenergetics and growth offish. Pacific Cooperative Water Pollution and Fisheries Research Laboratories, Agricultural Experiment Station, Oregon State University.

Weber, T. E. and Bosworth, B. G. (2005). Effects of 28 day exposure to cold temperature or feed restriction on growth, body composition, and expression of genes related to muscle growth and metabolism in channel catfish. Aquaculture 246, 483-492.

Weber, T. E., Small, B. C. and Bosworth, B. G. (2005). Lipopolysaccharide regulates myostatin and MyoD independently of an increase in plasma cortisol in channel catfish (Ictalurus punctatus). Domest. Anim. Endocrinol. 28, 64-73.

Weintraub, H., Davis, R., Tapscott, S., Thayer, M., Krause, M., Benezra, R., et al. (1991). The myoD gene family: nodal point during specification of the muscle cell lineage. Science 251, 761-766.

Wolowczuk, I., Verwaerde, C., Viltart, O., Delanoye, A., Delacre, M., Pot, B., et al. (2008). Feeding our immune system: impact on metabolism. Clin. Dev. Immunol. 2008, 639803.

Won, E. T., Douros, J. D., Hurt, D. A. and Borski, R. J. (2016). Leptin stimulates hepatic growth hormone receptor and insulin-like growth factor gene expression in a teleostfish, the hybrid striped bass. Gen. Comp. Endocrinol. 229, 84-91.

Wong, A. O. L., Chang, J. P. and Peter, R. E. (1993). Dopamine functions as a growth hormone-releasing factor in the goldfish, Carassius auratus. Fish Physiol. Biochem. 11, 77-84.

Wong, A. O. L., Zhou, H., Jiang, Y. and Ko, W. K. W. (2006). Feedback regulation of growth hormone synthesis and secretion infish and the emerging concept of intrapituitary feedback loop. Comp. Biochem. Physiol. A Mol. Integr. Physiol. 144, 284-305.

Wood, A. W., Duan, C. and Bern, H. A. (2005). Insulin-like growth factor signaling infish. Int. Rev. Cytol. 243, 215-285.

Wu, S., Ji, G., Liu, J., Zhang, S., Gong, Y. and Shi, L. (2015). TBBPA induces developmental toxicity, oxidative stress, and apoptosis in embryos and zebrafish larvae (Danio rerio). Environ. Toxicol. Available at: http://onlinelibrary.wiley.com/doi/10.1002/tox.22131/abstract.

Xie, Z., Nair, U. and Klionsky, D. J. (2008). Atg8 controls phagophore expansion during autophagosome formation. Mol. Biol. Cell. 19, 3290-3298.

Xu, C., Wu, G., Zohar, Y. and Du, S.-J. (2003). Analysis of myostatin gene structure, expression and function in zebrafish. J. Exp. Biol. 206, 4067-4079.

Yabu, T., Imamura, S., Mizusawa, N., Touhata, K. and Yamashita, M. (2012). Induction of autophagy by amino acid starvation infish cells. Marine Biotechnol. 14, 491-501.

Yu, J., Iwashita, M., Kudo, Y. and Takeda, Y. (1998). Phosphorylated insulin-like growth factor (IGF) -binding protein-1 (IGFBP-1) inhibits while non-phosphorylated IGFBP-1 stimulates IGF-I-induced amino acid uptake by cultured trophoblast cells. Growth Horm. IGF Res. 8, 65-70.

Zhu, Z., Zeng, X., Lin, X., Xu, Z. and Sun, J. (2015). Effects of ration levels on growth and reproduction from larvae tofirst-time spawning in the female Gambusia affinis. Int. J. Mol. Sci. 16, 5604-5617.

第 6 章　渗透压应激的稳态（体内平衡）反应

鱼类的血液和它们生活的水环境的渗透压差异很大，因此，鱼类一直处于一定程度的渗透应激下。另外，生理应激源及相关的反应可能对水盐平衡有较大的影响（Schreck，Tort，2016；本书第 1 章）。鱼类和它所处水环境的渗透压梯度由渗透压传感器感知并通过细胞内信号通路传导给渗透压调节器官中的效应器，进而从快速的转录后修饰到长期的基因转录调节，在不同时间范围引发整体的和渗透特异的适应性反应。前者包括转运蛋白和渗透物生产酶活性以及细胞骨架组成、血液运输和代谢的改变。近来基于全基因组、分子生理手段以及新兴模式物种的研究已鉴定出在渗透压调节适应中起关键作用的新的转录因子和效应分子。在本章，我们将综述近年来关于不同生物学组织水平（从基因到个体）以及不同时间尺度（从急性到慢性的反应）上鱼类渗透压反应的数据，以着重阐明现在已知的信息并提出下一步的研究领域。

6.1　导　言

鱼类能够主动调节体液渗透压和离子水平，使之不同于它们所生存的淡水和海水环境（Edwards，Marshall，2012）。渗透压和离子水平可能被一系列的应激源影响（Schreck，Tort，2016；本书第 1 章）（Winberg 等，2016；本书第 2 章）。为了保持体液平衡，交感神经系统和内分泌系统下丘脑-脑垂体-肾间腺轴为应激源产生的初级反应所激活，导致儿茶酚胺和皮质醇分别从嗜铬细胞和肾上腺皮质腺的类固醇生成细胞中释放。（Wendelaar Bonga，1997）（Gorissen，Flik，2016；本书第 3 章）。这些激素作用于鳃以调节离子运输细胞中离子转运蛋白的活性，且能改变呼吸动脉-小动脉分流和渗透压调节动脉-静脉的相对流通（Olson，2002），这些改变可以显著影响离子和水的平衡（Redding 等，1984；Takei，Loretz，2006）。皮质醇还能通过重组渗透调节器官促进鱼类对淡水和海水环境的长期适应，尽管在这两种不同水环境中离子和水的运输需求是相反的（Wendelaar Bonga，1997）。另外，激素导致的血糖变化能够改变鳃和其他渗透压调节器官中 ATP 驱动的离子泵的活性（Bickler，Buck，2007）（见 Sadoul，Vijayan，2016；本书第 5 章；本章第 6.5 节）。

正如 Schreck 和 Tort（2016；本书第 1 章）所说，本书旨在针对不同刺激的生理反应提供一个整体的综述，而不仅限于任何特殊刺激。在这点上，McDonald 和 Milligan（1997）综述了不同应激源对淡水大麻哈鱼水盐平衡主要的应激反应的影响。其他文章也描述了非渗透压应激源的影响：

（1）真鲷（*Pagrus pagrus*）生长速度和血浆渗透压（Vargau-Chacoff 等，2011）；

（2）舌齿鲈（*Dicentrarchus labrax*）血浆离子浓度（Sinha 等，2015）；

（3）软骨鱼不同渗透压调节器官水盐和尿素平衡的氧气供应调节（Skomal, Mandelman, 2012）;

（4）淡水刺魟（*Potamotrygon* cf. *histrix*）组织水通透性和钠钾离子丢失（Brinn 等，2012）。

然而关于一般应激源对渗透压调节影响的研究是比较少的，近年来对于包括鱼类在内的一系列动物渗透压反应的认知取得了迅速而深入的进展。就相关组织和分子而言，似乎保持水和离子平衡的机制在不同的应激源当中是相似的。因此，在本章，我们主要关注鱼类在对渗透压应激的反应中如何保持体液的稳态。从渗透压感知到细胞体积调节的信号级联在非鱼类模式物种中已经被研究，并且其基本的机制似乎和鱼类是非常相似。因此，我们也综述了非鱼类模型的文献，以期为鱼类今后的研究方向提供参考。

6.1.1　鱼类渗透压调节

为了更好地了解不同应激源对鱼类的影响，在此，我们简单描述一下渗透压调节的基本原理。我们还会简单讨论一下酸碱平衡，因为它和离子运输细胞功能密切相关，并且应激反应可能包含 pH 的改变。考虑到未来的研究需要，这些信息也是很有用的。关于这个课题的详细信息可见 McCormick 等（2013）。体液平衡的保持是一个复杂的过程，参与的是不同的渗透压调节器官，主要是鳃、消化道和肾（图 6.1）。这些器官之中，鳃是主要作用器官，能够直接感知外部渗透压改变，并能根据环境水型积极吸收或分泌单价离子（如 Na^+、K^+ 和 Cl^-），以保持血浆渗透压在小范围内变化。脑并不是一种渗透压调节器官，但是由于它能够控制饮水活动，所以对于渗透压调节也很重要。在海水鱼中，吞进去的海水连同 Na^+ 和 Cl^- 一起被肠道吸收（图 6.1），但是肠道在淡水鱼渗透压调节中的作用较小。在海水鱼中，肾脏是关键的二价离子（Mg^{2+}、Ca^{2+} 和 SO_4^{2-}）分泌器官（Beyenbach, 2004），然而在淡水鱼中，它能够分泌大量的稀释尿液以拮抗体表丰富的水流。

构成渗透压调节器官的上皮细胞暴露于不同的渗透应激中，这取决于它们面对的是顶面还是细胞基底外侧面；这一点在不同组织中也是不同的。例如，鳃上皮细胞顶膜直接接触外部淡水或海水环境，然而肠细胞顶膜面临着受外界媒介影响而不断变化的离子浓度。在海水鱼中，吞进去的海水在消化道被逐级加工，单价离子浓度降低而二价离子浓度升高（图 6.1）。肾小管细胞与过滤的血浆接触，在渗透应激中，管液渗透压的改变是缓慢而微小的（图 6.1）。因此，相比于鱼类其他渗透压调节组织以及暴露于高浓度尿素和其他渗透物的哺乳动物肾髓质细胞，鱼类肾脏就是无渗透性应激的组织（Beuchat, 1996）。与顶膜相反，上皮细胞基底外侧膜受到间质流体的冲洗而在渗透应激中改变微小。很显然，鳃面临的刺激是最猛烈的，其次是肠道，最后是肾脏（图 6.1）。

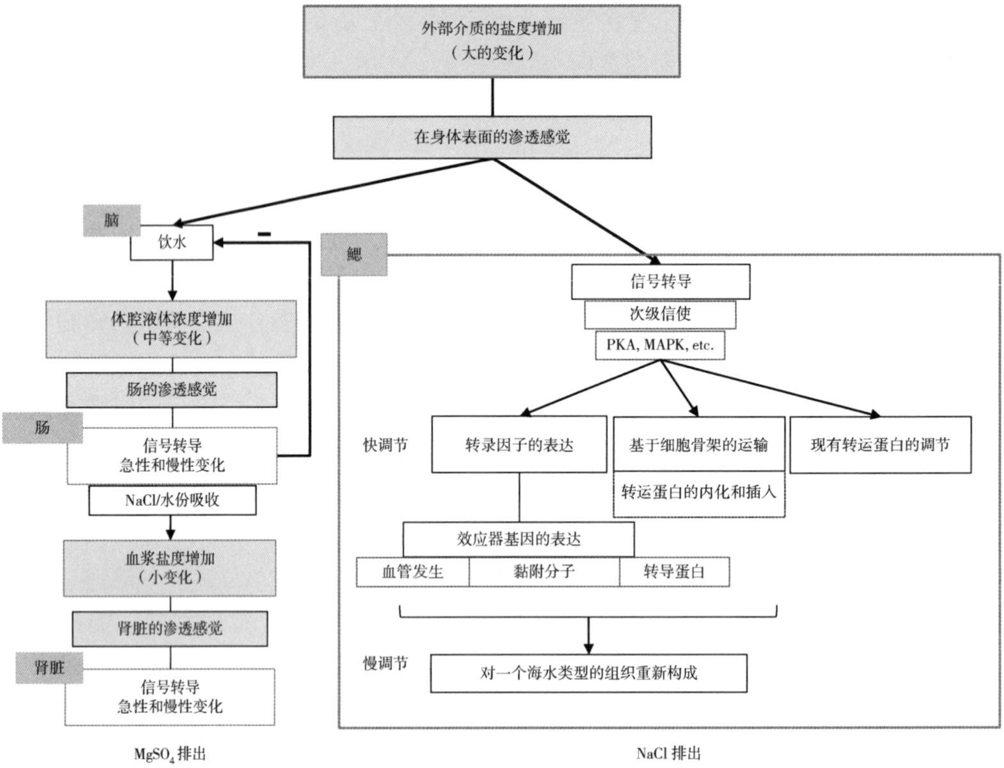

图 6.1　鱼类不同渗透压调节组织在渗透应激反应过程中随时间而进行的连续活动（这里以海水的挑战为例证）

注：MAPK，有丝分裂激活蛋白激酶；PKA，蛋白激酶 A。

细胞顶膜和基底外侧膜有不同的离子转运蛋白吸收或分泌以保持细胞外液平衡。例如，淡水鱼的鳃离子运输细胞通过不同类型的细胞从缺少离子的环境中吸收 NaCl，借助 Na^+-Cl^- 协同转运蛋白（NCC）、Na^+/H^+ 交换器（NHE）或者其他位于顶膜的 Na^+ 结合转运蛋白，这些蛋白均由基底外侧 Na^+/K^+ - ATPase（NKA）提供能量（Evans 等，2005；Hwang 等，2011）。海水鱼离子运输细胞主要通过顶端囊性纤维化运输调节（CFTR）氯离子通道、$Na^+-K^+-2Cl^-$ 协同转运蛋白 1（NKCC1）以及 NKA 来分泌 NaCl（Evans 等，2005；Hwang 等，2011）。在受到渗透压应激后，在渗透压调节组织的所有上皮细胞中，这些转运蛋白活性都会发生变化。

渗透压应激反应可以临时分为两个阶段，即急性阶段和慢性阶段（图 6.2）。例如，一旦从淡水转移到海水中，鳃上皮细胞能够逆转离子运输从吸收变成分泌。这种快速的转变可能会在去除淡水转运蛋白或将海水转运蛋白从细胞内募集到细胞膜后发生（图 6.1）。渗透压传感器的信号转导可能会调控不同激酶或磷酸酶系统，使细胞膜上淡水型酶失活而激活海水型酶。然而，在鱼类中，这种急性阶段的调控信息我们还知之甚少。在慢性阶段，渗透压调节器官在渗透应激中能够通过旧细胞凋亡以及祖先细胞分化成新细胞而重组成具有相反功能的新组织（图 6.2）。举例来说，鱼类在转入海水一周后，消化道上皮细胞会变得薄而透明，同时，由于血管的发育，细胞表面变得微红（Yuge 等，2003）。这些变化有利于水和离子的吸收，是海水鱼肠道的重要功能。和鳃离子运输细胞的广泛研究一起（Evans 等，2005；Hwang 等，2011；Hiroi，McCormick，2012；Hwang，Lin，2014；Takei 等，2014），相比于急性变化，这些慢性改变在细胞和分子水平已经得到很好的描述。

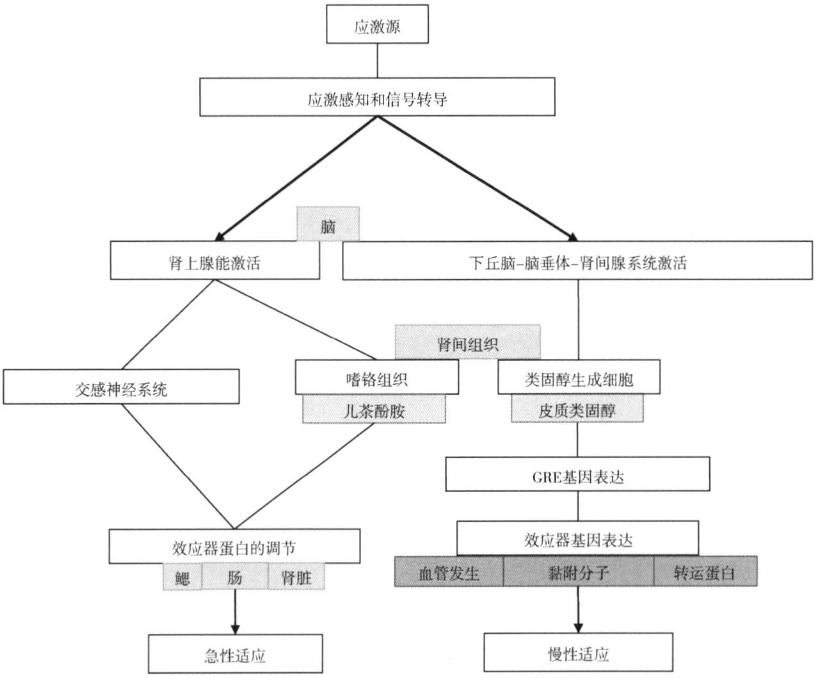

图6.2 鱼类在短期和长期适应新环境时对应激源（如渗透压）的激素的和神经的反应

注：其他参与渗透压应激的激素在文中介绍。对普通应激反应的详细级联通路的描述见 Schreck 和 Tort（2016；本书第1章）。GRE，糖皮质激素反应元件。

6.1.2 海水和淡水中的 pH 调节

在环境的应激中（海洋酸化或污染）或者内部代谢导致的 pH 波动下，鱼类需要保持体液平衡，因为 CO_3^- 和 H^+ 调节与其他离子密切相关。鱼类和哺乳动物相似，主要通过三种补偿机制保持酸碱平衡：

（1）碳酸氢盐或非碳酸氢盐缓冲液的快速理化缓冲；

（2）开放性 CO_2-HCO_3^- 缓冲系统的呼吸调节；

（3）动物和环境之间酸碱相关分子的净运输（Evans 等，2005）。

相比在空气中呼吸的哺乳动物，在水中呼吸的鱼类由于存在呼吸的限制（较低的氧溶解度和较高的水黏度），主要采用第三种机制来完成酸碱平衡调节（Claiborne 等，2002；Evans 等，2005）。与肾脏和肠道相比，鳃被普遍认为控制大多数的酸碱移动（90%或更高）（Claiborne 等，2002）。由于海水中 Na^+ 和 Cl^- 浓度高，目前研究者普遍认为海水鱼通过鳃细胞顶膜的 NHE 和 HCO_3^-/Cl^- 交换器（阴离子交换器，

AE)，依据离子梯度分泌产生代谢的酸和碱。不同类型的酸碱分泌离子运输细胞已经在海水和广盐性鱼类中被鉴别出来。酸分泌离子运输细胞具备顶膜 NHE2/3 和基底外侧 NKA，而碱分泌离子运输细胞具备顶膜钠离子依赖性氯/碘化物溶质运输蛋白(SLC26A4) 和基底外侧 H^+-ATPase (HA)，它们都作用于体液平衡的调节 (Evans 等，2005；Hwang 等，2011；Hwang，Lin，2014)。对 NHE 表达离子运输细胞及其表达与功能已经有很详尽的描述。钠离子依赖性氯/碘化物溶质运输蛋白 (SLC26A4) 表达离子运输细胞在软骨鱼中研究广泛，而硬骨鱼的相关信息比较少。

与海水鱼相比，淡水鱼鳃中 H^+ 和 HCO_3^- 的转运机制似乎更加复杂多样，或许是因为热力学问题，外界低浓度的 Na^+、Cl^- 和 H^+ 梯度不利于相关转运蛋白的运输。电中性 NHE3 或者其他亚型的作用是有争议的，尤其是涉及低 Na^+ 或 pH 的环境 (Avella，Bornancin，1989；Parks 等，2008)。近年来，氨转运 Rh 蛋白的鉴定帮助我们理解其中包含的机制 (Nakada 等，2007a，b)。随后的分子生理学研究提出了一个模型，顶膜 Rhcg1 能介导细胞内 NH_4^+ 去质子化、顶膜酸捕获以及随后 H^+ 梯度驱动 NHE3 以完成 H^+ 分泌和 Na^+ 吸收的功能 (Wu 等，2010)。另一方面，在一些物种如斑马鱼中发现，是顶膜 HA 而不是 NHE 在酸分泌和 Rhcg1 介导的 $NH_4{}^+$ 分泌中发挥主要作用 (Guh 等，2015；Lin 等，2006；Shih 等，2008，2012)。为了完成跨膜 H^+ 分泌，离子运输细胞中的碳酸酐酶 (细胞溶质 CA2 和膜型 CA15) 和基底外侧 AE1 (不是 Na^+ - HCO_3^- 协同转运蛋白，NBC) 必须和顶膜 NHE 或 HA 共同发挥作用 (Gilmour，2012；Hwang，Chou，2013；Lin 等，2008；Hsu 等，2014；Lee 等，2011)。

6.1.3 细胞体积调节

渗透压反应的细胞体积调节已经在脊椎动物细胞如红细胞以及单细胞有机体如酵母菌中得到广泛的研究。渗透压感知和信号转导系统的基本机制在脊椎动物和酵母菌中似乎是相同的，但是后来累积的相关认识将更为广泛。因此，这些非鱼类的数据未来将被引入到恰当的鱼类细胞相关研究中。

细胞暴露于高渗透压应激中后会皱缩，进而导致细胞膜上的大分子拥挤，细胞内离子浓度增加，并对细胞骨架产生机械刺激。这些改变可能会使细胞感知到其体积变化并启动远端信号作用，最终导致细胞膨胀或体积恢复。举例来说，瞬时受体电位香草酸型 (TRPV) 通道是一种可能的传感器分子，能够改变细胞内 Ca^{2+} 水平以调节一定量的激酶或磷酸酶活性，最终导致目标转运蛋白的磷酸化状态改变。激活的蛋白随后可能改变细胞内的渗透压以恢复细胞体积。也有一种可能，细胞骨架在收缩蛋白修饰将细胞质囊中的转运蛋白募集到细胞膜这一过程中发挥一定的作用 (Fletcher 和 Mullins，2010)。一些转运蛋白在高渗透压应激下会被卷入到离子流入过程；NKCC1 (SLC12A2) 能够一次吸收四种渗透物进入细胞，然而 NHE1 (SLC9A1) 和 AE (或许是 SLC4 或 SLC26 家族) 一起分别完成 Na^+ 和 Cl^- 吸收及 H^+ 和 HCO_3^- 的交换。带电的 H^+ 和 HCO_3^- 能够产生 H_2O 和 CO_2 以减少介质渗透压。NaCl 流入由 NKA 供能以保持细胞质中低的 Na^+ 浓度。另一个转运系统的作用取决于细胞类型和皱缩模式。各种各样的激酶被认为能够磷酸化转运蛋白并控制它们的活性，包括应激反应激酶（OSR1）或不含赖氨酸的 Ste20p-相关脯氨酸丙氨酸激酶复合体（SPAK）和肌球蛋白轻链激酶（Hoffmann 等，2009）。其他可能的转运系统包括有机渗透物转运蛋白（如牛磺酸转运蛋白）和高渗诱导的阴离子通道。

在低渗的应激中，细胞体积增加和随后的体积恢复可能是由于膨胀反应激活或张力控制的通过 Cl^- 和 K^+ 分泌离子通道进行的离子流出导致的（Hoffmann 等，2009）。Cl^- 通道可能是 ClC 或 Ca^{2+} 激活的氯化物通道家族成员，但是目前并未得到清晰的阐述。一种膨胀激活的 K^+ 流出通道已经在多种细胞中被论证，在不同的细胞类型中还有一些不同的 K^+ 通道也参与离子流出。现行的通道包括双孔结构域 K^+ 通道家族或者漏电通道家族成员如 TASK2/KCNK5、TREK1/KCNK2 和 TRAAK/KCNK4。

在鳉科鱼类底鳉（*Fundulus heteroclitus*）中，渗透压应激能够控制海水型离子运输细胞中 NaCl 分泌（Marshall，2011）。如果发生低渗休克，离子运输细胞会通过失活基底外侧 NKCC1 来减少 Cl^- 分泌。失活是由黏着斑激酶导致的去磷酸化作用引发的，这和渗透压传感器整合素 β1 相关联（Marshall 等，2008）。通过使用不同的激酶抑制剂，在鱼类中已经证明信号级联中可能涉及的远端信号激酶。这些激酶包括已经在酵母菌中发现的 p38 细胞分裂素激活的蛋白激酶（MAPK）、OSR1、SPAK 等。在鳉科鱼类中也已经研究低渗休克对其他离子运输细胞转运蛋白的影响（Duranton 等，2000；Avella 等，2009；Marshall 等，2009），并且体积调节机制在鳉科鱼类（Marshall 等，2002）和鳗鱼（Lionetto 等，2005）的肠道细胞中也有报道。然而，关于鱼类细胞体积调节机制仍需要开展大量的研究。

6.2　对高渗透压应激的反应

高渗海水环境对离子或渗透压调节器来说是一种应激源（Schreck，Tort，2016；本书第 1 章），并且鱼类会通过活跃地保持水分并分泌离子对它产生一定的生理反应（McCormick 等，2013）。应激激素在保持内稳态中发挥关键作用（Gorissen，Flik，2016；本书第 3 章）。因为海水保持着高浓度的单价和二价离子，所以体内平衡反应要分开两种离子类型来讨论（图 6.1）。

6.2.1　单价离子的应激源

在海水硬骨鱼类血浆中 Na^+ 浓度高于 Cl^-（分别为 170 mM Na^+ 和 130 mM Cl^-），但是在海水中是相反的（分别为 450 mM Na^+ 和 520 mM Cl^-）。这表示 Cl^- 相比 Na^+ 受到更活跃的调节。鱼类从淡水转到海水的离子调节器主要是 Cl^- 而不是 Na^+。鳗鱼转入海水后会在 Cl^- 而不是 Na^+ 的驱动下快速饮水，这一事实已经得到证明：鳗鱼暴露于 $MgCl_2$ 和氯化胆碱溶液中会开始吞水，而暴露在 $NaNO_3$ 和 Na_2SO_4 溶液中不会（Hirano，1974）。并且，原位灌注实验表明肠道管腔液 Cl^- 浓度能抑制海水鳗的过量饮水（Ando，Nagashima，1996）。因此，Cl^- 传感器可能存在于体表和肠腔表面以调节饮水活动。因为肠腔与外部环境是连续的，所以在饮入海水后传感器能够检测外部环境中 Cl^- 离子的改变。无疑的是，鱼类 Cl^- 感知机制是一个值得进一步研究的领域。

进入食道的海水会沿着消化道被逐渐加工，并且高达 80% 的海水在到达食道末端的时候会被吸收（Takei，Loretz，2011；Grosell，2011）。海水硬骨鱼的这种吸收能力是令人惊讶的，因为陆生动物在饮用海水后会失水。水的吸收是通过 NKCC2、NCC 和 NHE + AE 通道驱动的高效的 NaCl 吸收而完成的，主要是通过水通道蛋白（AQPs）促进水的平行运输（Grosell，2011；Madsen 等，2014；Ando 和 Takei，2014）。同样地，大量的 HCO_3^- 分泌到肠腔会使 $Ca/MgCO_3$ 沉淀（镁方解石），这将减少肠腔液渗透压并进一步加强水的吸收（Wilson 等，2009；Grosell 等，2009）。因此，除了海水鳗是暴露于高浓度的单价离子 HCO_3^-（120 mM）条件下（Tsukada，Takei，2006），其他鱼肠道上皮细胞的肠腔侧均暴露于高浓度的二价离子（Mg^{2+}，200 mM；Ca^{2+}，30 mM；SO_4^{2-}，100 mM）和低浓度的单价离子（Na^+，4 mM；Cl^-，30 mM）条件下。由于 HCO_3^- 浓度高，肠腔液 pH 会大于 9.0（Gorsell 等，2009）。因此，肠道上皮细胞的肠腔侧暴露于独特的渗透应激中。鱼类转入海水后肠道上皮细胞的分子转换机制已经很好地阐述并被大量地报道（Grosell，2011）。

当鱼在海水中时，Na^+ 和 Cl^- 顺着浓度梯度通过鳃和肠道被动进入鱼体。当广盐性鱼类从淡水进入到海水后，大量的 NaCl 会快速进入鱼体，因为当鱼在淡水环境中时，鳃和肠道的调节系统是直接作用于离子吸收的。由渗透压传感器检测到高渗透压应激后就会立即抑制渗透压调节器官上皮细胞的吸收系统并激活分泌系统（图 6.1）。然后，通过吸收细胞凋亡及分化出新的分泌细胞，渗透压调节器官被激素重新构成（Takei，McCormick，2013）。

离子分泌的分子和细胞机制随后将会以鳃离子运输细胞为例进行详细描述。海水鱼的离子运输细胞是含有丰富线粒体的细胞，专门用于 NaCl 分泌，并且在基底外侧膜上有丰富的 NKA 结合的 K^+ 通道（Kir5.1）和 NKCC1，在顶膜上有 CFTR（Suzuki 等，1999；Evans 等，2005；Marshall，2001；Hwang，Lin，2014）。离子运输细胞的静息电位是非常低的，因为激活的 NKA 将 3 个 Na^+ 排出细胞之外换取 2 个 K^+，而多余的 K^+ 根据浓度梯度通过 Kir5.1 而离开细胞。另外，细胞溶质中的 Cl^- 浓度由于 NKCC1 通道活化的 $2Cl^-$ 离子流入可能会变得异常高。低的细胞电位和高的 Cl^- 浓度会通过顶膜 CFTR 将 Cl^- 排到海水中，尽管海水中 Cl^- 浓度大于 500 mM。

6.2.2 二价离子的应激源

相比于海水，驯化于海水的鳗鱼血浆中二价离子浓度比例（Mg^{2+}，50 mM vs 1.6 mM；Ca^{2+}，10 mM vs 1.2 mM；SO_4^{2-}，35 mM vs 0.8 mM）一般要比单价离子浓度低（Na^+，450 mM vs 170 mM；Cl^-，520 mM vs 120 mM）（Watanabe，Takei，2012a）。这种在鱼体和环境之间极度的二价离子浓度梯度可能会对鱼类形成应激源；因此，海水鱼能够防止这些二价离子进入身体中。至于 SO_4^{2-}，通过研究食管结扎后的海鳗中 $^{35}SO_4^{2-}$ 的流入差异，环境中的 SO_4^{2-} 以 1.55 μmol/kg 每小时的速度进入海水鳗鱼体内，其中85%通过体表，主要是鳃，15%通过消化道。肠道几乎不会吸收二价离子，因此，腔液内SO_4^{2-}浓度的增加被用来作为水分吸收的标记，但是大量的SO_4^{2-}似乎会被消化道吸收以帮助吸收水分。在海水鳗鱼中，显著的二价离子流入可以通过其尿液中高浓度的二价离子来推断（116 mM Mg^{2+}和45 mM SO_4^{2-}）（Watanabe，Takei，2012a）。因此，二价离子的主要分泌部分是肾脏（Beyenbach，2004；Kurita 等，2008；Watanabe，Takei，2011），而鳃主要分泌单价离子（图6.1）。淡水中的鳗鱼可以从含有3 mM SO_4^{2-}的淡水中以每小时1.76 μmol/kg的速度逆向浓度梯度吸收SO_4^{2-}，然而在海水中的鳗鱼，所有的SO_4^{2-}流入均可以解释为海水的被动扩散（Watanabe，Takei，2012a）。因此，在脊椎动物中，鱼类是一种研究Mg^{2+}和SO_4^{2-}调节的极好模型。

当鳗鱼在淡水中时，肾小球滤液中几乎所有的SO_4^{2-}都是近端小管通过顶膜SLC13A1和基底外侧SLC26A1（Kurita 等，2008）重吸收的，在转移到海水中后，为了分泌SO_4^{2-}，顶膜SLC13A1会被SLC26A6替代（Watanabe，Takei，2011）。值得注意的是，近端小管从吸收转变到分泌并不是受海水中SO_4^{2-}的驱动，而是受Cl^-的驱动，因为水环境中的$MgCl_2$对这种转变是有效的，而$MgSO_4$是无效的（Watanabe，Takei，2012b）。另外，Cl^-离子主要通过硬骨鱼的肾脏来感知盐度的压力。

6.2.3　激素参与

典型的应激激素如皮质醇和肾上腺素能够影响海水中的水和离子平衡（Gorissen，Flik，2016；本书第3章）。另外，一些快速作用的激素如血管紧缩素Ⅱ（AngⅡ）和心钠素，能够刺激鱼类应激激素的产生（McCormick，2001；Takei，Loretz，2006；Takei，McCormick，2013）。除此之外，慢作用激素分泌，如皮质醇和生长激素可以被快速作用的激素激发。慢作用激素能够缓慢上调和长期海水适应相关的基因转录。现已知皮质醇和生长激素协同刺激硬骨鱼类鳃离子运输细胞的分化（Madsen 等，2009；Armesto 等，2014）以及弹涂鱼消化上皮细胞的重构（Takahashi 等，2006）。已经报道外周GH/IGF-I系统在鱼类生理中的作用包括渗透压调节（Reindl，Sheridan，2012）。其他应激相关的渗透压调节激素包括甲状腺激素（Peter，2011）和瘦素（Baltzegar 等，2014）。信号系统和靶标渗透压调节基因将是理解应激激素作用下一步的研究内容。

6.3　对低渗透压应激的反应

与高渗透压的海水相比，低渗透压的淡水缓冲能力较低，并且影响鱼类体液、离子与酸碱平衡的离子组成和pH水平等因素的变化较大。在此，考虑到这些参数在分子水平和细胞水平已经得到了广泛研究，我们主要关注不同水平的Na^+、Cl^-、Ca^{2+}和低pH的影响。

6.3.1 离子组成的应激源

淡水环境相对于鱼体液的离子水平较低，一般会导致鱼体中离子被动丢失，所以，为保持离子平衡，必须要通过积极的吸收来进行补偿（Mcdonald，Rogano，1986；Perry，Laurent，1989；Chou 等，2002）。人工淡水 Na^+、Cl^-或 Ca^{2+}水平相比于天然淡水会比较低，这似乎并不会显著影响被动的离子流出，但是在虹鳟和罗非鱼快速转入稀释的水环境时会使相关离子的吸收减少（Perry，Laurent，1989；Chou 等，2002）。由于严重的离子丢失导致的离子失衡，通过上调吸收通路和/或同时抑制被动流出等方式可以减弱或恢复，这取决于应激源和鱼的种类（Perry，Laurent，1989；Hwang 等，1996；Chou 等，2002；Chen 等，2003）。

有些离子如 Ca^{2+}的绝对值，可能会影响其他离子的调控。虹鳟急性暴露在含低浓度 Ca^{2+} 的淡水中时会增加 Na^+ 被动流出，而对 Na^+ 流入没有显著影响（McWilliams，1982），然而慢性的高 Ca^{2+}应激几天后会导致 Na^+净丢失（McDonald 等，1980）。已经阐述的可能机制是通过进一步操控水中 Ca^{2+}水平。添加额外 10 mM $CaCl_2$到正常淡水中，一开始会导致离子细胞浓度减少，之后伴随虹鳟鳃 Na^+流入的加强，离子细胞会有补偿性增加（Avalla 等，1987）。持续数周高浓度环境 Ca^{2+}（和 NaCl）的应激显示和低 NaCl 相似的影响，能刺激离子吸收和离子细胞功能（Laurent 等，1985；Chang 等，2001）。

离子细胞的离子转运功能的补偿调节对鱼类处理淡水环境中离子缺乏的应激是关键而重要的。根据免疫组化的连续在体观察和无创扫描离子选择电极技术（SIET），Cl^-吸收功能的激活是由活化的 NCC 表达罗非鱼离子细胞导致的。这些似乎是源于先存的但是在急性低 Cl^-应激中细胞增殖和分化均未发生改变的离子细胞转运蛋白的表达和功能的修饰（Chang 等，2003；Lin 等，2004；Horing 等，2009a）。在渗透压应激慢性阶段的补偿和提高过程后，调节活动都包含相关离子细胞的增殖和分化（Hwang，Lin，2014；Cuh 等，2015）。

有关运输相关离子的特定转运蛋白的鉴定和功能分析方面的分子证据，近来已经可以用来证明鱼类暴露于低渗压应激下的补偿机制。低 Na^+应激会促进鳃或胚胎卵黄囊上皮细胞中 NHE3（罗非鱼和斑马鱼）、Rhcg1、AE1b 和膜型 CA15a（斑马鱼）的表达（Yan 等，2007；Lin 等，2008；Inokuchi 等，2009；Lee 等，2011；Shih 等，2012）。同样，斑马鱼 NCC 和 SCL26 型 AE 以及位于鳟鱼和斑马鱼鳃上皮细胞中的钙离子通道（ECaC）的表达会分别受到低 Cl^-和低 Ca^{2+}应激的刺激（Bayaa 等，2009；Liao 等，2007；Pan 等，2005；Perry 等，2009；Shahsavarani 等，2006；Wang 等，2009）。这些近年来在模式物种中的发现将来一定会延伸到其他的物种。

6.3.2 低 pH 的应激源

环境中急性酸应激破坏了体液中的酸碱平衡和其他离子渗透压平衡，而且鱼类的这些反应会在几分钟到几小时内启动（Evans 等，2005）。考虑到近年来分子和细胞生理学研究方法的进展，这种在慢性和急性阶段的反应和相关的补偿方式需要在一些物种中做更详细的探讨。使用 SIET 技术（Lin 等，2006），斑马鱼胚胎表皮中具有 H^+-ATP 酶富含离子细胞（酸分泌主要细胞），其顶膜 H^+ 分泌活动在低 pH（pH 4）刺激下几小时后相比于对照组会被激活（Furukawa 等，2015）。增强的酸分泌活动可能包含转录后翻译的调节，但既不包含 HA 的转录调节也不包含翻译调节，因为在 pH 4 条件下，它的 mRNA 表达直到第四天也未发生改变（Chang 等，2009）。在 pH 4 应激的慢性阶段，另一方面，HR 离子细胞的补偿酸分泌功能似乎是受到细胞发生或基因组修饰。根据详细的 SIET 和免疫组化分析，这些调节的实现不仅是通过增加 HR 离子细胞的数量，还通过加强单细胞的酸分泌功能（Horng 等，2009b）。而且，HR 细胞的增加源于失去 2（GCM2）-标记的离子细胞前体细胞的神经胶质细胞分化和上皮细胞 p63 标记的干细胞增殖（Chang 等，2009；Horng 等，2009b）。为应对低 pH 应激而引起的细胞发生和基因组的补偿调节活动在其他物种中也有报道（Hwang 等，2011；Hwang，Lin，2014）。

酸性应激会对除了酸碱调节以外的许多生理机制产生影响。在这些机制中，对体液中 Na^+ 平衡影响的研究最充分。根据 Kwong 等（2014）的综述，通过增加被动 Na^+ 流出并同时抑制 Na^+ 流入，外界酸性一般会导致 Na^+ 净丢失。在酸性应激中，显著的被动 Na^+ 流出或许与紧密结合蛋白介导的增大的细胞旁通透性有关（Kwong 等，2014）。同样地，淡水种类表现出显著的全身性 Na^+ 丢失，这主要是在急性酸性应激中，由增加的被动 Na^+ 流出和受损的 Na^+ 吸收所造成。在随后的慢性阶段，斑马鱼体液 Na^+ 平衡的补偿调节似乎依赖于增强的 Na^+ 吸收，而不是依赖于 Na^+ 流出的改变（Kwong 等，2014）。斑马鱼 Na^+ 吸收的补偿主要源于 NCC（但不是 NHE3b）的功能刺激和通过 NCC 表达离子细胞分化的上调（Chang 等，2013），因为 NHE3b 表达由于不恰当的热动力学而受外界的酸性抑制（Yan 等，2007）。

酸性环境也会影响 Cl^- 和 Ca^{2+} 的调节。在酸性应激中，受损的体液 Cl^- 平衡是由于血液 HCO_3^- 的减少并伴随着 Cl^- 通过 SLC26A4 被吸收而导致的（Kwong 等，2014）。至于斑马鱼在酸性应激中 Ca^{2+} 吸收机制的补偿调节，pH 4 应激在开始时会抑制 Ca^{2+} 的吸收功能（机体 Ca^{2+} 含量因而减少）（Horng 等，2009b；Kumai 等，2015），而在暴露于这种应激中几天后，Ca^{2+} 吸收机能会上调（Horng 等，2009b；Kumai 等，2015）。

6.3.3 激素参与

渗透压应激既能引发鱼类的一般反应，也能影响体液离子或酸碱平衡，和这些反应相关的激素活动已经在本书第 2、3、4 章描述。在鱼类中，激素能像在哺乳动物中一样通过调节离子和渗透压调节机能而调节体液的离子和酸碱平衡，而且这些激素活动对由于渗透压或离子应激而导致机体平衡紊乱的补偿调节是非常关键的。Evans 等（2005）全面评述了许多激素在淡水鱼类离子和渗透压调节中的调节作用；然而，多数被提到的详细通路与精确目标的离子转运功能、离子转运蛋白与离子细胞（亚型，细胞阶段）相关的调节活动仍然是不清楚的。只有当明确的离子细胞亚型和它们相关的离子转运蛋白在这些模式物种如斑马鱼、青鳉和罗非鱼中被鉴定出来的时候，才能准确地阐明这些调节通路。此外，激素和受体同源的或内源水平的基因击倒和过表达技术，并且和更多传统的药理学方法相结合，就能够对激素信号通路有更加详细的阐述（Hwang，Chou，2013；Guh 等，2015）。我们在图 6.3 中总结了已有的离子细胞功能的相关模型，该模型能帮助我们更好地理解激素在渗透压应激刺激激导致的体液平衡紊乱的恢复过程中调节离子和酸碱转运的作用。

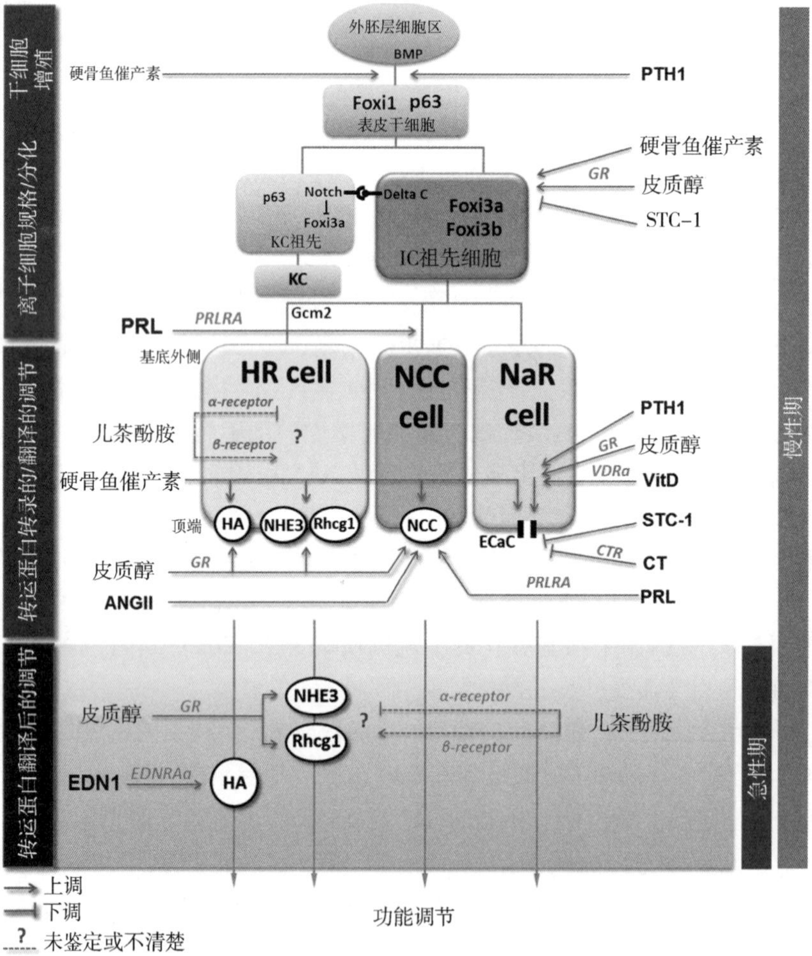

图 6.3 激素活动对低渗透压应激下斑马鱼离子调节过程的不同影响

注：不同的激素分别控制不同离子细胞数量（细胞增殖、特化和分化）、转运蛋白丰富性（mRNA 和蛋白表达）和活性（转录后调节）。关于离子特化和分化，可参考 Hwang 和 Chou（2013）。CA，碳酸酐酶；ClC，氯化物通道；CT，降钙素；CTR，降钙素受体；ECaC，上皮细胞钙离子通道；EDN1，内皮素 1；EDNRAa，内皮素受体 Aa；GCM2，星形胶质细胞 2；GR，糖皮质激素受体；HA，H^+-ATPase；HR，H^+-ATP 酶-rich；NaR，Na^+/K^+-ATPase-rich；NBC，Na^+-HCO_3^- 协同转运蛋白；NCC，Na^+-Cl-协同转运蛋白；NHE，Na^+/H^+ 交换器；NKA，Na^+/K^+-ATP 酶；PRL，催乳素；PRLRA，催乳素受体 A；PTH，副甲状腺激素；RAS，血管紧张素系统；Rhcg，Rh 家族 C 型糖蛋白；STC，司腺钙蛋白；VDRa，维生素 D 受体 a；VitD，维生素 D。

皮质醇、维生素 D（VitD）和甲状旁腺激素曾被报道对鱼类的 Ca^{2+} 吸收发挥积极作用（Evans 等，2005）。近年关于斑马鱼的研究证明皮质醇-糖皮质激素受体（cortisol-GR）、VitD-receptor（VDRa）和 PTH1 对 Ca^{2+} 吸收产生影响，主要是通过顶膜 ECaC 表达的调节而不是基底外侧膜转运蛋白 PMCA2 和 NCX1b（Lin 等，2011，2012，2014）。这一观点支持了假说：在低浓度 Ca^{2+} 暴露下，ECaC 是 Ca^{2+} 调节的靶基因，而 PMCA2 和 NCX1b 不是（Hwang，Lin，2014），这一点与哺乳类动物相同（Hoenderop 等，2005）。采用传统激素注射和/或手术切除激素合成腺体等方法开展相关研究得到的结论认为，降钙激素，如司腺钙蛋白（STC）和降钙素（CT）是鱼类在渗透压和离子应激刺激下其体液中 Ca^{2+} 调节的重要成分。STC-1 和 CT 的表达在外界环境高 Ca^{2+} 浓度应激下会发生上调（Tseng 等，2009a；Lafont 等，2011；Lin 等，2014）。和早期高钙血症激素的描述相似，STC-1 和 CT 在它们的调控作用中也会靶向 ECaC。

在高渗透压应激和低渗透应激下，皮质醇都能够调节水盐平衡（McCormick，2001；Evans 等，2005；Takei 等，2014）。皮质醇-糖皮质激素受体介导的斑马鱼 Na^+ 吸收活动有助于对酸性应激的耐性（Kumai 等，2012a）。另一方面，催乳素（PRL）一直被认为是一种淡水适应的激素；然而准确的目标蛋白仍然没有很好地了解（Breves 等，2014）。对于罗非鱼和斑马鱼，PRL 对 NCC 转录和翻译的作用已经被清晰地阐述，而对 NHE3 和 ECaC 的作用还不清楚（Breves 等，2013）。血管紧张素系统能够参与鱼类体液水平和血压的调控（Takei，Tsuchida，2000），并且近来在斑马鱼的手工操作中还被认为能够刺激 Na^+ 和 Cl^- 活动（Kumai 等，2014）。在神经末梢或嗜铬细胞释放的儿茶酚胺被认为能够通过 α-和 β-肾上腺素能受体对鱼类在淡水的高渗透压调节中产生影响（Evans 等，2005）。在近期的一项斑马鱼研究中，证明 β-肾上腺素能受体有利于酸性应激下 Na^+ 吸收的补偿性增加（Kumai 等，2012b）。

SIET 分析证明了皮质醇对斑马鱼 H^+ 分泌的作用，而基于外源皮质醇对 FOXI3a（一种控制离子运输细胞分化的转录因子）表达的刺激影响和 HR 细胞数量的相关研究，认为这种作用至少在一定程度上是通过调节 HR 离子细胞分化实现的（Cruz 等，2013a，b；Guh 等，2015）。内皮素 1（EDN1）和它的一个受体 EDNRAa 也参与斑马鱼酸分泌的相关机能中，尤其是在急性酸性应激中，并且这似乎是通过 HA 在酸胁迫后的转录介导的（Guh 等，2014）。同样地，EDN1-EDNRAa 和皮质醇-GR 可能分别在酸应激的急性和慢性阶段调节鱼类的酸分泌。

对付低渗透应激的体液离子和酸碱平衡的调节，不仅可以通过转运蛋白活性的调整达到，还可以通过细胞遗传学作用对离子运输细胞的影响实现。离子水平的调节在慢性渗透压刺激阶段是非常重要的，并且受到激素的紧密控制。催产素、皮质醇、PTH1、正调节蛋白和 STC-1，一种负调节蛋白，近来都被鉴定为斑马鱼离子转运蛋白功能的控制因子，主要通过离子运输细胞在慢性刺激阶段的增殖和分化来实现（Chou 等，2011；Cruz 等，2013a，b；Trayer 等，2013；Kwong，Perry，2015）（图 6.3）。这些细胞遗传作用在其他鱼类中是否也比较发达，是非常重要且具有挑战性的问题。

6.4 对稳态的应激感知

渗透压传感器能够检测介质中的盐度变化，从而启动快速的预期的反应，减缓盐度和水平衡的剧烈变化，这在鳗鱼转入海水中1分钟内即产生 Cl^- 反应并开始吞水的现象中得到例证（Hirano，1974）。然而，似乎存在着多个对渗透压应激的传感器负责检测细胞外液体积或特定离子浓度的变化，进而启动多种生理的和行为的反应，以应对应激。在这里，我们总结了渗透压感知领域的新发现，主要是从传感器分子水平到非基因组和基因组水平等方面诱导体液平衡的反应（图6.1）。特别着重于近期通过转录组方法鉴定的渗透压感知转录因子和它们的靶标基因。

6.4.1 渗透压传感器

尽管近期在本系列丛书的一卷中已经综述了渗透压的感知机制（Kültz，2013），但对鱼类渗透压传感器相关的认识还很少。在脊椎动物中，相似的分子很可能都可以作为渗透压传感器，所以，为了未来的鱼类研究，我们将会介绍在非鱼类物种中渗透压传感器已有的相关知识。涉及盐度敏感的高血压的盐度感受机制在哺乳动物中已进行大量研究（Orlov，Mongin，2007），揭示出至少存在四种不同的感知机制：①用于TRPV家族成员控制的加压素释放的脑脊液（CSF）渗透压（Clapham等，2001）；②通过未能鉴定的体积调节阴离子通道而由神经胶质细胞调节的渗透压（Bres等，2000）；③CSF中的非典型电压门 Na^+ 通道Nax控制的 Na^+ 浓度（Noda，2007）；④肾脏致密斑细胞肾小管液中 Cl^- 对肾小管-肾小球中的反馈调节和NKCC2驱动的肾素释放（Bell等，2003）。钙感知受体（CaSR）也是一种盐度传感器（Loretz，2008）。外界渗透压的改变导致细胞体积改变能通过细胞骨架张力的变化而检测出来。

TRPV是一类机械、热力和化学敏感的无选择性阴离子通道，其中，TRPV1和TRPV4被认为是渗透压传感器，能对内在的（血浆）渗透压改变产生反应。这种反应由细胞膜延伸的变化而驱动，并且影响因素包括不同的转运蛋白、渗透压调节的激素释放和参与渗透压调节的不同基因的表达（Orlov，Mongin，2007；Sladek，Johnson，2013）。缺少N端结构域的TRPV1位于哺乳动物产生加压素的大细胞神经元和感觉室周器官如穹窿下器（SFO）和终板血管器（OVLT）。TRPV1似乎作为一种高渗透压传感器作用于神经垂体中的加压素释放及调节饮水。事实上，*trpv1* 敲除的小鼠在脑室内高渗NaCl缓冲液和缺水条件下，和同窝对照组小鼠相比会表现出加压素分泌减少并且饮水减少（Naeini等，2006）。*Trpv4* 也在这些脑中区域表达，并且它的敲除能够破坏小鼠正常的加压素分泌和饮水活动，导致血容量减少。这些敲除的小鼠在高渗透压力下还表现为加压素分泌减少及饮水活动减少（Liedtke，Friedman，2003）。因此，在正常动物中，TRPV4被认为是一种低渗透压传感器，而TRPV1和TRPV4可能形成异质二聚体以感知高渗透压应激。

研究者通过芯片方法已经对TRPV家族成员的分子进化模式做了研究，并且在红鳍东方鲀（*Takifugu rubripes*）中已经鉴定出两种 TRPV1 和一种 TRPV4（Saito，Shingai，2006）。TRPV1 曾在南亚野鲮（*Labeo rohita*）的精子细胞膜中发现（Majhi 等，2013），在斑马鱼的神经元细胞膜中亦有发现（Gau 等，2013）。然而，这些研究都是检测其作为热力传感器的功能，但是在鱼类中其作为渗透压传感器的功能还需要研究。在莫桑比克罗非鱼脑垂体细胞中，TRPV4 在低渗诱导的催乳素分泌过程中也发挥重要作用（Seale 等，2012）。因为 *trpv* 基因广泛存在于已研究的物种中，它们作为渗透压传感器的功能是一个有意义的研究方向。

Nax，一种可能的 Na^+ 传感器，已经在哺乳类中被鉴定出来（Noda，2007）。Nax 能够监控细胞外液 Na^+ 浓度，并且水缺失导致的 Nax 增加能通过 SFP 或者 OVLT 抑制 Na^+ 吸收。在 $Nax^{-/-}$ 小鼠中，水缺失会诱导较大的神经元活动，并不能抑制 Na^+ 吸收（Watanabe 等，2000）。因为 Nax 免疫活性集中在神经胶质的加工过程，因此，从神经胶质细胞到神经元一定会有抑制性信号转导的作用。另外，上皮细胞 Na^+ 通道已经在大细胞的加压素神经元中发现，进一步说明 Nax 是一种 Na^+ 传感器（Teruyama 等，2012）。尽管 Nax 同源物已经在鱼类中被鉴定出来，但是它作为 Na^+ 传感器的功能是一个值得关注的主题。然而，所有先前提及的实验都表明高渗的 NaCl 或水缺失都会导致细胞外液 Na^+ 和 Cl^- 增加。因此，在这些室周器官中还可能存在 Cl^- 传感器，这和认为它们存在于鱼类体表的观点相似（见第 6.2.1 节）。

亨利环厚升支（TAHL）中的致密斑能够感知腔液中的 Cl^- 而不是 Na^+，并且能够调节哺乳动物肾小球过滤速率和肾素释放（Kotchen 等，1978；Wilcox，1983）。Cl^- 传感器被认为是 NKCC2，它在 TAHL 上皮细胞顶膜上游大量地分布（Bell 等，2003；Edwards 等，2014）。通过 NKCC2 流入的四种渗透物诱导的细胞体积增加能够诱导 ATP 和其他血管收缩剂的释放，并通过肾小球来减少血液流动和肾素释放（Boudreault，Grygorczyk，2002）。NKCC2 作为 Cl^- 传感器的作用在硬骨鱼类中是个有意义的课题，因为它们的肠道有顶膜 NKCC2，这与 TAHL 相似（Grosell，2011）。正如之前提到的，用含 Cl^- 的缓冲液灌注肠腔能够抑制饮水活动，因此，NKCC2 可能在鱼类中也能感知 Cl^-。

CaSR 是 G 蛋白受体家族 C 的其中一员，包括化学受体如信息素受体和嗅觉受体。在低毫摩尔范围内结合无机二价阳离子 Ca^{2+} 和 Mg^{2+} 之后，CaSR 能够通过磷脂酶或 MAPK 激活细胞内信号级联通路。因为 CaSR 位于体表，如鳃和皮肤，并且海水中包含高浓度 Ca^{2+}（10 mM）和 Mg^{2+}（50 mM），可以合理地推测鱼类细胞外的 CaSR 可以作为一种渗透压传感器感知二价阳离子（Nearing 等，2002；Loretz，2008）。事实上，低溶质的盐浓度在莫桑比克罗非鱼（Loretz 等，2004）和白斑角鲨（Fellner，Parker，2004）中能够显著增加 CaSR 对 Ca^{2+} 的亲和力。CaSR 对 Ca^{2+} 的亲和力比对 Mg^{2+} 的亲和力要高 3 倍，然而海水中 Mg^{2+} 浓度要比 Ca^{2+} 高 5 倍。因此，在海水中，CaSR 可能像感知 Ca^{2+} 一样地感知 Mg^{2+}。

6.4.2 传感器的信号转导

渗透压传感器感受到渗透压刺激后通过细胞内信号级联能够启动一系列活动来保

持体液平衡。然而在鱼类中相关的认识较少，研究多集中于单细胞，包括酵母菌（Lopertz-Maury，2008；de Nadal 等，2011），其相似的机制或许可以应用于鱼类细胞的在体研究。

当细胞处于高渗 NaCl 缓冲液中时，它们会快速失水并且又逐渐恢复。为了抵抗皱缩，细胞能够积累不干扰酶活性的小分子有机物（Yancy 等，1982），以平衡溶质中的渗透压。近年来已在鳗鱼和罗非鱼中揭示肌醇生物合成通道产生的渗透物的调节作用（Kalujnala 等，2013；Sacchi 等，2013）。巧合的是，细胞能通过不同细胞质蛋白磷酸化和去磷酸化作用来应对快速变化，如细胞骨架重组、囊泡运输、代谢调节和其他机能，以保持细胞的功能（de Nadal 等，2011）。另外，渗透压调节基因的转录调节发生在渗透压应激的慢性反应阶段，其中的信号级联通路已在酵母菌中得到全面阐述（Saito，Takahashi，2004）。在感知到渗透压增加后，应激激活蛋白激酶信号通路（SPAK）被激活以进行信号传导。该通路主要的责任分子是 p38 相关的 MAPK，它能够通过连续的激酶磷酸化作用而被激活（Capaldi 等，2008），进一步通过转录因子磷酸化作用使渗透压调节相关基因表达（Alepus 等，2001）。

相似的渗透压反应可能在哺乳动物肾髓质细胞（Sheikh-Hamad，Gustin，2004）和鱼鳃细胞中出现。SAPK 通路的主要分子还是 p38 MAPK，它不仅能激活渗透压调节基因表达，还能调节 mRNA 的稳定性和翻译（Cuadrado，Nebreda，2010）。在能直接感知渗透压的鱼鳃中，至少有两条路线可能用来应对应激（图 6.1）。最快的机制是利用细胞骨架网络；渗透压力通过闭合或张开连接外部介质的顶膜来诱导离子转运细胞机械的或形态的改变。关于哺乳动物加压素对水通道蛋白 2（AQP2）的影响以及胰岛素对葡萄糖转运蛋白 4 影响的研究都证明细胞渗透压反应能通过胞吐作用从细胞内募集新的转运蛋白，并通过胞吞作用将老的转运蛋白去除并移到细胞质内（Leto，Saltiel，2012；Möller，Fention，2012）。这种转运蛋白标记的囊泡交换通过由激酶调节的细胞骨架活动介导。然而，囊泡交换在鱼类中还未被证实，部分原因是加压催产素不是一个抗利尿激素，并且胰岛素在鱼类中也不是低血糖激素。MAPKs 和其他激酶在渗透压调节上皮细胞中的作用是鱼类研究的下一个目标。

除了细胞骨架活动，渗透压感受器信号启动一系列基因的表达，从最早基因表达开始，多数可能是渗透压敏感的转录因子；由于研究表明紧张性增强子结合蛋白对哺乳动物的陆生适应起作用（Lam 等，2004），因而推测这些转录因子能够控制对高渗透压应激的适应。这些基因也叫渗透压反应元件结合蛋白或活化 T 细胞中的核因子。ORE 的共有序列已经被阐明（Ferraris 等，1999），并在哺乳动物 AQP2 和尿素转运蛋白 A1 基因启动子区域中被发现。研究结果表明，在鱼类中，对渗透压应激反应的最早基因可能调控参与渗透压调节的基因表达。

6.4.3 细胞内信号级联通路靶标

靶标效应器的活性主要受到两种机制的调节：发生最快的基于细胞骨架的机制和渗透压敏感转录因子的转录调节。所有的细胞都有细胞骨架，一种丝状聚合物和调节蛋白连通的网络（Fletcher，Mullins，2010）。累积的证据表明细胞骨架在细胞体积感知和细胞内物质运输中都发挥重要的作用。在弹涂鱼的皮肤中，NaCl 分泌离子运输细胞被外周细胞环绕，并且只有顶膜通过裂口暴露在外部介质中。鱼类转移到低渗环境后，裂口会由于外周细胞表面延伸而消失，这可能会覆盖离子细胞以致终止在海水时的离子分泌（Sakamoto 等，2000）。使用尤斯房室（Ussingchamber）方法开展的一项体外研究表明，鳉科鱼类在暴露于淡水后会立刻增加跨上皮抗性，这可能是由于外周细胞关闭了裂口所导致（Daborn 等，2001）。这种闭合是由环绕裂口的表皮扁平细胞形态改变而引起，因为它会被细胞色素氧化酶 D 抑制。

当广盐性鱼类从淡水进入到海水中时，离子吸收的离子细胞一定会将其功能转换为分泌，并且这种转换会是立刻的，使体内离子平衡保持在一个小范围内。最容易且最快的方式就是利用具有分泌功能的转运蛋白替换离子吸收转运蛋白，这可以通过膜上吸收转运蛋白内部化（胞吞作用）和通过细胞内交换的方式募集细胞囊泡中的分泌转运蛋白到细胞膜上（胞吐作用）。这种交换调节是通过骨架中压缩蛋白磷酸化来实现（Valenti 等，2005），但是目前对于鱼类膜交换方面的信息还知之甚少。

在渗透压应激中，鱼类渗透压调节器官中最早的基因转录调节发生在几个小时之内。这类最早基因大多可能是能够调节渗透调节效应基因表达的转录因子。首次在鱼类中报道的相关基因是渗透转录因子 1（OSTF-1），它是人类 TSC22 家族蛋白 3 的一种同源基因，TSC22 在人类中被鉴定为一种亮氨酸拉链蛋白，能够作为一种转录调节器（Fiol，Kültz，2004）。在渗透压刺激 2 小时后，罗非鱼鳃的这些基因表达会上调。由于 OSTF-1 基因在其他广盐性鱼类如鳗鱼和青鳉鳃中也会受到海水刺激后上调表达（Tse 等，2008），因此，它成为一种广受关注的研究海水适应的主要基因。然而，在体 OSTF-1 基因的击倒对青鳉的海水适应没有影响（Tse 等，2011），这表明 OSTF-1 在海水适应中的作用有限。另一项通过差异显示技术的研究在大西洋鲑中检测到一个值得关注的富含甘氨酸的 RNA 结合蛋白，这种蛋白和哺乳动物低温诱导 RNA 结合蛋白相似，在高渗透压应激 48 小时后，鳃瓣的该基因表达会发生上调（Pan 等，2004）。这个基因的表达在低温应激中并未上调，并且在高渗透压应激中，它在肝脏，肾脏和心脏中的表达亦都没有上调。

渗透压应激反应的相关基因采用微阵列方法已经在经历淡水到海水以及海水到淡水转换 12 小时后的长颌姬虾虎鱼鳃中鉴定出来（Evans，Somero，2008，2009）。以基因本体论数据库为基础的差异表达基因的功能分类结果表明它们被区分为细胞信号、能源生存、转录调节、细胞外基质、细胞骨架、蛋白交换、离子稳态、有机渗透物和其他功能分类。在所有注释基因中，FK506 结合蛋白 51（FKBP-51）和转录控制肿瘤蛋白是皮质醇及 NKA 活动的主要调节对象，而皮质醇和 NKA 在渗透压适应调节中都起重要作用。激素相关基因如胰岛素、催乳素和 GH（IGF-1 和生长激素抑制素）会对渗透压刺激产生反应。在淡水和海水中“回家”的红大麻哈鱼（*Oncorhynchus nerka*），鳃和肝脏的转录组分析已经完成，但是其中差异表达基因是很少的（Evans 等，2011）。他们认为，渗透压基因转录调节的改变是预期的，正如“回家”的大麻哈鱼（*Oncorhynchus keta*）在从海水进入淡水前，海水型离子细胞会消失，而淡水型离子细胞出现（Uchida 等，1997）。基于微阵列的比较转录组测序也在广盐性鳉科鱼类——雨点卢氏鳉（*Lucania parva*）的海水种群和淡水种群以及狭盐性淡水鱼蓝鳍卢氏鳉（*Lucania goodie*）的鳃器官中进行；检测到许多与离子运输和细胞黏附相关的差异表达基因（Kozak 等，2013）。另外，在不同渗透压环境下，白鲑（*Coregonus lavaretus*）幼鱼的大规模蛋白组测序分析中发现了细胞因子和 Ca^{2+} 与 Na^+ 转运蛋白是渗透压敏感蛋白（Papakostas 等，2012）。

适应淡水、海水及高盐度海水的广盐性罗非鱼鳃和消化道的转录组测序分析已经完成（Li 等，2014）；在高渗透压溶质中，鳃中典型的 NaCl 分泌运输蛋白（NKAα1 和 α3，CFTR，NKCC1a）和肠道中 NaCl 吸收蛋白（NKAα1 和 α3，NKCC2）表达均会上调。另外，NKCC1b 和 CFTR 基因在罗非鱼肠道中上调，表明在硬骨鱼肠道中有分泌型上皮细胞的存在。比较转录组分析还被用于广盐性莫桑比克罗非鱼和狭盐性尼罗罗非鱼肠道研究（Ronkin 等，2015），结果表明，在高渗透压应激中，血管紧张素转化酶（AQP8）、SLAC34A2（Na^+-磷酸盐共转运蛋白）、SLC43 A2（不依赖 Na^+ 的氨基酸转运蛋白）在莫桑比克罗非鱼中上调，而在尼罗罗非鱼中下调。这些基因可能参与莫桑比克罗非鱼对海水高盐耐受性的调节。在青鳉（*Oryzias latipes*）中，高渗透压应激后几个小时内 19 个基因的表达会发生显著上调（Wong 等，2014）。然而，其中只有 5 个对渗透压应激的反应是特异的，其余 14 个基因在淡水-淡水转移时也会发生反应，表明它们是对人为操作应激产生的反应。这些应激特异性基因包括 CCAAT/增强子结合蛋白（CEBP）β 和 δ，DNA 修复和重组蛋白 RAD54-like 2（RAD54L2），低氧诱导因子 3α（HIF3A）和 LIM 结构域结合蛋白 1（LDB1）。CEBP 结合位点已经在血管紧张素基因和转运蛋白基因如 NKCC2 和 CFTR 中鉴定出来。鱼类中其他参与渗透压反应的基因仍然需要探索。渗透压敏感转录因子如血清/糖皮质激素调节激酶 1（SKG1）和 TSC22D3（OSTF-1）都包含在对渗透压和手工操作应激引起反应的基因之中。它们也处在其他渗透压调节基因的蛋白作用网络中，这表明它们都参与渗透压的调节（Wong 等，2014）。

6.5 对渗透压应激反应的能量代谢

能量供应对于耗能生理过程如渗透压反应来说是限速因子。为了应对渗透压应激，鱼类需要激活和新合成转运蛋白、酶，或者其他相关蛋白，以完成补偿性调节过程，这些过程都需要额外的能量。这可能影响到鱼类的生长、发育和生殖（Sadoul，Vijayan，2016；见本书第 5 章），需要对鱼类应激产生的代谢活动进行一个全面的讨论。

6.5.1 氧消耗

传统的测量渗透压相关的能量消耗方式是在不同盐度处理下测量鱼整个有机体的

氧消耗速率的变化。测量氧气消耗速率通常在鱼类渗透压研究中均有开展（Boeuf，Payan，2001；Ern 等，2014；Morgan，Iwama，1991）。盐度增加伴随着氧气消耗的改变，鲑科鱼类的生长速度会下降。然而，通过比较不同的物种发现，氧气消耗和环境盐度之间并没有正相关性（Morgan，Iwama，1991）。一个物种的自然史可能决定了它面对盐度改变后的代谢反应类型，也就是说，最低代谢速率和一个物种主要栖息的环境相关。另一方面，包括大量物种的近期一篇综述认为，在正常生活史的盐度中或者等渗性的盐度中，最低氧气吸收没有明显的趋向（Ern 等，2014）。Boeuf 和 Payan（2001）认为，为了应对渗透压应激，许多鱼类会消耗能量预算的 20% ～ 50%。Morgan 和 Iwama（1991）认为，通过决定整个生物体氧气消耗来量化渗透压应激后的能量消耗，可能会和其他代谢通路混淆。为了克服在整个鱼体研究中存在的内在问题，分离鳃（或其他组织）的代谢速率可用来估算特定渗透压调节器官的能量消耗（Tseng，Hwang，2008）。在分离的虹鳟鱼鳃中，乌本苷（一种 NKA 抑制剂）敏感的或巴弗洛霉素（一种 HA 抑制剂）敏感的氧气消耗和 ATP 或离子流驱动的氧气消耗一致（Kirschner，1995；Morgan，Iwama，1991）。鳃细胞的氧气消耗速率似乎和全鱼的氧气消耗相比要小些（小 4% ～ 20%）（McCormick 等，1989；Morgan，Iwama，1991）。

一些鱼类在自然界高盐水中生活得很好；它们的能量代谢如何进行而使鱼类能够应对这种严酷的环境，在鱼类渗透压调节中是一个重要的课题。在近期的综述中，Gonzalez（2002）提出一个普遍的策略，能抑制鳃组织的水通透性并增加鳃 NKA 表达活性和离子细胞的大小和密度，而又不出现增加代谢速率的明显趋势。在高盐水中代谢速率相对于海水中的变化较小，这主要是由于减少鳃的氧气通透性、降低常规活动和其他因素导致的氧气吸收减少；然而，目前还没有结论性的答案（Gonzalez，2012）。同样地，氧气消耗确实提供了和渗透压相关的一些重要的基本信息。然而，解析这些和应对渗透压能量消耗相关的数据，需要考虑到物种、生活阶段、适应状态以及生理参数的测量，而最重要的是所采用的研究方法。

6.5.2　代谢修饰

除了评估代谢的成本，不同代谢燃料和代谢组织重组的作用对于了解鱼类如何对付渗透应激也是很重要的。氧氮熵（O∶N）源于氧气消耗速率和氨分泌速率的比例，并且当所消耗的底物从脂质到蛋白质时会减少。从淡水转到海水中的 15 ～ 30 天之间，小锯盖鱼（*Centropomus parallulus*）有利用脂质（碳水化合物）替换蛋白作为能量底物的趋势（da Silva Rocha 等，2005）。相似地，减少使用蛋白作为能量底物趋势也在锯盖鱼（*Centropomus undecimalis*）从淡水转到海水 1 ～ 2 周的时间内出现（Gracia-Lopez 等，2006）。这些发现表明，鱼类为应对渗透压应激产生的代谢活动会在不同的中间代谢通路中发生转变，提示在高盐情况下，保留细胞溶质中游离氨基酸的策略和在较高盐度中细胞体积的调节相关（da Silva Rocha 等，2005；Gracia-Lopez 等，2006）；许多物种在高渗透压刺激后组织氨基酸水平增加进一步证实了上述观点

（Assem，Hanke，1983；Bystriansky 等，2007；Chang 等，2007a；Fiess 等，2007；Tok 等，2009）。

对能量底物和相关酶的直接研究也支持了为应对渗透压应激伴随的代谢重组和转变的观点。黄鳝（*Monopterus albus*）在从淡水转到海水后，肌肉和肝脏游离氨基酸水平（尤其是谷氨酰胺）和谷氨酰胺合成酶表达增加，并且伴随着胺分泌减少。这表明氨基酸分解代谢减少，并且一些非必需氨基酸的合成增加，以便调节面对渗透压应激时的细胞体积（Tok 等，2009）。渗透压应激导致的代谢重组似乎会临时发生改变。糖原分解可能提供从淡水到海水中急性渗透压应激所需要的糖类，并且来自氨基酸的糖原形成也相当重要（Baltzegar 等，2014）。肝糖原含量在渗透压应激 24 小时内就会变得稳定（Chang 等，2007b），之后糖原合成被激发，然后进一步合成脂肪（Baltzegar 等，2014）。血糖和氨基酸水平以及肝脏氨基转移酶活性在一些鱼类，包括罗非鱼、北极红点鲑和海鲷在受到盐胁迫 96 小时（Bystriansky 等，2007）后被激发。

渗透压应激导致的代谢重组也有组织特异性。海鲷从淡水转到海水后，肝脏糖原分解加剧而糖分解潜能下降，鳃和脑中糖酵解活动加强（Sangiao-Alvarellos 等，2003）。这些发现表明，通过糖原合成，肝脏是供应葡萄糖的主要器官，由它供应到鳃和脑中的葡萄糖进一步驱动离子转运机能和其他细胞过程以应对渗透压应激。

渗透压应激后脂代谢相关的信息是零碎的且部分是有矛盾的。北极红点鲑的鳃、肝脏和红肌肉中肉碱棕榈酰转移酶、三羟基酰基辅酶 A 脱氢酶（HOAD）和苹果酸脱氢酶活性在高渗透压应激 96 小时后没有改变。这些影响可能与血浆中未酯化脂肪酸水平比较恒定有关（Bystriansky 等，2007）。然而，海鲷从淡水转到海水 96 小时后血浆甘油三酯水平增加（Sangiao-Alcarellos 等，2003）。北极红点鲑在转移到海水中 2 个月内，葡萄糖 6 磷酸脱氢酶开始增加，从而使肝脏用于脂肪合成的等价物减少并且激活 HOAD 活性。这说明进入海水后肝脏脂质代谢能力增强（Aas-Hansen 等，2005）。我们认为这些都是鱼类在转入高渗环境时缓解渗透压应激以保持渗透压稳定的作用。

前面的讨论提到糖异生在渗透压应激和急性酸应激过程中对于能量代谢非常重要。根据 PCK1 敲除、葡萄糖供应和 SIET 等实验结果来看（Furukawa 等，2015），磷酸烯醇式丙酮酸 1（PCK1）介导的糖异生在斑马鱼中由酸性胁迫诱导，主要通过 HR 离子细胞中 HA 通道较高速率的酸分泌活动。而且，谷氨酰胺和谷氨酸似乎是柠檬酸循环中间物的主要补充物质，它会被 PCK 的活动削减。在近期的一项研究中（Zikos 等，2014），罗非鱼对高渗环境的适应导致鳃中 NKCC、细胞色素氧化酶Ⅳ（线粒体电子运输链中一个重要的酶）、糖原磷酸化酶（GP）、过氧化氢酶体增殖激活受体 γ 共激活子 1α（PGC-1α，一种线粒体的生物调节蛋白）等的表达加强。另外，标准代谢速率也上升。所有这些数据说明，为应对低渗透压应激，线粒体生物合成参与较多的能量消耗（Zikos 等，2014）。同样地，近期的研究对与离子和酸碱调节相关的复杂能量代谢通路在细胞或亚细胞水平提供了较为深入的阐述，这些还需要在其他物种中做进一步探讨。

6.5.3 代谢产物运输

肝脏通过糖酵解为其他组织输出葡萄糖以满足它们（包括渗透压调节器官）在生理和细胞水平需求的各种能量消耗。脑和肌肉也储存了一定数量的糖原，但是和肝脏相比则非常少。在哺乳动物的脑中，糖原主要储存于星形胶质细胞和星形细胞中，在神经元中没有（Pfeiffer-Guglielmi 等，2003），糖原会降解为乳酸并能够从星形胶质细胞穿梭到高耗能的神经元中，尤其在能量缺乏的时候（Brown 等，2003；Pfeiffer-Guglielmi 等，2007）。近来在罗非鱼和斑马鱼鳃中已发现相同的模式（Hwang 等，2011）（图6.4）。在罗非鱼中，一种新型鳃细胞，糖原丰富（GLR）细胞，被认为是一个能量库，能够在急性盐应激下及时为鳃离子细胞提供能量（Tseng 等，2007）。GLR 细胞靠近离子运输细胞，富含糖原，并且都表达 GP（鳃特异型）和糖原合成酶基因（Chang 等，2007b；Tseng 等，2007）。鳃 GR 细胞的糖异生作用在受到海水刺激1～3小时内即被激活，用来为邻近的离子运输细胞快速提供能量并可能用于提升离子分泌能力。随后（海水刺激6～12小时）通过糖异生提供的能量可能来源于肝

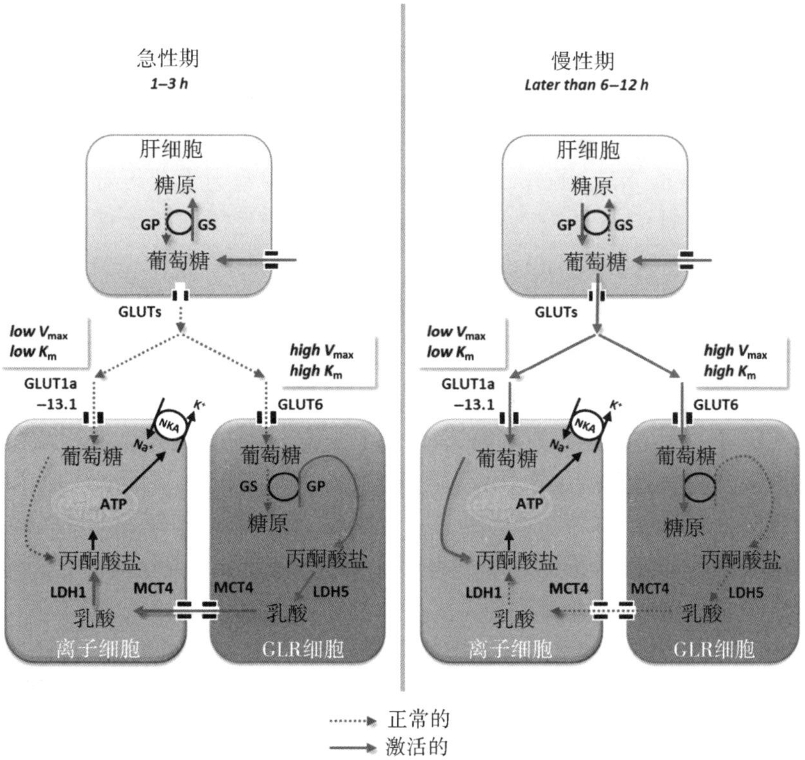

图6.4 渗透压应激时离子细胞的碳水化合物能量供应模型

注：详细参考文中第6.5.3节。蓝色连接线和红色线分别表明正常状态和激活状态。GLUT，糖转运蛋白；GLR，富糖原；GP，糖原磷酸化酶；GS，糖原合成酶；LDH，乳酸脱氢酶；MCT，单羧酸转运蛋白。（见书后彩图）

脏（Chang 等，2007b；Tseng 等，2007）。同样地，肝脏是身体中核心的糖类储藏所，然而在急性盐胁迫下，鳃似乎能够作为一个局部能源供应地（图 6.4）。GLR 细胞与离子细胞之间，肝细胞和离子细胞之间的能量迁移的时间和空间关系和哺乳动物星形胶质细胞与神经元之间以及肝细胞与神经元之间的关系是相似的。

相比其他代谢物（丙氨酸和油酸），罗非鱼鳃细胞能够氧化更多的葡萄糖和乳酸盐（Perry，Walsh，1989）。这些能源底物如何被运输到燃料离子细胞中去应对渗透压应激是另一个重要问题（Tseng 等，2009b）。许多 GLUTs 已经在鱼类中被鉴别出来，并且不止一个在渗透压感受器官（鳃、肾脏、肠道等）有表达（Hall 等，2014；Polakof 等，2012；Tseng 等，2009b；Tseng，Hwang，2008）。令人惊讶的是，几乎没有关于鱼类面对渗透压应激后 GLUT 在能量转移中起作用的相关信息（Balmaceda-Aguilera 等，2012；Tseng 等，2009b）。金头鲷（*Sparus aurata*）在适应不同的盐浓度时，鳃和脑中 GLUT1 基因的 mRNA 表达会增加（Balmaceda-Aguilera 等，2012），表明加强葡萄糖运输能够为鳃抵抗渗透压应激产生的离子调节提供能量。因为鳃通常能够表达多个同种类型的 GLUT，并且已经产生不同类型的有着相应离子转运蛋白和功能的离子细胞，因此，不鉴定出 GLUT 在每种鳃细胞中的特定亚型，就很难描述渗透压适应过程中能量转移的准确细胞和分子机制。在斑马鱼体内鉴定出的 18 个 GLUT 亚型中，GLUT-1a 和 GLUT-13.1 在不同类型的鳃和皮肤的离子细胞中表达，而 GLUT6 在 GLR 细胞中特异表达（Tseng 等，2009b）。

这些 GLUTs 的葡萄糖转运功能在特异离子缺失的应激中会被激活以便为相应的离子细胞供能（Tseng 等，2009b，2001）。深入的传递动力学分析着重说明在渗透压应激下鳃离子细胞和 GLR 细胞之间的空间关系对于能量供应的生理重要性。在斑马鱼中，GLUT-1a（Km = 3.4 mM）和 GLUT-13.1（1.5 mM）使离子细胞能够有效地对抗已糖吸收，以便能够调节由于渗透压应激而产生过多能量消耗所导致的低血糖情况下的离子运输（Tseng 等，2011）。与之相反，GLUT6（Km = 17.5 mM）使 GLR 细胞能够吸收来自餐后血液中过多的葡萄糖用于能量储存。当急性渗透压应激已经引发鱼鳃中离子运输调节和急性能量需求时，上述过程就是非常重要的。但还需要在其他物种中进行更多的研究来支持这个观点。

6.6 结束语和展望

在本章中，我们想要综述关于鱼类在不同组织中以不同时间范围面对各种渗透压应激如何保持水和离子平衡的最近研究进展。我们主要关注渗透压应激的主题，因为该领域已被深入阐述，并且在不同的应激源中，离子和水调节的基本离子机制是相似的。对于鱼类如何感知应激、如何通过细胞内信号通路传递信息，以及最终如何通过不同的基因组和非基因组水平的机制应对应激的相关数据都正在快速积累。需要在未来研究中关注的主题都已在每一节中进行了讨论，但是重要的主题是：

（1）直接和环境水接触的上皮细胞的体积调节，如皮肤和鳃，需要参考酵母菌中的机制做进一步阐述。

（2）二价离子应对高渗透应激的调节机制，尤其是 Mg^{2+} 和 SO_4^{2-}，与单价离子一样重要（Na^+ 和 Cl^-）。

（3）SLC26A4 介导的碱分泌机制比起 NHE 介导的酸分泌机制相关认识还比较肤浅，需要在硬骨鱼类中做进一步研究。

（4）顶端 H^+-ATP 酶介导的酸分泌机制已经在斑马鱼中得到阐述，由于硬骨鱼类机制变化多样（Takei 等，2014），所以还需要在其他淡水鱼中进一步确认。

（5）转运蛋白和离子细胞在高渗透压或酸应激下的补偿反应，需要在除模式物种（斑马鱼、罗非鱼和鲑鱼）之外的其他鱼类中做进一步研究。

（6）斑马鱼中出现的适应渗透压应激的如催产素、皮质醇、PTH1 和 STC-1 的细胞遗传作用，需要在其他物种中研究。

（7）鱼类中已经开展一些关于渗透压感受器和信号转导通路的研究，但是基于哺乳动物和酵母菌获得的数据还需要做进一步分析。

（8）环境的渗透压应激似乎主要通过位于口腔和消化道腔上皮细胞上的 Cl^- 感受器，而感受器分子的一致性需要在广盐性鱼类中做进一步阐述。

（9）和渗透压相关的能量代谢详细通路的近期研究进展需要在其他物种中做进一步探讨。

（10）在盐度应激下鳃细胞和肝脏细胞之间能量移动的时间和空间关系需要在除罗非鱼之外的其他鱼类中做更多研究来证实。

Y. 竹井祥郎　P. –P. 黄　著
张　勇　译
林浩然　校

参考文献

Aas-Hansen, O., Vijayan, M. M., Johnsen, H. K., Cameron, C. and Jorgensen, E. H. (2005). Resmoltification in wild, anadromous Arctic char (Salvelinus alpinus): a survey of osmoregulatory, metabolic, and endocrine changes preceding annual seawater migration. Can. J. Fish. Aquat. Sci. 62, 195-204.

Alepuz, P. M., Jovanovic, A., Reiser, V. and Ammerer, G. (2001). Stress-induced MAP kinase Hog 1 is part of transcription activation complexes. Mol. Cell. 7, 767-777.

Ando, M. and Nagashima, K. (1996). Intestinal Na^+ and Cl^- levels control drinking behavior in the seawater-adapted eel, Anguilla japonica. J. Exp. Biol. 199, 711-716.

Ando, M. and Takei, Y. (2014). Intestinal absorption of salts and water. In Eel Physiology (eds. F. Trischitta, Y. Takei and P. Sébert), pp. 160-177. Boca Raton: CRC Press.

Armesto, P. , Campinho, M. A. , Rodriguez-Rua, A. , Cousin, X. , Power, D. M. , Manchado, M. , et al. (2014). Molecular characterization and transcriptional regulation of the Na^+/K^+ ATPase a subunit isoforms during development and salinity challenge in a teleost fish, the Senegalese sole (Solea senegalensis). Comp. Biochem. Physiol. 175B, 23-38.

Assem, H. and Hanke, W. (1983). The significance of the amino-acids during osmotic adjustment in teleost fish. I. Changes in the euryhaline Sarotherodon mossambicus. Comp. Biochem. Physiol. 74A, 531-536.

Avella, M. and Bornancin, M. (1989). A new analysis of ammonia and sodium transport through the gills of the freshwater rainbow trout (Salmo gairdneri). J. Exp. Biol. 142, 155-175.

Avella, M. , Masoni, A. , Bornancin, M. and Mayergostan, N. (1987). Gill morphology and sodium influx in the rainbow-trout (Salmo gairdneri) acclimated to artificial freshwater environments. J. Exp. Zool. 241, 159-169.

Avella, M. , Ducoudret, O. , Pisani, D. F. and Poujeol, P. (2009). Swelling-activated transport of taurine in cultured gill cells of sea bass: physiological adaptation and pavement cell plasticity. Am. J. Physiol. 296, R1149-R1160.

Balmaceda-Aguilera, C. , Martos-Sitcha, J. A. , Mancera, J. M. and Martinez-Rodriguez, G. (2012). Cloning and expression pattern of facilitative glucose transporter 1 (GLUT1) in gilthead sea bream Sparus aurata in response to salinity acclimation. Comp. Biochem. Physiol. 163A, 38-46.

Baltzegar, D. A. , Reading, B. J. , Douros, J. D. and Borski, R. J. (2014). Role for leptin in promoting glucose mobilization during acute hyperosmotic stress in teleostfishes. J. Endocrinol. 220, 61-72.

Bayaa, M. B. , Vulesevic, A. , Esbaugh, M. , Braun, M. E. , Ekker, M. and Perry, S. F. (2009). The involvement of SLC26 anion transporters in chloride uptake in zebrafish (Danio rerio) larvae. J. Exp. Biol. 212, 3283-3295.

Bell, P. D. , Lapointe, J. Y. and Peti-Peterdi, J. (2003). Macula densa cell signaling. Annu. Rev. Physiol. 65, 481-500.

Beuchat, C. A. (1996). Structure and concentrating ability of the mammalian kidney: correlations with habitat. Am. J. Physiol. 271, R157-R179.

Beyenbach, K. W. (2004). Kidney sans glomeruli. Am. J. Physiol. 286, F811-F827.

Bickler, P. E. and Buck, L. T. (2007). Hypoxia tolerance in reptiles, amphibians, andfishes: life with variable oxygen availability. Annu. Rev. Physiol. 69, 145-170.

Boeuf, G. and Payan, P. (2001). How should salinity influence fish growth? Comp. Biochem. Physiol. 130C, 411-423.

Boudreault, F. and Grygorczyk, R. (2002). Cell swelling-induced ATP release and gadoliniumsensitive channels. Am. J. Physiol. 282, C216-C226.

Bres, V., Hurbin, A., Duboid, A., Orcel, H., Moos, F. C., Rabie, A., et al. (2000). Pharmacological characterization of voltage-sensitive, taurine permeable anion channels in rat supraoptic glial cells. Br. J. Pharmacol. 130, 1976-1982.

Breves, J. P., Serizier, S. B., Goffin, V., McCormick, S. D. and Karlstrom, R. O. (2013). Prolactin regulates transcription of the ion uptake Na^+/Cl^- cotransporter (ncc) gene in zebrafish gill. Mol. Cell. Endocrinol. 369, 98-106.

Breves, J. P., McCormick, S. D. and Karlstrom, R. O. (2014). Prolactin and teleostionocytes: new insights into cellular and molecular targets of prolactin in vertebrate epithelia. Gen. Comp. Endocrinol. 203, 21-28.

Brinn, R. P., Marcon, J. I., McComb, D. M., Gomes, L. C., Abreu, J. S. and Baldisseroto, B. (2012). Stress responses of the endemic freshwater cururu stingrey (Potamotrygon cf. histrix) during trasnportation in the Amazon region of the Rio Negro. Comp. Biochem. Physiol. 162A, 139-145.

Brown, A. M., Tekkok, S. B. and Ransom, B. R. (2003). Glycogen regulation and functional role in mouse white matter. J. Physiol. 549, 501-512.

Bystriansky, J. S., Frick, N. T. and Ballantyne, J. S. (2007). Intermediary metabolism of Arctic char Salvelinus alpinus during short-term salinity exposure. J. Exp. Biol. 210, 1971-1985.

Capaldi, A. P., Kaplan, T., Liu, Y., Habib, N., Regev, A., Friedman, N., et al. (2008). Structure and function of a transcriptional network activated by the MAPK Hog 1. Nat. Genet. 40, 1300-1306.

Chang, I. C., Lee, T. H., Yang, C. H., Wei, Y. Y., Chou, F. I. and Hwang, P. P. (2001). Morphology and function of gill mitochondria-rich cells infish acclimated to different environments. Physiol. Biochem. Zool. 74, 111-119.

Chang, I. C., Wei, Y. Y., Chou, F. I. and Hwang, P. P. (2003). Stimulation of Cl uptake and morphological changes in gill mitochondria-rich cells in freshwater tilapia (Oreochromis mossambicus). Physiol. Biochem. Zool. 76, 544-552.

Chang, E. W., Loong, A. M., Wong, W. P., Chew, S. F., Wilson, J. M. and Ip, Y. K. (2007a). Changes in tissue free amino acid contents, branchial Na^+/K^+-ATPase activity and bimodal breathing pattern in the freshwater climbing perch, Anabas testudineus (Bloch), during seawater acclimation. J. Exp. Zool. 307, 708-723.

Chang, J. C., Wu, S. M., Tseng, Y. C., Lee, Y. C., Baba, O. and Hwang, P. P. (2007b). Regulation of glycogen metabolism in gills and liver of the euryhaline tilapia (Oreochromis mossambicus) during acclimation to seawater. J. Exp. Biol. 210, 3494-3504.

Chang, W. J., Horng, J. L., Yan, J. J., Hsiao, C. D. and Hwang, P. P. (2009). The transcription factor, glial cell missing 2, is involved in differentiation and functional regulation of H^+-ATPase-rich cells in zebrafish (Danio rerio). Am. J. Physiol. 296, R1192-1201.

Chang, W. J., Wang, Y. F., Hu, H. J., Wang, J. H., Lee, T. H. and Hwang, P. P. (2013). Compensatory regulation of Na + absorption by Na^+/H^+ exchanger and Na^+-Cl^- cotransporter in zebrafish (Danio rerio). Front. Zool. 10, 46.

Chen, Y. Y., Lu, F. I. and Hwang, P. P. (2003). Comparisons of calcium regulation infish larvae. J. Exp. Zool. 295, 127-135.

Chou, M. Y., Yang, C. H., Lu, F. I., Lin, H. C. and Hwang, P. P. (2002). Modulation of calcium balance in tilapia larvae (Oreochromis mossambicus) acclimated to low-calcium environments. J. Comp. Physiol. B 172, 109-114.

Chou, M. Y., Hung, J. C., Wu, L. C., Hwang, S. P. and Hwang, P. P. (2011). Isotocin controls ion regulation through regulating ionocyte progenitor differentiation and proliferation. Cell. Mol. Life Sci. 68, 2797-2809.

Chou, M. Y., Lin, C. H., Chao, P. L., Hung, J. C., Cruz, S. A. and Hwang, P. P. (2015). Stanniocalcin-1 controls ion regulation functions of ion-transporting epithelium other than calcium balance. Int. J. Biol. Sci. 11, 122-132.

Claiborne, J. B., Edwards, S. L. and Morrison-Shetlar, A. I. (2002). Acid-base regulation infishes: cellular and molecular mechanisms. J. Exp. Zool. 293, 302-319.

Clapham, D. E., Runnels, L. W. and Strubing, C. (2001). The TRP ion channel family. Nat. Rev. Neurosci. 2, 387-396.

Cruz, S. A., Chao, P. L. and Hwang, P. P. (2013a). Cortisol promotes differentiation of epidermal ionocytes through Foxi3 transcription factors in zebrafish (Danio rerio). Comp. Biochem. Physiol. 164A, 249-257.

Cruz, S. A., Lin, C. H., Chao, P. L. and Hwang, P. P. (2013b). Glucocorticoid receptor, but not mineralocorticoid receptor, mediates cortisol regulation of epidermal ionocyte development and ion transport in zebrafish (Danio rerio). PLoS One 8, e77997.

Cuadrado, A. and Nebreda, A. R. (2010). Mehanisms and functions of p38 MAPK signalling. Biochem. J. 429, 405-417.

Daborn, K., Cozzi, R. R. F. and Marshall, W. S. (2001). Dynamics of pavement cell-chloride cell interactions during abrupt salinity change in Fundulus heteroclitus. J. Exp. Biol. 204, 1889-1899.

da Silva Rocha, A. J., Gomes, V., Van Ngan, P., de Arruda Campos Rocha Passos, M. J. and Rios Furia, R. (2005). Metabolic demand and growth of juveniles of Centropomus parallelus as function of salinity. J. Exp. Mar. Biol. Ecol. 316, 157-165.

deNadal, E. , Ammerer, G. and Posas, F. (2011). Controlling gene expression in response to stress. Nat. Rev. 12, 833-845.

Duranton, C. , Mikulovic, E. , Tauc, M. , Avella, M. and Poujeol, P. (2000). Potassium channel in primary culture of seawaterfish gill cells II. Channel activation by hypotonic shock. Am. J. Physiol. 279, R1659-R1670.

Edwards, A. , Castrop, H. , Laghmani, K. , Vallon, V. and Layton, A. T. (2014). Effects of NKCC2 isoform regulation on NaCl transport in thick ascending limb and macula densa: amodeling study. Am. J. Physiol. 307, F137-F146.

Edwards, S. L. and Marshall, W. S. (2012). Principles and Patterns of Osmoregulation and Euryhalinity in Fishes. In Fish Physiology-Euryhaline Fishes, Vol. 32 (eds. S. C. McCormick, A. P. Farrell and C. J. Brauner), pp. 1-44. London: Academic Press.

Ern, R. , Huong, D. T. , Cong, N. V. , Bayley, M. and Wang, T. (2014). Effect of salinity on oxygen consumption infishes: a review. J. Fish. Biol. 84, 1210-1220.

Evans, D. H. , Piermarini, P. M. and Choe, K. P. (2005). The multifunctionalfish gill: dominant site of gas exchange, osmoregulation, acid-base regulation, and excretion of nitrogenous waste. Physiol. Rev. 85, 97-177.

Evans, T. G. and Somero, G. N. (2008). A macroarray-based transcriptomic time-course of hyper-and hypo-osmotic stress signaling events in the euryhalinefish, Gillichthys mirabilis: osmosensors to effectors. J. Exp. Biol. 211, 3636-3649.

Evans, T. G. and Somero, G. N. (2009). Protein-protein interactions enable rapid adaptive response to osmotic stress infish gills. Commun. Integr. Biol. 2, 94-96.

Evans, T. G. , Hammil, E. , Kaukinen, K. , Schulze, A. D. , Patterson, D. A. , English, K. K. , et al. (2011). Transcriptomics of environmental acclimatization and survival in wild adult Pacific sockeye salmon (Oncorhynchus nerka) during spawning migration. Mol. Ecol. 20, 4472-4489.

Fellner, S. K. and Parker, L. (2004). Ionic strength and the polyvalent cation receptor of shark rectal gland and artery. J. Exp. Zool. 301A, 235-239.

Ferraris, J. D. , Williams, C. K. , Ohtaka, A. and GarcÍa-Pérez, A. (1999). Functional consensus for mammalian osmotic response elements. Am. J. Physiol. 276, C667-673.

Fiess, J. C. , Kunkel-Patterson, A. , Mathias, L. , Riley, L. G. , Yancey, P. H. , Hirano, T. , et al. (2007). Effects of environmental salinity and temperature on osmoregulatory ability, organic osmolytes, and plasma hormone profiles in the Mozambique tilapia (Oreochromis mossambicus). Comp. Biochem. Physiol. 146A, 252-264.

Fiol, D. F. and Kültz, D. (2004). Rapid hyperosmotic coinduction of two tilapia

(Oreochromis mossambicus) transcription factors in gill cells. Proc. Natl. Acad. Sci. U. S. A. 102, 927-932.

Fletcher, D. A. and Mullins, R. D. (2010). Cell mechanics and the cytoskeleton. Nature 463, 485-492.

Furukawa, F., Tseng, Y. C., Liu, S. T., Chou, Y. L., Lin, C. C., Sung, P. H., et al. (2015). Induction of phosphoenolpyruvate carboxykinase (PEPCK) during acute acidosis and its role in acid secretion by V-ATPase-expressing ionocytes. Int. J. Biol. Sci. 11, 712-725.

Gau, P., Poon, J., Ufret-Vincenty, C., Snelson, C. D., Gordon, S. E., Raible, D. W., et al. (2013). The zebrafish ortholog of TRPV1 is required for heat-induced locomotion. J. Neurosci. 33, 5249-5260.

Gilmour, K. M. (2012). New insights into the many functions of carbonic anhydrase infish gills. Respir. Physiol. Neurobiol. 184, 223-230.

Gonzalez, R. J. (2012). The physiology of hyper-salinity tolerance in teleostfish: a review. J. Comp. Physiol. B 182, 321-329.

Gorissen, M. and Flik, G. (2016). Endocrinology of the Stress Response in Fish. In Fish Physiology-Biology of Stress in Fish, Vol. 35 (eds. C. B. Schreck, L. Tort, A. P. Farrell and C. J. Brauner), San Diego, CA: Academic Press.

Gracia-Lopez, V., Rosas-Vazquez, C. and Brito-Perez, R. (2006). Effects of salinity on physiological conditions in juvenile common snook Centropomus undecimalis. Comp. Biochem. Physiol. 145A, 340-345.

Grosell, M. (2011). The role of the gastrointestinal tract in salt and water balance. In Fish Physiology-The Multifunctional Gut of Fish, Vol. 30 (eds. M. Grosell, A. P. Farrell and C. J. Brauner), pp. 135-164. San Diego: Academic Press.

Grosell, M., Mager, E. M., Williams, C. and Taylor, J. R. (2009). High rates of HCO_3^- secretion and Cl^- absorption against adverse gradients in the marine teleost intestine: the involvement of an electrogenic anion exchanger and H^+-pump metabolon?. J. Exp. Biol. 212, 1684-1696.

Guh, Y. J., Tseng, Y. C., Yang, C. Y. and Hwang, P. P. (2014). Endothelin-1 regulates H^+ ATPase-dependent transepithelial H^+ secretion in zebrafish. Endocrinology 155, 1728-1737.

Guh, Y. J., Lin, C. H. and Hwang, P. P. (2015). Osmoregulation in zebrafish: ion transport mechanisms and functional regulation. EXCLI J. 14, 627-659.

Hall, J. R., Clow, K. A., Short, C. E. and Driedzic, W. R. (2014). Transcript levels of class I GLUTs within individual tissues and the direct relationship between GLUT1 expression and glucose metabolism in Atlantic cod (Gadus morhua). J. Comp. Physiol. B 184, 483-496.

Hirano, T. (1974). Some factors regulating water intake in the eel, Anguilla japonica. J. Exp. Biol. 61, 171-178.

Hiroi, J. and McCormick, S. D. (2012). New insights into gill ionocyte and ion transporter function in euryhaline and diadromousfish. Respir. Physiol. Neurobiol. 184, 257-268.

Hoenderop, J. G., Nilius, B. and Bindels, R. J. (2005). Calcium absorption across epithelia. Physiol. Rev. 85, 373-422.

Hoffmann, E. K., Lambert, I. H. and Pedersen, S. F. (2009). Physiology of cell volume regulation in vertebrates. Physiol. Rev. 89, 193-277.

Horng, J. L., Hwang, P. P., Shih, T. H., Wen, Z. H., Lin, C. S. and Lin, L. Y. (2009a). Chloride transport in mitochondrion-rich cells of euryhaline tilapia (Oreochromis mossambicus) larvae. Am. J. Physiol. 297, C845-854.

Horng, J. L., Lin, L. Y. and Hwang, P. P. (2009b). Functional regulation of H^+-ATPase-rich cells in zebrafish embryos acclimated to an acidic environment. Am. J. Physiol. 296, C682-692.

Hsu, H. H., Lin, L. Y., Tseng, Y. C., Horng, J. L. and Hwang, P. P. (2014). A new model forfish ion regulation: identification of ionocytes in freshwater-and seawater-acclimated medaka (Oryzias latipes). Cell. Tissue. Res. 357, 225-243.

Hwang, P. P. and Chou, M. Y. (2013). Zebrafish as an animal model to study ion homeostasis. Pflugers. Arch. 465, 1233-1247.

Hwang, P. P. and Lee, T. H. (2007). New insights intofish ion regulation and mitochondrionrich cells. Comp. Biochem. Physiol. 148A, 479-497.

Hwang, P. P. and Lin, L. Y. (2014). Gill ion transport, acid-base regulation and nitrogen excretion. In The Physiology of Fishes (eds. D. Evans, J. B. Claiborne and S. Currie), pp. 205-233. Boca Raton: CRC Press.

Hwang, P. P., Tung, Y. C. and Chang, M. H. (1996). Effect of environmental calcium levels on calcium uptake in tilapia larvae (Oreochromis mossambicus). Fish Physiol. Biochem. 15, 363-370.

Hwang, P. P., Lee, T. H. and Lin, L. Y. (2011). Ion regulation infish gills: recent progress in the cellular and molecular mechanisms. Am. J. Physiol. 301, R28-47.

Inokuchi, M., Hiroi, J., Watanabe, S., Hwang, P. P. and Kaneko, T. (2009). Morphological and functional classification of ion-absorbing mitochondria-rich cells in the gills of Mozambique tilapia. J. Exp. Biol. 212, 1003-1010.

Kalujnala, S., Gellatly, S. A., Hazon, N., Villasenor, A., Yancy, P. H. and Cramb, G. (2013). Seawater acclimation and inositol monophosphatase isoform expression in the Enropean eel (Anguilla anguilla) and Nile tilapia (Oreochromis niloticus). Am. J. Physiol. 305, R369-R384.

Kirschner, L. B. (1995). Energetics of osmoregulation in fresh-water vertebrates. J. Exp. Zool. 271, 243-252.

Kotchen, T. A., Galla, J. H. and Luke, R. G. (1978). Contribution of chloride to the inhibition of plasma renin by sodium chloride in rat. Kidney Int. 13, 201-207.

Kozak, G. M., Brennan, R. S., Berdan, E. L., Fuller, R. C. and Whitehead, A. (2014). Functional and population genomic divergence within and between two species of killifish adapted to different osmotic niches. Evolution 68 (1), 63-80.

Kültz, D. (2012). Osmosensing. In Fish Physiology-Euryhaline Fishes, Vol. 32 (eds. S. C. McCormick, A. P. Farrell and C. J. Brauner), pp. 45-68. London: Academic Press.

Kumai, Y., Nesan, D., Vijayan, M. M. and Perry, S. F. (2012a). Cortisol regulates Na + uptake in zebrafish, Danio rerio, larvae via the glucocorticoid receptor. Mol. Cell. Endocrinol. 364, 113-125.

Kumai, Y., Ward, M. A. and Perry, S. F. (2012b). beta-Adrenergic regulation of Na^+ uptake by larval zebrafish Danio rerio in acidic and ion-poor environments. Am. J. Physiol. 303, R1031-1041.

Kumai, Y., Bernier, N. J. and Perry, S. F. (2014). Angiotensin-II promotes Na^+ uptake in larval zebrafish, Danio rerio, in acidic and ion-poor water. J. Endocrinol. 220, 195-205.

Kumai, Y., Kwong, R. W. and Perry, S. F. (2015). A role for transcription factor glial cell missing 2 in Ca^{2+} homeostasis in zebrafish, Danio rerio. Pflugers Arch. 467, 753-765.

Kurita, Y., Nakada, T., Kato, A., Doi, H., Mistry, A. C., Chang, M. H., et al. (2008). Identification of intestinal bicarbonate transporters involved in formation of carbonate precipitates to stimulate water absorption in marine teleost fish. Am. J. Physiol. 294, R1402-R1412.

Kwong, R. W. and Perry, S. F. (2015). An essential role for parathyroid hormone in gill formation and differentiation of ion-transporting cells in developing zebrafish. Endocrinology en20141968.

Kwong, R. W., Kumai, Y. and Perry, S. F. (2014). The physiology offish at low pH: the zebrafish as a model system. J. Exp. Biol. 217, 651-662.

Lafont, A. G., Wang, Y. F., Chen, G. D., Liao, B. K., Tseng, Y. C., Huang, C. J., et al. (2011). Involvement of calcitonin and its receptor in the control of calcium-regulating genes and calcium homeostasis in zebrafish (Danio rerio). J. Bone Miner. Res. 26, 1072-1083.

Lam, A. K. M., Ko, B. C. B., Tam, S., Morris, R., Yang, J. Y., Chung, S. K., et al. (2004). Osmotic response element-binding protein (OREBP) is an essential regulator of the urine concentrating mechanism. J. Biol. Chem. 279, 48048-48054.

Laurent, P., Hobe, H. and Dunelerb, S. (1985). The role of environmental sodium-chloride relative to calcium in gill morphology of fresh-water salmonidfish. Cell Tissue Res. 240, 675-692.

Lee, Y. C., Yan, J. J., Cruz, S. A., Horng, J. L. and Hwang, P. P. (2011). Anion exchanger 1b, but not sodium-bicarbonate cotransporter 1b, plays a role in transport functions of zebrafish H^+-ATPase-rich cells. Am. J. Physiol. 300, C295-307.

Leto, D. and Saltiel, A. R. (2012). Regulation of glucose transport by insulin: traffic control of GLUT4. Nat. Rev. 13, 383-396.

Li, Z., Lui, E. Y., Wilson, J. M., Ip, Y. K., Lin, Q., Lam, T. J., et al. (2014). Expression of key ion transporters in the gill and esophageal-gastrointestinal tract of euryhaline Mozambique tilapia, Oreochromis mossambicus, acclimated to fresh water, seawater and hypersaline water. PLoS One 9, e87591.

Liao, B. K., Deng, A. N., Chen, S. C., Chou, M. Y. and Hwang, P. P. (2007). Expression and water calcium dependence of calcium transporter isoforms in zebrafish gill mitochondrionrich cells. BMC Genomics 8, 354.

Liedtke, W. and Friedman, J. M. (2003). Abnormal osmotic regulation in trpv4/ mice. Proc. Natl. Acad. Sci. U. S. A. 100, 13698-13703.

Lin, C. H., Huang, C. L., Yang, C. H., Lee, T. H. and Hwang, P. P. (2004). Time-course changes in the expression of Na, K-ATPase and the morphometry of mitochondrion-rich cells in gills of euryhaline tilapia (Oreochromis mossambicus) during freshwater acclimation. J. Exp. Zool. 301, 85-96.

Lin, C. H., Tsai, I. L., Su, C. H., Tseng, D. Y. and Hwang, P. P. (2011). Reverse effect of mammalian hypocalcemic cortisol infish: cortisol stimulates Ca^{2+} uptake via glucocorticoid receptor-mediated vitamin D3 metabolism. PLoS One 6, e23689.

Lin, C. H., Su, C. H., Tseng, D. Y., Ding, F. C. and Hwang, P. P. (2012). Action of vitamin D and the receptor, VDRa, in calcium handling in zebrafish (Danio rerio). PLoS One 7, e45650.

Lin, C. H., Su, C. H. and Hwang, P. P. (2014). Calcium-sensing receptor mediates Ca^{2+} homeostasis by modulating expression of PTH and stanniocalcin. Endocrinology 155, 56-67.

Lin, L. Y., Horng, J. L., Kunkel, J. G. and Hwang, P. P. (2006). Proton pump-rich cell secretes acid in skin of zebrafish larvae. Am. J. Physiol. 290, C371-378.

Lin, T. Y., Liao, B. K., Horng, J. L., Yan, J. J., Hsiao, C. D. and Hwang, P. P. (2008). Carbonic anhydrase 2-like a and 15a are involved in acid-base regulation and Na^+ uptake in zebrafish H^+-ATPase-rich cells. Am. J. Physiol. 294,

C1250-1260.

Lionetto, M. G. , Giordano, M. E. , de Nuccio, F. , Nicolardi, G. , Hoffmann, E. K. and Schettino, T. (2005). Hypotonicity induced K^+ and anion conductive pathways activation in eel intestinal epithelium. J. Exp. Biol. 208, 749-760.

Lópetz-Maury, L. , Marguerat, S. and Bähler, J. (2008). Tuning gene expression to changing environments: from rapid responses to evolutionary adaptation. Nat. Rev. 9, 583-593.

Loretz, C. A. (2008). Extracellular calcium-sensing receptors infish. Comp. Biochem. Physiol. 149A, 225-245.

Loretz, C. A. , Pollina, C. , Hyodo, S. , Takei, Y. , Chang, W. and Shoback, D. (2004). cDNA cloning and functional expression of a Ca^{2+}-sensing receptor with truncated carboxyterminal tail from the Mozambique tilapia (Oreochromis mossambicus). J. Biol. Chem. 279, 53288-53297.

Madsen, S. S. , Kiilerich, P. and Tipsmark, C. K. (2009). Multiplicity of expression of Na^+, K^+-ATPase a-subunit isoforms in the gill of Atlantic salmon (Salmo salar): cellular localisation and absolute quantification in response to salinity change. J. Exp. Biol. 212, 78-88.

Madsen, S. S. , Bujak, J. and Tipsmark, C. K. (2014). Aquaporin expression in the Japanese medaka (Oryzias latipes) in freshwater and seawater: challenging the paradigm of intestinal water transport?. J. Exp. Biol. 217, 3108-3121.

Majhi, R. K. , Kumar, A. , Yadav, M. , Swain, N. , Kumari, S. , Saha, A. , et al. (2013). Thermosensitive ion channel TRPV1 is endogenously expressed in the sperm of a fresh water teleostfish (Labeo rohita) and regulates sperm motility. Channels 7, 483-492.

Marshall, W. S. (2011). Mechanosensitive signalling infish gill and other ion transporting epithelia. Acta Physiol. 202, 487-499.

Marshall, W. S. , Howard, J. A. , Cozzi, R. R. F. and Lynch, E. M. (2002). NaCl andfluid secretion by the intestine of the teleost Fundulus heteroclitus: involvement of CFTR. J. Exp. Biol. 205, 745-758.

Marshall, W. S. , Katoh, F. , Main, H. P. , Sers, N. and Cozzi, R. R. F. (2008). Focal adhesion kinase and b1 integrin regulation of Na^+, K^+, $2Cl^-$ cotransporter in osmosensing ion transporting cells of killifish, Fundulus heteroclitus. Comp. Biochem. Physiol. 150A, 288-300.

Marshall, W. S. , Watters, K. D. , Hovdestad, L. R. , Cozzi, R. R. F. and Katoh, F. (2009). CFTR Cl^- channel functional regulation by phosphorylation of focal adhesion kinase at tyrosine 407 in osmosensitive ion transporting mitochondria rich cells of euryhaline killifish. J. Exp. Biol. 212, 2365-2377.

McCormick, S. D. (2001). Endocrine control of osmoregulation in teleostfish. Am. Zool. 41.

McCormick, S. D., Moyes, C. D. and Ballantyne, J. S. (1989). Influence of salinity on the energetics of gill and kidney of Atlantic salmon (Salmo salar). Fish Physiol. Biochem. 6, 243-254.

McCormick, S. C., Farrell, A. P. and Brauner, C. J. (2012). Fish Physiology-Euryhaline Fishes, Vol. 32. London: Academic Press.

McDonald, D. G. and Rogano, M. S. (1986). Ion regulation by the rainbow-trout, Salmo gairdneri, in ion-poor water. Physiol. Zool. 59, 318-331.

McDonald, D. G., Hobe, H. and Wood, C. M. (1980). The influence of calcium on the physiological responses of the rainbow trout, Salmo gairdneri, to low environmental pH. J. Exp. Biol. 88, 109-131.

McDonald, G. and Milligan, L. (1997). Ionic, osmotic and acid-base regulation instress. In Fish Stress and Health in Aquaculture (eds. G. K. Iwama, A. D. Pickering, J. P. Sumpter and C. B. Schreck), pp. 119-144. Cambridge: Cambridge University Press.

McWilliams, P. G. (1982). The effects of calcium on sodiumfluxes in the brown trout, Salmo trutta, in neutral and acid water. J. Exp. Biol. 96, 439-442.

Möller, H. B. and Fenton, R. A. (2012). Cell biology of vasopressin-regulated aquaporin-2 trafficking. Pflugers. Arch. 464, 133-144.

Morgan, J. D. and Iwama, G. K. (1991). Effects ofsalinity on growth, metabolism, and ion regulation in juvenile rainbow and steelhead trout (Oncorhynchus mykiss) and fall chinook salmon (Oncorhynchus tshawytscha). Can. J. Fish. Aquat. Sci. 48, 2083-2094.

Nakada, T., Hoshijima, K., Esaki, M., Nagayoshi, S., Kawakami, K. and Hirose, S. (2007a). Localization of ammonia transporter Rhcg1 in mitochondrion rich cells of yolk sac, gill, and kidney of zebrafish and its ionic strength-dependent expression. Am. J. Physiol. 293, R1743-1753.

Nakada, T., Westhoff, C. M., Kato, A. and Hirose, S. (2007b). Ammonia secretion fromfish gill depends on a set of Rh glycoproteins. FASEB J. 21, 1067-1074.

Nearing, J., Betka, M., Quinn, S., Hentschel, H., Elger, M., Baum, M., et al. (2002). Polyvalent cation receptor proteins (CaRs) are salinity sensor infish. Proc. Natl. Acad. Sci. U. S. A. 99, 9231-9236.

Noda, M. (2007). Hydromineral neuroendocrinology: mechanism of sensing sodium levels in the mammalian brain. Exp. Physiol. 92, 513-522.

Olson, K. R. (2002). Vascular anatomy of thefish gill. J. Exp. Zool. 293, 214-231.

Orlov, S. N. and Mongin, A. A. (2007). Salt sensing mechanisms in blood pressure regulation and hypertension. Am. J. Physiol. 293, H2039-H2053.

Pan, F., Zarate, J., Choudhury, A., Rupprecht, R. and Bradley, T. M. (2004). Osmotic stress of salmon stimulates upregulation of a cold inducible RNA binding protein (CIRP) similar to that of mammals and amphibians. Biochemie 86, 451-461.

Pan, T. C., Liao, B. K., Huang, C. J., Lin, L. Y. and Hwang, P. P. (2005). Epithelial Ca^{2+} channel expression and Ca^{2+} uptake in developing zebrafish. Am. J. Physiol. 289, R1202-1211.

Papakostas, S., Vasemagi, A., Vaha, J.-P., Himberg, M., Peil, L. and Primmer, C. R. (2012). A proteomics approach reveals divergent molcular responses to salinity in populations of European whitefish (Coregonus lavaretus). Mol. Ecol. 21, 3516-3530.

Parks, S. K., Tresguerres, M. and Goss, G. G. (2008). Theoretical considerations underlying Na^+ uptake mechanisms in freshwater fishes. Comp. Biochem. Physiol. 148C, 411-418.

Perry, S. F. and Laurent, P. (1989). Adaptational responses of rainbow-trout to lowered external NaCl concentration-contribution of the branchial chloride cell. J. Exp. Biol. 147, 147-168.

Perry, S. F., Vulesevic, B. and Bayaa, M. (2009). Evidence that SLC26 anion transporters mediate branchial chloride uptake in adult zebrafish (Danio rerio). Am. J. Physiol. 297, R988-997.

Perry, S. F. and Walsh, P. J. (1989). Metabolism of isolatedfish gill cells: contribution of epithelial chloride cells. J. Exp. Biol 144, 507-520.

Peter, M. C. S. (2011). The role of thyroid hormones in stress response offish. Gen. Comp. Endocrinol. 172, 198-210.

Pfeiffer-Guglielmi, B., Fleckenstein, B., Jung, G. and Hamprecht, B. (2003). Immunocytochemical localization of glycogen phosphorylase isozymes in rat nervous tissues by using isozyme-specific antibodies. J. Neurochem. 85, 73-81.

Pfeiffer-Guglielmi, B., Francke, M., Reichenbach, A. and Hamprecht, B. (2007). Glycogen phosphorylase isozymes and energymetabolism in the rat peripheral nervous system-an immunocytochemical study. Brain Res. 1136, 20-27.

Polakof, S., Panserat, S., Soengas, J. L. and Moon, T. W. (2012). Glucose metabolism infish: a review. J. Comp. Physiol. B 182, 1015-1045.

Redding, J. M., Schreck, C. B., Barks, E. K. and Ewing, R. D. (1984). Cortisol and its effects on plasma thyroid hormone and electrolyte concentrations in fresh water and during seawater acclimation in yearing coho salmon, Oncorhynchus kisutch. Gen. Comp. Endocrinol. 56, 146-155.

Reindl, K. M. and Sheridan, M. A. (2012). Peripheral regulation of the growth hormoneinsulin-like growth factor system infish and other vertebrates. Comp. Biochem. Physiol. 163A, 231-245.

Ronkin, D., Seroussi, E., Nitzan, T., Dron-Faigenboim, A. and Cnaani, A. (2015). Intestinal transcriptome analysis revealed differential salinity adaptation between two tilapiine species. Comp. Biochem. Physiol. 13D, 35-43.

Sacchi, R., Li, J., Villarreal, F., Gardell, A. M. and Kultz, D. (2013). Salinity-induced regulation of the myo-inositol biosynthetic pathway in tilapia gill epithelium. J. Exp. Biol. 216, 4626-4638.

Sadoul, B. and Vijayan, M. M. (2016). Stress and Growth. In Fish Physiology-Biology of Stress in Fish, Vol. 35 (eds. C. B. Schreck, L. Tort, A. P. Farrell and C. J. Brauner), San Diego, CA: Academic Press.

Saito, H. and Takahashi, K. (2004). Regulation of the osmoregulatory HOC MAPK cascade in yeast. J. Biochem. 136, 267-272.

Saito, S. and Shingai, R. (2006). Evolution of thermoTRP ion channel homologs in vertebrates. Physiol. Genomics 27, 219-230.

Sakamoto, T., Yokota, S. and Ando, M. (2000). Rapid morphological oscillation of mitochondrion-rich cell in estuarine mudskipper following salinity changes. J. Exp. Zool. 286, 666-669.

Sangiao-Alvarellos, S., Laiz-Carrion, R., Guzman, J. M., Martin del Rio, M. P., Miguez, J. M., Mancera, J. M., et al. (2003). Acclimation of S. aurata to various salinities alters energy metabolism of osmoregulatory and nonosmoregulatory organs. Am. J. Physiol. 285, R897-907.

Schreck, C. B. and Tort, L. (2016). The Concept of Stress in Fish. In Fish Physiology-Biology of Stress in Fish, Vol. 35 (eds. C. B. Schreck, L. Tort, A. P. Farrell and C. J. Brauner), San Diego, CA: Academic Press.

Seale, A. P., Watanabe, S. and Grau, E. G. (2012). Osmoreception: perspectives on signal transduction and environmental modulation. Gen. Comp. Endocrinol. 176, 354-360.

Shahsavarani, A., McNeill, B., Galvez, F., Wood, C. M., Goss, G. G., Hwang, P. P., et al. (2006). Characterization of a branchial epithelial calcium channel (ECaC) in freshwater rainbow trout (Oncorhynchus mykiss). J. Exp. Biol. 209, 1928-1943.

Sharif Naeini, R., Witty, M. F., Seguera, P. and Bourque, C. W. (2006). An N-terminal variant of Trpv1 channelis required for osmosensory transduction. Nat. Neurosci. 9, 93-98.

Sheikh-Hamad, D. and Gustin, M. C. (2004). MAP kinases and the adaptive response to hypertonicity: functional preservation from yeast to mammals. Am. J. Physiol. 287,

F1102-F1110.

Shih, T. H., Horng, J. L., Hwang, P. P. and Lin, L. Y. (2008). Ammonia excretion by the skin of zebrafish (Danio rerio) larvae. Am. J. Physiol. 295, C1625-1632.

Shih, T. H., Horng, J. L., Liu, S. T., Hwang, P. P. and Lin, L. Y. (2012). Rhcgl and NHE3b are involved in ammonium-dependent sodium uptake by zebrafish larvae acclimated to lowsodium water. Am. J. Physiol. 302, R84-93.

Sinha, A. K., Rasoloniriana, R., Dasan, A. F., Pipralia, N., Blust, R. and De Boeck, G. (2015). Interactive effect of high environmentalammonia and nutritional status on ecophysiological performance of European sea bass (Dicentrarchus labrax) acclimated to reduced seawater salinities. Aquat. Toxicol. 160, 39-56.

Skomal, G. B. and Mandelman, J. W. (2012). The physiological response to anthropogenic stressors in marine elasmobranchfishes: a review with a focus on the secondary response. Comp. Biochem. Physiol. 162A, 146-155.

Sladek, C. D. and Johnson, A. K. (2013). Integration of thermal and osmotic regulation of water homeostasis: the role of TRPV channels. Am. J. Physiol. 305, R669-R678.

Suzuki, Y., Itakura, M., Kashiwagi, M., Nakamura, N., Matsuki, T., Sakuta, H., et al. (1999). Identification by differential display of a hypertonicity-inducible inward rectifier potassium channel highly expressed in chloride cells. J. Biol. Chem. 274, 11376-11382.

Takahashi, H., Sakamoto, T. and Narita, K. (2006). Cell proliferation and apoptosis in the anterior intestine of an amphibious, euryhaline mudskipper (Periophthalmus modestus). J. Comp. Physiol. 176B, 463-468.

Takei, Y. and Loretz, C. A. (2006). Endocrinology. In The Physiology of Fishes (eds. D. H. Evans and J. B. Claiborne), third ed., pp. 271-318. Boca Raton: CRC Press.

Takei, Y. and Loretz, C. A. (2010). The gastrointestinal tract as an endocrine/ neuroendocrine/ paracrine organ: organization, chemical messengers and physiological targets. In Fish Physiology-The Multifunctional Gut of Fish, Vol. 30 (eds. M. Grosell, A. P. Farrell and C. J. Brauner), pp. 261-317. London: Academic Press.

Takei, Y. and McCormick, S. D. (2012). Hormonal control offish euryhalinity. In Fish Physiology-Euryhaline Fishes, vol. 32 (eds. S. D. McCormick, A. P. Farrell and C. J. Brauner), pp. 69-123. San Diego: Academic Press.

Takei, Y. and Tsuchida, T. (2000). Role of the renin-angiotensin system in drinking of seawateradapted eels Anguilla japonica: a reevaluation. Am. J. Physiol. 279, R1105-R1111.

Takei, Y., Hiroi, J., Takahashi, H. and Sakamoto, T. (2014). Diverse mechanisms for bodyfluid regulation in teleost fishes. Am. J. Physiol. 307, R778-R792.

Teruyama, R., Sakuraba, M., Wilson, L. L., Wandrey, L. E. and Armstrong, W. E. (2012). Epithelial Na^+ sodium channels in magnocellular cells of the rat supraoptic and paraventricular nuclei. Am. J. Physiol. 302, E273-285.

Tok, C. Y., Chew, S. F., Peh, W. Y., Loong, A. M., Wong, W. P. and Ip, Y. K. (2009). Glutamine accumulation and up-regulation of glutamine synthetase activity in the swamp eel, Monopterus albus (Zuiew), exposed to brackish water. J. Exp. Biol. 212, 1248-1258.

Trayer, V., Hwang, P. P., Prunet, P. and Thermes, V. (2013). Assessment of the role of cortisol and corticosteroid receptors in epidermal ionocyte development in the medaka (Oryzias latipes) embryos. Gen. Comp. Endocrinol. 194, 152-161.

Tse, K. F. W., Chow, S. C. and Wong, C. K. C. (2008). The cloning of eel osmotic stress transcription factor and the regulation of its expression in primary gill cell culture. J. Exp. Biol. 211, 1964-1968.

Tse, K. F. W., Lai, K. P. and Takei, Y. (2011). Medaka osmotic stress transcription factor 1b (Ostf1b/TSC22D3-2) triggers hyperosmotic responses of different ion transporters in medaka gill and human embryonic kidney cells via the JNK signaling pathway. Int. J. Biochem. Cell. Biol. 43, 1764-1775.

Tseng, Y. C. and Hwang, P. P. (2008). Some insights into energy metabolism for osmoregulation infish. Comp. Biochem. Physiol. 148C, 419-429.

Tseng, Y. C., Huang, C. J., Chang, J. C., Teng, W. Y., Baba, O., Fann, M. J., et al. (2007). Glycogen phosphorylase in glycogen-rich cells is involved in the energy supply for ion regulation infish gill epithelia. Am. J. Physiol. 293, R482-491.

Tseng, D. Y., Chou, M. Y., Tseng, Y. C., Hsiao, C. D., Huang, C. J., Kaneko, T., et al. (2009a). Effects of stanniocalcin 1 on calcium uptake in zebrafish (Danio rerio) embryo. Am. J. Physiol. 296, R549-557.

Tseng, Y. C., Chen, R. D., Lee, J. R., Liu, S. T., Lee, S. J. and Hwang, P. P. (2009b). Specific expression and regulation of glucose transporters in zebrafish ionocytes. Am. J. Physiol. 297, R275-290.

Tseng, Y. C., Lee, J. R., Lee, S. J. and Hwang, P. P. (2011). Functional analysis of the glucose transporters-1a, -6, and-13.1 expressed by zebrafish epithelial cells. Am. J. Physiol. 300, R321-329.

Tsukada, T. and Takei, Y. (2006). Integrative approach to osmoregulatory action of atrial natriuretic peptide in seawater eels. Gen. Comp. Endocrinol. 147, 31-38.

Uchida, K., Kaneko, T., Yamaguchi, A., Ogasawara, T. and Hirano, T. (1997). Reduced hypoosmoregulatory ability and alteration infill chloride cell distribution in mature chum salmon (Oncorhynchus keta) migrated upstream for spawining. Marine Biol. 129, 247-253.

Valenti, G., Procino, G., Tamma, G., Carmosino, M. and Svelto, M. (2005). Minireview: aquaporin 2 trafficking. Endocrinology 146, 5063-5070.

Vargau-Chacoff, L., Calvo, A., Ruiz-Jarabo, I., Villarroel, F., Munoz, J. L., Tinoco, A. B., et al. (2011). Growth performance, osmoregulatory and metabolic modifications in red porgy fry, Pagrus pagrus, under different environmental salinities and stocking densities. Aquac. Res. 42, 1269-1278.

Wang, Y. F., Tseng, Y. C., Yan, J. J., Hiroi, J. and Hwang, P. P. (2009). Role of SLC12A10. 2, a Na-Cl cotransporter-like protein, in a Cl uptake mechanism in zebrafish (Danio rerio). Am. J. Physiol. 296, R1650-1660.

Watanabe, E., Fujikawa, A., Matsunaga, H., Yasoshima, Y., Sako, N., Yamamoto, T., et al. (2000). Nav2/NaG channel is involved in control of salt-intake behavior in the CNS. J. Neurosci. 20, 7743-7751.

Watanabe, T. and Takei, Y. (2011). Molecular physiology and functional morphology of sulfate excretion by the kidney of seawater-adapted eels. J. Exp. Biol. 214, 1783-1790.

Watanabe, T. and Takei, Y. (2012a). Vigorous SO_4^{2-} influx via the gills is balanced by enhanced SO_4^{2-} excretion by the kidney in eels after seawater adaptation. J. Exp. Biol. 215, 1775-1781.

Watanabe, T. and Takei, Y. (2012b). Environmental factors responsible for switching on the SO_4^{2-} excretory system in the kidney of seawater eels. Am. J. Physiol. 301, R402-R411.

Wendelaar Bonga, S. E. (1997). The stress response infish. Physiol. Rev. 77, 591-625.

Wilcox, C. S. (1983). Regulation of renal bloodflow by plasma chloride. J. Clin. Invest. 71, 726-735.

Wilson, R. W., Millero, F. J., Taylor, J. R., Walsch, P. J., Christensen, V., Jennings, S., et al. (2009). Contribution offish to marine inorganic carbon cycle. Science 323, 359-362.

Winberg, S., Höglund, E. and Øverli, Ø. (2016). Variation in the Neuroendocrine Stress Response. In Fish Physiology-Biology of Stress in Fish, Vol. 35 (eds. C. B. Schreck, L. Tort, A. P. Farrell and C. J. Brauner), San Diego, CA: Academic Press.

Wong, M. K. S., Ozaki, H., Suzuki, Y., Iwasaki, W. and Takei, Y. (2014). Discovery of osmotic sensitive transcription factors infish intestine via a transcriptomic approach. BMC Genomics 15, 1134.

Wu, S. C., Horng, J. L., Liu, S. T., Hwang, P. P., Wen, Z. H., Lin, C. S., et al. (2010). Ammonium-dependent sodium uptake in mitochondrion-rich cells of medaka (Oryzias latipes) larvae. Am. J. Physiol. 298, C237-250.

Yan, J. J., Chou, M. Y., Kaneko, T. and Hwang, P. P. (2007). Gene expression of Na^+/H^+ exchanger in zebrafish H^+-ATPase-rich cells during acclimation to low-Na^+ and acidic environments. Am. J. Physiol. 293, C1814-1823.

Yancy, P. H., Clark, M. E., Hand, S. C., Bowlus, R. D. and Somero, G. N. (1982). Living with water stress: evolution of osmolytes systems. Science 217, 1214-1222.

Yuge, S., Inoue, K., Hyodo, S. and Takei, Y. (2003). A novel guanylin family (guanylin, uroguanylin and renoguanylin) in eels: possible osmoregulatory hormones in intestine and kidney. J. Biol. Chem. 278, 22726-22733.

Zikos, A., Seale, A. P., Lerner, D. T., Grau, E. G. and Korsmeyer, K. E. (2014). Effects of salinity on metabolic rate and branchial expression of genes involved in ion transport and metabolism in Mozambique tilapia (Oreochromis mossambicus). Comp. Biochem. Physiol. 178A, 121-131.

第 7 章　应激和运动缓解应激的效应：心血管、新陈代谢和骨骼肌的调节

鱼类的运动方式是游动，很多鱼类通过持续的游动进行摄食、迁徙、繁殖和逃避捕食者。在不同的水生环境（如远洋、底栖、溯河）中，不同鱼类的生活方式千差万别，这反映在它们不同的游动能力上。鱼类游动会增加骨骼肌和心呼吸系统的活动，这就需要增加代谢能量的产生，以维持鱼类游动过程中和游动之后体内的平衡。无论游动能力如何，应激反应的共同特征是刺激心血管系统，增加氧气输送和组织对氧气的摄取。鱼类在应激状态下或者是游动过程中，必须优先进行氧气的输送，以提高代谢状态。能量的产生和利用决定生理行为，鱼类游动既可以看作潜在的生理应激源，也可以看作一种减轻应激的机制。根据强度和持续时间，鱼类游动通常分为突发性的和持续性的。鱼类突发性游动会对机体多个生理系统造成巨大的压力，引起代谢、酸碱、渗透压和电解质平衡的紊乱。相比之下，鱼类长时间的持续游动并不会导致血液循环中皮质醇和儿茶酚胺的含量发生显著变化，但却能产生有益于机体的生理反应，并增强随后对应激源的抵抗能力。总的来说，在活动力强的鱼类中，持续的游动可以促进生长，提高有氧代谢能力，降低皮质醇水平，减少相互的攻击行为。因此，在水产养殖或实验研究中，对鱼类自由游动的限制可能使其丧失游动产生的有益生理作用，因而可能对鱼体产生压力。在本章中，我们将鱼类的游动行为作为一种潜在的应激源和一种有助于维持体内平衡与增强其适应能力的减压行为来探讨。

7.1　导　言

应激是由外部或内部的刺激（可称为应激源或应激物）而产生的破坏机体稳态的状态（Schreck，2010；Schreck，Tort，2016；见本书第 1 章）。鱼类暴露在自然环境、水产养殖和实验室环境中的各种应激源中。鱼类的应激反应是应对应激源和促进体内平衡的正常反应。这种反应始于儿茶酚胺和皮质醇迅速地释放到血液循环中，并能改变一系列和血液循环的底物与能量代谢有关的生化反应。突发的应激反应对盐水平衡、呼吸和心血管功能有显著影响。我们可以认为，应激反应所产生的主要结果是激素介导的储存能量调动和神经体液增强氧气输送到需要能量的组织中。应激反应可以是行为上的，也可以是生理上的，机体平衡状态被破坏之后，其恢复到应激前水平的速度可能会影响个体的生长和繁殖等生命过程，甚至会决定其是否可以存活。由应激导致的能量消耗的增加最终可能会影响到一个或多个生理功能。鱼类在游动过程中代谢率会较高，在不同强度和持续时间下，研究鱼类游动与应激诱导的行为改变之间的相互作用是很重要的。有许多研究是关于鱼类游动的限制因素和游动能力的，但很

少有研究涉及低强度游动会诱导生理适应机制以及将游动行为视为一种潜在的减压行为。

鱼类游动是通过增加骨骼肌和心呼吸系统的运动，这就需要增加代谢能量的产生，以维持游动过程中和游动过程后体内稳态的平衡。与陆生动物不同的是，许多鱼类为了觅食、迁移、躲避敌害而不间歇地进行游动。鱼类游动的能力在不同鱼类之间差异非常大，鱼类游动通常根据强度和持续时间分为爆发型和持续型。对鱼类竭力游动和游动后恢复情况的研究，由于获得许多生理知识以及引起竭力游动与持续几小时恢复时期的许多应激活动，是很有价值的。例如，和水产养殖有关的鱼类捕捞和释放、生物学调查的电捕鱼、运输以及各种手工操作应激都可以刺激鱼类进行高强度的游动。骨骼肌和血液乳酸的迅速增加以及血液氧含量和 pH 值相应地降低是急剧的应激反应的生化反应标志（Wendelaar Bonga，1997）。和应激相关的鱼类竭力游动甚至可能超过机体可维持的稳态，并在竭力游动数小时后导致鱼类死亡（Black，1958；Wood 等，1983；Van Ginneken 等，2008）。假设一个应激反应没有超出机体可维持稳态的能力，一些研究表明集群（Veiseth 等，2006）或竭力游动（Pagnotta 等，1994；Milligan，1996）后，在机体恢复期间，鱼类积极游动可以加快血液皮质醇水平、酸碱平衡、水代谢平衡和能量代谢平衡的恢复。然而，这并不是一个普遍性的发现（Meyers，Cook，1996；Powell，Nowak，2003；Kieffer 等，2011），游动是否影响机体恢复的速度，可能取决于应激源、研究的鱼种类和恢复时的游动强度。

一般来说，慢性应激对鱼类是有害的（Schreck，Tort，2016；见本书第 1 章）。与鱼类突发游动或仅几秒到几分钟的竭力游动相反，在特定条件下持续游动数小时[强度低于临界游动速度（U_{crit}）]，不会引起血液循环中应激激素的峰值（Hughes 等，1988；Wood，1991；Perry，Bernier，1999），也不会刺激骨骼肌的无氧能量代谢。例如，成年红大麻哈鱼可以在 15℃的环境中以 75% 的临界游动速度游 4 个小时以上，并且不会增加血液乳酸水平（Steinhausen 等，2008）。当鲑科鱼类接近临界游动速率时，它们会改变游动状态，加快糖酵解速率，快速抽搐白肌，增加血糖乳酸含量并降低血液的 pH 值（Brauner 等，2000）。如果鱼类在较低或中等强度下持续较长时间游动，并在低于神经内分泌阈值进行应激反应，机体会通过刺激作用（hormesis）过程产生适应性，并引起有益于机体的生理反应和随后对应激源的抵抗能力。因此，根据鱼类的种类、游动的类型和持续时间，游动可以认为是一种重要的应激源和应激缓解因子。事实上，有大量的文献报道，很多鱼类持续地进行游动，对机体会产生积极的生理作用（Davison，1997；Palstra，Planas，2013；见本章第 7.3 节）。据报道，在应对应激源时，中等强度下持续游动可以降低血液皮质醇水平（Boesgaard 等，1993；Postlethwaite，McDonald，1995），提高鱼的应激反应能力，促进鱼类健康和福利（Jorgensen，Jobling，1993；Ward，Hilwig，2004；Veiseth 等，2006；Castro 等，2011；Larsen 等，2012；McKenzie 等，2012）。相反，有一个假设，限制鱼类游动能力和降低游动潜在的有益生理作用（如在水产养殖中经常发生的那样），会对鱼类产生应激的作用。

我们的目的是综述心脏呼吸系统和骨骼肌系统对鱼类急剧游动的生理反应，并将这些反应与鱼类普遍出现的应激反应联系起来。我们还将着重说明最近的一些发现，即鱼类长期游动可能对其生理适应能力起着一种重要的刺激作用，包括减弱对应激源的反应程度。我们大量阅读了60多年以来的同行评议文献，并参考了本书其他章节的内容和以前的综述。大多数关于鱼类游动效应的研究都集中在血液、骨骼肌（因为骨骼肌组织决定了鱼类游动行为并受到疲劳的影响）和心呼吸系统。需要指出的是，在实验室环境下，大多数关于鱼类生理应激反应和游动的研究只针对少数几种鱼类——主要是硬骨鱼类和淡水鲑科鱼类。因此，读者应该注意，已有的研究发现及其研究趋势可能不会扩展到自然环境中庞大而多样的脊椎动物群体所有成员。

7.2 游动的生理需求和应激连续体

7.2.1 引言

本节的目的是提出鱼类游动的定义，并探明鱼类游动所产生相关生理紊乱/反应。我们对鱼类游动行为的大部分了解都是在实验室里通过实验得出的。鱼类高强度游动和竭力游动（包括厌氧的白肌持续活动长达20秒）会对多个生理系统造成重大的应激。这些应激包括代谢平衡、酸碱平衡、渗透压平衡和电解质平衡方面的破坏（Wood，1991；Kieffer，2000）。而在另一种情况下，“持续游动”会以较低的速度持续较长时间（>200分钟），并且其消耗的能量主要通过有氧代谢产生。然而，基于最初机体激素和代谢情况的改变，即使鱼类以中等强度和低强度游动，在刚开始时游动对机体产生的影响也类似于急性的应激（Nielsen等，1994）。因此，鱼类持续游动是否会成为机体的应激源将取决于神经内分泌的反应和时间。目前认为，鱼类和哺乳动物一样，运动可以促进其生理适应的能力，影响游动行为以及对各种应激源的反应。

7.2.2 应激和运动的神经内分泌方面

Gorissen和Flik对内分泌应激轴和应激反应的调控进行了综述（Gorissen，Flik，2016；见本书第3章）。应激激素的一个基本作用是为参与搏斗、逃逸（游动）或应对的系统提供能量。简单地说，对各种物理的和环境的应激源，血液中儿茶酚胺、肾上腺素和去甲肾上腺素的水平会提高。这些应激源包括外部低氧、高碳酸血症、暴露在空气中、代谢性酸中毒、暴露在软水中以及进行竭力游动（Randall，Perry，1992）。鱼类游动期间，应激激素的快速释放主要依赖于鱼类游动的强度和持续时间，而在较低强度的游动过程中，血液中儿茶酚胺水平基本保持不变（Hughes等，1988；Wood，1991；Perry，Bernier，1999）。竭力游动后，血液中的肾上腺素和去甲肾上腺素浓度立刻升到最高，肾上腺素浓度比去甲肾上腺素浓度高2～3倍（Milligan，1996）。在游动停止后大约2小时内，血液儿茶酚胺的浓度迅速恢复到静息状态。如前所述，鱼类积极游动可能会改善血液循环的皮质醇水平，促进竭力游动后的恢复（Pagnotta 等，1994；Milligan，1996）。

应激反应也可以通过肾上腺素能的神经支配直接介导到相关的器官，如心脏（尽管不在盲鳗和板鳃鱼类中）和鳃与体循环的血管。总的来说，儿茶酚胺直接或间接地介导能量代谢、氧气供应、增加提供能量底物以及输送给各个组织的氧气量（Wendelaar Bonga，1997；Pankhurst，2011）。儿茶酚胺其他的作用包括增加血液葡萄糖水平（Wright 等，1989）、改变鳃的扩散能力、增加心脏血液的输出量和鳃的通氧、通过刺激红细胞从脾脏释放以提高血液运输氧的能力，以及提高红细胞 pH 值（见下文和图 7.1）。儿茶酚胺也可以通过增加心脏输出量而间接增加鳃的离子交换，从而增加和水环境接触的鳃片表面积。

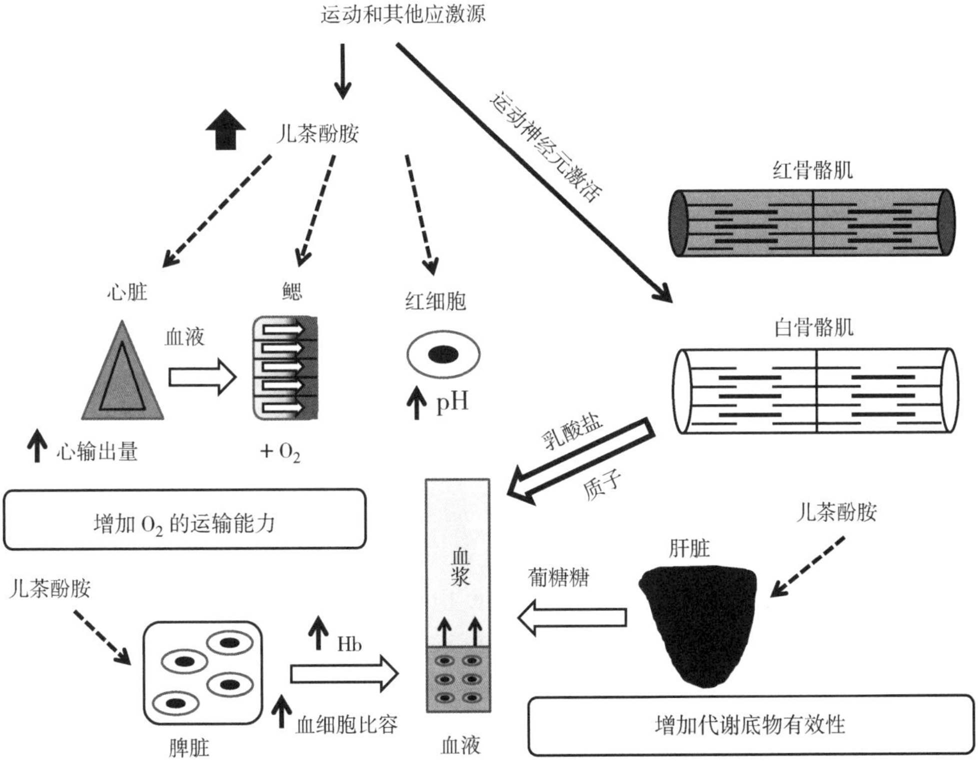

图 7.1　鱼类在游动和面对各种应激源期间，对儿茶酚胺和肌肉激活的血液循环、呼吸和代谢反应

注：本图着重表示在应激、运动和运动后恢复期间肾上腺素能对心脏、鳃、红细胞和脾脏的影响以增强血液 O_2 的运输作用。肝脏通过糖原分解增加血液的葡萄糖浓度。在竭力运动期间，在厌氧的白肌中贮存的能量可能分解为乳酸和质子而释放到血液中，造成酸中毒。

皮质醇影响多种生理功能，包括能量代谢、氨基酸代谢、离子调节和免疫功能。硬骨鱼类皮质醇的主要作用对象包括肝脏和骨骼肌。血液皮质醇含量增加使蛋白质转换速率、氨的产量、糖异生和脂质分解速率增加（Mommsen 等，1999）。皮质醇通过增加 Na^+/K^+ ATP 酶活性和 Na^+ 的转运对鳃的渗透压调节也起重要的作用

（McCormick，1995）。总而言之，在应激反应中，皮质醇促进能量底物的分解和水平衡的恢复。说明皮质醇在鱼类剧烈游动中作用的研究很少；然而，和严重的有害影响不同，血液皮质醇对鱼类摄食行为、生长、身体状况和生存以及需氧的游动等，似乎不受慢性应激的影响（Gregory，Wood，1999）。鱼类游动后血液皮质醇含量的升高可能会抑制骨骼肌糖原的合成，从而延长恢复的时间（Milligan，2003）。

7.2.3 鱼类应激和游动过程中的能量代谢

在本节中，我们将比较并联系鱼类应激反应时和游动过程中的代谢反应。Sadoul 和 Vijayan（2016；见本书第 5 章）将应激对代谢率、能量平衡和能量底物分解的反应都归纳为生理应激源。“新陈代谢”是指生物体中所有化学反应的总和，包括合成代谢和分解代谢过程。其中细胞最重要的代谢过程是三磷酸腺苷（ATP）的有氧和无氧合成及其水解为二磷酸腺苷（ADP）、无机磷酸盐（Pi）和 H^+。应激反应是一个需要能量的过程，能量是所有生理活动的限制因素。应激源可以促进能量储备的快速调动和利用。应激源也能改变耗氧量、水平衡和酸碱平衡、心血管的功能以及直接影响机体活动范围和生理行为的骨骼肌活动。

7.2.3.1 游动对水和离子稳态的影响

Takei 和 Hwang（2016；见本书第 6 章）提出了一个详细的关于渗透压调节、水平衡以及对渗透压生理反应的观点。众所周知，鱼体内的渗透压对水平衡和离子平衡有重要影响，并会增加离子运输机制的能量需求。由于不同的水中含盐量和采用不同的实验方法，以往的研究表明鱼类鳃约占其总耗氧量的 4%（Morgan，Iwama，1999）、7%（Mommsen，1984）和 20%（McCormick 等，1989），应激反应时增加的耗氧量主要用于渗透压调节和离子调节（Wendelaar Bonga，1997）。鱼类在不同强度、不同持续时间内游动，保持水平衡所需的能量还有待确定。

呼吸-渗透压调节在气体交换和渗透负担之间的调整已经了解清楚（Wood，Randall，1973；Nilsson，Sundin，1998），并已通过急性应激和应激激素水平的升高来说明。肾上腺素通过以下几种方式促进环境中氧气的摄取：①通过提高心脏排出量和腹主动脉血压来增加鳃的气体交换，促进摄氧量；②降低鳃部血管阻力；③增加鳃上皮细胞紧密接合的通透性；④增加鳃叶数量，从而增加有效的呼吸表面积，以便气体、水和离子的扩散交换。因此，增加离子被动丢失和水分子吸取与较高的耗氧率和血液循环中儿茶酚胺浓度有关。急性应激反应和游动过程中的任何变化都可能影响流经鳃部的被动离子流。然而，激素对离子调节反应的控制有些不确定性。虽然对禁闭应激的渗透调节反应和有氧运动相似，但血液循环的激素作用型式却不同（Postlethwaite，McDonald，1995）。

7.2.3.2 肌肉活动和运动

鱼类游动过程中生理反应的比较研究对于了解运动肌的活动、神经内分泌反应、代谢补偿和心呼吸能力应对应激的状态等都是非常有价值的。游动对于鱼类来说能量消耗非常高，而这些能量大部分来自于骨骼肌收缩（Gerry，Ellerby，2014）。毫不奇怪，所有脊椎动物在运动和其他应激反应中都会提高代谢的速率。鱼类游动所耗费的能量直接和肌肉组织收缩的强度与持续时间相关，并且由有氧代谢和无氧代谢这两种方式提供能量（见下文）。

鱼类在水中游动时，身体会受到阻力。阻力与鱼类游动速度的平方成正比。因此，氧的消耗同增加的速度呈指数增长（Brett，1964），耗氧能力和输送氧气能力（呼吸代谢）的上限限制了游动速度（Bennett，1991）。应激源可以引发鱼类行为的改变并干扰其生理活动。例如，在高的养殖密度下（Anras，Lagardere，2004），在运输过程中（Chandroo 等，2005），人工养殖的虹鳟游动活性较高。然而，即使超过 48 小时的休息时间，相对于没有经过运输的对照组的鱼，经过运输的虹鳟需要 50 分钟才能平静下来（Chandroo 等，2005）。相反，尽管皮质醇的长期升高对摄食行为和生长有负面影响，但皮质醇升高本身不会影响虹鳟（Gregory，Wood，1999）和鲤鱼（Liew 等，2013）的游动行为。其他受可能影响能量代谢的应激因素影响的鱼类行为包括逃避敌害、摄食和温度调节（Schreck 等，1997）。如果幼年虹鳟鱼经历了竭力游动，那么游动行为会减弱至少 6 小时，并且长时间游动带来的不适会增加被捕食的风险并损害其觅食能力（Jenkins 等，2007）。总之，迄今为止的证据表明，应激源会影响鱼类的游动能力，鱼类竭力游动还会影响到很多其他行为。

了解各种肌肉的特征是掌握鱼类游动的能量需求和游动能力限制的基础。鱼类的骨骼肌主要分为两类，即白肌纤维和红肌纤维，它们具有不同的代谢能力和收缩特性（Driedzic，Hochachka，1978）。白肌的血管密度有限，线粒体少，肌红蛋白少，主要依靠厌氧代谢产生能量。白肌占游动相关肌肉组织的绝大部分（80%～95%），为高强度、短时间的爆发游动提供能量。这两种类型的肌纤维具有不同的底物偏好，产生 ATP 的代谢方式也不同（见下文）。随着鱼类从低速游动向高速游动的转变，乳酸盐积累在白肌中，白肌纤维会快速地补充（Johnston，Goldspink，1973）。红肌纤维具有广泛的维管联结，线粒体和肌红蛋白的含量很高，比白肌纤维的氧耗高得多。在鱼类缓慢长时间的游动时，红肌纤维利用有氧代谢产生的能量，而在高速游动时，参与糖酵解的红肌纤维和白肌纤维都会利用起来（Rome 等，1993；Johnson 等，1994；Richards 等，2002c）。

尽管很多研究者都是通过让鱼类进行高强度游动来研究其生理能力和生理恢复能力，但在自然条件下，鱼类可能很少会经历这种行为。也有例外，例如人类的干预，包括以休闲或者商业目的用钩和线钓鱼和用网网鱼。急性应激源（如捕捉、手工操作、运输、被强迫游动、低氧、渗透压和温度刺激、拥挤或暴露于水中污染物）都已证明会提高肌肉和血液中的乳酸水平，降低血液 pH 值，增加氧耗（Wendelaar Bonga，1997）。自由游动的鱼通常会选择较低速度的游动，这样可以持续较长的时间（Krohn，Boisclair，1994）。Driedzic 和 Hochchka（1978）、Moyes 和 West（1995）、Richards 等（2002c）已经提出了关于鱼类游动期间能量底物利用率的综合性观点。

7.2.3.3 能量的产生和利用

生理应激反应的一个标志是动物代谢率的增加以及恢复体内平衡提供细胞需要的循环底物的增加（Wendelaar Bonga，1997；Sadoul，Vijayan，2016；见本书第5章）。Fry介绍了“代谢活动范围”，并且把它定义为最大代谢率和标准代谢率两者之间的差异。这个概念仅限于有氧代谢，但它强调了生理活动的能量限制和对各种环境的行为反应（Fry，1947）。细胞有多种生化通路和能量底物可供选择，用于再生ATP（生物合成、离子泵和机械工作）。儿茶酚胺刺激糖原分解和无氧糖酵解，而皮质醇刺激有氧和无氧代谢、氮排泄和糖异生作用。任何动物的代谢速率都是有限的，应激源能导致能量短缺以及限定动物是否能承受某种特定的应激源。通过生物能量方程可以得出能量消耗和同时执行多项作用的能力（Schreck，Tort，2016；见本书第1章）。很明显，保持能量平衡是适应应激源和维持活动的基本要求，比如游动（Schreck，2010）。然而，不同的环境和人为应激源会降低鱼类越过生理局限和行为障碍进行活动的能力，最终可能导致死亡（Portz等，2006）。

1. 磷酸原

在鱼类游动期间和之后，代谢率的升高需要能量供需的匹配以维持体内平衡。ATP是肌肉收缩和其他细胞功能能量的直接来源，在高强度的肌肉收缩中，内源性ATP供应会受到限制，并很快就会用尽。相反，细胞内的磷酸肌酸（Crp）和肌酸激酶可以快速利用ADP和H^+重新合成ATP，在没有O_2的情况下帮助维持细胞内ATP的水平。这种机制（即动员磷酸盐）在组织（条状平滑肌和神经组织）中具有重要意义。在高强度运动开始时，糖酵解速率提高之前，白肌阻止ATP浓度发生大的变化，ADP通过磷酸肌酸（Crp）的磷酸化是特别重要的（Newsholme，Leech，1983；Hochachka，Somero，2002；Richards等，2002a）。与其他组织相比，鱼类白肌纤维磷酸肌酸和ATP的静息浓度非常高，白肌中CP的恢复是非常快的，通常在竭力游动后的1小时之内就可恢复（van Ginneken等，2008）。因此，Crp在维持细胞能量状态中起重要作用，特别是在高强度、不可持续的游动活动中（例如，U_{crit}的90%）。根据组织、一种应激源或多重应激源的特定ATP需求和细胞内肌酸的原始水平，Crp系统可以作为保护机制防止自由能的临界减少。

2. 糖酵解和糖异生

葡萄糖是一种基本的细胞能量底物。一些鱼类组织（如大脑、鳃、心脏和血液细胞）主要依靠葡萄糖进行能量代谢（Polakof等，2011）。在应激期间，血液皮质醇和儿茶酚胺浓度的升高会刺激储存在肝脏的肝糖原分解，从而提高血液循环中葡萄糖水平并增加能量底物的供应，以满足更高的能量需求（Sadoul，Vijayan，2016；见本书第5章）。值得注意的是，与哺乳动物相比，一些鱼类（虹鳟）在低氧和拥挤饲养等应激中血糖变化很大（Polakof等，2011）。

糖酵解很重要，因为它是细胞内葡萄糖代谢的重要通路。剧烈运动时白肌收缩所需的能量主要通过局部糖原的无氧代谢转换成乳酸（Richards等，2002a）。糖原（而不是葡萄糖）转化为乳酸盐具有50%的能量优势（3：2 ATP）（Newsholme，Leech，

1983）。因为糖酵解与细胞有氧代谢相比是一种非常低效的产生 ATP 的方式，所以在剧烈运动中代谢速率很高时，它必须以很高的速率发生，而这样会导致乳酸和质子的积累。乳酸，反过来可以作为有氧代谢组织的能量来源和葡萄糖糖异生作用的底物（Suarez，Mommsen，1987；Gladden，2004）。在竭力游动后，乳酸盐被保留在白肌中，这一过程延长了极限运动后的恢复时间（Van Raaij 等，1996）。

细胞内的糖酵解受到应激激素、离子、腺苷酸和关键酶的调节。众所周知，儿茶酚胺会影响鱼类局部和全身碳水化合物的代谢（Fabbri 等，1998）。例如，糖原磷酸化酶（glycogen phospylase，GP）对白色骨骼肌的作用，在催化糖原转化为乳酸生成的步骤中，由肾上腺素通过 β-肾上腺素感受器和第二信使环腺苷酸（cAMP）激活。在肌肉收缩的过程中发生的肌质 Ca^{2+} 浓度升高也可以激活 GP，导致细胞内糖原的分解和 ATP 的再生。在运动过程中，Pi、游离 ADP、游离 AMP、磷酸肌苷（IMP）、NH_3等在细胞内的积累也可以激活虹鳟白肌细胞中糖酵解以产生 ATP（Richards 等，2002a）。在爆发运动的前 10 秒，ATP 的高转化率主要是由 Crp 的水解和糖酵解支持（Richards 等，2002a）。此外，竭力运动会大量增加血液 NH_3 的含量（Mommsen，Hochachka，1988），这种代谢物可能会促进糖酵解产生 ATP。

3. 氧化磷酸化

细胞再生 ATP 的第三种方式是通过氧气和一系列复杂的通道进行的（线粒体中包含的三羧酸或柠檬酸循环、电子转移和氧化磷酸化）。氧气向新陈代谢旺盛的细胞中运输，特别是在鱼类游动过程中骨骼肌的收缩，依赖于心血管和呼吸反应的协调作用。虽然利用磷酸盐（ATP 和 Crp）和无氧糖酵解进行短期的能量生产主要依赖于细胞内的燃料，但在运动或应激反应期间，持续的氧化磷酸化也可以利用血液循环中的燃料供应。因此，在运动和应激反应过程中，心血管系统为需要能量的细胞提供血液循环燃料起着至关重要的作用。

需要不断供应 ATP 进行生物合成、离子转运和/或收缩活动的器官和组织，会表现出较高耗氧量并且细胞内线粒体会得到补充（Hochachka，1994）。考虑到在应激反应和持续运动中，线粒体功能对于细胞能量的平衡和耗氧量的增加至关重要，因此，氧化能力是机体最大代谢率的决定因素，决定机体代谢能力的范围。尽管在竭力运动和运动后恢复过程中，脂质似乎不是白肌重要的能量燃料，但有证据表明，在运动后的恢复过程中，多种鱼类会激活氧化磷酸化和脂质分解代谢，从而在该组织中合成 ATP（Richards 等，2002b）。

关于儿茶酚胺和皮质醇对鱼类线粒体和氧化代谢的直接影响，我们知之甚少。由埋植物引起的皮质醇升高并不会影响有氧的游动（Gregory，Wood，1999），但它确实增加了动物在应对手工操作应激（Davis，Schreck，1997）以及皮质醇刺激后的氧耗（Morgan，Iwama，1996；De Boeck 等，2001）。虽然在鱼类中还没有得到证实，但最近的研究表明，线粒体中含有糖皮质激素和儿茶酚胺受体，因而能够感知以及对细胞的应激并产生反应（Manoli 等，2007；Picard 等，2014）。许多线粒体的功能以及功能失调是由糖皮质激素引起的，这些激素可以增加线粒体的生物发生（biogenesis），从而增加线粒体产生 ATP 的能力，满足应激反应和应激反应后恢复阶段升高的能量需求。

4. 耗氧量的测量

代谢率的变化可以通过耗氧量和呼吸计的间接量热法来估算。鱼类有氧代谢率是高度可变的，并且对许多因素敏感，如手工操作、进食时间和温度等（Fry，1971）。许多研究都发现，在鱼类进行急性应激反应和游动时，会刺激能量代谢和摄氧量。急性的应激反应会使幼年虹鳟的耗氧量增加 1 倍，并且耗氧量和血液皮质醇水平的升高呈正相关（Barton，Schreck，1987）。这些研究者估计，鱼类多达 25% 的活动可能是应对轻微和短暂的干扰。持续 5 分钟的竭力游动会导致运动后耗氧量增加 2. 0 ～2. 5 倍，几乎会耗尽全身糖原、ATP 和 CP 储存，乳酸含量会增加 5 倍之多（Scarabello 等，1991）。

动物整体耗氧量的测量不能用来测定特定器官或生理过程的能量消耗。虽然暴露于不同的应激源中（Barton，Schreck，1987）可以观察到代谢率（O_2消耗）的增加，但在应激期间，代谢能量的分布却知之甚少。在应激状态下提高能量需求可能包括渗透调节、蛋白质合成和骨骼肌和心肌的收缩活动。此外，与应激反应相关的能量需求增加不会随着特定应激源的终止而停止。特别是短暂的手工操作和暴露于空气中的应激，根据不同的物种、生命阶段和应激源的激烈程度，可将 1 小时的耗氧量（Davis 和 Schreck，1997）提高到 3 ～4 天的耗氧量（Korovin 等，1982）。运动后的耗氧量升高，称为运动后的过量耗氧量（EPOC），可用于评估鱼类的代谢恢复时间和游动的无氧能量消耗（Lee 等，2003）。经过竭力游动，虹鳟的 EPOC 和血液循环中的乳酸可以坚持 6 小时，而 CP、ATP 的恢复和糖原存储则分别可以持续 5 分钟、1 小时和至少 6 小时（Scarabello 等，1991）。幼年鲑鱼在其他的应激源影响下，如急性手工操作（乳酸含量升高 2 ～8 小时，依营养状况而定）（Vijayan，Moon，1992）以及装载和运输之后（持续运输 0. 5 ～4. 5 小时，血液乳酸含量升高 48 小时以上）（Iversen 等，1998），血液乳酸含量会持续升高。

7. 2. 4 对应激和运动的心血管和呼吸调节

有效应激反应的一个关键因素是对氧气转运的刺激和组织对氧气的吸收（Wendelaar Bonga，1997）。鱼类必须优先输送氧气，并根据代谢速率的升高而调整心呼吸活动（Hicks，Bennett，2004），如鱼类在游动过程中和/或应激中。在应激反应期间，氧和能量底物可能从一个或多个需要能量的功能中重新分配，以支持最紧迫的生存需求。这种反应包括血液循环中的儿茶酚胺和神经释放的儿茶酚胺，以及心呼吸功能的改变（Randall，Perry，1992）。然而，目前我们对鱼类心呼吸功能相关的皮质醇非基因组效应的了解非常有限。

菲克原理表明耗氧量（机体或器官的 MO_2）等于血流或心脏血液输出量和血液中氧提取量的乘积（Fick，1870）。因此，动物机体耗氧量的增加会增加心脏血液输出量，提高动脉氧含量（CaO_2）和/或降低静脉氧含量（CvO_2）。心脏血液输出量由心搏量和心率决定。应激反应、自发性活动和失血都会增加静息状态心脏血液的输出量

（Farrell，Jones，1992）。CaO_2-CvO_2代表动静脉氧含量差异。增加 CaO_2-CvO_2的能力主要取决于动脉氧含量和血红蛋白浓度（Hb），虽然鱼类在持续的游动活动中，CvO_2的降低是正常的（Kiceniuk，Jones，1977）。血液中 Hb 的浓度取决于 Hb 的总循环质量和血浆的容积。在应激状态下，心脏血液输出量和血液氧含量的重要性直接关系到有氧能量代谢，为支持细胞稳态和应对应激源而使用更多的氧气来产生能量。

鱼类从平稳状态游动达到临界游动速度（U_{crit}）期间，氧气输送系统能力已经被确定，我们称之为有氧代谢能力。根据鱼的大小和水温的不同，鲑科鱼类氧的输送和组织的耗氧量可以增加基础水平（标准代谢率）的 5～15 倍（Brett，1972）。在 U_{crit}时，虹鳟 MO_2的增加能比静息水平高 7.5 倍（Kiceniuk，Jones，1977）。鲑鱼在 U_{crit}时，动脉血液中仍然保持完全饱和的氧气，而静脉血中的氧气被活跃组织最大限度地耗尽（Farrell，2007）。对持续运动的综合反应包括：①增加鳃的通气量；②由于心脏输出量增加，腹主动脉血压升高，导致血液流经鳃的量增加；③将血液从不活跃的组织重新分配到收缩的肌肉组织。心脏输出量随游动速度的增加而增加，接近 U_{crit}的最大值。与菲克原理相一致，通过提高心率和心搏量来增加心脏输出量，并且增加血液中 O_2的提取（降低 CvO_2），从而增加 O_2的运输量。最后，鱼类增加 O_2的运输有一个生理限制，并且鱼鳃的通气量、心血管功能或血液特征的局限都可能对机体形成应激并损害生理功能（见下文）。

7.2.4.1　心血管的调节

调节心脏和血管系统的自主神经能促进循环系统的平衡，向需要能量的组织增加氧气输送量。和其他脊椎动物相似，硬骨鱼类心脏的双重神经分布模式已由 Nilsson（1983）和 Santer（1985）广泛研究。心脏肾上腺素能神经分布在大多数硬骨鱼类中都有研究，但在圆口类、板鳃鱼类和鲽科类中还没有。硬骨鱼心脏会接收肾上腺素能纤维，它可以释放肾上腺素和去甲肾上腺素。根据种类不同，儿茶酚胺与心肌和起搏组织的肾上腺素受体结合（α 或 β），并且可以通过 β-肾上腺素受体起到正的变时效应（Nilsson，1983），或通过 α-肾上腺素受体发挥负的变时效应（Tirri，Ripatti，1982）。除了盲鳗，迷走神经纤维（类胆碱的）也可以调节鱼类的心脏，并和肾上腺素能纤维一起决定心率。儿茶酚胺增加心脏对预负荷的敏感，从而增加心搏量（Farrell 等，1986）。大多数鱼类肾上腺素能神经可以刺激血管系统，儿茶酚胺可以增加血压并在肠道受应激时诱导血管收缩（通过 α-肾上腺素受体）（Axelsson 等，1989）。在应激状态下，释放的儿茶酚胺和皮质醇也会降低脱氧血液的黏度，从而降低红细胞在血液中的流动速度（Sorensen，Weber，1995）。

鱼类在低强度游动时血液循环的儿茶酚胺调节血液流动的作用可能并不显著，因为在轻度或中度应激中，血液循环的儿茶酚胺水平一般不会改变，因而不会促进心血管反应。只有在急性应激时，才会有较多的儿茶酚胺释放到血液中（Perry，Bernier，1999）。有氧运动时的心率加速可能是通过减少对心脏迷走神经（胆碱类）输入，而不是额外的肾上腺素刺激决定的（Farrell，Jones，1992）。相反，鱼类突发性游动经常导致心率、心排血量和血压下降，而随后恢复期间心血管功能提升（Farrell，

1982）。由于肌肉和血液之间的液体流动，竭力的游动会减少血浆容量。这种血液浓缩情况，连同脾脏收缩，可能有助于增加动脉血含氧量（Randall，Perry，1992）。

潜在的应激源，如水温升高和氧气含量降低，会引起和游动不同的心血管反应。由于对心脏起搏细胞的直接刺激作用，温度的急剧升高会加快心率和增加心脏血液输出量，但不会增加心脏的每搏输出量（Harper 等，1995）。然而，这种反应和心脏泵动能力是有限的，可以确定热耐受性的上限。也有迹象表明，肾上腺素能的成分对于鱼类在接近致死温度时的热耐受性和心脏活动性能都很重要（Ekstrom 等，2014）。另一方面，中度至重度的低氧会导致心率降低，原因是心脏胆碱能紧张性的增强，以及心搏量的补偿性增加，以维持一些种类的心脏血液输出量（Farrell，Jones，1992）。低氧也增加了鳃的血管阻力，心动过缓和动脉高血压并不会提高虹鳟的动脉氧含量，也没有促进气体的转运（Perry，Desforges，2006）。

与哺乳动物和两栖动物的研究一致，皮质醇可以通过非基因组机制快速调节鱼类的心脏功能。具体地说，在体外实验中，虹鳟在氧气充沛的条件下，皮质醇的生理水平会快速（在 3 ～ 10 分钟内）增加心肌收缩力（Farrar，Rodnick，2004）。皮质醇的积极作用是对肾上腺素的补充，可能涉及多胺和一氧化氮信号级联反应等机制。需要进一步研究来确定鱼类心血管对皮质醇的反应，并评估应激诱导的皮质醇含量变化是否有助于维持升高的心脏功能水平，以及可能影响整个动物机体。

7.2.4.2 呼吸调节

在轻微和中度应激刺激期间，鱼类会增加鳃的水流交换速度（过度通氧），并提高心脏的血液输出量（鳃部血液的流通加快）以加强气体交换。在持续的游动过程中，鳃气体交换速度的加快是由于呼吸速率的微小增加和心搏量的较大变化所致（Jones，Randall，1978）。然而，在爆发速度下，鳃的气体交换可能完全停止（Satchell，1991）。通氧增强可以通过减少鳃叶表面的边缘层来增强氧气通过鳃的传递。虽然血液循环的儿茶酚胺在调节气体交换中的作用尚不清楚，并且仍然是一个有争议的问题，但是鳃气体交换的调节涉及外部和内部的氧分压。此外，应激状态下血液中 O_2 含量的降低似乎是硬骨鱼释放儿茶酚胺的一种接近的刺激（Randall，Perry，1992）。然而，在应激状态下，刺激气体的交换可能会超出维持正常血氧水平的能力。例如，尽管动脉血氧水平正常，但鱼类在竭力游动后的恢复过程中会长时间过度通气（Wood，1991）。在氧气代谢需求之外增加气体交换的好处包括维持体内酸碱平衡、二氧化碳的排泄和氨排泄（Zhang，Wood，2009）。鱼类经过竭力游动，在整个恢复过程中，细胞内和细胞外氨含量都会增加（Wang 等，1994）。

由于心脏输出的血液必须通过鳃，鳃血流量的增加和运动引起的心脏血液输出量的增加成正比。儿茶酚胺通过提高心脏的输出量和主动脉背侧血压而间接提高鳃对氧气的摄取，从而增加血液流经鳃叶的量，并使每片鳃叶血液分布更均衡（Randall，Perry，1992）。由于心搏量的增加而增强的脉搏压力将使更多血液流经每片鳃叶，特别是在鳃丝的远端（Farrell，1980；Farrell 等，1980）。应激诱导的儿茶酚胺还会通

过 β-肾上腺素能激活红细胞 Na^+/H^+ 交换，以及随后细胞 pH 的升高来抵消由代谢性酸中毒引起的 Hb-O_2亲和力降低（Nikinmaa，1992），从而影响血氧含量。鱼类剧烈游动的过程中，儿茶酚胺通过 α-肾上腺感受器诱导脾脏收缩，导致红细胞进入血液循环，从而使血液循环中血细胞比容、CaO_2和 Hb 浓度升高（Nilsson，1983）。图7.1表示一些关于心血管、呼吸和代谢对鱼类游动的反应。

7.2.5　游动和应激的极限

长时间处于应激状态下，最终鱼类会精疲力竭，造成体内平衡机制无法弥补过度应激对机体的不利状况。关于这方面的一个重要的但未引人注意的例子是，在实验室中，鱼类游动后恢复不好的情况下，竭力游动的鱼死亡率非常高（Wood 等，1983；van Ginneken 等，2008）。在淡水鱼类的产卵迁移过程中，以无氧呼吸游动为主的野生鲑鱼也会死亡（Burnett 等，2014）。如前所述，即使鱼能存活下来，筋疲力尽的鱼的游动能力在很长一段时间也会下降（Lee-Jenkins 等，2007）。已知的竭力游动后的致死原因包括：由于应激激素过量产生导致的高速率能量消耗、储存能量的消耗、磷酸化程度降低和严重的酸中毒（Wood 等，1983；van Ginneken 等，2008）。鱼类突发的游动也会破坏渗透压和电解质平衡（Wood，1991；Milligan，1996；Kieffer，2000）。对这些挑战的生理反应涉及多个组织，在恢复过程中需要大量的能量产生，但还可能无法维持稳态。最后，鱼类竭力游动后血液的酸中毒、高钾血症和低氧会限制心脏的泵血能力（Driedzic，Gesser，1994），因此在恢复过程中使向组织输送的氧气量减少。很明显，心脏血液输出量不能支持过量的代谢需求。虽然肾上腺素能的刺激可以减轻酸中毒、高钙血症和低氧引起的心律失常，但在高温下肾上腺素的有益作用会降低（Farrell 等，1996）。因此，取决于环境条件，鱼类可能最终会调理生理的局限，以维持氧气的运输和能量储备的恢复、酸碱平衡和离子平衡，以便在高强度游动和其他与鱼类养殖过程以及休闲渔业有关的严重应激源影响下能够完成恢复。

7.3　与游动和应激相关的生理适应

7.3.1　引言

如前所述，游动是硬骨鱼类生存不可或缺的一部分。游动也可以被认为是一种重要的生理行为，因为它和鱼类生存所必需的许多行为，如繁殖（如迁徙、求爱）、捕食、躲避捕食、环境（如温度、盐度、光照、深度）偏好、社会行为等密切相关。根据鱼类游动的强度和持续时间的不同，游动可以被看作是一种机体应激源，能够引起显著的应激反应。因此，鱼类的游动能力、行为表现和活力是决定其机体健康的重要因素。

野生鱼类的大多数游动行为，包括摄食、繁殖迁移、觅食活动和探索活动，都可以归类为有应激反应迹象的持续游动。这种游动方式不同于高强度爆发式游动，后者会导致应激激素的迅速释放。这种游动方式可以维持很长一段时间（如小时、天、月），直到鱼类精疲力竭（Beamish，1979），超过机体维持稳态的能力。持续游动是需氧运动，也就是说，它由营养物质的氧化提供燃料，这些营养物质为骨骼肌和心肌的持续收缩活动提供了必要的ATP。这也和由Crp和厌氧糖酵解支持的突发（即逃逸反应、捕食）或爆发式的游动行为相反。鱼类持续游动的有氧呼吸部分，其能量消耗可以通过耗氧量来估算（Tudorache等，2013）。持续游动是指以低代谢（即需氧）速率进行的游动。从实验中得出，在一定距离内，鱼类以持续游动的速度游动所耗费的能量最低，因此效率最高，这种游动速度被称为最佳游动速度（U_{opt}）。最佳游动速度会随着不同的物种和同一物种的不同发育阶段而变化。值得注意的是，以最佳游动速度游动的鱼表现出了明显的适应性反应，例如，鱼类的生长速度加快、食物转换效率提高、骨骼肌肥大、营养物质利用率增加。因此，以最佳游动速度进行有氧游动可能在能量消耗、生理和行为方面都是有益的，它们在水产养殖中增加鱼类产量、促进鱼类健康的潜力是显著的（Davison，1997；Palstra，Planas，2011）。本节综述了有氧游动对心血管系统的各方面影响，以及对摄食、消化、代谢以及对骨骼肌生长和鱼类健康的影响。本节的主要目的是研究和应激反应相关的游动训练的表型适应性，并且提供增强鱼类行为表现能力和减轻应激能力的机制。

7.3.2 游动训练对心血管系统的影响

脊椎动物横纹肌的可塑性较强，运动训练可以增强哺乳动物和鱼类的氧运输系统。因此，对于一些特定的鱼类，心血管系统的形态和性能的变化可能会增强其游动能力，并且随后会改变其对应激源的反应。成年鲑鱼的游动训练会诱导心脏大小、心脏最大输血量和机体的耗氧量、血细胞比容、动脉氧含量和组织氧摄取量等微小但明显的增加（Gamperl，Farrell，2004）。游动训练也会增加鲤科鱼类（Sanger，Potscher，2000）、虹鳟（Davie等，1986）和斑马鱼（*Danio rerio*）（Pelster等，2003）的轴肌毛细血管（血管化），从而减少红细胞和附近的肌细胞之间的扩散距离。这些适应性行为在发育早期就会发生，并可能提高其对环境应激源的耐受性。例如，经过耐力训练的斑马鱼幼鱼游动消耗的氧气比未经训练的幼鱼少，而且经过训练的幼鱼对低氧的耐受性更强（Bagatto等，2001）。游动训练显然在细胞和分子水平上对心脏有重要影响，通过过度生长和血浆生成的反应而增强心脏的活动性（Ellison等，2012）。然而，尽管有研究表明游动可以诱导心脏的生长速度加快（Takle，Castro，2013），但导致这种效应原因的证据还几乎没有。在Castro等（2013）最近的一项研究中报道了大西洋鲑鱼（*Salmo salar*）游动训练对心室心肌细胞增殖和过度生长的标志物（如PCNA、MEF2C和ACTA1）以及血管生成标志物（如VEGF、VEGF-R2和红细胞生成素）的表达有刺激作用。这些结果表明，鲑鱼的游动训练可能通过激活哺乳动物已知的参与心脏生长调节的一组特定基因，诱导心脏肥大和过度增生。进而，心脏的生长潜在地增加了心脏的输血量，在应激状态下，促进更多氧气和营养物质的传递。研究表明，鲑鱼的心肌肥大和应激反应后的皮质醇水平呈正相关（Johansen等，2011），在游动训练中，皮质醇也可能促进心肌基因表达和重构。最终，研究确定运动诱导的适应性变化的分子靶点和通路，如鱼类心脏增大和血管化增强，将有助于提高我们对心血管可塑性和对应激源缓解反应的理解。

7.3.3　游动训练对摄食和能量代谢的影响

游动是一项能量需求很高的活动，因此，它对鱼类的营养和代谢过程有着深刻的影响。据报道，持续的游动训练可以刺激鱼类摄取食物，最重要的是，可以提高一些鱼类的食物转换效率（Davison，1997；Magnoni 等，2013b，c），有助于促进生长作用。众所周知，在水产养殖环境中，应激会降低机体的生长速率（Sadoul，Vijayan，2016；见本书第 5 章）。从新陈代谢的角度来看，骨骼肌是鱼类最重要的组织之一，因为在游动时骨骼肌对机体活动有很大的作用。之前的研究主要是用鲑科鱼类进行（Magnoni 等，2013b，c），持续游动会促进营养物质的转运，这是通过刺激脂类的利用，以碳水化合物作为燃料以及蛋白质在骨骼肌的储存来实现的。最近的研究已经将通过持续游动刺激骨骼肌能量底物的利用扩展到非鲑科鱼类物种，例如海鲷（Martin-Perez 等，2012；Felip 等，2013）、鲤鱼（He 等，2013）、中华倒刺鲃（*Spinibarbus sinensis*）（Zhao 等，2012）。几项关于持续游动对骨骼肌转录组和蛋白质组影响的研究，已经可以说明在持续游动时骨骼肌代谢适应性变化的分子基础。目前已有证据表明，在一些鱼类的蛋白质和基因表达水平上，对游动的应激反应，骨骼肌中碳水化合物代谢的增加涉及葡萄糖 GLUT4-转运通道的刺激及其细胞内磷酸化，以及通过糖酵解增加代谢（Martin-Perez 等，2012；Magnoni 等，2013，2014；Palstra 等，2014）。因此，鱼类游动诱导的肌肉收缩有降糖效果（Felip 等，2012），这是由于骨骼肌细胞 GLUT4-依赖的葡萄糖摄取增加，直接刺激 GLUT4 基因的转录（Marin-Juez 等，2013），以及离体的骨骼肌细胞葡萄糖吸收的电脉冲刺激（Magnoni 等，2014）。有意义的是，已经证明持续的游动可以提高含有高水平碳水化合物食物的利用率，例如用于水产养殖的饲料，并改善血糖控制能力（Felip 等，2012；Martin-Perez 等，2012）。因此，持续的游动可以看作是一种代谢促进剂，可以优化鱼类食物的利用。此外，在骨骼肌中，脂质的主要作用是作为能量底物支持持续性游动（Magnoni，Weber，2007），有证据表明，和脂蛋白分解代谢（如脂蛋白脂肪酶）、脂肪酸吸收（如 CD36）以及线粒体脂肪酸运输（CPT1）相关基因的表达增加，会增强鱼类游动过程中骨骼肌的 β-氧化（Magnoni 等，2013a，2014；Palstra 等，2014）。这些观察结果，再加上鱼类游动过程中刺激骨骼肌脂肪酸分解酶（如 3-OH 酰基辅酶 A 脱氢酶）的活性（Johnston，Moon，1980a，b），清楚表明持续游动可以通过 β-氧化刺激脂质代谢，增强骨骼肌的有氧代谢能力。游动训练通过改变个体应激和应对反应，是否能增强鱼肌肉有氧代谢的生化能力，还有待证实。

如前所述，持续游动过程中骨骼肌的一个重要代谢结果是增加蛋白质代谢，由于蛋白质合成的增加，产生一种蛋白质节约效果为肌肉收缩供应蛋白质，从而导致净蛋白质沉积（Felip 等，2012；Martin-Perez 等，2012）。虹鳟、海鲷、斑马鱼蛋白组学和转录组学研究已经证明，持续的游动可以增加骨骼肌中参与蛋白质合成的蛋白和基因家族的表达（Martin-Perez 等，2012；Magnoni 等，2013a；Palstra 等，2013，2014）。与哺乳动物一样，信号通路 IGF-1/PI3K/mTOR 是鱼体内蛋白质合成增加的可能机制之一，因为在斑马鱼白肌中该信号通路基因表达量是通过斑马鱼游动而增加的。这一信号通路也和斑马鱼骨骼肌块增加有关（Palstra 等，2014）。值得关注的是，鱼类游动过程中骨骼肌的转录研究表明，游动诱导的蛋白质合成速率增加和蛋白质降解速率增加同时发生。具体来说，游动提高了参与泛素-蛋白酶体系统、凋亡调控和蛋白质水解系统（如钙蛋白酶 calpain，组织蛋白酶）的基因表达（Magnoni 等，2013a；Palstra 等，2013，2014），这可能是生长状态下组织重构的必要条件。因此，这些结果提供了充分的证据，表明由持续游动引起的骨骼肌蛋白质的高水平转换，导致了蛋白质沉积的净增加，这是游动诱导鱼类生长的基础（见本章第 7.3.4 节）。通过促进蛋白质沉积和骨骼肌块的增加，游动也可以增加一些需求能量行为（如迁徙和性成熟）的能量储备，从而提高鱼类的整体健康水平。

7.3.3.1 鱼类游动过程中骨骼肌能量传感机制：AMPK

Sadoul 和 Vijayan（2016；见本书第 5 章）概述了能量传感器及其在应激状态下的能量传感机制。这里我们着重介绍一种特殊的异三聚体酶，即磷酸腺苷依赖的蛋白激酶（AMPK），它既是能量传感器，又是维持细胞能量稳态的能量代谢主开关（Hardie，Sakamoto，2006）。AMPK 的激活是通过细胞能量状态的降低（即 AMP：ATP 比率的增加）来实现的，因而可以通过应激源加速 ATP 的利用（如运动、升高温度和渗透不平衡）或限制 ATP 合成（如低氧）来实现（Weber，2011）。AMPK 作为一种糖原传感器（McBride 等，2009），可能通过抑制合成代谢、能量需求过程，如糖原合成、糖异生、脂质合成和蛋白质合成，在应激反应中进一步增强能量的供应（Weber，2011）。在骨骼肌中，肌肉纤维的收缩是通过 ATP 水解提供能量的。AMPK 通路的激活最重要的一个作用是刺激与产生 ATP 相关的代谢通路，如骨骼肌中游离脂肪酸和葡萄糖的摄取和氧化（Richter，Ruderman，2009）。在鱼类中，很少有人研究 AMPK 是不是鱼类游动诱导骨骼肌肌肉纤维收缩影响细胞代谢的潜在介质。Magnoni 等（2012）报道了虹鳟骨骼肌细胞 AMPK 异四聚体的性质和特征，以及合成 AMPK 的激活剂，如 AICAR 和二甲双胍。此外，鳟鱼骨骼肌细胞 AMPK 的药物激活导致 GLUT4-介导的葡萄糖的摄取和利用增加（Magnoni 等，2012）。随后，有研究证明 AMPK 活性在虹鳟白肌纤维和红肌纤维增加以应对持续游动，并且电刺激培育的虹鳟肌小管可以诱导 AMPK 活性（Magnoni 等，2014）。这项研究还表明，电刺激对葡萄糖摄取和利用的作用需要 AMPK 活性。有趣的是，药物激活鳟鱼肌肉细胞 AMPK 能增加转录调节因子 PGC-1α 的表达（Magnoni 等，2012），而在哺乳动物中已经知道能诱导线粒体的发生和刺激有氧代谢，从而产生 ATP（Lin 等，2005）。这些研究清楚地表明 AMPK 是鱼类骨骼肌中的一个重要因子，它将肌肉纤维收缩的激活转变为能量稳态和持续功能的适应性代谢反应。因此，AMPK 的重要性还应该延伸到应激反应中，包括鱼的游动和游动后的恢复。

除了打开 ATP 的再生途径，AMPK 的激活还会关闭 ATP 的消耗（即合成代谢）途径，如蛋白的合成，主要是为骨骼肌收缩保留能量（Hardie，Sakamoto，2006）。然而，该模型存在的问题之一是，鱼类游动时骨骼肌蛋白合成量和质量的增加（Felip 等，2012）很难和骨骼肌 AMPK 活性的增加相协调（Magnoni 等，2014）。在鱼类游动过程中，AMPK 如何刺激分解代谢以再生 ATP，而同时又增加蛋白质合成是一个问题，这甚至在哺乳动物中都未解决。

7.3.4 游动对训练骨骼肌生长的影响

游动是骨骼肌的收缩功能和骨骼肌发生重大变化的生理刺激（图 7.3）。脊椎动物骨骼肌受生理环境的影响很大，随着使用的增加，骨骼肌可以改变其收缩能力、兴奋性、代谢能力和质量（Egan，Zierath，2013）。因此，骨骼肌被认为是一种独特的可塑性和适应性较强的组织，可以增强游动能力，调节对应激源的行为反应。如前所述（见本章第 7.2.3 节），人们普遍认为鱼的持续游动主要是由慢肌纤维或红肌纤维的收缩完成的，而爆发式游动主要是由快肌纤维或白肌纤维的收缩完成的。然而，仅靠红肌纤维的收缩，很难维持持久的游动，例如长时间的繁殖或摄食洄游（如鳗鲡、鲑科鱼类、金枪鱼类）。最近，关于白肌纤维几乎专一地参与高强度游动（引发明显的应激反应）的观点受到了质疑（Videler，2011）。事实上，有证据表明，在低强度的持续游动中，白肌纤维也会收缩（Moyes，West，1995；Johnston，1999；Coughlin，2002）。

7.3.4.1 有氧游动及其对骨骼肌质量的调节

机体生长是由多个变量调控的综合反应：摄食、肠胃的消化吸收、血液循环激素和局部激素以及所受到的应激（Sadoul，Vijayan，2016；见本书第 5 章）。已有研究证明持续的游动可以促进多种鱼类生长，特别是当鱼类以 U_{opt} 或者接近 U_{opt} 持续游动时（Davison，Herbert，2013）。对于那些在自然环境中持续的游动速度接近于实验得出的 U_{opt}（大多数鲑科鱼）的物种来说尤其如此。游动也促进了一些非鲑科鱼类的生长，包括条纹狼鲈（*Morone saxatilis*）（Young，Cech，1993）、亚马逊石脂鲤（*Brycon amazonicus*）（Arbelaez-Rojas，Moraes，2009，2010）、黄尾鰤（*Seriola lalandi*）（Yogata，Oku，2000；Brown 等，2011）、海鲷（Ibarz 等，2011）、中华倒刺鲃（Li 等，2010）、巨鳍鲤（*Piaractus mesopotamicus*）（Da Silva Nunes 等，2013）、斑马鱼（Pelster 等，2003；van der Meulen 等，2006；Palstra 等，2010）。在研究过的鱼类中，游动诱导的生长与骨骼肌白肌纤维和骨骼肌红肌纤维的肥大有关（Johnston 和 Moon，1980a，b；Davison，1997；Bugeon 等，2003），这支持了白肌纤维可能确实在持续游动中发挥重要作用的观点。在细胞水平上，由鱼类游动引起的白肌纤维的肥厚变化（包括肌纤维横截面积和肌纤维周长等形态学参数的增加）导致其向较大纤维的较高丰度的方向转移（Johnston，Moon，1980；Ibarz 等，2011；Palstra 等，2014），这最终将导致肌肉质量的增加。因此，我们认为，在鱼体内，白肌纤维肥大（白色骨骼肌占体重的50%以上）是鱼类游动促进生长作用的机制之一。

鱼类通过游动刺激白肌纤维生长的机制已经可以通过转录组来解释。最近的研究指出，鱼类持续游动引起的骨骼肌肌纤维肥大可以解释为骨骼肌转录组发生了重大的变化，如成年虹鳟和斑马鱼的研究所显示（Magnoni 等，2013；Palstra 等，2013，2014）。在分子水平上，游动诱导的骨骼肌收缩活动导致编码结构的基因表达增强，以及对骨骼肌的收缩和细胞外基质的组成部分进行调节（Magnoni 等，2013；Palstra 等，2014）。具体而言，鱼类游动导致参与激活神经-肌肉联系、兴奋-收缩耦合、肉瘤细胞收缩力向肌膜传导、细胞-直肠传导、细胞外基质传递和肌肉结构维持的基因表达增加（Magnoni 等，2013a；Palstra 等，2013，2014）。这些研究表明，持续的游动可以促进白肌纤维的肌原纤维发生，并有力地支持了鱼类在持续游动下增生性的肌肉生长在很大程度上是由于肌原纤维蛋白增加的说法（Johnston 等，2011）。支持这一观点的证据来自观察到的鳟鱼白肌中蛋白质沉积的增加是对游动的反应（Felip 等，2012；见第 7.3.3 节）。对运动过程中鱼白骨骼肌的转录分析也首次提供了调控因子的证据，这些因子可能涉及将收缩信号转化为激活转录的程序，从而增加肌原纤维的发生。在这方面，这些调控因子在斑马鱼中的作用很明显，它们是鱼骨骼肌质量的潜在调节因子，并且是在游动诱导的收缩活动中转录激活的（Palstra 等，2010，2014）。这些研究表明，在硬骨鱼中，由于持续游动导致收缩活动，白骨骼肌发生肥大生长，而一些肌肉来源的可能刺激肥大生长的因子已被确定。重要的是，这些因子可以作为白骨骼肌生长和功能的潜在标记，也可以作为肌肉对应激反应的有价值的指标。

到目前为止，还没有关于鱼类游动对骨骼肌增生刺激作用的结论性报道。一些研究报告称，鱼类在游动时，白肌纤维密度或数量没有变化（Ibarz 等，2011；Rasmussen 等，2011；Palstra 等，2014）。然而，也有报道表明游动可能刺激斑马鱼骨骼肌细胞的增殖（vander Meulen 等，2006；Palstra 等，2014），并证明在斑马鱼成年后这些细胞会保持其增殖能力（Zhang，Anderson，2014）。无论游动诱导肌肉生长的机制如何，收缩作用的扩展都有可能增强鱼类的游动能力，并对捕食、躲避敌害、迁徙和繁殖等行为产生直接影响。

7.3.4.2 骨骼肌通过游动训练获得有氧呼吸表型

细胞内细胞器、膜通道和骨骼肌血管网络扩张的生理适应和生化适应，可能有助于应对应激并增强鱼类行为能力。鱼类游动的速度在 U_{opt} 或接近于 U_{opt} 时会增强骨骼肌的有氧代谢能力。这在红肌（一种具有高线粒体氧化酶活性的组织）中表现得尤为明显。一些研究也表明，持续游动增加了柠檬酸合成酶的活性和 mRNA 的表达水平（Johnston，Moon，1980a，b；van der Meulen 等，2006；LeMoine 等，2010；Magnoni 等，2013a）以及线粒体的密度（Pelster 等，2003）。有意义的是，通过持续的游动，白肌的氧化能力似乎也会增加，如同线粒体氧化标记物和肌原蛋白表达量的增加所表现的那样（Johnston，Moon，1980；McClelland 等，2006；van der Meulen 等，2006；LeMoine 等，2010；Magnoni 等，2013）。此外，鱼类持续游动过程中，红肌和白肌需氧表现型的增加会伴随组织血管化的增加（见下文）。

7.3.4.3　需氧表现型的增加与游动训练后的恢复情况有关

考虑到鱼类游动诱导的骨骼肌收缩的重大影响，其相关的问题是游动训练是否会影响游动的行为表现。有几项研究已经解决了这个问题，很多鱼类在持续游动速度下进行游动训练后，其游动性能显著提高（Pearson 等，1990；Davison，1997；Holk，Lykkeboe，1998；McDonald 等，1998；Liu 等，2009；Li 等，2010；Anttila 等，2011；Zhao 等，2012）。在大西洋鲑、褐鳟（*Salmo trutta*）和白鲑（*Coregonus lavaretus*）中，提高游动能力与二氢吡啶和阿诺碱受体的密度增加相关，并且在白肌细胞与红肌细胞内和兴奋-收缩偶联机制相关的细胞内钙离子的释放中起重要作用（Anttila 等，2008，2011）。此外，鱼类白肌和红肌氧化能力的增加伴随着游动能力的增加，它是由线粒体酶丰度和活性（Farrell 等，1991；Anttila 等，2006，2011；McClelland 等，2006）以及氧化通道转录水平的上调所决定的（McClelland 等，2006；LeMoine 等，2010；Magnoni 等，2013；Palstra 等，2013，2014）。重要的是，已经证实鱼类游动可以增加白肌和红肌中的毛细血管数量（Davie 等，1986；Pelster 等，2003；Ibarz 等，2011；Palstra 等，2014），这可能会增强血液和骨骼肌之间气体和代谢物的交换。通过上述观察得出结论：在持续游动速度下进行游动训练可能会诱导肌肉的适应性反应，增加肌肉摄取氧的能力和代谢物的扩散交换。因此，游动训练对游动能力的提高很可能是骨骼肌收缩能力提高的结果，部分原因是氧化磷酸化和细胞内钙离子的循环增加 ATP 的生成。虽然没有直接进行实验，但我们可以推测游动训练后的肌肉适应性可能有助于确定应激反应的可能发生及其量度，或者和应激反应有关的能量学与恢复情况。

7.3.5　游动训练对应激的影响：行为、健康和福利

鱼类在蓄养和高密度饲养（即水产养殖）中的一个重要问题是对它们先天性游动行为的破坏。在目前的水产养殖条件下，鱼的游动活动水平无法与野生鱼相比，因此，鱼是在比自然环境更加静息的环境中饲养的。禁闭强迫游动行为的改变可能意味着对自然行为的偏离，因此，它可以认为是一种应激的来源。禁闭和减少与游动有关的活动也可能剥夺鱼游动的有益生理作用，从而构成额外的应激来源。尽管禁闭诱导的游动行为中断的实际影响非常重要，但迄今为止，游动活动与行为、鱼的健康以及它们在应激反应中的影响之间的关系得到的关注相对较少。然而，一些研究已经测试这种设想：研究通过游动训练（即强迫游动）恢复游动行为是否可以增强鱼类的健康和福利状况？

7.3.5.1　游动训练对行为和应激的影响

研究表明，随着水流速度的增加，鲑鱼的游动强度会增加，这主要是通过诱导群

体行为来改变蓄养鱼的行为，如北极红点鲑和虹鳟（Grunbaum 等，2008；Larsen 等，2012）。鱼对适度水流速度的趋流性反应减少了自发活动，促进了鱼群的聚集。除了水流速度，移动的光刺激可以诱导鱼的视动反应，已有研究证明，这可以增加蓄养的大西洋鲑鱼的集群行为和游动活动（Herbert 等，2011）。利用诱导视动反应来操纵大型网箱内鱼的游动行为，改变鱼在网箱中的游动速度和游动深度，这对于积极游动鱼类的大规模生产具有潜在价值。

除了提高新陈代谢效率（Herskin，Steffensen，1998），游动训练诱导的集群会减少鱼的攻击性行为（Christiansen 等，1989；Adams 等，1995；Davison，1997；Brannas，2009），可能是由于游动的能量消耗。在缓流条件下，游动训练破坏了个体的社会性等级的关系，减少了互相攻击的发生率，特别是在摄食期间（Brannas，2009）。有研究表明，鱼类被迫游动和集群活动是为了减少应激，这可能是相互攻击减少的结果。游动训练能降低血液皮质醇的水平（Woodward，Smith，1985；Boesgaard 等，1993；Postlethwaite，McDonald，1995；Herbert 等，2011；Arbelaez-Rojas 等，2013），并且在应激后加快了皮质醇水平恢复到基础的非应激水平（Veiseth 等，2006）。因此，持续的游动可以减轻鱼的应激，同时减少应激恢复时间，从而提高游动性能和生长速度（Milligan 等，2000；Veiseth 等，2006；McKenzie 等，2012）（图 7.2 和图 7.3）。然而，其他一些研究并没有显示游动对急性应激后代谢恢复有明显影响（Kieffer 等，2011）。虽然游动对减轻应激的效应是确信无疑的，但游动训练方式、鱼的种类、营养状况以及发育阶段的差异可能会造成一些和文献报道相互矛盾之处。

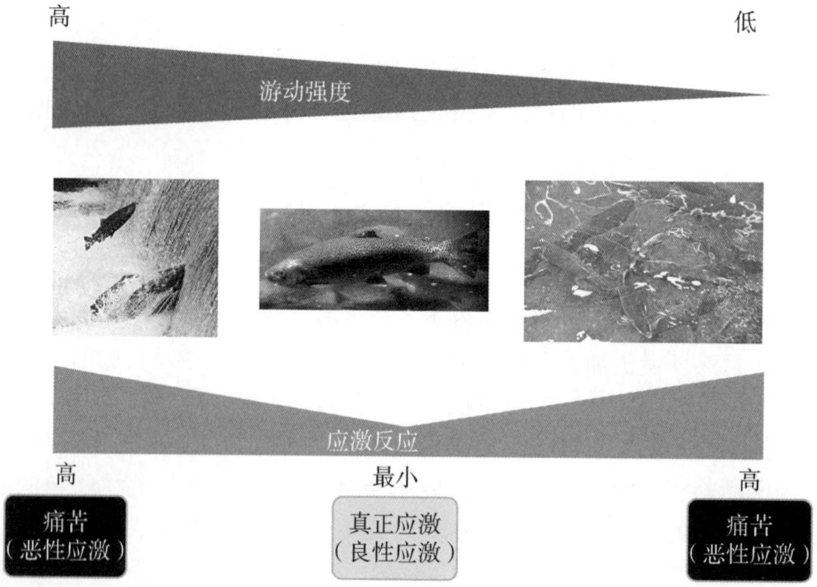

图 7.2　概念式的模型描述在不同的环境状态中游动强度和应激反应之间的关系

注：在鲑科鱼类中，应激反应在低和高的游动强度中实现。中等的或最适的游动速度减少应激反应；如果游动坚持超过延长的时间，可能引起几种不同的适应性。

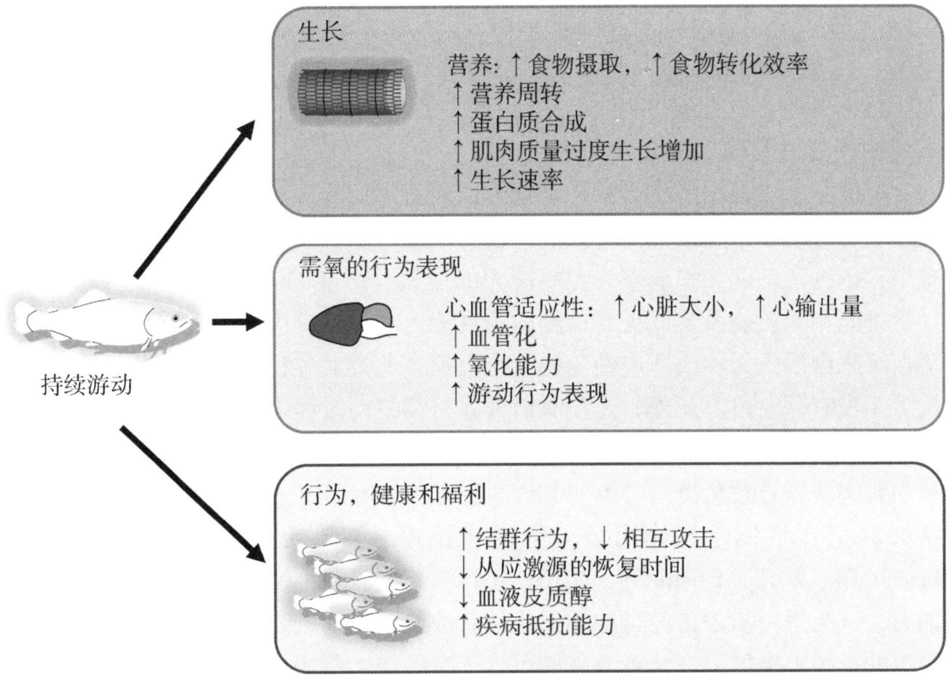

图 7.3　持久游动对生理的与行为的度量正面影响的总结

注：这些影响能引起行为表现增强和提高对随后应激源的抵抗能力。

7.3.5.2　游动训练对抗病性的影响

众所周知，人类锻炼对预防许多疾病具有重要的作用，特别是对心血管病和代谢疾病的预防。一直有人致力于研究游动对鱼类抵抗病原体产生的有益作用，特别是考虑到在水产养殖中鱼类疾病的发病率相对较高。据推测，不能强迫鱼类以它们持续的速度游动，加上水产养殖中鱼群的密度较高，这可能造成鱼类更易感染疾病。这一观点得到了一项研究的支持，即在持续的游动条件下饲养的大西洋鲑鱼的死亡率低于静息的鱼类（Totland 等，1987）。依照这个思路，其他研究人员推测，鱼类持续的有氧游动会提高它们对病原体的抵抗力。最近两项关于大西洋鲑鱼的研究证实了这一假设。在第一项研究中，研究人员将鱼分为两组，一组是间隔训练，另一组是连续 6 周的有氧游动训练。经过 6 周的训练期之后，用传染性胰腺坏死病毒（IPNV）对鱼进行攻毒实验。结果表明，经过间歇性有氧游动训练的鱼比未经训练的鱼对 IPNV 的抵抗能力要强，尽管效果不够显著（Castro 等，2011）。在第二项研究中，在对抗 IPNV（Castro 等，2013b）之前，大西洋鲑鱼再次接受连续的游动训练（在三个不同的游动强度下）或间隔性的游动训练（在两个不同的持续时间下），共 10 周，然后用传染性胰腺坏死病毒（IPNV）对鱼进行攻毒实验。在这项研究中，在游动训练中接受中等强度训练的鱼对 IPNV 的抵抗能力高于其他组，包括未进行游动训练的鱼，但没有统计学上的显著差异。然而，根据游动能力来筛选鱼，把游动能力强的和游动能力弱的鱼分开，随后用 IPNV 对鱼进行攻毒实验，游动能力强的鱼对 IPNV 的抗性明显高于游动能力弱的鱼（Castro 等，2013b）。这些研究的结果表明，持续游动训练会提高对致命病毒感染的抵抗能力。即使无显著影响，这些结果应该会鼓励更多的研究利用游动训练来提高鱼类抵抗各种疾病的能力以及确定特异性的减压机制。

有趣的是，当根据鱼先天的游动能力强弱将它们分开时，游动训练的效果更为显著，这表明游动能力越强可能对疾病抵抗力也越强。从实际应用的角度来看，在水产养殖中，依据游动能力强弱作为免疫能力的度量是可能的。值得注意的一个方面是，游动是否能调节免疫系统以提高鱼类的免疫能力和免疫反应能力。到目前为止，通过游动训练来激活先天性或适应性免疫系统以提高鱼类对各种生理的和其他的应激源的抵抗力还没有研究过。此外，还没有研究在持续游动过程中，肌肉收缩如何在增强鱼的免疫系统中发挥作用。然而，与运动的哺乳动物（Wiendl 等，2005）类似，游动诱导的肌肉收缩导致鱼类白骨骼肌和红骨骼肌中免疫相关分子的基因表达发生改变，包括抗原呈递机制、补体因子、免疫球蛋白、细胞因子、趋化因子及其受体（Magnoni 等，2013；Palstra 等，2013，2014）。经常锻炼鱼类骨骼肌可以明显提高免疫能力，这为进一步评估经过游动训练的鱼或游动能力较强的鱼对病原体和其他各种应激源的抵抗力提供了一个有意义的可行方案。综上所述，长期的游动活动可以调节免疫系统应对潜在生物应激源的生理反应，并最终确定鱼对特殊病原体的耐受能力。在鱼游动的过程中，血液循环的皮质醇减少和免疫抑制能力的消失是不是维持免疫能力的相关介质，还有待证实。

7.3.5.3 游动训练对应激和鱼类福利的影响

如 Schreck 和 Tort 所述（2016；见本书第 1 章），野生的或人为的外界环境是多变的，这会直接影响机体对应激源的生理反应。据报道，游动训练对游动行为的刺激、血液皮质醇水平的降低、从急性应激中恢复以及对致病性感染的抵抗能力的加强，显著地表明游动训练可以减少应激反应并改善鱼的福利（图 7.3）。Huntingford 和 Kadri（2013）在最近的一篇综述中认为，持续的游动除了可以改善鱼的生理状态，还可以通过增加鱼类集群行为、减少相互的攻击性行为以及促进神经发生和相关的学习能力来改善和应激相关的行为过程。即使通过鱼类游动来促进神经发生和学习能力仍然需要证实，但是研究已经证明游动可以增加大脑多巴胺能活性（Ortiz，Lutz，1995）。其他研究也表明多巴胺能活性的增加与鱼类游动相关（Johansson 等，2004 年）。总的来说，持续游动引起的生理和行为反应似乎遵从福利和减压的功能定义。然而，准确地确定每一个物种及其不同发育阶段持续游动的最佳速度和游动对其生理与行为的益处是非常有意义的，特别对水产养殖的鱼类，这是应用游动训练以减轻外界压力、提高生长速度和增强抗病性的前提。

7.4　总结及未来展望

虽然关于鱼类游动对其生理的影响已有大量的研究，但还没有尝试建立个体应激反应的多变性（Winberg 等，2016；见本书第 2 章）和诱导游动的生理反应之间的关系。假设个体持续地对应激物反应，如反应前的个体在一端而反应中的个体在同一范围的另一端，那么相关的问题是游动的应对方式是否基于鱼类游动能力。几项研究表明，不同鱼类个体的游动能力可能存在差异，这使得研究人员可以区分游动能力不同的鱼类（Claireaux 等，2005；Castro 等，2013）。例如，根据在游动测试中的表现，大西洋鲑鱼被归类为游动能力强的鱼类，据报道，它们对病毒感染的抵抗能力高于游动能力弱的鱼类（Castro 等，2013b），这表明个体游动能力和免疫能力之间可能存在联系。显然还需要进一步的研究来确定持续游动是否可以通过激活免疫系统来改善鱼的健康状况和增强其对付应激的能力。

鱼类持续游动还有其他益处，特别是生长速度的增加、健康状况的改善，以及相互攻击性行为和应激反应的减少。因此，很明显，持续游动训练有利于鱼类养殖（即水产养殖）。如今在水产养殖中实施鱼类游动训练在技术上还有很大的限制：我们如何让鱼在人工环境中以预期的速度游动？未来这一领域的研究很可能为水产养殖产业生产健壮的鱼类创造必要条件。

最后，我们可能对鱼类游动对行为和负责记忆获取与恢复以及空间与社会学习的神经过程的影响感兴趣。众所周知，应激会影响学习能力和行为方式改变（Noakes，Jones，2016；见本书第 9 章）。最近对哺乳动物的研究发现，运动与认知功能之间存在联系，通过运动可以诱发肌肉衍生因子（即鸢尾素）的产生，而这种肌肉衍生因子可能增强大脑中脑衍生神经营养因子（BDNF）的表达（Wrann 等，2013）。有趣的是，已知的哺乳动物运动的抗抑郁作用是和骨骼肌收缩改变肌肉组织色氨酸代谢的能力有关，从而促进动物的认知和情绪（Agudelo 等，2014）。这些研究阐明了哺乳动物骨骼肌和大脑之间化学信号的交叉，并为鱼类生理学家提供重要的未解之谜。游动会对鱼的大脑产生类似的影响吗？关于鱼类游动、福利和减压之间的关系，游动对鱼来说是一种积极和有益的行为吗？将鱼类生理的与行为的变化以及持续游动和各种应激源结合起来进行比较研究，最终将加深我们对游动机理，以及游动和野生或蓄养鱼类应激反应之间相互联系的理解。

K. J. 罗德尼克　J. V. 普拉纳斯　著
张　勇　译
林浩然　校

参考文献

Adams, C. E., Huntingford, F. A., Krpal, J., Jobling, M. and Burnett, S. J. (1995). Exercise, agonistic behaviour and food acquisition in Arctic charr Salvelinus alpinus. Environ. Biol. Fishes 43, 213-218.

Agudelo, L. Z., Femenia, T., Orhan, F., Porsmyr-Palmertz, M., Goiny, M., Martinez-Redondo, V., et al. (2014). Skeletal muscle PGC-1 alpha1 modulates kynurenine metabolism and mediates resilience to stress-induced depression. Cell 159, 33-45.

Anras, M.-L. B. and Lagardère, J. P. (2004). Measuring cultured fish swimming behaviour: first results on rainbow trout using acoustic telemetry in tanks. Aquaculture 240, 175-186.

Anttila, K., Järvilehto, M. and Mänttäri, S. (2008). The swimming performance of brown trout and whitefish: the effects of exercise on Ca^{2+} handling and oxidative capacity of swimming muscles. J. Comp. Physiol. B 178, 465-475.

Anttila, K., Jokikokko, E., Erkinaro, J., Jaervilehto, M. and Maenttaeri, S. (2011). Effects of training on functional variables of muscles in reared Atlantic salmon Salmo salar smolts: connection to downstream migration pattern. J. Fish. Biol. 78, 552-566.

Anttila, K., Mänttäri, S. and Järvilehto, M. (2006). Effects of different training protocols on Ca^{2+} handling and oxidative capacity in skeletal muscle of Atlantic salmon (Salmo salar L.). J. Exp. Biol. 209, 2971-2978.

Arbelaez-Rojas, G. A., Hackbarth, A., Inoue, L. A. K. A., Dias de Moraes, F. and Moraes, G. (2013). Sustained swimming mitigates stress in juvenilefish Brycon amazonicus. J. Appl. Aquac. 25, 271-281.

Arbelaez-Rojas, G. A. and Moraes, G. (2009). Sustained swimming and stocking density iteraction in the performance and body composition of matrinxa Brycon amazonicus juveniles. Cienc. Rural 39, 201-208.

Arbelaez-Rojas, G. A. and Moraes, G. (2010). Optimization of sustaining swimming speed of matrinxa Brycon amazonicus: performance and adaptive aspects. Sci. Agr. 67, 253-258.

Axelsson, M., Driedzic, W. R., Farrell, A. P. and Nilsson, S. (1989). Regulation of cardiacoutput and gut blood-flow in the sea raven, Hemitripterus americanus. Fish Physiol. Biochem. 6, 315-326.

Bagatto, B., Pelster, B. and Burggren, W. W. (2001). Growth and metabolism of larval zebrafish: effects of swim training. J. Exp. Biol. 204, 4335-4343.

Barton, B. A. andSchreck, C. B. (1987). Metabolic cost of acute physical stress in juvenile steelhead. Trans. Am. Fish. Soc. 116, 257-263.

Beamish, F. W. H. (1979). Swimming capacity. In Fish Physioogy-Locomotion, Vol. 7 (eds. W. S. Hoar and D. J. Randall), pp. 101-187. New York, NY: Academic Press.

Bennett, A. F. (1991). The evolution of activity capacity. J. Exp. Biol. 160, 1-23.

Black, E. C. (1958). Hyperactivity as a lethal factor infish. J. Fish. Res. Bd. Can. 15, 573-586.

Boesgaard, L., Nielsen, E. N. and Rosenkilde, P. (1993). Moderate exercise decreases plasma cortisol levels in Atlantic salmon (Salmo salar). Comp. Biochem. Physiol. A 106, 641-643.

Brannas, E. (2009). The effect of moderate exercise on growth and aggression depending on social rank in groupsof Arctic charr (Salvelinus alpinus L.). Appl. Anim. Behav. Sci. 119, 115-119.

Brauner, C. J., Thorarensen, H., Gallaugher, P., Farrell, A. P. and Randall, D. J. (2000). The interactions between O_2 and CO_2 exchange in rainbow trout during graded sustained exercise. Resp. Physiol. 119, 83-96.

Brett, J. R. (1964). The respiratory metabolism and swimming performance of young sockeye salmon. J. Fish. Res. Bd. Can. 21, 1183-1226.

Brett, J. R. (1972). Metabolic demand for oxygen infish, particularly salmonids, and a comparison with other vertebrates. Resp. Physiol. 14, 151-170.

Brown, E. J., Bruce, M., Pether, S. and Herbert, N. A. (2011). Do swimmingfish always grow fast? Investigating the magnitude and physiological basis of exercise-induced growth in juvenile New Zealand yellowtail kingfish, Seriola lalandi. Fish Physiol. Biochem. 37, 327-336.

Bugeon, J., Lefevre, F. and Fauconneau, B. (2003). Fillet texture and muscle structure in brown trout (Salmo trutta). Aquac. Res. 34, 1287-1295.

Burnett, N. J., Hinch, S. G., Braun, D. C., Casselman, M. T., Middleton, C. T., Wilson, S. M., et al. (2014). Burst swimming in areas of highflow: delayed consequences of anaerobiosis in wild adult sockeye salmon. Physiol. Biochem. Zool. 87, 587-598.

Castro, V., Grisdale-Helland, B., Helland, S. J., Kristensen, T., Joergensen, S. M., Helgerud, J., et al. (2011). Aerobic training stimulates growth and promotes disease resistance in Atlantic salmon (Salmo salar). Comp. Biochem. Physiol. A 160, 278-290.

Castro, V., Grisdale-Helland, B., Helland, S. J., Torgersen, J., Kristensen, T., Claireaux, G., et al. (2013a). Cardiac molecular-acclimation mechanisms in response to swimming-induced exercise in Atlantic salmon. PLoS One 8, e55056.

Castro, V., Grisdale-Helland, B., Jorgensen, S. M., Helgerud, J., Claireaux, G.,

Farrell, A. P. , et al. (2013b). Disease resistance is related to inherent swimming performance in Atlantic salmon. BMC Physiol. 13, 1.

Chandroo, K. P. , Cooke, S. J. , McKinley, R. S. and Moccia, R. D. (2005). Use of electromyogram telemetry to assess the behavioural and energetic responses of rainbow trout, Oncorhynchus mykiss (Walbaum) to transportation stress. Aquac. Res. 36, 1226-1238.

Christiansen, J. S. , Ringo, E. and Jobling, M. (1989). Effects of sustained exercise on growth and body composition offirst-feeding fry of Arctic charr, Salvelinus alpinus L. Aquaculture 79, 329-335.

Claireaux, G. , McKenzie, D. J. , Genge, A. G. , Chatelier, A. , Aubin, J. and Farrell, A. P. (2005). Linking swimming performance, cardiac pumping ability and cardiac anatomy in rainbow trout. J. Exp. Biol. 208, 1775-1784.

Coughlin, D. J. (2002). Aerobic muscle function during steady swimming infish. Fish Fisher. 3, 63-78.

Da Silva Nunes, C. , Moraes, G. , Fabrizzi, F. , Hackbarth, A. and Arbelaez-Rojas, G. A. (2013). Growth and hematology of pacu subjected to sustained swimming and fed different protein levels. Pesq. Agropec. Bras. 48, 645-650.

Davie, P. S. , Wells, R. M. and Tetens, V. (1986). Effects of sustained swimming on rainbow trout muscle structure, blood oxygen transport, and lactate dehydrogenase isozymes: evidence for increased aerobic capacity of white muscle. J. Exp. Zool. 237, 159-171.

Davis, L. E. and Schreck, C. B. (1997). The energetic response to handling stress in juvenile coho salmon. Trans. Am. Fish. Soc. 126, 248-258.

Davison, W. (1997). The effects of exercise training on teleostfish, a review of recent literature. Comp. Biochem. Physiol. A 117, 67-75.

Davison, W. and Herbert, N. A. (2013). Swimming-enhanced growth. In Swimming Physiology of Fish (eds. A. P. Palstra and J. V. Planas), pp. 177-202. Heidelberg: Springer.

De Boeck, G. , Alsop, D. and Wood, C. (2001). Cortisol effects on aerobic and anaerobic metabolism, nitrogen excretion, and whole-body composition in juvenile rainbow trout. Physiol. Biochem. Zool. 74, 858-868.

Driedzic, W. R. and Gesser, H. (1994). Energy metabolism and contractility in ectothermic vertebrate hearts-hypoxia, acidosis, and low temperature. Physiol. Rev. 74, 221-258.

Driedzic, W. R. and Hochachka, P. W. (1978). Energy metabolism during exercise. In Fish Physiology-Locomotion, Vol. 7 (eds. W. S. Hoar and D. J. Randall), pp. 503-543. New York, NY: Academic Press.

Egan, B. and Zierath, J. R. (2013). Exercise metabolism and the molecular regulation of skeletal muscle adaptation. Cell Metab. 17, 162-184.

Ekström, A., Jutfelt, F. and Sandblom, E. (2014). Effects of autonomic blockade on acute thermal tolerance and cardioventilatory performance in rainbow trout, Oncorhynchus mykiss. J. Therm. Biol. 44, 47-54.

Ellison, G. M., Waring, C. D., Vicinanza, C. and Torella, D. (2012). Physiological cardiac remodelling in response to endurance exercise training: cellular andmolecular mechanisms. Heart 98, 5-10.

Fabbri, E., Capuzzo, A. and Moon, T. W. (1998). The role of circulating catecholamines in the regulation offish metabolism: an overview. Comp. Biochem. Physiol. C120, 177-192.

Farrar, R. S. and Rodnick, K. J. (2004). Sex-dependent effects of gonadal steroids and cortisol on cardiac contractility in rainbow trout. J. Exp. Biol. 207, 2083-2093.

Farrell, A. P. (1980). Vascular pathways in the gill of ling cod, Ophiodon elongatus. Can. J. Zool. 58, 796-806.

Farrell, A. P. (1982). Cardiovascular changes in the unanaesthetized lingcod (Ophiodon elongatus) during short-term progressive hypoxia and spontaneous activity. Can. J. Zool. 60, 933-941.

Farrell, A. P. (2007). Cardiorespiratory performance during prolonged swimming tests with salmonids: a perspective on temperature effects and potential analytical pitfalls. Philos. T. Roy. Soc. B 362, 2017-2030.

Farrell, A. P., Gamperl, A. K., Hicks, J. M. T., Shiels, H. A. and Jain, K. E. (1996). Maximum cardiac performance of rainbow trout (Oncorhynchus mykiss) at temperatures approaching their upper lethal limit. J. Exp. Biol. 199, 663-672.

Farrell, A. P., Johansen, J. A. and Suarez, R. K. (1991). Effects of exercise-training on cardiac performance and muscle enzymes in rainbow trout, Oncorhynchus mykiss. Fish Physiol. Biochem. 9, 303-312.

Farrell, A. P. and Jones, D. R. (1992). The heart. In Fish Physiology-The Cardiovascular System, Vol. 12A (eds. W. S. Hoar, D. J. Randall and A. P. Farrell), pp. 1-88. San Diego, CA: Academic Press.

Farrell, A. P., Macleod, K. R. and Chancey, B. (1986). Intrinsic mechanical-properties of the perfused rainbow-trout heart and the effects of catecholamines and extracellular calcium under control and acidotic conditions. J. Exp. Biol. 125, 319-345.

Farrell, A. P., Sobin, S. S., Randall, D. J. and Crosby, S. (1980). Intralamellar blood-flow patterns in fish gills. Am. J. Physiol. 239, R428-R436.

Felip, O., Blasco, J., Ibarz, A., Martin-Perez, M. and Fernandez-Borras, J.

(2013). Beneficial effects of sustained activity on the use of dietary protein and carbohydrate traced with stable isotopes 15N and 13C in gilthead sea bream (Sparus aurata). J. Comp. Physiol. B 183, 223-234.

Felip, O., Ibarz, A., Fernández-Borrás, J., Beltrán, M., MartÍn-Pérez, M., Planas, J. V., et al. (2012). Tracing metabolic routes of dietary carbohydrate and protein in rainbow trout (Oncorhynchus mykiss) using stable isotopes ([13C] starch and [15N] protein): effects of gelatinisation of starches and sustained swimming. Br. J. Nutr. 107, 834-844.

Fick, A. (1870). ? ber die Messung den Blutquantums in der Herzventrikeln. Sitzungb. Phys. Med. Ges. Würzburg 16.

Fry, F. E. J. (1947). Effects of environment on animal activity. Univ. Toronto Stud. Biol. Ser. 55, 1-62.

Fry, F. E. J. (1971). The effect of environmental factors on the physiology offish. In Fish Physiology-Environmental Relations and Behavior, Vol. 6 (eds. W. S. Hoar and D. J. Randall), pp. 1-98. New York, NY: Academic Press.

Gamperl, A. K. and Farrell, A. P. (2004). Cardiac plasticity infishes: environmental influences and interspecific differences. J. Exp. Biol. 207, 2539-2550.

Gerry, S. P. and Ellerby, D. J. (2014). Resolving shifting patterns of muscle energy use in swimmingfish. PLoS One 9, e106030.

Gladden, L. B. (2004). Lactate metabolism: a new paradigm for the third millennium. J. Physiol. Lond. 558, 5-30.

Gorissen, M. and Flik, G. (2016). Endocrinology of the Stress Response in Fish. In Fish Physiology-Biology of Stress in Fish, Vol. 35 (eds. C. B. Schreck, L. Tort, A. P. Farrell and C. J. Brauner), San Diego, CA: Academic Press.

Gregory, T. R. and Wood, C. M. (1999). The effects ofchronic plasma cortisol elevation on the feeding behaviour, growth, competitive ability, and swimming performance of juvenile rainbow trout. Physiol. Biochem. Zool. 72, 286-295.

Grünbaum, T., Cloutier, R. and Le Franc-ois, N. R. (2008). Positive effects of exposure to increased water velocity on growth of newly hatched Arctic charr, Salvelinus alpinus L. Aquac. Res. 39, 106-110.

Hardie, D. and Sakamoto, K. (2006). AMPK: a key sensor of fuel and energy status in skeletal muscle. Physiology 21, 48-60.

Harper, A. A., Newton, I. P. and Watt, P. W. (1995). The effect of temperature on spontaneous action-potential discharge of the isolated sinus venosus from winter and summer plaice (Pleuronectes platessa). J. Exp. Biol. 198, 137-140.

He, W., Xia, W., Cao, Z.-D. and Fu, S.-J. (2013). The effect of prolonged exercise training on swimming performance and the underlying biochemical mechanisms in juvenile common carp (Cyprinus carpio). Comp. Biochem. Physiol. A 166, 308-315.

Herbert, N. A., Kadri, S. and Huntingford, F. A. (2011). A moving light stimulus elicits a sustained swimming response in farmed Atlantic salmon, Salmo salar L. Fish Physiol. Biochem. 37, 317-325.

Herskin, J. and Steffensen, J. F. (1998). Energy savings in sea bass swimming in a school: measurements of tail beat frequency and oxygen consumption at different swimming speeds. J. Fish. Biol. 53, 366-376.

Hicks, J. W. and Bennett, A. F. (2004). Eat and run: prioritization of oxygen delivery during elevated metabolic states. Resp. Physiol. Neurobiol. 144, 215-224.

Hochachka, P. W. (1994). Muscles as Molecular and Metabolic Machines. Boca Raton, FL: CRC Press.

Hochachka, P. W. and Somero, G. N. (2002). Biochemical Adaptation. Mechanisms and Process in Physiological Evolution. New York, NY: Oxford University Press.

Holk, K. and Lykkeboe, G. (1998). The impact of endurance training on arterial plasma K^+ levels and swimming performance of rainbow trout. J. Exp. Biol. 201, 1373-1380.

Hughes, G. M., Le-Bras-Pennec, Y. and Pennec, J.-P. (1988). Relationships between swimming speed, oxygen consumption, plasma catecholamines and heart performance in rainbow trout (S. gairdneri R.). Exp. Biol. 48, 45-49.

Huntingford, F. A. and Kadri, S. (2013). Exercise, stress and welfare. In Swimming Physiology of Fish (eds. A. P. Palstra and J. V. Planas), pp. 161-176. Heidelberg: Springer.

Ibarz, A., Felip, O., Fernández-Borràs, J., Martín-Pérez, M., Blasco, J. and Torrella, J. (2011). Sustained swimming improves muscle growth and cellularity in gilthead sea bream. J. Comp. Physiol. B 181, 209-217.

Iversen, M., Finstad, B. and Nilssen, K. J. (1998). Recovery from loading and transport stress in Atlantic salmon (Salmo salar L.) smolts. Aquaculture 168, 387-394.

Johansen, I. B., Lunde, I. G., Rosjo, H., Chrisyensen, G., Nilsson, G. E., Bakken, M., et al. (2011). Cortisol response to stress is associated with myocardial remodeling in salmonidfishes. J. Exp. Biol. 214, 1313-1321.

Johansson, V., Winberg, S., Jonsson, E., Hall, D. and Bjornsson, B. T. (2004). Peripherally administered growth hormone increases brain dopaminergic activity and swimming in rainbow trout. Horm. Behav. 46, 436-443.

Johnson, T. P., Syme, D. A., Jayne, B. C., Lauder, G. V. and Bennett, A. F. (1994). Modeling red muscle power output during steady and unsteady swimming in largemouth bass. Am. J. Physiol. 267, R481-R488.

Johnston, I. A. (1999). Muscle development and growth: potential implications forflesh quality in fish. Aquaculture 177, 99-115.

Johnston, I. A., Bower, N. I. and Macqueen, D. J. (2011). Growth and the regulation of myotomal muscle mass in teleostfish. J. Exp. Biol. 214, 1617-1628.

Johnston, I. A. and Goldspink, G. (1973). Study of swimming performance of crucian carp Carassius carassius (L.) in relation to effects of exercise and recovery on biochemical changes in myotomal muscles and liver. J. Fish. Biol. 5, 249-260.

Johnston, I. A. and Moon, T. W. (1980a). Endurance exercise training in the fast and slow muscles of a teleostfish (Pollachius virens). J. Comp. Physiol. B 135, 147-156.

Johnston, I. A. and Moon, T. W. (1980b). Exercise training in skeletal muscle of brook trout (Salvelinus fontinalis). J. Exp. Biol. 87, 177-194.

Jones, D. R. and Randall, D. J. (1978). The respiratory and circulatory systems during exercise. In Fish Physiology-Bioenergetics and Growth, Vol. 8 (eds. W. S. Hoar and D. J. Randall), pp. 425-501. New York, NY: Academic Press.

Jorgensen, E. H. and Jobling, M. (1993). The effect of exercise on growth, food utilization and osmoregulatory capacity of juvenile Atlantic salmon, Salmo salar. Aquaculture 116, 233-246.

Kiceniuk, J. W. and Jones, D. R. (1977). Oxygen transport system in trout (Salmo gairdneri) during sustained exercise. J. Exp. Biol. 69, 247-260.

Kieffer, J. D. (2000). Limits to exhaustive exercise infish. Comp. Biochem. Physiol. A 126, 161-179.

Kieffer, J. D., Kassie, R. S. and Taylor, S. G. (2011). The effects of low-speed swimming following exhaustive exercise on metabolic recovery and swimming performance in brook trout (Salvelinus fontinalis). Physiol. Biochem. Zool. 84, 385-393.

Korovin, V. A., Zybin, A. S. and Legomin, V. B. (1982). Response of pondfish young to stress factors caused by transfers in geo thermal fish farming. Vopr. Ikhtiol. 22, 280-284.

Krohn, M. M. and Boisclair, D. (1994). Use of a stereo video system to estimate the energy expenditure of free-swimmingfish. Can. J. Fish. Aquat. Sci. 51, 1119-1127.

Larsen, B. K., Skov, P. V., McKenzie, D. J. and Jokumsen, A. (2012). The effects of stocking density and low level sustained exercise on the energetic efficiency of rainbow trout (Oncorhynchus mykiss) reared at 191C. Aquaculture 324, 226-233.

Lee, C. G., Farrell, A. P., Lotto, A., Hinch, S. G. and Healey, M. C. (2003). Excess post-exercise oxygen consumption in adult sockeye (Oncorhynchus nerka) and coho (O. kisutch) salmon following critical speed swimming. J. Exp. Biol. 206,

3253-3260.
Lee-Jenkins, S. S. Y., Binder, T. R., Karch, A. P. and McDonald, D. G. (2007). The recovery of locomotory activity following exhaustive exercise in juvenile rainbow trout (Oncorhynchus mykiss). Physiol. Biochem. Zool. 80, 88-98.

LeMoine, C. M. R., Craig, P. M., Dhekney, K., Kim, J. J. and McClelland, G. B. (2010). Temporal and spatial patterns of gene expression in skeletal muscles in response to swim training in adult zebrafish (Danio rerio). J. Comp. Physiol. B 180, 151-160.

Li, X.-M., Cao, Z.-D., Peng, J.-L. and Fu, S.-J. (2010). The effect of exercise training on the metabolic interaction between digestion and locomotion in juvenile darkbarbel catfish (Peltebagrus vachelli). Comp. Biochem. Physiol. A 156, 67-73.

Liew, H. J., Chiarella, D., Pelle, A., Faggio, C., Blust, R. and De Boeck, G. (2013). Cortisol emphasizes the metabolic strategies employed by common carp, Cyprinus carpio at different feeding and swimming regimes. Comp. Biochem. Physiol. A 166, 449-464.

Lin, J., Handschin, C. and Spiegelman, B. M. (2005). Metabolic control through the PGC-1 family of transcription coactivators. Cell Metab. 1, 361-370.

Liu, Y., Cao, Z. D., Fu, S.-J., Peng, J.-L. and Wang, Y. X. (2009). The effect of exhaustive chasing training and detraining on swimming performance in juvenile darkbarbel catfish (Peltebagrus vachelli). J. Comp. Physiol. B 179, 847-855.

Magnoni, L. J., Crespo, D., Ibarz, A., Blasco, J., Fernández-Borràs, J. and Planas, J. V. (2013a). Effects of sustained swimming on the red and white muscle transcriptome of rainbow trout (Oncorhynchus mykiss) fed a carbohydrate-rich diet. Comp. Biochem. Physiol. A 166, 1-12.

Magnoni, L. J., Felip, O., Blasco, J. and Planas, J. V. (2013b). Metabolic fuel utilization during swimming: optimizing nutritional requirements for enhanced performance. In Swimming Physiology of Fish (eds. A. P. Palstra and J. V. Planas), pp. 203-235. Heidelberg: Springer.

Magnoni, L. J., Marquez-Ruiz, P., Palstra, A. P. and Planas, J. V. (2013c). Physiological consequences of swimming-induced activity in trout. In Trout: From Physiology to Conservation (eds. S. Polakof and T. W. Moon), pp. 321-350. New York, NY: Nova Science Publishers.

Magnoni, L. J., Palstra, A. P. and Planas, J. V. (2014). Fueling the engine: induction of AMPactivated protein kinase in trout skeletal muscle by swimming. J. Exp. Biol. 217, 1649-1652.

Magnoni, L. J., Vraskou, Y., Palstra, A. P. and Planas, J. V. (2012). AMP-activated protein kinase plays an important evolutionary conserved role in the regulation of glucose metabolism infish skeletal muscle cells. PLoS One 7, e31219.

Magnoni, L. and Weber, J. M. (2007). Endurance swimming activates trout lipoprotein lipase: plasma lipids as a fuel for muscle. J. Exp. Biol. 210, 4016-4023.

Manoli, I., Alesci, S., Blackman, M. R., Su, Y. A., Rennert, O. M. and Chrousos, G. P. (2007). Mitochondria as key components of the stress response. Trends. Endocrinol. Metab. 18, 190-198.

Marín-Juez, R., Díaz, M., Morata, J. and Planas, J. V. (2013). Mechanisms regulating GLUT4 transcription in skeletal muscle cells are highly conserved across vertebrates. PLoS One 8, e80628.

Martin-Perez, M., Fernandez-Borras, J., Ibarz, A., Millan-Cubillo, A., Felip, O., de Oliveira, E., et al. (2012). New insights intofish swimming: a proteomic and isotopic approach in gilthead sea bream. J. Proteome. Res. 11, 3533-3547.

McBride, A., Ghilagaber, S., Nikolaev, A. and Hardie, D. G. (2009). The glycogen-binding domain on the AMPK beta subunit allows the kinase to act as a glycogen sensor. Cell Metab. 9, 23-34.

McClelland, G. B., Craig, P. M., Dhekney, K. and Dipardo, S. (2006). Temperature-and exercise-induced gene expression and metabolic enzyme changes in skeletal muscle of adult zebrafish (Danio rerio). J. Physiol. 577, 739-751.

McCormick, S. D. (1995). Hormonal control of gill Na^+, K^+-ATPase and chloride cell function. In Cellular and Molecular Approaches to Fish Ionic Regulation (eds. C. M. Wood and T. J. Shuttleworth), pp. 285-315. San Diego, CA: Academic Press.

McCormick, S. D., Moyes, C. D. and Ballantyne, J. S. (1989). Influence of salinity on the energetics of gill and kidney of Atlantic salmon (Salmo salar). Fish. Physiol. Biochem. 6, 243-254.

McDonald, D. G., Milligan, C. L., McFarlane, W. J., Croke, S., Currie, S., Hooke, B., et al. (1998). Condition and performance of juvenile Atlantic salmon (Salmo salar): effects of rearing practices on hatcheryfish and comparison with wild fish. Can. J. Fish. Aquat. Sci. 55, 1208-1219.

McKenzie, D. J., Hoglund, E., Dupont-Prinet, A., Larsen, B. K., Skov, P. V., Pedersen, P. B., et al. (2012). Effects of stocking density and sustained aerobic exercise on growth, energetics and welfare of rainbow trout. Aquaculture 338, 216-222.

Meyer, W. F. and Cook, P. A. (1996). An assessment of the use of low-level aerobic swimming in promoting recovery from handling stress in rainbow trout. Aquac. Int. 4, 169-171.

Milligan, C. L. (1996). Metabolic recovery from exhaustive exercise in rainbow trout. Comp. Biochem. Physiol. A 113, 51-60.

Milligan, C. L. (2003). A regulatory role for cortisol in muscle glycogen metabolism in rainbow trout Oncorhynchus mykiss Walbaum. J. Exp. Biol. 206, 3167-3173.

Milligan, C. L., Hooke, G. B. and Johnson, C. (2000). Sustained swimming at low velocity following a bout of exhaustive exercise enhances metabolic recovery in rainbow trout. J. Exp. Biol. 203, 921-926.

Mommsen, T. P. (1984). Metabolism of thefish gill. In Fish Physiology-Gills: Ion and Water Transfer, Vol. 10B (eds. W. S. Hoar and D. J. Randall), pp. 203-238. San Diego, CA: Academic Press.

Mommsen, T. P. and Hochachka, P. W. (1988). The purine nucleotide cycle as 2 temporally separated metabolic units-a study on trout muscle. Metabolism 37, 552-556.

Mommsen, T. P., Vijayan, M. M. and Moon, T. W. (1999). Cortisol in teleosts: dynamics, mechanisms of action, and metabolic regulation. Rev. Fish Biol. Fish. 9, 211-268.

Morgan, J. D. and Iwama, G. K. (1996). Cortisol-induced changes in oxygen consumption and ionic regulation in coastal cutthroat trout (Oncorhynchus clarki clarki) parr. Fish Physiol. Biochem. 15, 385-394.

Morgan, J. D. and Iwama, G. K. (1999). Energy cost of NaCl transport in isolated gills of cutthroat trout. Am. J. Physiol. 277, R631-R639.

Moyes, C. D. and West, T. G. (1995). Exercise metabolism offish. In Biochemistry and Molecular Biology of Fishes, vol. 4 (eds. P. W. Hochachka and T. P. Mommsen), pp. 368-392. Amsterdam: Elseiver Science.

Newsholme, E. A. and Leech, A. R. (1983). Biochemistry for the Medical Sciences. Chichester: Wiley and Sons.

Nielsen, M. E., Boesgaard, L., Sweeting, R. M., McKeown, B. A. and Rosenkilde, P. (1994). Plasma levels of lactate, potassium, glucose, cortisol, growth hormone and triiodo-Lthyronine in rainbow-trout (Oncorhynchus mykiss) during exercise at various levels for 24 h. Can. J. Zool. 72, 1643-1647.

Nikinmaa, M. (1992). How does environmental pollution affect red cell function infish. Aquat. Toxicol. 22, 227-238.

Nilsson, S. (1983). Autonomic Nerve Function in Vertebrates. Berlin: Springer

Verlag.

Nilsson, S. and Sundin, L. (1998). Gill bloodflow control. Comp. Biochem. Physiol. A 119, 137-147.

Noakes, D. L. G. and Jones, K. M. M. (2016). Cognition, Learning, and Behavior. In Fish Physiology-Biology of Stress in Fish, Vol. 35 (eds. C. B. Schreck, L. Tort, A. P. Farrell and C. J. Brauner), San Diego, CA: Academic Press.

Ortiz, M. and Lutz, P. L. (1995). Brain neurotransmitter changes associated with exercise and stress in a teleostfish (Sciaenops ocellatus). J. Fish. Biol. 46, 551-562.

Pagnotta, A., Brooks, L. and Milligan, L. (1994). The potential regulatory roles of cortisol in recovery from exhaustive exercise in rainbow trout. Can. J. Zool. 72, 2136-2146.

Palstra, A. P., Beltran, S., Burgerhout, E., Brittijn, S. A., Magnoni, L. J., Henkel, C. V., et al. (2013). Deep RNA sequencing of the skeletal muscle transcriptome in swimmingfish. PLoS One 8, e53171.

Palstra, A. P. and Planas, J. V. (2011). Fish under exercise. Fish Physiol. Biochem. 37, 259-272.

Palstra, A. P. and Planas, J. V. (2013). Swimming Physiology of Fish: Towards Using Exercise to Farm a Fit Fish in Sustainable Aquaculture. Heidelberg: Springer.

Palstra, A. P., Rovira, M., Rizo-Roca, D., Torrella, J. R., Spaink, H. P. and Planas, J. V. (2014). Swimming-induced exercise promotes hypertrophy and vascularization of fast skeletal musclefibres and activation of myogenic and angiogenic transcriptional programs in adult zebrafish. BMC Genomics 15, 1136.

Palstra, A. P., Tudorache, C., Rovira, M., Brittijn, S. A., Burgerhout, E., VandenThillart, G. E., et al. (2010). Establishing zebrafish as a novel exercise model: swimming economy, swimmingenhanced growth and muscle growth marker gene expression. PLoS One 5, e14483.

Pankhurst, N. W. (2011). The endocrinology of stress infish: an environmental perspective. Gen. Comp. Endocrinol. 170, 265-275.

Pearson, M. P., Spriet, L. L. and Stevens, E. D. (1990). Effect of sprint training on swim performance and white muscle metabolism during exercise and recovery in rainbow trout (Salmo gairdneri). J. Exp. Biol. 149, 45-60.

Pelster, B., Sa ¨nger, A. M., Siegele, M. and Schwerte, T. (2003). Influence of swim training on cardiac activity, tissue capillarization, and mitochondrial density in muscle tissue of zebrafish larvae. Am. J. Physiol. 285, R339-R347.

Perry, S. F. and Bernier, N. J. (1999). The acute humoral adrenergic stress response infish: facts and fiction. Aquaculture 177, 285-295.

Perry, S. F. and Desforges, P. R. (2006). Does bradycardia or hypertension enhance gas transfer in rainbow trout (Oncorhynchus mykiss)?. Comp. Biochem. Physiol. A 144, 163-172.

Picard, M., Juster, R. P. and McEwen, B. S. (2014). Mitochondrial allostatic load puts the "gluc" back in glucocorticoids. Nat. Rev. Endocrinol. 10, 303-310.

Polakof, S. M., Mommsen, T. P. and Soengas, J. L. (2011). Glucosensing and homeostasis: fromfish to mammals. Comp. Biochem. Physiol. B 160, 123-149.

Portz, D. E., Woodley, C. M. and Cech, J. J., Jr. (2006). Stress-associated impacts of short-term holding onfishes. Rev. Fish Biol. Fish. 16, 125-170.

Postlethwaite, E. K. and McDonald, D. G. (1995). Mechanisms of Na + and Cl regulation in freshwater-adapted rainbow trout (Oncorhynchus mykiss) during exercise and stress. J. Exp. Biol. 198, 295-304.

Powell, M. D. and Nowak, B. F. (2003). Acid-base and respiratory effects of confinement in Atlantic salmon affected with amoebic gill disease. J. Fish. Biol. 62, 51-63.

Randall, D. J. and Perry, S. F. (1992). Catecholamines. In Fish Physiology-The Cardiovascular System, Vol. 12B (eds. W. S. Hoar, D. J. Randall and A. P. Farrell), pp. 255-300. San Diego, CA: Academic Press.

Rasmussen, R. S., Heinrich, M. T., Hyldig, G., Jacobsen, C. and Jokumsen, A. (2011). Moderate exercise of rainbow trout induces only minor differences in fatty acid profile, texture, white muscle fibres and proximate chemical composition of fillets. Aquaculture 314, 159-164.

Richards, J. G., Heigenhauser, G. J. F. and Wood, C. M. (2002a). Glycogen phosphorylase and pyruvate dehydrogenase transformation in white muscle of trout during high-intensity exercise. Am. J. Physiol. 282, R828-R836.

Richards, J. G., Heigenhauser, G. J. F. and Wood, C. M. (2002b). Lipid oxidation fuels recovery from exhaustive exercise in white muscle of rainbow trout. Am. J. Physiol. 282, R89-R99.

Richards, J. G., Mercado, A. J., Clayton, C. A., Heigenhauser, G. J. F. and Wood, C. M. (2002c). Substrate utilization during graded aerobic exercise in rainbow trout. J. Exp. Biol. 205, 2067-2077.

Richter, E. A. and Ruderman, N. B. (2009). AMPK and the biochemistry of exercise: implications for human health and disease. Biochem. J. 418, 261-275.

Rome, L. C., Swank, D. and Corda, D. (1993). Howfish power swimming. Science

261, 340-343.

Sadoul, B. and Vijayan, M. M. (2016). Stress and Growth. In Fish Physiology-Biology of Stress in Fish, Vol. 35 (eds. C. B. Schreck, L. Tort, A. P. Farrell and C. J. Brauner), San Diego, CA: Academic Press.

Sänger, A. M. and Pötscher, U. (2000). Endurance exercise training affects fast white axial muscle in the cyprinid species Chalcalburnus chalcoides mento (Agassiz, 1832), cyprinidae, teleostei. Basic Appl. Myol. 10, 297-300.

Santer, R. M. (1985). Morphology and Innervation of the Fish Heart, vi. Berlin: Springer Verlag.

Satchell, G. H. (1991). Physiology and Form of Fish Circulation. Cambridge: Cambridge University Press.

Scarabello, M., Heigenhauser, G. J. F. and Wood, C. M. (1991). The oxygen debt hypothesis in juvenile rainbow trout after exhaustive exercise. Resp. Physiol. 84, 245-259.

Schreck, C. B. (2010). Stress andfish reproduction: the roles of allostasis and hormesis. Gen. Comp. Endocrinol. 165, 549-556.

Schreck, C. B., Olla, B. L. and Davis, M. W. (1997). Behavior responses to stress. In Fish Stress and Health in Aquaculture vol. Society of Experimental Biology Seminar Series 62 (eds. G. K. Iwama, A. D. Pickering, J. P. Sumpter and C. B. Schreck), pp. 145-170. Cambridge: Cambridge University Press.

Schreck, C. B. and Tort, L. (2016). The Concept of Stress in Fish. In Fish Physiology-Biology of Stress in Fish, Vol. 35 (eds. C. B. Schreck, L. Tort, A. P. Farrell and C. J. Brauner), San Diego, CA: Academic Press.

Sorensen, B. and Weber, R. E. (1995). Effects of oxygenation and the stress hormones adrenaline and cortisol on the viscosity of blood from the trout Oncorhynchus mykiss. J. Exp. Biol. 198, 953-959.

Steinhausen, M. F., Sandblom, E., Eliason, E. J., Verhille, C. and Farrell, A. P. (2008). The effect of acute temperature increases on the cardiorespiratory performance of resting and swimming sockeye salmon (Oncorhynchus nerka). J. Exp. Biol. 211, 3915-3926.

Suarez, R. K. and Mommsen, T. P. (1987). Gluconeogenesis in teleostfishes. Can. J. Zool. 65, 1869-1882.

Takei, Y. and Hwang, P.-P. (2016). Homeostatic Responses to Osmotic Stress. In Fish Physiology-Biology of Stress in Fish, Vol. 35 (eds. C. B. Schreck, L. Tort, A. P. Farrell and C. J. Brauner), San Diego, CA: Academic Press.

Takle, H. and Castro, V. (2013). Molecular adaptive mechanisms in the cardiac muscle of exercisedfish. In Swimming Physiology of Fish (eds. A. P. Palstra and J. V. Planas), pp. 257-274. Heidelberg: Springer.

Tirri, R. and Ripatti, P. (1982). Inhibitory adrenergic control of heart-rate of perch (Percafluviatilis) in vitro. Comp. Biochem. Physiol. C 73, 399-401.

Totland, G. K., Kryvi, H., Jødestøl, K. A., Christiansen, E. N., Tangeras, A. and Slinde, E. (1987). Growth and composition of the swimming muscle of adult Atlantic salmon (Salmo salar L.) during long-term sustained swimming. Aquaculture 66, 299-313.

Tseng, Y. C. and Hwang, P. P. (2008). Some insights into energy metabolism for osmoregulation infish. Comp. Biochem. Physiol. C 148, 419-429.

Tudorache, C., De Boeck, G. and Claireaux, G. (2013). Forced and preferred swimming speeds offish: a methodological approach. In Swimming Physiology of Fish (eds. A. P. Palstra and J. V. Planas), pp. 81-108. Heidelberg: Springer.

van der Meulen, T., Schipper, H., van den Boogaart, J. G. M., Huising, M. O., Kranenbarg, S. and van Leeuwen, J. L. (2006). Endurance exercise differentially stimulates heart and axial muscle development in zebrafish (Danio rerio). Am. J. Physiol. 291, R1040-R1048.

van Ginneken, V., Coldenhoff, K., Boot, R., Hollander, J., Lefeber, F. and van den Thillart, G. (2008). Depletion of high energy phosphates implicates post-exercise mortality in carp and trout; an in vivo ^{31}P-NMR study. Comp. Biochem. Physiol. A 149, 98-108.

VanRaaij, M. T. M., VandenThillart, G., Vianen, G. J., Pit, D. S. S., Balm, P. H. M. and Steffens, B. (1996). Substrate mobilization and hormonal changes in rainbow trout (Oncorhynchus mykiss, L.) and common carp (Cyprinus carpio, L.) during deep hypoxia and subsequent recovery. J. Comp. Physiol. B 166, 443-452.

Veiseth, E., Fjaera, S. O., Bjerkeng, B. and Skjervold, P. O. (2006). Accelerated recovery of Atlantic salmon (Salmo salar) from effects of crowding by swimming. Comp. Biochem. Physiol. B 144, 351-358.

Videler, J. J. (2011). An opinion paper: emphasis on white muscle development and growth to improve farmedfish flesh quality. Fish Physiol. Biochem. 37, 337-343.

Vijayan, M. M. and Moon, T. W. (1992). Acute handling stress alters hepatic glycogen metabolism in food-deprived rainbow trout (Oncorhynchus mykiss). Can. J. Fish. Aquat. Sci. 49, 2260-2266.

Wang, Y. X., Heigenhauser, G. J. F. and Wood, C. M. (1994). Integrated responses to exhaustive exercise and recovery in rainbow trout white muscle-acid-base, phosphogen, carbohydrate, lipid, ammonia, fluid volume and electrolyte metabolism. J. Exp. Biol. 195, 227-258.

Ward, D. L. and Hilwig, K. D. (2004). Effects of holding environment and exercise conditioning on swimming performance of southwestern nativefishes. N. Am. J. Fish Manage. 24, 1083-1087.

Weber, J. M. (2011). Metabolic fuels: regulatingfluxes to select mix. J. Exp. Biol. 214, 286-294.

Wendelaar Bonga, S. E. (1997). The stress response infish. Physiol. Rev. 77, 591-625.

Wiendl, H., Hohlfeld, R. and Kieseier, B. (2005). Immunobiology of muscle: advances in understanding an immunological microenvironment. Trends. Immunol. 26, 373-380.

Winberg, S., Höglund, E. and Øverli, Ø. (2016). Variation in the Neuroendocrine Stress Response. In Fish Physiology-Biology of Stress in Fish, Vol. 35 (eds. C. B. Schreck, L. Tort, A. P. Farrell and C. J. Brauner), San Diego, CA: Academic Press.

Wood, C. M. (1991). Acid-base and ion balance, metabolism, and their interactions, after exhaustive exercise infish. J. Exp. Biol. 160, 285-308.

Wood, C. M. and Randall, D. J. (1973). Influence of swimming activity on sodium balance in rainbow trout (Salmo gairdneri). J. Comp. Physiol. 82, 207-233.

Wood, C. M., Turner, J. D. and Graham, M. S. (1983). Why dofish die after severe exercise? J. Fish. Biol. 22, 189-201.

Woodward, J. J. and Smith, L. S. (1985). Exercise training and the stress response in rainbow trout, Salmo gairdneri Richardson. J. Fish. Biol. 26, 435-447.

Wrann, C. D., White, J. P., Salogiannnis, J., Laznik-Bogoslavski, D., Wu, J., Ma, D., et al. (2013). Exercise induces hippocampal BDNF through a PGC-1alpha/FNDC5 pathway. Cell Metab. 18, 649-659.

Wright, P. A., Perry, S. F. and Moon, T. W. (1989). Regulation of hepatic gluconeogenesis and glycogenolysis by catecholamines in rainbow trout during environmental hypoxia. J. Exp. Biol. 147, 169-188.

Yogata, H. and Oku, H. (2000). The effects of swimming exercise on growth and whole-body protein and fat contents of fed and unfedfingerling yellowtail. Fish. Sci. 66, 1100-1105.

Young, P. S. and Cech, J. J. (1993). Improved growth, swimming performance, and muscular development in exercised-conditioned young-of-the-year striped bass (Morone saxatilis). Aquat. Sci. 50, 703-707.

Zhang, H. and Anderson, J. E. (2014). Satellite cell activation and populations on single musclefiber cultures from adult zebrafish (Danio rerio). J. Exp. Biol. 217, 1910-1917.

Zhang, L. and Wood, C. M. (2009). Ammonia as a stimulant to ventilation in rainbow trout Oncorhynchus mykiss. Resp. Physiol. Neurobiol. 168, 261-271.

Zhao, W.-W., Pang, X., Peng, J.-L., Cao, Z.-D. and Fu, S.-J. (2012). The effects of hypoxia acclimation, exercise training and fasting on swimming performance in juvenile qingbo (Spinibarbus sinensis). Fish Physiol. Biochem. 38, 1367-1377.

第 8 章　生殖和发育

应激通常会对雌雄鱼的生殖行为有抑制作用，但是在少数情况下也可能有促进作用。抑制作用包括抑制卵巢和精巢发育，抑制排卵和产卵，以及产出较小的卵粒和幼体。而对后代的长期影响仍缺乏研究。对内分泌的影响包括抑制下丘脑、垂体和性腺类固醇激素，其中对性腺雌雄激素的合成影响最大。事实上，应激调节的激素改变能够对生殖系统产生系统的和直接的影响，并且实验范式通常无法区别这两种影响，因此，了解应激干扰生殖活动的机理是比较复杂的。在这个前提下，有证据表明应激内分泌轴的所有层次对生殖都有抑制作用，尤其是关于皮质类固醇作用的证据最有力，值得注意的是，现有大量的研究皮质醇影响的文献，这也部分反映出类固醇激素的检测相对容易。这些影响包括系统代谢影响、基因组糖皮质激素受体介导的影响以及通过一些非基因组过程的直接影响，其中可能包括对类固醇转换酶和结合蛋白对底物的争夺。多数应激相关的研究包含对捕获的或养殖的鱼类群体的实验室评估，然而应激影响自由生存的野生鱼类方面的研究资料还很少。现已清楚生殖过程可以保持在大范围的皮质类固醇浓度下，但是越来越多的证据表明生殖过程的社会性调节可能也是通过应激过程介导的。在生殖功能的所有层次上，对应激-生殖相互作用的理解还需进一步深入研究。

8.1　导　言

在高等脊椎动物中已经证实生理应激对生殖有较大的抑制作用，尤其是在哺乳动物中研究历史最长（Pottinger，1999）。最开始，人们对于了解低等脊椎动物尤其是鱼类生殖过程的兴趣并不大，John Sumpter 和他的同事在 1987 年写道：“尽管有一些偶然的证据能够从自然鱼类和实验室鱼类相关研究中获得，但是直接的证据几乎没有。并且，应激可能影响鱼类生殖轴的机理仍未知。”（Sumpter 等，1987）考虑到许多物种在蓄养或养殖过程中性成熟持续失败，人们对内分泌调控作为一种管理策略的兴趣日益浓厚，这多少是意料之外的（Fostier，Jalabert，1982）。一系列对鲑科鱼类的开拓性研究建立了应激和生殖之间的关系（Pickering 等，1987；Sumpter 等，1987；Carragher 等，1989；Carragher，Sumpter，1990；Pottinger，Pickering，1990；Pottinger 等，1991），并开始探索其中可能包含的机制，也为之后的大量研究打下基础（Pankhurst，Van Der Kraak，1997；Schreck 等，2001；Milla 等，2009；Leatherland 等，2010；Schreck，2010；Fuzzen 等，2011）。这些研究详细评估了应激对许多物种的影响，拓展了我们对于应激影响相关机制的理解，并且研究了在人工生殖条件下如何规避它们。

近年来，分子技术的进步为重新评价和拓展对现存范式的理解（Maruska，2014），以及在转录物组水平上探索确定更为复杂的相互作用的存在和关系提供了很好的机会（Aluru，Vijayan，2009）。目前，许多研究都关注应激对近期捕获或养殖种群的影响，只有较少的研究例外，但是几乎没有关于这些应激因素作用于自然种群方面的认识（Pankhurst，2011）。对于野生种群和人工养殖种群来说，还有一个悬而未决的问题，即应激对生殖健康的长期影响，以及这些影响在多大程度上是跨越世代的。关于鱼类的信息仍很少，只有一些来源于高等脊椎动物的明确提示（Chand，Lovejoy，2011；Lane，2014）和对有毒物质在鱼类的长期个体和群体影响中后成说（epigenesis）作用的新见解（Bhandari 等，2015；Kamstra 等，2014）。这些问题为下一阶段鱼类应激和生殖的研究确定了议题，并将在目前的理解和知识差距的背景下加以探讨。

8.2 生殖调节

8.2.1 生殖模式和环境调节

鱼类的生殖模式是高度可变的，了解它们的共性及差异对于评估应激对生殖的影响是重要的先决条件。配体发育模式、性腺生长和发育，以及产卵已经由 Pankhurst（1998a）做了详细的阐述，在此进行简单总结。终身一次生殖物种（仅产一次卵，如鳗鱼和太平洋鲑鱼）雌性个体卵巢滤泡直到产卵时才到达到输卵管（鲑鱼体腔内）仅成熟一次，而雄性个体则将或多或少的全部成熟配子发育为精子，这称为同步配子发育。多次生殖物种能够多次产卵，并被分为每个生殖季节单次排卵（其他的鲑科鱼类，配子同步发育）和每个生殖季节能进行多次排卵和产卵，即配子发育不同步。配子发育不同步在硬骨鱼类中最常见，即同一时间点性腺中存在多种发育时期的配子。雌性排卵和雄性排精后，配子通过产卵活动（有时被错误用于描述排卵过程）被释放到水中然后开始受精作用，少数卵胎生物种则是通过交配受精。产卵模式也有很大的变化，从成对或成群的中上层浮游性或底层性产卵到有些种类在底质中产卵受精和保护巢穴。

生殖周期受到环境变化的同步和调节，广泛的季节性周期通常受到光周期的影响，而较精细的季节内调整受到温度的强烈影响，也取决于环境、降雨、月周期、营养和社会地位等一系列其他因素的影响（Pankhurst，Porter，2003）。和其他脊椎动物相同，鱼类的性别主要由遗传决定；然而，与其他脊椎动物高度保守的遗传系统不同，鱼类的性别决定基因有很多（Devlin，Nagaham，2002；Martinez 等，2014）。更复杂的是，许多鱼类还能定期进行性别转换，或在青春期，或在生殖成熟期，有些鱼出生为雄性而在某个阶段变成雌性（先雄后雌），而有些则出生为雌性然后变为雄性（先雌后雄），在少数物种中还存在双向性别转换（Frisch，2004）。

8.2.2 生殖的内分泌调控

环境信号通过激活下丘脑-垂体-性腺轴被转换成内分泌信号（图 8.1），最开始是通过来源于下丘脑的促性腺激素释放激素（GnRH，一种十肽）的合成与释放。所有的脊椎动物包括鱼类都能够表达2种以上的GnRH。截至目前，在硬骨鱼类中已经鉴定出8种形式的GnRH（Zohar等，2010）。GnRHs能够在多个组织中表达，但是在下丘脑产生的GnRH1在HPG轴激活中发挥主要作用，主要通过突触刺激垂体中的促性腺激素细胞。在部分物种中，垂体促性腺激素细胞还会受到多巴胺生成神经元的抑制作用，而多巴胺和GnRH的平衡决定了促性腺激素合成和释放的速度和浓度（Dufour等，2010）。另外，GnRH的合成与释放调节主要通过最近发现的亲吻激动素（Kisspeptin）及其受体系统，该系统被认为是外源和内源环境对生殖系统产生影响的主要途径（Zohar等，2010；Shahjahan，2013），但其作用机理仍未知。据推断，光诱导的松果体褪黑素分泌的改变也能调节下丘脑-垂体轴（Migaud等，2010），但其作用机理也尚未阐明。

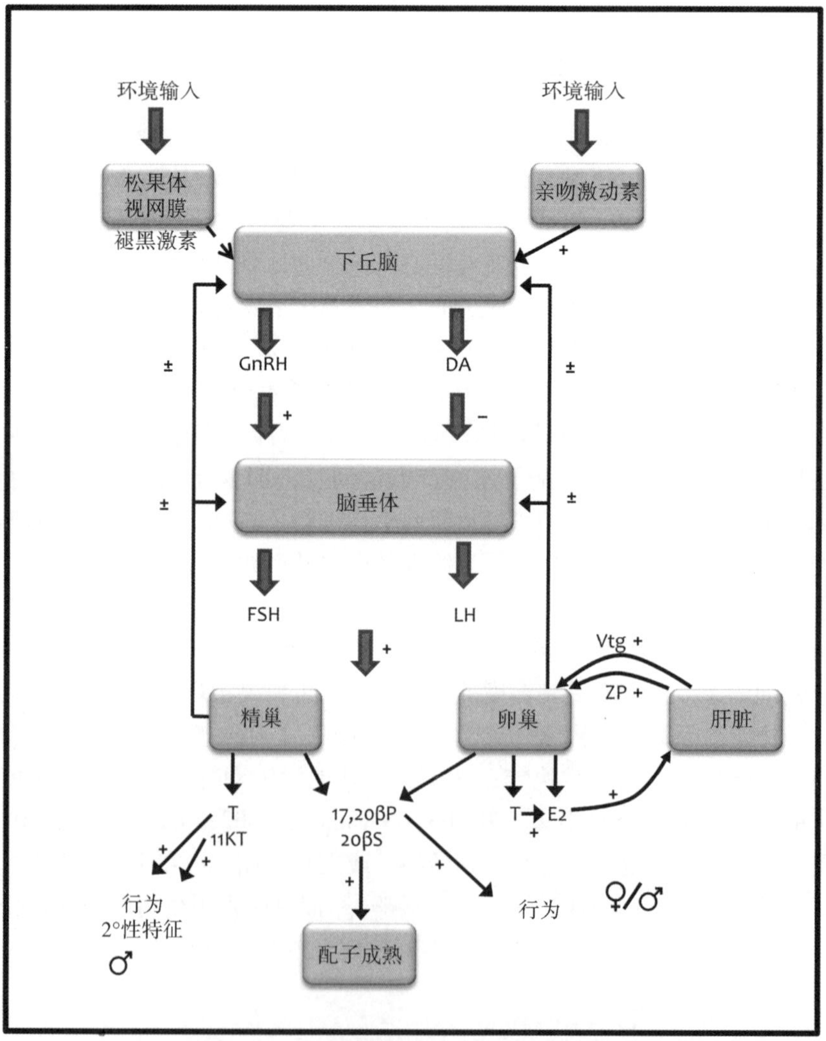

图8.1 鱼类下丘脑垂体性腺轴调节的概述

注：文中列出了缩写。+，刺激效应；−，抑制作用；虚线表示可能的但未经确认的调节输入。

垂体受到 GnRH 刺激后合成并释放卵泡刺激素（FSH）和黄体生成素（LH）（Planas，Swanson，2008；Levavi-Sivan 等，2010）（这与之前研究中提到的 GTH-I 和 GTH-II 同义）。FSH 参与促进早期配子形成，而 LH 调节卵巢和精巢的晚期成熟，两者均通过膜结合的 G 蛋白偶联受体激活（Planas，Swanson，2008）。配体受体结合后激活腺苷酸环化酶，随后细胞内环化腺苷酸酶 cAMP 浓度增加，并激活类固醇激素合成相关的蛋白激酶级联通路并促进胆固醇分子的降解（Pankhurst，2008）。级联反应的第一阶段就是将胆固醇传递到线粒体内膜上的降解部位，该过程主要通过类固醇急性调节蛋白（StAR）控制，这是降解通路中的主要限速步骤。从类固醇细胞核中系列酶调控的碳原子裂解以及活性基团的加入，使 21 C 孕激素（和肾间组织中的皮质激素）以及较短的 19 C 雄激素和 18 C 雌激素的裂解顺序升高。

广泛的中间物和潜在的活性类固醇产物由性腺组织产生，但是主要的活性黄体酮是 17, 20β-二羟基-4-孕烯-3-酮（17, 20βP）和 17, 20β, 21-三羟基-4-孕烯-3-酮（20βS）（雌雄），雄激素睾酮（T）（雌雄），酮基睾酮（11KT）（主要是雄）以及雌二醇（E2）（主要是雌）（Pankhurst，2008）。黄体酮能够调节配体的最终成熟，并参与一些生殖行为。雄激素，在雌性体内是雌激素的前提，在雄性中参与雄性精巢成熟以及第二性征的发育，并参与占有领地和攻击行为（Schulz 等，2010；Munakata，Kobayashi，2010）。雌激素主要负责雌激素受体介导的卵黄前体卵黄蛋白原（VTG）的产生，卵黄蛋白原主要由肝细胞产生，随后在卵黄蛋白原生成过程中吸收进入卵细胞，另外，卵壳、透明带（ZP）蛋白等也由肝细胞产生（Pankhurst，2008）。有趣的是，雌激素亦参与精巢基因表达的调节，并被认为能够参与促进精原细胞增殖（Schulz 等，2010）。在先雄后雌的性逆转物种中，在雄性阶段主要产生 11KT，在兼性阶段 11KT 含量降低而 E2 含量增加，而在先雌后雄的物种中则相反（Frisch，2004）。类固醇则可以通过细胞内受体的经典（基因组）作用，在 HPG 轴的较高水平上产生正负反馈（Pankhurst，2008），并通过膜结合的类固醇受体产生短期的调节作用（Thomas，2003）。HPG 轴及其控制的复杂性为生殖系统应激的潜在作用提供了丰富的画面。

8.3　应激对生殖的影响

8.3.1　对生殖行为的影响

首次关于鉴别应激对生殖影响的研究是 De Montalembert 等（1978）对捕获的白斑狗鱼（*Esox lucius*）卵巢闭锁发育的观察报道，以及环境应激对内华达鳉（*Cyprinodon nevadensis*）影响的讨论（Gerking，1982）。随后的研究表明，人为刺激（捕获、禁闭和养殖）会对生殖行为产生抑制作用，而对非鲑科鱼类物种如卵巢不同步发育物种和捕捞的野生鱼类的影响更为明显。

持续 9 个月，每周将虹鳟（*Oncorhynchus mykiss*）暴露在干扰应激下会导致雌鱼延迟排卵，卵粒体积和重量减少，雄鱼精子密度下降（Campbell 等，1992）。将虹鳟短暂禁闭两周也会带来相同的影响，而对褐鳟（*Salmo trutta*）没有类似的影响（Campbell 等，1994）。相反，在生殖发育或最后成熟时期，连续两周每日将虹鳟暴露在扰动应激中会促进其排卵，并且在早期卵黄蛋白形成阶段施加应激刺激会导致卵粒较小（Contreras-Sanchez 等，1998）。Pottinger 和 Carrick（2000）报道对于高皮质醇反应的虹鳟个体，一周的禁闭会使其卵死亡率增加，而北极红点鲑在高浓度鱼虱的应激影响下排卵会延迟，但是卵粒质量和存活率没有改变（Tveiten 等，2010）。

应激对非鲑科鱼类生殖行为的影响似乎更深刻。金赤鲷（*Pagrus auratus*）在被捕获 3 天后就会终止排卵（Carragher，Pankhurst，1991），在被捕获 7 天后会导致卵巢闭锁（Cleary 等，2000）。绿鳍鱼（*Chelidonichthys kumu*）在被捕获 48 小时后卵巢闭锁程度非常高（图 8.2）；条纹婢鳊（*Latris lineata*）在每日取样 9 天后，较大的卵黄囊泡会丢失（Morehead，1998）；在红腹罗非鱼（*Tilapia zillii*）中发现，拥挤会导致其卵巢闭锁程度增加，产卵受到抑制（Coward 等，1998）。对金赤鲷（*Pagrus auratus*）（Cleary 等，2002）、布氏棘鲷（*Acanthopagrus butcheri*）（Haddy，Pankhurst，2000）和克林雷氏鲇（*Rhamdia quelen*）（Soso 等，2008）的研究发现，被捕获后 1 天，GnRH 类似物和人绒毛膜促性腺激素（hCG）的促排卵反应就会被抑制。

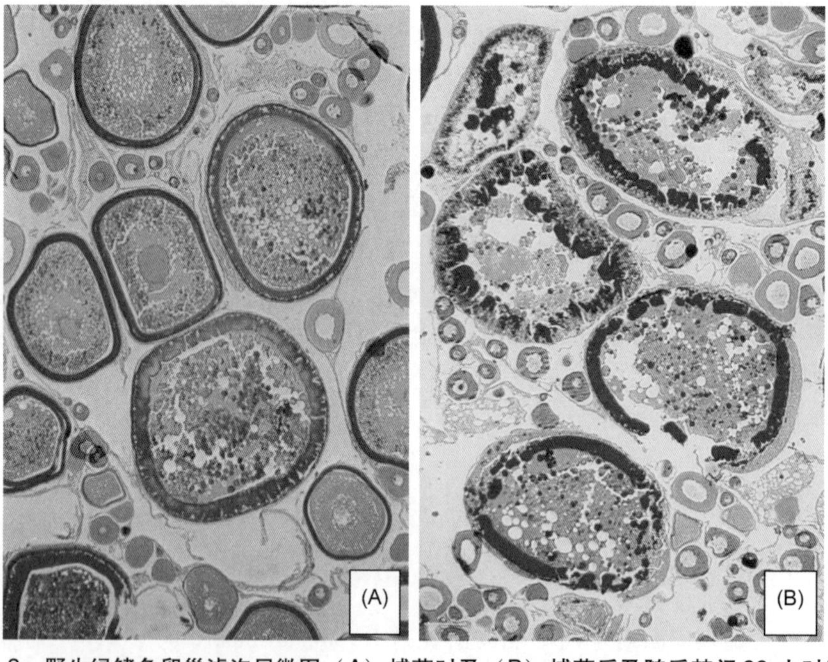

图 8.2　野生绿鳍鱼卵巢滤泡显微图（A）捕获时及（B）捕获后及随后禁闭 96 小时卵黄蛋白发生的滤泡晚期闭锁发展情况

注：见书后彩图。

有限的可用数据表明应激影响生殖行为，并且能够影响子代。在应激下，孵化率、上浮稚鱼以及孵化后 28 天虹鳟幼鱼存活率均下降（Campbell 等，1992，1994），但是，另有研究表明在卵黄蛋白形成时期持续 3 个月的不同应激刺激对存活率没有影响（Contreras-Sánchez 等，1998）。受到应激影响的实验室驯养的大西洋鳕鱼（*Gadus morhua*），其产卵、卵粒产量及生育力和受低应激刺激的鱼相似，但是异常发育的幼鱼比例比较高（Morgan 等，1999）。雀鲷（*Pomacentrus amboinensis*）雌鱼在面对同种个体数量增加以及食卵天敌增加的情况下，其幼鱼孵化时个体较小，用皮质醇处理产卵雌鱼也会产生相同的影响（McCormick，1998，1999，2006）。用皮质醇处理的雌鱼产出的幼鱼的形态变异也会较大。每天捕获和追逐雌性丽鱼（*Neolamprologus pulcher*）会使其产卵的间隔时间增加且产出卵粒较小（Mileva 等，2011），而大西洋鲑鱼（*Salmo salar*）在受精前 6 天用外源皮质醇处理，其卵死亡率较高，孵化时鱼苗的体积较小，并且异常幼鱼的比例较高（Eriksen 等，2006，2007）。

实际上，应激对子代生活史行为产生的可能影响是未知的，也没有任何关于应激对世代影响的信息。Auperin 和 Geslin（2008）证明应激刺激受精后 14～55 天（在受精后 32 天孵化）虹鳟的卵或幼体会抑制 5 月龄个体受到应激后皮质醇的增加，而用外源皮质醇处理也能得到相同的影响。据推测，受到母源应激产生的皮质醇刺激的卵粒可能也有相同的影响。哺乳动物的研究以及鱼类多代毒理学研究也为应激可能的跨世代影响提供了线索。对表观遗传学修饰过程的理解正在不断加深，环境通过一些作用过程影响遗传表达，包括核酸碱基主要是胞嘧啶的甲基化和组氨酸的转录后修饰（Chand，Lovejoy，2011；Siklenkia 等，2015）。这反过来能够干扰转录因子和基因启动子的结合能力。甲基化改变是非常稳定的并且能够跨世代保留。

在哺乳动物中，卵细胞中的许多表观遗传标记在受精时会被消除，但是一些位点的表观遗传学改变是不确定的，可能会跨世代遗传（Lane 等，2014）。在雄鱼中，表观遗传学改变包括 DNA 甲基化以及参与调节基因转录的非编码小 RNA 的改变。对哺乳动物的研究发现，毒素、内分泌干扰物、吸烟和肥胖等应激在人类和啮齿动物中能引发表观遗传改变（Lane 等，2014）。还没有关于应激对鱼类产生表观遗传影响的证据，但是，暴露于有毒物质中的确能够对大量基因造成表观遗传改变，包括 DNA 甲基化、组蛋白修饰和小 RNA 影响（Bhandari 等，2015；Kamstra 等，2014），还有一些跨世代影响证据，例如胖头鲅（*Pimephales promelas*）暴露于外源性雌激素条件下会导致 F1 代雌性生殖力下降（Staples 等，2011）。可以合理地设想，应激对成年鱼的影响通过应激介导的表观遗传可以在子代中表现出来，甚至可能影响更远的后代。

8.3.2　应激对生殖内分泌系统的影响

由于化学结构保守且易于测量，多数研究最初都会检测应激对类固醇激素合成的影响，而对分析系统可供应用的少数种类还会检测其垂体激素。近年来，分子生物技术的快速发展大大扩展了可能的查询范围。早期对雄性褐鳟的研究表明，急性搬运应激会导致血浆皮质醇和促肾上腺皮质激素（ACTH）升高，血浆睾酮和 11KT 在刺激 1～4 小时内下降，并且在长时间慢性应激下对激素的影响也会一直保持（Pickering 等，1987；Sumpter 等，1987）。禁闭雌性褐鲑和雄性虹鳟 2 周会使其血浆皮质醇升高、血浆睾酮下降，但是血浆雌激素含量在两者中均不受影响，而血浆 VTG 含量在雌性虹鳟中会下降（Campbell 等，1994）。相反，据 Pottinger 和 Carrick（2000）报道，强制禁闭雌性虹鳟 2～24 小时，会导致其血浆雌激素含量下降，而 Pankhurst 和 Dedual（1994）也发现，禁闭正在迁徙的野生虹鳟 24 小时后，其血浆睾酮和雌激素含量下降，皮质醇含量升高。养殖的红大麻哈鱼（*Oncorhynchus nerka*）被禁闭 15～30 分钟，雌雄个体睾酮含量均下降，雄性个体 11KT 含量也下降（Kubokawa 等，1999），并且北极红点鲑暴露于高浓度的海水虱寄生虫环境中，其雌雄个体血浆雄激素含量下降，雄性个体血浆 11KT 以及雌性个体血浆雌激素含量也下降，而血浆皮质醇含量升高（Tveiten 等，2010）。

非鲑科鱼类也存在相似的影响，被捕获 60 分钟后，野生雄性云纹犬牙石首鱼（*Cynoscion nebulosus*）血浆雄激素和 11KT 含量下降（Safford，Thomas，1987）；金赤鲷被捕获禁锢后，在 1～6 小时内，其皮质醇含量上升，雌激素下降，雄激素在捕获 6 小时后也下降，而 17,20 βP 在两个时间点均上升（Carragher，Pankhurst，1991）；而野生黑鲷在被捕获后 15 分钟，皮质醇含量升高，在被捕获后 30 分钟，雄激素下降，1 小时后雌激素下降，6 小时后 11KT 下降，而最初血浆 17,20 βP 在雌雄个体中均升高（Haddy，Pankhurst，1999；图 8.3）。同样地，野生雌性绿鳍鱼在被捕获 24 小时后，皮质醇缓慢升高，雄雌激素下降（Clearwater，Pankhurst，1997）；野生雌性条纹婢鳊血浆皮质醇也会持续上升，雌雄激素下降，而血浆 17,20 βP 没有改变（Morehead，1998）；橙线雀鲷（*Acanthochromis polyacanthus*）在被捕获后 3 小时和 6 小时雌鱼雄激素下降，雌激素在被捕获后 6 小时也下降，而雄鱼在被捕获 6 小时后雄激素和 11KT 含量均下降（Pankhurst，2001）。拟鲤（*Rutilus rutilus*）通常体内的皮质醇水平非常高，而在应激下，其皮质醇水平会有更高水平的提升，而在急性和慢性应激下，血浆雌激素水平下降（Pottinger 等，1999）。相似地，白亚口鱼（*Catostomus commersoni*）被禁闭后，雄性血浆雄激素和 11KT 水平下降，雌鱼血浆雄激素和雌激素水平也下降（Van Der Kraak 等，1992；Jardine 等，1996）。在克林雷氏鲶中的表达模式则不同，对养殖雌性个体每日进行网捕干扰，其雄激素水平并没有下降，但雌激素含量减少（Soso，2008）。条纹狼鲈（*Morone saxatilis*）受到重复网捕时，其雄激素和 11KT 水平会降低，但是反应的强度与个体的应激反应状态有关，皮质醇含量低的个体其血浆雄激素水平会受到更明显的抑制（Castranova 等，2005）。总体而言，以上研究表明，生殖系统在面对应激时，群体和物种内不同个体的应激反应存在一定差异。也有证据表明不同生殖阶段对应激的敏感性也不同。小口黑鲈（*Micropterus dolomieu*）雄性个体在栖息巢穴被捕获后血浆雄激素水平下降，但是这种影响在亲代抚育条件下有所减轻（O'Connor 等，2011）。在其他种类中，应激反应会随着生殖阶段而改变（Pankhurst，2011）。还不清楚的是，这种变化是否扩大了应激对生殖的影响。

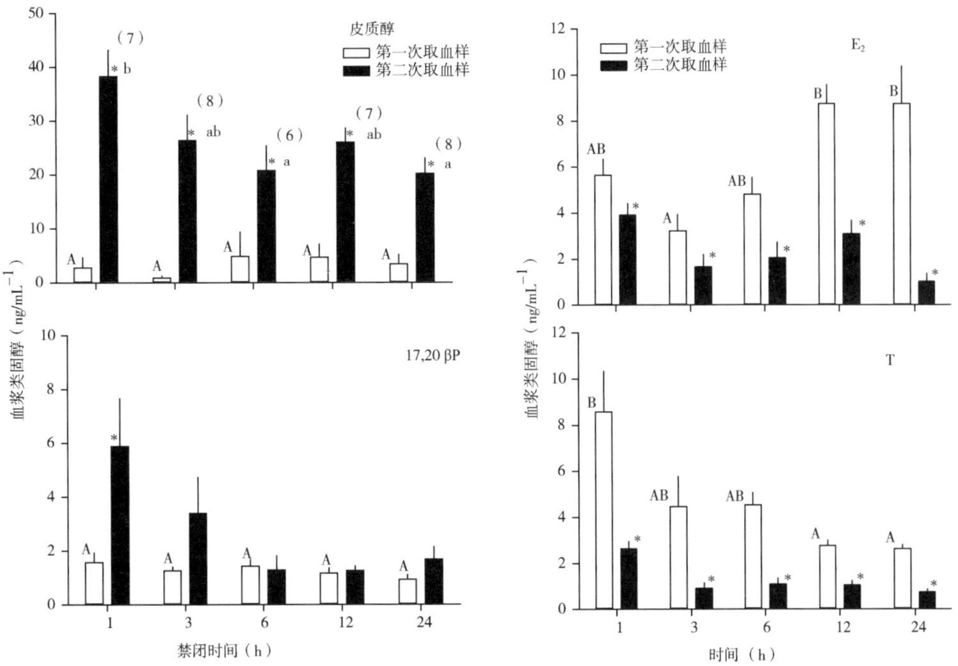

图 8.3　野生雌性布氏棘鲷在被捕获后不同的禁闭期对血浆皮质醇，17, 20 βP，T 和 E2 水平的影响

注：平均值 + SE（n）；大写字母和小写字母分别表示第一次（在捕获时-清晰条形图）和第二次（在禁闭后-填满条形图后）取血的采样时间之间的差异；每次禁闭时第一次和第二次取血之间有显著差异的值用星号标记。转载自 Haddy J A 和 Pankhurst N W（1999）。应激诱导布氏棘鲷血浆性激素浓度的变化。*J. Fish. Biol.* 55，1304-1316。

在高等脊椎动物中，应激能够抑制 GnRH 的合成与释放、抑制 LH 和 FSH 释放以及释放脉冲的频率，并与鱼类相同，能够抑制性腺类固醇的合成（Pottinger，1999；Fuzzen 等，2011）。关于应激对 HPG 轴较高层次的影响的认识还很少。早期关于应激对垂体功能的影响也是比较模糊的。在雄性褐鳟中，急性应激会导致其血浆 LH 升高（Pickering 等，1987；Sumpter 等，1987）；白亚口鱼在一系列取样处理后，雄性血浆 LH 水平降低而雌性没有变化（Van Der Kraak 等，1992）。相反，在排卵期前的虹鳟中，急性低水压应激能够提高血浆皮质醇含量而抑制睾酮含量，但对血浆 LH 含量没有影响（Pankhurst，Van Der Kraak，2000）。日本竹筴鱼（*Trachurus japonicus*）被捕获后，尽管 GnRH1 被抑制，但是处于相同生殖阶段的野生个体和捕获个体之间 FSHβ 和 LHβ 基因表达并没有差异，且血浆雌激素水平也没有差异（Imanaga 等，2014）。处于从属地位的非领地内的雄性非洲伯氏妊丽鱼（*Astatotilapia burtoni*）垂体 LHβ 和 FSHβ mRNA，以及血浆 FSH、LH、T 和 11KT 含量相对于占有领地的雄鱼要低一些，这种影响可以理解为是一种社会行为介导的应激反应（Maruska，2014）。非领地内的雄鱼脑中视前区（POA）促肾上腺皮质激素释放因子受体（CRFR1）mRNA 含量增加，GnRH1 mRNA 含量减少，垂体中 CRF 和 CRFR1 mRNA 增加，而 GnRH1-

R1 mRNA 含量减少。先前的研究也发现了类似的现象，非领地内雄鱼视前区产生 GnRH 的神经元较小（Fox 等，1997）。

在一些硬骨鱼类中，性腺芳香化酶（cyp19a1）和脑芳香化酶（cyp19b1）基因中发现糖皮质激素受体（GR）反应元件表明皮质类固醇可能参与性逆转的相关基因调节（Gardner 等，2005）。然而，应激诱导的改变在成年鱼性逆转中所起作用相关的证据比较模糊。去掉一雄多雌社会群体特性的雄性蓝带虾虎鱼（*Lythrypnus dalli*）会导致优势雌鱼发生快速的行为改变，并且随后转变为雄性（Black 等，2011）。雌鱼攻击行为增加和脑芳香化酶活性显著降低相联系，但伴随而来的不是驱动行为的改变。社会性优势关系可能导致从属地位鱼类血浆皮质醇水平升高，但这一般只在建立优势地位期间发生（Sloman，Armstrong，2002）；并且考虑到应激对雌激素合成有抑制作用，似乎应激相关的社会性优势效应并不能解释雌性先熟雌鱼保持的社会地位。应激通过抑制雄性激素的产生来维持雌性地位的证据也不是特别有力。为了检测对雄激素、皮质类固醇转换酶、11β 羟化酶（11βH）、11β 羟化类固醇脱氢酶（11βHSD）的底物争夺是否会抑制 11KT 的合成，研究人员用硅胶包埋的皮质醇去处理雌性先熟的圆拟鲈（*Parapercis cylinderica*）（Frisch 等，2007）。11βH 能介导雄烯二酮（A）和睾酮（T）转变成相关的 11β 羟基化衍生物，并且 11βHSD 既参与羟基蒽醌到肾上腺雄甾酮（11KA）（11KT 合成的前体）的转变过程，又参与 11β 羟色胺酸转变为 11KT 的过程。11βH 也能介导脱氧皮质醇到皮质醇的转变，并且 11βHSD 能够介导皮质醇转变为低活性代谢物可的松（Perry，Grober，2003）。激素处理 21 天的鱼和对照组鱼以相同的速度发生性逆转，但对血浆雄激素和 11KT 水平没有影响。这显然是一个有相当大的进一步研究余地的领域。

8.3.3 热应激：一个特例？

在性别决定的阶段，升高温度对幼鱼和未成熟鱼的生殖过程以及成熟鱼的性成熟和产卵影响很大，并且越来越多的证据表明性别决定可能会部分受到应激的调控。在许多物种中，性别决定是热依赖性的，较高的温度能够驱动性别向雄性发展。这是由于高温对芳香化酶产生抑制作用，进而减少雄激素向雌激素的转变（Devlin，Nagahama，2002；Guiguen 等，2010）。近期的研究认为应激可能参与该过程的调控。银汉鱼（*Odontesthes bonariensis*）幼鱼在 29℃养殖条件下饲养，100% 为雄性，并且皮质醇、雄激素和 11KT 含量比在 17℃养殖条件下的鱼都要高（Hattori 等，2009）。升高温度对生育的影响可以通过用皮质醇、皮质类固醇以及地塞米松处理来实现。用青鳉来做类似的实验，32℃～34℃温度下养殖的遗传雌性的个体当中会有表型为雄性的个体产生，并且其体内皮质醇升高，而 FSH 受体 mRNA 表达减少（Hayashi 等，2010）。同样，这种影响也可以通过皮质醇处理和改良皮质醇合成抑制剂甲吡酮来实现，因此，可以得出结论，升高温度能够诱导皮质醇分泌增加，进而抑制生殖细胞增殖和 FSHR 表达。随后关于青鳉的研究表明，异常高温和皮质醇处理都能抑制 cyp19a1 和性腺体细胞衍生的生长因子的表达（Kitano 等，2012）。这些影响可以通过雌激素处理被消除，说明高温对生育力的影响是通过抑制芳香化酶表达来调节。遗传雌性的牙鲆（*Paralichthys olivaceus*）在正常温度范围可以发育为雌性，而在 27℃养殖条件下则会被诱导发育为雄性，并且在高温条件下鱼的皮质醇含量会比较高。在体的实验表明，高温诱导的雄性化过程可以用甲吡酮处理来阻止，而皮质醇能够抑制离体的 cyp19a1 的基因表达（Yamaguchi 等，2010）。在该研究中，皮质醇对 FSHR 没有影响，而在体外，重组的鲷鱼 FSH 能够终止皮质醇对 cyp19a1 的抑制作用。在 27℃而不是 18℃ 的条件下，GR 能够结合到 cyp19a1 启动子上的 cAMP 反应元件。Fernandino 等（2013）提出了一种应激的附加效应，认为雄性表型可能是受雄激素驱动的而不仅仅因为没有雌激素。应激诱导的皮质醇增加对性腺 11βHSD 的上调影响既增加了 11KT 的合成，又通过转变为可的松而使皮质醇失活。这些作者认为，这些机制都不是相互排斥的，可能是多种效应共同作用的结果。

应激在调节温度对成鱼生殖影响中的作用还不清楚。温度升高对生殖过程有强烈的抑制作用，并且这种影响发生在内分泌通路的多个环节（Pankhurst，Munday，2011）。异常升温会抑制亲吻激动素（kisspeptin）及其受体的表达（Shahjahan 等，2013），减少 GnRH、GnRHR、LHβ（Okuzawa 等，2003；Okuzawa，Gen，2013）和 FSHR（Soria 等，2008）的表达，增加多巴胺能对垂体 LH 释放的抑制作用（Gillet 和 Breton，2009），抑制类固醇转换酶，尤其是芳香化酶（Lim 等，2003；Anderson 等，2012a），并能减少血浆类固醇、VTG 和 ZP 基因的表达（Tveiten，Johnsen，2001；Soria 等，2008；Pankhurst，King，2010；Anderson 等，2012b；Okuzawa，Gen，2013）。还伴随着延迟成熟、卵巢闭锁水平增加、延迟排卵等一些与应激抑制生殖过程很相似的影响。然而，有限的证据表明，皮质类固醇激素调节机制可能并不参与这一过程。

一项关于适度升高（正常范围）温度对培养大西洋鲑鱼的影响的广泛研究表明，在抑制 cyp19a1 的同时，对卵黄发生和卵巢发育以及 VTG、ZP 基因表达具有一致的抑制作用，并能够降低血浆 T、E2、VTG 水平，延迟排卵以及影响 GnRH 类似物诱导的 17,20 βP 的产生（Pankhurst，King，2010；Pankhurst 等，2011；Anderson 等，2012a）。类似的作用在北极红点鲑亦表现为卵泡对 LH 的反应性降低，17,20 βP 产量降低，卵母细胞在较高温度下对 17,20 βP 的敏感性降低（Gillet 等，2011）。在大西洋鲑鱼的实验中，它们的血浆皮质醇水平对热刺激的反应没有差异，而这一结果在驯养的群体中仍然保持在正常水平的范围内（Pankhurst 等，2011）。叉头转录因子（forkhead factor，foxl2）可能是对 cyp19a1 表达进行热调控的较强候选因子，它参与 cyp19a1 的转录调控，在高温下会受到热损伤（Nakamoto 等，2006；Wang 等，2007；Yamaguchi 等，2007）。cyp19a1 基因的 GR 反应元件（Gardner 等，2005）不妨碍皮质醇在热应激的鱼体内起作用，但许多抑制温度升高的变化似乎发生在鱼类能够维持正常稳态的范围内，而热对下丘脑-脑垂体-肾间腺轴（HPI）的激活可能不起作用。相比之下，与温度快速下降相关的热冲击确实会激活 HPI 轴。雄性幼鲤暴露在从 25℃到 11℃ 的快速温度变化中，血浆皮质醇水平升高，性腺随之减小（Goos，

Consten，2002）。温度迅速而大幅度地上升可能会引起类似的影响。

8.3.4 低氧影响

低氧是一种能够引起内分泌应激反应并对生殖内分泌过程有抑制作用的环境因素。河流和河口的物种似乎已经进化，以应对自然发生的低氧时期（Dabrowski 等，2003；Landry 等，2007；Thomas 等，2007）。但是，许多水生生境温度升高，以及海洋沿岸和地表水由于营养负荷而缺氧的情况日益频繁，表明现在有更广泛的物种面临低氧应激的危险。实验研究表明，暴露在低氧环境下，生殖受到一定程度的抑制。淡水白鲳（*Piaractus brachypomus*）暴露于低氧环境 3 天后，表现为雄性血浆 T 和 11KT 水平降低，雌性血浆 T 和 E2 水平降低，幼鱼存活率降低，畸形患病率增加（Dabrowski 等，2003）。同样，缺氧 12 周后，鲤鱼（*Cyprinus carpio*）的性腺指数（GSI）降低，雌性鲤鱼的 T 和 E2 水平降低，雄性睾酮、精子和精子活力降低，卵子的生殖力、孵化成功率和幼鱼存活率降低（Wu 等，2003）。在稍后的一项研究中，缺氧的鲤鱼血浆 LH 水平降低，最后成熟延缓，但性腺激素没有受到影响（Wang 等，2008）。暴露于缺氧环境 1 个月的大底鳉（*Fundulus grandis*）雌雄 GSI 均降低，雌鱼血浆 E2 降低，雄鱼血浆 E2 降低，产卵量减少，产卵开始较晚（Landry 等，2007）。

通过一系列的实验，将大西洋绒须石首鱼（*Micropogonias undulatus*）暴露于长时间（长达 10 周）的低氧环境中，雌性 GSI 降低，卵母细胞发育迟缓，血浆 E2 和 VTG 降低，ER 基因表达降低，雄性睾丸中 GSI、血浆 11KT 水平和精子比例降低（Thomas 等，2006，2007）。在河口低氧处直接取样的鱼类中也发现了类似的效应。经实验室处理的鱼的血浆 LH 水平、大脑 GnRH mRNA 水平和下丘脑 5-羟色胺水平（5-羟色胺对大西洋绒须石首鱼的 GnRH 释放有刺激作用）也有所下降（Thomas 等，2007）。在稍后对大西洋绒须石首鱼的研究中，4 周的缺氧导致 GSI 降低，血浆 20βS 水平降低，卵母细胞对 hCG 和 20βS 的体外成熟反应降低（Thomas，Rahman，2009）。

最近对斑马鱼（*Danio rerio*）的实验表明，低氧会在多个水平上影响基因的表达（Lu 等，2014）。在低氧环境中暴露 3 周，会导致雌雄鱼血浆 E2 和 T 的水平降低，并能导致雌鱼 sGnRH、FSHβ、cyp19a1、垂体和性腺 ER 和 VTG 表达下降，雄鱼 LHβ、LHR、StAR、20βHSD 和 11βH 基因表达下降。这些都是基因特异性作用，而不是基因表达普遍下调的结果。遗传型雌青鳉（*O. latipes*）的受精卵暴露在缺氧环境下 12 天会发育为雄性表型并且 vasa 基因（负责原代生殖细胞增殖及卵巢分化）表达下调，睾丸发育相关的 Y 染色体（DMY）基因上 DNA 结合基序的上调（Cheung 等，2014），证明低氧是影响性别决定的另一个环境变量，其影响似乎是基因特异性的。

和热效应一样，还有一个问题是：低氧是否对内分泌级联有直接影响，或者至少通过激活 HPI 轴来发挥其部分作用？急性严重低氧可导致鱼类血浆皮质醇水平升高（Perry，Gilmour，1999），但目前尚不清楚这是否也发生在对较不急性或持续低氧的反应中。更复杂的是，在高密度养鱼生产的不良水质（包括低氧）条件下，皮质醇对禁闭应激的反应实际上是受到抑制的（Pickering，Pottinger，1987）。低氧对生殖的抑制作用最初被解释为应激反应的一部分（Dabrowski 等，2003），但是低氧会降低大西洋绒须石首鱼体内 5HT 的水平（在大西洋绒须石首鱼体内，5HT 会随着应激而升高）（Thomas 等，2007），而基因特异性的效应证明（Cheung 等，2014；Lu 等，2014），低氧对鱼类生殖过程的作用不需要激活 HPI 轴。

8.3.5　应激对生殖的刺激作用

虽然大多数研究报道了应激对生殖的抑制作用，但似乎至少在某些条件下，皮质类固醇激素可能在生殖过程中发挥着许可作用或刺激作用，而这些作用大多见于雌性的排卵周期（Milla 等，2009）。大量研究表明，皮质类固醇激素对体外卵母细胞成熟具有增强或直接刺激作用（Goetz，1983；Greeley 等，1986；Upadhyaya，Haider，1986；Patino，Thomas，1990）。但皮质类固醇激素通常比孕激素的直接成熟效力低得多，进而 11-氧合皮质类固醇激素的效力也不如 11-脱氧皮质类固醇激素。然而，在其他生殖发育研究中，皮质醇对离体的卵母细胞成熟无增强作用（Pankhurst，1985）。也有证据表明皮质类固醇具有直接作用，其中 11-脱氧类固醇在体外具有最强的成熟潜力，并且也存在于大西洋胸棘鲷（*Hoplostethus atlanticus*）（Pankhurst，1987）和塔氏油白鱼（*Chalcalburnus tarichi*）（Unal 等，2008）最后成熟的血浆中。塔氏油白鱼在 hCG 处理后血浆 11-脱氧类固醇水平也升高，有证据表明卵巢（Kime 等，1992）和睾丸（Lee 等，2000）组织都可以通过放射性标记的类固醇前体合成皮质类固醇激素。皮质类固醇激素在配子成熟过程中是否具有直接的功能作用，或者是否由于具有模仿羟化孕激素作用的功能基因（如 21-羟基化）而表现出成熟活性，目前尚不清楚。

如前一节所述，应激对体内成熟类固醇水平的典型影响要么是维持应激水平，要么是血浆浓度增加。至少在鲑科鱼类中，成熟类固醇似乎也能刺激肾间腺皮质醇的产生（Barry 等，1997），这与一些物种的皮质类固醇周期性增加是一致的。在排卵期对雄性的影响似乎都是负面的，几乎没有证据表明皮质类固醇激素在睾丸的发育过程中起着促进或支持睾丸成熟的作用（Milla 等，2009）。

8.4　应激反应的机制

应激影响生殖内分泌轴的评估机制是复杂的，因为 HPI 轴的激素产物对生殖性能的直接作用和 HPI 轴对行为、代谢和生长的调节作用所产生的间接影响都有其作用范围（Leatherland 等，2010）。在这个前提下，一个合理的假设是，就像在哺乳动物中一样，鱼类也会在内分泌级联的多个层面上受到影响，其中许多影响将由皮质类固醇的作用来调节。GRs 已确定在生殖轴多个位点包括端脑尾部的 GnRH 神经元和脑的 POA、脑垂体的 FSH 和 LH 分泌细胞（Teitsma 等，1999）、肝脏的肝细胞（Lethimonier 等，2002）、卵巢和睾丸组织（Takeo 等，1996；Milla 等，2008）。Borski 等（2001）也报道了皮质醇对莫桑比克罗非鱼垂体催乳素分泌细胞的非基因组效应，对哺乳动物的研究已经确认了膜结合的皮质类固醇受体在很多组织中介导非基因组效应（Tasker 等，2006）。这导致了对皮质醇调节应激影响鱼类生殖作用的最初关注。

8.4.1 皮质醇的作用：在体方案

在体内使用皮质醇通常模拟应激反应中的抑制作用。褐鳟在植入皮质醇颗粒 18 天后，卵巢和睾丸重量下降，T 以及雌性 E2 和 VTG 水平降低，雌雄垂体 LH 含量均降低，而雄性和雌性的血浆 11KT 水平没有影响。雄性虹鳟血浆 LH 下降，但对其他生殖参数没有影响，而未成熟雌性虹鳟血浆 VTG 下降（Carragher 等，1989）。莫桑比克罗非鱼（*Oreochromis mossambicus*）植入皮质醇 18 天后，同样显示雌性体重、GSI、血浆 T 和 E2 水平下降，雄性血浆 T 下降。雌性滤泡闭锁的发病率也有所增加（Foo，Lam，1993a，b）。在未成熟雌性虹鳟体内植入皮质醇，4 周后肝脏 E2 结合位点减少，但血浆 E2 结合能力增加，这说明皮质醇可能是抑制 VTG 生成的机制之一（Pottinger，Pickering，1990）。在稍后的一项研究中也报道了类似的效应，在褐鳟体内植入皮质醇会导致血浆 E2、VTG 下降和肝脏 E2 结合能力降低，并增加血浆 E2 的结合能力（Pottinger 等，1991）。卵黄发生前的虹鳟植入皮质醇 15 天，肝脏 ER mRNA 水平降低，而用地塞米松体处理离体的肝细胞可抑制 E2 对 ER mRNA 的自刺激（Lethimonier 等，2002）。在布氏棘鲷中发现应激对类固醇结合蛋白（SBP）能力的反向作用，禁闭应激导致黑鲷在应激 6 小时后 SBP 能力下降，但对虹鳟的 SBP 结合能力或亲和力均无影响（Hobby 等，2000）。在相伴结合的研究中，加入 100 倍性类固醇激素浓度的皮质醇能够有效替换结合位点上的 E2，这表明，皮质醇效应可能来自 SBP 能力的降低，以及在正常性类固醇激素滴度低于应激诱导的皮质醇水平的物种中皮质醇的直接竞争性结合效应（Hobby 等，2000）。

一系列雄性鲤鱼幼鱼实验（Consten 等，2000，2001a，b；Goos，Consten，2002）表明，鲤鱼喂食含糖皮质激素的食物长达 160 天，会导致脑中 GnRH 含量降低，垂体 FSH 和 LH mRNA、血浆 T、11KT、11KA、GSI 下降，精子发生延迟。这种效应需要一段时间才会出现，并伴有生长抑制，因此很难将皮质醇的效应归因于 HPG 轴上的直接作用。用 11KA 共处理的鱼，恢复的 11KT 水平并不能抵消对性腺生长的抑制作用，说明抑制作用不受血浆雄激素下降的介导（Goos，Consten，2002）。然而，正如前面提到的应激效应，并不是所有的皮质醇效应都是抑制的。每周注射皮质醇，4 周后导致欧洲鳗鲡（*Anguilla anguilla*）LHβ 表达上调，垂体 LH 升高（Huang 等，1999）。

为了避免皮质醇长期处理所产生效应的模糊性，Pankhurst 和 Van Der Kraak（2000）研究了雌性虹鳟注射皮质醇后血浆皮质醇短暂升高的影响。在卵黄生成中期处理对血浆 T 和 E2 水平均无影响，而排卵前的鱼在注射后 1 小时和 3 小时 T 水平受到抑制，在 3 小时和 6 小时 E2 水平受到抑制。对血浆 LH 无影响，表明皮质醇的类固醇激素抑制作用可通过直接作用以及在内分泌级联中增长，介于 LH-LHR 结合和 T 的合成之间。在油船中给野生雄性大口黑鲈（*Micropterus salmoides*）注射皮质醇，短期内血浆皮质醇水平升到非常高的水平，也会影响它们的生殖行为，在接受皮质醇处理后，占有领地的雄鱼遗弃巢穴的概率增加。然而，糖皮质激素处理的鱼也显示出细菌感染的发生率增加，其行为效应可以解释为皮质醇对新陈代谢和健康状况的系统性影响（O'Connor 等，2009）。这一结论得到野生雄性雀鲷在短时间内（24 ～ 48 h）接受水下皮质醇注射后筑巢行为没有任何变化的实验结果所支持（Pankhurst，2001）。

8.4.2　皮质醇的作用：离体方案

对皮质醇在体给药效应的解释是设想它们通过直接作用而产生。这一假设得到了离体实验的支持，该实验用皮质醇培养虹鳟的卵巢滤泡。在皮质醇的生理剂量下，基础 T 和 E2 的分泌受到剂量依赖性的抑制，皮质醇水平低至 5 ng/mL 时，影响 T 的产生（Carragher，Sumpter，1990）。在随后的研究中，对类似实验的广泛重复中发现了不太一致的效果，在皮质醇浓度为 100 ～ 1000 ng/mL 的 20 个实验中，只有 4 个实验抑制了基础的 E2，皮质醇对 17-羟孕酮（17 P）或 hCG 刺激的 E2 产生均无影响（Pankhurst，1998b）。既没有关于结合性类固醇产生的任何一致性效果，亦没有将类固醇吸收到卵母细胞中而使培养基中游离类固醇含量降低。通过对体外类固醇产量通常低得多的非鲑科鱼类采用相同的程序进行研究，鳟鱼卵泡在培养过程中产生的高水平类固醇可能掩盖了较为细微的影响。浓度高达 1000 ng/mL 的皮质醇对金鱼（*Carassius auratus*）、鲤鱼或澳大拉西亚笛鲷（Pankhurst，1995）卵巢卵泡中 hCG 或鲤鱼 LH 的甾体前体 25-羟基胆固醇的转化、基础 T 和 E2 的生成，以及对 hCG 或鲤鱼 LH 的类固醇反应均无影响（Pankhurst 等，1995）。由此可以得出结论，皮质醇在卵巢水平上的直接作用不足以解释应激施加影响后血浆类固醇水平有时急剧下降的原因。

最近，Alsop 等（2009）也没有发现皮质醇对斑马鱼卵泡 hCG 刺激的 E2 分泌有任何影响。进一步检测虹鳟卵泡（Reddy 等，1999）发现，皮质醇在 100 ng/mL 时对卵黄发生后期的基础的和 LH 刺激导致的 T 和 E2 的产生具有抑制作用，而对排卵期前的滤泡没有影响。结论是，皮质醇的抑制作用是存在的，但只有在 T 和 E2 的基础水平较高时才会出现，尽管所报道的水平与早期研究中发现的水平相似或更低（Pankhurst，1998b）。稍后的一项使用虹鳟卵黄生成中期卵泡的研究显示，在含有 10 ng/mL 皮质醇的孵育中，基础的和 cAMP 刺激的 T 和 E2 分泌均被抑制，但皮质醇对体外 T 转化为 E2 没有影响。皮质醇处理还抑制 StAR 和 P450 侧链裂解酶（负责介导胆固醇转化为孕烯醇酮）mRNA 水平的表达（Barkataki 等，2011）。结论是皮质醇的抑制作用是由于它抑制胆固醇向线粒体内膜的传递和转化为孕烯醇酮。

同样的效果也出现在捕获后 24 小时受到应激的雀鲷获取的卵泡中，在这些卵泡中血浆类固醇水平降低，但卵泡仍将 17 P 转化为 T，T 转化为 E2，但对 hCG 没有反应（Pankhurst，2001）。尚不清楚的是，离体卵泡接触皮质醇产生的不一致影响是否能够说明皮质醇在卵巢层面没有一种强有力的抑制作用，或者是因为建立的共同决定血浆类固醇循环水平的动态分泌、运输、代谢和清除过程这一体外系统的不足。应激对血浆类固醇水平的影响可以非常迅速，这一事实表明，如果皮质醇是诱因，那么至少这种影响的早期阶段需要通过非基因组机制才能发生。

对睾丸组织的影响似乎较为一致。用 LH 刺激鲤鱼精巢，体外 11KT 的产量在喂食皮质醇长达 160 天的鱼体内较低，但这种效果只在青春期发育的早期出现（Consten 等，2000）。采用高浓度皮质醇（2000 ng/mL）孵育精巢会导致 11KA 浓度降低（Consten 等，2002）。其机制是通过底物与皮质醇竞争 11βHSD，如前所述，11β-羟雄烯二酮转化为 11KA，皮质醇转化为可的松。也有离体的证据表明皮质醇在生殖轴的其他部位有直接作用。Lethimonier 等（2002）报道了 GR 介导的抑制虹鳟肝细胞中 E2 诱导的 ER 表达上调，其机制涉及皮质醇抑制转录因子 C/EPBβ 与 ER 基因启动子区域的结合。

8.4.3 其他应激因素的影响

大多数研究都检测了皮质醇的影响，但针对其他应激相关的内分泌因子的研究较少。黑皮质素受体 2（MC2R）在斑马鱼的卵巢和精巢组织中均有强烈表达，MC2R 配体 ACTH 可抑制 hCG 刺激的游离卵巢卵泡中 E2 的分泌，但不抑制 E2 的基础分泌（Alsop 等，2009）。ACTH 的抑制作用可以通过与环 AMP 激活剂毛喉素和 8-溴环 AMP 共同作用而被消除，说明其作用水平高于腺苷酸环化酶的激活水平。在较长的孵育时间内，这种抑制作用也消失了，这表明没有检测到 ACTH 对褐鳟卵泡中 LH 刺激的 E2 产生有任何影响（Sumpter 等，1987），或者虹鳟卵泡产生 T 和 E2 的基础产物（Fiztgibbon，Pankhurst，未发表的数据）可能上升，或者是由于在鲑科鱼类中缺乏类似的效应，或者是使用的孵育参数的选择不同。也有证据表明其他垂体源性因子在生殖应激抑制中的作用。在莫桑比克罗非鱼中，β-内啡肽给药 22 天降低了垂体远端促性腺细胞中 LHβ 亚基的免疫组化染色（Ganesh，Chabbi，2013）。这种效应可通过多次日常手工操作和网捕活动而重现，并通过与阿片样物质受体拮抗剂纳曲酮（naltrexone）的协同处理而减弱，表明该效应是阿片样物质受体介导的。

CRF 也参与了与应激相关效应的调节，但这方面也存在一些不确定性，即这些效应是直接的还是通过 HPI 内分泌轴中较低的激素介导的？每天腹腔注射含有 CRF 的莫桑比克罗非鱼，持续 22 天，血浆皮质醇水平升高，垂体 LHβ 染色强度降低，GSI 降低，卵巢卵黄滤泡丢失，滤泡闭锁增加（Chabbi，Ganesh，2013）。网捕活动和手工操作应激会产生类似的效应，而这些效应可以通过与甲吡酮的共同处理来改善，这表明这些效应是皮质醇介导的，而不是通过 CRF 的直接作用。CRF 确实对一系列行为有直接影响，包括抑制食欲和进食，减少觅食和增加运动能力（Bernier，2006；Lowry，Moore，2006），并且有证据表明，CRF 也可能在调节生殖行为中发挥作用。如前所述，非占有领地的非洲丽鱼类的脑和垂体 CRFR 升高；垂体 CRF mRNA 升高，GnRH 神经元变小，GnRHR、FSHβ、LHβ mRNA 降低；血浆 LH、FSH、雄激素水平降低（Maruska，2014）。非占有领地的鱼进入空出的领地所引起的社会地位变化导致迅速的行为变化（10 分钟）；30 分钟内血浆 LH、FSH、雄激素升高；长期（5 ～ 7 天）

GnRH 神经元的体积增加。由此推断，CRF 在抑制处于从属社会地位雄鱼的生殖发育中起着关键作用。社会地位变化的短时间状态表明，这些影响要么是由于 CRF 的直接作用，要么是皮质激素介导的，于是，它们肯定是通过非基因组过程发生的。

急性应激反应最直接的组成部分包括中性介导儿茶酚胺中的去甲肾上腺素（NA）和肾上腺素（AD）从肾脏的嗜铬组织释放进入血液循环，产生的影响包括增加血红蛋白对氧的亲和力、刺激心脏、升高动脉血压、释放肝糖原以增加血浆葡萄糖水平（Pankhurst，2011）。现有的证据表明，血浆儿茶酚胺水平的急性升高主要是刺激作用而不是抑制作用，有助于从应激中全面恢复，并保护心脏等器官（Eliason 等，2011）。生殖过程似乎也是如此。Yu 和 Peter（1992）报道了 NA 对离体金鱼下丘脑薄片 GnRH 释放的刺激作用，用 α1 肾上腺素能受体激动剂处理可复制该作用，而用 α1 受体拮抗剂可将它阻断。在哺乳动物中，NA 对卵巢类固醇生成具有典型的容许或刺激作用（Aguado，2002），用 NA 和 α 肾上腺素能受体激动剂体外孵育虹鳟和雀鲷的卵泡，结果是没有变化或者增加基础 E2 的生成（Fitzgibbon，Gonzalez-Reynoso，Pankhurst，未发表的数据）。AD 还会刺激美洲红点鲑（*Salvelinus fontinalis*）卵泡收缩和排卵，这种效应为 α 肾上腺素能受体拮抗剂阻断（Goetz 和 Bradley，1994）。儿茶酚胺还可以对舌齿鲈（*Dicentrarchus labrax*）产生间接的抑制作用，其中 NA 和 AD 通过 β 肾上腺素能受体刺激肾间组织产生皮质类固醇（Rotllant 等，2006）。

有记载的应激对生殖性能的影响及其可能机制的概要见图 8.4。

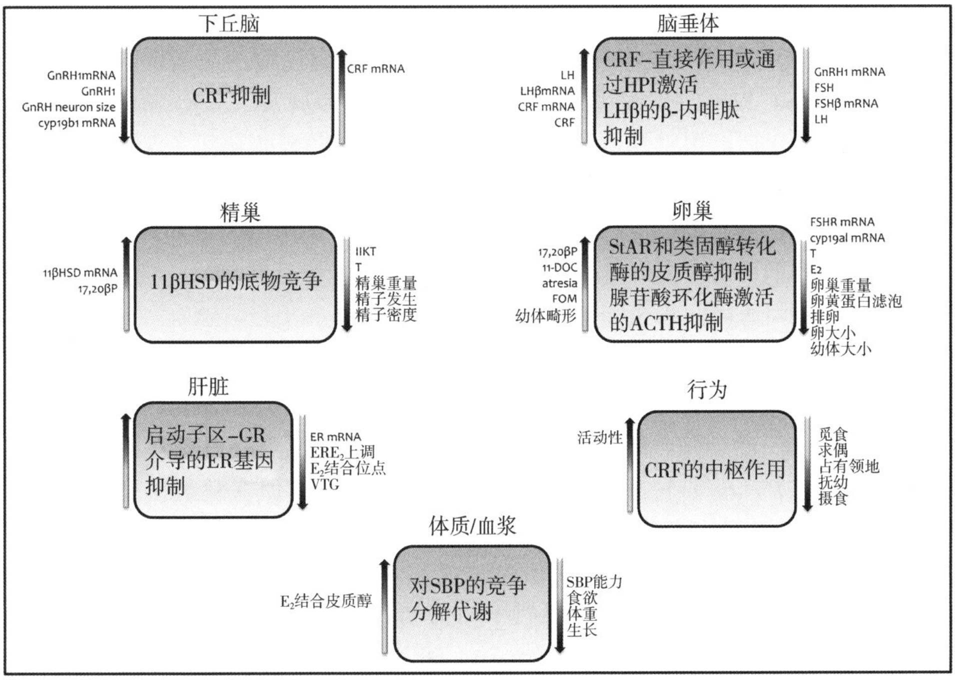

图 8.4　应激对下丘脑-垂体-性腺轴（HPG）组成部分影响的概述

注：向上箭头表示刺激作用；向下箭头表示抑制作用。文本框概括了可能的作用机制。

8.5 自然环境中的应激对生殖的影响

在考虑野生动物是否会在正常的活动范围和条件下承受应激的问题时，Pankhurst（2011）认为答案通常是否定的。我们对应激生理的理解往往受到文献的支配，这些文献主要研究的是在非自然条件下蓄养和养殖的鱼类，它们通过行为改变来避免或减轻应激的能力有限。对野生鱼类的进一步了解受到了少量研究的限制，这些研究针对的是自由生活鱼类的表现，而可能被预测为应激的自然现象（如暴风雨和洪水）通常都未能进行观察和取样（Pankhurst，2011）。对鸟类的实地研究表明，在可预测的条件下（即动物通常善于应对），环境的严酷程度和生殖的需求都不会产生应激（Wingfield，1994）。也有一些来自鱼类的证据表明，可预测的应激源产生的应激反应不如不可预测的应激反应那样显著（Galhardo 等，2011）。在足够的挑战性条件下提高皮质类固醇（对于鸟类，典型的是风暴或一段时间的高温），在行为上有一个从筑巢到觅食或寻求庇护的转变，这有助于减轻应激，并在条件允许的情况下支持体能恢复以及生殖行为的恢复（Romero 等，2000）。一般的观点是，像迁徙和生育这样的事件是需要的，但不一定是应激的，Schreck（2010）指出，如果对应激源的抵抗在能量上是昂贵的，那么这可能反映为分配支持生殖的能量资源的能力下降。

有限的野外数据显示，在自然种群中，个体间的皮质醇水平差异相当大，而这似乎和对生殖的任何负面影响无关。在水下和捕获取样的占有领地的成年雀鲷，血浆皮质醇水平从小于 1 ng/mL 到 42 ng/mL，但雌性皮质醇血浆 T 和 E2 水平以及雄性 T 和 11KT 水平之间无相关性（如前所述，在实验室中，性类固醇的应激抑制在两性中都存在）（Pankhurst，2011）。同样，雄性蓝鳃太阳鱼（*Lepomis macrochirus*）在亲代抚育过程中，即使皮质醇水平高达 125 ng/mL，仍然表现和产卵和卵子保护行为相联系的血浆 T 和 11KT 的正常周期（Magee 等，2006）。正如 Pankhurst（2011）所指出的，这些研究得出的结论是生殖活动可以在皮质类固醇的一定范围内得到维持，野生鱼类血浆皮质醇水平的适度升高并不意味着受到应激，或者这种升高与应激有关，但持续时间不长，不会对生殖产生负面影响。

营养状况显然对鱼类应对或处理应激的能力有显著影响，从而维持正的能量平衡，使其能够得到持续支持或进行生殖过程（Schreck，2010）。来自热带雀鲷自然种群的证据支持这一观点。珊瑚覆盖式样和质量引起的生境变化对刺雀鲷或鹿角雀鲷（*Amblyglyphidodon curacao*）的血浆皮质醇水平没有影响，但在鱼类密度较低的生境中，卵巢状况最好，从而减少了获取浮游生物食物资源的竞争（Pankhurst 等，2008a，b）。相反，在珊瑚覆盖和鱼类密度较高的生境中，卵巢状况最差，有趣的是，在人类眼中最美丽的栖息地它们生殖的成功率比较高。双色雀鲷（*Stegastes partitus*）在不同的栖息地表现出不同的行为，生活在珊瑚碎石中的鱼比在珊瑚中的鱼显示较强的攻击行为、较频繁的求爱和隐蔽行为，尽管它们静息的血浆皮质醇水平没有差异，并且在两种栖息地中，皮质醇在应激时都有增加（Schrandt，Lema，2011）。

然而，应激刺激下 CRF 和硬骨鱼紧张素 I（UI）mRNA 表达的性别差异都是栖息地特异性的，但这些差异的意义尚不清楚。

一个有趣的自然鱼类行为是，应激对生殖的调节存在于有社会性调节生殖状态的物种之中。这一点在丽鱼类身上得到了强烈的体现，在丽鱼种群中，只有少数雄性是优势地位的、占有领地的性活跃的鱼类。而大多数从属地位的、非占有领地鱼类的性行为不活跃，生殖内分泌活动水平较低，同时脑垂体 CRF 和 CRFR1 mRNA 表达升高（Maruska，2014）。由此推断，从属地位鱼类的这种效应是对社会优势地位的应激反应，但这究竟是一种经典的应激反应，还是 CRF 普遍的神经调节作用的表达，目前尚不清楚。最近的研究表明，这种影响不是皮质类固醇激素介导的，在群体间转移莫桑比克罗非鱼优势地位雄鱼诱导社会关系的不稳定，导致血浆 T 和 11KT 增加，但对优势地位或从属地位雄鱼血浆皮质醇水平均无影响（Almeida 等，2014）。

8.6　未来的方向

尽管这里讨论了大量的研究工作，并且自 Sumpter 和他的同事揭示出我们对应激和生殖的影响所知甚少以来，我们在这些方面取得了非常重大的进展，但仍有很多问题需要探讨（Sumpter 等，1987）。我们在了解应激对子代在发育的最初阶段之后所产生的影响方面几乎没有取得实质性进展，在了解应激对子代生殖性能的影响或应激的多个世代影响方面也没有取得任何进展。在性腺水平上，应激的影响得到了一致的证明，但在 HPG 轴的较高水平上，对应激的影响和相互作用的理解仍然很不完整。我们对产生这些影响的机制的理解同样是不完整的。关于皮质类固醇的作用，目前还不是很清楚；当然也有长期的系统效应，可能是在较短时间内直接的基因组效应，也可能是由非基因组机制介导的快速的近端效应，但这些效应怎样以及何时起作用尚不清楚。对于 CRF 在 HPG 轴直接调节中的作用及其通过 ACTH、UI、β 内啡肽和皮质类固醇等下游作用的影响，我们的了解也非常有限。目前也没有关于应激对 kisspeptin-kissR 系统可能影响的信息，对儿茶酚胺在调节生殖过程中可能发挥作用的了解仍然相对有限。分子生物学研究技术的普及提供了一个机会，可以迅速扩展到以前由于无法获得适当的参数测量而受到限制的领域，但在这里，这个过程也不太可能是简单易做的。Chand 和 Lovejoy（2011）归纳的问题是：描述发生在脊椎动物的这种（应激反应激活对生殖功能的抑制）分子机制仍然是一个重大挑战。情况仍然如此。

目前还不清楚热应激广泛表现出的抑制作用和同样深刻的低氧作用是否通过 HPI 轴被调节。这对于我们如何预测气候和环境变化对自然种群的影响，或者如何管理热应激对养殖鱼类种群的影响具有重要意义。改善策略应该倾向于关注 HPG 轴的内分泌调节（Pankhurst，King，2010）。应激可能是热损伤和生殖抑制之间的中介，这一可能性表明，通过操纵 HPI 轴，可能有尚未探索的和更有效的保护热损伤的潜力。最后，仍然有人强烈地怀疑我们关于应激在自然栖息地和种群中的作用的看法，认为我们受到从实验室实验中学到的应激过程的过度影响。我们仍然不知道自然条件是否

（或何时）有足够的应激来引发在实验室实验中所观察到的那种生殖的抑制作用，也不知道应激在自由生活群体的正常调控和生殖组织中发挥了什么作用。在鱼类中，社会性因素对生殖的调控与 HPI 轴的激活联系一起被证明能够使应激发生变化，但似乎不太可能维持这种状态。总而言之，研究应激对生殖的影响，从工作平台到自然环境，并不缺少研究的底质，但也要有一个坚实的基础来进行研究。

N. W. 潘克赫斯特　著
张　勇　译
林浩然　校

参 考 文 献

Aguado, L. I. (2002). Role of the central and peripheral nervous system in the ovarian function. Microscopy Res. Tech. 59, 462-473.

Almeida, O., Gonçalves-de-Freitas, E., Lopes, J. S., Oliveira, R. F. (2014). Social instability promotes hormone-behavior associated patterns in a cichlid fish. Horm. Behav. 66, 369-382.

Alsop, D., Ings, J. S., Vijayan, M. M. (2009). Adrenocorticotropic hormone suppresses gonadotropin-stimulated estradiol release from zebrafish ovarian follicles. PLoS ONE. 4, e6463.

Aluru, N., Vijayan, M. M. (2009). Stress transcriptomics in fish: a role for genomic cortisol signalling. Gen. Comp. Endocrinol. 164, 142-150.

Anderson, K., Swanson, P., Pankhurst, N., King, H., Elizur, A. (2012). Effect of thermal challenge on plasma gonadotropin levels and ovarian steroidogenesis in female and repeat spawning Tasmanian Atlantic salmon (*Salmo salar*). Aquaculture. 334, 205-212.

Anderson, K., King, H., Pankhurst, N., Ruff, N., Pankhurst, P., Elizur, A. (2012). Effect of elevated temperature on estrogenic induction of vitellogenesis and zonagenesis in juvenile Atlantic salmon (*Salmo salar*). Mar. Freshwat. Behav. Physiol. 45, 1-15.

Auperin, B., Geslin, M. (2008). Plasma cortisol response to stress in juvenile rainbow trout is influenced by their life history during early development and by egg cortisol content. Gen. Comp. Endocrinol. 158, 234-239.

Barkataki, S., Aluru, N., Li, M., Lin, L., Christie, H., Vijayan, M. M. (2011). Cortisol inhibition of 17β-estradiol secretion by rainbow trout ovarian follicles involves modulation of *Star* and *P450scc* gene expression. J. Aquac. Res. Devel. S2, 001.

Barry, T. P., Riebe, J. D., Parrish, J. J., Malison, J. A. (1997). Effects of 17, 20β-dihydroxy-4-pregnen-3-one on cortisol production by rainbow trout interrenal tissue in vitro. Gen. Comp. Endocrinol. 107, 172-181.

Bernier, N. J. (2006). The corticotrophin-releasing factor system as a mediator of the appetite-suppressing effects of stress in fish. Gen. Comp. Endocrinol. 146, 45-55.

Bhandari, R. K., Deem, S. L., Holliday, D. K., Jandegian, C. M., Kassotis, C. D., Nagel, S. C. (2015). Effects of the environmental estrogenic contaminants bisphenol A and 17α-ethinyl estradiol on sexual development and adult behaviors in aquatic wildlife species. Gen. Comp. Endocrinol. 214, 195-219.

Black, M. P., Balthazart, J., Baillien, M., Grober, M. S. (2011). Rapid increase in aggressive behavior precedes the decrease in brain aromatase activity during socially mediated sex change in *Lythrypnus dalli*. Gen. Comp. Endocrinol. 170, 119-124.

Borski, R. J., Hyde, G. N., Fruchtman, S., Tsai, W. S. (2001). Cortisol suppresses prolactin release through a non-genomic mechanism involving interactions with the plasma membrane. Comp. Biochem. Physiol. 129B, 533-541.

Campbell, P. M., Pottinger, T. G., Sumpter, J. P. (1992). Stress reduces the quality of gametes produced by rainbow trout. Biol. Reprod. 47, 1140-1150.

Campbell, P. M., Pottinger, T. G., Sumpter, J. P. (1994). Preliminary evidence that chronic confinement stress reduces the quality of gametes produced by brown and rainbow trout. Aquaculture. 120, 151-169.

Carragher, J. F., Pankhurst, N. W. (1991). Stress and reproduction in a commercially important marine fish, *Pagrus auratus* (Sparidae). In: Scott, A. P., Sumpter, J. P., Kime, D. E., Rolfe, M. S.,. FishSymp, Sheffield, 253-255.

Carragher, J. F., Sumpter, J. P. (1990). The effect of cortisol on the secretion of sex steroids from cultured ovarian follicles of rainbow trout. Gen. Comp. Endocrinol. 77, 403-407.

Carragher, J. F., Sumpter, J. P., Pottinger, T. G., Pickering, A. D. (1989). The deleterious effects of cortisol implantation on reproductive function in two species of trout, *Salmo trutta* L. and *Salmo gairdneri* Richardson. Gen. Comp. Endocrinol. 76, 310-321.

Castranova, D. A., King, W., Woods, L. C. (2005). The effects of stress on androgen production, spermiation response and sperm quality in high and low cortisol responsive domesticated male striped bass. Aquaculture. 246, 413-422.

Chabbi, A., Ganesh, C. B. (2013). Glucocorticoid synthesis inhibitor metyrapone blocks stress-induced suppression along luteinizing hormone secreting cells-ovary axis in the fish *Oreochromis mossambicus*. J. Exp. Zool. 321A, 125-134.

Chand, D., Lovejoy, D. A. (2011). Stress and reproduction: controversies and challenges. Gen. Comp. Endocrinol. 171, 253-257.

Cheung, C. H. Y., Chiu, J. M. Y., Wu, R. S. S. (2014). Hypoxia turns female medaka fish into phenotypic males. Ecotoxicology. 23, 1260-1269.

Clearwater, J. C., Pankhurst, N. W. (1997). The response to capture and confinement stress of plasma cortisol, plasma sex steroids and vitellogenic oocytes in the marine teleost, red gurnard (*Chelidonichthys kumu*) (Triglidae). J. Fish. Biol. 50, 429-441.

Cleary, J. J., Pankhurst, N. W., Battaglene, S. C. (2000). The effect of capture and handling stress on plasma steroid levels and gonadal condition in wild and farmed snapper, *Pagrus auratus* (Sparidae). J. World Aquacult. Soc. 31, 558-569.

Cleary, J. J., Battaglene, S. C., Pankhurst, N. W. (2002). Capture and handling stress affects the endocrine and ovulatory response to exogenous hormone treatment in snapper, *Pagrus auratus* (Bloch & Schneider). Aquacult. Res. 33, 1-10.

Consten, D., Lambert, J. G. D., Goos, H. J. Th. (2000). Inhibitory effects of cortisol on in vivo and in vitro androgen secretion in male common carp, *Cyprinus carpio*. In: Norberg, B., Kjesbu, O. S., Taranger, G. L., Andersson, E., Stefansson, S. O., Proceedings of the 6th International Symposium on the Reproductive Physiology of Fish. John Grieg AS, Bergen, 192.

Consten, D., Bogerd, J., Komen, J., Lambert, J. G. D., Goos, H. J. Th. (2001). Long-term cortisol treatment inhibits pubertal development in male common carp, *Cyprinus carpio* L. Biol. Reprod. 64, 1063-1071.

Consten, D., Lambert, J. G. D., Goos, H. J. Th. (2001). Cortisol affects testicular development in male common carp, *Cyprinus carpio* L., but not via an effect on LH secretion. Comp. Biochem. Physiol. 129B, 671-677.

Consten, D., Keuning, E. D., Terlou, M., Lambert, J. G. D., Goos, H. J. Th. (2002). Cortisol effects on the testicular androgen synthesizing capacity in common carp, *Cyprinus carpio* L. Fish Physiol. Biochem. 25, 91-98.

Contreras-Sánchez, W. M., Schreck, C. B., Fitzpatrick, M. S., Pereira, C. B. (1998). Effects of stress on the reproductive performance of rainbow trout (*Oncorhynchus mykiss*). Biol. Reprod. 58, 439-447.

Coward, K., Bromage, N. R., Little, D. C. (1998). Inhibition of spawning and associated suppression of sex steroid levels during confinement in the substrate-spawning *Tilapia zillii*. J. Fish. Biol. 52, 152-165.

Dabrowski, K., Rinchard, J., Ottobre, J. S., Alcantara, F., Padilla, P., Ciereszko, A. (2003). Effect of oxygen saturation in water on reproductive performance of pacu *Piaractus brachypomus*. J. World Aquacult. Soc. 34, 441-449.

De Montalembert, G., Jalabert, B., Bry, C. (1978). Precocious induction of maturation and ovulation in northern pike (*Esox lucius*). Ann. Biol. Ann. Bioch.

Biophys. 18, 969-975.

Devlin, R. H., Nagahama, Y. (2002). Sex determination and sex differentiation in fish: an overview of genetic, physiological and environmental influences. Aquaculture. 208, 191-364.

Dufour, S., Sebert, M.-E., Weltzein, F.-A., Rousseau, K., Pasqualini, C. (2010). Neuroendocrine control by dopamine of teleost reproduction. J. Fish. Biol. 76, 129-160.

Eliason, E. J., Clark, T. D., Hague, M. J., Hanson, L. M., Gallagher, Z. S., Jeffries, K. M. (2011). Differences in thermal tolerance among sockeye salmon populations. Science. 332, 109-112.

Eriksen, M. S., Bakken, M., Espmark, Å., Braastad, B. O. (2006). Prespawning stress in farmed Atlantic salmon *Salmo salar*: maternal cortisol exposure and hyperthermia during embryonic development affect offspring survival, growth and incidence of malfunctions. J. Fish. Biol. 69, 114-129.

Eriksen, M. S., Espmark, Å., Braastad, B. O., Salte, R., Bakken, M. (2007). Long-term effects of maternal cortisol exposure and mild hyperthermia during embryogeny on survival, growth and morphological anomalies in farmed Atlantic salmon *Salmo salar* offspring. J. Fish. Biol. 70, 462-473.

Fernandino, J. I., Hattori, R. S., Moreno Acosta, O. D., Strüssman, C. A., Somoza, G. M. (2013). Environment stress-induced testis differentiation: androgen as a by-product of cortisol inactivation. Gen. Comp. Endocrinol. 192, 36-44.

Foo, J. T. W., Lam, T. J. (1993). Retardation of ovarian growth and depression of serum steroid levels in the tilapia *Oreochromis mossambicus*, by cortisol implantation. Aquaculture. 115, 133-143.

Foo, J. T. W., Lam, T. J. (1993). Serum cortisol response to handling stress and the effect of cortisol implantation on testosterone level in the tilapia *Oreochromis mossambicus*. Aquaculture. 115, 145-158.

Fostier, A., Jalabert, B. (1982). Physiological basis of practical means to induce ovulation in fish. In: Richter, C. J. J., Goos, H. J. Th, Proceedings of the International Symposium on the Reproductive Physiology of Fish 1982. Pudoc, Wageningen, 164-173.

Fox, H. E., White, S. A., Kao, M. H. F., Fernald, R. D. (1997). Stress and dominance in a social fish. J. Neurosci. 17, 6463-6469.

Frisch, A. (2004). Sex-change and gonadal steroids in sequentially-hermaphroditic teleost fish. Rev. Fish Biol. Fish. 14, 481-489.

Frisch, A. J., Walker, S. P. W., McCormick, M. I., Solomon-Lane, T. K. (2007). Regulation of protogynous sex change by competition between corticosteroids and androgens: an experimental test using sandperch, *Parapercis cylindrica*. Horm. Behav. 52, 540-545.

Fuzzen, M. L. M., Bernier, N. J., Van Der Kraak, G. (2011). Stress and reproduction. In: Norris, D. O., Lopez, K. H., Hormones and Reproduction of Vertebrates. Volume 1: Fishes. Elsevier, Amsterdam, 103-117.

Galhardo, L., Vital, J., Oliveira, R. F. (2011). The role of predictability in the stress response of a cichlid fish. Physiol. Behav. 102, 367-372.

Ganesh, C. B., Chabbi, A. (2013). Naltrexone attenuates stress-induced suppression of LH secretion in the pituitary gland in the cichlid fish *Oreochromis mossambicus*: evidence for the opioidergic mediation of reproductive stress response. Fish Physiol. Biochem. 39, 627-636.

Gardner, L., Anderson, T., Place, A. R., Dixon, B., Elizur, A. (2005). Sex change strategy and aromatase genes. Steroid Biochem. Mol. Biol. 94, 395-404.

Gerking, S. D. (1982). The sensitivity of reproduction in fishes to stressful environmental conditions. In: Richter, C. J. J., Goos, H. J. Th, Proceedings of the International Symposium on the Reproductive Physiology of Fish 1982. Pudoc, Wageningen, 224-228.

Gillet, C., Breton, B. (2009). LH secretion and ovulation following exposure of Arctic charr to different temperature and photoperiod regimes: responsiveness of females to a gonadotropin-releasing hormone analogue and a dopamine antagonist. Gen. Comp. Endocrinol. 162, 210-218.

Gillet, C., Breton, B., Mikolajczyk, T., Bodinier, P., Fostier, A. (2011). Disruption of the secretion and action of 17,20β-dihydroxy-4-pregnen-3-one in response to a rise in temperature in the Arctic charr, *Salvelinus alpinus*. Consequences on oocyte maturation and ovulation. Gen. Comp. Endocrinol. 172, 392-399.

Goetz, F. W. (1983). Hormonal control of oocyte final maturation and ovulation in fishes. In: Hoar, W. S., Randall, D. J., Donaldson, E. M., . Academic Press, New York, NY, 117-170.

Goetz, F. W., Bradley, J. A. (1994). Stimulation of in vitro ovulation and contraction of brook trout (*Salvelinus fontinalis*) follicles by adrenaline through α-adrenoreceptors. J. Reprod. Fertil. 100, 381-385.

Goos, H. J. Th, Consten, D. (2002). Stress adaptation, cortisol and pubertal development in the male common carp, *Cyprinus carpio*. Mol. Cell. Endocrinol. 197, 105-116.

Greeley, M. S., Calder, D. R., Taylor, M. H., Hols, H., Wallace, R. A. (1986). Oocyte maturation in the mummichog (*Fundulus heteroclitus*): effects of steroids on germinal vesicle breakdown of intact follicles in vitro. Gen. Comp. Endocrinol. 62, 281-289.

Guiguen, Y., Fostier, A., Piferrer, F., Chang, C.-F. (2010). Ovarian aromatase and estrogens: a pivotal role for gonadal sex differentiation and sex change in fish. Gen. Comp. Endocrinol. 165, 352-366.

Haddy, J. A., Pankhurst, N. W. (1999). Stress-induced changes in concentrations of plasma sex steroids in black bream. J. Fish. Biol. 55, 1304-1316.

Haddy, J. A., Pankhurst, N. W. (2000). The efficacy of exogenous hormones in stimulating changes in plasma steroids and ovulation in wild black bream *Acanthopagrus butcheri* is improved by treatment at capture. Aquaculture. 191, 351-366.

Hattori, R. S., Fernandino, J. I., Kishii, A., Kimura, H., Kinno, T., Oura, M. (2009). Cortisol-induced masculinization: does thermal stress affect gonadal fate in pejerrey, a teleost fish with temperature-dependent sex determination?. PLoS One. 4, e6548.

Hayashi, Y., Kobira, H., Yamaguchi, T., Shiraishi, E., Yazawa, T., Hirai, T. (2010). High temperature causes masculinization of genetically female medaka by elevation of cortisol. Mol. Reprod. Dev. 77, 679-686.

Hobby, A. C., Pankhurst, N. W., Haddy, J. A. (2000). The effect of short term confinement stress on binding characteristics of sex steroid binding protein (SBP) in female black bream (*Acanthopagrus butcheri*) and rainbow trout (*Oncorhynchus mykiss*). Comp. Biochem. Physiol. 125A, 85-94.

Huang, Y.-S., Rousseau, K., Sbaiti, M., Le Belle, N., Schmitz, M., Dufour, S. (1999). Cortisol selectively stimulates pituitary gonadotropin β-subunit in a primitive teleost, *Anguilla anguilla*. Endocrinology. 140, 1228-1235.

Imanaga, Y., Nyuji, M., Amano, M., Takahashi, A., Kitano, H., Yamaguchi, A. (2014). Characterization of gonadotropin-releasing hormone and gonadotropin in jack mackerel (*Trachurus japonicus*): comparative gene expression analysis with respect to reproductive dysfunction in captive and wild fish. Aquaculture. 428-429, 226-235.

Jardine, J. J., Van Der Kraak, G. J., Munkittrick, K. R. (1996). Capture and confinement stress in white sucker exposed to bleached kraft pulp mill effluent. Ecotoxicol. Environ. Saf. 33, 287-298.

Kamstra, J. H., Aleström, P., Kooter, J. M., Legler, J. (2014). Zebrafish as a model to study the role of DNA methylation in environmental toxicology. Environ. Sci. Pollut. Res.

Kime, D. E., Scott, A. P., Canario, A. V. M. (1992). In vitro biosynthesis of steroids, including 11-deoxycortisol and 5α-pregnane-3β, 7α, 20β-tetrol, by ovaries of the goldfish *Carassius auratus* during the stage of oocyte final maturation. Gen. Comp. Endocrinol. 87, 375-384.

Kitano, T., Hayashi, Y., Shiraishi, E., Kamei, Y. (2012). Estrogen rescues masculinization of genetically female medaka by exposure to cortisol or high temperature. Mol. Reprod. Dev. 79, 719-726.

Kubokawa, K., Watanabe, T., Yoshioka, M., Iwata, M. (1999). Effects of acute stress on plasma cortisol, sex steroid hormone and glucose levels in male and female sockeye salmon during the breeding season. Aquaculture. 172, 335-349.

Landry, C. A., Steele, S. L., Manning, S., Cheek, A. O. (2007). Long term hypoxia suppresses reproductive capacity in the estuarine fish, *Fundulus grandis*. Comp. Biochem. Physiol. 148A, 317-323.

Lane, M., Robker, R. L., Robertson, S. A. (2014). Parenting from before conception. Science. 345, 756-760.

Leatherland, J. F., Li, M., Barkataki, S. (2010). Stressors, glucocorticoids and ovarian function in teleosts. J. Fish. Biol. 76, 86-111.

Lee, S. T. L., Lam, T. J., Tan, C. H. (2000). Corticosteroid biosynthesis in vitro by testes of the grouper (*Epinephelus coioides*) after 17α-methyltestosterone-induced sex inversion. J. Exp. Zool. 287, 453-457.

Lethimonier, C., Flouriot, G., Kah, O., Ducouret, B. (2002). The glucocorticoid receptor represses the positive autoregulation of the trout estrogen receptor gene by preventing the enhancer effect of a C/EBPβ-like protein. Endocrinology. 143, 2961-2974.

Levavi-Sivan, B., Bogerd, J., Mañanós, E. L., Gómez, A., Lareyre, J. J. (2010). Perspectives on fish gonadotropins and their receptors. Gen. Comp. Endocrinol. 165, 412-437.

Lim, B.-S., Kagawa, H., Gen, K., Okuzawa, K. (2003). Effects of water temperature on the gonadal development and expression of steroidogenic enzymes in the gonad of juvenile red seabream, *Pagrus major*. Fish Physiol. Biochem. 28, 161-162.

Lowry, C. A., Moore, F. L. (2006). Regulation of behavioural responses by corticotrophin releasing factor. Gen. Comp. Endocrinol. 146, 19-27.

Lu, X., Yu, R. M. K., Murphy, M. B., Lau, K., Wu, R. S. S. (2014). Hypoxia disrupts gene modulation along the brain-pituitary-gonad (BPG) -liver axis. Ecotoxicol. Environ. Saf. 102, 70-78.

Magee, S. E., Neff, B. D., Knapp, R. (2006). Plasma levels of androgen and cortisol in relation to breeding behaviour in parental male bluegill sunfish, *Lepomis macrochirus*. Horm. Behav. 49, 598-609.

Martinez, P. , Viñas, A. M. , Sánchez, L. , Díaz, N. , Ribas, L. , Piferrer, F. (2014). Genetic architecture of sex determination in fish: applications to sex ratio control in aquaculture. Front. Genet. 5, 340.

Maruska, K. P. (2014). Social regulation of reproduction in male cichlid fishes. Gen. Comp. Endocrinol. 207, 2-12.

McCormick, M. I. (1998). Behaviorally induced maternal stress in a fish influences progeny quality by a hormonal mechanism. Ecology. 79, 1873-1883.

McCormick, M. I. (1999). Experimental test of the effect of maternal hormones on larval quality of a coral reef fish. Oecologia. 118, 412-422.

McCormick, M. I. (2006). Mothers matter: crowding leads to stressed mothers and smaller offspring in marine fish. Ecology. 87, 1104-1109.

Migaud, H. , Davie, A. , Taylor, J. F. (2010). Current knowledge on the photoneuroendocrine regulation of reproduction in temperate fish species. J. Fish. Biol. 76, 27-68.

Mileva, V. R. , Gilmour, K. M. , Balshine, S. (2011). Effects of maternal stress on egg characteristics in a cooperatively breeding fish. Comp. Biochem. Physiol. 158A, 22-29.

Milla, S. , Terrien, X. , Sturm, A. , Ibrahim, F. , Giton, F. , Fiet, J. (2008). Plasma 11-deoxycorticosterone (DOC) and mineralocorticoid receptor testicular expression during rainbow trout *Oncorhynchus mykiss* spermiation: implication with 17α, 20β-dihydroxyprogesterone on the milt fluidity?. Reprod. Biol. Endocrinol. 6, 19.

Milla, S. , Wang, N. , Mandiki, S. N. M. , Kestemont, P. (2009). Corticosteroids: friends or foes of teleost reproduction?. Comp. Biochem. Physiol. 153A, 242-251.

Morehead, D. T. (1998). Effect of capture, confinement and repeated sampling on plasma steroid concentrations and oocyte size in female striped trumpeter *Latris lineata* (Latrididae). Mar. Freshwat. Res. 49, 373-377.

Morgan, M. J. , Wilson, C. E. , Crim, L. W. (1999). The effect of stress on reproduction in Atlantic cod. J. Fish. Biol. 54, 477-488.

Munakata, A. , Kobayashi, M. (2010). Endocrine control of sexual behaviour in teleost fish. Gen. Comp. Endocrinol. 165, 456-468.

Nakamoto, M. , Matsuda, M. , Wang, D. -S. , Nagahama, Y. , Shibata, N. (2006). Molecular cloning and analysis of gonadal expression of foxl2 in the medaka, *Oryzias latipes*. Biochem. Biophys. Res. Comm. 344, 353-361.

O'Connor, C. M. , Gilmour, K. M. , Arlinghaus, R. , Van Der Kraak, G. , Cooke, S. J. (2009). Stress and parental care in a wild teleost fish: insights from exogenous supraphysiological cortisol implants. Physiol. Biochem. Zool. 82, 709-719.

O'Connor, C. M. , Yick, C. Y. , Gilmour, K. M. , Van Der Kraak, G. , Cooke, S. J. (2011). The glucocorticoid stress response is attenuated but unrelated to reproductive investment during parental care in a teleost fish. Gen. Comp. Endocrinol. 170, 215-221.

Okuzawa, K. , Gen, K. (2013). High water temperature impairs ovarian activity and gene expression in the brain-pituitary-gonadal axis in female red seabream during the spawning season. Gen. Comp. Endocrinol. 194, 24-30.

Okuzawa, K. , Kumakura, N. , Gen, K. , Yamaguchi, S. , Lim, B. -S. , Kagawa, H. (2003). Effect of high water temperature on brain-pituitary-gonad axis of the red seabream during its spawning season. Fish Physiol. Biochem. 28, 439-440.

Pankhurst, N. W. (1985). Final maturation and ovulation of oocytes of the goldeye, *Hiodon alosoides* (Rafinesque), in vitro. Can. J. Zool. 63, 1003-1009.

Pankhurst, N. W. (1987). In vitro steroid production by ovarian follicles of orange roughy (*Hoplostethus atlanticus* Collett), from the continental slope off New Zealand. In: Idler, D. R. , Crim, L. W. , Walsh, J. M. , Reproductive Physiology of Fish 1987. Memorial University of Newfoundland, St John's, 266.

Pankhurst, N. W. (1998). Reproduction. In: Black, K. D. , Pickering, A. D. , Biology of Farmed Fish. Sheffield Academic Press, Sheffield, 1-26.

Pankhurst, N. W. (1998). Further evidence of the equivocal effects of cortisol on in vitro steroidogenesis by ovarian follicles of rainbow trout *Oncorhynchus mykiss*. Fish Physiol. Biochem. 19, 315-323.

Pankhurst, N. W. (2001). Stress inhibition of reproductive endocrine processes in a natural population of the spiny damselfish *Acanthochromis polyacanthus*. Mar. Freshwat. Res. 52, 753-761.

Pankhurst, N. W. (2008). Gonadal steroids: functions and patterns of change. In: Rocha, M. J. , Arukwe, A. , Kapoor, B. G. , Fish Reproduction. Science Publishers, Enfield, NH, 67-111.

Pankhurst, N. W. (2011). Stress in fish: an environmental perspective. Gen. Comp. Endocrinol. 170, 265-275.

Pankhurst, N. W. , Dedual, M. (1994). Effects of capture and recovery on plasma levels of cortisol, lactate and gonadal steroids in a natural population of rainbow trout, *Oncorhynchus mykiss*. J. Fish. Biol. 45, 1013-1025.

Pankhurst, N. W. , King, H. R. (2010). Temperature and salmonid reproduction: implications for aquaculture. J. Fish. Biol. 76, 69-85.

Pankhurst, N. W. , Munday, P. L. (2011). Effects of climate change on fish reproduction and early life history stages. Mar. Freshwat. Res. 62, 1015-1026.

Pankhurst, N. W. , Porter, M. J. R. (2003). Cold and dark or warm and light:

variations on the theme of environmental control of reproduction. Fish Physiol. Biochem. 28, 385-389.

Pankhurst, N. W., Van Der Kraak, G. (1997). Effects of stress on growth and reproduction. In: Iwama, G. K., Pickering, A. D., Sumpter, J. P., Schreck, C. B., Fish Stress and Health in Aquaculture. Cambridge University Press, Cambridge, 73-93.

Pankhurst, N. W., Van Der Kraak, G. (2000). Evidence that acute stress inhibits ovarian steroidogenesis in rainbow trout in vivo, through the action of cortisol. Gen. Comp. Endocrinol. 117, 225-237.

Pankhurst, N. W., Van Der Kraak, G., Peter, R. E. (1995). Evidence that the inhibitory effects of stress on reproduction in teleost fish are not mediated by the action of cortisol on ovarian steroidogenesis. Gen. Comp. Endocrinol. 99, 249-257.

Pankhurst, N. W., Fitzgibbon, Q. P., Pankhurst, P. M., King, H. R. (2008). Habitat-related variation in reproductive endocrine condition in the coral reef damselfish *Acanthochromis polyacanthus*. Gen. Comp. Endocrinol. 155, 386-397.

Pankhurst, N. W., Fitzgibbon, Q., Pankhurst, P., King, H. (2008). Density effects on reproduction in natural populations of the staghorn damsel *Amblyglyphidodon curacao*. CYBIUM. 32 (Suppl. 2), 297-299.

Pankhurst, N. W., King, H. R., Anderson, K., Elizur, A., Pankhurst, P. M., Ruff, N. (2011). Thermal impairment of reproduction is differentially expressed in maiden and repeat spawning Atlantic salmon. Aquaculture. 316, 77-87.

Patino, R., Thomas, P. (1990). Induction of maturation of Atlantic croaker oocytes by 17α, 20β, 21-trihydroxy-4-pregnen-3-one in vitro-consideration of some biological and environmental variables. J. Exp. Zool. 255, 97-109.

Perry, A. N., Grober, M. S. (2003). A model for the social control of sex change: interactions of behaviour, neuropeptides, glucocorticoids, and sex steroids. Horm. Behav. 43, 31-38.

Perry, S. F., Gilmour, K. M. (1999). Respiratory and cardiovascular systems during stress. In: Balm, P. H. M., Stress Physiology in Animals. Sheffield Academic Press, Sheffield, 52-107.

Pickering, A. D., Pottinger, T. G. (1987). Poor water quality suppresses the cortisol response of salmonid fish to handling and confinement. J. Fish. Biol. 30, 363-374.

Pickering, A. D., Pottinger, T. G., Carragher, J., Sumpter, J. P. (1987). The effects of acute and chronic stress on the levels of reproductive hormones in the plasma of mature male brown trout, *Salmo trutta* L. Gen. Comp. Endocrinol. 68, 249-259.

Planas, J. V., Swanson, P. (2008). Physiological function of gonadotropins in fish.

In: Rocha, M. J., Arukwe, A., Kapoor, B. G., Fish Reproduction. Science Publishers, Enfield, NH, 37-66.

Pottinger, T. G. (1999). The impact of stress on animal reproductive activities. In: Balm, P. H. M., Stress Physiology in Animals. Sheffield Academic Press, Sheffield, 130-177.

Pottinger, T. G., Carrick, T. R. (2000). Indicators of reproductive performance in rainbow trout *Oncorhynchus mykiss* (Walbaum) selected for high and low responsiveness to stress. Aquacult. Res. 31, 367-375.

Pottinger, T. G., Pickering, A. D. (1990). The effect of cortisol administration on hepatic and plasma estradiol-binding capacity in immature female rainbow trout (*Oncorhynchus mykiss*). Gen. Comp. Endocrinol. 80, 264-273.

Pottinger, T. G., Campbell, P. M., Sumpter, J. P. (1991). Stress-induced disruption of the salmonid liver-gonad axis. In: Scott, A. P., Sumpter, J. P., Kime, D. E., Rolfe, M. S., Reproductive Physiology of Fish 1991. FishSymp 91, Sheffield, 114-116.

Pottinger, T. G., Yeomans, W. E., Carrick, T. R. (1999). Plasma cortisol and 17β-oestradiol levels in roach exposed to acute and chronic stress. J. Fish. Biol. 54, 525-532.

Reddy, P. K., Renaud, R., Leatherland, J. F. (1999). Effects of cortisol and triiodo-l-thyronine on the steroidogenic capacity of rainbow trout ovarian follicles at two stages of oocyte maturation. Fish Physiol. Biochem. 21, 129-140.

Romero, L. M., Reed, J. M., Wingfield, J. C. (2000). Effects of weather on corticosterone responses in wild free-living passerine birds. Gen. Comp. Endocrinol. 118, 113-122.

Rotllant, J., Ruane, N. M., Dinis, M. T., Canario, A. V. M., Power, D. M. (2006). Intra-adrenal interactions in fish: catecholamine stimulated cortisol release in sea bass (*Dicentrarchus labrax* L.). Comp. Biochem. Physiol. 143A, 375-381.

Safford, S. E., Thomas, P. (1987). Effects of capture and handling on circulating levels of gonadal steroids and cortisol in the spotted seatrout, *Cynoscion nebulosus*. In: Idler, D. R., Crim, L. W., Walsh, J. M., Reproductive Physiology of Fish 1987. Memorial University of Newfoundland, St John's, 312.

Schrandt, M. N., Lema, S. C. (2011). Habitat-associated intraspecific variation in behavior and stress responses in a demersal coral reef fish. Mar. Ecol. Prog. Ser. 443, 153-166.

Schreck, C. B. (2010). Stress and fish reproduction: the roles of allostasis and hormesis. Gen. Comp. Endocrinol. 165, 549-556.

Schreck, C. B., Conteras-Sánchez, W., Fitzpatrick, M. S. (2001). Effects of stress on fish reproduction, gamete quality and progeny. Aquaculture. 197, 3-24.

Schulz, R. W., de Franca, L. R., Lareyre, J.-J., Le Gac, F., Chiarini-Garcia, H., Nobrega, R. H. (2010). Spermatogenesis in fish. Gen. Comp. Endocrinol. 165, 390-411.

Shahjahan, Md, Kitahashi, T., Ogawa, S., Parhar, I. S. (2013). Temperature differentially regulates the two kisspeptin systems in the brain of zebrafish. Gen. Comp. Endocrinol. 193, 78-85.

Siklenkia, K. (2015). Disruption of histone methylation in developing sperm impairs offspring health transgenerationally. Science. 350, 651.

Sloman, K. A., Armstrong, J. D. (2002). Physiological effects of dominance hierarchies: laboratory artefacts or natural phenomena?. J. Fish. Biol. 61, 1-23.

Soria, F. N., Strüssman, C. A., Miranda, L. A. (2008). High water temperatures impair the reproductive ability of the pejerrey fish *Odontesthes bonariensis*: effects on the hypophyseal-gonadal axis. Physiol. Biochem. Zool. 81, 898-905.

Soso, A. B., Barcellos, L. J. G., Ranzani-Paiva, M. J., Kreutz, L. C., Quevedo, R. M., Lima, M. (2008). The effects of stressful broodstock handling on hormonal profiles and reproductive performance of *Rhamdia quelen* (Quoy & Gaimard) females. J. World Aquacult. Soc. 39, 835-841.

Staples, C. A., Tilghman Hall, A., Friederich, U., Caspers, N., Klecka, G. M. (2011). Early life-stage and multigeneration toxicity study with bisphenol A and fathead minnows (*Pimephales promelas*). Ecotoxicol. Environ. Saf. 74, 1548-1557.

Sumpter, J. P., Carragher, J. F., Pottinger, T. G., Pickering, A. D. (1987). Interaction of stress and reproduction in trout. In: Idler, D. R., Crim, L. W., Walsh, J. M., Reproductive Physiology of Fish 1987. Memorial University of Newfoundland, St Johns, 299-302.

Takeo, J., Hata, J. H., Segawa, C., Toyohara, H., Yamashita, S. (1996). Fish glucocorticoid receptor with splicing variants in the DNA binding domain. FEBS Lett. 389, 244-248.

Tasker, J. G., Di, S., Malcher-Lopes, R. (2006). Minireview: rapid glucocorticoid signalling via membrane-associated receptors. Endocrinology. 147, 5549-5556.

Teitsma, C. A., Anglade, I., Lethimonier, C., Le Dréan, G., Saligaut, D., Ducouret, B. (1999). Glucocorticoid receptor immunoreactivity in neurons and pituitary cells implicated in reproductive functions in rainbow trout: a double immunohistochemical study. Biol. Reprod. 60, 642-650.

Thomas, P. (2003). Rapid, nongenomic steroid actions initiated at the cell surface: lessons from studies with fish. Fish Physiol. Biochem. 28, 3-12.

Thomas, P., Rahman, M. S. (2009). Chronic hypoxia impairs gamete maturation in

Atlantic croaker induced by progestins through nongenomic mechanisms resulting in reduced reproductive success. Environ. Sci. Technol. 43, 4175-4180.

Thomas, P., Rahman, M. S., Kummer, J. A., Lawson, S. (2006). Reproductive endocrine dysfunction in Atlantic croaker exposed to hypoxia. Mar. Env. Res. 62, S249-S252.

Thomas, P., Rahman, M. S., Khan, I. A., Kummer, J. A. (2007). Widespread endocrine disruption and reproductive impairment in an estuarine fish population exposed to seasonal hypoxia. Proc. R. Soc. B. 274, 2693-2701.

Tveiten, H., Johnsen, H. K. (2001). Thermal influences on temporal changes in plasma testosterone and oestradiol-17β concentrations during gonadal recrudescence in female common wolfish. J. Fish. Biol. 59, 175-178.

Tveiten, H., Bjørn, P. A., Johnson, H. K., Finstad, B., McKinley, R. S. (2010). Effects of the sea louse *Lepeophtheirus salmonis* on temporal changes in cortisol, sex steroids, growth and reproductive investment in Arctic charr *Salvelinus alpinus*. J. Fish. Biol. 76, 2318-2341.

Ünal, G., Erdoğan, E., Oağuz, A. R., Kaptaner, B., Kankaya, E., Elp, M. (2008). Determination of hormones inducing oocyte maturation in *Chalcalburnus tarichi* (Pallas, 1811). Fish Physiol. Biochem. 34, 447-454.

Upadhyaya, N., Haider, S. (1986). Germinal vesicle breakdown in oocytes of catfish, *Mystus vittatus* (Bloch): Relative in vitro effectiveness of estradiol-17β, androgens, corticosteroids, progesterone, and other pregnene derivatives. Gen. Comp. Endocrinol. 63, 70-76.

Van Der Kraak, G. J., Munkittrick, K. R., McMaster, M. E., Portt, C. B., Chang, J. P. (1992). Exposure to bleached kraft pulp mill effluent disrupts the pituitary-gonadal axis of white sucker at multiple sites. Toxicol. Appl. Pharmacol. 115, 224-233.

Wang, D.-S., Kobayashi, T., Zhou, L.-Y., Paul-Prasanth, B., Ijiri, S., Sakai, F. (2007). Foxl2 up-regulates aromatase gene transcription in a female-specific manner by binding to the promoter as well as interacting with Ad4 binding protein/steroidogenic factor 1. Mol. Endocrinol. 21, 712-725.

Wang, S., Yuen, S. S. F., Randall, D. J., Hung, C. Y., Tsui, T. K. N., Poon, W. L. (2008). Hypoxia inhibitis fish spawning via LH-dependent final oocyte maturation. Comp. Biochem. Physiol. 148C, 363-369.

Wingfield, J. C. (1994). Modulation of the adrenocortical response to stress in birds. In: Davey, K. G., Peter, R. E., Tobe, S. S., Perspectives in Comparative Endocrinology. National Research Council of Canada, Ottawa, 520-528.

Wu, R. S. S., Zhou, B. S., Randall, D. J., Woo, N. Y. S., Lam, P. K. S.

(2003). Aquatic hypoxia is an endocrine disruptor and impairs fish reproduction. Environ. Sci. Technol. 37, 1137-1141.

Yamaguchi, T., Yamaguchi, S., Hirai, T., Kitano, T. (2007). Follicle-stimulating hormone signalling and foxl2 are involved in transcriptional regulation of aromatase gene during gonadal sex differentiation in Japanese flounder, *Paralichthys olivaceus*. Biochem. Biophys. Res. Comm. 359, 935-940.

Yamaguchi, T., Yoshinaga, N., Yazawa, T., Gen, K., Kitano, T. (2010). Cortisol is involved in temperature-dependent sex determination in the Japanese flounder. Endocrinology. 151, 3900-3908.

Yu, K. L., Peter, R. E. (1992). Adrenergic and dopaminergic regulation of gonadotropin-releasing hormone release from goldfish preoptic-anterior hypothalamus and pituitary in vitro. Gen. Comp. Endocrinol. 85, 138-146.

Zohar, Y., Muñoz-Cueto, J. A., Elizur, A., Kah, O. (2010). Neuroendocrinology of reproduction in teleost fish. Gen. Comp. Endocrinol. 165, 438-455.

第 9 章　认知、学习和行为

我们综述鱼类中行为与应激之间的关联性及因果关系。我们以应激下的生态与进化角度进行观察与实验研究为基础，将应激生理作用和鱼类行为的研究关联起来。鱼类行为的理论模型，包括实验与比较心理学，可帮助我们理解生理以何种方式影响或指导行为，相反地，行为也会影响生理。目前研究最有成效的领域是进行生理、应激和行为的整合研究，包括多种内容，如觅食和摄食行为、迁移、学习、育幼与社会行为以及生活史模式。对于其他模式鱼类的更广泛的研究加深了我们对分子遗传学机制和进化生态学推论的认识。我们考虑最优化概念的中心作用及如何将应激的生理和行为联系起来。生理学的最优化指的是近期的因果关系。而对行为来说，最优化更多是指最终的、进化的结果。我们会展示最优化如何综合考虑应激的近期影响和最终作用。我们提出可能的未来研究方向，这些研究方向会继续增强我们对鱼类行为和应激近期影响及最终作用的理解。

9.1　应激如何影响行为和行为如何影响应激

鱼类①必须应对复杂的不断变化的外部环境。一些环境的物理参数，如全球大洋中的深海海域的周年变化可能要比北半球温带池塘的浅水，或是热带珊瑚礁海域，又或是淡水溪流的周年变化要小得多（图 9.1）。尽管如此，生物环境一直在变化，而且经常是不可预测的，如捕食者是否存在、食物是否充裕及同一物种是否临近（Jones，2005，2007）。同时，它们必须面对复杂的内部环境及在耐受范围内调节无数的生理参数（Hoar，1966）。如果可能的话，鱼类也必须要处理内外环境、生存、繁殖等所有方面（Dawkins，1976）。考虑鱼类通过什么方式反映这些外部和内部环境变化来对行为进行操作性的定义是必要的。

鱼类行为是一个复杂的命题（Noakes，Baylis，1990），但是我们采用了一种与生理和应激有关的可预测的方法。我们对于行为研究的方法主要基于下述假说，我们在鱼类中观察到行为反应是适应性的，因为它们是通过自然选择形成的。这与 Dobzhansky（1973）的名言“如果不从进化论的角度分析问题，生物学的一切都毫无意义”是一致的，而这种基本的假说已经用于解释一些动物行为的大范围复杂性（Hamilton，1998；Ridley，2003）。

现今存活的鱼都有一个完整的展示其如何生存和繁衍而来的进化路径，因此，我

---

① 我们在后述中使用“鱼”来代表一种鱼中的一个或多个个体，而“鱼类”则代表多种鱼中的一个或多个个体。

们认为当今我们观察到的行为一定是反映其成功的一面（如，行为是适应性的）。总结一个特定物种的生活史策略就像是做简要的概览。当然现存鱼类肯定是多样性的，它们和已知化石形式记录的鱼类一起展示出可能出现的许多种策略（Dawkins，2009）。

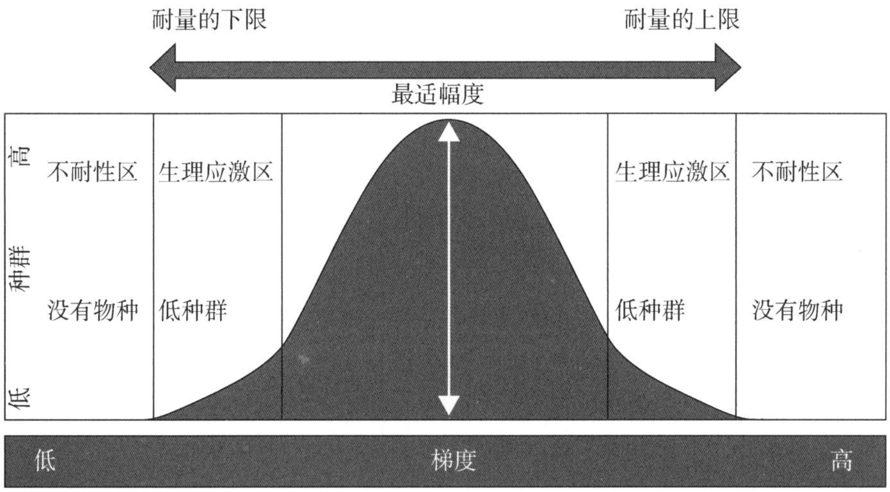

图 9.1 假定鱼类处在一个特定的环境梯度，以温度为例的耐受范围、最适范围和应激范围的示意图

注：修改自 McFarland D（1999）. Animal Behaviour. New York，NY：Prentice-Hall，with permission。所表现的鱼个体数以及群体数在耐受极限时最少，而在最适值时最多。

传统上都认为动物的行为是生理的结果，例如，睾酮水平上升是因，好斗的行为是果（Fernald，1976）；甲状腺素水平升高是因，运动能力提高是果（Hoar，1953）。我们认识到这种因果关系，此外，我们将会考虑到行为能够影响生理机能这个观点，而且我们将会强调行为如何作为我们通常认为的应激反应的表现形式。

我们将要综述鱼类中应激、认知、学习及行为的关系。我们的结论就是如果我们把重点放在行为的适应性本质这一背景条件下，我们就可以对鱼类中应激、认知、学习及行为的关系有更好的理解（Spagnoli 等，2016）。我们与其试图将可能引起应激的所有条件归类，还不如对大量文献进行综述并考虑到对应激源的反应行为是适应性。再者就是，我们得出结论，由于任何鱼类都有着许多互相冲突的需求，尽管鱼类的行为有着很明显的适应性，但任何鱼类无一例外都会受到应激的影响。

9.2 最优化、偏好性及决策

大量文献中，有许多对应激进行了定义（Schreck，Tort，2016；见本书第1章）。"应激源（应激物）是任何倾向于扰乱鱼内部生理状态体内平衡（稳态）的事物"，这是我们认定的应激源的一般定义（McFarland，1999）。内部或外部应激源导致这种扰乱的产生。鱼类生理应激反应的详细内容已在本书的其他章节描述，我们在此不再赘述。然而，我们确实应考虑到不良应激（痛苦）与良性应激（真正应激）间的显著差别，这一差异是由各不同个体的行为变化而体现出来的更加全面的内容。我们认为鱼类的耐受极限和外部与内部的波动（实际的和潜在的应激源）是密切相关的。一些人为的外部应激源会导致某些鱼类的死亡（Paetzold 等，2009）

我们假设以温度作为单个物理影响因子。近年来有大量文献报道关于温度对鱼类生理与行为的影响（Coutant，1997；Jobling，1981）。多年来，我们一直意识到这一点，部分原因是我们可以持续地对温度进行监测（Beamish，1970），但更重要的是因为温度对任何鱼的新陈代谢都有非常重大的影响（Brett，1971；Wood，MacDonald，1997；Allan 等，2015；Lawrence 等，2015）。我们假定对于一尾特定的鱼来说，它所处环境的外部温度可在致死低温到致死高温之间变化。已经明确鱼不能在高于致死高温或低于致死低温环境下存活（图 9. 2）。在耐受范围内，在致死低温和致死高温之间，一尾单个的鱼会有最适生存温度（Brett，1952），而这一最适生存温度会受热经历的影响（即驯化）。一尾鱼所选择的温度会受到应激的影响。研究表明，应激的斑马鱼会对温度过热做出反应（Rey 等，2015），而且一些物种的生长反应也会受到应激的影响而发生改变（Eldridge 等，2015）。

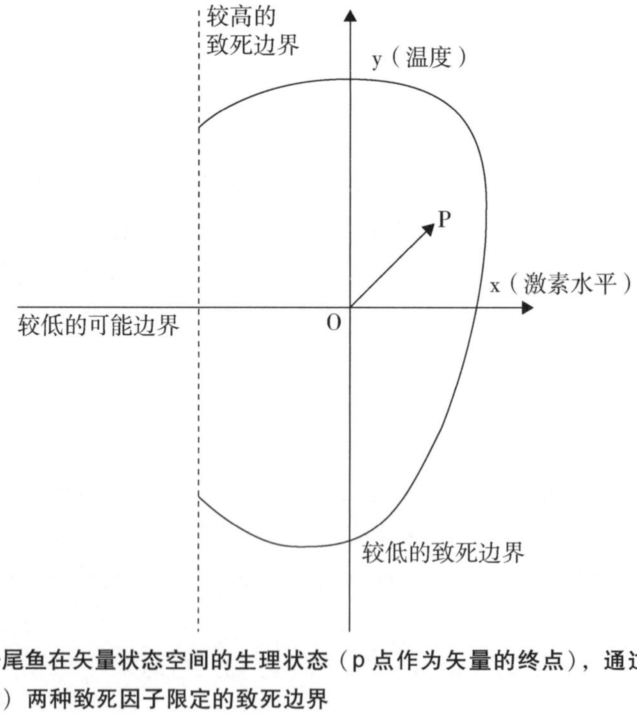

图 9. 2 一尾鱼在矢量状态空间的生理状态（p 点作为矢量的终点），通过温度（y 轴）与激素水平（x 轴）两种致死因子限定的致死边界

注：O 值代表温度和激素水平共同的最优值。生理状态可能是这两种因子致死边界内的任何位置。根据 McFarland D（1999）. Animal Behaviour. New York，NY：Prentice-Hall 修改，允许引用。

考虑到生理学和行为学的需要，最优化这一概念在近年来已经被大量提及。在鱼类中，我们认为对行为和应激进行综合考虑的核心就是最优化。我们意识到鱼类在生理方面或者是顺应型（如，身体内部条件依随着外部条件）或者是调节型（如，在一定的极限范围内，它们维持内部条件相对独立于外部条件）（Hoar，1966）。一些鱼类可在一定范围内调节它们的体温，这不会影响我们所描述事物的一般性质（Carey 等，1971；Block，Finnerty，1994；Wegner 等，2015）。实际上，最适温度肯定是在一个相当狭窄的温度范围内（Jobling，1981；McCauley，Casselman，1981）。在最适温度下，鱼的生理表现在活力范围内将会是最优的（Fry，1947；Beamish，1970）。在鱼的最适温度范围以上或以下，它的生理应激范围也是随之增加的（Iwama 等，2011）。那些在生理上或者行为上被认为是恶劣的条件，是因为它们都会导致应激反应。

所谓的“恶劣”，意味着鱼类会或者应该试图避免这些状况。作为这种观点的一部分，经常做出的假设是一尾鱼偏好的温度应该是相当于最适温度（Hazel，Prosser，1974；Beitinger，Fitzpatrick，1979；McCauley，Casselman，1981）。如果温度仅稍微低于或高于最适温度，鱼类个体可能会通过生理反应来进行补偿（通常称为稳态反应）（Hazel，Prosser，1974；Beitinger，Fitzpatrick，1979；Ott 等，1980；Pankhurst，1997，2011）。但生理的补偿也会有一定的限度（McFarland，1999）。这种反应可能和血液流向不同组织模式的改变或新陈代谢活性的改变有关（Beitinger，Fitzpatrick，1979）。

这些生理反应和行为反应或补偿效应相结合将会使鱼的温度恢复到最适状态（图9.3）。例如，鱼可能会迁移到水体中更接近最适环境的水域（行为的温度调节，温度偏好性）（Coutant，1987）。如果我们现在考虑两种外部环境因子（例如，温度和 pH），我们就可以开始了解任何鱼需要维持最优内部生理状态所面对的复杂状况。一尾鱼可能调节它的内部温度至接近最适温度，但如果环境中 pH 出现时空上的变化，那么鱼亦将要对这一改变做出反应。

如果将我们考虑的范围扩展到包括在任何时间内影响一尾鱼的许多可能的因素或者内部和外界刺激（如温度、溶解氧、pH、血液激素水平、捕食者的存在、附近的食物、盐度等），很明显，个体要实现最优的体内生理状态面临的是一个复杂的困难局面（Takei，Hwang，2016；见本书第6章）。例如，行为能立刻驱使动物靠近最适温度条件而同时会从最适条件移到有遮蔽物的地方。然而尽管未经训练的观察者可能会声称，“不做任何事情”从来不是一条鱼的选择。它一定经常会“做一些事情”，这些事情就是我们所认为的行为。鱼在任何一个时刻所做的事情都可以认为是与动物状态相关的行为决策的后果（Houston，McNamara，1999）。因此，鱼在任何时间点上的行为几乎都不可避免地是为了达到最佳条件而在相互冲突的需求方面妥协。（Sneddon 等，2016；见本书第12章）。当动物有机会在互不相容的活动中进行改变时，我们就有可能观察到动物的行为反应（Noakes，1986a，b）。动物对行为活动优先级别的评估排序是基于动物从事某个特定活动需要付出多少努力这一标准，而这一标准也更具有说服力（Dawkins，2003）。

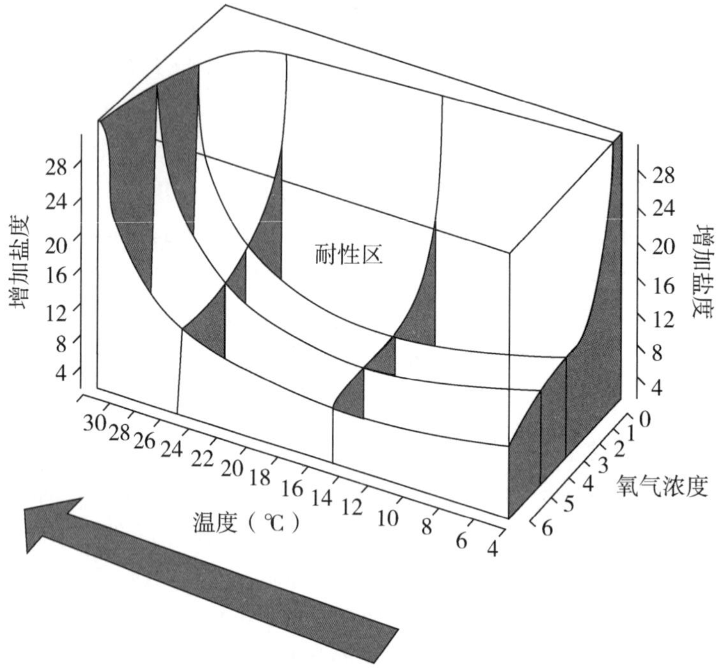

图 9.3　一尾鱼受到盐度、氧气和温度的组合因素产生的生理反应表面示意图

注：修改自 McLeese（1956）。通过三维表面界定鱼对这些环境因子组合的耐受力。反应表面外的状况对鱼是致死的。

鱼类的行为将会约束在所谓的最终普通行为通道中（McFarland，1999）。除极少数情况外，一尾鱼一次仅能做一件事。例如，一尾雄性刺鱼（stickleback）可能会摄食，但不可能同时保卫自己的领土、筑巢、求偶（Noakes，1986a，b）。一尾鱼的行为可能是作用于降低一种外部（或内部）因素所引起的应激，但是它可能不会同时发生行为去降低同一时间由任何众多的其他内部或外部因素所引起的应激。

我们也注意到，在任何时间点，鱼类个体的行为都是一个决策的后果，而这个决策应该受到动物年龄和发育状态（如未成熟幼鱼与性成熟成鱼），及内部与外部环境条件相互作用的影响。动物内部状态发生的变化，取决于先前的经历（参照我们先前所讨论的各种认知），还有发育的条件。一个明显的例子就是鲑科鱼类稚鱼（smolts）的洄游活动，从淡水洄游到海水环境（Seals Price，2003）。鱼类从淡水到海水的洄游是非常有意义的生态的、生理的和行为的转变（McCormick 等，2013）。有很多这种鱼类，以及鲑鱼类（鲑属；大麻哈鱼属）已经被作为对象研究稚鱼从淡水到海水的运动，以及成鱼到由海水洄游到淡水产卵的反向运动（Boeuf，1993；Romer 等，2013；Thompson 等，2015）。鲑鱼的溯河洄游是个体根据其发育状态、年龄、外部环境因素（一年中的时间、水温、水流）和内部条件（性别、大小、生长史）综合做出的主动、自愿的行为决定（Thompson 等，2015）。

还有许多其他鱼类的例子，这些鱼类的行为反应是内部因素作用的结果，包括亲本的影响（Allan 等，2014），还结合到外部环境因子的作用（Domenici 等，2014；Donelson 等，2014；Allan 等，2015；Ferrari 等，2015）。这些例子中有些我们认为是属于自然状态的，而其他的则被认为是预示着海洋条件的改变（如酸化和变暖）。这些情况中我们所观察到的行为都是相同过程的结果。

在极端情况下，个体应对特定刺激的行为会是一种很明显和突然的方式。传统的例子就是在许多研究中的应急状态（搏斗或逃逸）（Hertz 等，1982；Jansel 等，1995；Dawkins，2012；Adamo，2014；Baciadonna，McElligott，2015；Domenici，Ruxton，2015；Wiseman，2015）。这些行为的其他选择必定是明显的，并且可能的情况就是在这一形势下该个体要么没有替代的选择，要么替代选择的受益（代价）是最终的（如保护幼者、逃离捕食者）。我们把这些作为一个极端连续性的例子，稍后会对其进行阐述。

行为是鱼类与环境相互作用所采取的一种重要的、直接的方式。很明显，现在任何一尾活鱼一定是已成功适应一系列早期环境条件的祖先的后裔。鱼类对内部或外部环境变化的反应通过长期自然选择已经形成。有一段混乱的命名历史和鱼类对它们环境的潜在反应相关（Reebs，2001；Keenleyside，1997）。鱼的行为反应可以方便地认为是一种行为模式，有时被认为是物种的习性（Manning，Dawkins，2012）。

有些人将习性定义为先天的，或遗传决定的，或是鱼类天生的行为活动。很清楚的是，自然选择一定会影响鱼类的行为，就像自然选择会影响鱼类的生理或形态那样，把行为简单地认为是先天的或者学习得到的是严重的误导。先天与学习二分法的谬误早已被认识并已烟消云散（Dawkins，2004）。

最值得我们关注的重要差别是，当我们考虑到行为的时候，我们一定要想到行为差异的产生机制。两尾鱼之间在某一行为上的差异，例如是迁移到海洋而不是停留在淡水中，是这些个体间遗传差异导致的，还是它们所处的环境差异导致的呢？在更为现实的情况下，个体间的差异很可能是遗传和环境差异相互作用的结果（Manning，Dawkins，2012）。我们认为一尾鱼的典型行为，就是当一个物种处在通常的生存条件下，是其祖先在自然选择压力下的适应性反应的表达（Charnov，1993；Houston，McNamara，1999；Dawkins，2009）。根据定义，行为策略决定鱼在每个时间点的行为（Houston，McNamara，1999）。

有时有一种倾向，认为动物对我们所定义或认为的应激刺激或环境的反应是不良适应的（Iwama 等，2011；Careau 等，2014）。目前有大量关于鱼类中引起反应的应激源及应激（和不良应激）的文献（Noakes，Leatherland，1977；Young 等，2010；Laursen 等，2011；Careau 等，2014）。对生理性应激与行为之间的关系已经做过多次综述（Schreck 等，1997；Yue 等，2006）。长时间或极端应激条件会给一尾鱼带来负面的影响，这个观点是肯定正确的。所以，我们可能同样会认为应激中的行为反应是适应性的，因为它们能够减轻鱼的应激。

人们普遍接受的是应激对鱼的影响是负面的，无论是水族馆驯养的鱼还是自然环境下的野生鱼类（Ejike，Schreck，1980；Conte，2004；Cachat 等，2011；Galhardo

等，2011；Cook 等，2014；Monaghan，Spencer，2014；Pavlidis 等，2015）。但至关重要的是，区分应激与不良应激（Selye，1975，1976a，b，c；Szabo 等，2012）。更确切地说，是要区分不良应激（痛苦）与良性应激（真正应激）（Selye，1978；LeFever 等，2003；Hargrove 等，2015）。Selye 的话最能说明这一点，“应激中没有不良应激”（Selye，1976c）。我们认为例子中的极端情况是不良应激对个体产生的负面影响。从个体的角度讨论，良性应激的影响是正面的，这部分将在其他章节综述。良性应激是长期的，因为个体会一直受到有可能出现冲突的外部与内部的影响，这也将会影响任何时间点的行为（Selye，1976c）。毫无疑问，任何一种极端的或长期的、物理的或生理的条件都会引起鱼的损伤和疾病。例如，暴露在有毒化学物质下会引起鱼类的生理性的不良应激（Paetzold 等，2009；Grassie 等，2013）。我们没有将这些情况及它们对鱼的影响纳入应激所产生的适应性反应的范畴当中。我们同意不良应激对鱼一定是负面的，而且应该避免或尽量减少不良应激。现在还不明确的是应激、短时的不良应激对鱼的影响是否必然是负面的，以及是否至少有一些理论可以预测某些应激在任何时间都是不可避免的。我们将把鲑科鱼类作为模式物种来详细考虑这种可能性。

9.3 模式动物：鲑鱼

我们把鲑科鱼类的行为作为一种模型是基于很多原因的。目前有大量关于鲑科鱼类的生物学、行为学、生态学和进化方面的文献（Magnan 等，2002；Quinn，2005；Piccolo 等，2014；Woo，Noakes，2014）。文献涵盖我们所要考虑的物种自然环境下全面的生理学与行为学。为方便起见，我们希望使用鲑科鱼类生活史，但也要承认在方便的同时，不同鱼在属、种和当地种群中的差异是显著的。我们所使用的鲑科鱼类生活史是溯河产卵的，终生一胎的，生活在北半球固有的范围内（Fraser 等，2011）。

鲑鱼生物学要比考虑人工孵化的鲑鱼和野生的鲑鱼间的差异以及导致这种差异的原因可能更加没有异议（Fagerlund 等，1981；Noakes，Corrarino，2010；Noakes，2014）。有大量描述性的研究记录野生鲑鱼与人工孵化鲑鱼的差异，包括和应激源与应激反应有关的差异（Fenderson 等，1968；Woodward，Strange，1987；Ruzzante，1994；Jonsson，Jonsson，2006；Matsuda，2013；Goetz 等，2015）。然而，这些差异是定量的表现而非定性的，因此，我们将野生鲑鱼与人工孵化鲑鱼对应激源的行为反应作为例子涵盖在其中。

如果说鲑科鱼类的一生毫无疑问要去面对许多应激源，那么大量鱼类溯河产卵的迁移一定在应激源排名中名列前茅。激素一直被认为和鲑鱼的生理性应激有关，这些激素可参与这些物种正常生活史中的行为事件（Carruth 等，2012）。迁徙的鲑鱼一定会遭遇的自然环境改变包括盐度、温度、水流及捕食者的出现（Hodgson 等，2006；Hayes 等，2012；Romer 等，2013；Kavanagh，Olson，2014；Melnychuk 等，2014）。此外，在鱼类迁徙过程中有大量人为因素导致应激的例子（Danley 等，2002）。当然，太平洋鲑鱼（大麻哈鱼属）生活史的典型特征就是它们的几乎自杀式的终生一胎行为（Quinn，2005）。并不使我们感到意外的是，有大量关于这些鲑科鱼类应激反应的文献（Contreras-Sanchez 等，1998；Schreck，2000；Iwama 等，2011）。

9.4 鱼类中学习与应激的关系

9.4.1 学习、可塑性与问题解决

如果从长期来看鱼一生中的改变（应激源）是可预测的，那么就会导致进化上的适应性。若个体生活中相对较短时期内的环境变化是可预测的，那么鱼将会以我们认为的适应性方式——学习来做出反应（Hollis，1982；Houston，McNamara，1999；Braithwaite，2010）。在我们考虑应激对鱼类学习的影响之前，我们首先需要对学习做一个定义，这一定义要适用于大范围内各种不同的生态和生理条件下的多个不同物种。我们这里将使用 W. E. Thorpe 和其他个体生态学家提出的学习的定义（Manning，Dawkins，2012）。我们也注意到其他关于行为的观点包括不同的可使用的学习定义和行为的其他适应性方面（Houston，McNamara，1999）。我们所使用的定义：学习是个体行为的适应性改变，是经历所产生的结果。这一定义的第一个关键在于，行为的改变是适应性的，换句话说，动物会因学习而获得一些优势（短期的或长期的）。在这里，鱼的学习将会降低应激反应。第二个关键在于，行为的改变是个体发生改变（而非群体，或是连续的世代）。第三个关键在于，行为的改变是经历所产生的结果（而非单纯生长或成熟）。

学习经常通过相对简单的分类系统进行分类，这种分类方法并不能真实反映这种现象的复杂性。为了方便，我们还是使用这套简单的分类系统，而且它也体现了绝大多数鱼类中学习反应的研究结果。我们用来描述学习的术语如下：习惯化、经典条件反射、工具条件反射或印记。

习惯化经常被认为是最简单的学习形式。习惯化很容易被定义为鱼对反复的中性刺激的反应下降，中性刺激是指初始的能激发鱼的反应（Manning，Dawlins，2012）。中性刺激可以是鱼从视觉、嗅觉或触觉中探测到的几乎任何刺激。习惯化很明显是适应性的，因为动物停止对没有明显影响的刺激做出反应（Nilsson 等，2012）。起初，鱼对我们所讨论的应激源（例如，一些环境特征的突然改变，如经过的阴影，或低频声音）的刺激产生反应。鱼的反应可能是生理反应或行为反应。于是可能出现心率提高、激素突然释放、远离刺激源等反应。然而，当刺激和鱼无关时，鱼将对频繁发生的这一刺激（应激源）停止反应。习惯化是频繁的和普遍的，甚至我们考虑以它们作为学习的例子。我们认为鱼“习惯于”任何一系列突发的环境刺激或改变。然而，习惯化对于鱼的应激反应来说又是极其重要的，可能比其他任何一种形式的学习都要重要。事实上，外部环境中能被一尾鱼检测到的万事万物都可能产生显著影响，这些事物或者是应激源，或者将会成为应激源的一些

事物的指标。但来自外部环境的大部分事物对鱼没有影响，至少在大多数时间是这样的。

实质上，习惯化是我们和其他动物区分有意与无意的相关关系的方法。可能是因果关系，但一般来说这不是关键的问题。我们需要知道所有特定的环境信号或线索是否都和应激源相关。如果有关，想必运作的规则（策略）是通过生理的或行为的反应进行恰当的反应以最小化应激源的影响，并让个体的生理状态恢复到最优条件。如果没有关联，运作规则会忽略这种信号或线索。较为普遍的现象是已被大家所了解的刺激过滤，过滤刺激是动物选择性反应相关刺激的方法，这些相关刺激来自非常大量的潜在信息，动物个体对这些海量信息是无法一一处理的。鉴于一尾鱼可检测到来自环境的一些事物，习惯化将会是限制动物对相关信息的反应的方法。在公共水族馆的鱼对水族馆附近大量的人类活动缺乏反应就是一个很明显的例子。

经典条件反射，有时指的是巴甫洛夫的条件反射，亦即有明显的适应性，但不像习惯性，这种条件涉及动物将两种先前无任何关联的刺激联系起来（Bitterman，1964；Hollis 等，1989）。一个和动物无关的初始刺激或信号却产生了对动物有意义的后果（这里指一个应激源）（Sneddon 等，2016；见本书第 12 章）。巴甫洛夫摇响铃铛，一只狗听到铃声并可能显示出对声音的一些反应（如转向声音、心率加速、远离声音），随后很快就会习惯（停止响应）重复的铃声。然而，如果铃响跟随着一个相关事件（如有肉碎在口中），狗很快就会通过预期的获得肉碎（分泌唾液）而对重复铃声进行反应。很快狗就会单独对铃声响应，分泌唾液。一尾鱼一旦发现头顶有阴影就会寻找遮蔽物，伴随着心率改变，或其他潜在对应激源的反应。如果将头顶出现阴影频繁与在渔场获得食物联系起来，那么很快鱼就会改变自己的反应方式，在头顶出现阴影的时候游到水面觅食。然而，如果头顶出现阴影的后果是捕食的鸟类潜入水中攻击鱼的话（Penaluna，2015），那么头顶阴影的频繁出现将会导致鱼游向遮蔽物，并出现一系列预期的对潜在捕食者攻击的应激反应。许多作者表明，这种特别的例子可以解释人工孵化的鲑鱼和野生鲑鱼之间的显著差异，以及当人工孵化的鲑鱼被释放到开放水域时所受的伤害（Thorpe，Morgan，1978；Pickering 等，1987；Wiley 等，1993；Olla 等，1994；Berejikian 等，2001；Brown 等，2003；Malavasi 等，2008）。

用温和的电击刺激心率这一经典条件反射已经被频繁地用来检测鱼类的视觉、嗅觉及其他外部刺激的感知能力（Offitt，1971；Beauchamp，Rowe，1977；Beauchamp 等，1979；Holland，Teeter，1981；Hawryshyn，McFarland，1987；Hawryshyn 等，1989；Morin 等，1989；Hawryshyn，1992，2000；Rodriguez 等，2005）。这甚至可能展示一些鱼类在经典条件反射下的竞争性反应（Thompson，Sturm，1965）。

Karen Hollis 和她的同事们给我们展示了大量鱼类在经典条件反射下对竞争或求爱活动产生预期反应的简明例子（Hollis，1997，1982；Hollis 等，1989，1995，1997）。鱼已经将一些其他的中性刺激（如闪电）与随后因闪电出现的领土入侵者联系在一起，并比入侵的鱼更快启动竞争性反应。结果就是，它们与入侵的懵懂个体相比处于更有优势的地位。毫无疑问，这种学习对鱼的行为具有非常深刻的影响，尤其对应激源的反应。我们还要强调的是，有一种微妙的可能性，在考虑鱼类对频繁的及可预测的事件产生生理的及行为的反应时，野生鱼和驯养的鱼将会有一定差异。

这种情况被描述为一种条件补偿反应（Siegel 和 Ramos，2002）。我们知道，稳态作为一种复杂的内部生理反应，可恢复（趋向）动物的生理状态到最优值（稳定的生理状态）（Gross，1998；McEwen，2005；Cooper，2008）。如果一只动物暴露在外部应激源下，那么它的稳态反应将中和应激源的反应作为补偿反应。例如，如果应激源可增加心率，那么稳态反应将会降低心率。如果特殊的应激源是重复的，尤其是以可预测形式重复时，那么动物可能会通过经典条件反射方式生理感应预知应激源。伴随该应激源重复出现次数的增加，作为预期补偿反应的结果，这一动物的应激反应将会逐步减弱（Siegel，Ramos，2002）。我们可能也会遇到自相矛盾的情况，就是在预知的情况下将动物重复暴露于一个可预知的应激源，可能会使动物不产生应激。例如，在一段时间内的每天同一时间规律地注射药物，最初我们会在动物中检测到明显的应激反应，但超过一定时间后应激反应会下降，以至于最终我们可能检测不到显著的反应。我们可能认为出现这种情况是因为动物对应激源的重复暴露产生了耐受性。但如果相同应激源以不可预测的方式或不可预测的时间出现，那么我们预期的动物应激反应和初始暴露时候的反应一样大。很明显我们通过动物在自然状态下的反应准确预测了条件补偿反应的趋势。这种关键特征再次作为应激源的可预知性，与之相伴的是经典条件反射所产生的后果（Ferno 等，2011）。因此，学习，至少以经典条件反射的形式，在短期和长期情况下都能显著地降低应激。

工具条件反射（有时是指操作条件反射）是一种名为 Skinner 或 Skinner box 的学习形式（Mackintosh，1969；Ferno 等，2011；Agrillo 等，2012），尽管首先使用谜盒作为实验手段的是 Thordike（Burnham，1972）。有大量例子表明鱼类拥有学习执行一些活动的能力，例如推杆、游过一根管子、穿越光束而获得食物奖励、控制环境温度，乃至获得探访同伴的机会（Attia 等，2012）。利用由鱼操作的自动进料器已成为许多水产养殖系统的标准做法（Alanara，1992，1996；Boujard 等，1996；Millot 等，2011；Patton，Braithwaite，2015）。这些都是工具条件反射教科书式的范例（Sneddon 等，2016）。

对鱼类已有大量关于学习方面的研究，研究利用的是模式物种（一般用的是金鱼 *Carassius auratus*）。这些研究的目的通常是比较鱼的表现，尤其是学习方面，并和鸟类、哺乳类或其他模式物种进行比较（Mackintosh，1969；Ferno 等，2011）。虽然这些研究已经记录了鱼类的表现，但研究人员很少研究应激或有压力的环境对鱼类学习的影响。甚至有一些研究比较了不同鱼类在移动栖息地或者空间移动方面的学习能力（McAroe 等，2015）。这些比较可能会告诉我们一些关于比较意义中的学习状况，但这些比较不能直接解答一些和鱼类生活史生态学相关的特定的学习问题。然而，工具条件反射很可能会导致鱼类应激下降，因为它给动物提供环境改变的机会

(McAroe 等，2015)。

研究显示，应激与学习之间存在着大量可能的因果关系（Askey 等，2006；Ferrari 等，2012；Roche 等，2012；Eaton 等，2015；Mesquita 等，2015)。不出意外，这种影响是短暂的，但也有证据显示这种作用会延续给后代。例如，与未遭受应激影响亲本的子一代相比，受应激影响亲本的子一代三刺鱼（*Gasterosteus aculeatus*）的学习能力会降低（Mommer，Bell，2013)。类似的，雌性网纹花鳉（*Poecilia reticulata*）子代的学习反应和自发攻击会受到其母本是否受到温和应激处理的影响（Eaton 等，015)。这种应激与学习的关系是复杂的，在某种情况下可能是与直觉相悖的。例如，在有些情况下应激会增强学习（Larra 等，2014)，而在其他情况下应激会减损或阻碍学习（Zoladz 等，2011；Dominique，McGaugh，2014)。在大多数这样的例子中，我们尚未了解其作用的生理机制，但应激与学习的关系既不简单也不一致则是非常明确的。

一些鱼类对从受伤害的同类中释放的化学物质（Schreckstoff 信息素）的典型反应称为预警反应（Frisch，1942；Smith，1992)。预警反应是鱼对应激产生的一种反应。这种反应已经被作为标准方法检测鱼类应激对学习的影响（Smith，1992；Manek 等，2012)。饲养或蓄养条件可明显地影响鱼类（包括鲤科鱼类和鲑科鱼类）的预警反应。在一些情况下，预警反应的改变是鱼学习的结果，或在早期发育中至少在一段较长时间里起着环境影响的作用（Berejikian 等，1999；Manek 等，2012)。

在鱼生活史中，有一些例子提供大量关于学习的详细信息。关于印记的研究是阐述得最为清晰的。印记是和学习相联系的，而这种学习局限于鱼的生活早期一个较短的时期（敏感时期)，对该个体生活史的后期也有延续的影响。可能最好的例子就是大西洋鲑鱼和太平洋鲑鱼（大麻哈鱼属）的行为。我们通过对鲑科鱼类早期学习阶段的研究已经了解到在物种自然历史背景下的学习（Suboski，Templeton，1989)。了解得比较清晰的是早期经历的化学的（可能是嗅觉）和地磁的信号可能会导致大麻哈鱼行为的持续改变。在鱼类生活中学习最著名的例子之一可能是来自于对鲑鱼嗅觉印记和归巢的研究（Hasler，1966；Hasler 等，1978；Nordeng，1977，1983，2009；Johnsen，Hasler，1980；Nordeng，Bratland，2006)。尽管细节仍然被继续调查研究，但这种现象如此广泛被接受就是因为它为许多地区标准孵化场的实践奠定了基础（Dittman 等，2015)。“气味是著名的记忆触发器”（Dawkins，2013)。

基于这个例子的背景，有必要简要总结一下鲑鱼的生活周期。大西洋鲑鱼与太平洋鲑鱼的生活周期有着明显的差异。俗名的使用不是一直都有用。事实上，有些人已经提到早期欧洲探索者在北美将鲑鱼的俗名强加于太平洋物种引起了不必要的混乱（Dymond，1932)。然而，溯河产卵鲑鱼生活史的重要特征就是性成熟的成鱼要回到它们出生的河流中产卵。这些后代在它们重新迁徙到它们觅食、生长和性成熟的海洋之前，需要花费大量时间（许多个月或许多年）在它们出生的河流中，然后再次迁徙回到它们出生的河流产卵而完成一个生命史周期。鲑鱼洄游到它们各自的出生溪流是这些鱼的显著特征（Hansen，Quinn，1998；Quinn，2005；Waples，Hendry，

2008)。

早期有些关于鲑鱼在不同流域间移居的研究提供了它们精确的归巢行为是学习结果的第一个证据(Dittman, Quinn, 1996)。作用机制的研究提供很多关于这种现象的详细信息。嗅觉印记的生理关联已被详细地了解。淡水中幼鱼甲状腺素和皮质醇会发生波动,当它们性成熟的时候,鲑鱼的嗅觉感受器对其家乡溪流所特有的化学物质就会敏感(Hasler, 1966; Hasler 等, 1978; Stabell, 1984; Carruth 等, 2002; Lema, Nevitt, 2014)。这些化学物质可能来自同伴,或来自溪流中的其他来源,但鲑鱼会形成化学物质信号和位置之间强烈的相互关系,当它们回来产卵时,就能返回它们幼鱼期所处的水域(Brett, Groot, 1963; Hasler 等, 1978; Fisknes, Doving, 1982; Quinn, Hara, 1986; Brannon, Quinn, 1990; Dittman, Quinn, 1996; Ueda, 2014)。

有些人可能会提出异议,严格地说,鲑科鱼类早期的化学印记应该和应激无关。原因可能是嗅觉印记的发生是这些鱼正常生活史的一部分,而应激则是鱼所经历的非正常部分。我们理解这个观点,但我们认为应激的概念应该包括在鱼类正常生活史中的任何一个事件或场合(见本书其他章节)。因此,我们认为甲状腺素、皮质醇以及其他可能参与鲑鱼幼鱼的卵黄吸收和下游迁徙的激素,以及随后成鱼上游洄游的激素水平的升高,都是鱼的生理应激的表现(Schreck, 1990; Dittman, Quinn, 1996; Schreck 等, 1997; Congleton 等, 2000)。另外,我们知道,格外特别的记忆就是人们可以将其跟生命中特别有压力的事件联系在一起。对我们在哪里,我们所看到的,以及我们在 1963 年 11 月 22 日或者 2001 年 9 月 11 日所听到的有特别详细的回忆,这些例子对我们某些人来说是很熟悉的。然而我们并不能记住 1963 年 11 月 21 日或者 2001 年 9 月 10 日的任何事情。

太平洋鲑鱼甲状腺素水平的提高能使嗅觉受体对起始的学习和随后对化学刺激的反应更加敏感(Cooper, Scholz, 1976; Carruth 等, 2002; Lema, Nevitt, 2004)。这对于鲑科鱼类和其他鱼类的其他激素和其他学习例子都是可以相比拟的。丽鱼科鱼类精巧的育幼行为,特别是它们对自己的幼鱼甚至不同的发育阶段都具有特异的识别能力,涉及这种学习(Noakes, 1999; Galhardo 等, 2011; Morandini 等, 2014; Ramallo 等, 2014)。在丽鱼科鱼类复杂社会关系中的多数时间,学习和相关应激是紧密联系的(Francis, 1988; Maruska 等, 2013)。学习,以经典条件反射(Hollis 等, 1995; Hollis 等, 1997)或工具条件反射(Sneddon 等, 2016; 见本书第 12 章)的形式能够调节鱼类的行为和生理状态(Fernald, 2015)。在一些鱼类中,母本的应激亦可能会影响它们后代的学习能力(Eaton 等, 2015)。

在鱼类行为生态学广泛范围内亦有关于学习的重要研究(Charnov, 1976; Dill, 1983; Drew, Gumm, 2015; Furtbauer 等, 2015)。鱼类显示出在觅食、配偶选择、栖息地选择和生存以及生活史相关的活动方面相当大的可塑性及适应性(Charnov, 1976, 1978, 1993; Godin, 1999; Conrad 等, 2011; Davies 等, 2012; Mittelbach 等, 2014; Bond 等, 2015)。一些作者报道这些行为的可变性可以描述及定义为一致性行为表型(McLaughlin 等, 1999; Pham 等, 2012; Oswald 等, 2013; Thornqvist 等,

2015）。这些研究涉及的种类繁多，从鲑科鱼类到斑马鱼，所处的环境从景观生态到高度受控和人为设计的实验室测试环境（Grant，Noakes，1978a，b；Dittman，Quinn，1996；Brown，Laland，2001；Brown 等，2003；Askey 等，2006；Oswald 等，2013；Tudorache 等，2015）。其中包括一些已知的鱼类最复杂的学习例子。我们可以有把握地推测，实质上所有的例子也包括至少一定程度的鱼类生理应激（甚至可能是不良应激）。

栖息地和栖息地选择是在自然环境中鱼类行为最基础的方面（Nakano，1994；Hughes，2000；Brockmark 等，2010；Jones 等，2014）。鱼类对栖息地特征的反应，以及栖息地选择的偏好性可能会受到鱼生理条件的影响。对栖息地特征的行为反应当然是鱼类在自然条件中最重要的亦是研究得最频繁的方面（Fausch，Northcote，1992；Hughes，2000；Hughes，Grand，2000；Piccolo 等，2007；Berejikian 等，2013；Bunt 等，2013；White，Brown，2015）。正如前所述，和栖息地特征相关的行为包括对遮蔽物、温度、水质、捕食者和同伴以及其他的反应。其中对猎物的反应是研究最为频繁的，并且提供了大量关于最优觅食的文献（Charnov，1976；Green，1987；Kamil 等，1987；Stephens，Krebs，1987）。毫无疑问，这些关于最优化和行为决策的例子我们先前已经讨论过了。

鱼类的觅食行为已经通过各种理论与技术方法广泛研究了许多年。一种广泛应用的基本技术用来鉴定自然环境中鱼个体的胃中成分（Chiasson 等，1995；Kawaguchi，Nakano，2001；Nakano，Murakami，2001）。很难从这些数据中总结出任何关于学习或应激的结论。更为复杂的技术，例如稳定同位素分析会告诉我们更多关于这些个体的长期摄食行为。然而，尽管这些数据可以帮助我们推断长期摄食习性，但它们能告诉我们关于野生鱼类觅食和摄食行为相关应激或学习的信息却很少（Church 等，2008；Gao 等，2008）。因此，基于驯养鱼类和野生鱼类的研究所获得的引人注目的证据表明，学习对于食物选择、觅食地点和觅食活动都有重要影响（Dill，1983；Grant，Noakes，1987a，b；Nielsen，1992；McLaughlin 等，1999）。

有直接的实验证据以及相关的数据显示，鲑科鱼类和其他鱼类能很快根据猎物多少调整它们的觅食（Piccolo 等，2014）。单个个体能发展一致性的捕猎偏好性（Grant，Noakes，1987a，b；McLaughlin 等，1994），这或者是基于鱼的内在本质特征，或者是经验的结果（个体发生或学习）。我们从养殖鱼类摄食的详细研究中得知，摄食反应会受到食物颗粒大小、质地、活动状态、颜色、组成和出现在鱼面前的方式所影响（Hobling，1985；Noakes，Grant，1992；Brown 等，2003；Ferno 等，2011；Attia 等，2012；Nilsson 等，2012）。在上述例子的大多数情况下，即使不是全部，鱼也在学习对食物特征与食物投给方式的反应。

9.5　一些关键的知识空白

区分信息和知识（或理解力）是很重要的。综合、假说和原理是科学研究的对象。有时这些是指通俗文化中的本土法则。这在很大程度上是对19世纪统治世界的自然法观点的坚持。关键在于这样的法则或规则成为我们真正理解的基础，而不是简单的事实或观察的持续累积。

基于我们在这方面的文献综述以及鱼类行为的相关领域，我们提出三个重要的分析和信息领域，以便为鱼类行为和应激研究提供关键信息。这三个富有成效而活跃的研究领域之间都有各自的相交处进行相互作用，它们对我们理解应激和行为的机制有很大帮助。

第一个领域是基因调节与行为之间的相交处。我们相信这是一个关键的知识空白，这空白可以作为研究热点的基础课题。目前有大量关于基因调节研究的结果（Brakefield，2011；Bourret 等，2013；Oswald 等，2013；McKinney 等，2015）。模式鱼系统当然是斑马鱼，但鲑科鱼类的研究也有相当的潜力，尤其是对一些生活史差异有联系的物种。那些如“基因要么存在，要么不存在”以及“存在和行为相关的基因”的说法早已过时。鲑科鱼类中养殖硬头鳟和野生硬头鳟会出现基因活化的差异，这种差异可能是因为养殖场中对鱼的驯化所导致，这也为我们提供了典型的例子（Araki 等，2007，2008；Christie 等，2012）。这些研究所使用的方法会让我们对基因与行为间因果关系有更深入的理解。同时也为我们提供了通过外部环境人为操作基因活化而产生我们所期望的行为表型鱼的相关知识。生产具有特定行为表型的鱼类，尤其是鲑科鱼，长期以来一直是渔业经营管理的长期目标，涵盖了从保护和恢复到充实和支持的收获计划。

我们提出的第二个领域是未来研究应着眼于应激对行为的跨世代影响。有时是指母体的非遗传效应或表观遗传效应（Leblanc 等，2011）。这个领域可以认为是基因调节与行为领域的一个分支，而跨世代影响无疑是基因调节差异的结果。然而，我们在这里提出这个研究领域作为未来研究的特殊需求，是因为我们意识到了解鱼类生活中应激作用的重要性（Sloman，2010）。对养殖硬头鳟（*Oncorhynchus mykiss*）跨世代影响的详细研究证明了我们能预测的模式（Araki 等，2007，2009；Christie 等，2012，2014）。开展该领域的研究工作将能揭示和分析鱼的应激反应和一系列与应激反应相关的内外部环境因子之间的大量潜在关系（Sneddon 等，2016；见本书第12章）。对于寻求联系应激与因果关系因子的一般通路，这些研究会是最有效的。

我们提出的第三个领域是根据鱼（行为的）-应激相关的行为预测模型做出决策。这个研究领域是以最优模型的建立为前提。因为我们在本章已经概述，我们的观点就是鱼类对应激反应可能最好地被认为是适应性的。它们的反应将通过自然选择塑造。然而，这需要我们定义什么状态是最优化。Houston 和 McNamara（1999）已经提出了解行为的最全面、最简明的方法之一。可能他们提出的模型亦必须包括生理、生态、进化、生活史、性别差异及它们与应激关系的研究。这个研究领域一定具有挑战性，但最后可能有利于对鱼类生活中所遇到的应激有更全面的理解。

9.6 致 谢

我们感谢 Carl Schreckon 和 Lluis Tort 邀请我们参与本章编写，亦感谢他们对文稿提出建设性的评论和建议。我们也要特别感谢 Victoria Braithwaite 所给予的富有洞察力的综述与评论，还要真挚地感谢 Lynn Bouvier 对修订稿进行了无数细节的修改，提供了宝贵的帮助。我们还要感谢许多年来众多和我们分享他们想法、观点和研究的同事、学生和助理们。我们试图通过直接引用他们发表的论文来认同这些观点。我们也要感谢所有支持我们研究活动和项目的机构与组织，以及我们现在所在的学院和研究所为我们的写作提供的知识分子家园。

D. L. G. 诺阿克斯　K. M. M. 琼斯　著
卢丹琪　译
林浩然　校

参 考 文 献

Adamo, S. A. (2014). The effects of stress hormones on immune function may be vital for the adaptive reconfiguration of the immune system during fight-or-flight behavior. Integr. Comp. Biol. 54, 419-426.

Agrillo, C., Petrazzini, M. E. M., Piffer, L., Dadda, M., Bisazza, A. (2012). A new training procedure for studying discrimination learning in fish. Behav. Brain Res. 230, 343-348.

Alanärä, A. (1992). Demand feeding as a self-regulating feeding system for rainbow trout (*Oncorhynchus mykiss*) in net-pens. Aquaculture. 108, 347-356.

Alanärä, A. (1996). The use of self-feeders in rainbow trout (*Oncorhynchus mykiss*) production. Aquaculture. 145, 1-20.

Allan, B. J., Miller, G. M., McCormick, M. I., Domenici, P., Munday, P. L. (2014). Parental effects improve escape performance of juvenile reef fish in a high-CO_2 world. Proc. R. Soc. Lond. B Biol. Sci. 281, 20132179.

Allan, B. J., Domenici, P., Munday, P. L., McCormick, M. I. (2015). Feeling the heat: the effect of acute temperature changes on predator-prey interactions in coral reef fish. Conserv. Physiol. 3, cov011.

Araki, H., Cooper, B., Blouin, M. S. (2007). Genetic effects of captive breeding cause a rapid, cumulative fitness decline in the wild. Science. 318, 100-103.

Araki, H., Berejikian, B. A., Ford, M. J., Blouin, M. S. (2008). Fitness of hatchery-reared salmonids in the wild. Evol. Appl. 1, 342-355.

Araki, H., Cooper, B., Blouin, M. S. (2009). Carry-over effect of captive breeding reduces reproductive fitness of wild-born descendants in the wild. Biol. Lett. 5, 621-624.

Askey, P. J., Richards, S. A., Post, J. R., Parkinson, E. A. (2006). Linking angling catch rates and fish learning under catch-and-release regulations. N. Am. J. Fish. Manage. 26, 1020-1029.

Attia, J., Millot, S., Di-Poï, C., Bégout, M.-L., Noble, C., Sanchez-Vazquez, F. J. (2012). Demand feeding and welfare in farmed fish. Fish Physiol. Biochem. 38, 107-118.

Baciadonna, L., McElligott, A. (2015). The use of judgement bias to assess welfare in farm livestock. Anim. Welfare. 24, 81-91.

Beamish, F. W. H. (1970). Influence of temperature and salinity acclimation on temperature preferenda of the euryhaline fish Tilapia nilotica. J. Fish. Res. Board Can. 27, 1209-1214.

Beauchamp, R. D., Rowe, J. S. (1977). Goldfish spectral sensitivity: a conditioned heart rate measure in restrained or curarized fish. Vision. Res. 17, 617-624.

Beauchamp, R., Rowe, J., O'Reilly, L. (1979). Goldfish spectral sensitivity: identification of the three cone mechanisms in heart-rate conditioned fish using colored adapting backgrounds. Vision Res. 19, 1295-1302.

Beitinger, T. L., Fitzpatrick, L. C. (1979). Physiological and ecological correlates of preferred temperature in fish. Am. Zool. 19, 319-329.

Berejikian, B. A., Smith, R. J. F., Tezak, E. P., Schroder, S. L., Knudsen, C. M. (1999). Chemical alarm signals and complex hatchery rearing habitats affect antipredator behavior and survival of Chinook salmon (*Oncorhynchus tshawytscha*) juveniles. Can. J. Fish. Aquat. Sci. 56, 830-838.

Berejikian, B., Tezak, E., Riley, S., LaRae, A. (2001). Competitive ability and social behaviour of juvenile steelhead reared in enriched and conventional hatchery tanks and a stream environment. J. Fish. Biol. 59, 1600-1613.

Berejikian, B. A., Campbell, L. A., Moore, M. E. (2013). Large-scale freshwater habitat features influence the degree of anadromy in eight Hood Canal *Oncorhynchus mykiss* populations. Can. J. Fish. Aquat. Sci. 70, 756-765.

Bitterman, M. (1964). Classical conditioning in the goldfish as a function of the CS-UCS interval. J. Comp. Physiol. Psychol. 58, 359.

Block, B. A., Finnerty, J. R. (1994). Endothermy in fishes: a phylogenetic analysis of constraints, predispositions, and selection pressures. Environ. Biol. Fish. 40, 283-302.

Boeuf, G. (1993). Salmonid smolting: a pre-adaptation to the oceanic environment. In: Rankin, J. C., Jensen, F. B., Fish ecophysiology. Chapman and Hall, London, 105-135.

Bond, M. H., Miller, J. A., Quinn, T. P. (2015). Beyond dichotomous life histories in partially migrating populations: cessation of anadromy in a long-lived fish. Ecology. 96 (7), 1899-1910.

Boujard, T., Jourdan, M., Kentouri, M., Divanach, P. (1996). Diel feeding activity and the effect of time-restricted self-feeding on growth and feed conversion in European sea bass. Aquaculture. 139, 117-127.

Bourret, V., Kent, M. P., Primmer, C. R., Vasemägi, A., Karlsson, S., Hindar, K. (2013). SNP-array reveals genome-wide patterns of geographical and potential adaptive divergence across the natural range of Atlantic salmon (*Salmo salar*). Mol. Ecol. 22, 532-551.

Braithwaite, V. (2010). Do Fish Feel Pain?. Oxford University Press, Oxford.

Brakefield, P. M. (2011). Evo-devo and accounting for Darwin's endless forms. Philos. Trans. R. Soc. B Biol. Sci. 366, 2069-2075.

Brannon, E., Quinn, T. (1990). Field test of the pheromone hypothesis for homing by pacific salmon. J. Chem. Ecol. 16, 603-609.

Brett, J. R. (1952). Temperature tolerance in young Pacific salmon, genus *Oncorhynchus*. J. Fish. Res. Board Can. 9, 265-323.

Brett, J. R. (1971). Energetic responses of salmon to temperature. A study of some thermal relations in the physiology and freshwater ecology of sockeye salmon (*Oncorhynchus nerka*). Am. Zool. 11, 99-113.

Brett, J. R., Groot, C. (1963). Some aspects of olfactory and visual responses in Pacific salmon. J. Fish. Res. Board Can. 20, 287-303.

Brockmark, S., Adriaenssens, B., Johnsson, J. I. (2010). Less is more: density influences the development of behavioural life skills in trout. Proc. R. Soc. B Biol. Sci. 277, 3035-3043.

Brown, C., Laland, K. (2001). Social learning and life skills training for hatchery reared fish. J. Fish. Biol. 59, 471-493.

Brown, C., Davidson, T., Laland, K. (2003). Environmental enrichment and prior experience of live prey improve foraging behaviour in hatchery-reared Atlantic salmon. J. Fish. Biol. 63, 187-196.

Bunt, C. M., Mandrak, N. E., Eddy, D. C., Choo-Wing, S. A., Heiman, T. G., Taylor, E. (2013). Habitat utilization, movement and use of groundwater seepages by larval and juvenile Black Redhorse, Moxostoma duquesnei. Environ. Biol. Fish. 96, 1281-1287.

Burnham, J. C. (1972). Thorndike's puzzle boxes. J. Hist. Behav. Sci. 8, 159-167.

Cachat, J. M., Canavello, P. R., Elegante, M. F., Bartels, B. K., Elkhayat, S. I., Hart, P. C. (2011). Modeling stress and anxiety in zebrafish. In: Kalueff, A. V., Cachat, J. M., Zebrafish models in neurobehavioral research. Springer, Dordrecht, 73-88.

Careau, V., Buttemer, W. A., Buchanan, K. L. (2014). Developmental stress can uncouple relationships between physiology and behaviour. Biol. Lett. 10, 20140834.

Carey, F. G., Teal, J. M., Kanwisher, J. W., Lawson, K. D., Beckett, J. S. (1971). Warm-bodied fish. Am. Zool. 11, 137-143.

Carruth, L. L., Jones, R. E., Norris, D. O. (2002). Cortisol and Pacific salmon: a new look at the role of stress hormones in olfaction and home-stream migration. Integr. Comp. Biol. 42, 574-581.

Charnov, E. L. (1976). Optimal foraging, the marginal value theorem. Theor. Popul. Biol. 9, 129-136.

Charnov, E. L. (1978). Evolution of eusocial behavior: offspring choice or parental parasitism?. J. Theor. Biol. 75, 451-465.

Charnov, E. L. (1993). Life History Invariants. Oxford University Press, Oxford, UK.

Chiasson, P., Beamish, F. W. H., Noakes, D. L. G. (1995). Benthic invertebrates and stomach contents of juvenile sturgeon in the Moose river basin, Ontario. Can. J. Fish. Aquat. Sci. 54, 2866-2871.

Christie, M. R., Marine, M. L., French, R. A., Blouin, M. S. (2012). Genetic adaptation to captivity can occur in a single generation. Proc. Natl. Acad. Sci. 109, 238-242.

Christie, M. R., French, R. A., Marine, M. L., Blouin, M. S. (2014). How much does inbreeding contribute to the reduced fitness of hatchery-born steelhead (*Oncorhynchus mykiss*) in the wild?. J. Hered. 105, 111-119.

Church, M. R., Ebersole, J. L., Rensmeyer, K. M., Couture, R. B., Barrows, F. T., Noakes, D. L. (2008). Mucus: a new tissue fraction for rapid determination of fish diet switching using stable isotope analysis. Can. J. Fish. Aquat. Sci. 66, 1-5.

Congleton, J. L., LaVoie, W. J., Schreck, C. B., Davis, L. E. (2000). Stress indices in migrating juvenile Chinook salmon and steelhead of wild and hatchery origin before and after barge transportation. Trans. Am. Fish. Soc. 129, 946-961.

Conrad, J. L., Weinersmith, K. L., Brodin, T., Saltz, J., Sih, A. (2011). Behavioural syndromes in fishes: a review with implications for ecology and fisheries management. J. Fish. Biol. 78, 395-435.

Conte, F. (2004). Stress and the welfare of cultured fish. Appl. Anim. Behav. Sci. 86, 205-223.

Contreras-Sanchez, W., Schreck, C., Fitzpatrick, M., Pereira, C. (1998). Effects of stress on the reproductive performance of rainbow trout (*Oncorhynchus mykiss*). Biol. Reprod. 58, 439-447.

Cook, K. V., Crossin, G. T., Patterson, D. A., Hinch, S. G., Gilmour, K. M., Cooke, S. J. (2014). The stress response predicts migration failure but not migration rate in a semelparous fish. Gen. Comp. Endocrinol. 202, 44-49.

Cooper, S. J. (2008). From claude bernard to walter cannon. Emergence of the concept of homeostasis. Appetite. 51, 419-427.

Cooper, J. C., Scholz, A. T. (1976). Homing of artificially imprinted steelhead (rainbow) trout, *Salmo gairdneri*. J. Fish. Res. Board Can. 33, 826-829.

Coutant, C. C. (1977). Compilation of temperature preference data. J. Fish. Board Can. 34, 739-745.

Coutant, C. C. (1987). Thermal preference: when does an asset become a liability?. Environ. Biol. Fish. 18, 161-172.

Danley, M. L., Mayr, S. D., Young, P. S., Cech, J. J. (2002). Swimming performance and physiological stress responses of splittail exposed to a fish screen. N. Am. J. Fish. Manage. 22, 1241-1249.

Davies, N. B., Krebs, J. R., West, S. A. (2012). An Introduction to Behavioural Ecology. John Wiley & Sons, Oxford.

Dawkins, R. (1976). The Selfish Gene. Oxford University Press, Oxford.

Dawkins, M. S. (2003). Behaviour as a tool in the assessment of animal welfare. Zoology. 106, 383-387.

Dawkins, R. (2004). Extended phenotype-but not too extended. A reply to Laland, Turner and Jablonka. Biol. Philos. 19, 377-396.

Dawkins, R. (2009). The Greatest Show on Earth. The Evidence for Evolution. Simon & Schuster, New York, NY.

Dawkins, M. S. (2012). Why Animals Matter: Animal Consciousness, Animal Welfare, and Human Well-being. Oxford University Press, Oxford.

Dawkins, R. (2013). An Appetie for Wonder. Harper Collins, New York, NY.

Dill, L. M. (1983). Adaptive flexibility in the foraging behavior of fishes. Can. J. Fish. Aquat. Sci. 40, 398-408.

Dittman, A., Quinn, T. (1996). Homing in Pacific salmon: mechanisms and ecological basis. J. Exp. Biol. 199, 83-91.

Dittman, A. H., Pearsons, T. N., May, D., Couture, R. B., Noakes, D. L. (2015). Imprinting of hatchery-reared salmon to targeted spawning locations: a new embryonic imprinting paradigm for hatchery programs. Fisheries. 40, 114-123.

Dobzhansky, T. (1973). Nothing in biology makes sense except in the light of evolution. Am. Biol. Teach. 35, 125-129.

Domenici, P., Ruxton, G. D. (2015). 8 Prey behaviors during fleeing: escape trajectories, signaling, and sensory defensesEscaping From Predators: An Integrative View of Escape Decisions. Cambridge University Press, Cambridge, 199.

Domenici, P., Allan, B. J., Watson, S. -A., McCormick, M. I., Munday, P. L. (2014). Shifting from right to left: the combined effect of elevated CO_2 and temperature on behavioural lateralization in a coral reef fish. PLoS One. 9 (1), e87969.

Dominique, J. -F., McGaugh, J. L. (2014). Stress and the regulation of memory: from basic mechanisms to clinical implications. Neurobiol. Learn. Mem. 112, 1.

Donelson, J. M., McCormick, M. I., Booth, D. J., Munday, P. L. (2014). Reproductive acclimation to increased water temperature in a tropical reef fish. PLoS One. 9 (5), e97223.

Drew, J. A., Gumm, J. M. (2015). Learning and behavior in reef fish: fuel for microevolutionary change?. Ethology. 121, 2-7.

Dymond, J. R. (1932). The Trout and Other Game Fishes of British Columbia. Department of Fisheries, Ottawa, ON.

Eaton, L., Edmonds, E., Henry, T., Snellgrove, D., Sloman, K. (2015). Mild maternal stress disrupts associative learning and increases aggression in offspring. Horm. Behav. 71, 10-15.

Ejike, C., Schreck, C. B. (1980). Stress and social hierarchy rank in coho salmon. Trans. Am. Fish. Soc. 109, 423-426.

Eldridge, W. H., Sweeney, B. W., Law, J. M. (2015). Fish growth, physiological stress, and tissue condition in response to rate of temperature change during cool or warm diel thermal cycles. Can. J. Fish. Aquat. Sci. 72, 1527-1537.

Fagerlund, U. H. M., McBride, J. R., Stone, E. T. (1981). Stress-related effects of hatchery rearing density on coho salmon. Trans. Am. Fish. Soc. 110, 644-649.

Fausch, K. D., Northcote, T. G. (1992). Large woody debris and salmonid habitat in a small Coastal British Columbia stream. Can. J. Fish. Aquat. Sci. 49, 682-693.

Fenderson, O. C., Everhart, W. H., Muth, K. M. (1968). Comparative agonistic and feeding behavior of hatchery-reared and wild salmon in aquaria. J. Fish. Res. Board Can. 25, 1-14.

Fernald, R. (1976). The effect of testosterone on the behavior and coloration of adult male cichlid fish (Haplochromis burtoni, Günther). Horm. Res. Paediatr. 7, 172-178.

Fernald, R. D. (2015). Social behaviour: can it change the brain?. Anim. Behav. 103, 259-265.

Fernö, A., Huse, G., Jakobsen, P. J., Kristiansen, T. S., Nilsson, J. (2011). Fish behaviour, learning, aquaculture and fisheriesFish Cognition and Behavior. Wiley-Blackwell, Oxford, 359-404.

Ferrari, M. C., Manassa, R. P., Dixson, D. L., Munday, P. L., McCormick, M. I., Meekan, M. G. (2012). Effects of ocean acidification on learning in coral reef fishes. PLoS One. 7, e31478.

Ferrari, M., Munday, P. L., Rummer, J. L., McCormick, M. I., Corkill, K., Watson, S. A. (2015). Interactive effects of ocean acidification and rising sea temperatures alter predation rate and predator selectivity in reef fish communities. Glob. Change Biol. 21, 1848-1855.

Fisknes, B., Døving, K. (1982). Olfactory sensitivity to group-specific substances in Atlantic salmon (*Salmo salar* L.). J. Chem. Ecol. 8, 1083-1092.

Francis, R. C. (1988). Socially mediated variation in growth rate of the Midas cichlid: the primacy of early size differences. Anim. Behav. 36, 1844-1845.

Fraser, D. J., Weir, L. K., Bernatchez, L., Hansen, M. M., Taylor, E. B. (2011). Extent and scale of local adaptation in salmonid fishes: review and meta-analysis. Heredity. 106, 404-420.

Frisch, K. V. (1942). Über einen schreckstoff der cischhaut und seine biologische bedeutung. Zeit. Vergl. Physiol. 29, 46-145.

Fry, F. E. J. (1947). Effects of the environment on animal activity. University of Toronto Studies, Biological Series No. 55. Publications of the Ontario Fisheries Research Laboratory No. 68, 62.

Fürtbauer, I., Pond, A., Heistermann, M., King, A. J. (2015). Personality, plasticity and predation: linking endocrine and behavioural reaction norms in stickleback fish. Funct. Ecol. 29, 931-940.

Galhardo, L., Vital, J., Oliveira, R. F. (2011). The role of predictability in the stress response of a cichlid fish. Physiol. Behav. 102, 367-372.

Gao, Y., Bean, D., Noakes, D. L. (2008). Stable isotope analyses of otoliths in identification of hatchery origin of Atlantic salmon (*Salmo salar*) in maine. Environ. Biol. Fish. 83, 429-437.

Godin, J.-G. J. E. (1999). Behavioural Ecology of Teleost Fishes. Oxford University Press, New York, NY.

Goetz, F. A., Jeanes, E., Moore, M. E., Quinn, T. P. (2015). Comparative migratory behavior and survival of wild and hatchery steelhead (*Oncorhynchus mykiss*) smolts in riverine, estuarine, and marine habitats of Puget Sound, Washington. Environ. Biol. Fish. 98, 357-375.

Grant, J. W. A., Noakes, D. L. G. (1987). Movers and stayers: foraging tactics of young-of-the-year brook charr, *Salvelinus fontinalis*. J. Anim. Ecol. 56, 1001-1013.

Grant, J. W. A., Noakes, D. L. G. (1987). A simple model of optimal territory size for drift-feeding fish. Can. J. Zool. 65, 270-276.

Grassie, C., Braithwaite, V. A., Nilsson, J., Nilsen, T. O., Teien, H.-C., Handeland, S. O. (2013). Aluminum exposure impacts brain plasticity and behavior in Atlantic salmon (*Salmo salar*). J. Exp. Biol. 216 (Pt 16), 3148-3155.

Green, R. F. (1987). Stochastic models of optimal foraging. In: Kamil, A. C., Krebs, J. R., Pulliam, H. R., Foraging behavior. Springer, Dordrecht, 273-302.

Gross, C. G. (1998). Claude bernard and the constancy of the internal environment. Neuroscientist. 4, 380-385.

Hamilton, W. (1998). Narrow roads of gene land: The collected papers of W. D. Hamilton Volume 1: Evolution of social behaviour. Oxford University Press, Oxford.

Hansen, L. P., Quinn, T. P. (1998). The marine phase of the Atlantic salmon (*Salmo salar*) life cycle, with comparisons to Pacific salmon. Can. J. Fish. Aquat. Sci. 55, 104-118.

Hargrove, M. B., Becker, W. S., Hargrove, D. F. (2015). The HRD eustress model generating positive stress with challenging work. Hum. Resour. Dev. Rev. 14, 279-298.

Hasler, A. (1966). Underwater Guideposts. *University Wisconsin Press, Madison*, 155.

Hasler, A. D., Scholz, A. T., Horrall, R. M. (1978). Olfactory imprinting and homing in salmon. Am. Sci. 66, 347-355.

Hawryshyn, C. W. (1992). Polarization vision in fish. Am. Sci., 164-175.

Hawryshyn, C. W. (2000). Ultraviolet polarization vision in fishes: possible mechanisms for coding e-vector. Philos. Trans. R. Soc. Lond. *B Biol. Sci.* 355, 1187-1190.

Hawryshyn, C. W., McFarland, W. N. (1987). Cone photoreceptor mechanisms and the detection of polarized light in fish. J. Comp. Physiol. A. 160, 459-465.

Hawryshyn, C. W., Arnold, M. G., Chaisson, D. J., Martin, P. C. (1989). The ontogeny of ultraviolet photosensitivity in rainbow trout (*Salmo gairdneri*). Vis. Neurosci. 2, 247-254.

Hayes, S. A., Hanson, C. V., Pearse, D. E., Bond, M. H., Garza, J. C., MacFarlane, R. B. (2012). Should I stay or should I go? The influence of genetic origin on emigration behavior and physiology of resident and Anadromous juvenile *Oncorhynchus mykiss*. N. Am. J. Fish. Manage. 32, 772-780.

Hazel, J. R., Prosser, C. L. (1974). Molecular mechanisms of temperature compensation in poikilotherms. Physiol. Rev. 54, 620-677.

Hertz, P. E. , Huey, R. B. , Nevo, E. (1982). Fight versus flight: body temperature influences defensive responses of lizards. Anim. Behav. 30, 676-679.

Hoar, W. S. (1953). Control and timing of fish migration. Biol. Rev. 28, 437-452.

Hoar, W. S. (1966). General and Comparative Physiology. Prentice Hall, New York, NY.

Hodgson, S. , Quinn, T. P. , Hilborn, R. A. Y. , Francis, R. C. , Rogers, D. E. (2006). Marine and freshwater climatic factors affecting interannual variation in the timing of return migration to fresh water of sockeye salmon (*Oncorhynchus nerka*). Fish. Oceanogr. 15, 1-24.

Holland, K. N. , Teeter, J. H. (1981). Behavioral and cardiac reflex assays of the chemosensory acuity of channel catfish to amino acids. Physiol. Behav. 27, 699-707.

Hollis, K. L. (1982). Pavlovian conditioning of signal-centered action patterns and autonomic behavior: a biological analysis of function. Adv. Study Behav. 12, 1-64.

Hollis, K. L. (1997). Contemporary research on Pavlovian conditioning: a "new" functional analysis. Am. Psychol. 52, 956.

Hollis, K. L. , Cadieux, E. L. , Colbert, M. M. (1989). The biological function of Pavlovian conditioning: a mechanism for mating success in the blue gourami (*Trichogaster trichopterus*). J. Comp. Psychol. 103, 115.

Hollis, K. L. , Dumas, M. J. , Singh, P. , Fackelman, P. (1995). Pavlovian conditioning of aggressive behavior in blue gourami fish (*Trichogaster trichopterus*): winners become winners and losers stay losers. J. Comp. Psychol. 109, 123.

Hollis, K. L. , Pharr, V. L. , Dumas, M. J. , Britton, G. B. , Field, J. (1997). Classical conditioning provides paternity advantage for territorial male blue gouramis (*Trichogaster trichopterus*). J. Comp. Psychol. 111, 219.

Houston, A. I. , McNamara, J. M. (1999). Models of Adaptive Behaviour. Cambridge University Press, Cambridge, UK.

Hughes, N. F. (2000). Testing the ability of habitat selection theory to predict interannual movement patterns of a drift-feeding salmonid. Ecol. Freshw. Fish. 9, 4-8.

Hughes, N. , Grand, T. (2000). Physiological ecology meets the ideal-free distribution: predicting the distribution of size-structured fish populations across temperature gradients. Environ. Biol. Fishes. 59, 285-298.

Iwama, G. K. , Pickering, A. , Sumpter, J. , Schreck, C. (2011). Fish Stress and Health in Aquaculture. Cambridge University Press, Cambridge.

Jansen, A. S. , Van Nguyen, X. , Karpitskiy, V. , Mettenleiter, T. C. , Loewy, A. D. (1995). Central command neurons of the sympathetic nervous system: basis of the fight-or-flight response. Science. 270, 644-646.

Jobling, M. (1981). Temperature tolerance and the final preferendum—rapid methods for the assessment of optimum growth temperatures. J. Fish. Biol. 19, 439-455.

Jobling, M. (1985). Physiological and social constraints on growth of fish with special reference to Arctic charr, *Salvelinus alpinus* L. Aquaculture. 44, 83-90.

Johnsen, P. B., Hasler, A. D. (1980). The use of chemical cues in the upstream migration of coho salmon, *Oncorhynchus kisutch* Walbaum. J. Fish. Biol. 17, 67-73.

Jones, K. (2005). The effect of territorial damselfish (family Pomacentridae) on the space use and behaviour of the coral reef fish, *Halichoeres bivittatus* (Bloch, 1791) (family Labridae). J. Exp. Mar. Biol. Ecol. 324, 99-111.

Jones, K. M. M. (2007). Distribution of behaviours and species interactions within home range contours in five Caribbean reef fish species (Family Labridae). Environ. Biol. Fishes. 80, 35-49.

Jones, K. M. M., McGrath, P. E., Able, K. W. (2014). White perch morone americana (Gmelin, 1789) habitat choice and movements: comparisons between phragmites-invaded and spartina reference marsh creeks based on acoustic telemetry. J. Exp. Mar. Biol. Ecol. 455, 14-21.

Jonsson, B., Jonsson, N. (2006). Cultured Atlantic salmon in nature: a review of their ecology and interaction with wild fish. ICES J. Marine Sci. J. du Conseil. 63, 1162-1181.

Kamil, A. C., Krebs, J. R., Pulliam, H. R. (1987). Foraging Behavior. Plenum Press, New York, NY, 676.

Kavanagh, M., Olson, D. E. (2014). The effects of rearing density on growth, fin erosion, survival, and migration behavior of hatchery winter steelhead. N. Am. J. Aquac. 76, 323-332.

Kawaguchi, Y., Nakano, S. (2001). Contribution of terrestrial invertebrates to the annual resource budget for salmonids in forest and grassland reaches of a headwater stream. Freshw. Biol. 46, 303-316.

Keenleyside, M. H. (2012). Diversity and Adaptation in Fish Behaviour. Springer Science & Business Media, .

Larra, M. F., Schulz, A., Schilling, T. M., deSá, D. S. F., Best, D., Kozik, B. (2014). Heart rate response to post-learning stress predicts memory consolidation. Neurobiol. Learn. Mem. 109, 74-81.

Laursen, D. C., Olsén, H. L., de Lourdes Ruiz-Gomez, M., Winberg, S., Höglund, E. (2011). Behavioural responses to hypoxia provide a non-invasive method for distinguishing between stress coping styles in fish. Appl. Anim. Behav. Sci. 132, 211-216.

Lawrence, D. J., Beauchamp, D. A., Olden, J. D. (2015). Life-stage-specific physiology defines invasion extent of a riverine fish. J. Anim. Ecol. 84, 879-888.

Leblanc, C. A. L., Benhaïm, D., Hansen, B. R., Kristjánsson, B. K., Skúlason, S. (2011). The importance of egg size and social effects for behaviour of Arctic charr juveniles. Ethology. 117, 664-674.

Le Fevre, M., Matheny, J., Kolt, G. S. (2003). Eustress, distress, and interpretation in occupational stress. J. Manage. Psychol. 18, 726-744.

Lema, S. C., Nevitt, G. A. (2004). Evidence that thyroid hormone induces olfactory cellular proliferation in salmon during a sensitive period for imprinting. J. Exp. Biol. 207, 3317-3327.

Mackintosh, N. (1969). Comparative studies of reversal and probability learning: rats, birds and fish. In: Gilbert, R. M., Sutherland, N. S., Animal Discrimination Learning. Academic Press, New York, London, 137-162.

Magnan, P., Audet, C., Legault, M., Rodríguez, M. A., Taylor, E. B. (2002). Ecology, behaviour and conservation of the charrs, genus *Salvelinus*. Springer, Dordrecht.

Malavasi, S., Georgalas, V., Mainardi, D., Torricelli, P. (2008). Antipredator responses to overhead fright stimuli in hatchery-reared and wild European sea bass (*Dicentrarchus labrax* L.) juveniles. Aquac. Res. 39, 276-282.

Manek, A. K., Ferrari, M. C., Sereda, J. M., Niyogi, S., Chivers, D. P. (2012). The effects of ultraviolet radiation on a freshwater prey fish: physiological stress response, club cell investment, and alarm cue production. Biol. J. Linn. Soc. 105, 832-841.

Manning, A. W. G., Dawkins, M. S. (2012). Introduction to Animal Behaviour. Cambridge University Press, Cambridge.

Maruska, K. P., Zhang, A., Neboori, A., Fernald, R. D. (2013). Social opportunity causes rapid transcriptional changes in the social behaviour network of the brain in an African cichlid fish. J. Neuroendocrinol. 25, 145-157.

Matsuda, K. (2013). Regulation of feeding behavior and psychomotor activity by corticotropin-releasing hormone (CRH) in fish. Front. Neurosci. 7.

McAroe, C. L., Craig, C. M., Holland, R. A. (2015). Place versus response learning in fish: a comparison between species. Anim. Cogn., 1-9.

McCauley, R., and Casselman, J. (1981). The final preferendum as an index of the temperature for optimum growth in fish. In: *Proceedings of the World Symposium on Aquaculture in Heated Effluents and Recirculation Systems*. Vol. 2, pp. 81-93, Stavanger Norway.

McCormick, S. D., Farrell, A. P., Brauner, C. J. (2013). Fish Physiology: Euryhaline Fishes, *Vol.* 32. *Academic Press*. Academic Press, London.

McEwen, B. S. (2005). Stressed or stressed out: what is the difference?. J. Psychiatry Neurosci. 30, 315.

McFarland, D. (1999). Animal Behaviour. Prentice-Hall, New York, NY.

McKinney, G. J., Hale, M. C., Goetz, G., Gribskov, M., Thrower, F. P., Nichols, K. M. (2015). Ontogenetic changes in embryonic and brain gene expression in progeny produced from migratory and resident *Oncorhynchus mykiss*. Mol. Ecol. 24 (8), 1792-1809.

McLaughlin, R. L., Grant, J. W. A., Kramer, D. L. (1994). Foraging movements in relation to morphology, water-column use, and diet for recently emerged brook trout (*Salvelinus fontinalis*) in still-water pools. Can. J. Fish. Aquat. Sci. 51, 268-279.

McLaughlin, R. L., Ferguson, M. M., Noakes, D. L. G. (1999). Adaptive peaks and alternative foraging tactics in brook charr: evidence of short-term divergent selection for sitting-and-waiting and actively searching. Behav. Ecol. Sociobiol. 45, 386-395.

McLeese, D. W. (1956). Effects of temperature, salinity and oxygen on the survival of the American lobster. J. Fish. Res. Bd. Can. 13, 247-272.

Melnychuk, M. C., Korman, J., Hausch, S., Welch, D. W., McCubbing, D. J., Walters, C. J. (2014). Marine survival difference between wild and hatchery-reared steelhead trout determined during early downstream migration. Can. J. Fish. Aquat. Sci. 71, 831-846.

Mesquita, F. O., Borcato, F. L., Huntingford, F. A. (2015). Cue-based and algorithmic learning in common carp: a possible link to stress coping style. Behav. Processes. 115, 25-29.

Millot, S., Péan, S., Chatain, B., Bégout, M.-L. (2011). Self-feeding behavior changes induced by a first and a second generation of domestication or selection for growth in the European sea bass, *Dicentrarchus labrax*. Aquat. Living Resour. 24, 53-61.

Mittelbach, G. G., Ballew, N. G., Kjelvik, M. K., Fraser, D. (2014). Fish behavioral types and their ecological consequences. Can. J. Fish. Aquat. Sci. 71, 927-944.

Mommer, B. C., Bell, A. M. (2013). A test of maternal programming of offspring stress response to predation risk in threespine sticklebacks. Physiol. Behav. 122, 222-227.

Monaghan, P., Spencer, K. A. (2014). Stress and life history. Curr. Biol. 24, R408-R412.

Morandini, L., Honji, R. M., Ramallo, M. R., Moreira, R. G., Pandolfi, M. (2014). The interrenal gland in males of the cichlid fish Cichlasoma dimerus:

relationship with stress and the establishment of social hierarchies. Gen. Comp. Endocrinol. 195, 88-98.

Morin, P.-P., Dodson, J. J., Doré, F. Y. (1989). Thyroid activity concomitant with olfactory learning and heart rate changes in Atlantic salmon *Salmo salar*, during smoltification. Can. J. Fish. Aquat. Sci. 46, 131-136.

Nakano, S. (1994). Variation in agonistic encounters in a dominance hierarchy of freely interacting red-spotted masu salmon (*Oncorhynchus masou ishikawai*). Ecol. Freshw. Fish. 3, 153-158.

Nakano, S., Murakami, M. (2001). Reciprocal subsidies: dynamic interdependence between terrestrial and aquatic food webs. Proc. Natl. Acad. Sci. 98, 166-170.

Nielsen, J. L. (1992). Microhabitat-specific foraging behavior, diet, and growth of juvenile coho salmon. Trans. Am. Fish. Soc. 121, 617-634.

Nilsson, J., Stien, L. H., Fosseidengen, J. E., Olsen, R. E., Kristiansen, T. S. (2012). From fright to anticipation: reward conditioning versus habituation to a moving dip net in farmed Atlantic cod (*Gadus morhua*). Appl. Anim. Behav. Sci. 138, 118-124.

Noakes, D. L. (1986). When to feed: decision making in sticklebacks, *Gasterosteus aculeatus*. In: Simensted, C. A., Cailliet, G. M., Contemporary studies on fish feeding: the proceedings of GUTSHOP'84. Springer, Dordrecht, 95-104.

Noakes, D. L. G. (1986). Genetic basis of fish behaviour. In: Pitcher, T. J., The behaviour of teleost fishes. Croom Helm, London, 3-22.

Noakes, D. L. G. (1999). Onogeny of behaviour. In: Keenleyside, M. H. A., Behaviour of Cichlid Fishes. Croom-Helm, London.

Noakes, D. L. G. (2014). Behavior and genetics of salmon. In: Wood, P. T., Noakes, D. J., Salmon: Biology, Ecological Impacts and Economic Importance. Nova, New York, NY, 195-222.

Noakes, D. L. G., Baylis, J. R. (1990). Fish behavior. In: Schreck, C. B., Moyle, P. B., Methods in Fish Biology. American Fisheries Society, Bethesda, MD, 553-585.

Noakes, D. L. G., Corrarino, C. (2010). The oregon hatchery research center: an experimental laboratory in a natural setting. World Aquac. 41, 33-37.

Noakes, D. L., Grant, J. (1992). Feeding and social behaviour of brook and lake charrFeeding Behavior and Culture of Salmonid Fishes. World Aquaculture Society, The Oregon Hatchery Research Center, Orlando, 13-20.

Noakes, D. L., Leatherland, J. F. (1977). Social dominance and interrenal cell activity in rainbow trout, *Salmo gairdneri* (Pisces, Salmonidae). Environ. Biol. Fishes. 2, 131-136.

Nordeng, H. (1977). A pheromone hypothesis for homeward migration in anadromous salmonids. Oikos. 28, 155-159.

Nordeng, H. (1983). Solution to the "Char Problem" based on Arctic char (*Salvelinus alpinus*) in Norway. Can. J. Fish. Aquat. Sci. 40, 1372-1387.

Nordeng, H. (2009). Char ecology. Natal homing in sympatric populations of anadromous Arctic char *Salvelinus alpinus* (L.): roles of pheromone recognition. Ecol. Freshw. Fish. 18, 41-51.

Nordeng, H., Bratland, P. (2006). Homing experiments with parr, smolt and residents of anadromous Arctic char *Salvelinus alpinus* and brown trout *Salmo trutta*: transplantation between neighbouring river systems. Ecol. Freshw. Fish. 15, 488-499.

Offitt, G. C. (1971). Response of the tautog (Tautoga onitis, teleost) to acoustic stimuli measured by classically conditioning the heart rate. Cond. Reflex. 6, 205-214.

Olla, B., Davis, M., Ryer, C. (1994). Behavioural deficits in hatchery-reared fish: potential effects on survival following release. Aquac. Fish. Manage. 25 (Suppl. 1), 19-34.

Oswald, M. E., Singer, M., Robison, B. D. (2013). The quantitative genetic architecture of the bold-shy continuum in zebrafish, *Danio rerio*. PLoS One. 8, e68828.

Ott, M. E., Heisler, N., Ultsch, G. R. (1980). A re-evaluation of the relationship between temperature and the critical oxygen tension in freshwater fishes. Comp. Biochem. Physiol. A Physiol. 67, 337-340.

Paetzold, S. C., Ross, N. W., Richards, R. C., Jones, M., Hellou, J., Bard, S. M. (2009). Up-regulation of hepatic ABCC2, ABCG2, CYP1A1 and GST in multixenobiotic-resistant killifish (*Fundulus heteroclitus*) from the Sydney Tar Ponds, Nova Scotia, Canada. Mar. Environ. Res. 68, 37-47.

Pankhurst, N. (1997). Temperature effects on the reproductive performance of fish. Cambridge University Press, Cambridge, 159.

Pankhurst, N. W. (2011). The endocrinology of stress in fish: an environmental perspective. Gen. Comp. Endocrinol. 170, 265-275.

Patton, B. W., Braithwaite, V. A. (2015). Changing tides: ecological and historical perspectives on fish cognition. Wiley Interdiscip. Rev. Cogn. Sci. 6, 159-176.

Pavlidis, M., Theodoridi, A., Tsalafouta, A. (2015). Neuroendocrine regulation of the stress response in adult zebrafish, *Danio rerio*. Prog. Neuropsychopharmacol. Biol. Psychiatry. 60, 121-131.

Penaluna, B. E., Dunham, J. B., Noakes, D. L. (2015). Instream cover and shade mediate avian predation on trout in semi-natural streams. Ecol. Freshw. Fish. 25,

405-411.

Pham, M., Raymond, J., Hester, J., Kyzar, E., Gaikwad, S., Bruce, I. (2012). Assessing social behavior phenotypes in adult zebrafish: shoaling, social preference, and mirror biting tests. In: Kalueff, A. V., Stewart, A. M., Zebrafish protocols for neurobehavioral research. Dordrecht: Humana Press, Springer, , 231-246.

Piccolo, J. J., Hughes, N. F., Bryant, M. D. (2007). The effects of water depth on prey detection and capture by juvenile coho salmon and steelhead. Ecol. Freshw. Fish. 16, 432-441.

Piccolo, J., Noakes, D. L., Hayes, J. W. (2014). Preface to the special drift foraging issue of environmental biology of fishes. Environ. Biol. Fishes. 97, 449-451.

Pickering, A., Griffiths, R., Pottinger, T. (1987). A comparison of the effects of overhead cover on the growth, survival and haematology of juvenile Atlantic salmon, *Salmo salar* L., brown trout, *Salmo trutta* L., and rainbow trout, *Salmo gairdneri* Richardson. Aquaculture. 66, 109-124.

Quinn, T. P. (2005). The Behavior and Ecology of Pacific Salmon and Trout. American Fisheries Society, Bethesda, MD.

Quinn, T. P., Hara, T. J. (1986). Sibling recognition and olfactory sensitivity in juvenile coho salmon (*Oncorhynchus kisutch*). Can. J. Zool. 64, 921-925.

Ramallo, M. R., Morandini, L., Alonso, F., Birba, A., Tubert, C., Fiszbein, A. (2014. The endocrine regulation of cichlids social and reproductive behavior through the eyes of the chanchita, *Cichlasoma dimerus* (Percomorpha; Cichlidae). J. Physiol. Paris. 108, 194-202.

Reebs, S. G. (2001). Fish Behaviour in the Aquarium and in the Wild. Cornell University Press, Ithaca, NY.

Rey, S., Huntingford, F. A., Boltana, S., Vargas, R., Knowles, T. G., Mackenzie, S. (2015). Fish can show emotional fever: stress-induced hyperthermia in zebrafish. Proc. Biol. Sci. 282.

Ridley, M. (2003). Evolution. Ridley, 2003 Ridley, M., Evolutionthird ed2003Wiley, New York, NY, Wiley, New York, NY.

Roche, D. P., McGhee, K. E., Bell, A. M. (2012). Maternal predator-exposure has lifelong consequences for offspring learning in threespined sticklebacks. Biol. Lett. 8, 932-935.

Rodriguez, F., Duran, E., Gomez, A., Ocana, F., Alvarez, E., Jimenez-Moya, F. (2005). Cognitive and emotional functions of the teleost fish cerebellum. Brain Res. Bull. 66, 365-370.

Romer, J. D., Leblanc, C. A., Clements, S., Ferguson, J. A., Kent, M. L.,

Noakes, D. (2013). Survival and behavior of juvenile steelhead trout (*Oncorhynchus mykiss*) in two estuaries in Oregon, USA. Environ. Biol. Fishes. 96, 849-863.

Ruzzante, D. E. (1994). Domestication effects on aggressive and schooling behavior in fish. Aquaculture. 120, 1-24.

Schreck, C. B. (1990). Physiological, behavioral, and performance indicators of stress. Am. Fish. Soc. Symp. 8, 29-37.

Schreck, C. (2000). Accumulation and long-term effects of stress in fish. Biol. Anim. Stress. , 147-158.

Schreck, C. , Olla, B. , Davis, M. , Iwama, G. , Pickering, A. , Sumpter, J. (1997). Behavioral responses to stress. Cambridge University Press, Cambridge, 145-170.

Schreck, C. B. , Tort, L. (2016). The Concept of Stress in Fish. In: Schreck, C. B. , Tort, L. , Farrell, A. P. , Brauner, C. J. , . Academic Press, San Diego, CA.

Seals Price, C. C. B. S. (2003). Stress and saltwater entry behavior of juvenile chinook salmon (*Oncorhynchus tshawytscha*): conflict in physiological motivation. Can. J. Fish. Aquat. Sci. 60, 910-918.

Selye, H. (1975). Confusion and controversy in the stress field. J. Hum. Stress. 1, 37-44.

Selye, H. (1976). Forty years of stress research: principal remaining problems and misconceptions. Can. Med. Assoc. J. 115, 53.

Selye, H. (1976). The stress concept. Can. Med. Assoc. J. 115, 718.

Selye, H. (1976). Stress without distress. Signet, New York, NY.

Selye, H. (1978). On the real benefits of eustress. Psychol. Today. 11, 60-70.

Siegel, S. , Ramos, B. (2002). Applying laboratory research: drug anticipation and the treatment of drug addiction. Exp. Clin. Psychopharmacol. 10, 162.

Sloman, K. (2010). Exposure of ova to cortisol pre-fertilisation affects subsequent behaviour and physiology of brown trout. Horm. Behav. 58, 433-439.

Smith, R. J. F. (1992). Alarm signals in fishes. Rev. Fish Biol. Fish. 2, 33-63.

Sneddon, L. U. , Wolfenden, D. C. C. , Thomson, J. S. (2016). Stress Management and Welfare. In: Schreck, C. B. , Tort, L. , Farrell, A. P. , Brauner, C. J. , . Academic Press, San Diego, CA.

Spagnoli, S. , Kent, M. L. , Lawrence, C. (2016). Stress in laboratory fishes. In: Schreck, L. T. C. B. , Stress in Fishes. Academic Press, New York, NY.

Stabell, O. B. (1984). Homing and olfaction in salmonids: a critical review with special reference to the Atlantic salmon. Biol. Rev. 59, 333-388.

Stephens, D. W. , Krebs, J. R. (1987). Forgaing Theory. Princeton University Press,

Princeton, NJ.

Suboski, M. D. , Templeton, J. J. (1989). Life skills training for hatchery fish: social learning and survival. Fish. Res. 7, 343-352.

Szabo, S. , Tache, Y. , Somogyi, A. (2012). The legacy of Hans Selye and the origins of stress research: a retrospective 75 years after his landmark brief "Letter" to the Editor# of nature. Stress. 15, 472-478.

Takei, Y. , Hwang, P. -P. (2016). Homeostatic Responses to Osmotic Stress. In: Schreck, C. B. , Tort, L. , Farrell, A. P. , Brauner, C. J. , . Academic Press, San Diego, CA.

Thompson, T. , Sturm, T. (1965). Classical conditioning of aggressive display in Siamese fighting fish. J. Exp. Anal. Behav. 8, 397-403.

Thompson, N. F. , Leblanc, C. A. , Romer, J. D. , Schreck, C. B. , Blouin, M. S. , Noakes, D. L. (2015). Sex-biased survivorship and differences in migration of wild steelhead (*Oncorhynchus mykiss*) smolts from two coastal Oregon rivers. Ecol. Freshw. Fish.

Thörnqvist, P. -O. , Höglund, E. , Winberg, S. (2015). Natural selection constrains personality and brain gene expression differences in Atlantic salmon (*Salmo salar*). J. Exp. Biol. 218, 1077-1083.

Thorpe, J. E. , Morgan, R. I. G. (1978). Parental influence on growth rate, smolting rate and survival in hatchery reared juvenile Atlantic salmon, *Salmo salar*. J. Fish. Biol. 13, 549-556.

Tudorache, C. , ter Braake, A. , Tromp, M. , Slabbekoorn, H. , Schaaf, M. J. (2015). Behavioral and physiological indicators of stress coping styles in larval zebrafish. Stress. 18, 1-8.

Ueda, H. (2014). Homing ability and migration success in Pacific salmon: mechanistic insights from biotelemetry, endocrinology, and neurophysiology. Mar. Ecol. Progr. Ser. 496, 219-232.

Waples, R. S. , Hendry, A. P. (2008). Evolutionary perspectives on salmonid conservation and management. Evol. Appl. 1, 183-188.

Wegner, N. C. , Snodgrass, O. E. , Dewar, H. , Hyde, J. R. (2015). Whole-body endothermy in a mesopelagic fish, the opah, *Lampris guttatus*. Science. 348, 786-789.

White, G. , Brown, C. (2015). Microhabitat use affects goby (Gobiidae) cue choice in spatial learning task. J. Fish. Biol. 86 (4), 1305-1318.

Wiley, R. W. , Whaley, R. A. , Satake, J. B. , Fowden, M. (1993). An evaluation of the potential for training trout in hatcheries to increase poststocking survival in streams. N. Am. J. Fish. Manage. 13, 171-177.

Wiseman, S. M. (2015). Soundscape response in animals. J. Acoust. Soc. Am. 138.

Woo, P. T. K., Noakes, D. J. (2014). Salmon: Biology, Ecological Impacts and Economic Importance. Nova Science, New York, NY.

Wood, C. M., MacDonald, D. G. (1997). Global warmingSociety for Experimental Biology Seminar Series (No. 61). Cambridge University Press, Cambridge.

Woodward, C. C., Strange, R. J. (1987). Physiological stress responses in wild and hatchery-reared rainbow trout. Trans. Am. Fish. Soc. 116, 574-579.

Young, P. S., Swanson, C., Cech, J. J. (2010). Close encounters with a fish screen III: behavior, performance, physiological stress responses, and recovery of adult delta smelt exposed to two-vector flows near a fish screen. Trans. Am. Fish. Soc. 139, 713-726.

Yue, S., Duncan, I., Moccia, R. D. (2006). Do differences in conspecific body size influence social stress in domestic rainbow trout?. Environ. Biol. Fishes. 76, 425-431.

Zoladz, P. R., Clark, B., Warnecke, A., Smith, L., Tabar, J., Talbot, J. N. (2011). Pre-learning stress differentially affects long-term memory for emotional words, depending on temporal proximity to the learning experience. Physiol. Behav. 103, 467-476.

第 10 章　应激和疾病防御：免疫系统和免疫内分泌的相互作用

鱼类的内分泌与免疫的关系，其中一部分与应激反应相关，主要是由激素和细胞因子的紧密相互作用所介导的。事实上，应激能够抑制某些免疫系统组分，使得鱼类易受感染和疾病的侵害。本章总结了应激源对抗病能力和免疫系统的影响，同时更新了鱼类内分泌对免疫系统调节作用的相关知识，包括系统和局部水平的影响，以及应激条件下免疫应答的组织，特别强调了激素、它们的受体以及系统相互作用的影响。基本上，低水平的应激刺激（真正应激或良性应激）可能会导致免疫力的增强，而较严重的应激刺激则会造成免疫抑制。应激源所引起的免疫应答是由中枢和外周的内分泌系统介导的。

10.1　导　言

像在所有动物中一样，先前遭受过应激刺激的鱼类会出现典型的疾病。只要影响了与能量供应或者与代谢途径相关的免疫分子，各种类型的应激源都可能会影响免疫力。应激源显著地激活了激素并且直接影响了受体介导的与免疫系统分子的相互作用，导致免疫力和抗病力的丧失。已知有如下这些免疫抑制效应的应激源，如养殖管理方式、行为应激、环境污染、饮食习惯相关的改变。适应负荷由一个能降低免疫机制有效性的应激源引起，允许病原体对宿主产生更大的毒力作用。例如，已经表明温度的急剧改变（从27℃至19℃～23℃或者31℃～35℃）会降低罗非鱼（*Oreochromis mossambicus*）对病原体的抗病能力且削弱宿主的免疫应答（Ndong 等，2007）。多变的水温增加了舌齿鲈（*Dicentrarchus labrax*）对病原体（如诺达病毒）的易感性，并使舌齿鲈血浆中的皮质醇浓度、渗透度、IgM 水平、体重急剧改变（Varsamos 等，2006）。相似地，越冬综合征会令金头鲷（*Sparus aurata*）更易感染鳗败血假单胞菌（*Pseudomonas anguilliseptica*），金头鲷的血浆皮质醇水平升高并且产生严重的免疫抑制效应（Tort 等，1998）。行为应激，例如建立社会等级的过程，会导致占优势地位的富攻击性的北极红点鲑（*Salvelinus alpinus*）之间明显的打斗，造成抗病力明显下降，包括消除已黏附在肠道上的益生菌和起保护作用的黏膜（Ringo 等，2014）。斑点叉尾鮰经低水压处理（Small，Bilodeau，2005），或者在应激源刺激或者糖皮质醇处理后暴露于病原原生生物时，病原体易感性和死亡数增加（Davis 等，2003）。

本章综述应激源对免疫应答的影响，包括感受应激源的中枢和局部机制、细胞和分子反应的产生，以及所导致的促进和抑制效应。我们也强调激素作为内分泌系统和免疫系统之间信息的传递者，包括下丘脑激素、垂体激素、肾间激素和生长激素，同时也考虑环境应激源对鱼类免疫的影响。

10.2　应激源对免疫应答的影响

发生在应激鱼类的多种免疫变化都可以认为是对免疫过程的抑制（图 10.1）。尽管一些应激会增强或激活免疫应答，这些应激中的大部分对于免疫力是有害的，我们将会在后续的内容中继续讨论。值得一提的是，长期的应激源通常会产生抑制效应。拥挤的海鲈免疫力下降，表现为细胞毒性减弱及化学发光活性降低（Vazzana 等，2002）。相似地，高密度养殖的真鲷（Montero 等，1999；MacKenzie 等，2006a，b）和赤鲷（*Pagrus pagrus*）（Rotllant，Tort，1997）的补体蛋白水平会降低，而皮质醇水平会升高。高密度饲养的金头鲷（*Sparus aurata*）头肾白细胞的补体蛋白水平和吞噬活性会受到抑制，并且能够促进细胞向血液中迁移（Ortuno 等，2001）。反复追逐 3 天以上也会造成真鲷补体活性、溶菌酶活性、血凝集活性和抗体效价等水平的降低（Sunyer，Tort，1995）。

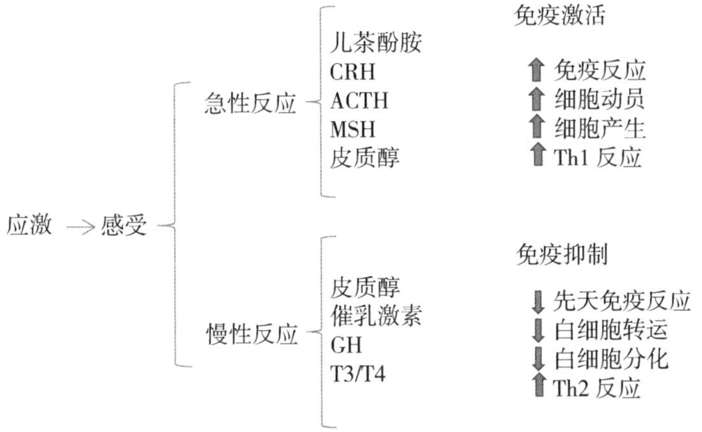

图 10.1　应激源和免疫激活及免疫抑制之间的关系

注：图表反映了应激源引起的主要的免疫应答和所参与的激素。急性反应是由激活机制鉴别的，尤其是在头肾中细胞的产生和动员及 Th1 淋巴细胞的促炎症反应。慢性反应通常呈现抑制性，导致先天免疫应答的削弱，抑制白细胞分化和将 Th2 淋巴细胞介导的促炎反应向抑制转变。CRH，促肾上腺皮质激素释放激素；ACTH，促肾上腺皮质激素；MSH，促黑激素；GH，生长激素；T3，三碘甲腺原氨酸；T4，甲状腺素；TH1，I 型辅助性 T 淋巴细胞；TH2，II 型辅助性 T 淋巴细胞；向上的箭头 = 激活；向下的箭头 = 抑制。

在基因组水平上同样能够观察到这些类型的反应，因为在应激的鱼类中，免疫相关的基因可能会显著上调。例如，虹鳟（*Oncorhynchus mykiss*）在面对重金属铜暴露或细菌脂多糖等应激源时，免疫相关基因如 IL-1β、IL-6 和 TNF-α 的表达会出现显著变化（Teles 等，2011）。在真鲷的饲料中，用蔬菜油来代替鱼油也会诱导促炎细胞因子表达的改变，同时会抑制免疫应答（Montero 等，2003，2010）。大西洋鳕鱼

（*Gadus morhua*）受持续的手工操作应激后会增加 IL-1β 的本底表达，而杀鲑气单胞菌（*Aeromonas salmonicida*）感染会降低大西洋鲑鱼（*Salmo salar*）的白细胞活性（Fast 等，2008）。低氧的河鲈（*Perca fluviatilis*）减少了 C3 补体蛋白的表达并诱导转铁蛋白的过度表达（Douxfils 等，2014）。

10.2.1 抑制作用与增强作用

虽然应激源趋向于引发对机体的整体抑制或不利的影响，一些应激源引起的单个免疫应答也会出现适应反应（adaptive responses）。因此，一个复杂的状态如整体应激反应将会取决于所处的时间、反应本身、应激源的强度和持续的时间。虽然特异性的增强反应有时候会出现，但是大多数观察到的反应是抑制的（Miller 等，1995；Dhabhar，2002，2008）。因此，改变淋巴细胞向皮肤和其他外周位置转运，在这些部位应激源会引起免疫应答的增强（Dhabhar，2000），同时，这些反应也许会在其他部位被延迟。先前的应激暴露会影响免疫系统回应的能力以及回应的方式。因此，调节系统的有效性需要克服应激源的影响，包括免疫活性的激活。目前对于这些效应的理解主要在细胞和分子水平，但仍不能够有效地预测任何特定的应激源对于特定的免疫应答或促炎性免疫应答的影响（Pruett，2003）。关于鱼类曾经历应激源处理的影响、如何预测以及各种反应的变化，可以参阅本书的其他章节（见本书第 2、3、9 章）。

应激源引起的免疫应答可分为两个阶段。首先，主要的反应是激活大量的和挑战直接相关的应答机制，刺激受体和引发快速反应。值得一提的是，这个活化阶段和急性期蛋白的产生相关，释放短暂储存或者产生的激素和肽类，并且动员已有的资源。一些活化反应包括促炎性标记物的增加、溶菌酶活性的增加、增加补体蛋白 C3 的表达（Sunyer，Tort，1995；Demers，Bayne，1997），还包含适应性免疫应答的增强。MacArthur 等（1984）向欧洲鲽（*Pleuronectes platessa*）腹腔注射内毒素和皮质醇后，发现腹腔灌洗中的髓系白细胞增加。短期手工操作应激也会促使银大麻哈鱼（*Oncorhynchus kisutch*）每个头肾白细胞的糖皮质激素受体位点大量增加（Maulet 和 Schreck，1991）。短期的拥挤应激使得促炎性因子 IL-1β 和 IL-8 表达上调，同时抗菌基因和 g-型溶菌酶的量也会增加（Caipang 等，2008）。

如果应激源持续存在，第二阶段通常会引起抑制。这个抑制是由两个主要因素造成的：缺乏资源（异型稳态负荷）以维持细胞分裂和增殖的蛋白合成的高需求，同时相关介体的作用能够引起免疫机制效能下调或降低。抑制效应主要和 HPI 轴释放的最终应激激素（尤其是皮质醇）有关；然而，部分反应也可由调节激素介导，如儿茶酚胺或促肾上腺皮质激素释放激素。这个双重的反应方式，如前文所述，取决于反应的时间过程、应激源的持续性和机体曾经的经历。

用来预测激活或抑制反应的指标很难确定，这是由于受到应激条件和时间的限制。在进行抗原性刺激之前给予一个急性应激处理，会显著增强皮肤中的细胞免疫并促进皮肤的白细胞重新调动，而在敏化作用之前给予慢性应激处理，皮肤免疫应答会出现抑制（Dhabhar，2008）。这表明应激对皮肤所介导的细胞免疫具有双向影响。总体来说，我们必须理解的是，目前还没有一个标记物能够确定应激所导致的是激活还是抑制作用，以及预测应激对免疫系统影响的严重性。

10.2.2　免疫刺激后的应激感受：全身与局部反应

应激源被一个或者多个感知系统（如神经传感器、激素或者免疫受体）感知后会激活神经内分泌反应。后续免疫机制的活化是由激素介导的，这些激素（主要是皮质醇）在所有细胞中都有相应的受体，随后激活大量的通道（图 10.2）。免疫细胞上也有 GRs，因而能够直接反应循环系统中的皮质醇（Gorissen，Flik，2016；见本书第 3 章）。皮质类固醇激素对免疫系统的作用将在 10.4 节中综述。

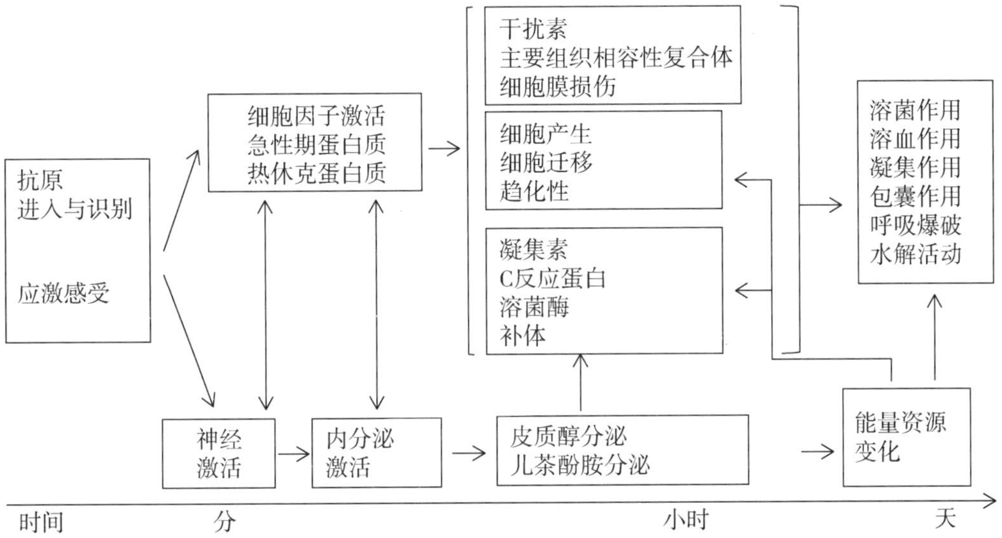

图 10.2　对应激源和抗原性分子的免疫和神经内分泌反应之间共激活的时间进程

注：箭头表示各因子之间的相互作用。

然而，伴随着这些激素介导的免疫应答的是与之相关的局部反应。外部刺激，例如水中的物理化学变化、皮肤损伤，以及食物诱导的肠道环境变更等等，都可以作为应激源在鱼类特异的交互表面（通常是鳃、皮肤或肠）而检测到。已经表明局部反应所涉及机制不仅是应激源特异的，而且也和预警反应变化（包括免疫通道）有关。因此，器官和全身的区室（头肾、胸腺和脾脏）负责白细胞的产生、T 细胞的增殖和抗原捕获，而局部反应涉及先天性和获得性免疫媒介，包括细胞免疫和体液免疫，包含帮助组织免受病原体侵害的因子有溶菌酶、凝集素、蛋白酶和补体蛋白等。局部反应发生在黏膜表面，而中枢器官中并未发生反应的预激活。这是预警状态总体反应的一个关键点，因此，局部和全身性反应会结合起来产生显著的效果。

尽管全身性免疫内分泌反应的研究相对来说较为深入，但最近已对局部反应的重要性进行了研究，基本上都是在侵入口（如鳃、皮肤、肠道）发生反应，特别是黏膜上发生的反应。发生在黏膜上的免疫应答不仅在病原体的识别上起着“第一道防线”的作用，而且它也参与了各种免疫应答机制、吞噬作用、抗体反应、抗菌分子，甚至更显著的，能够广泛识别大范围的其他类型的应激源（Salinas 等，2011；Caipang 等，2011；Esteban 等，2012）（图 10.3）。

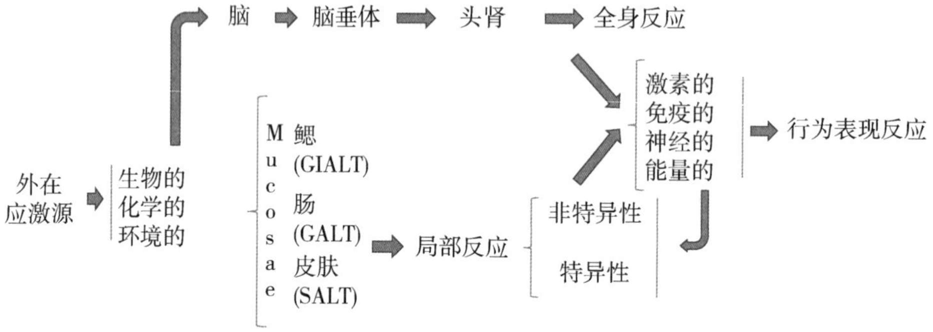

图 10.3　外部刺激所引起的黏膜和全身反应模式

注：局部及全身反应影响总体的表现和它们之间的相互作用。GIALT，鳃相关的淋巴组织；GALT，肠道相关的淋巴组织；SALT，皮肤相关的淋巴组织。

黏膜表面持续从环境中获取信息、识别变化并处理这些信息以便鱼类能够适应环境，从而维持稳态并存活下来。这些黏膜表面的功能是多种多样的，从种内和种间的交流、营养吸收、气体交换到微生物的识别（Salinas 等，2011）。在所有脊椎动物和许多无脊椎动物中，作为免疫位点的黏膜能够在病原体入侵后触发激烈的免疫应答（Gomez 等，2013；Zhang 等，2014）。尽管物种间存在形态学和生理学差异，但所有脊椎动物都拥有这些黏膜免疫组织或者黏膜相关的淋巴组织（MALT），它们能够控制黏膜位点的免疫应答。在硬骨鱼中，4 个 MALT 已经被详细描述：鼻腔相关的淋巴组织（NALT）、皮肤相关的淋巴组织（SALT）、鳃相关的淋巴组织（GIALT）和肠道相关的淋巴组织（GALT）。这些黏膜组织不仅能够产生一种精准的免疫应答，而且能够对非免疫的应激源做出反应。例如抗菌活性、抗病毒反应、细胞因子的产生、应激反应和抗凋亡活性相关的基因都能够在大西洋鳕鱼的皮肤中找到，由此证明皮肤是免疫活化位点，尤其是鱼类的腹部皮肤（Caipang 等，2011）。

目前没有关于黏膜表面免疫细胞类群在应激后结果的报道。目前已知鳟鱼皮肤中的皮质醇能降低和 B 细胞活化相关的基因的表达（Krasnov 等，2012）。重要的是，分泌型 IgA，一种哺乳动物的黏膜免疫球蛋白，它在黏膜表面的变化取决于应激源的性质，被用作应激的标记分子（Volkmann，Weekes，2006）。在硬骨鱼类中，作为黏膜特异的免疫球蛋白 sIgT/Z，很可能在应激后修饰其分泌活动。一般而言，应激源能够激活免疫应答进程，无论该应激源是不是对免疫的刺激。值得一提的是，这样就可能可以界定应激反应的一些特性，包括免疫系统的抑制，如改变胸腺、脾脏和机体其他淋巴组织的结构，改变白细胞的数量和分布，或者消化道呈现出血和溃疡的症状（Tort，2011）。

10.2.3　应激与细胞及体液免疫应答

应激能够改变组织和器官中免疫细胞的数量和招募速率以增加有效的免疫应答，这包括白细胞在身体各区室内的差异化分布。急性应激刺激头肾中血细胞的产生，包括红细胞和白细胞，这就导致循环系统中的白细胞数量增加，有利于分布到靶组织中。这是由交感神经系统的活化和儿茶酚胺的释放导致的。血液中白细胞类型改变的频率是由急性应激造成的。在应激期间，淋巴细胞和单核细胞的百分比显著降低，而噬中性粒细胞的数量和百分比显著增加（Wojtaszek 等，2002）。遭受急性应激处理的动物，血液中有更多活化的巨噬细胞，同时 T 细胞的活化加强并且增加监视 T 细胞招募到皮肤中。因此，急性应激能够显著提升白细胞亚群进入免疫激活位点的动力学和幅度（Dhabhar，2002）。这些发现表明，先天性免疫（巨噬细胞）和适应性免疫（监视 T 细胞）是重要的相互作用的组分，是应激引起的初级反应（如儿茶酚胺分泌）增强的重要结果。细胞运输是以趋化特异的方式进行的，在这种情况下，组织损伤或抗原/病原体产生的化学诱导物和急性应激协同作用，以进一步确定招募到的特异的白细胞亚群（Dhabhar，2002）。在鱼类中，慢性应激引起相似效应的研究鲜有报道。由于持续应激会造成下丘脑-垂体-肾上腺轴的持续激活，血液白细胞亚群最初的调动和分布情况会因应激强度和持续时间而造成血液白细胞数量的普遍降低。白细胞主要集中在受影响的器官中，因此，中枢器官和血液中都出现细胞数量降低的情况（Dhabhar，2002；Tort，2011）。

免疫应答中的一个关键的细胞类型是 B 淋巴细胞或 B 细胞。硬骨鱼类的 B 淋巴细胞具有先天性免疫和适应性免疫特征。先天性免疫的特征包括天然抗体的分泌、细胞因子的产生和吞噬能力（Sunyer，2013），这些都和预警反应的发生有关。硬骨鱼表达 3 种不同的免疫球蛋白：IgM、IgT（在一些物种中也称作 IgZ）和 IgD。IgM 是血清和皮肤黏膜上最常见的免疫球蛋白，也是全身性免疫应答的关键成员，而 IgT 则是肠黏膜组织主要的免疫球蛋白（Salinas 等，2011）。尽管哺乳动物和鱼类的 B 淋巴细胞都具有这些先天性免疫的能力，它们在机体防御的重要性方面有时是不可比拟的。例如，已经证实哺乳动物的吞噬 B 细胞存在于腹腔并占 B 细胞总数的 10% ～ 15%（Parra 等，2012），而在硬骨鱼中，吞噬 B 细胞存在于全身各个区室（包括血液、脾脏和头肾），并且占 B 细胞总数的 60% （Li 等，2006；Zhang 等，2010）。因此，B 细胞在鱼类先天性免疫中的作用似乎比在哺乳动物中重要得多，同时，鱼类 B 细胞连接先天性免疫和适应性免疫的功能在硬骨鱼类中是最重要的（Zhu 等，2014）。应激会减少免疫后的鲤鱼体内血液循环 B 淋巴细胞的数量和降低抗体反应（Stolte 等，2009）。应激也会诱发海鲈 IgM 水平的降低并增加海鲈对诺达病毒的易感性（Varsamos 等，2006）。总的来说，这表明皮质醇或其他应激介导因子会引起对鱼类免疫力，尤其是对 B 细胞的抑制效应。

用皮质醇处理鱼会导致免疫球蛋白分泌减少或者 IgM 表达量降低。另外，也有关于皮质醇处理引起抗体分泌细胞数量的减少的报道（Hou 等，1999；Saha 等，2004）。鲤鱼注射皮质醇后会诱导其 B 细胞凋亡并减少增殖（Weyts 等，1998b）。尽管皮质醇的在体和离体研究已经广泛用于建立研究硬骨鱼类应激进程的模型，但其他激素也可能参与到应激反应中。例如，鲤鱼在遭受突然的温度变化后皮质醇水平会升高，但抗体的产生并没有减少，抗体分泌细胞的数量也没有变化（Saha 等，2002），这说明在应激状态下，其他组分参与了免疫系统的活化。因此，虽然皮质醇是参与免疫相关应激变化的主要激素，但神经内分泌系统对免疫的调节作用不是单纯利用皮质醇释放和传导信号这么简单。

应激源还会引起鱼类体液免疫应答的一系列变化，首先表现为与非特异性免疫应答相关的大量蛋白被激活，其次是细胞和组织中释放大量介导免疫应答的细胞因子。许多研究表明，应激源可引起溶菌酶、补体蛋白或抗菌肽等先天免疫因子的改变（Montero 等，2009；Saurabh，Sahoo，2008；Mauri 等，2011；Costas 等，2011；Teles 等，2012）。皮质醇可以抑制炎症消退过程中促炎细胞因子的释放。哺乳动物中活化的应激激素可通过抑制 Th1 细胞因子（TNF-α、干扰素 γ、IL-2、IL-12）的产生，诱导其他 Th 亚群的细胞因子［IL-4、IL-10、IL-13、转化生长因子（TGF）-β］的产生来抑制全身性 Th1/炎症反应以及促进向 Th2 反应的转变（Elenkov，Chrousos，1999）。另外，糖皮质类固醇调节适应性的免疫介体，抑制 IL-12 和 IFN 的合成或上调 IL-4、IL-10，并诱导 TGF-β 的分泌（Borghetti 等，2009）。

有证据表明，即使不存在病原体，若干初级免疫应答机制也会受到刺激；这是非抗原应激源在起作用。在分析应激期间免疫应答基因上调后的整体基因表达情况时，这种情况可以清晰观察到（Mackenzie 等，2005；Teles 等，2012；Xia 等，2013；Zhu 等，2013）。尖吻鲈（*Lates calcarifer*）在受到 LPS 处理、哈维氏弧菌感染、高盐环境或禁食等多种应激源胁迫时，肠道 cDNA 文库测序显示，其整体表达谱相似，而无论涉及哪种应激源，在所有应激条件下均有 37 个基因差异表达。这些数据表明，在这些不同的应激条件下，基因调控能够总体协调和微调（Xia 等，2013）。

10.3 应激后免疫应答的组织结构：神经免疫内分泌和头肾的作用

脊椎动物的神经、免疫和内分泌系统协同作用，处理和反应外部或内部的信息输入，特别是在快速和/或强有力的反应的应激情况下。神经系统是调节网络的主要组成部分，调节整体的非特异性即时反应和修饰激素的活性。应激期间释放的应激激素可能有助于免疫系统对潜在的挑战（如感染、伤口修复、组织变化）做好准备，而当应激信号被大脑感知后则可能被视作预警信号。这种联系在所有脊椎动物中都存在，和应激、CRH、肾上腺皮质激素（ACTH）、糖皮质醇、生物胺和细胞因子相关的分子在不同物种和类群的进化过程中都是相似的并被完好地保留下来（Ottaviani，Franceschi，1996）。细胞因子和神经肽在神经内分泌系统和免疫系统中都起作用。这就解释了为什么应激会有免疫和神经内分泌细胞以及信使分子的全身性参与。因此，明显的是，多系统间相互作用的调节和微调可以避免初级反应的过度活化，并允许进一步作用机制的组织以适应特定的挑战。这些互动机制不仅适用于特定的应激时期，对适应和准备应对未来的挑战也很有用。

头肾在组织应激反应中起核心作用，这涉及头肾中 3 个调节系统间的密切沟通（Verburg-Van Kemenade 等，2009；Tort，2011；Gorissen，Flik，2016；见本书第 3 章）。鱼类头肾的功能在脊椎动物中是独一无二的，因为在这个器官中所有 3 个调节系统都发挥着重要的作用：儿茶酚胺（肾上腺素和去甲肾上腺素）是通过特异的自主纤维激活嗜铬组织而释放的；ACTH 刺激肾间细胞分泌皮质醇；免疫细胞的产生和处理是为了后续的成熟和早期分化；制造红细胞是为了向组织输送氧气。当必须组织应激反应时，头肾所有的这些功能就都可以动用起来：调节代谢、渗透压和离子交换，调节免疫应答，为能量代谢提供血细胞和氧气。这样，鱼类红细胞作为有核的细胞也可以在全身性的免疫应答中发挥作用，从而能够合成相关的免疫分子。事实上，鳟鱼红细胞表达和调节特定模式的识别受体 mRNA，并能够识别特定病原体相关分子模式（PAMP），这是先天免疫应答的核心（Morera 等，2011）。不同 PAMPs 的体外刺激导致免疫受体和应答因子 IFN-α 等特异性 mRNA 的从头合成（Morera 等，2011）。总的来说，头肾是组织和整合应激反应时的关键器官，包括几个过程，如血细胞的合成和动员以及导致代谢变化的激素释放。此外，它也是神经免疫内分泌相互作用最有可能发生的地方，从而为多系统相互交叉调节提供了机会。

在这 3 个调节系统之间存在的双向关系中，内分泌-免疫系统的关系得到最为广泛的研究，尤其是皮质醇与免疫应答的相互作用。尽管如此，哺乳动物的内分泌系统和免疫系统之间的相互作用也发生在肾上腺切除的动物身上，这表明各种神经递质、神经肽和下丘脑肽 CRH 也可能会参与此过程（Leonard，2005）。总的来说，目前被评估的任何一种激素都可能对免疫因子或机制产生影响，尽管其中一些激素在调节免疫应答方面发挥着主要作用（Gorissen，Flik，2016；见本书第 3 章）。对于异型稳态负荷，应激时的能量消耗需求很高；其他的激素轴，如主要支持代谢的生长激素（GH）参与的促生长轴（somatotropic axis），也会对免疫功能产生影响。除了免疫系统的能量需求增加外，在免疫激活期间，细胞因子、糖皮质激素和促生长激素之间复杂的相互作用可能会导致代谢的改变，从而减少用于生长的资源。此外，食物摄取量、能量平衡、细胞代谢和免疫功能都是相互关联的，很可能是依赖于中枢（脑、脑垂体）和头肾间的相互作用（Yada 等，2002，2005；Tort，2011）（见下一节）。

10.4 激素在免疫系统中的作用

基本上，本节讨论的所有激素在某种程度上都和应激有关，要么直接作为应激反应的一部分，要么通过改变或扰乱功能而间接发挥作用。此外，免疫刺激激素很可能调节和应激相关的免疫抑制。激素的多能性是鱼类内分泌系统的特点之一，激素在应激过程中所起的作用是非常复杂的。

10.4.1 下丘脑激素

在鱼类的应激反应中，下丘脑控制头肾功能的重要性在本书的其他章节已进行了深入的讨论（特别是 Gorissen，Flik，2016；见本书第 3 章）。CRH 被认为是（肾上腺）皮质类固醇激素类中最重要的一种促分泌物，可作为鱼类 HPI 轴的发动分子；下丘脑 CRH 刺激 ACTH 的分泌，皮质醇分泌受血液循环系统中 ACTH 水平的控制。除了在 HPI 轴中起促分泌作用外，一些下丘脑激素还直接或间接地影响鱼类的免疫系统。

下丘脑神经末梢形成垂体后叶的神经垂体，释放神经垂体激素，如黑色素聚集激素（MCH）。MCH 对鱼类免疫系统的影响与背景颜色的适应有关（Harris，Bird，2000）。利用虹鳟头肾分离出的白细胞进行离体研究，发现 MCH 可直接刺激虹鳟白细胞的吞噬作用（Harris，Bird，1998）。促性腺激素释放激素（GnRH）主要产生于下丘脑，刺激脊椎动物腺垂体释放促性腺激素（Van Der Kraak，2009；Tsutsui 等，2010；Gopurappilly 等，2013）。GnRH 是性腺激素最重要的刺激因子之一，在鱼类性成熟过程中参与免疫抑制作用。GnRH 的作用不局限于下丘脑-垂体-性腺（HPG）轴，还被认为和内源性的生长因子及细胞因子起相互作用（Marchetti 等，2001）。在虹鳟中，GnRH 可促进细胞增殖、吞噬过程中超氧化物的生成，上调白细胞中促炎细胞因子 TNF-α 的 mRNA 水平（Yada，2012）。另外有两种下丘脑肽通过 GnRH 调节生殖，亲吻激动素（kisspeptin）通过 GnRH 神经元的作用刺激促性腺激素的分泌，而促性腺激素抑制激素（GnIH）则起抑制作用（Tsutsui 等，2010；Gopurappilly 等，2013）。这些 HPG 轴的刺激和抑制因子可能在性成熟过程中起免疫调节的作用。

10.4.2 垂体激素

垂体激素是调节鱼类免疫功能的重要因子。早前关于垂体切除对硬骨鱼影响的研究表明垂体激素在维持有效的免疫系统中起重要作用（Slicher，1961；Pickford，1971；Yada，Nakanishi，2002）。注射生长激素可以在许多方面增强鱼类免疫的特异性和非特异性防御（Balm，1997；Weyts 等，1999；Yada，2007；Tort，2011）。Calduke-Giner 等（1995）揭示了真鲷头肾淋巴细胞中存在生长激素结合位点。目前已从几种鱼类中分离并测序了鱼类生长激素的受体（Calduch-Giner 等，2000；Fukada 等，2001；Lee 等，2001；Shved 等，2009，2011）。和上述的刺激作用相反，长期使用生长激素会抑制鱼类的免疫功能（Cuesta 等，2005，2006）。生长激素转基因对鱼类免疫功能的影响是模棱两可的，有的显示刺激作用，有的则显示有抑制作用或没有作用（Jhingan 等，2003；Wang 等，2006；Mori 等，2007）。生长激素对罗非鱼细胞因子 TNF-α 生成的调节作用也不明确（Shved 等，2011）。这些矛盾的结果可能和不同研究系统在发育和能量状态上的差异有关。

3 个垂体肽类，生长激素、催乳素（PRL）和生长促乳素（SL），被认为是来自同一个祖分子（Kaneko，1996；Kawauchi 等，2009）。PRL 对鱼类免疫系统的影响和 GH、PRL 增强虹鳟吞噬作用、白细胞增殖作用十分相似（Sakai 等，1996a，b；Yada 等，2004）。PRL 受体 mRNA 在几种鱼类的肾脏和肠道中表达（Sandra 等，2000；Prunet 等，2000；Tseet 等，2000；Higashimoto 等，2001；Santos 等，2001）。从尼罗罗非鱼外周血液和头肾分离的白细胞中也观察到 PRL 受体的表达，说明 PRL 对鱼的免疫系统有直接作用（Sandra 等，2000）。在海水驯化的罗非鱼的头肾白细胞中，PRL 受体 mRNA 水平升高，同时呼吸爆发作用增强（Yada 等，2002）。生长促乳素对鱼类免疫功能没有明显影响（Sakai 等，1996a）。但在虹鳟一个缺失大部分垂体中间部的突变体中，血浆皮质醇水平和溶菌酶活性要比正常鱼低。生长促乳素产生的细胞位于垂体中间部的神经垂体组织，而这个虹鳟突变体血液循环中的生长促乳素水平非常低。生长促乳素可能通过其他免疫调节因子间接参与鱼类免疫功能的调节。

ACTH 和阿片促黑素皮质激素原（POMC）来源的肽类除了在 HPI 轴中发挥作用外，还直接影响鱼类的免疫功能，这与其在高等脊椎动物中的功能相似。ACTH 受体样分子在海鲈的胸腺和脾脏中表达（Mola 等，2005）。在体外使用 ACTH、黑色素细胞刺激激素和 POMC N-端肽，能增强鳟鱼和鲤鱼的几种先天性免疫功能（Bayne，Levy，1991a，b；Harris，Bird，1997，1998，2000；Harris 等，1998；Takahashi 等，2000；Sakai 等，2001）。此外，ACTH 还能调节金头鲷白细胞中一些促炎细胞因子基因的表达，抑制 IL-1β，刺激 IL-6、TNF-α 和 TGF-β1（Castillo 等，2009）。β-内啡肽是另一种 POMC 来源的肽，它直接刺激鳟鱼和鲤鱼的几种免疫功能（Watanuki 等，1999，2000；Takahashi 等，2000）。阿片样肽调节哺乳动物的免疫功能，而阿片样物质对鱼类炎症的调节作用可以由阿片样物质受体拮抗剂纳洛酮和金鱼头肾细胞特异性结合的定位而得到支持（Chadzinska 等，1997）。

10.4.3　肾间激素

内分泌对鱼类免疫功能调控的研究大多都关注于机体通过增加肾间激素分泌来抑制免疫应答，以应对环境的应激源。然而，肾间激素，如肾上腺素和皮质醇，对鱼类免疫系统的几个部位具有抑制作用和刺激作用。

10.4.3.1　下丘脑-交感神经-嗜铬细胞轴的免疫调节作用

下丘脑-交感神经-嗜铬细胞轴是介导儿茶酚胺释放所引起应激反应的重要调节系统（Randall，Perry，1992）。嗜铬细胞中儿茶酚胺的释放主要受节前交感神经的调节。内分泌和非内分泌因子，如血浆离子水平的波动，也参与儿茶酚胺释放的调节（Randall，Perry，1992；Wendelaar Bonga，1997）。交感神经对鱼类免疫系统的自发调节似乎有抑制作用，至少对特异性免疫是如此。Flory（1989）证明化学切除交感神经可增强银大麻哈鱼的抗体反应。使用肾上腺素和去甲肾上腺素能降低翠鳢（*Channa punctatus*）的吞噬作用（Roy，Rai，2008）。此外，肾上腺素体外给药降低了金头鲷白细胞中促炎细胞因子 mRNA 水平（Castillo 等，2009）。

利用受体的激动剂和拮抗剂是研究儿茶酚胺对鱼类免疫功能影响的一种方法，然而，结果似乎会出现矛盾。β-肾上腺素能激动剂异丙肾上腺素醇对鱼类免疫系统的抑制作用已经被反复观察到（Flory，1990；Bayne，Levy，1991a，b；Flory，Bayne，1991 年；Finkenbine 等，1997）。相比之下，α_1肾上腺素能激动剂苯肾上腺素能增强虹鳟白细胞的抗体反应和化学发光（Flory，1990；Flory，Bayne，1991）。α_2激动剂氯硝定对抗体反应的刺激作用可被α_2拮抗剂育亨宾阻断，而不被α_1拮抗剂普拉唑嗪阻断，提示存在两种不同类型的肾上腺素能受体（Flory，1990）。在体外使用苯肾上腺素也能刺激虹鳟吞噬细胞的呼吸爆发作用（Bayne，Levy，1991a，b）。然而，同一个物种的吞噬指数，即被黏附细胞吞噬的酵母细胞数量，可以被上述的同种激动剂苯肾上腺素抑制（Narnaware 等，1994；Narnaware，Baker，1996）。

不同的肾上腺素能和胆碱能受体可以在几种鱼类的淋巴器官中找到（Józefowski 等，1995；Nickerson 等，2003；Owen 等，2007）。应激能诱导虹鳟红细胞的 β 肾上腺素能受体的动力学改变，但尚未发现白细胞发生变化（Reid，Perry，1991；Reid 等，1993）。在哺乳动物中，和神经递质相关的受体特异性信号转导通路以及受体刺激后下游基因的表达变化已有详细报道（Sanders 等，2001）。神经系统肾上腺素能受体功能表达方面的分子研究表明，α_1和α_2受体在鱼类中均存在（Svensson 等，1993；Yasuoka 等，1996）。为了阐明儿茶酚胺在鱼类免疫功能中的作用，需要对肾上腺素能受体和胆碱能受体进行功能分析。

神经递质和神经肽通常和 G 蛋白偶联受体（GPCRs）结合，进而激活信号通路，如蛋白激酶，它们在高等脊椎动物也是由免疫介体激活的信号。肾上腺素降低了体外培养的真鲷巨噬细胞促炎症细胞因子 IL-1β 的表达，但不影响 TNF-α、TGF-β 和 IL-6 细胞因子的表达（Castillo 等，2009）。此外，当肾上腺素与脂多糖联合使用时，它可以降低脂多糖诱导的 IL-1β 和 TNF-α 的表达（Castillo 等，2009）。

10.4.3.2 皮质类固醇的免疫调节

皮质醇抑制鱼类免疫系统的许多方面，如抗体产生、白细胞有丝分裂和吞噬作用（Balm，1997；Wendelaar Bonga，1997；Weyts 等，1999；Yada，Nakanishi，2002；Tort，2011）。虹鳟巨噬细胞的增殖受到皮质醇的调节（Pagniello 等，2002）。相反，一些研究表明皮质醇能显著增强免疫功能。White 和 Fletcher（1985）观察到给鲽鱼（*Pleuronectes platessa*）施用皮质醇后能刺激 C 反应蛋白的产生，而 C 反应蛋白是参与炎症过程的急性期蛋白。皮质醇刺激炎症的能力（以及其他免疫相关的参数）和皮质醇在高等脊椎动物中承担的伤口愈合和组织修复的作用一致（Buckingham 等，

1996）。使用微阵列基因表达谱的分析显示，和炎症过程相关的大部分基因表达都会因皮质醇的处理而降低（MacKenzie 等，2006a，b）。上述是利用虹鳟白细胞所得到的研究成果。然而，在金头鲷的头肾白细胞中，一些促炎性细胞因子的基因表达也受到皮质醇的抑制（Castillo 等，2009）。

皮质醇对淋巴细胞的凋亡具有双相调节作用（Weyts 等，1998a，b；Wojtaszek 等，2002）。皮质醇对从鲤鱼头肾、脾脏和外周血液分离的 B 细胞的凋亡和增殖有着不同的影响（Verburg-Van Kemenade 等，1999）。皮质醇对白细胞亚型之间的不同影响似乎和应激的鲶鱼 B 细胞对嗜中性粒细胞的比值波动相一致（Ellsaesser，Clem，1986；Ainsworth 等，1991）。皮质醇对白细胞亚型细胞凋亡和有丝分裂的不同影响似乎与白细胞总数的变化有关。在哺乳动物中，儿茶酚胺似乎和皮质类固醇具有协同作用，通过凋亡机制调节白细胞的群体（Boomershine 等，2001）。有证据表明，皮质类固醇激素并不抑制鱼的整体免疫功能，但在应激反应过程中调节淋巴细胞在机体不同位置的重新分布（Weyts 等，1999；Dhabhar，Mc Ewen，2001；Schreck，Maule，2001）。

糖皮质类固醇激素抑制 NF-kB 信号通路，导致 IL-1、IL-6、TNF 等促炎细胞因子的生成减少（Sternberg，2006）。皮质醇抑制鱼类炎症细胞因子的表达（Saeij 等，2003）。皮质醇和 LPS 协同刺激头肾吞噬细胞 IL-1 mRNA 的表达（Engelsma 等，2003）。皮质醇抑制细胞因子的表达（TNF、TGF-β、IL-6），当皮质醇和脂多糖同时给予时，真鲷细胞因子的诱导作用下调。此外，皮质醇可能激活一些巨噬细胞的功能而导致炎症的缓解（Castro 等，2011）。

10.4.4　应激反应时鱼类免疫中受体介导的皮质醇的作用

鱼类免疫系统中由受体介导的激素作用机制是通过受体特异性激动剂和拮抗剂，或者通过标记配体进行结合分析来开展的。不同的皮质类固醇激素受体 GR-1、GR-2 和盐皮质类固醇激素受体（MR）已经在虹鳟中鉴定（Ducouret 等，1995；Colombe 等2000；Bury，2003）。这些不同类型的皮质类固醇激素受体已经在几种鱼类中鉴定，这意味着不同器官之间皮质激素类固醇受体基因表达的差异调节，是表达不同的功能（Prunet 等，2006；Stolte 等，2006，2008a，b；McCormick 等，2008；Flores 等，2012）。

10.4.4.1　在体实验

放射受体测定法已经证实在银大麻哈鱼几种淋巴组织中分离的白细胞中，皮质醇的特异性结合有一定差异（Maule，Schreck，1990，1991）。不同的白细胞亚型之间，皮质醇敏感性的差异可能和皮质激素受体的动力学特性有关。圆鳍雅罗鱼（*Leuciscus cephalus*）鳃的胞液中皮质醇的低结合亲和力似乎和血浆高的皮质醇水平有关，而不亚于虹鳟，圆鳍雅罗鱼是一种抗应激鱼类（Pottinger 等，2000）。GRs 已在几种鱼类中克隆并测序，并在淋巴器官特别是脾脏中检测到其 mRNA 的分布（Ducouret 等，1995；Takeo 等，1996；Tagawa 等，1997）。然而，对鱼类免疫系统中 GR 基因表达调控的详细研究很少。在大麻哈鱼（*Oncorhynchus keta*）的鳃中检测到 GR mRNA，其表达水平受到环境盐度的影响（Uchida 等，1998）。这些研究结果进一步支持了鱼类可能使用 GR 作为食物代谢和渗透调节介体的发现（Ducouret 等，1995）。另一方面，Colombe 等（2000）克隆了鱼类 MR，并证明它和高等脊椎动物 MR 有明显的同源性，但鱼类 MR 的重组蛋白优先结合皮质醇。有一种可能性，皮质醇诱导的鱼类免疫调节可能同时由 GR 和 MR 来介导。

急性应激后，鳟鱼和鲤鱼白细胞中的皮质激素受体 mRNA 水平出现短暂升高，随后出现了免疫抑制（Yada 等，2007；Stolte 等，2008a，b）。LPS 注射免疫刺激后，在金头鲷包括脾脏在内的几个组织中都发现皮质激素受体基因的表达增加（Acerete 等，2007）。然而，在虹鳟中，皮质激素受体 mRNA 水平短暂提升后也出现了持续的下降（Yada 等，2007）。将缓释的皮质醇埋植在虹鳟体内，可使头肾、大脑、鳃和性腺中的皮质类固醇受体的 mRNA 水平提升；脾脏、肝脏和肌肉中的 mRNA 水平则会降低（Teles 等，2013）。

鱼类免疫系统中的皮质类固醇受体在皮质醇分泌升高后会出现两种表达模式（图 10.4）。皮质激素受体的快速上调似乎与免疫抑制相关。相反，皮质激素受体表达的慢性下调可能导致对皮质醇刺激的免疫功能脱敏（Yada 等，2007；Stolte 等，2008a，b；Teles 等，2013）。与淡水驯化的鳟鱼相比，盐水驯化的硬头鳟白细胞中皮质激素受体 mRNA 水平也显著降低（Yada 等，2008）。鱼类应激反应对免疫系统的调节作用是通过皮质激素受体介导的，这一过程应该在对应激源脱敏以及对环境变化的适应过程中进行研究。

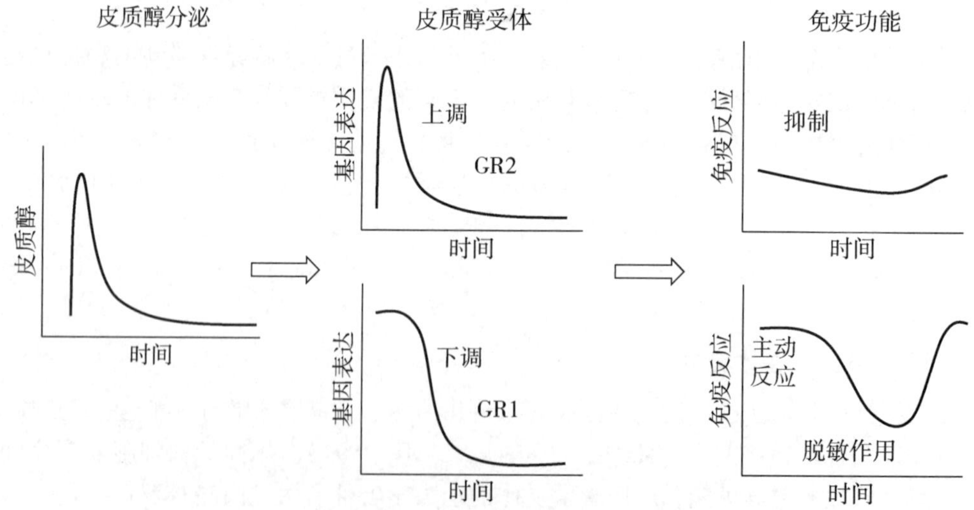

图 10.4　受体介导的皮质醇对免疫功能影响的概念图

注：皮质醇分泌调节特定的 GR 受体，导致免疫应答的抑制或脱敏。GR，糖皮质激素受体。

10.4.4.2　离体实验

一项利用鲤鱼外周血液白细胞进行的结合研究显示，鲤鱼外周血液白细胞中存在一类皮质醇的结合位点，皮质激素受体拮抗剂可以抑制皮质醇减少中性粒细胞凋亡的影响（Weyts 等，1998a，c）。皮质醇诱导的细胞凋亡和抑制增殖的效应在脂多糖激活后的鲤鱼白细胞中更为明显（Overberg-van Lemonade 等，1999）。然而，皮质醇对脂多糖刺激的鲤鱼肾脏白细胞中的 GR 和 MR mRNA 水平则无显著影响（Stolte 等，2008a，b）。

热休克蛋白家族（HSPs）的成员是普遍表达的、保守的蛋白质，按照分子质量分为 HSP30、HSP70 和 HSP90。它们也被认为皮质激素受体介导的胞内信号传导的关键分子。糖皮质激素受体和 90 kDa HSP（HSP90）的结合被认为可以增加其稳定性，促进同源二聚体的形成以结合靶标基因启动子区的保守序列（Buckingham 等，1996；Pratt，Tot，1997 年；Richter，Buchner，2001）。此外，已知 HSP90 的转录调控和一些细胞因子的相互作用（Auphan 等，1995；Scheinman 等，1995；Stephanou，Latchman，1999）。在鱼类中，HSP 的表达可被内分泌系统调节。热休克诱导肝细胞或鳃中表达 HSP30、HSP70、HSP90，这一效应会因皮质醇的引入而受到抑制，虽然皮质醇在没有热休克的情况下对 HSP 的基础水平并没有显著影响（Deane 等，1999；Iwama 等，1999；Ackerman 等，2000；Basu 等，2001；Sathiyaa 等，2001）。体外施用肾上腺素可增加虹鳟肝细胞 HSP70 水平，β 受体阻断剂普萘洛尔能抵消该效应（Ackerman 等，2000）。体内给以 GH 和 PRL 可降低平鲷（*Sparus sarba*）肝脏 HSP70 的表达（Deane 等，1999）。虽然 HSP 参与鱼类免疫抑制的情况尚不清楚，但 HSP 介导了皮质激素受体水平上对应激反应的内分泌调控。

10.4.5　促生长轴和鱼类免疫系统

和高等脊椎动物一样，GH 对鱼类的促生长作用在很大程度上是由胰岛素样生长因子（IGF）-I 介导的（Duan，1998）。在哺乳动物中，IGF-I 也被认为是对免疫系统许多方面，包括特异性和非特异性防御，有刺激作用的重要因子（Clark，1997；Dorshkind，Horseman，2000；Venters 等，2001）。和哺乳动物中的相关报告一致，IGF-I 在鱼类中同样具有免疫调节功能（Yada，2007，2009）。

在几种鱼类的淋巴组织中检测到 IGF-I 基因的表达（Duguay 等，1992，1996；Shamblott，Chen，1993；Loffing-Cueni 等，1998；Yada 等，2002；Shved 等，2009）。罗非鱼吞噬细胞在体外可表达 IGF-I 基因，分泌大量 IGF-I；然而，培养基中添加 GH 并没有使 IGF-I 分泌量发生显著变化（Yada，2007）。这和 GH 处理人淋巴细胞的研究结果一致（Clayton 等，1994）。用 GH 处理可增加大鼠脾脏中生成 IGF-I 的白细胞数量（Baxter 等，1991）。小鼠淋巴瘤细胞转染 GH 过表达的载体后，其 IGF-I 蛋白和 IGF-I mRNA 水平也高于对照细胞（Weigent，Arnold，2005）。这些看似矛盾的发现可能是源于物种间或细胞制备方法间的差异。在广盐性鱼类中，IGF-I 也是重要的电解质平衡调节因子，GH 直接刺激鳃中 IGF-I 基因的表达，而鳃是海水驯化过程中一个主要的渗透调节器官（Sakamoto，McCormick，2006）。综上所述，这些结果表明 GH/IGF-I 轴在硬骨鱼免疫系统中起着重要的调节作用。下一步的研究必须阐明 GH/IGF-I 轴在不同环境条件下对不同鱼类的免疫系统所起的作用。

在鱼白细胞中进行 GH 基因的垂体外表达，结果提示 GH 的自分泌或旁分泌起着调节免疫功能的作用（Yada，Nakanishi，2002；Yada 等，2005）。生长素释放肽是一种饥饿的信号激素，最初在大鼠胃中发现，被认为是生长激素分泌型受体的内源性配体（Kojima 等，1999）。生长素释放肽在脊椎动物中被发现，它的 GH 分泌作用也在几种鱼类中得到证实（Unniappan，Peter，2005；Kaiya 等，2012）。使用生长素释放肽可增加虹鳟吞噬细胞中超氧化物的生成，而 GH 被免疫中和后（通过添加抗鲑鱼 GH 的血清到培养基中）就会阻断生长素释放肽的刺激作用（Yada 等，2006）。这些结果表明，生长素释放肽刺激鱼的白细胞作用，至少有部分是通过白细胞分泌的生长激素来实现的。

GnRH 被认为是一种强有力的刺激物，不仅对促性腺激素有刺激作用，而且对几种鱼类（包括虹鳟）的 GH 分泌也有刺激作用（Holloway，Leatherland，1997；Melamed 等，1998；Canosa 等，2007；Gahete 等，2009）。如前所述，GnRH 刺激虹鳟的几种免疫功能（Yada，2012）。与生长素释放肽的作用相比，GnRH 对白细胞中 GH mRNA 的表达水平没有明显的影响，GH 被免疫中和后并没有阻断 GnRH 的作用，表明 GnRH 没有通过鱼类白细胞 GH 的旁分泌表达而是直接作用于鱼类免疫（Yada，2012）。

影响鱼类免疫功能的其他可能的生长因子是甲状腺激素。然而，只有少数研究检测了这类激素对鱼类免疫功能的影响。放射性碘注射引起的底鳉（*Fundulus heteroclitus*）甲状腺功能减退并导致循环白细胞数量显著减少（Slicher，1961）。Ball 和 Hawkins（1976）观察到，给以甲状腺素或者哺乳动物甲状腺刺激激素（TSH）可以恢复垂体切除的帆鳍花鳉（*Poecilia latipinna*）血液循环中白细胞的数量。考虑到垂体-甲状腺轴在调控高等脊椎动物免疫系统中以及在个体发育早期（Marsh，Erf，1996；Dorshkind，Horseman，2000）和随后变态中的重要性，需要进一步的研究以阐明甲状腺激素在鱼类免疫系统中的作用。例如，在牙鲆（*Paralichthys olivaceus*）变态过程中，甲状腺激素刺激红细胞从未成熟期向成熟期转变（Miwa，Inui，1991），但这种激素对白细胞类群或淋巴组织的影响尚未阐明。甲状腺激素也被认为是控制鲑科鱼类一龄幼鲑向二龄稚鲑下海洄游的转化，以及随后包括免疫在内的各种形态和生理变化（McCormick，1995）的关键因子，下面将进一步详细地叙述。

10.5　环境应激源和鱼类免疫

10.5.1　环境盐度

Takei 和 Hwang（2016；见本书第 6 章）提到环境盐度变化所造成的应激反应也应考虑。这里我们关注这些应激反应如何影响鱼类的免疫。虹鳟急性暴露在海水中会升高血浆皮质醇水平，降低抗体水平，但延长在海水中的暴露并不影响免疫功能（Betoulle 等，1995）。事实上，对海水的适应提高了虹鳟的非特异性免疫（Yada 等，2001，2012）。褐鳟（*Salmo trutta*）在海水环境下还可活化血浆溶菌酶活性和头肾白细胞的吞噬活性（Marc 等，1995）。在真鲷中，环境盐度不影响吞噬活性（Narnaware 等，1997，1998）。适应海水的莫桑比克罗非鱼，吞噬作用活化后出现呼吸爆发的增强，而血浆溶菌酶水平不变（Yada 等，2002）。相反，环境盐度较高时，罗非鱼血浆溶菌酶活性和免疫球蛋白水平升高（Dominguez 等，2004，2005）。我们认为，不同物种或不同实验条件下，盐度对免疫功能影响的差异可能与机体对高渗透环境的适应性有关。广盐性鱼类是一种能耐受盐度变化引起的应激的鱼类，它们在适应高渗环境的同时，能够通过内分泌调节的变化来促进免疫功能。然而，在广盐性鱼类（包括鲑科鱼类、罗非鱼和真鲷）中，每种免疫功能对盐度变化的反应似乎存在着物种特异性的差异。鱼类的渗透调节研究还需要进一步比较盐度对不同鱼类（包括狭盐性鱼类）免疫功能的影响。

在海水驯化过程中观察到，一些广盐性鱼类免疫功能的激活似乎和 GH 的渗透调节作用有关。在海水驯化过程中，褐鳟免疫功能增强和血浆 GH 水平提高呈正相关关系（Marc 等，1995）。除了促进身体生长和新陈代谢，GH 还能帮助广盐性硬骨鱼适应高渗环境；环境盐度能刺激内源性 GH 的分泌（Sakamoto 等，1993；Yada，1994；McCormick，1995；Bjornsson，1997；Shepherd 等，1997）。在海水驯化的广盐性鱼类中，GH 分泌的增加不仅可以促进渗透调节，还可以增强免疫功能。海水渗透调节作用对 PRL 的抑制已为我们所熟知（Sakamoto，McCormick，2006）。然而，PRL 对免疫功能的刺激作用甚至在海水驯化后出现，同时伴随着其受体的表达（Yada 等，2002）。PRL 的免疫调节作用似乎与它的渗透调节并不相关，但参与鱼类免疫系统中激素的局部表达。

10.5.2　温度和季节性

水温是影响鱼类免疫功能的主要环境因素之一（Fries，1986；Bly，Clem，1992；Manning，Nakanishi，1996；Schreck，1996；Bly 等，1997；Le Morvan 等，1998；Hernandez，Tort，2003）。当温度低于最佳免疫应答发生的范围，但仍在生理范围内时，这种情况会抑制细胞和体液的特异性免疫功能（Manning，Nakanishi，1996）。前期研究表明，辅助 T 细胞（而不是记忆 T 细胞或 B 细胞）对低温敏感，这是基于分裂原对淋巴细胞的增殖反应，混合白细胞反应和抗体产生反应的结果（Bly，Clem，1992；Manning，Nakanishi，1996；Le Morvan 等，1998）。辅助性 T 细胞的热敏感度是由质膜中脂肪酸和/或糖类成分的特性决定的（Clem，1992；Le Morvan 等，1998）。相比之下，非特异性免疫（如吞噬活性和非特异性细胞毒性）往往比特异性免疫更能耐受低温（Ainsworth 等，1991；Dexiang，Ainsworth，1991；Collazos 等，1994；Kurata 等，1995）。当然，适当的水温能比过热或过冷的环境产生更高水平的特异性和非特异性免疫功能（Dominguez 等，2004，2005）。众所周知，温度波动除了直接影响免疫功能外，还通过其他机制（如应激）影响免疫功能。对于非特异性免疫功能，温度波动（而不是稳定低温）可能是应激源（Elliott，1981；Le Morvan 等，1998）。冷应激可能由下丘脑-垂体-甲状腺轴介导，影响哺乳动物的免疫系统（Davis，1998）。然而，如前所述，甲状腺激素对鱼类免疫系统的作用尚不清楚。

性成熟通常会伴有免疫抑制的症状。在鲑科鱼类的性成熟期间，杀菌活性下降和感染频率增加（Richards，Pickering，1978；Pickering，Christie，1980；Iida 等，1989）。鲑科鱼性成熟也与血浆溶菌酶活性降低、总白细胞和抗体产生细胞数量减少同时发生（Pickering，Pottinger，1987；Maule 等，1996）。虹鳟、金鱼和褐菖鲉（*Sebastiscus marmoratus*）在繁殖过程中血浆免疫球蛋白水平呈下降趋势（Nakanishi，1986；Suzuki 等，1996，1997）。在许多鱼类繁殖期间也观察到血液循环中的皮质醇水平升高，表明这种应激激素介导了免疫抑制（Schreck，Maule，2001）。促性腺激素和性激素分泌的变化以及对这些激素的反应是脊椎动物性成熟过程中值得注意的内分泌现象（Bentley，1998；Blazquez 等，1998）。性类固醇激素对鱼的免疫功能也有直接作用，雄激素和雌激素有着不同的影响。在高等脊椎动物中，雌性和雄性在免疫应答上的区别已有详细报道（Chapman 等，1996；Lin 等，1996；Marchetti 等，2001）。对鱼类免疫功能的性别差异也有报道，这意味着性类固醇激素在调节免疫系统中的重要性（Tatner，1996；Yano，1996）。已经知道应激诱导的皮质醇分泌会影响鱼类的繁殖过程（Barton，Iwama，1991；Pankhurst，Van Der Kraak，1997；Wendelaar Bonga，1997；Schreck，Maule，2001）。鱼类在性成熟期间免疫应答的变化似乎是由皮质醇和性激素之间的相互作用引起的。

溯河洄游的鲑科鱼类从幼鲑到初次由河入海的小鲑转变过程，涉及形态的、行为的和生理的变化，这些变化为初次由河入海的小鲑的转变过程做好了向海洋迁移和在海洋中生活的准备（Hoar，1988；Barron，1986；Dickhoff 等，1997）。循环系统中的激素如甲状腺激素、皮质醇和 GH 普遍升高（Hoar，1988；Dickhoff 等，1997），而给幼鲑注射这些激素能促进幼鲑为下海洄游做准备的各种特征的出现（Bjornsson，1997；McCormick，2013）。在溯河洄游的鲑科鱼类由一龄幼鲑向二龄鲑的转变过程中观察到免疫应答的短暂下降（Schreck，1996）。转变过程中可观察到物种之间，甚至同一物种在不同条件下血浆激素水平的不同型式（Bjornsson，1997；Dickhoff 等，1997）。Olsen 等（1993）发现，在自然光照状况下，大西洋鲑鱼由一龄幼鲑向二龄鲑的转变过程中，血浆溶菌酶活性没有发生显著变化。转变过程中环境的调节作用，尤其是由光照引起的环境调节是众所周知的（Hoar，1988）。在虹鳟的两个品系虹鳟和硬头鳟中，白细胞和脾脏中表达的皮质激素受体 mRNA 水平的季节性变化和春季血浆中甲状腺激素与皮质醇水平的升高相一致（Yada 等，2014）。光照或昼夜节律被认为是免疫系统的环境调节剂之一（Esteban 等，2006）。

10.6　未来的方向

有很多重要的研究课题是关于应激源对免疫系统的影响以及其他调节系统如神经和内分泌系统等之间的关系。一个重要的方面是，应激源通常会引起一些反应，这些反应表现为免疫应答，即使应激源不是病原体或"非自身"的生物体，例如化学或物理应激源。这些反应为何被激活以及如何被激活是一个关键的未知领域。另一个至关重要的信息空白是黏膜表面的免疫应答，黏膜表面是各种不同应激刺激的入口，以及应激、黏膜反应和局部微生物群之间相互作用的位置。虽然这一领域目前正在深入开展研究，但还需要更多的关注。关于相互作用，还有许多研究工作要做，以便了解激素、细胞因子和神经肽是如何相互联系，以及不同的组织是否有这些分子的受体来对付应激刺激。特别是神经肽和/或儿茶酚胺与免疫应答之间的相互作用需要阐明，因为对皮质类固醇等其他介质以及其免疫效应和所涉及的通道已经了解比较多。关于特异性免疫应答，鱼类如何根据行为特征或基于个体对刺激的感知或应对策略产生不同的免疫反应还有待研究。此外，我们对鱼类免疫系统的一些基础知识还相当缺乏，而这些知识在建立保护鱼体健康和抵抗病原体特别是应激源存在时的解决方案时，是必需的。最后，确定不同种、属和科的鱼类在应激条件下的免疫策略是否存在系统发生的差异也很有意义。

T. 雅达　L. 托特　著
卢丹琪　译
林浩然　校

参 考 文 献

Acerete, L., Balasch, J. C., Castellana, B., Redruello, B., Roher, N., Canario, A. V. (2007). Cloning of the glucocorticoid receptor (GR) in gilthead seabream (*Sparus aurata*). Differential expression of GR and immune genes in gilthead seabream after an immune challenge. Comp. Biochem. Physiol. 148B, 32-43.

Ackerman, P. A., Forsyth, R. B., Mazur, C. F., Iwama, G. K. (2000). Stress hormones and the cellular stress response in salmonids. Fish Physiol. Biochem. 23, 327-336.

Ainsworth, A. J., Dexiang, C., Waterstrat, P. R., Greenway, T. (1991). Effect of temperature on the immune system of channel catfish (*Ictalurus punctatus*) -I. Leucocyte distribution and phagocyte function in the anterior kidney at 10℃. Comp.

Biochem. Physiol. 100A, 907-912.

Auphan, N., DiDonato, J. A., Rosette, C., Helmberg, A., Karin, M. (1995). Immunosuppression by glucocorticoids: inhibition of NF-kB activity through induction of IlB synthesis. Science. 270, 286-290.

Ball, J. N., Hawkins, E. F. (1976). Adrenocortical (interrenal) responses to hypophysectomy and adenohypophysial hormones in the teleost *Poecilia latipinna*. Gen. Comp. Endocrinol. 28, 59-70.

Balm, P. H. M. (1997). Immune-endocrine interactions. In: Iwama, G. K., Pickering, A. D., Sumpter, J. P., Schreck, C. B., Fish Stress and Health in Aquaculture. Cambridge University Press, Cambridge, 195-221.

Barron, M. G. (1986). Endocrine control of smoltification in anadromous salmonids. J. Endocrinol. 108, 313-319.

Barton, B. A., Iwama, G. K. (1991). Physiological changes in fish from stress in aquaculture with emphasis on the response and effects of corticosteroids. Ann. Rev. Fish Dis. 1, 3-26.

Basu, N., Nakano, T., Grau, E. G., Iwama, G. K. (2001). The effects of cortisol on heat shock protein 70 levels in two fish species. Gen. Comp. Endocrinol. 124, 97-105.

Baxter, J. B., Blalock, J. E., Weigent, D. A. (1991). Characterization of immunoreactive insulin-like growth factor-I from leukocytes and its regulation by growth hormone. Endocrinology. 129, 1727-1734.

Bayne, C. J., Levy, S. (1991). Modulation of the oxidative burst in trout myeloid cells by adrenocorticotropic hormone and catecholamines: mechanisms and action. J. Leukoc. Biol. 50, 554-560.

Bayne, C. J., Levy, S. (1991). The respiratory burst of rainbow trout, *Oncorhynchus mykiss* (Walbaum), phagocytes is modulated by sympathetic neurotransmitters and the 'neuro' peptide ACTH. J. Fish Biol. 38, 609-619.

Bentley, P. J. (1998). Comparative Vertebrate Endocrinology. Bentley, 1998 Bentley, P. J., Comparative Vertebrate Endocrinologythird ed1998Cambridge University Press, Cambridge, Cambridge University Press, Cambridge.

Betoulle, S., Troutaud, D., Khan, N., Deschaux, R. (1995). Résponse anticorps, cortisolémie et prolactinémie chez la truite arc-en-ciel. C. R. Acad. Sci. Paris. 318, 677-681.

Björnsson, B. Th. (1997). The biology of salmon growth hormone: from daylight to dominance. Fish Physiol. Biochem. 17, 9-24.

Blázquez, M., Bosma, P. T., Fraser, E. J., Van Look, K. J. W., Trudeau, V. L. (1998). Fish as models for the neuroendocrine regulation of reproduction and growth. Comp. Biochem. Physiol. 119C, 345-364.

Bly, J. E. , Clem, L. W. (1992). Temperature and teleost immune functions. Fish Shellfish Immunol. 2, 159-171.

Bly, J. E. , Quiniou, S. M. -A. , Clem, L. W. (1997). Environmental effects of fish immune mechanisms. Dev. Biol. Stand. 90, 33-43.

Boomershine, C. S. , Wang, T. , Zwilling, B. S. (2001). Neuroendocrine regulation of macrophage and neutrophil function. In: Ader, R. , Felten, D. L. , Cohen, N. , . Academic Press, San Diego, 289-300.

Borghetti, P. , Saleri, R. , Mocchegiani, E. , Corradi, A. , Martelli, P. (2009). Infection, immunity and the neuroendocrine response. Vet. Immunol. Immunopathol. 130, 141-162.

Buckingham, J. C. , Christian, H. C. , Gillies, G. E. , Philip, J. G. , Taylor, A. D. (1996). The hypothalamo-pituitary-adrenocortical immune axis. In: Marsh, J. A. , Kendall, M. D. , The Physiology of Immunity. CRC Press, Boca Raton, 331-354.

Bury, N. R. , Sturm, A. , Le Rouzic, P. , Lethimonier, C. , Ducouret, B. , Guiguen, Y. (2003). Evidence for two distinct functional glucocorticoid receptors in teleost fish. J. Mol. Endocrinol. 31, 141-156.

Caipang, C. M. A. , Brinchmann, M. F. , Berg, I. (2008). Changes in selected stress and immune-related genes in Atlantic cod, *Gadus morhua*, following overcrowding. Aquac. Res. 39, 1533-1540.

Caipang, C. M. A. , Lazado, C. C. , Brinchmann, M. F. , Rombout, J. H. W. M. , Kiron, V. (2011). Differential expression of immune and stress genes in the skin of Atlantic cod (*Gadus morhua*). Comp. Biochem. Physiol. D. 6, 158-162.

Calduch-Giner, J. A. , Sitjà-Bobadilla, A. , Alvarez-Pellitero, P. , Pérez-Sánchez, J. (1995). Evidence for a direct action of GH on haemopoietic cells of a marine fish, the gilthead sea bream (*Sparus aurata*). J. Endocrinol. 146, 459-467.

Calduch-Giner, J. A. , Duval, H. , Chesnel, F. , Boeuf, G. , Prérez-Sánchez, J. , Boujard, D. (2000). Fish growth hormone receptor: molecular characterization of two membrane-anchored forms. Endocrinology. 142, 3269-3273.

Canosa, L. F. , Chang, J. P. , Peter, R. E. (2007). Neuroendocrine control of growth hormone in fish. Gen. Comp. Endocrinol. 151, 1-26.

Castillo, J. , Teles, M. , Mackenzie, S. , Tort, L. (2009). Stress-related hormones modulate cytokine expression in the head kidney of gilthead seabream (*Sparus aurata*). Fish Shellfish Immunol. 27, 493-499.

Castro, R. , Zou, J. , Secombes, C. J. , Martin, S. A. M. (2011). Cortisol modulates the induction of inflammatory gene expression in a rainbow trout macrophage cell line. Fish Shellfish Immunol. 30, 215-223.

Chadzinska, M., Józefowski, S., Bigaj, J., Plytycz, B. (1997). Morphine modulation of thioglycollate-elicited peritoneal inflammation in the goldfish, *Carassius auratus*. Arch. Immunol. Ther. Exp. 45, 321-327.

Chapman, J. C., Despande, R., Michael, S. D. (1996). Estrogen-mediated interactions between the immune and female reproductive systems. In: Marsh, J. A., Kendall, M. D., The Physiology of Immunity. CRC Press, Boca Raton, 239-261.

Clark, R. (1997). The somatogenic hormones and insulin-like growth factor-1: stimulators of lymphopoiesis and immune function. Endocr. Rev. 18, 157-179.

Clayton, P. E., Day, R. N., Silva, C. M., Hellmann, P., Day, H. K., Thorner, M. O. (1994). Growth hormone induces tyrosine phosphorylation but does not alter insulin-like growth factor-I gene expression in human IM9 lymphocytes. J. Mol. Endocrinol. 13, 127-136.

Collazos, M. E., Ortega, E., Barriga, C. (1994). Effect of temperature on the immune system of a cyprinid fish (*Tinca tinca*, L.). Blood phagocyte function at low temperature. Fish Shellfish Immunol. 4, 231-238.

Colombe, L., Fostier, A., Bury, N., Pakdel, F., Guiguen, Y. (2000). A mineralcorticoid-like receptor in the rainbow trout, *Oncorhynchus mykiss*: cloning and characterization of its steroid binding domain. Steroids. 65, 319-328.

Costas, B., Conceição, L., Aragão, C., Martos, J. A., Ruiz-Jarabo, I., Mancera, J. M. (2011). Physiological responses of Senegalese sole (Solea senegalensis Kaup, 1858) after stress challenge: effects on non-specific immune parameters, plasma free amino-acids. Aquaculture. 316, 68-76.

Cuesta, A., Laiz-Carrión, R., Del Río, M. P., Meseguer, J., Mancera, J. M., Esteban, M. A. (2005). Salinity influences the humoral immune parameters of gilthead seabream (*Sparus aurata* L.). Fish Shellfish Immunol. 18, 255-261.

Cuesta, A., Arjona, R. L.-C. F., del Río, M. P. M., Meseguer, J., Mancera, J. M., Esteban, M. Á. (2006). Effect of PRL, GH and cortisol on the serum complement and IgM levels in gilthead seabream (*Sparus aurata* L.). Fish Shellfish Immunol. 20, 427-432.

Davis, K. B., Griffin, B. R., Gray, W. L. (2003). Effect of dietary cortisol on resistance of channel catfish to infection by *Ichthyophthirius multifiliis* and channel catfish virus disease. Aquaculture. 218, 121-130.

Davis, S. L. (1998). Environmental modulation of the immune system via the endocrine system. Dom. Anim. Endocrinol. 15, 283-289.

Deane, E. E., Kelly, S. P., Lo, C. K., Woo, N. Y. (1999). Effects of GH, prolactin and cortisol on hepatic heat shock protein 70 expression in a marine teleost *Sparus sarba*. J. Endocrinol. 161, 413-421.

Demers, N. E., Bayne, C. J. (1997). The immediate effects of stress on hormones and plasma lysozyme in rainbow trout. Dev. Comp. Immunol. 21, 363-373.

Dexiang, C., Ainsworth, A. J. (1991). Effect of temperature on the immune system of channel catfish (*Ictalurus punctatus*) -II. Adaptation of anterior kidney phagocytes to 10℃. Comp. Biochem. Physiol. 100A, 913-918.

Dhabhar, F. S. (2000). Acute stress enhances while chronic stress suppresses skin immunity. The role of stress hormones and leukocyte trafficking. Ann. N. Y. Acad. Sci. 9178, 76-93.

Dhabhar, F. S., McEwen, B. S. (2001). Bidirectional effects of stress and glucocorticoid hormones on immune function: Possible explanations for paradoxical observations. In: Ader, R., Felten, D. L., Cohen, N.,. Academic Press, San Diego, 301-338.

Dhabhar, F. S. (2002). Stress-induced augmentation of immune function—the role of stress hormones, leukocyte trafficking, and cytokines. Brain Behav. Immun. 16, 785-798.

Dhabhar, F. S. (2008). Enhancing versus suppressive effects of stress on immune function: implications for immunoprotection versus immunopathology. Allergy Asthma Clin. Immunol., 2-11.

Dickhoff, W. W., Beckman, B. R., Larsen, D. A., Duan, C., Moriyama, S. (1997). The role of growth in endocrine regulation of salmon smoltification. Fish Physiol. Biochem. 17, 231-236.

Dominguez, M., Takemura, A., Tsuchiya, M., Nakamura, S. (2004). Impact of different environmental factors on the circulating immunoglobulin levels in the Nile tilapia, *Oreochromis niloticus*. Aquaculture. 241, 491-500.

Dominguez, M., Takemura, A., Tsuchiya, M. (2005). Effects of changes in environmental factors on the non-specific immune response of Nile tilapia, *Oreochromis niloticus* L. Aquac. Res. 36, 391-397.

Dorshkind, K., Horseman, N. D. (2000). The roles of prolactin, growth hormone, insulin-like growth factor-I, and thyroid hormones in lymphocyte development and function: insights from genetic models of hormone and hormone receptor deficiency. Endocr. Rev. 21, 292-312.

Douxfils, J., Lambert, S., Mathieu, C., Milla, S., Mandiki, S. N. M., Henrotte, E. (2014). Influence of domestication process on immune response to repeated emersion stressors in Eurasian perch (*Perca fluviatilis*, L.). Comp. Biochem. Physiol. A. 173, 52-60.

Duan, C. (1998). Nutritional and developmental regulation of insulin-like growth factors in fish. J. Nutr. 128, 306S-314S.

Ducouret, B. , Tujague, M. , Ashraf, J. , Mouchel, N. , Servel, N. , Valotaire, Y. (1995). Cloning of a teleost fish glucocorticoid receptor shows that it contains a deoxyribonucleic acid-binding domain different from that of mammals. Endocrinology. 136, 3774-3783.

Duguay, S. J. , Park, L. K. , Samadpour, M. , Dickhoff, W. W. (1992). Nucleotide sequence and tissue distribution of three insulin-like growth factor I prohormones in salmon. Mol. Endocrinol. 6, 1202-1210.

Duguay, S. J. , Lai-Zhang, J. , Steiner, D. F. , Funkenstein, B. , Chan, S. J. (1996). Developmental and tissue-regulated expression of IGF-I and IGF-II mRNAs in *Sparus aurata*. J. Mol. Endocrinol. 16, 123-132.

Elenkov, I. J. , Chrousos, G. P. (1999). Stress hormones, Th1/Th2 patterns, pro/anti-inflammatory cytokines and susceptibility to disease. Trends Endocrinol. Metab. 10, 359-368.

Elliott, J. M. (1981). Some aspects of thermal stress on freshwater teleosts. In: Pickering, A. D. , Stress and Fish. Academic Press, London, 209-245.

Ellsaesser, C. F. , Clem, L. W. (1986). Haematological and immunological changes in channel catfish stressed by handling and transport. J. Fish Biol. 28, 511-521.

Engelsma, M. Y. , Hougee, S. , Nap, D. , Hofenk, M. , Rombout, J. H. , van Muiswinkel, W. B. (2003). Multiple acute temperature stress affects leucocyte populations and antibody responses in common carp, *Cyprinus carpio* L. Fish Shellfish Immunol. 15, 397-410.

Esteban, M. Á. , Cuesta, A. , Rodríguez, A. , Meseguer, J. (2006). Effect of photoperiod on the fish innate immune system: a link between fish pineal gland and the immune system. J. Pineal Res. 41, 261-266.

Fast, M. D. , Hosoya, S. , Johnson, S. C. , Afonso, L. O. (2008). Cortisol response and immune-related effects of Atlantic salmon (*Salmo salar* Linnaeus) subjected to short-and long-term stress. Fish Shellfish Immunol. 24, 194-204.

Finkenbine, S. S. , Gettys, T. W. , Brunett, K. G. (1997). Direct effects of catecholamines on T and B cell lines of the channel catfish, *Ictalurus punctatus*. Dev. Comp. Immunol. 21, 155.

Flores, A. -M. , Shrimpton, J. M. , Patterson, D. A. , Hills, J. A. , Cooke, S. J. , Yada, T. (2012). Physiological and molecular endocrine changes in maturing wild sockeye salmon, *Oncorhynchus nerka*, during ocean and river migration. J. Comp. Physiol. B. 182, 77-90.

Flory, C. M. (1989). Automatic innervation of the spleen of the coho salmon, *Oncorhynchus kisutch*: a histochemical demonstration and preliminary assessment of its immunoregulatory role. Brain Behav. Immun. 3, 331-344.

Flory, C. M. (1990). Phylogeny of neuroimmunoregulation: effects of adrenergic and cholinergic agents on the in vitro antibody response of the rainbow trout, *Oncorhynchus mykiss*. Dev. Comp. Immunol. 14, 283-294.

Flory, C. M., Bayne, C. J. (1991). The influence of adrenergic and cholinergic agents on the chemiluminescent and mitogenic responses of leukocytes from the rainbow trout, *Oncorhynchus mykiss*. Dev. Comp. Immunol. 15, 135-142.

Fries, C. R. (1986). Effects of environmental stressors and immunosuppressants on immunity in *Fundulus heteroclitus*. Am. Zool. 26, 271-282.

Fukada, H., Ozaki, Y., Adachi, S., Yamauchi, K., Hara, A. (2001). GeneBank accession number AB071216.

Gahete, M. D., Durán-Prado, M., Luque, R. M., Martínez-Fuentes, A. J., Quintero, A., Gutiérrez-Pascual, E. (2009.) Understanding the multifactorial control of growth hormone release by somatotropes. Ann. N. Y. Acad. Sci. 1163, 137-153.

Gomez, D., Sunyer, J. O., Salinas, I. (2013). The mucosal immune system of fish: the evolution of tolerating commensals while fighting pathogens. Fish Shellfish Immunol. 35, 1729-1739.

Gopurappilly, R., Ogawa, S., Parhar, I. S. (2013). Functional significance of GnRH and kisspeptin, and their cognate receptors in teleost reproduction. Front. Endocrinol. 4, 24.

Gorissen, M., Flik, G. (2016). Endocrinology of the Stress Response in Fish. In: Schreck, C. B., Tort, L., Farrell, A. P., Brauner, C. J.,. Academic Press, San Diego, CA.

Harris, J., Bird, D. J. (1997). The effects of α-MSH and MCH on the proliferation of rainbow trout (*Oncorhynchus mykiss*) lymphocytes in vitro. In: Kawashima, S., Kikuyama, S., Advances in Comparative Endocrinology. Monduzzi Editore, Bologna, 1023-1026.

Harris, J., Bird, D. J. (1998). Alpha-melanocyte stimulating hormone (α-MSH) and melanin-concentrating hormone (MCH) stimulate phagocytosis by head kidney leucocytes of rainbow trout (*Oncorhynchus mykiss*) in vitro. Fish Shellfish Immunol. 8, 631-638.

Harris, J., Bird, D. J. (2000). Supernatants from leucocytes treated with melanin-concentrating hormone (MCH) and α-melanocyte stimulating hormone (α-MSH) have a stimulatory effect on rainbow trout (*Oncorhynchus mykiss*) phagocytes in vitro. Vet. Immunol. Immunopathol. 76, 117-124.

Harris, J., Bird, D. J., Yeatman, L. A. (1998). Melanin-concentrating hormone (MCH) stimulates the activity of rainbow trout (*Oncorhynchus mykiss*) head kidney phagocytes in vitro. Fish Shellfish Immunol. 8, 639-642.

Hernández, A., Tort, L. (2003). Annual variation of complement, lysozyme and haemagglutinin levels in serum of the gilthead sea bream *Sparus aurata*. Fish Shellfish Immunol. 15, 479-481.

Higashimoto, Y., Nakao, N., Ohkubo, T., Tanaka, M., Nakashima, K. (2001). Structure and tissue distribution of prolactin receptor mRNA in Japanese flounder (*Paralichthys olivaceus*): conserved and preferential expression in osmoregulatory organs. Gen. Comp. Endocrinol. 123, 170-179.

Hoar, W. S. (1988). The physiology of smolting salmonids. In: Hoar, W. S., Randall, D. J. Academic Press, San Diego, 275-343.

Holloway, A. C., Leatherland, J. F. (1997). The effects of *N*-methyl-d, l-aspartate and gonadotropin-releasing hormone on in vitro growth hormone release in steroid-primed immature rainbow trout, *Oncorhynchus mykiss*. Gen. Comp. Endocrinol. 107, 32-43.

Hou, Y., Suzuki, Y., Aida, K. (1999). Effects of steroids on the antibody producing activity of lymphocytes in rainbow trout. Fish Sci. 65, 850-855.

Iida, T., Takahashi, K., Wakabayashi, H. (1989). Decrease in the bactericidal activity of normal serum during the spawning period of rainbow trout. Bull. Jpn. Soc. Sci. Fish. 55, 463-465.

Iwama, G. K., Vijayan, M. M., Forsyth, R. B., Ackenrian, P. A. (1999). Heat shock proteins and physiological stress in fish. Am. Zool. 39, 901-909.

Jhingan, E., Devlin, R. H., Iwama, G. K. (2003). Disease resistance, stress response and effects of triploidy in growth hormone transgenic coho salmon. J. Fish Biol. 63, 806-823.

Józefowski, S., Gruca, P., Józkowicz, A., Plytycz, B. (1995). Direct detection and modulatory effects of adrenergic and cholinergic receptors on goldfish leukocytes. J. Mar. Biotechnol. 3, 171-173.

Kaiya, H., Hosoda, H., Kangawa, K., Miyazato, M. (2012). Determination of nonmammalian ghrelin. In: Kojima, M., Kangawa, K., Ghrelin. Academic Press, San Diego, 75-87.

Kaneko, T., 1996. Cell biology of somatolactin. Int. Rev. Cytol. 169, 1-24.

Kaneko, T., Kakizawa, S., Yada, T. (1993). Pituitary of "cobalt" variant of the rainbow trout separated from the hypothalamus lacks most pars intermedial and neurohypophysial tissue. Gen. Comp. Endocrinol. 92, 31-40.

Kaneko, T., Kakizawa, S., Yada, T., Hirano, T. (1993). Gene expression and intercellular localization of somatolactin in the pituitary of rainbow trout. Cell Tissue Res. 272, 11-16.

Kawauchi, H., Sower, S. A., Moriyama, S. (2009). The neuroendocrine regulation of prolactin and somatolactin secretion in fish. In: Bernier, N. J., van der Kraak, G., Farrell, A. P., Brauner, C. J., Fish Neuroendocrinology. Elsevier Academic Press, San Diego, 197-234.

Kojima, M., Hosoda, H., Date, Y., Nakazato, M., Matsuo, H., Kangawa, K. (1999). Ghrelin is a growth-hormone-releasing acylated peptide from stomach. Nature. 402, 656-660.

Krasnov, A., Skugor, S., Todorcevic, M., Glover, K. A., Nilsen, F. (2012). Gene expression in Atlantic salmon skin in response to infection with the parasitic copepod *Lepeophtheirus salmonis*, cortisol implant, and their combination. BMC Genomics. 13, 130.

Kurata, O., Okamoto, N., Suzumura, E., Sano, N., Ikeda, Y. (1995). Accommodation of carp natural killer-like cells to environmental temperature. Aquaculture. 129, 421-424.

Le Morvan, C., Troutaud, D., Deschaux, P. (1998). Differential effects of temperature on specific and nonspecific immune defences in fish. J. Exp. Biol. 201, 165-168.

Lee, L. T. O., Nong, G., Chan, Y. H., Tse, D. L. Y., Cheng, C. H. K. (2001). Molecular cloning of a teleost growth hormone receptor and its functional interaction with human growth hormone. Gene. 270, 121-129.

Leonard, B. E. (2005). The HPA and immune axes in stress: the involvement of the serotonergic system. Eur. Psychiatry. 20 (Suppl. 3S), 302-306.

Li, J., Barreda, D. R., Zhang, Y. A., Boshra, H., Gelman, A. E., Lapatra, S. (2006). B lymphocytes from early vertebrates have potent phagocytic and microbicidal abilities. Nat. Immunol. 7, 1116-1124.

Lin, T.-M., Lustig, R. H., Chang, C. (1996). The role of androgens-androgen receptor in immune system activity. In: Marsh, J. A., Kendall, M. D., The Physiology of Immunity. CRC Press, Boca Raton, 263-276.

Loffing-Cueni, D., Schmid, A. C., Graf, H., Reinecke, M. (1998). IGF-I in the bony fish *Cottus scorpius*: cDNA, expression and differential localization in brain and islets. Mol. Cell. Endocrinol. 141, 187-194.

MacArthur, J. I., Fletcher, T. C., Pirie, B. J. S., Davidson, R. J. L., Thomson, A. W. (1984). Peritoneal inflammatory cells in plaice, *Pleuronectes platessa* L.: effects of stress and endotoxin. J. Fish Biol. 25, 69-81.

MacKenzie, S., Iliev, D., Liarte, C., Koskinen, H., Planas, J. V., Goetz, F. W. (2006). Transcriptional analysis of LPS-stimulated activation of trout (*Oncorhynchus mykiss*) monocyte/macrophage cells in primary culture treated with cortisol. Mol.

Immunol. 43, 1340-1348.

MacKenzie, S., Liarte, D. I., Koskinen, H., Planas, J. V., Goetz, F. W., Mölsä, H. (2006). Transcriptional analysis of LPS-stimulated activation of trout (*Oncorhynchus mykiss*) monocyte/macrophage cells in primary culture treated with cortisol. Mol. Immunol. 43, 1340-1348.

Manning, M. J., Nakanishi, T. (1996). The specific immune system: cellular defenses. In: Iwama, G., Nakanishi, T., The Fish Immune System: Organism, Pathogen, and Environment. Academic Press, San Diego, 159-205.

Marc, A. M., Quentel, C., Severe, A., Le Bail, P. Y., Boeuf, G. (1995). Changes in some endocrinological and non-specific immunological parameters during seawater exposure in the brown trout. J. Fish Biol. 46, 1065-1081.

Marchetti, B., Morale, M. C., Gallo, F., Lomeo, E., Testa, N., Tirolo, C. (2001). The hypothalmo-pituitary-gonadal axis and the immune system. In: Ader, R., . Elsevier Academic Press, San Diego, 363-389.

Marsh, J. A., Erf, G. F. (1996). Interactions between the thyroid and the immune system. In: Marsh, J. A., Kendall, M. D., The Physiology of Immunity. CRC Press, Boca Raton, 211-235.

Maule, A. G., Schreck, C. B. (1990). Glucocorticoid receptors in leukocytes and gill of juvenile coho salmon (*Oncorhynchus kisutch*). Gen. Comp. Endocrinol. 77, 448-455.

Maule, A. G., Schreck, C. B. (1991). Stress and cortisol treatment changed affinity and number of glucocorticoid receptors in leukocytes and gill of coho salmon. Gen. Comp. Endocrinol. 84, 83-93.

Maule, A. G., Schrock, R., Slater, C., Fitzpatrick, M. S., Schreck, C. B. (1996). Immune and endocrine responses of adult chinook salmon during freshwater migration and sexual maturation. Fish Shellfish Immunol. 6, 221-233.

Mauri, I., Romero, A., Acerete, L., MacKenzie, S., Roher, N., Callol, A. (2011). Changes in complement responses in Gilthead seabream (*Sparus aurata*) and European seabass (*Dicentrarchus labrax*) under crowding stress, plus viral and bacterial challenges. Fish Shellfish Immunol. 30, 182-188.

McCormick, S. D. (1995). Hormonal control of gill Na^+, K^+-ATPase and chloride cell function. In: Wood, C. M., Shuttleworth, T. J., Cellular and Molecular Approaches to Fish Ionic Regulation. Academic Press, San Diego, 285-315.

McCormick, S. D. (2013). Smolt physiology and endocrinology. In: McCormick, S. D., Farrell, A. P., Brauner, C. J., Fish Physiology Vol. 32 Euryhaline Fishes. Academic Press, Amsterdam, 200-251.

McCormick, S. D., Regish, A., O'Dea, M. F., Shrimpton, J. M. (2008). Are we missing a mineralocorticoid in teleost fish? Effects of cortisol, deoxycorticosterone and aldosterone on osmoregulation, gill Na^+, K^+-ATPase activity and isoform mRNA levels in Atlantic salmon. Gen. Comp. Endocrinol. 157, 35-40.

Melamed, P., Rosenfeld, H., Elizur, A., Yaron, Z. (1998). Endocrine regulation of gonadotropin and growth hormone gene transcription in fish. Comp. Biochem. Physiol. 119C, 325-338.

Miwa, S., Inui, Y. (1991). Thyroid hormone stimulates the shift of erythrocyte populations during metamorphosis of the flounder. J. Exp. Zool. 259, 222-228.

Mola, L., Gambarelli, A., Pederzoli, A., Ottaviani, E. (2005). ACTH response to LPS in the first stages of development of the fish *Dicentrarchus labrax* L. Gen. Comp. Endocrinol. 143, 99-103.

Montero, D., Marrero, M., Izquierdo, M. S., Robaina, L., Vergara, J. M., Tort, L. (1999). Effect of Vitamin E and C and dietary supplementation on some immune parameters of gilthead seabream (*Sparus aurata*) juveniles subjected to crowding stress. Aquaculture. 171, 269-278.

Montero, D., Kalinowski, T., Obach, A., Robaina, L., Tort, L., Caballero, M. J. (2003). Vegetable lipid sources for gilthead seabream (*Sparus aurata*): effects on fish health. Aquaculture. 225, 353-370.

Montero, D., Lalumera, G., Izquierdo, M. S., Caballero, M. J., Saroglia, M., Tort, L. (2009). Establishment of dominance relationships in gilthead sea bream (*Sparus aurata*) juveniles during feeding: effects on feeding behaviour, feed utilization, and fish health. J. Fish Biol. 74, 1-16.

Montero, D., Mathlouthi, F., Tort, L., Afonso, J. M., Torrecillas, S. (2010). Replacement of dietary fish oil by vegetable oils affects humoral immunity and expression of pro-inflammatory cytokines genes in gilthead sea bream *Sparus aurata*. Fish Shellfish Immunol. 29, 1073-1081.

Morera, D., Roher, N., Ribas, L., Balasch, J. C., Doñate, C., Callol, A. (2011). RNA-Seq reveals an integrated immune response in nucleated erythrocytes. PloS One. 6 (10), e26998.

Mori, T., Hiraka, I., Kurata, Y., Kawachi, H., Mano, N., Devlin, R. H. (2007). Changes in hepatic gene expression related to innate immunity, growth and iron metabolism in GH-transgenic amago salmon (*Oncorhynchus masou*) by cDNA subtraction and microarray analysis, and serum lysozyme activity. Gen. Comp. Endocrinol. 151, 42-54.

Nakanishi, T. (1986). Seasonal changes in the humoral immune response and the lymphoid tissues of the marine teleost, *Sebastiscus marmoratus*. Vet. Immunol. Immunopathol. 12, 213-221.

Narnaware, Y. K. , Baker, B. I. (1996). Evidence that cortisol may protect against the immediate effects of stress on circulating leukocytes in the trout. Gen. Comp. Endocrinol. 103, 359-366.

Narnaware, Y. K. , Baker, B. I. , Tomlinson, M. G. (1994). The effect of various stresses, corticosteroids and adrenergic agents on phagocytosis in the rainbow trout *Oncorhynchus mykiss*. Fish Physiol. Biochem. 13, 31-40.

Narnaware, Y. K. , Kelly, S. P. , Woo, N. Y. S. (1997). Effect of injected growth hormone on phagocytosis in silver sea bream (*Sparus sarba*) adapted to hyper-and hypo-osmotic salinities. Fish Shellfish Immunol. 7, 515-517.

Narnaware, Y. K. , Kelly, S. P. , Woo, N. Y. S. (1998). Stimulation of macrophage phagocytosis and lymphocyte count by exogenous prolactin administration in silver sea bream (*Sparus sarba*) adapted to hyper-and hypo-osmotic salinities. Vet. Immunol. Immunopathol. 61, 387-391.

Ndong, D. , Chen, Y. Y. , Lin, Y. H. , Vaseeharan, B. , Chen, J. C. (2007). The immune response of tilapia *Oreochromis mossambicus* and its susceptibility to *Streptococcus iniae* under stress in low and high temperatures. Fish Shellfish Immunol. 22, 686-694.

Nickerson, J. G. , Dugan, S. G. , Drouin, G. , Perry, S. F. , Moon, T. W. (2003). Activity of the unique-adrenergic Na^+/H^+ exchanger in trout erythrocytes is controlled by a novel β_3-AR subtype. Am. J. Physiol. Regul. Integr. Comp. Physiol. 285, R526-R535.

Olsen, Y. A. , Reitan, F. J. , Roed, K. H. (1993). Gill Na^+, K^+-ATPase activity, plasma cortisol level, and non-specific immune response in Atlantic salmon (*Salmo salar*) during parr-smolt transformation. J. Fish Biol. 43, 559-573.

Ortuño, J. , Esteban, M. A. , Meseguer, J. (2001). Effects of short-term crowding stress on the gilthead seabream (*Sparus aurata* L.) innate immune response. Fish Shellfish Immunol. 11, 187-197.

Ottaviani, E. , Franceschi, C. (1996). The neuroimmunology of stress from invertebrates to man. Prog. Neurobiol. 48, 421-440.

Owen, S. F. , Giltrow, E. , Huggett, D. B. , Hutchinson, T. H. , Saye, J. -A. , Winter, M. J. (2007). Comparative physiology, pharmacology and toxicology of β-blockers: mammals versus fish. Aquat. Toxicol. 82, 145-162.

Pagniello, K. B. , Bols, N. C. , Lee, L. E. (2002). Effect of corticosteroids on viability and proliferation ofthe rainbow trout monocyte/macrophage cell line, RTS11. Fish Shellfish Immunol. 13, 199-214.

Pankhurst, N. W. , Van Der Kraak, G. (1997). Effects of stress on reproduction and growth of fish. In: Iwama, G. K. , Pickering, A. D. , Sumpter, J. P. , Schreck, C.

B. , Fish Stress and Health in Aquaculture. Cambridge University Press, Cambridge, 73-93.

Parra, D. , Rieger, A. M. , Li, J. , Zhang, Y. A. , Randall, L. M. , Hunter, C. A. (2012). Pivotal advance: peritoneal cavity B-1 B cells have phagocytic and microbicidal capacities and present phagocytosed antigen to CD4 + T cells. J. Leukoc. Biol. 91, 525-536.

Pickering, A. D. , Christie, P. (1980). Sexual differences in the incidence and severity of ectoparasitic infestation of the brown trout, *Salmo trutta* L. J. Fish Biol. 16, 669-683.

Pickering, A. D. , Pottinger, T. G. (1987). Lymphocytopenia and interrenal activity during sexual maturation in the brown trout, *Salmo trutta* L. J. Fish Biol. 30, 41-50.

Pickford, G. E. , Srivastave, A. K. , Slicher, A. M. , Pang, P. K. T. (1971). The stress response in the abundance of circulating leucocytes in the killifish, *Fundulus heteroclitus*. I. The cold-shock sequence and the effects of hypophysectomy. J. Exp. Zool. 177, 89-96.

Pottinger, T. G. , Carrick, T. R. , Appleby, A. , Yeomans, W. E. (2000). High blood cortisol levels and low cortisol receptor affinity: is the chub, *Leuciscus cephalus*, a cortisol-resistant teleost?. Gen. Comp. Endocrinol. 120, 108-117.

Pratt, W. B. , Toft, D. O. (1997). Steroid receptor interactions with heat shock protein and immunophilin chaperones. Endocr. Rev. 18, 306-360.

Pruett, S. B. (2003). Stress and the immune system. Pathophysiology. 9, 133-153.

Prunet, P. , Sandra, O. , Le Rouzic, P. , Marchand, O. , Laudet, V. (2000). Molecular characterization of the prolactin receptor in two fish species, tilapia *Oreochromis niloticus* and rainbow trout, *Oncorhynchus mykiss*: a comparative approach. Can. J. Physiol. Pharmacol. 78, 1086-1096.

Prunet, P. , Sturm, A. , Milla, S. (2006). Multiple corticosteroid receptors in fish: from old ideas to new concepts. Gen. Comp. Endocrinol. 147, 17-23.

Randall, D. J. , Perry, S. F. (1992). Catecholamines. In: Hoar, W. S. , Randall, D. J. , Farrell, A. P. Academic Press, San Diego, 255-300.

Reid, S. D. , Perry, S. F. (1991). The effects and physiological consequences of raised levels of cortisol on rainbow trout (*Oncorhynchus mykiss*) erythrocyte β-adrenoreceptors. J. Exp. Biol. 158, 217-240.

Reid, S. D. , Lebras, Y. , Perry, S. F. (1993). The in vitro effect of hypoxia on the trout erythrocyte β-adrenergic signal transduction system. J. Exp. Biol. 176, 103-116.

Richards, R. H. , Pickering, A. D. (1978). Frequency and distribution patterns of *Saprolegnia* infection in wild and hatchery-reared brown trout *Salmo trutta* L. and char *Salvelinus alpinus* (L.). J. Fish Dis. 1, 69-82.

Richter, K. , Buchner, J. (2001). HSP90: chaperoning signal transduction. J. Cell. Physiol. 188, 281-290.

Ringø, E. , Zhou, Z. , He, S. , Erik, R. (2014). Effect of stress on intestinal microbiota of Arctic charr, Atlantic salmon, rainbow trout and Atlantic cod: a review. Afr. J. Microbiol. Res. 8, 609-618.

Rotllant, J. , Tort, L. (1997). Cortisol and glucose responses after acute stress by net handling in the sparid red porgy previously subjected to crowding stress. J. Fish Biol. 51, 21-28.

Roy, B. , Rai, U. (2008). Role of adrenoceptor-coupled second messenger system in sympatho-adrenomedullary modulation of splenic macrophage functions in live fish *Channa punctatus*. Gen. Comp. Endocrinol. 155, 298-306.

Saeij, J. P. , Verburg-van Kemenade, L. B. , van Muiswinkel, W. B. , Wiegertjes, G. F. (2003). Daily handling stress reduces resistance of carp to *Trypanoplasma borreli*: in vitro modulatory effects of cortisol on leukocyte function and apoptosis. Dev. Comp. Immunol. 27, 233-245.

Saha, N. R. , Usami, T. , Suzuki, Y. (2002). Seasonal changes in the immunr activities of common carp (*Cyprinus carpio*). Fish Physiol. Biochem. 26, 379-387.

Saha, N. R. , Usami, T. , Suzuki, Y. (2004). In vitro effects of steroid hormones on IgM-secreting cells and IgM secretion in common carp (*Cyprinus carpio*). Fish Shellfish Immunol. 17, 149-158.

Sakai, M. , Kobayashi, M. , Kawauchi, H. (1996). In vitro activation of fish phagocytic cells by GH, PRL and somatolactin. J. Endocrinol. 151, 113-118.

Sakai, M. , Kobayashi, M. , Kawauchi, H. (1996). Mitogenic effect of growth hormone and prolactin on chum salmon *Oncorhynchus keta* leukocytes in vitro. Vet. Immunol. Immunopathol. 53, 185-189.

Sakai, M. , Yamaguchi, T. , Wananuki, H. , Yasuda, A. , Takahashi, A. (2001). Modulation of fish phagocytic cells by *N*-terminal peptides of proopiomelanocortin (NPP). J. Exp. Zool. 290, 341-346.

Sakamoto, T. , McCormick, S. D. (2006). Prolactin and growth hormone in fish osmoregulation. Gen. Comp. Endocrinol. 147, 24-30.

Sakamoto, T. , McCormick, S. D. , Hirano, T. (1993). Osmoregulatory actions of growth hormone and its role of action in salmonids: a review. Fish Physiol. Biochem. 11, 155-164.

Salinas, I. , Zhang, Y. A. , Sunyer, J. O. (2011). Mucosal immunoglobulins and B cells of teleost fish. Dev. Comp. Immunol. 35, 1346-1365.

Sanders, V. M. , Kasprowicz, D. J. , Kohm, A. P. , Swanson, M. A. (2001). Neurotransmitter receptors on lymphocytes and lymphoid cells. In: Ader, R. ,

Felten, D. L. , Cohen, N. , . Academic Press, San Diego, 161-196.

Sandra, O. , Le Rouzic, P. , Cauty, C. , Edery, M. , Prunet, P. (2000). Expression of the prolactin receptor (tiPRL-R) gene in tilapia *Oreochromis niloticus*: tissue distribution and cellular localization in osmoregulatory organs. J. Mol. Endocrinol. 24, 215-224.

Santos, C. R. A. , Ingleton, P. M. , Cavaco, J. E. B. , Kelly, P. A. , Edery, M. , Power, D. M. (2001). Cloning, characterization, and tissue distribution of prolactin receptor in the sea bream (*Sparus aurata*). Gen. Comp. Endocrinol. 121, 32-47.

Sathiyaa, R. , Campbell, T. , Vijayan, M. M. (2001). Cortisol modulates HSP90 mRNA expression in primary cultures of trout hepatocytes. Comp. Biochem. Physiol. 129B, 679-685.

Saurabh, S. , Sahoo, P. K. (2008). Lysozyme: an important defence molecule of fish innate immune system. Aquac. Res. 39, 223-229.

Scheinman, R. I. , Cogwell, P. C. , Lofquist, A. K. , Baldwin, A. S. (1995). Role of transcriptional activation of IkBα in mediation of immunosuppression by glucocorticoids. Science. 270, 283-286.

Schreck, C. B. (1996). Immunomodulation: endogenous factors. In: Iwama, G. , Nakanishi, T. Academic Press, New York, NY, 311-337.

Schreck, C. B. , Maule, A. G. (2001). Are the endocrine and immune systems really the same thing?. In: Goos, H. J. Th, Rastogi, R. K. , Vaudry, H. , Pierantoni, R. , Perspective in Comparative Endocrinology: Unity and Diversity. Monduzzi Editore, Bologna, 351-357.

Schreck, C. B. , Tort, L. (2016). The Concept of Stress in Fish. In: Schreck, C. B. , Tort, L. , Farrell, A. P. , Brauner, C. J. Academic Press, San Diego, CA.

Shamblott, M. J. , Chen, T. T. (1993). Age-related and tissue-specific levels of five forms of insulin-like growth factor mRNA in a teleost. Mol. Mar. Biol. Biotechnol. 2, 351-361.

Shepherd, B. S. , Ron, B. , Burch, A. , Sparks, R. , Richman, N. H. , Shimoda, S. K. (1997). Effects of salinity, dietary level of protein and 17α-methyltestosterone on growth hormone (GH) and prolactin ($tPRL_{177}$ and $tPRL_{188}$) levels in the tilapia, *Oreochromis mossambicus*. Fish Physiol. Biochem. 17, 279-288.

Shved, N. , Berishvili, G. , Häusermann, E. , D'Cotta, H. , Baroiller, J. -F. , Eppler, E. (2009). Challenge with 17α-ethinylestradiol (EE2) during early development persistently impairs growth, differentiation, and local expression of IGF-I and IGF-II in immune organs of tilapia. Fish Shellfish Immunol. 26, 524-530.

Shved, N. , Berishvili, G. , Mazel, P. , Baroiller, J. -F. , Eppler, E. (2011).

Growth hormone (GH) treatment acts on the endocrine and autocrine/paracrine GH/IGF-axis and on TNF-α expression in bony fish pituitary and immune organs. Fish Shellfish Immunol. 31, 944-952.

Slicher, A. M. (1961). Endocrinological and hematological studies in *Fundulus heteroclitus* (*Linn.*). Bull. Bingham Ocean Coll. 17, 3-55.

Small, B. C., Bilodeau, A. L. (2005). Effects of cortisol and stress on channel catfish (*Ictalurus punctatus*) pathogen susceptibility and lysozyme activity following exposure to *Edwardsiella ictaluri*. Gen. Comp. Endocrinol. 142, 256-262.

Stephanou, A., Latchman, D. S. (1999). Transcriptional regulation of the heat shock protein genes by STAT family transcription factors. Gene Expr. 7, 311-319.

Sternberg, E. M. (2006). Neural regulation of innate immunity: a coordinated nonspecific host response to pathogens. Nat. Rev. Immunol. 6, 318-328.

Stolte, E. H., Verburg van Kemenade, B. M. L., Savelkoul, H. F. J., Flik, G. (2006). Evolution of glucocorticoid receptors with different glucocorticoid sensitivity. J. Endocrinol. 190, 17-28.

Stolte, E. H., de Mazon, A. F., Leon-Koosterziel, K. M., Jęsiak, M., Bury, N. R., Strum, A. (2008). Corticosteroid receptors involved in stress regulation in common carp, *Cyprinus carpio*. J. Endocrinol. 198, 403-417.

Stolte, E. H., Nabuurs, S. B., Bury, N. R., Sturm, A., Flik, G., Savelkoul, H. F. (2008). Stress and innate immunity in carp: corticosteroid receptors and pro-inflammatory cytokines. Mol. Immunol. 46, 70-79.

Sunyer, J. O. (2013). Fishing for mammalian paradigms in the teleost immune system. *Nat. Immunol.* 14, 320-326.

Sunyer, J. O., Tort, L. (1995). Natural hemolytic and bactericidal activities of sea bream *Sparus aurata* serum are effected by the alternative complement pathway. Vet. Immunol. Immunopathol. 45, 333-345.

Suzuki, Y., Orito, M., Iigo, M., Kezuka, H., Kobayashi, M., Aida, K. (1996). Seasonal changes in blood IgM levels in goldfish, with special reference to water temperature and gonadal maturation. Fish Sci. 62, 754-759.

Suzuki, Y., Otaka, T., Sato, S., Hou, Y. Y., Aida, K. (1997). Reproduction related immunoglobulin changes in rainbow trout. Fish Physiol. Biochem. 17, 415-421.

Svensson, S. P., Bailey, T. J., Pepperl, D. J., Grundström, N., Ala-Uotila, S., Scheinin, M. (1993). Cloning and expression of a fish β_2-adrenoceptor. Br. J. Pharmacol. 110, 54-60.

Tagawa, M., Hagiwara, H., Takemura, A., Hirose, S., Hirano, T. (1997). Partial cloning of the hormone-binding domain of the cortisol receptor in tilapia,

Oreochromis mossambicus, and changes in the mRNA level during early embryonic development. Gen. Comp. Endocrinol. 108.

Takahashi, A., Takasaka, T., Yasuda, A., Amemiya, Y., Sakai, M., Kawauchi, H. (2000). Identification of carp proopiomelanocortin-related peptides and their effects on phagocytes. Fish Shellfish Immunol. 10, 273-284.

Takei, Y., Hwang, P.-P. (2016). Homeostatic Responses to Osmotic Stress. In: Schreck, C. B., Tort, L., Farrell, A. P., Brauner, C. J., . Academic Press, San Diego, CA.

Takeo, J., Hara, S., Segawa, C., Toyokawa, H., Yamashita, S. (1996). Fish glucocorticoid receptor with splicing variants in the DNA binding domain. FEBS Lett. 389, 244-248.

Tatner, M. F. (1996). Natural changes in the immune system of fish. In: Iwama, G., Nakanishi, T., The Fish Immune System: Organism, Pathogen, and Environment. Academic Press, San Diego, 255-287.

Teles, M., Mackenzie, S., Boltaña, S., Callol, A., Tort, L. (2011). Gene expression and TNF-alpha secretion profile in rainbow trout macrophages following exposures to copper and bacterial lipopolysaccharide. Fish Shellfish Immunol. 30, 340-346.

Teles, M., Tridico, R., Callol, A., Fierro-Castro, C., Tort, L. (2013). Differential expression of the corticosteroid receptors GR1, GR2 and MR in rainbow trout organs with slow release cortisol implants. Comp. Biochem. Physiol. 164A, 506-511.

Tort, L. (2011). Stress and immune modulation in fish. Dev. Comp. Immunol. 35, 1366-1375.

Tort, L., Padros, F., Rotllant, J., Crespo, S. (1998). Winter syndrome in the gilthead sea bream *Sparus aurata*. Immunological and histopathological features. Fish Shellfish Immunol. 8, 37-47.

Tse, D. L. Y., Chow, B. K. C., Chan, C. B., Lee, L. T. O., Cheng, C. H. K. (2000). Molecular cloning and expression of a prolactin receptor in goldfish (*Carassius auratus*). Life Sci. 66, 593-605.

Tsutsui, K., Bentley, G. E., Kriegsfeld, L. J., Osugi, T., Seong, J. Y., Vaudry, H. (2010). Discovery and evolutionary history of gonadotrophin-inhibitory hormone and kisspeptin: new key neuropeptides controlling reproduction. J. Neuroendocrinol. 22, 716-727.

Uchida, K., Kaneko, T., Tagawa, M., Hirano, T. (1998). Localization of cortisol receptor in branchial chloride cells in chum salmon fry. Gen. Comp. Endocrinol. 109,

175-185.

Unniappan, S. , Peter, R. E. (2005). Structure, distribution and physiological functions of ghrelin in fish. Comp. Biochem. Physiol. 140A, 396-408.

Van Der Kraak, G. (2009). The GnRH system and the neuroendocrine regulation of reproduction. In: Bernier, N. J. , Van Der Kraak, G. , Farrell, A. P. , Brauner, C. J. , Fish Neuroendocrinology. Academic Press, London, 113-149.

Varsamos, S. , Flik, G. , Pepin, J. F. , Bonga, S. E. W. , Breuil, G. (2006). Husbandry stress during early life stages affects the stress response and health status of juvenile sea bass, *Dicentrarchus labrax*. Fish Shellfish Immunol. 20, 83-96.

Vazzana, M. , Cammarata, M. , Cooper, E. L. , Parrinello, N. (2002). Confinement stress in sea bass (*Dicentrarchus labrax*) depresses peritoneal leukocyte cytotoxicity. Aquaculture. 210, 231-243.

Venters, H. K. , Dantzer, R. , Freund, G. G. , Broussard, S. R. , Kelley, K. W. (2001). Growth hormone and insulin-like growth factor as cytokines in the immune system. In: Ader, R. , Felten, D. L. , Cohen, N. , . Academic Press, San Diego, 339-362.

Verburg-Van Kemenade, B. M. L. , Nowak, B. , Engelsma, M. Y. , Weyts, F. A. A. (1999). Differential effects of cortisol on apoptosis and proliferation of carp B-lymphocytes from head kidney, spleen and blood. Fish Shellfish Immunol. 9, 405-415.

Verburg-Van Kemenade, B. M. L. , Stolte, E. H. , Metz, J. R. , Chadzinska, M. (2009). Neuroendocrine-immune interactions in teleost fish. Fish Neuroendocrinology. , 313-364.

Volkmann, E. R. , Weekes, N. Y. (2006). Basal SIgA and cortisol levels predict stress-related health outcomes. Stress Health. 22, 11-23.

Wang, W. -B. , Wang, Y. -P. , Hu, W. , Li, A. -H. , Cai, T. -Z. , Zhu, Z. -Y. (2006). Effects of the "all-fish" growth hormone transgene expression on non-specific immune functions of common carp, *Cyprinus carpio* L. Aquaculture. 259, 81-87.

Watanuki, H. , Gushiken, Y. , Takahashi, A. , Yasuda, A. , Sakai, M. (2000). In vitro modulation of fish phagocytic cells by β-endorphin. Fish Shellfish Immunol. 10, 203-212.

Watanuki, N. , Takahashi, A. , Yasuda, A. , Sakai, M. (1999). Kidney leucocytes of rainbow trout, *Oncorhynchus mykiss*, are activated by intraperitoneal injection of β-endorphin. Vet. Immunol. Immunopathol. 71, 89-97.

Weigent, D. A. , Arnold, R. E. (2005). Expression of insulin-like growth factor-1 and insulin-like growth factor-1 receptors in EL4 lymphoma cells overexpressing growth hormone. Cell Immunol. 234, 54-66.

Wendelaar Bonga, S. E. (1997). The stress response in fish. Physiol. Rev. 77, 591-625.

Weyts, F. A. A., Flik, G., Verburg-Van Kemenade, B. M. L. (1998). Cortisol inhibits apoptosis in carp neutrophilic granulocytes. Dev. Comp. Immunol. 22, 563-572.

Weyts, F. A. A., Flik, G., Rombout, J. H. W. M., Verburg-Van Kemenade, B. M. L. (1998). Cortisol induces apoptosis in activated B cells, not in other lymphoid cells of common carp, *Cyprinus carpio*. Dev. Comp. Immunol. 22, 551-562.

Weyts, F. A. A., Verburg-Van Kemenade, B. M. L., Flik, G. (1998). Characterization of glucocorticoid receptors in peripheral blood leukocytes of carp, *Cyprinus carpio* L. Gen. Comp. Endocrinol. 111, 1-8.

Weyts, F. A. A., Cohen, N., Flik, G., Verburg-Van Kemenade, B. M. L. (1999). Interactions between the immune system and the hypothalamo-pituitary-interrenal axis in fish. Fish Shellfish Immunol. 9, 1-20.

White, A., Fletcher, T. C. (1985). The influence of hormones and inflammatory agents on C-reactive protein, cortisol and alanine aminotransferase in the place (*Pleuronectes platessa* L.). Comp. Biochem. Physiol. 80C, 99-104.

Wojtaszek, J., Dziewulska-Szwajkowska, D., Lozińska-Gabska, M., Adamowicz, A., Dzugaj, A. (2002). Hematological effects of high dose of cortisol on the carp (*Cyprinus carpio* L.): cortisol effect on the carp blood. Gen. Comp. Endocrinol. 125, 176-183.

Yada, T. (2007). Growth hormone and fish immune system. Gen. Comp. Endocrinol. 152, 353-358.

Yada, T. (2009). Effects of insulin-like growth factor-I on non-specific immune functions in rainbow trout. Zool. Sci. 26, 338-343.

Yada, T. (2012). Effect of gonadotropin-releasing hormone on phagocytic leucocytes of rainbow trout. Comp. Biochem. Physiol. 155C, 375-380.

Yada, T., Nakanishi, T. (2002). Interaction between endocrine and immune systems in fish. Int. Rev. Cytol. 220, 35-92.

Yada, T., Hirano, T., Grau, E. G. (1994). Changes in plasma levels of the two prolactins and growth hormone during adaptation to different salinities in the euryhaline tilapia, *Oreochromis mossambicus*. Gen. Comp. Endocrinol. 93, 214-223.

Yada, T., Azuma, T., Takagi, Y. (2001). Stimulation of non-specific immune functions in seawater-adapted rainbow trout, *Oncorhynchus mykiss*, with reference to the role of growth hormone. Comp. Biochem. Physiol. 129B, 695-701.

Yada, T., Uchida, K., Kajimura, S., Azuma, T., Hirano, T., Grau, E. G. (2002). Immunomodulatory effects of prolactin and growth hormone in the tilapia, *Oreochromis mossambicus*. J. Endocrinol. 173, 483-492.

Yada, T., Misumi, I., Muto, K., Azuma, T., Schreck, C. B. (2004). Effects of prolactin and growth hormone on proliferation and survival of cultured trout leucocytes. Gen. Comp. Endocrinol. 136, 298-306.

Yada, T., Muto, K., Azuma, T., Hyodo, S., Schreck, C. B. (2005). Cortisol stimulates growth hormone gene expression in rainbow trout leucocytes in vitro. Gen. Comp. Endocrinol. 142, 248-255.

Yada, T., Muto, K., Azuma, T., Fukamachi, S., Kaneko, T., Hirano, T. (2006). Effects of acid water exposure on plasma cortisol, ion balance, and immune functions in the "cobalt" variant of rainbow trout. Zool. Sci. 23, 707-713.

Yada, T., Azuma, T., Hyodo, S., Hirano, T., Grau, E. G., Schreck, C. B. (2007). Differential expression of corticosteroid receptor genes in trout immune system in response to acute stress. Can. J. Fish. Aquat. Sci. 64, 1382-1389.

Yada, T., Hyodo, S., Schreck, C. B. (2008). Effects of seawater acclimation on mRNA levels of corticosteroid receptor genes in osmoregulatory and immune systems in trout. Gen. Comp. Endocrinol. 156, 622-627.

Yada, T., McCormick, S. D., Hyodo, S. (2012). Effects of environmental salinity, biopsy, and GH and IGF-I administration on the expression of immune and osmoregulatory genes in the gills of Atlantic salmon (*Salmo salar*). Aquaculture. 362-363, 177-183.

Yada, T., Miyamoto, K., Miura, G., Munakata, A. (2014). Seasonal changes in gene expression of corticoid receptors in anadromous and non-anadromous strains of rainbow trout *Oncorhynchus mykiss*. J. Fish Biol. 85, 1263-1278.

Yano, T. (1996). The nonspecific immune system: humoral defense. In: Iwama, G., Nakanishi, T., The Fish Immune System: Organism, Pathogen, and Environment. Academic Press, San Diego, 105-157.

Yasuoka, A., Abe, K., Arai, S., Emori, Y. (1996). Molecular cloning and functional expression of the 1α-adrenoceptor of medaka fish, *Oryzias latipes*. Eur. J. Biochem. 235, 501-507.

Zhang, T., Qiu, L., Sun, Z., Wang, L., Zhou, Z., Liu, R. (2014). The specifically enhanced cellular immune responses in Pacific oyster (*Crassostrea gigas*) against secondary challenge with *Vibrio splendidus*. Dev. Comp. Immunol. 45, 141-150.

Zhang, Y. A., Salinas, I., Li, J., Parra, D., Bjork, S., Xu, Z. (2010). IgT, a primitive immunoglobulin class specialized in mucosal immunity. Nat. Immunol. 11, 827-835.

Zhu, L. Y., Lin, A. F., Shao, T., Nie, L., Dong, W. R., Xiang, L. X. (2014). B cells in teleost fish act as pivotal initiating APCs in priming adaptive immunity: an evolutionary perspective on the origin of the B-1 cell subset and B7 molecules. J. Immunol. 192, 2699-2714.

第 11 章　鱼类的应激指标

鱼池里的鱼在被捕获前为渔网追逐，然后离开了水，再被安置在附近测试场地。这会令它应激吗？我们怎么知道？我们检测的指标可靠吗？鱼类应激的量化已经从最初使用放射免疫测定法检测血浆中的皮质醇，发展到使用基于基因组开发的检测方法，其指标范围涵盖了从细胞内到整个生物体水平。热休克蛋白（HSPs）的表达量、代谢相关酶的活性，可以和能直接观察到的条件反射、存活率相结合。传统的指标和新兴的指标都各有其优缺点，使用时是有组织特异性和特定条件特异性的。在选择、检测和解读应激指标的时候，必须考虑到生态学的、生物学的和方法上的各种因素的影响。对应激源生理反应在种内和种间、性别、生命周期不同阶段和时间上的差异可能被混淆到一个应激状态中。尽管有很多类型的指标，但我们对于怎样将指标的绝对水平和应激的严重程度与恢复情况之间联系起来，至今依旧理解有限。在暴露于应激源的天然野生种群中，指标能否准确地反映应激的特性，仍是个有待进一步讨论的问题。研究领域的整合，利益相关者和用户群体的参与，以及将个体水平的指标转化成群体水平或者生态水平的过程，都将有助于填补这些知识空白。

11.1　我们为什么要检测应激？

正如 Schreck 和 Tort 所阐述的那样，应激是包括鱼类在内的所有脊椎动物生命活动中与生俱来的组成部分。应激的检测结果告诉我们，当面对有害的刺激时，鱼如何有效地抵抗死亡及恢复机体平衡状态。这些信息随后转化为进化和生态理论，以了解动物如何适应或能够适应将来的应激源。从根本上说，检测机体应激反应有助于我们了解携带效应（如 O'Connor 等，2010）、亲本效应（如 Sopinka 等，2014）、个体性（如 Aubin-Horth 等，2012）和生活史变化（如 Pottinger，Carrick，2001；Ricklefs，Wikelski，2002）。如果不对应激反应指标进行分类、量化、细化和解读，就不能清楚了解鱼类对外部和内部环境变化所产生反应的重要性和影响（Schulte，2014）。例如：如果没有一个明确的应激指标，某个动物的反应能被认定为应激反应吗？没有受过处理的动物能被描述成实验的对照组吗？此外，环境的改变能被归类为某个应激源吗？应激的程度超过生物的组织结构水平，检测应激有助于将个体反应（Calow，Forbes，1998；Fefferman，Romero，2013）与群体水平和生态健康（Dale 和 Beyeler，2001）联系起来。

从应用的前景来看，检测应激是非常必要的，用来确认鱼类健康福利是如何受到与人类互相作用的影响的。例如：评估蓄养鱼群（如孵化场、养殖场、水族馆）的应激状态通常是为了减少应激，使鱼群能最大化地增长和存活。事实上，鱼类应激的实验式研究有了一定的基础，继续扩展研究内容，在水产养殖中的相关研究是检测接触、饲养、运输（Barton 等，1980；Portz 等，2006）及麻醉等（Iwama 等，1989；Trushenski 等，2010）操作对养殖鱼类的健康和繁殖能力的影响。应激量化基本的机制研究将持续在驯化的鱼群中（如虹鳟、罗非鱼）开展。现在已将应激指标视为“鱼类福利”的客观指标（Iwama，2007），用来分析休闲渔业（如 Morrissey 等，2005；Landsman 等，2015）、经济渔业（Marcalo 等，2006；Raby 等，2015）以及实验用鱼（Brydges 等，2009；Eaton 等，2015）的鱼类行为表现的影响。人们对使用优化的应激指标以评估野生种群状态的兴趣越来越大（Madliger，Love，2014）。这种将基础研究与应用结合并促进应激反应的检测特别有价值。将应激的进化与生态基础和有关工业与环保工作人员对应激的检测结合起来，可以为鱼类生物学、人类生活与文化方面有效的管理策略提供指导。

在这里，我们提供了从细胞到个体水平的应激指标的总结（本章第 11.3 节）。接着，我们归纳了检测和解读这些指标时需要考虑的重要事项（本章第 11.4 节），讨论了将个体水平指标扩展到种群和生态系统水平的过程（本章第 11.5 节），并且总结研究途径和新的指标，为进一步的研究提供基础保障（本章第 11.6 节）。我们感谢 S. Marshall Adam 所著《Biological Indicators of Stress in Aquatic Ecosystems》一书中对水生生态系统中生物应激指标所做的开创性综述，特别是 Barton 等人（Barton 等，2002）关于鱼类生理的和状态相关的应激指标的章节。我们鼓励读者去参考这些文章以便进一步了解鱼类的应激指标。

11.2 应激的量化

根据靶标的不同指标，鱼类应激量化方法也不尽相同。量化的主要区别在于检测的是不是应激反应的本身（即初级反应/HPI 活性）（Gorissen，Flik，2016；见本书第 3 章），或其他生理、行为或生活史特征（即次级和三级检测）（Mazeaud 等，1977；见本章第 11.3 节）是与刺激 HPI 轴同时发生的改变，还者是因 HPI 轴受刺激所导致的结果。评估一个个体暴露在急性或慢性应激源时所产生的反应大小，通常需要在一段时间内重复采样。必须建立尽可能精确的理想指标的基准线/静息水平，并作为一系列检测的首个时间点。检测整个时间段里关键的主要参数（如儿茶酚胺、皮质醇）的变化有助于确定应激反应的特性（见 11.3.2 部分）。当慢性应激被量化后，主要指标的静息水平是在应激源暴露停止时检测（在慢性应激源暴露后由急性应激源所引发的水平）。已经建立的应激反应量化方法虽然得到了广泛的重复验证和认可，但也只是代表整个动物对应激源的反应的一部分（见 11.3.3 部分）。

应激反应的实验评估并非没有挑战。由于身体规格的限制（如鱼非常大或非常小）使得无法获得重复性样品的个体，基于预定采样间隔而使量化的指标峰值出现非故意的误差，在组织取样过程中的捕捉和手工操作引发了第二次应激反应（Baker，Vynne，2014），所有这些都限制了统计和解读数据的准确性。个体暴露在慢性应激源中的适应性也会影响解读（见 11.4.4 部分）。最终没有适度的测定或替代的检测指标，这样的机制研究将限制其扩展到群体水平上的应用中（Calow，Forbes，1998；见本章第11.5节）。然而，机制的研究是构建应激知识基础的关键，对生理的与行为的过程和应激相结合的研究设计具有推动和指导作用，并简化未来对应激的验证（例如，针对单个应激源后的时间点，以确定研究动物的应激反应）。

基于建立应激源暴露、HPI活化程度以及一系列个体反应之间的关系（Wendelarr-Bonga，1997；Iwama 等，1997；Mommsen 等，1999；Schreck，2010；Barton，2002），应激的定量可以参考HPI轴激活后引发的初级、次级和三级反应的量化（如基因表达、免疫功能、代谢、生长、生殖、性能、行为；见本章第11.3节）。典型的反应量化需要检测指标在应激物前和应激物后的水平。通常情况下，检测后者的反应是在应激源暴露后检测。观察到的应激源前和应激源后的差异可以视为应激指标。抽样方式可提供一个快速的指引，说明动物暴露在应激源中性状如何变化，但不能反映捕获过程中最大的反应或反应后的恢复。一般来说，当反应被定量时，与HPI轴功能直接相关的一个初级指标（如血浆皮质醇），也要在和应激源接触前以及和应激源接触后的一个时间点上被检测。研究物种和特异性的应激源时，关于初级、次级和三级应激指标的时间点，理想情况下应该基于最初时间过程中抽样（Pickering 等，1982；Donaldson 等，2014）。当指标的取样时间点是任意选择的，会引起一系列的混乱。例如，两条鱼在取样时间点上都有相同的升高应激指标，但一条鱼最后死了，另一条鱼却恢复了。此外，根据暴露在应激源中是连续的还是断续的（见11.4.4部分），都有可能使得一个应激源处理后的指标水平不稳定（Schreck，Tort，2016；见本书第1章，图1.5）。因此，在应激源和其他次级/三级检测之间建立明确的机制联系并非易事。然而，我们对应激过程中（包括恢复过程）HPI轴和生物有机体行为表现如何变化还了解得有限，但是Schreck和Tort对其进行了概念性的探讨（2016；见本书第1章，图1.5和图1.6）。

研究应激反应的定量和适应性特征之间的联系将有助于对生活史生理基础的深入理解。例如，应激源所引起的血浆糖皮质激素和适应性之间的联系（Breuner 等，2008）在鸟类（Mac Dougall-Shackleton 等，2009）和爬行动物（Romero，Wikelski，2001）中都已经检测到，现在在鱼类（Cooket 等，2014）中也发现了这样的联系。这种将应激反应本身（或其组成部分）和下游适应性变化相结合的综合量化方法促进了生理学家和生态学家之间的合作，推动了生态生理学或环境保护生理学等学科间领域的发展（Wikelski，Cooke，2006；Cooke 等，2013；Boonstra，2013a）。

11.3 鱼类应激的检测

需要注意的是，虽然应激量化被广泛研究并且有丰富的指标去鉴定一尾应激的鱼（表11.1至表11.3），但我们掌握的绝大多数应激指标都是初步的，特别是在非实验环境中（例如，如何了解生活在野外的鱼是否处在应激反应中?）。指标水平升高可能预示着鱼处在应激中，但指标水平下降并不等同于鱼的应激减少或不应激。我们希望读者在调查和选择检测鱼类应激方法时考虑到这一点。

表 11.1　细胞的和分子的应激指标

指标	抽样和分析注意事项	指标优点	指标缺点
氧化应激			
●代谢途径产生活性氧（ROS）作为一种天然副产品（Costantini, 2008） ●活性氧会破坏生物分子，尤其是脂质、蛋白质、RNA 和 DNA ●抗氧化剂可以通过阻止活性氧的形成，或者通过消除活性氧氧化应激来防止活性氧造成的损害 ●当活性氧产生压倒抗氧化剂的均衡能力时，生物分子发生损伤（Lesser，2006）	●由于氧化应激可以由 ROS 的过量产生或抗氧化剂不足引起，氧化应激可以通过（1）测量 ROS 来量化；（2）测定抗氧化水平；（3）测量生物分子的损伤 ●测量活性氧、抗氧化剂和生物分子损伤有多种标记（Lesser，2006） ●活性氧往往是不稳定的，因此，测量抗氧化剂或生物分子损伤是更常见的标志 ●根据所测物质的不同，标记物通过各种比色法测定，不同的标记物更容易在血浆、血清、尿液、组织匀浆或细胞培养中测定（Valavandis 等，2006）	●氧化应激是代谢过程中不可避免的副产物，因而可视为生命代价 ●氧化应激是在生态环境中作为繁殖成本（如 Alonso-Alverz 等，2004）、免疫反应成本（如 Torres and Velando，2007）或高强度能量消耗成本（如迁移成本）（Rankin，Burchsted，1992）来测量的。 ●氧化应激也是暴露在具有挑战性的环境下的结果，比如受到严重污染的区域（如 Bacanskas 等，2004）	●由于氧化应激产生于复杂的过程，其结果很难解释 ●测量氧化应激通常需要专门的设备，而且可能相对昂贵，尽管商业试剂盒正变得越来越普遍
端粒长度			
●环境应激可引起氧化应激，氧化应激如果不被抵消，可导致端粒缩短，加速细胞（可能还有机体）衰老（Monaghan 等，2009） ●环境应激如心理应激（Epel 等，2004）或生殖能力的提高（Kotrschal 等，2007）与端粒缩短有关	●为了确定相对端粒长度，定量 PCR 可以测量 DNA 样本与参考 DNA 样本在端粒重复拷贝数与单拷贝基因拷贝数的比值上的差异（Caethon，2002）	●潜在的强大指标，可以跨越环境应激、氧化应激源和机体衰老之间的差距	●环境应激源、细胞应激、端粒缩短和机体衰老之间的联系在很大程度上仍未得到验证（Monaghan，Haussmann，2006）。为了有效地利用端粒长度作为应激指标，还需要进行更多的研究来了解与端粒缩短相关的三级结果

续表 11.1

指标	抽样和分析注意事项	指标优点	指标缺点
热休克蛋白（HSPs）			
●HSPs，在热休克因子 1（HSF1）的控制下，指示细胞应激反应和 HSP 表达增加，以维持细胞内稳态（Iwama 等，2004） ●大多数 HSPs 是蛋白质折叠、修复和降解的分子伴侣（Moseley，1997）	●利用实时荧光定量 PCR（qRT-PCR）可以检测 HSP 的表达，需要分离基因组 DNA，提取全 RNA，逆转录酶 PCR 扩增（Fangue 等，2006） ● Hsp70 以前已被 ELISA、BIAcore 和以 bead 为基础的流式细胞仪定量；BIAcore 和 FACS 比 ELISA 更敏感，需要的样品更少 ●HSP 表达水平依赖于邻近序列，这意味着需要建立基线表达水平（Tomanek，Somero，1999）	●热休克蛋白对一系列应激源敏感（例如，温度快速变化、盐度挑战、手工操作）（Palmisano 等，2000；Donaldson 等，2008） ●广泛研究，功能了解	●HSPs 的表达与环境有关，因为它们对应激源的大小和持续时间很敏感（Iwama 等，2004），也能适应以前的应激源（Somero，Hoffman，1996）
即时早期基因和转录因子			
● 即时早期基因（IEGs）在细胞应激反应后几分钟内被诱导（Hughes，Dragunow，1995） ●通常测量的 IEGs 包括转录因子 c-fos、fosB、c-jun、JUNB、c-myc、egr-1（Inuzuka 等，1999） ●低氧诱导因子 1（HIF-1a）和 NUPR1 等转录因子通常在应激过程中被激活（Semenza，1998；Momoda 等，2007）	● IEG 的表达可用 qRT-PCR 或微阵列法测定 ●可在多种组织（如心脏、肝脏、鳃）中测量 ●可采用非致死的鳃活检，便于综合研究（如生理遥测）（Miller 等，2011）	●应激和恢复的敏感指标（Momoda，2007；Donaldson 等，2014）	●大多数研究都集中在哺乳动物身上，但是在更广泛的动物类群中解释 IEG 的激活将有助于识别转录应激反应的上游调节因子（Kassahn 等，2009） ●到目前为止，对鱼类的研究通常集中在不同的时间进程（Krasnov 等，2005）、物种、组织（Kassahn 等，2009）、技术（Prunet 等，2008）和感兴趣的基因上 ●未来的研究需要了解跨物种的功能作用和下游效应

续表 11.1

指标	抽样和分析注意事项	指标优点	指标缺点
胞内酶			
●细胞内的酶，如丙氨酸转氨酶（ALT）、天冬氨酸转氨酶（AST）、乳酸脱氢酶（LDH），或肌酸激酶（CK），是由细胞损伤或死亡释放的 ●因此，血浆中这些酶的存在是组织损伤发生的有用指标（Henry，1996） ●许多酶是组织特异性的，因而可以提供关于已经发生的组织损伤类型的信息（Wagner，Congleton，2004）	●组织损伤指标可通过血浆比色法测定 ●商业实验室也经常使用自动分析仪来测量血浆中的酶	●血浆中细胞内酶的存在是损伤的良好指标 ●商业实验室的可用性使得测量组织损伤的指标相对便宜，而且在等离子体中易于测量，而无须投资专门的设备	●这些指标通常不会升高，除非发生了身体损伤，因此作为不包括组织损伤的应激源的指标就不那么有用了（Wagner，Congleton，2004）

表 11.2　初级和次级的生理应激指标

指标	抽样和分析注意事项	指标优点	指标缺点
儿茶酚胺			
●当个体面临挑战时，生理反应首先是从嗜铬细胞中立即释放儿茶酚胺激素肾上腺素和去甲肾上腺素（Reid 等，1998；Gallo，Civinini，2003） ●肾上腺素和去甲肾上腺素的释放与经典的应激反应有关	●儿茶酚胺通常是通过色谱和电化学检测在血浆中测定的（Woodward，1982）	●儿茶酚胺对各种应激源都有反应（Reid 等，1998；Pottinger，2008），因此，儿茶酚胺的测定能够在很短的时间内提供关于急性应激源反应的非常准确的信息	●儿茶酚胺对捕捉和处理的反应非常迅速（即在几秒钟内），因此，如果没有专门的设备和实验室饲养的动物，很难量化儿茶酚胺的水平

续表 11.2

指标	抽样和分析注意事项	指标优点	指标缺点
类固醇激素			
●儿茶酚胺释放后，应激反应以下丘脑-垂体-肾间轴（HPI）激活为特征。HPI 轴激活涉及一系列复杂的相互作用，最终导致糖皮质激素的产生和释放（Mommsen 等，1999；Barton，2002；Pottinger，2008） ●对循环糖皮质激素的测量是一个个体是否正在经历应激的指标 ●促肾上腺皮质激素释放激素（corticotropin hormone，CRH）和促肾上腺皮质激素（肾上腺皮质激素，ACTH）是参与 HPI 轴活化的中间激素，也是常用的测量指标	●类固醇激素通常通过放射免疫测定法或酶联免疫测定法在血浆或组织匀浆中测定（Pottinger，2008；Sheriff 等，2011） ●糖皮质激素也可以在尿液和粪便中测量，因此，也可以在水样中提取和测量（Ellis 等，2004；Pottinger，2008） ●由于糖皮质激素面对挑战会迅速变化，糖皮质激素的测量通常在暴露于应激源之前和之后进行，以获得应激反应的测量；基线和应激后水平，或应激反应性，提供了关于个体状态的信息（Breuner 等，2008；Bonier 等，2009）	●应激后糖皮质激素水平可以提供有关个体如何受特定刺激影响的信息（例如，捕捉和手工操作应激、不同的保持条件、急性暴露）（Sapolsky 等，2000；Barton，2002） ●基线糖皮质激素水平可以提供关于动物是否正经历慢性环境应激的信息，在某些情况下还可以预测未来的表现和生存（Bonier 等，2009） ●在某些情况下，应激反应可以预测未来的表现和生存（Breuner 等，2008） ●循环的糖皮质激素水平也与生活史特征和权衡有关（Wingfield 等，1998；Ricklefs，Wikelski，2002）	●循环的糖皮质激素对捕捉和手工操作反应迅速（通常在 3～5 分钟内）（Romero，Reed，2005），因此，往往很难获得野生动物的基线水平 ●基线（Bonier 等，2009）和应激诱导（Breuner 等，2008）的糖皮质激素水平与未来表现和生存之间的关系是环境和物种特异性的，结果可能难以解释
代谢物			
●一旦糖皮质激素产生并进入循环，它们与一系列帮助动物生存和从挑战中恢复的次级反应相关联（Sapolsky 等，2000），包括动员储存的葡萄糖（Barton，2002） ●厌氧应激源（如剧烈运动）产生乳酸等厌氧代谢产物（Wood 等，1983）	●葡萄糖和乳酸可通过血浆或组织匀浆中的比色法测定 ●乳酸和葡萄糖也可以用为糖尿病患者或运动训练设计的便携式测量仪在全血中测量（Wells，Pankhurst，1999；Beecham 等，2006；Stoot 等，2014）	●代谢物对于评估特定应激源的急性反应非常有用（Barton 等，2002），尤其是乳酸盐情况下的运动应激源（Wood 等，1983） ●易于获得的便携式测量仪使乳酸和葡萄糖易于测量，价格低廉，使用非常小的血样，没有专门的设备，这使它们成为评估急	●由于葡萄糖和乳酸盐都受到应激反应之外的一般代谢过程的影响，因此，基线结果可能难以解释，而这些指标作为衡量对特定应激源的急性反应最为有用

续表 11.2

指标	抽样和分析注意事项	指标优点	指标缺点
		性应激源时在现场条件下测量的良好参数（Wells，Pankhurst，1999；Beecham 等，2006）	
渗透压和离子浓度			
●当鱼类受到急性应激时，肾上腺素的升高会引起血管收缩和心输出量增加（Mazeaud，Mazeaud，1981），这又会由于鳃瓣灌注增加而增加鳃的扩散能力（Randall，Perry，1992） ●这种扩散能力的增加会引起鳃部离子转移的增加，以及随后等离子体渗透压的变化，特别是在血液循环中 Na^+ 和 Cl^- 的浓度变化（McDonald，Milligan，1997）	●等离子体渗透压是用渗透压计测量的 ●离子可以用分光光度法在血浆中测量 ●市面上有一些测量血浆中一些常见离子（如 Na^+）的仪器 ●商业实验室也经常使用自动分析仪来测量血浆中的离子	●整体渗透压或离子平衡的变化是急性应激的良好指标 ●商业实验室的可用性使得离子测量相对便宜，而且在血浆中容易测量，不需要专门的设备	●这些指标是急性应激的有用指标，但在长期暴露的情况下往往难以解释，因为它们与环境有关，并受到多种内外因素的影响（McDonald，Milligan，1997）
营养指标			
●血浆中的营养指标（如总蛋白、总胆固醇、甘油三酯）提供了有关可用于燃料活动的调动能量储存当前水平的信息（Wagner，Congleton，2004；Congleton，Wagner，2006）	●血浆中的营养指标通常通过比色法测定 ●商业实验室也经常使用自动分析仪来测量血浆中的营养指标	●血浆中的营养指标可以提供鱼类最近摄食史的信息（Congleton，Wagner，2006） ●商业实验室的可用性使得测量营养指标相对便宜，而且不需要专门的设备就可以很容易地在血浆中测量	●营养指标并没有显示出对禁食和应激源的一致反应，其结果与物种和环境有关。营养指标有时很难解释为一般健康的指标（Wagner，Congleton，2004；Congleton，Wagner，2006；O'Connor 等，2011）

续表 11.2

指标	抽样和分析注意事项	指标优点	指标缺点
生物能学			
●能量是生命的货币，所以理解它在身体过程中的分配可以作为机体应激的敏感指标（Beyers 等，2002） ●能量储存和脂质含量与生存、繁殖和生活史策略有关（Henderson，Tocher，1987；Adams，1999）	●糖源是一种长期的能量储备，通常用水解和酶解法在组织匀浆中测定 ●组织匀浆中的磷酸肌酸（PCr）和三磷酸腺苷（ATP）通常通过比色法测定 ●身体邻近成分（即脂质、蛋白质、水、有机灰分占身体的比例）和分析可以用来确定能量如何在各个区室之间分配 ●Ω 弹量热法提供了关于组织总能量密度的信息 ●也可以评估脂质成分（如胆固醇、脂肪酸、甘油三酯），尽管这些很少被用作应激指标 ●商业试剂盒可用来测量组织匀浆中的能量储存	●商业实验室的可用性使得测量营养指标相对便宜，而且不需要专门的设备就可以很容易地在血浆中测量 ●提供组织状态的长期指示 ●信息可以被纳入生物能量学模型（特别是与代谢率信息相结合时），以置于更广阔的背景下（Beyers 等，2002） ●可以使用非致命电子设备（如手持微波能量计，Crossin，Hinch，2005；生物电阻抗分析，Kushner，1992）	●由于一些能量储备指标在急性运动应激（如 PCr 或 ATP）下变化非常迅速；为了解释结果，了解这种动物的近期经历是很重要的 ●一些措施，如邻近身体成分分析，需要相对大量的组织，因此往往是致命的 ●不同性别、不同体型的鱼类，其能量具有内在的差异性，需要对这些因素进行控制 ●非致命性取样工具需要校准 ●一般对急性应激源没有反应（Schreck，2000）
白细胞			
●白细胞，或称白血细胞，是血液中在免疫防御和炎症中起重要作用的细胞集群 ●大多数脊椎动物的白细胞有 5 种：嗜碱性细胞、嗜酸性细胞、淋巴细胞、单核细胞和中性粒细胞（鸟类和爬行动物的嗜异细胞）	●白细胞谱通常是通过对染血涂片中的 100 个白细胞进行光镜查看得到的（Davis 等，2008）	●白细胞谱可以预测未来的行为表现和生存能力，如对感染的易感性（Al-Murrani 等，2002）、增长率（Moreno 等，2002）和生存率（Lobato 等，2005；Kilgas 等，2006）	●白细胞谱受疾病、感染以及应激的影响，因此很难解释白细胞谱的变化（Davis 等，2008）

续表 11.2

指标	抽样和分析注意事项	指标优点	指标缺点
●每种 WBC 类型的相对比例都受到应激源的影响（Dhabhar 等，1996），从而为动物健康和应激暴露提供了一个有用的测量方法 ●最常见的测量方法是中性粒细胞或嗜异细胞与淋巴细胞的比值（N：L 或 H：L 比值）（Davis 等，2008）		●血液涂片相对便宜，而且很容易从捕获的野生动物身上获得 ●白细胞对捕获和手工处理的反应相对较慢（即在数小时或数天内）（Davis 等，2008），因此，白细胞谱可以方便地用于测量野生动物的基线应激水平 ●白细胞的反应在不同的分类群中是保守的，因此，从一个分类群中获得的结果应该是广泛适用的（Davis 等，2008）	
血细胞比容			
●红细胞，是血液中的输氧细胞 ●血细胞比容是血液中红细胞的体积百分比	●血细胞比容最常用的测量方法，是收集细胞体积（PCV），它是通过在毛细管中离心整个血液样本得到的，毛细管将血液分层 ●收集的红细胞的体积除以血样的总体积就得到了 PCV	●血细胞比容通常随着暴露于应激源而增加，相对便宜，测量简单，不需要专门的检测	●血细胞比容在某些情况下会因刺激而增加或减少，这取决于具体的刺激，结果可能难以解释

表 11.3　机体整体的应激指标

指标	抽样和分析注意事项	指标优点	指标缺点
反射			
●简单的反射指标，如直立翻身的能力 ●描述鱼类对外界刺激或自主神经系统功能的神经反应的特征越来越流行（Davis，2010）	●反射可以单个地评估（存在或不存在），也可以作为一个复合而得出一个分数（Davis，2010） ●需要验证每个鱼种的反射，但与大多数鱼种相关的一些常见反射是右反射（即把鱼翻过来，3 秒后看它是否恢复直立方向）和抓尾反射（即抓住鱼尾，看鱼是否试图挣脱） ●不需要任何专业设备，并对鱼类活动提供立即（＜20 s）的测定	●实验室中和野外多种鱼类死亡率的预测（Davis，2010；Raby 等，2012） ●快速、简单、廉价地评估没有观察者偏差的反射 ●由于不需要任何设备或科学技能，能培训利益相关者（如垂钓者、渔业观察员） ●由于传统生理测量方法不能预测死亡率而发展起来的（Davis 等，2001） ●不依赖于鱼的大小、活动状态，或适应情况（Davis，2010）	●需要验证所有物种的反射，因为并非所有鱼类的反射都相同 ●相对较新的方法，所以发表的例子相对较少 ●反射预测死亡率的确切机制尚不清楚 ●不适当的反射选择或解释可能是含糊不清的
行为			
●运动、觅食、社会互动、捕食-猎物动力学、栖息地选择和其他方面的行为 ●行为是一个生态相关的指标，需要整合各种生理系统（Schreck 等，1997）	●可以在实验室和野外测量 ●使用生物遥测和生物记录仪（如肌电图［EMG］遥测）远程研究自由游动鱼类行为的工具越来越多（Cooke 等，2004） ●动作摄像机为研究鱼类在水中的行为提供了机会（Struthers 等，2015）	●考虑到生理和行为是内在联系的（Cooke 等，2014），整合其他测量方法的能力，包括那些与机体生理学更直接相关的测量方法（Scott，Sloman，2004） ●考虑到许多行为与食物获取和捕食者躲避直接的生态相关（Schreck 等，1997） ●许多行为终点成本不高（例如，不需要分析成本），但需要昂贵的或专门技术的设备	●很难识别行为的具体机制取决于能力、动机、感觉敏锐度和反应性（Schreck 等，1997）。 ●有必要控制实验假象和观察者的影响 ●如果终点不清楚，可能是主观的（而不是客观的）

续表 11.3

指标	抽样和分析注意事项	指标优点	指标缺点
游泳行为表现			
●例子包括速度、强度、游泳时间 ●游泳需要许多生物系统的整合，因此可以认为是应激和整个生物体状态的敏感集合体（Hammer，1995）	●最常用的定量方法是在实验室环境中使用泳道/水槽/环形呼吸计（Ellerby，Herskin，2013）或牵引带（Nelson 等，2002），尽管一些泳道是可移动的，可以在野外环境中使用（Farrell 等，2003） ●多种形式的游泳（如突发、临界游泳速度、耐力）可以测量（Beamish，1978） ●重复游泳的表现方法考虑到个体间的差异，对应激源损害行为表现的评估是有用的（Jain 等，1998）	●如果在呼吸计中游泳，则可以与代谢率（耗氧量）等其他指标相结合（Farrell 等，2003） ●游泳行为表现可认为具有很强的生态相关性（Plaut，2001）	●游泳行为表现在一定程度上反映了鱼的动力（如成熟、适宜的环境信号），可以不受应激状态影响 ●与购买设备相关的初始成本可能相对较高 ●游泳性能可能受到能量状态（喂食对禁食）的影响（Gingerich 等，2010）。
代谢率			
●食物转化为能量时的耗氧量 ●指示维持生命所需的最低代谢率 ●代谢率与健康、生长或生存之间可能存在邻近依赖关系（Burton 等，2011）	●在实验室环境中，在不受外界刺激的情况下，在静息、吸收后动物身上收集的典型样本（Nelson，Chabot，2011） ●可界定标准代谢率、最大代谢率或需氧范围（最大代谢率与标准代谢率之差）	●代谢率升高可解释为相对于对照组的应激状态（Barton，Schreck，1987） ●能够将数据合并到生物能量模型中，对一些不同的过程进行推断	●需要专门的设备和标准化的程序来产生数据（即，动物必须是后吸收性的，与外界刺激隔离） ●压力处理、个体差异、社会地位、适应时间、营养状况等因素都会影响结果（Sloman 等，2000；Nelson，Chabot，2011） ●数据受个体大小的影响很大，这使得跨过个体大小等级的比较具有挑战性 ●不同研究小组的技术差异很大，这使得种内比较具有挑战性（Nelson，Chabot，2011）

续表 11.3

指标	抽样和分析注意事项	指标优点	指标缺点
通气率			
●大多数鱼通过积极的通气将水从鳃上移过，这种通气包括打开和关闭鳃盖，可作为呼吸作用的表现（Barreto，Volpato，2004）	●可以通过直接观察或录像用秒表计数（White 等，2008）。 ●可以通过放置在鱼附近水中的生物电传感器进行远程测量（Altimiras，Larsen，2000） ●可以遥测鳃盖活动（无线电或声学）来估计通气率（Oswald，1978）	●区分不同应激源的能力（Barreto，Volpato，2006） ●相对简单的指标，可无创测量，费用低	●通气率似乎不能反映应激源在所有物种中的严重程度（Barreto，Volpato，2004） ●单是速率可能还不够，而且可能还需要定量测定振幅
心脏活动			
●心脏对于循环和生命维持是必不可少的，因此，心脏活动（如心率、每搏输出量、心输出量）是整个机体应激的相关指标（Farrell，1991）	●心率/心电图的测量相对简单，但考虑到一些鱼类是体积调节者，测量心输出量往往更好（Farrell，1991） ●多普勒和超声袖带可用于测量腹主动脉血流（Farrell，1991） ●在野外自由游动的鱼可以远程测量心率（Priede，1983；Armstrong，1998）或者在实验室中使用 ECG 生物记录仪（Raby 等，2015）或者将生物电传感器放置在鱼附近的水中（Altimiras，Larsen，2000）	●许多物种应激的可靠指标 ●用于记录不同畜养条件的反应（Rabben，Furevik，1993）、环境条件（Claireaux 等，1995；Lefrancois 等，1998）和人为干扰（Anderson 等，1998）	●心脏活动的基线值是可变的，这使得很难确定鱼在应激源后何时恢复 ●虽然有许多测量心脏活动的工具，但它们都具有相当大的技术挑战性

续表 11.3

指标	抽样和分析注意事项	指标优点	指标缺点
成长与生活史			
●生长和繁殖只有在其他过程的能量需求得到满足后才会发生 ●应激会降低生长速度 ●鱼类的一些坚硬结构（如鳞片、耳石、骨）沉积生长环，能准确测定年龄	●可以使用一系列的硬结构，其中一些可以以非致命性的方式收集（例如，鳞片） ●可以用不同的性腺指数作为生殖指数	●个体大小与雌性的繁殖力和某些物种雄性的繁殖能力呈正相关（Suski，Philipp，2004） ●生长速率的降低是应激的一个很好的指标（Pankhurst，Van der Kraak，1997）	●可能需要致死样本来获得数据 ●可能需要对群体进行长时间的抽样调查，以了解增长趋势，但此时停止或改变增长轨迹可能为时已晚 ●仅凭大小/年龄可能不足以了解群体水平的趋势，可能需要群体数据（即生育能力、存活率） ●需要跨年龄的级别来验证生长（Beamish，McFarlane，1983）
条件指数			
●应激条件指数包括长度-重量关系、器官-躯体指数和尸体解剖等（Barton 等，2002）	●方法的范围从相对非侵入性（如对活鱼的简单测量）到致命性（如对器官躯体指数） ●器官躯体指数是指器官重量与体重的比值（如肝体指数［肝脏：体重，HSI］，性腺指数［性腺：体重，GSI］，脏体细胞［全脏器：体重，VSI］，脾体指数［脾：体重，SSI］）（Barton 等，2002） ●低于或高于正常值的值表明，器官的能量分配受到应激的影响（Kebus 等，1992）	●简单而廉价（Bolger，Connolly，1989） ●良好的鱼群状况指标（例如，可检测慢性应激）（Barton 等，2002） ●一些非致死性的选择（例如，长度-重量分析、肥满度、相对重量）	●对短期应激不太敏感（Bolger，Connolly，1989） ●存在一些关于使用条件指数的批评，因为它们可能导致基于各种方法固有局限性的不适当结论（Cone，1989） ●更多涉及的测量（器官躯体指数和尸体解剖）是致死的 ●条件指数受到季节、发展阶段、性成熟和疾病状态的影响 ●通常需要较大的效应结果来检测应激

续表 11.3

指标	抽样和分析注意事项	指标优点	指标缺点
	●基于尸体解剖的方法（例如，健康评估指数）需要对被牺牲的鱼进行尸体解剖，从而将内脏的状况与概述的正常器官公开标准进行比较（Adams 等，1993）		
不对称性波动			
●生物一侧结构的差异发育（Jagoe，Haines，1985） ●对称结构应该来自相同的遗传物质，所以偏离对称可以采用基因突变的应激形式或环境应激的形式表示（Leary，Allendorf，1989）	●量化不对称波动的存在可以表示种群内部的应激 ●可作为种群层次丰度下降或不利环境条件下降之前的早期预警指标（Jagoe，Haines，1985）	●当测量或计算数量形态度量的特征时，可以既便宜又直接	●许多数量特征（如鳍条）是可变的，这使得它们作为应激指标是不可靠的（Leary，Allendorf，1989）。 ●并不是所有的物种或特征都适于将波动不对称性作为应激指数进行研究 ●可能需要大量样本来识别趋势/模式（Jagoe，Haines，1985） ●界定不对称和应激之间的因果机制或联系可能具有挑战性（Jagoe，Haines，1985）
繁殖和健康			
●生殖和生殖产出是一个对个体和物种存活和持续生存至关重要的过程 ●应激会导致繁殖减少	●可以评估雌雄间性的程度作为环境应激的一个指标（Bortone，Davis，1994） ●应激的影响可以在一系列生殖相关因素中观察到，如配子质量和/或生殖定时（Schreck 等，2001）	●当把应激和种群水平参数联系起来时，在生态学上是非常相关的 ●测量一组性腺指数（质量、大小、卵期、激素等）能作为生殖的代表指标 ●能用已知的双亲进行人工杂交以确定后代的存活和生存力（Campbell 等，1994）	●可能只需要在每年的特定时间收集繁殖数据 ●应激源和生殖降低之间的因果关系很难确定

续表 11.3

指标	抽样和分析注意事项	指标优点	指标缺点
生存			
●对应激源最极端的反应是死亡，体内平衡无法维持（Wood 等，1983）	●死亡率可以通过将鱼放在网子、笼子、围栏或水箱中并简单地计算死亡数量来测量（Gutowsky 等，2015） ●通常有助于标记/标签单个的鱼，以确定哪些个体死亡和它们的历史 ●越来越多的工具（如生物遥测）可用来研究野生自由游动鱼类的死亡率	●通常是简单、便宜的方法 ●死亡率是绝对的，是与生态有关的健康指标 ●与揭示死亡率机制基础的其他指标相结合时，效果最为显著（Cooke 等，2006）	●可能很难获得动物护理和利用以死亡为终点的认可 ●受到评估死亡率方法（如净效应或网箱效应）的偏差影响（Gutowsky 等，2015） ●鱼类的死亡有很多原因（如衰老），因此必须有适当的对照组

11.3.1　细胞和分子指标

和更高层次的生物组织反应一致，细胞和分子的应激反应有助于暂时耐受应激源（Kultz，2005；见表 11.1）。皮质醇参与了应激源所引起的分子反应，通过与影响转录调控的热休克因子（HSFs）相互作用，刺激金属硫蛋白、泛素和 HSPs 的表达（Vamvakopoulos，Chrousos，1994；Kassahn 等，2009）。皮质醇还与糖皮质激素受体结合，与 AP-1（活化蛋白-1）转录因子的 c-Jun 组分相互作用（Iwama 等，2006）。皮质醇与糖皮质激素受体调控转录效应取决于组织类型和 HSP90 表达水平（Basu 等，2001；Vijayan 等，2003）。一系列应激源引发了一系列相同的反应，包括 DNA 和蛋白质的损失修复、细胞周期阻滞或凋亡、细胞和分子碎片清除以及反映从合成代谢到分解代谢状态转变的细胞代谢变化（Iwama 等，2004）。总之，这些细胞和分子反应是由真核生物最小应激蛋白单位（进化保守的蛋白质）所引发的（Kultz，2005）（Faught 等，2016；见本书第 4 章）。

应激源暴露导致活性氧（ROS）的产生，从而导致细胞内氧化应激。氧化应激导致蛋白质损伤水平升高。泛素标记蛋白的数量可以反映蛋白质损伤的程度（Iwama 等，1998；胞内酶亦反映细胞损伤和死亡，见表 11.1）。细胞内的受损和泛素化蛋白会引起热休克反应去修复蛋白损伤（Wu，1995）。热休克反应的程度取决于应激源的强度及持续时间和适应状态（Iwama 等，2004；Somero，Hofmann，1996）。HSP 的表达受到热休克因子 1（HSF1）的调控，该因子在下丘脑-垂体-肾上腺（HPI）轴活化后将 HSPs 与 HSF1 分离，然后 HSF1 迁移到细胞核，开始 HSPs 的转录（Kassahn

等，2009）。ROS还能加速氧化应激导致端粒长度的减少，端粒是染色体的帽状结构，对基因组的稳定性至关重要。当端粒加速减少时，端粒又会加速细胞衰老（Richter，von Zglinicki，2007）。端粒长度和细胞衰老、组织衰老有关的论点已成为一个新兴的研究领域（Ricklefs，2008；Monaghan等，2009）。一系列的环境应激源，包括心理压力（Epel等，2004）或优质生育力（Kotrschal等，2007）都能影响脊椎动物端粒长度。有证据表明，环境应激源可以导致氧化应激，如果应激持续，可能触发细胞衰老，最终导致生物个体衰老（Monaghan等，2009）。

基因表达的变化（即定量、定性和反应系数的变化）可能和一系列应激源有关（Krasnov等，2005）。基因组学工具，如微阵列和基因表达谱如今广泛用于了解鱼类对一系列应激源的反应。常见的应激源包括温度（Jeffries等，2012）、低氧（Gracey等，2001）、手工操作应激（Donaldson等，2014）和有毒物质（Williams等，2003）。cDNA微阵列可以同时筛选数千个基因，以鉴别和参与一系列反应的生化通道相关的不同表达基因群。相对于评估初级、次级和三级应激指标的文献，很少有关于评估与急性应激源暴露相关的基因表达（Caipang等，2008；Prunet等，2008）的研究。相反，功能基因组学研究的主要目标之一是扩展我们对于环境条件如何影响基因表达的了解（Buckley，2007；Miller等，2009）。使用普通活组织（如肌肉和鳃组织）切片的微阵列已经鉴定出与红大麻哈鱼迁徙中不健康特征相关的潜在基因，不能到达产卵地的个体特征的不健康指标，包括凝血因子、与有氧呼吸有关的基因和与免疫功能相关的基因的表达下调（Miller等，2011）。尽管如此，对这些基因及基因家族功能和意义的了解依然是具有挑战性的，而基因如何响应应激源的刺激还不清楚。

在鱼类中，许多基因被视作是多种应激源的潜在生物指标。红大麻哈鱼与细胞凋亡相关的基因，如细胞色素C和转录因子JUNB，在温度的升高时表达量上调（Jefferies等，2012）。虹鳟（Momoda等，2007）暴露在低水位和空气的应激源中时，JUNB亦是上调的。在虹鳟面对应激源时，参与调节细胞生长和凋亡的转录因子NUPR1（Mallo等，1997）能持续上调数小时（Momoda等，2007）。和手工操作应激相关的研究中观察到虹鳟肝脏中与炎症、蛋白降解和免疫反应有关的生物学通道基因的表达发生变化（Momoda等，2007；Wiseman等，2007）。Krasnov等（2005）亦在虹鳟中观察到在经过反复渔网捕捞的应激之后，脑内和免疫反应、细胞增殖与生长、细胞凋亡与蛋白合成相关的基因表达发生变化，以及肾脏中和细胞生化过程相关的基因也发生变化。肝脏中糖异生、糖源分解和能量代谢相关基因的表达可以证明启动应激后的恢复（Momoda等，2007；Wiseman等，2007）。Donaldson等（2014）发现红大麻哈鱼物种特异性和性别特异性的基因组反应和运动应激后的应激应答和恢复有关。了解这些基因在应激反应期间行为的一个复杂因素是，迄今对鱼类进行的研究通常都注重于不同的时间进程（Krasnov等，2005）、物种、组织（Kassahn等，2009）、技术（Prunet等，2008）和感兴趣的基因。

11.3.2　初级和次级生理指标

应激的生理指标包括细胞、分子水平和整个动物机体水平之间的所有反应（表 11.2）。应激反应涉及广泛的生理反应，包括 HPI 轴的激活（Schreck，Tort，2016；见本书第 1 章，图 1.6）。初级指标（如儿茶酚胺和应激激素）和次级指标（如葡萄糖、离子平衡、酸碱平衡、免疫功能或其他能力代谢指标的变化）都可用于评估鱼类应激。次级反应变化发生时间比初级反应要慢。不同的初级指标和次级指标在使用上各有优缺点，这取决于感兴趣的应激源以及物种、种群和个体的背景信息（表 11.2）。

在评估对特定的水产养殖或者手工操作的反应，或者场地的严重干扰时，特定的初级和次级应激指标的变化特别有用。儿茶酚胺提供的是对应激源反应最快的初级反应，但由于其对应激源反应迅速，因此难以检测（Reid，1998；Pottinger，2008）。儿茶酚胺在实验室条件下可作为适当和强有力的指标，但往往在实际场地检测时不适用，因为它对捕捞和手工操作有显著性反应。皮质醇是另一种衡量初级应激反应的指标，是最常用的应激指标之一（Mommsen 等，1999；Barton，2002；Pottinger，2008）。由于皮质醇对特异性应激源的反应比儿茶酚胺要慢，所以它可以用于实验室或野外环境中量化本底水平和应激后水平，只要动物能在捕获后几分钟内采集样品（Romero，Reed，2005）。通常检测的次级应激指标包括：①葡萄糖增多，这是由于应激后分解代谢和机体向血液中释放葡萄糖的结果（Barton，2002）；②乳酸增多，是缺氧和运动应激引起的缺氧代谢指标（Wood 等，1983）；③摩尔渗透压浓度或特异性离子，由于儿茶酚胺的增加以及随后心率升高和鳃通透性的增加而改变（McDonald，Milligan，1997）；④白细胞，它不仅能反映对急性应激源的反应，还能预测未来的生存或行为表现（Davis 等，2008）。

许多生理指标都是在血浆中检测的，它们能够反映激素和代谢循环水平（Barton，2002；Pottinger，2008）。对大多数鱼类来说，采集血浆（或红细胞以检测血容量等指标）的好处在于，收集样品的手段不是致命的。此外，由于血浆是检测生理参数的常规样品，目前已进行许多有助于解读数值的研究，而且市面上也有用于促进数据形成的测定技术或仪器（Wells，Pankhurst，1999；Beecham，2006）。血浆检测指标的主要缺陷在于血浆并不总是最相关的组织，对于一些指标来说血浆检测没有意义。例如，慢性应激暴露后检测的生物能量指标，如糖源肝储存不能只通过血浆样品来获得数据。同样，磷酸肌酸（PCr）和三磷酸腺苷（ATP）是急性运动应激的指标，这些指标最好在肌肉组织中检测。因此，特定组织的取样取决于研究的目的以及感兴趣的初级/次级生理指标。综合这些注意事项表明，初级/次级应激指标提供了关于个体如何感知和应对环境挑战的信息，能够分析动物在这些挑战中所承受的应激程度，并且可能预测未来的行为表现和生存性（Breuner 等，2008）。

11.3.3 机体整体的指标

机体整体（或三级）应对应激源反应包括许多鱼类表征行为的变化，如生长、个体状态、抗病能力、新陈代谢（Sadoul，Vijayan，2016；见本书第5章）、心脏活动、游泳动作、行为、健康，甚至生存状态（Wedemeyer 等，1990；表11.3）。这些反应能作为应激的指标，通常也认为具有生态相关性。例如，生长率可以直接影响种群模型（Power，2002）和生物能量模型（Beyers 等，2002），而健康和生存能影响种群统计过程（McCallum，2000）。与食物获取、躲避捕食者和选择栖息地相关的行为也和生理基础有直接的生态相关性（Godin，1997）。许多和条件相关的检测包括单独体重检测（通常是整个有机体和特定器官）、体长或两者的测量（Goede，Barton，1990）。鱼类状态（如相对重量、肥满度）的传统检测通常是简单的计数，但由于缺乏特异性和广度而受到批评（Bolger，Connolly，1989）。所谓的器官整体指数（如器官与身体的比例）由于简单易行，被广泛地用作应激和状态指标，但它们需要致死的取样。一些研究者已经提出各种和身体状况与健康相关的指标组合，最常用的可能是由 Adams 等（1993）首次提出的健康评估指标（HAI）。可以对 HAI 进行修改，以满足研究人员和他们研究问题的需求。

游泳行为表现是另一个三级的应激反应指标，功能简单而易于记录和解读，而且在生态上也有意义，但合适的检测需要设计游泳管道或者水槽，而建筑这些水槽或管道在技术上具有挑战性，或者价格昂贵。临界游泳速度是一个几十年来一直被测量的指标（Beamish，1978），尽管它现在被认为在应用上具有一定局限（Plaut，2001），但它在研究应激处理时对鱼游泳能力的相对比较确实有应用价值。在 Portz 的报道中（2007）表示，如果一项研究的目标仅仅是对游泳能力进行恢复水平的比较，简单地测定在圆形鱼缸中被夹着尾巴追赶的鱼消耗体力的时间，就会产生和比较正规测定的临界游泳速度相关的值数。Jain 等（1997）发现鱼类个体的游泳能力有内在差异，从而改进游泳方案，创造了一种新的方法，比较鱼在第一次游泳后不久进行第二次游泳的能力。这种方法被称为“恢复率”，它是在面对不同应激源时机体整体状态的一个灵敏指标（Jain 等，1998）。我们综合游泳后心脏活动能力（如心率、心输出血量）和代谢率（耗氧量）的检测（Webber 等，1998），可以多层次了解应激对生物有机体表现的生态相关指标的影响。事实上，在实验室衍生的代谢、心功能、游泳能力和水温之间的关系可以结合实验场地环境来推导死亡率（Farrell 等，2008）。总之，使用一系列可能的技术检测鱼类游泳能力，可以提供关于应激如何影响鱼类生态重要方面有价值的信息。

三级应激指标比初级或次级应激反应的数据更加多样化，这些数据几乎完全属于生理学家的工作范畴。收集与三级应激反应有关的数据，以及解读这些数据，可能需要行为学家、生态学家和生理学家的专业知识。尽管如此，对于三级应激指标确实需要更直接和扎实的生理学基础。与心脏活动和代谢相关的指标是生理学家的研究领域，但这些也与游泳（游泳行为表现及一般的运动）和其他行为成分有关。代谢和运动主要由能量驱使（Boisclair，Leggett，1989）。能量的预算限定了身体的生长速度、生殖和行为。从这个意义上说，许多三级指标与生物能量学有关（Beyers 等，2002）。人们对于利用反射损伤作为鱼体活力指标的兴趣越来越大，这亦和机体代谢与神经功能有关（Davis，2010）。

当我们从关注稳态的直接指标（即初级或次级应激指标）转移到更多的机体整体水平指标时，需要整合不同的作用机制（如：神经内分泌过程），推断的应激程度更富有挑战性，而且需要研究比较不同组合的差异（如，对照组对比应激源 A 组和应激源 B 组）。例如，一些直接的死亡是自然现象，死亡并不能说明是由于鱼受到了应激；衰老是一种自然过程，鱼的衰老可能会提前发生，但这与它是否处于应激状态无关。同样，鱼的进食行为减少并不意味着鱼正在经历应激过程，可能只是因为它不饿，也可能受到先前的进食情况、代谢需求、食物合适性、季节性甚至是遗传因素的影响。这不像初级和次级指标（如皮质醇或渗透压）有明确的参考范围或处在一个特定的阈值（如 X ng/mL 或 X mmol/L）那么有意义。另外，单个的鱼只是个体的血液循环皮质醇相对于参考范围可能会升高，但还没有机体水平的显著变化，这是否意味着这只鱼有应激反应？在从三级反应指标去推测应激时，必须要使用恰当的对照组进行比较，结合背景知识也很重要。

尽管存在挑战，但使用三级指标去界定应激的优势在于收集和解读都很简易，特别是与其他只能通过实验室工作来量化的应激生理指标有关。例如，在鉴定生物机体的整体应激时，反射指标（如，存在或不存在特异性反射）和生存指标（如，生物个体是生存还是死亡）这两个指标的优势在于简易、准确和相关性强。值得注意的是，简单的反射损伤已经被证明与生存相关（Davis，2010；Raby 等，2012），现在为研究团队广泛接受，用来确定该领域动物的生理状态并预测死亡率。然而，有趣的是，反射损伤的机制基础还不清楚。相似的是，很多应激的三级指标能在传统实验室之外采用，对野外自由游动的鱼进行检测。例如，一些检测项目，如精细的运动活动（Cooke 等，2004）、心脏活动（Cooke 等，2004；Clark 等，2010）和血糖（Endo 等，2009）可以遥控检测，或者传输（即生物遥感/生物传感器），或者储存在电子标签中以供日后分析（即生物标签）。反射损伤的检测也可以在野外现场进行，因为它们不需要任何专门的仪器设备（Raby 等，2012）。

11.4　检测和解读应激的注意事项

关于应激源、研究动物和应激指标，在量化和解读鱼类的应激时需要考虑很多因素（图 11.3）。下面将着重指出这些因素的选择情况。

11.4.1　种间差异

不同种类的鱼对同样的应激源具有不同的反应也许并不奇怪。由于进化、生态（捕食）、环境（温度）和生活史的差异导致物种的趋异，HPI 活性和其他应激指标容易发生变化。不同物种间的应激反应在很大程度上受到不同的应激源类型、严重程度和持续时间变化的限制，这些变化会显著影响 HPI 应激反应的强度（见第 11.4.4 节）。当暴露在相同的应激源中时，不同物种的血浆皮质醇（Barton，2000，2002；Pottinger，2010）和葡萄糖（Jentoft 等，2005）水平、基因表达（Jeffries 等，2014）、免疫功能（Cnaani 等，2004）、栖息地偏爱（Jacobsen 等，2014）和回避行为等（Hansen 等，1999）都会改变，包括密切相关的物种间（如细鳞大麻哈鱼和红大麻哈鱼）（Donaldson 等，2014）也会发生类似情况。然而，在某些情况下，不同物种会对相同的应激源表现出相似的反应。Campbell 等（1994）指出，虹鳟和褐鲑反复暴露在空气中会使子代的存活率降低到类似的程度。Ryer 等（2004）表明，身体健康的褐盖鱼（*Anoplopoma fimbria*）幼鱼在被捕捞后表现出和体质弱小的狭鳕（*Theragra chalcogramma*）相似的行为障碍。设计多个物种（或杂交种）的实验（Noga 等，1994），暴露在相同的应激源中并检测相同的指标，可以最大限度地收集数据，帮助了解驱动产生种类特异性应激差异的因素。

11.4.2 种内差异

在相同物种内检测应激也会存在差异。通常情况下，在单一种群中启动应激的检测。环境与生态因素和地理上不同种群的联系可能会导致应激反应的差异。的确，在纵向的梯度中，北部和南部的底鳉（*Fundulus heteroclitus*）种群应对应激源所引发的血浆皮质醇反应存在差异（DeKoning 等，2004）。通风率所引起的应激反应，在埃氏短棒鳉（*Brachyrhaphis episcopi*）（约束应激物）（Brown 等，2005）和三刺鱼（*Gasterosteus aculeatus*）（捕食应激源）（Bell 等，2010）的种内差异取决于鱼来源于低捕食生境还是高捕食生境。在其他情况下，应激生理参数（如血浆皮质醇、乳酸、葡萄糖）在种群中不存在差异，但健康状态（存活率）仍可能在暴露于应激源后发生改变（Donaldson 等，2012）。从进化的角度来看，应激反应中的种群特异性差异是耐人寻味的，在对应激源的灵敏度做出种族范围的结论时不能忽视。

性别、体型、社会地位和驯化也是促使应激指标发生变化的因素。性别差异在某些物种中尤其明显。成年雌性大麻哈鱼（*Oncorhynchus spp.*）在应对温度应激源时具有更高的血浆皮质醇水平（Donaldson 等，2014）以及更高的死亡率（Martins 等，2012）。舌齿鲈（*Dicentrarchus labrax*）在应激中所引发血浆皮质醇表达量达到峰值的时间受到体型的影响（Fatira 等，2014）。社会等级地位不同的虹鳟在面对应激源之后的 HPI 活性也存在差异（Jeffrey 等，2014a）。孵化场和野外的鳟鱼（Woodward，Strange，1987；Lepage 等，2000）和大麻哈鱼（Johnsson 等，2001）的检测应激影响亦存在差异，应综合考虑所有的种内差异，以及如何最大化的提高人工饲养或野外放生的鱼群行为表现，以增加养殖产量。

最后，在应激反应方面，对遗传和个体/群体间差异（Kittilsen 等，2009；Ponttinger，2010；Hori 等，2012），以及一个个体内部生理和行为应激指标偶联（即应对方式）（Castanheira 等，2015）的认识也在迅速发展。水产养殖利用这种差异已经超过 10 年，培育出在遗传上不同的虹鳟，它们在面对监禁应激源后 3 小时产生的血浆皮质醇存在浓度高低的差异（Pottinger，Carrick，1999）。这种可变性与多数应激行为指标的可变性相一致（Overli 等，2007）。大西洋鲑（*Salmo salar*）（Fevolden 等，1991）、金头鲷（*Sparus aurata*）（Tort 等，2001）、大西洋鳕鱼（*Gadus morhua*）（Hori 等，2012）和条纹鲈（*Morone saxatilis*）（Wang 等，2004）中也存在面对应激出现高低反应的特征。这些个体水平的差异也可能影响到个体对用于应激量化的抽样方法的耐受性（如，从导管手术中恢复过来）（Bry，Zohar，1980）。总之，种内差异有多个方面可以影响应激指标，解读和描述这些差异是需要有根据的。

11.4.3　背景-特异性差异

除了辅助分子，HPI 轴活性、生理的和整体的反应量度都可以根据个体暴露在应激源的环境和生态背景而改变。HPI 应该和背景保持一致。最终，任何对动物健康的威胁必须通过 HPI 介导的生理的和行为的变化来耐受和克服（Schreck，Tort，2016；见本书第 1 章）。个体面对捕食者、资源竞争、限制摄食、低氧、水体污染或水温升高时所产生的生理和行为过程变化可能会有所不同。

在所有的组织结构水平上都观察到应激检测有背景相关的变化。在神经内分泌水平（主要指标）上，三刺鱼脑内单胺（如去甲肾上腺素、羟色胺）浓度随应激源类型的不同（陌生的同种类型与捕食者）（Bell 等，2007）而有所变化。在 HPI 轴的后期，捕食者诱导冬季美洲拟鲽（*Pseudopleuronectes americanus*）的整体皮质醇反应会根据捕食者的种类而产生差异（Breves，Specker，2005）。更细微的应激源类型差异可能不会引起不同的血浆反应（种内对比种间的入侵者）（Ros，2014）。值得注意的是，在生态相关性和潜在激烈性方面，两种截然不同的应激源可以引发相类似的反应。植入弹性体标签能在三刺鱼身上产生类似于模拟捕食者攻击后的血浆皮质醇反应（Furtbauer 等，2015）。其次，生理过程也会受到应激环境的影响。在大斑南乳鱼（*Galaxias maculatus*）中，代谢变化程度取决于个体是否暴露在视觉或嗅觉捕食信号下（Milano 等，2010）。在机体的整体水平上，通风率的变化因应激源不同而有差异（例如，监禁 vs 同种 vs 电击）（Barreto，Volpato，2006）。很可能在决定选择哪个指标、预测指标的反应以及解读结果时，应激源的环境和生态背景是和应激源的激烈程度交织在一起的（见第 11.4.4 节）。

在评估应激时，动物暴露在多个应激源中是可论证的大多数生物学相关的背景。实验设计结合多种不同应激源（如水温升高、渔业捕捞、水体污染或免疫刺激）可以检测更多的应激指标（Marcogliese 等，2005；Jacobsen，2014）。鱼类不止一次暴露在同样的应激源中，而之前对一个特别应激源的经历和个体的学习能力可能会影响应激指标（Barcellos 等，2010；见第 11.4.4 节中关于驯化的论述）。暴露在应激源之前的生存环境（即隔离对群居）可能通过同种应激的嗅觉信号交流而影响应激指标（Giacomini 等，2015；Bracellos 等，2014a）。群体中单独测试的鱼可以受到该物种的教养和社会倾向的驱使。加入额外的实验变量用来增加生物相关性可能会带来其他逻辑上的挑战，例如需要更大的样本量。如果优先考虑的是应激指标的生物学相关性，那么应该考虑到这种生态和环境背景的细微差别。

在某些情况下，识别合适的指标是很直观的（如，检测模拟捕食者攻击后的通风频率和躲避行为）。其他场所允许使用许多合适的指标。暴露在毒理应激源中，会影响基因的表达（Jeffries 等，2015）、配子质量（Khan，Weis，1987）、跨越整个生命阶段的生理过程和行为（Scott，Sloman，2004；Sloman，McNeil，2012）。使用多种指标的方法（Woodley，Peterson，2003）为发现或排除特定应激源的指标提供机会。然而，有必要提醒，应激的初级指标和次级指标可能和动物的整体或三级指标无关（例如，热源刺激时，HSP 的表达升高而游泳性能依旧保持不变）。即使在一组指标中，反应也可能不一样（例如，乳酸浓度的增加表明对肌肉的供氧不足，但 HSP 表达没有变化表明没有蛋白质的损伤）。因此，在确定检测应激的合适指标时，必须考虑应激源或应激源的性质以及各指标之间的关系。

11.4.4 应激物的严重程度

如 Schreck 和 Tort（2016；见本书第 1 章）所论述的，存在于特定背景下的差异是应激源是否处于急性/单一、重复或慢性/长期的状态。一个急性或单次暴露可能以短暂的持续时间（秒到分钟）为特征，并与适应性的生理反应相关。急性应激源持续时间的变化可以逐步改变应激指标水平（Gesto 等，2013，2015）。急性应激源的强度也可能随着预期损伤水平的不同而改变（例如，2 分钟监禁对比 5 分钟的渔网追捕），并再次导致应激指标水平的变化（Geslin，Auperin，2004）。对于污染应激源，在剂量浓度与指标反应程度或死亡率（即 LC_{50}）之间通常都有明确的型式。然而，正如第 11.4.3 节所提到的，应激源的不同严重程度可以引起相似的反应，以及同一个应激源能表现不同的持续时间（Fatira 等，2014）。

关于暴露在慢性应激源中的组成的定义并不总是一致的；慢性应激可以重复、持续地暴露于一个急性应激源中（例如，持续数周的日常手工操作），或连续长时间的暴露于应激源（例如，数周内持续的水温升高）。对这些暴露反应的一个指标可能取决于它在即时（如躲避行为）和长期（如生长）对健康影响中的作用（Schreck，2000）。此外，对于一些物种来说，暴露于慢性应激源是现实的（Boonstra，2013b）。虽然指标可能暗示是慢性应激状态（如异型稳态负荷）（McEwen，Wingfield，2003），但反应可能是适应性的（Boonstra，2013b）。重复暴露在应激源中，应激源之间的时间间隔也会影响应激指标的反应（Schreck，2000）。因此，为了验证应激指标的变化确实是由于“被打乱的负反馈作用”（Romero 等，2009）或异型稳态负荷（Schreck，Tort，2016；见本书第 1 章，图 1.3、图 1.5 和图 1.6），就必须考虑到适应性。

Schreck（2000）、Cyr 和 Romero（2009）指出使用应激的激素指标从脱敏和疲劳中识别习惯化（适应性）。用表现/行为指标来定义习惯化可能不那么简单，需要检测伴随的生理参数（即 HPI 轴活性）。通过反复暴露于急性应激源中（如在数周内每天手工操作应激后立即检测），在中间阶段对一部分动物进行采样，以确认 HPI 轴和应激指标仍然对应激源有反应。在持续的应激源暴露过程中（如连续暴露在水温升高的环境中经过数周后的一次急性追逐的应激），确认 HPI 轴和目标应激指标仍然有效。在一些长期处于应激状态的物种中，可以检测到内分泌应激反应的减弱（Barton 等，1987；Jentoft 等，2005；Barcellos，2006）。Wingfield（2011）等提出阈值的存在，超过阈值后，进行的应激反应不再具有适应性。当超过这些阈值时，对应激源的抵抗电位通过应激反应的衰减而增加（Wingfield 等，2011）。在慢性暴露过程中多次取样可以确定临界点，在此临界点上，应激指标不再反应（复原），或回复到基线、应激前的水平（疲劳）。

表型可塑性也应加以考虑，特别是对居住环境处于波动情况下的野生种群进行应激量化时，这些波动情况也属于慢性应激源（如气候变化、水体污染）（Silvestre 等，2012；Crozier，Hutchings，2014）。基于实验室的应激指标对野生动物可能不是可靠的指标（Dickens，Romero，2013），这可能是由于应对策略的改变。或者，可以在野生种群中检测到基于实验室的慢性应激指标，但不会有种群水平的后果。来自入侵安大略湖高污染地区的圆虾虎鱼种群表现出毒理学的应激指标［如，内分泌损伤（Marentette 等，2010）；受损伤的行为（Sopinka 等，2010）］。然而，污染地区的圆虾虎鱼种群数量稳定，而在相关区域的种群数量下降（McCallum 等，2014）。这个例子说明了新的机体稳态是如何出现的，可以解读为应激，但不会影响到种群水平（见本章第 11.5 节）。

11.4.5　野外与实验室

实验室和野外的应激检测各有优缺点。由于获取样本所用设备的复杂性，一些应激指标必须在实验室中测量（如通过导管连续抽血；见图 11.1）。其他指标只能在实地检测，因为无法在实验室中复制这种行为（如应激诱导的迁移率变化）（Donaldson 等，2011）。实验室研究可以控制和应激指标检测相混淆的变量，而在野外则很难做到（如水温）。这种控制的程度对于研究应激源和检测指标之间的机制联系是至关重要和必要的。尽管如此，在实验室环境中模拟鱼类在野外遇到的应激源并不能真正涵盖动物在生态相关条件下所引起的全部的应激反应。例如，亚致死的应激指标的潜在影响（如在渔业捕捞破坏平衡的应激源之后的释放后捕食）（Danylchuk 等，2007）在实验室中没有得到恰当的考虑。根据研究的假设和目的，选择应激指标的测定在实验室条件下可能比在野外条件下更合适，反之亦然。

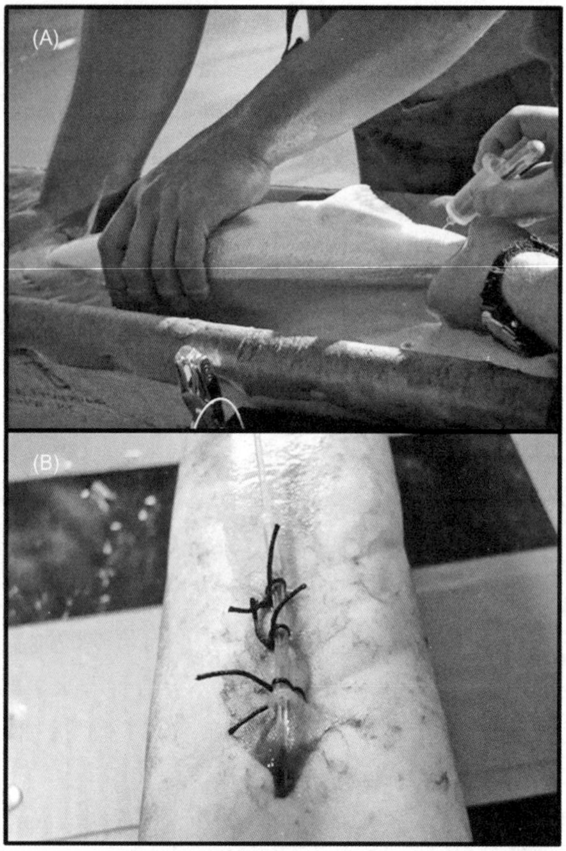

图 11.1　从活鱼体内抽取血样测定应激指标的方法

注：（A）使用注射器或 Vacutainer 从尾部血管抽取血液；或（B）在血管中（通常是主动脉）植入一导管，它还能够进行系列的取样。图片（A）由 Michael Donaldson 提供，图片（B）由 Michael Lawrence 提供。

远程传感设备的发展（Cooke 等，2004），以及对鱼类监护点设备的验证（Stoot 等，2014），都为曾经局限于实验室的指标检测对实地测量的支持。特别是，随着生物标记技术和生物遥感技术的进步，活性水平和能量（Burnett 等，2014）、觅食（Brownscombe 等，2014）和自由游动的鱼的心率（Clark 等，2008）的变化都可以在应激前、中、后检测到（Donaldson 等，2010；Raby 等，2015）。这种面向野外的机械式应激检测方法的局限性在于成本、标签植入的手术要求、检测效率和标签回收以及它们和研究的样本量密切相关的。测量野生鱼类自然暴露在不可预测的、不稳定的应激源（如，旋风，洪水）中的指标，可以令人难以置信地展现，但也是高度的机会主义（Wingfield，2013）。以野外研究为基础的整体协作方式，加上以作用机理为重点的补充实验室研究，可以提供最完整的应激反应评估。

11.4.6　时间方面

在研究动物应激时，可以考虑多个时间层次。首先，将在生命的哪个阶段测量应激指标？已经确定一些物种在发育早期会处于低反应期，例如虹鳟（Barry 等，1995）、大鳞大麻哈鱼（*Oncorhynchus tshawytscha*）（Feist 和 Schreck，2001）、黄鲈（*Perca flavescebs*）（Jentoft 等，2002）、湖鲟（*Acipenser fulvescens*）（Zubair 等，2012）在孵化前未能检测到应激诱导的内源性皮质醇生成。应激诱导的和静息的血浆皮质醇水平可以随着鲑鳟鱼类（Pottinger，Carrick，2000；Cook 等，2011）、鲶鱼（Barcellos 等，2014b）和抚育期的大口黑鲈（Jeffrey 等，2014b）的性成熟过程而发生变化。在二龄的大麻哈鱼中也存在反应过度的时期（Carey，McCormick，1998）。Koakoski 等（2012）的纵向研究发现，应激诱导的血浆皮质醇的浓度和峰值时间在小鱼、幼鱼和成年的克林雷氏鲇（*Rhamdia quelen*）之间存在差异。到目前为止，年龄效应主要集中在皮质醇作为应激指标（Schreck，Tort，2016；见本书第 1 章，图 1.4）。应激指标基底水平的生命阶段变化也可能混淆应激的量化。这一提醒也和其他生理的（如抗氧化防御的个体发生）（Otto，Moon，1996）、行为表现（如幼体珊瑚礁物种的快速生长）和行为（如逃避捕食者的个体发生）（Brown，1984）等方面的应激指标有关，它们在整个生命阶段中可能会有所不同。

其次，一天中选择什么时间段进行应激指标检测？静息心率（Aissaoui 等，2000）、血浆皮质醇（Cousineau 等，2014）和各种行为（如活动性）（Bayarri 等，2004）在昼夜周期中都会有波动。采集时间的不一致会造成数据的偏差，而不同采样时间的研究之间的比较会影响结论的准确性。然而，值得注意的是，有些物种可能在昼夜节律相关的特征上表现出可塑性（Reebs，2002）。

第三，什么时候测量应激后的应激指标？如前所述，时间-过程采样是最全面的方法，以确保获得上升、峰值和恢复等的指标数据。然而，对于不同的指标，诱导和恢复的时间是不同的（Gesto 等，2015）。儿茶酚胺（与其他交感神经系统的作用过程，如心率和通氧）会在瞬间升高，而皮质醇则需要更长的时间（几分钟到几小时）才会高于应激前的水平。应激源的暴露与 mRNA 丰度（即转录）变化之间的时间差是可以预期的。诱导和恢复时间也可能取决于应激源类型和严重程度而变化（见第 11.4.4 节）。确定和描述时间对应激反应影响本身就是应激生物学研究的热点之一；然而，处理水平（如，应激源类型或严重程度）的影响也是有意义的，时间的影响也应该考虑到。

11.5　从单个指标到生态系统健康

到目前为止，本章综述了个体水平的应激指标，这项工作的一个有意义的延伸在于个体应激如何扩展到种群和生态系统应激的进程。需要注意的是，本章中描述的分子、生理和个体的应激指标，相对于种群和整个生态系统对环境应激源的反应（月至年）而言，是在较短的时间（分钟到天）范围内具有反应性（Adams，Greeley，2000）。将个别的应激指标和下游的种群效应联系起来时，应激的持续效应和世代之间的组成部分显得尤为完整。严酷的越冬条件、资源利用率低的时期以及其他的环境应激源，都会对鱼类的表型产生持续的影响，即使应激指标指明正在恢复也不例外（O'Connor 等，2014；O'Connor，Cooke，2015）。应激对种群的潜在影响可能是由母系的匹配/错配来决定的（Sheriff，Love，2013；Love 等，2013），可能几个世代都不会表现出来。例如，Venturelli 等（2010）使用30年的数据库发现母系介导的对玻璃梭鲈（*Sander vitreus*）卵大小的影响，具有调节种群动态的能力。年龄较大、体型较大的雌性会产下较大的卵，而这些卵往往会生出存活率更高的后代。事实上，研究人员发现，在年龄较大的雌性数量较丰富的年份，种群的繁殖率较高（Venturelli 等，2010）。如果一种以老年雌性为目标的渔业应激源损害了这一群体的繁殖，种群的稳定性可以通过母体介导的机制（如卵的大小、数量或能量的含量）而发生波动。通过建立模型可以将单个应激指标和较大规模的过程联系起来（Calow，Forbes，1998；Fefferman，Romero，2013）或者将种群水平和生态系统水平的指标联系起来。

有许多常见的种群和生态系统水平的应激指标，包括种群丰度、栖息地利用、年龄和大小结构、性别比例和成熟年龄（Shuter，1990；Adams，Greeley，2000；Bartell，2006）。在生态系统层面，生物完整性或物种丰富度、食物链结构和生产力等指数的变化都可以反映环境的应激（Karr，1981；Odum，1985）。正如使用多个指标界定个体应激是一种可靠方法一样，将生物组织结构的多个层次结合起来以形成生态系统健康评估的方法可以提供最丰富的信息（Attrill，Depledge，1997；Adams，Greeley，2000；Bartell，2006；Yeom，Adams，2007）。例如，测量蓄养成年鲑鱼的热应激指标（Jeffries 等，2012）可以和自然经历较高水温的野生洄游鱼类的健康指标相联系（Martins 等，2012）。利用政府机构收集的种群评估数据，这些实验室和野外调查结果可以根据种群和物种的生存趋势进行调整。鲑鱼种群丰度的变化可以和其他类群的健康状况（Bryan 等，2013）以及生态系统水平的进程（Gende 等，2002）联系起来。将个体特征与生态系统联系起来是可能的，这需要将实验室记录的个体应激指标、野外获得的应激指标以及从一系列来源组合而形成的种群和生态系统属性进行配对。

然而，建立从个体到种群或生态系统的联系，确实需要纵向的数据集，包括单个水平的应激指标和基于现场的检测。当建立个体水平的应激指标和种群水平的变化相关联时，可以利用反应较快的应激指标作为即将对种群和生态系统影响的预警信号（Adams，Greeley，2000）。重视将岗哨的或者生态上重要的生物种类作为生物指标，可以通过评估生态系统的稳定性而明显有助于建立个体的应激型式（Adams，Greeley，2000）。这些努力充分显示了学者、政府和使用者团体之间合作的价值和必要性，以便将应激指标发展成为环境保护和管理的有用工具。

11.6　未来的应激指标

本章以及本书中概述的研究工作对鱼类如何检测和应对应激源提供了全面的理解。尽管现有的测量鱼类对应激源反应的工具非常广泛（图 11.2），而且可以回答的问题也非常多，但这一领域的研究仍有许多新的方向可以向前发展。随着人口的增长，对地球的影响预计将持续或加剧，提高我们对鱼类对应激源的反应，尤其是对多种应激源的反应的理解是至关重要的。我们认为，有 5 个主要领域是研究人员应该考虑作为未来工作目标的，希望能够开发新的应激指标和改进现有的应激指标，以便最大限度地提高效益和预测价值。

图 11.2　应用于鱼类的各种应激指标

注：应用于鱼类的各种应激指标包括一些相当新式的检测指标，如：（A）使用高流量（omic）的检测技术；（B）使用 point-of-care 手提式仪表在野外测量血液化学；（C）从小鱼整体抽提皮质醇；（D）评估反射状态；（E）评定游泳行为表现和代谢状态；（F）使用加速度计（biologger）评估自由游动鱼类的运动活动能力和能量学。图片提供者：（A）Katrina Cook，（B）Steven Cooke，（C）Julia Redfern，（D）Vivian Nguyen，（E）Zack Zucherman，（F）Jocob Brownscombe。

首先，也是最重要的一点，必须更好地将应激与干扰指标和生殖产出与健康指标联系起来。初次应激反应本身就是一个正在选用的特征（Wingfield 等，1998；Ricklefs，Wikelski，2002）。研究工作已经使用急性应激反应的量度作为种群可能经历下降风险的提示（Romero，Wikelski，2001）。尽管慢性应激与减少生殖之间的关系是确定的，但急性应激反应的激活和健康结果之间的联系是存在的，在某种程度上是相关关系，而不是因果机制。无论是急性还是慢性应激源暴露后，识别应激反应的激活，并不需要保证动物会比未暴露的动物经历体能下降。能够自信地将急性或慢性应激反应和健康状况下降或其他种群参数的负面变化联系起来，将是我们对应激反应重要性认识的巨大飞跃。同时，我们预测应激刺激的结果和使用皮质醇应激反应作为环境保护的早期预警系统的能力也会大大增强。与生殖有关的应激指标也可以和传递效应的概念相结合，使当前的应激源在多个生殖周期或世代之间产生影响。然后，与繁殖有关的指数可以和栖息地改变或气候变化等景观层面的应激源相联系，以识别这些广泛的挑战对种群的影响。

第二，就像应激和健康之间的联系一样，我们觉得有必要加强应激和鱼的行为表现之间的联系。鱼的行为表现是一个广泛的概念，包括游泳能力、有氧范围和应激反应范围等指标。这些不同指标和生存与食物获取等结果之间的联系在文献中得到了充分的证实（Plaut，2001；Farrellet 等，2008）。然而，应激和健康/繁殖也可能通过有机体行为表现的下降而间接联系在一起（例如，游泳行为表现的下降可能导致摄食减少和/或无法逃脱被捕食）。有许多行为表现的性能指标或反射障碍（如身体屈曲、呕吐反应）和应激呈正相关，可用于预测个体死亡率（Davis，2010）。由于这些反射损伤可以在野外轻松可靠地收集到，将这些行为表现的性能指标和生存之外的后果联系起来将是有益的，包括减少生殖投资或降低健康程度等概念。因此，了解应激在短期和长期内如何影响机体的行为表现，将有助于我们预测应激对个体的影响，以及最终对种群的影响。

第三，为了促进单个应激指标和种群水平进程之间的联系，我们需要进一步了解世代间影响的指标。卵子大小、受精成功率和胚胎存活率是亲代应激的指标。将这些指标与子代生理和行为特征的评价相结合，可以揭示世代间应激的潜在指标。目前仍然缺乏关于驱动子代表型变化的配子应激指标的知识。卵皮质醇水平升高被认为是一种可靠的应激指标，也是子代变化的机制（Gingerich，Suski，2011）。然而，母体的应激源暴露是否会改变卵子中的皮质醇水平仍是一个未知数（例如，Stratholt 等，1997；Sopinka 等，2014）。还需要进一步的实验来证实卵皮质醇（或其他激素，如甲状腺激素和性激素）的浓度是否是母体应激的可靠指标。此外，随着分子技术的进步（见下文），卵子、胚胎和精子转录组表观遗传学变化的量化（Cabrita 等，2014；Mommer，Bell，2014）已具备潜力成为一种有价值的应激指标。扩大以生殖为基础的应激指标将有助于预测应激从一代到下一代的连锁效应。

第四，最近的研究表明，应激激素可以在毛皮或羽毛等结构中沉积和存档（Bortolotti 等，2009；Sheriff 等，2011）。从毛皮或羽毛等结构中提取应激激素的能力提供了一种独特的、长期的、完整的应激轴活动史，可以作为动物生活中过去活动的目录。此外，这些结构可以以非致命性的方式收集，而且动物通常可以自由脱落。目前，我们知道有一项研究可以测量鱼鳞的应激激素（皮质醇）（Aerts 等，2015），这表明利用弹性鳞片作为鱼类应激生物标志物的潜力。我们鼓励对皮质醇沉积在鳞片上的研究，因为这可以提供一种宝贵的工具，既可以非致命性地确定自由游动动物的应激史，也可以将过去的应激状况和繁殖能力或健康状况有效地联系起来。

最后，过去十年出现了许多新技术，可用于量化动物对各种干扰的分子反应，包括转录组、基因表达、蛋白质生成等技术，这为量化生物与环境的相互作用提供了一个强有力的新途径（Evans，Hofmann，2012）。这些技术能提供可靠的应激指标，可以同时分析许多不同的生理系统，重要的是，能把基因表达的变化和生存与健康等生态相关的结果联系起来（Abzhanov 等，2006）。鱼类非常适合使用这些分子工具来进行研究，因为它们拥有许多不同的具核组织，这些组织可以采用非致命性的方式收集，例如，鳃（Jeffries 等，2014a）、红细胞（Dennis 等，2015）。当这些技术和整个机体的行为表现、世代间的效应（即表观遗传学）、景观层面的挑战或者人口统计的模式联系起来以确定种群层面的发展趋势，而不是简单地对动物的应激反应进行编目时，它们就显得特别有价值。因此，我们鼓励这些新工具的继续发展和提高。

10.7　结束语

鱼类暴露在应激源下会引起 HPI 轴的激活（主要反应）以及随后的次级和三级反应。HPI 轴的活性和整个动物的反应都可以量化以表示应激的状态（见本章第 11.3 节和表 11.1 至表 11.3）。在实验室和野外条件下对应激源进行定量是了解鱼类如何对其内外环境变化做出反应的基础。这方面的知识正在继续应用于野生种群的管理和保护以及维护捕获鱼类的福利。文献中报道的不同应激指标的数量是和不同因子的数量相匹配的，在量化和解释这些指标本身时必须考虑到（图 11.3）。无论是应激源的严重程度还是同时发生的特性，种群内和物种间反应的变化，或是在一天内测量指标的时间，我们对应激的理解都受到许多微妙而重要的变量的挑战。应激指标的未来需要结合到：①优化实验方法，以确保指标的可靠性和生态相关性；②研究个体内部各组织层面的应激（即分子到整体动物的反应），并将个体反应和种群与生态系统层面的应激指标结合起来；③解读与确定延续下来与世代之间影响有关的指标；④加强生物标记、遥测、基因组和内分泌应激型式无损伤评估等新技术；⑤继续加强学科间的合作。

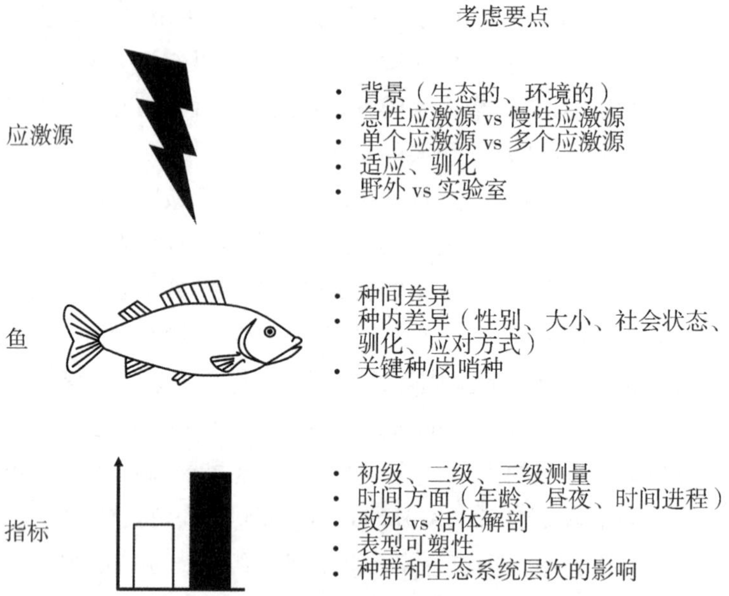

考虑要点
应激源
• 背景（生态的、环境的）
• 急性应激源 vs 慢性应激源
• 单个应激源 vs 多个应激源
• 适应、驯化
• 野外 vs 实验室
鱼
• 种间差异
• 种内差异（性别、大小、社会状态、驯化、应对方式）
• 关键种/岗哨种
指标
• 初级、二级、三级测量
• 时间方面（年龄、昼夜、时间进程）
• 致死 vs 活体解剖
• 表型可塑性
• 种群和生态系统层次的影响

图 11.3 量化和解读鱼类应激指标时的考虑要点

N. M. 索平卡 M. R. 唐纳森 C. M. 奥康纳 C. D. 苏斯基 S. J. 库克 著

卢丹琪 译

林浩然 校

参考文献

Abzhanov, A., Kuo, W. P., Hartmann, C., Grant, B. R., Grant, P. R., Tabin, C. J. (2006). The calmodulin pathway and evolution of elongated beak morphology in Darwin's finches. Nature. 442, 563-567.

Adams, S. M. (1999). Ecological role of lipids in the health and success of fish populations. In: Arts, M. T., Wainmann, B. C., *Lipids in* Freshwater *Ecosystems*. Springer, New York, NY, 132-160.

Adams, S. M. (2002). Biological Indicators of Aquatic Ecosystem Stress. American Fisheries Society, Maryland.

Adams, S. M., Greeley, M. S. (2000). Ecotoxicological indicators of water quality: using multi-response indicators to assess the health of aquatic ecosystems. Water Air Soil Poll. 123, 103-115.

Adams, S. M., Brown, A. M., Goede, R. W. (1993). A quantitative health assessment index for rapid evaluation of fish condition in the field. *Trans. Am. Fish. Soc*. 122, 63-73.

Aerts, J., Metz, J. R., Ampe, B., Decostere, A., Flik, G., De Saeger, S. (2015). Scales tell a story on the stress history of fish. PLoS One. 10, e0123411.

Aissaoui, A., Tort, L., Altimiras, J. (2000). Circadian heart rate changes and light-dependence in the Mediterranean seabream Sparus aurata. Fish. Physiol. Biochem. 22, 89-94.

Al-Murrani, W. K., Al-Rawi, I. K., Raof, N. M. (2002). Genetic resistance to *Salmonella typhimurium* in two lines of chickens selected as resistant and sensitive on the basis of heterophil/lymphocyte ratio. Brit. Poult. Sci. 43, 501-507.

Alonso-Alvarez, C., Bertrand, S., Devevey, G., Prost, J., Faivre, B., Sorci, G. (2004). Increased susceptibility to oxidative stress as a proximate cost of reproduction. Ecol. Lett. 7, 363-368.

Altimiras, J., Larsen, E. (2000). Non-invasive recording of heart rate and ventilation rate in rainbow trout during rest and swimming. Fish go wireless!. J. Fish Biol. 57, 197-209.

Anderson, W. G., Booth, R., Beddow, T. A., McKinley, R. S., Finstad, B., Økland, F. (1998). Remote monitoring of heart rate as a measure of recovery in angled Atlantic salmon, *Salmo salar* (L.). In: Lagardere, J. P., Begout Anras, M.-L., Claireaux, G., Advances in Invertebrates and Fish Telemetry. Springer, Netherlands, 233-240.

Armstrong, J. D. (1998). Relationships between heart rate and metabolic rate of pike: integration of existing data. J. Fish Biol. 52, 362-368.

Attrill, M. J., Depledge, M. H. (1997). Community and population indicators of ecosystem health: targeting links between levels of biological organisation. Aquat. Toxicol. 38, 183-197.

Aubin-Horth, N., Deschênes, M., Cloutier, S. (2012). Natural variation in the molecular stress network correlates with a behavioural syndrome. Horm. Behav. 61, 140-146.

Bacanskas, L. R., Whitaker, J., Di Giulio, R. T. (2004). Oxidative stress in two populations of killifish (*Fundulus heteroclitus*) with differing contaminant histories. Marine Environ. Res. 58, 597-601.

Baker, M. R., Vynne, C. H. (2014). Cortisol profiles in sockeye salmon: sample bias and baseline values at migration, maturation, spawning, and senescence. Fish Res. 154, 38-43.

Barcellos, L. J. G., Kreutz, L. C., Quevedo, R. M. (2006). Previous chronic stress does not alter the cortisol response to an additional acute stressor in jundiá

(*Rhamdia quelen*, Quoy and Gaimard) fingerlings. Aquaculture. 253, 317-321.

Barcellos, L. J. G., Ritter, F., Kreutz, L. C., Cericato, L. (2010). Can zebrafish *Danio rerio* learn about predation risk? The effect of a previous experience on the cortisol response in subsequent encounters with a predator. J. Fish Biol. 76, 1032-1038.

Barcellos, L. J., Koakoski, G., da Rosa, J. G., Ferreira, D., Barreto, R. E., Giaquinto, P. C. (2014). Chemical communication of predation risk in zebrafish does not depend on cortisol increase. Sci. Rep. 4, 5076.

Barcellos, L. J., Woehl, V. M., Koakoski, G., Oliveira, T. A., Ferreira, D., da Rosa, J. G. S. (2014). Puberty influences stress reactivity in female catfish *Rhamdia quelen*. *Physiol. Behav.* 128, 232-236.

Barreto, R. E., Volpato, G. L. (2004). Caution for using ventilatory frequency as an indicator of stress in fish. Behav. Process. 66, 43-51.

Barreto, R. E., Volpato, G. L. (2006). Ventilatory frequency of Nile tilapia subjected to different stressors. J. Exp. Anim. Sci. 43, 189-196.

Barry, T. P., Malison, J. A., Held, J. A., Parrish, J. J. (1995). Ontogeny of the cortisol stress response in larval rainbow trout. Gen. Comp. Endocrinol. 97, 57-65.

Bartell, S. M. (2006). Biomarkers, bioindicators, and ecological risk assessment-a brief review and evaluation. Environ. Bioindic. 1, 60-73.

Barton, B. A. (2000). Salmonid fishes differ in their cortisol and glucose responses to handling and transport stress. N. Am. J. Aquacul. 62, 12-18.

Barton, B. A. (2002). Stress in fishes: a diversity of responses with particular reference to changes in circulating corticosteroids. Integr. Comp. Biol. 42, 517-525.

Barton, B. A., Schreck, C. B. (1987). Metabolic cost of acute physical stress in juvenile steelhead. Trans. Am. Fish. Soc. 116, 257-263.

Barton, B. A., Peter, R. E., Paulencu, C. R. (1980). Plasma cortisol levels of fingerling rainbow trout (*Salmo gairdneri*) at rest, and subjected to handling, confinement, transport, and stocking. Can. J. Fish. Aquat. Sci. 37, 805-811.

Barton, B. A., Schreck, C. B., Barton, L. D. (1987). Effects of chronic cortisol administration and daily acute stress on growth, physiological conditions, and stress responses in juvenile rainbow trout. Dis. Aquat. Organ. 2, 173-185.

Barton, B. A., Morgan, J. D., Vijayan, M. M., Adams, S. M. (2002). Physiological and condition-related indicators of environmental stress in fish. In: Adams, S. M., Biological Indicators of Aquatic Ecosystem Stress. American Fisheries Society, Maryland, 111-148.

Basu, N., Nakano, T., Grau, E. G., Iwama, G. K. (2001). The effects of cortisol on heat shock protein 70 levels in two fish species. Gen. Comp. Endocrinol. 124, 97-105.

Bayarri, M. J., Munoz-Cueto, J. A., López-Olmeda, J. F., Vera, L. M., De Lama, M. R., Madrid, J. A. (2004). Daily locomotor activity and melatonin rhythms in Senegal sole (*Solea senegalensis*). Physiol. Behav. 81, 577-583.

Beamish, F. W. H. (1978). Swimming capacity. In: Hoar, W. S., Randall, D. J., . Academic Press, New York, NY, 101-187.

Beamish, R., McFarlane, G. A. (1983). The forgotten requirement for age validation in fisheries biology. Trans. Am. Fish. Soc. 112, 735-743.

Beecham, R. V., Small, B. C., Minchew, C. D. (2006). Using portable lactate and glucose meters for catfish research: acceptable alternatives to laboratory methods?. *N. Am. J.* Aquacul. 68, 291-295.

Bell, A. M., Backström, T., Huntingford, F. A., Pottinger, T. G., Winberg, S. (2007). Variable neuroendocrine responses to ecologically-relevant challenges in sticklebacks. Physiol. Behav. 91, 15-25.

Bell, A. M., Henderson, L., Huntingford, F. A. (2010). Behavioral and respiratory responses to stressors in multiple populations of three-spined sticklebacks that differ in predation pressure. J. Comp. Biochem. Physiol. B. 180, 211-220.

Beyers, D. W., Rice, J. A., Adams, S. M. (2002). Evaluating stress in fish using bioenergetics-based stressor-response models. In: Adams, S. M., Biological Indicators of Aquatic Ecosystem Stress. American Fisheries Society, Maryland, 289-320.

Boisclair, D., Leggett, W. C. (1989). The importance of activity in bioenergetics models applied to actively foraging fishes. Can. J. Fish. Aquat. Sci. 46, 1859-1867.

Bolger, T., Connolly, P. L. (1989). The selection of suitable indices for the measurement and analysis of fish condition. J. Fish Biol. 34, 171-182.

Bonier, F., Martin, P. R., Moore, I. T., Wingfield, J. C. (2009). Do baseline glucocorticoids predict fitness?. Trends Ecol. Evol. 24, 634-642.

Boonstra, R. (2013). The ecology of stress: a marriage of disciplines. Funct. Ecol. 27, 7-10.

Boonstra, R. (2013). Reality as the leading cause of stress: rethinking the impact of chronic stress in nature. Funct. Ecol. 27, 11-23.

Bortolotti, G. R., Marchant, T., Blas, J., Cabezas, S. (2009). Tracking stress: localisation, deposition and stability of corticosterone in feathers. J. Exp. Biol. 212,

1477-1482.

Bortone, S. A., Davis, W. P. (1994). Fish intersexuality as indicator of environmental stress. Bioscience., 165-172.

Breuner, C. W., Patterson, S. H., Hahn, T. P. (2008). In search of relationships between the acute adrenocortical response and fitness. Gen. Comp. Endocrinol. 157, 288-295.

Breves, J. P., Specker, J. L. (2005). Cortisol stress response of juvenile winter flounder (*Pseudopleuronectes americanus*, Walbaum) to predators. J. Exp. Mar. Biol. Ecol. 325, 1-7.

Brown, J. A. (1984). Parental care and the ontogeny of predator-avoidance in two species of centrarchid fish. Anim. Behav. 32, 113-119.

Brown, C., Gardner, C., Braithwaite, V. A. (2005). Differential stress responses in fish from areas of high-and low-predation pressure. J. Comp. Biochem. Physiol. B. 175, 305-312.

Brownscombe, J. W., Gutowsky, L. F., Danylchuk, A. J., Cooke, S. J. (2014). Foraging behaviour and activity of a marine benthivorous fish estimated using tri-axial accelerometer biologgers. Mar. Ecol. Prog. Ser. 505, 241-251.

Bry, C., Zohar, Y. (1980). Dorsal aorta catheterization in rainbow trout (*Salmo gairdneri*). II. Glucocorticoid levels, hematological data and resumption of feeding for five days after surgery. Reprod. Nutr. Dev. 20, 1825-1834.

Bryan, H. M., Darimont, C. T., Paquet, P. C., Wynne-Edwards, K. E., Smits, J. E. (2013). Stress and reproductive hormones in grizzly bears reflect nutritional benefits and social consequences of a salmon foraging niche. PloS One. 8, e80537.

Brydges, N. M., Boulcott, P., Ellis, T., Braithwaite, V. A. (2009). Quantifying stress responses induced by different handling methods in three species of fish. *Appl. Anim.* Behav. Sci. 116, 295-301.

Buckley, B. A. (2007). Comparative environmental genomics in non-model species: using heterologous hybridization to DNA-based microarrays. J. Exp. Biol. 210, 1602-1606.

Burnett, N. J., Hinch, S. G., Braun, D. C., Casselman, M. T., Middleton, C. T., Wilson, S. M. (2014). Burst swimming in areas of high flow: delayed consequences of anaerobiosis in wild adult sockeye salmon. Physiol. Biochem. Zool. 87, 587-598.

Burton, T., Killen, S. S., Armstrong, J. D., Metcalfe, N. B. (2011). What causes intraspecific variation in resting metabolic rate and what are its ecological consequences?. Proc. Roy. Soc. B. 278, 3465-3473.

Cabrita, E., Martínez-Páramo, S., Gavaia, P. J., Riesco, M. F., Valcarce, D. G., Sarasquete, C. (2014). Factors enhancing fish sperm quality and emerging tools for sperm analysis. Aquaculture. 432, 389-401.

Caipang, C. M. A., Brinchmann, M. F., Berg, I., Iversen, M., Eliassen, R., Kiron, V. (2008). Changes in selected stress and immune-related genes in Atlantic cod, *Gadus morhua*, following overcrowding. Aquac. Res. 39, 1533-1540.

Calow, P., Forbes, V. E. (1998). How do physiological responses to stress translate into ecological and evolutionary processes?. Comp. Biochem. Physiol. A. 120, 11-16.

Campbell, P. M., Pottinger, T. G., Sumpter, J. P. (1994). Preliminary evidence that chronic confinement stress reduces the quality of gametes produced by brown and rainbow trout. Aquaculture. 120, 151-169.

Carey, J. B., McCormick, S. D. (1998). Atlantic salmon smolts are more responsive to an acute handling and confinement stress than parr. Aquaculture. 168, 237-253.

Castanheira, M. F., Conceição, L. E., Millot, S., Rey, S., Bégout, M. L., Damsgard, B. (2015). Coping styles in farmed fish: consequences for aquaculture. Rev. Aquac. 7, 1-19.

Cawthon, R. M. (2002). Telomere measurement by quantitative PCR. Nucl. Acids Res. 30.

Claireaux, G., Webber, D., Kerr, S., Boutilier, R. (1995). Physiology and behaviour of free-swimming Atlantic cod (*Gadus morhua*) facing fluctuating salinity and oxygenation conditions. J. Exp. Biol. 198, 61-69.

Clark, T. D., Taylor, B. D., Seymour, R. S., Ellis, D., Buchanan, J., Fitzgibbon, Q. P. (2008). Moving with the beat: heart rate and visceral temperature of free-swimming and feeding bluefin tuna. Proc. R. Soc. B. 275, 2841-2850.

Clark, T. D., Sandblom, E., Hinch, S. G., Patterson, D. A., Frappell, P. B., Farrell, A. P. (2010). Simultaneous biologging of heart rate and acceleration, and their relationships with energy expenditure in free-swimming sockeye salmon (*Oncorhynchus nerka*). *J. Comp. Biochem. Physiol*. B. 180, 673-684.

Cnaani, A., Tinman, S., Avidar, Y., Ron, M., Hulata, G. (2004). Comparative study of biochemical parameters in response to stress in *Oreochromis aureus*, *O. mossambicus* and two strains of *O. niloticus*. Aquac. Res. 35, 1434-1440.

Cone, R. S. (1989). The need to reconsider the use of condition indices in fishery science. Trans. Am. Fish. Soc. 118, 510-514.

Congleton, J. L., Wagner, T. (2006). Blood-chemistry indicators of nutritional status in juvenile salmonids. J. Fish Biol. 69, 473-490.

Cook, K. V., McConnachie, S. H., Gilmour, K. M., Hinch, S. G., Cooke, S. J. (2011). Fitness and behavioral correlates of pre-stress and stress-induced plasma cortisol titers in pink salmon (*Oncorhynchus gorbuscha*) upon arrival at spawning grounds. Horm. Behav. 60, 489-497.

Cook, K. V., Crossin, G. T., Patterson, D. A., Hinch, S. G., Gilmour, K. M., Cooke, S. J. (2014). The stress response predicts migration failure but not migration rate in a semelparous fish. Gen. Comp. Endocrinol. 202, 44-49.

Cooke, S. J., Hinch, S. G., Wikelski, M., Andrews, R. D., Kuchel, L. J., Wolcott, T. G. (2004). Biotelemetry: a mechanistic approach to ecology. Trends Ecol. Evol. 19, 334-343.

Cooke, S. J., Hinch, S. G., Crossin, G. T., Patterson, D. A., English, K. K., Healey, M. C. (2006). Mechanistic basis of individual mortality in Pacific salmon during spawning migrations. Ecology. 87, 1575-1586.

Cooke, S. J., Sack, L., Franklin, C. E., Farrell, A. P., Beardall, J., Wikelski, M. (2013). What is conservation physiology? Perspectives on an increasingly integrated and essential science. Conserv. Physiol. 1, cot001.

Cooke, S. J., Blumstein, D. T., Buchholz, R., Caro, T., Fernández-Juricic, E., Franklin, C. E. (2014). Physiology, behavior, and conservation. Physiol. Biochem. Zool. 87, 1-14.

Costantini, D. (2008). Oxidative stress in ecology and evolution: lessons from avian studies. Ecol. Lett. 11, 1238-1251.

Cousineau, A., Midwood, J. D., Stamplecoskie, K., King, G., Suski, C. D., Cooke, S. J. (2014). Diel patterns of baseline glucocorticoids and stress responsiveness in a teleost fish (bluegill, *Lepomis macrochirus*). Can. J. Zool. 92, 417-421.

Crossin, G. T., Hinch, S. G. (2005). A nonlethal, rapid method for assessing the somatic energy content of migrating adult Pacific salmon. Trans. Am. Fish. Soc. 134, 184-191.

Crozier, L. G., Hutchings, J. A. (2014). Plastic and evolutionary responses to climate change in fish. Evol. App. 7, 68-87.

Cyr, N. E., Romero, L. M. (2009). Identifying hormonal habituation in field studies of stress. Gen. Comp. Endocrinol. 161, 295-303.

Dale, V. H., Beyeler, S. C. (2001). Challenges in the development and use of ecological indicators. Ecol. Indic. 1, 3-10.

Danylchuk, S. E. , Danylchuk, A. J. , Cooke, S. J. , Goldberg, T. L. , Koppelman, J. , Philipp, D. P. (2007). Effects of recreational angling on the post-release behavior and predation of bonefish (*Albula vulpes*): the role of equilibrium status at the time of release. J. Exp. Mar. Biol. Ecol. 346, 127-133.

Davis, M. W. (2010). Fish stress and mortality can be predicted using reflex impairment. Fish Fish. 11, 1-11.

Davis, M. W. , Olla, B. L. , Schreck, C. B. (2001). Stress induced by hooking, net towing, elevated sea water temperature and air in sablefish: lack of concordance between mortality and physiological measures of stress. J. Fish Biol. 58, 1-15.

Davis, A. K. , Maney, D. L. , Maerz, J. C. (2008). The use of leukocyte profiles to measure stress in vertebrates: a review for ecologists. Funct. Ecol. 22, 760-772.

DeKoning, A. L. , Picard, D. J. , Bond, S. R. , Schulte, P. M. (2004). Stress and interpopulation variation in glycolytic enzyme activity and expression in a teleost fish *Fundulus heteroclitus*. Physiol. Biochem. Zool. 77, 18-26.

Dennis, C. E. , Kates, D. F. , Noatch, M. R. , Suski, C. D. (2015). Molecular responses of fishes to elevated carbon dioxide. Comp. Biochem. Physiol. A. 187, 224-231.

Dhabhar, F. S. , Miller, A. H. , McEwen, B. , Spender, R. L. (1996). Stress-induced changes in blood leukocyte distribution-role of adrenal steroid hormones. J. Immunol. 157, 1638-1644.

Dickens, M. J. , Romero, L. M. (2013). A consensus endocrine profile for chronically stressed wild animals does not exist. Gen. Comp. Endocrinol. 191, 177-189.

Donaldson, M. R. , Cooke, S. J. , Patterson, D. A. , Macdonald, J. S. (2008). Cold shock and fish. J. Fish Biol. 73, 1491-1530.

Donaldson, M. R. , Clark, T. D. , Hinch, S. G. , Cooke, S. J. , Patterson, D. A. , Gale, M. K. (2010). Physiological responses of free-swimming adult coho salmon to simulated predator and fisheries encounters. Physiol. Biochem. Zool. 83, 973-983.

Donaldson, M. R. , Hinch, S. G. , Patterson, D. A. , Hills, J. , Thomas, J. O. , Cooke, S. J. (2011). The consequences of angling, beach seining, and confinement on the physiology, post-release behaviour and survival of adult sockeye salmon during upriver migration. Fish Res. 108, 133-141.

Donaldson, M. R. , Hinch, S. G. , Raby, G. D. , Patterson, D. A. , Farrell, A. P. , Cooke, S. J. (2012). Population-specific consequences of fisheries-related stressors on adult sockeye salmon. Physiol. Biochem. Zool. 85, 729-739.

Donaldson, M. R. , Hinch, S. G. , Jeffries, K. M. , Patterson, D. A. , Cooke, S. J. ,

Farrell, A. P. (2014). Species-and sex-specific responses and recovery of wild, mature pacific salmon to an exhaustive exercise and air exposure stressor. Comp. Biochem. Physiol. A. 173, 7-16.

Eaton, L., Edmonds, E. J., Henry, T. B., Snellgrove, D. L., Sloman, K. A. (2015). Mild maternal stress disrupts associative learning and increases aggression in offspring. Horm. Behav. 71, 10-15.

Ellerby, D. J., Herskin, J. (2013). Swimming flumes as a tool for studying swimming behavior and physiology: current applications and future developments. In: Palstra, A. P., Planas, J. V., Swimming Physiology of Fish. Springer Berlin Heidelberg, , 345-375.

Ellis, T., James, J. D., Stewart, C., Scott, A. P. (2004). A non-invasive stress assay based upon measurement of free cortisol released into the water by rainbow trout. J. Fish Biol. 65, 1233-1252.

Endo, H., Yonemori, Y., Hibi, K., Ren, H., Hayashi, T., Tsugawa, W. (2009). Wireless enzyme sensor system for real-time monitoring of blood glucose levels in fish. Biosens. Bioelectron. 24, 1417-1423.

Epel, E. S., Blackburn, E. H., Lin, J., Dhabhar, F. S., Adler, N. E., Morrow, J. D. (2004). Accelerated telomere shortening in response to life stress. Proc. Natl. Acad. Sci. U. S. A. 101, 17312-17315.

Evans, T. G., Hofmann, G. E. (2012). Defining the limits of physiological plasticity: how gene expression can assess and predict the consequences of ocean change. Philos. Trans. R. Soc. B. 367, 1733-1745.

Fangue, N. A., Hofmeister, M., Schulte, P. M. (2006). Intraspecific variation in thermal tolerance and heat shock protein gene expression in common killifish, *Fundulus heteroclitus*. J. Exp. Biol. 209, 2859-2872.

Farrell, A. P. (1991). From hagfish to tuna: a perspective on cardiac function in fish. Physiol. Zool. 64, 1137-1164.

Farrell, A. P., Lee, C. G., Tierney, K., Hodaly, A., Clutterham, S., Healey, M. (2003). Field-based measurements of oxygen uptake and swimming performance with adult Pacific salmon using a mobile respirometer swim tunnel. J. Fish Biol. 62, 64-84.

Farrell, A. P., Hinch, S. G., Cooke, S. J., Patterson, D. A., Crossin, G. T., Lapointe, M. (2008). Pacific salmon in hot water: applying aerobic scope models and biotelemetry to predict the success of spawning migrations. Physiol. Biochem. Zool. 81, 697-709.

Fatira, E. , Papandroulakis, N. , Pavlidis, M. (2014). Diel changes in plasma cortisol and effects of size and stress duration on the cortisol response in European sea bass (*Dicentrarchus labrax*). Fish. Physiol. Biochem. 40, 911-919.

Faught, E. , Aluru, N. , Vijayan, M. M. (2016). The Molecular Stress Response. In: Schreck, C. B. , Tort, L. , Farrell, A. P. , Brauner, C. J. , . Academic Press, San Diego, CA.

Fefferman, N. H. , Romero, L. M. (2013). Can physiological stress alter population persistence? A model with conservation implications. Conserv. Physiol. 1, cot012.

Feist, G. , Schreck, C. B. (2001). Ontogeny of the stress response in chinook salmon, Oncorhynchus tshawytscha. Fish. Physiol. Biochem. 25, 31-40.

Fevolden, S. E. , Refstie, T. , Røed, K. H. (1991). Selection for high and low cortisol stress response in Atlantic salmon (*Salmo salar*) and rainbow trout (*Oncorhynchus mykiss*). Aquaculture. 95, 53-65.

Fürtbauer, I. , King, A. J. , Heistermann, M. (2015). Visible implant elastomer (VIE) tagging and simulated predation risk elicit similar physiological stress responses in three-spined stickleback *Gasterosteus aculeatus*. J. Fish Biol. 86, 1644-1649.

Gallo, V. P. , Civinini, A. (2003). Survey of the adrenal homolog in teleosts. Inter. Rev. Cytol. 230, 89-187.

Gende, S. M. , Edwards, R. T. , Willson, M. F. , Wipfli, M. S. (2002). Pacific salmon in aquatic and terrestrial ecosystems. Bioscience. 52, 917-928.

Geslin, M. , Auperin, B. (2004). Relationship between changes in mRNAs of the genes encoding steroidogenic acute regulatory protein and P450 cholesterol side chain cleavage in head kidney and plasma levels of cortisol in response to different kinds of acute stress in the rainbow trout (*Oncorhynchus mykiss*). Gen. Comp. Endocrinol. 135, 70-80.

Gesto, M. , López-Patiño, M. A. , Hernάndez, J. , Soengas, J. L. , Míguez, J. M. (2013). The response of brain serotonergic and dopaminergic systems to an acute stressor in rainbow trout: a time course study. J. Exp. Biol. 216, 4435-4442.

Gesto, M. , López-Patiño, M. A. , Hernάndez, J. , Soengas, J. L. , Míguez, J. M. (2015). Gradation of the stress response in rainbow trout exposed to stressors of different severity: the role of brain serotonergic and dopaminergic systems. J. Neuroendocrinol. 27, 131-141.

Giacomini, A. C. V. V. , de Abreu, M. S. , Koakoski, G. , Idalêncio, R. , Kalichak, F. , Oliveira, T. A. (2015). My stress, our stress: blunted cortisol response to stress in isolated housed zebrafish. *Phys.* Behav. 139, 182-187.

Gingerich, A. J., Suski, C. D. (2011). The role of progeny quality and male size in the nesting success of smallmouth bass: integrating field and laboratory studies. Aquat. Ecol. 45, 505-515.

Gingerich, A. J., Philipp, D. P., Suski, C. D. (2010). Effects of nutritional status on metabolic rate, exercise and recovery in a freshwater fish. *J. Comp. Biochem. Physiol.* B. 180, 371-384.

Godin, J. G. J. (1997). Behavioural Ecology of Teleost Fishes. Oxford University Press, Oxford.

Goede, R. W., Barton, B. A. (1990). Organismic indices and an autopsy-based assessment as indicators of health and condition of fish. Am. Fish. Soc. Symp. 8, 93-108.

Gorissen, M., Flik, G. (2016). Endocrinology of the Stress Response in Fish. In: Schreck, C. B., Tort, L., Farrell, A. P., Brauner, C. J., . Academic Press, San Diego, CA.

Gracey, A. Y., Troll, J. V., Somero, G. N. (2001). Hypoxia-induced gene expression profiling in the euryoxic fish *Gillichthys mirabilis*. Proc. Natl. Acad. Sci. U. S. A. 98, 1993-1998.

Gutowsky, L. F., Aslam, W., Banisaeed, R., Bell, L. R., Bove, K. L., Brownscombe, J. W. (2015). Considerations for the design and interpretation of fishing release mortality estimates. Fish. Res. 167, 64-70.

Hammer, C. (1995). Fatigue and exercise tests with fish. Comp. Biochem. Physiol. A. 112, 1-20.

Hansen, J. A., Marr, J. C., Lipton, J., Cacela, D., Bergman, H. L. (1999). Differences in neurobehavioral responses of chinook salmon (*Oncorhynchus tshawytscha*) and rainbow trout (*Oncorhynchus mykiss*) exposed to copper and cobalt: behavioral avoidance. Environ. Toxicol. Chem. 18, 1972-1978.

Henderson, R. J., Tocher, D. R. (1987). The lipid composition and biochemistry of freshwater fish. Prog. Lipid Res. 26, 281-347.

Henry, J. B. (1996). Clinical Diagnosis and Management by Laboratory Methods. Saunders, Philadelphia.

Hori, T. S., Gamperl, A. K., Hastings, C. E., Vander Voort, G. E., Robinson, J. A. B., Johnson, S. C. (2012). Inter-individual and-family differences in the cortisol responsiveness of Atlantic cod (*Gadus morhua*). Aquaculture. 324, 165-173.

Hughes, P., Dragunow, M. (1995). Induction of immediate-early genes and the control of neurotransmitter-regulated gene expression within the nervous system. Pharmacol. Rev. 47, 133-178.

Inuzuka, H., Nanbu-Wakao, R., Masuho, Y., Mmrmatsu, M., Tojo, H., Wakao, H. (1999). Differential regulation of immediate early gene expression in preadipocyte cells through multiple signaling pathways. Biochem. Biophys. Res. Commun. 265, 664-668.

Iwama, G. K. (2007). The welfare of fish. Dis. Aquat. Organ. 75, 155-158.

Iwama, G. K., McGeer, J. C., Pawluk, M. P. (1989). The effects of five fish anaesthetics on acid-base balance, hematocrit, blood gases, cortisol, and adrenaline in rainbow trout. Can. J. Zool. 67, 2065-2073.

Iwama, G. K., Pickering, A. D., Sumpter, J. P., Schreck, C. B. (1997). Fish stress and health. In Aquaculture Society for Experimental Biology Seminar Series, vol. 62, Cambridge: Cambridge University Press.

Iwama, G. K., Thomas, P. T., Forsyth, R. B., Vijayan, M. M. (1998). Heat shock protein expression in fish. Rev. Fish Biol. Fish. 8, 35-56.

Iwama, G. K., Afonso, L. O. B., Todgham, A., Ackerman, P., Nakano, K. (2004). Are hsps suitable for indicating stressed states in fish?. J. Exp. Biol. 207, 15-19.

Iwama, G. K., Afonso, L. O. B., Vijayan, M. M. (2006). Stress in fishes. In: Evans, D. H., Claiborne, J. B., The Physiology of Fishes. Taylor & Francis, Boca Raton, 319-342.

Jacobsen, L., Baktoft, H., Jepsen, N., Aarestrup, K., Berg, S., Skov, C. (2014). Effect of boat noise and angling on lake fish behaviour. J. Fish Biol. 84, 1768-1780.

Jagoe, C. H., Haines, T. A. (1985). Fluctuating asymmetry in fishes inhabiting acidified and unacidified lakes. Can. J. Zool. 63, 130-138.

Jain, K. E., Hamilton, J. C., Farrell, A. P. (1997). Use of a ramp velocity test to measure critical swimming speed in rainbow trout (*Onchorhynchus mykiss*). Comp. Biochem. Physiol. A. 117, 441-444.

Jain, K. E., Birtwell, I. K., Farrell, A. P. (1998). Repeat swimming performance of mature sockeye salmon following a brief recovery period: a proposed measure of fish health and water quality. Can. J. Zool. 76, 1488-1496.

Jeffrey, J. D., Gollock, M. J., Gilmour, K. M. (2014). Social stress modulates the cortisol response to an acute stressor in rainbow trout (*Oncorhynchus mykiss*). Gen.

Comp. Endocrinol. 196, 8-16.

Jeffrey, J. D., Cooke, S. J., Gilmour, K. M. (2014). Regulation of hypothalamic-pituitary-interrenal axis function in male smallmouth bass (*Micropterus dolomieu*) during parental care. Gen. Comp. Endocrinol. 204, 195-202.

Jeffries, K. M., Hinch, S. G., Sierocinski, T., Clark, T. D., Eliason, E. J., Donaldson, M. R. (2012). Consequences of high temperatures and premature mortality on the transcriptome and blood physiology of wild adult sockeye salmon (*Oncorhynchus nerka*). Ecol. Evol. 2, 1747-1764.

Jeffries, K. M., Hinch, S. G., Gale, M. K., Clark, T. D., Lotto, A. G., Casselman, M. T. (2014). Immune response genes and pathogen presence predict migration survival in wild salmon smolts. Mol. Ecol. 23, 5803-5815.

Jeffries, K. M., Hinch, S. G., Sierocinski, T., Pavlidis, P., Miller, K. M. (2014). Transcriptomic responses to high water temperature in two species of Pacific salmon. Evol. Appl. 7, 286-300.

Jeffries, K. M., Brander, S. M., Britton, M. T., Fangue, N. A., Connon, R. E. (2015). Chronic exposures to low and high concentrations of ibuprofen elicit different gene response patterns in a euryhaline fish. Environ. Sci. Pollut. Res. 22 (22), 17397-17413.

Jentoft, S., Held, J. A., Malison, J. A., Barry, T. P. (2002). Ontogeny of the cortisol stress response in yellow perch (*Perca flavescens*). Fish. Physiol. Biochem. 26, 371-378.

Jentoft, S., Aastveit, A. H., Torjesen, P. A., Andersen, Ø. (2005). Effects of stress on growth, cortisol and glucose levels in non-domesticated Eurasian perch (*Perca fluviatilis*) and domesticated rainbow trout (*Oncorhynchus mykiss*). Comp. Biochem. Physiol. A. 141, 353-358.

Johnsson, J. I., Höjesjö, J., Fleming, I. A. (2001). Behavioural and heart rate responses to predation risk in wild and domesticated Atlantic salmon. Can. J. Fish. Aquat. Sci. 58, 788-794.

Karr, J. R. (1981). Assessment of biotic integrity using fish communities. Fisheries. 6, 21-27.

Kassahn, K. S., Crozier, R. H., Pörtner, H. O., Caley, M. J. (2009). Animal performance and stress: responses and tolerance limits at different levels of biological organisation. Biol. Rev. 84, 277-292.

Kebus, M. J., Collins, M. T., Brownfield, M. S., Amundson, C. H., Kayes, T. B., Malison, J. A. (1992). Effects of rearing density on the stress response and growth of rainbow trout. J. Aquat. Anim. Health. 4, 1-6.

Khan, A. T., Weis, J. S. (1987). Toxic effects of mercuric chloride on sperm and egg viability of two populations of mummichog, *Fundulus heteroclitus*. Environ. Pollut. 48, 263-273.

Kilgas, P., Tilgar, V., Mand, R. (2006). Hematological health state indices predict local survival in a small passerine bird, the great tit (*Parus major*). Physiol. Biochem. Zool. 79, 565-572.

Kittilsen, S., Ellis, T., Schjolden, J., Braastad, B. O., Øverli, Ø. (2009). Determining stress-responsiveness in family groups of Atlantic salmon (*Salmo salar*) using non-invasive measures. Aquaculture. 298, 146-152.

Koakoski, G., Oliveira, T. A., da Rosa, J. G. S., Fagundes, M., Kreutz, L. C., Barcellos, L. J. G. (2012). Divergent time course of cortisol response to stress in fish of different ages. Physiol. Behav. 106, 129-132.

Kotrschal, A., Ilmonen, P., Penn, D. J. (2007). Stress impacts telomere dynamics. Biol. Lett. 3, 128-130.

Krasnov, A., Koskinen, H., Pehkonen, P., Rexroad, C. E., Afanasyev, S., Mölsä, H. (2005). Gene expression in the brain and kidney of rainbow trout in response to handling stress. BMC Genomics. 6, 3.

Kultz, D. (2005). Molecular and evolutionary basis of the cellular stress response. Annu. Rev. Physiol. 67, 225-257.

Kushner, R. F. (1992). Bioelectrical impedance analysis: a review of principles and applications. J. Am. Coll. Nutr. 11, 199-209.

Landsman, S. J., Wachelka, H. J., Suski, C. D., Cooke, S. J. (2011). Evaluation of the physiology, behaviour, and survival of adult muskellunge (*Esox masquinongy*) captured and released by specialized anglers. Fish Res. 110, 377-386.

Leary, R. F., Allendorf, F. W. (1989). Fluctuating asymmetry as an indicator of stress: implications for conservation biology. Trends Ecol. Evol. 4, 214-217.

Lefrançois, C., Claireaux, G., Lagardere, J. P. (1998). Heart rate telemetry to study environmental influences on fish metabolic expenditure. In: Lagardere, J. P., Begout Anras, M.-L., Claireaux, G., *Advances in Invertebrates* and Fish Telemetry. Springer, Netherlands, 215-224.

Lepage, O., Overli, O., Petersson, E., Jarvi, T., Winberg, S. (2000). Differential stress coping in wild and domesticated sea trout. Brain Behav. Evol. 56, 259-268.

Lesser, M. P. (2006). Oxidative stress in marine environments: biochemistry and physiological ecology. Annu. Rev. Physiol. 68, 253-278.

Lobato, E., Moreno, J., Merino, S., Sanz, J. J., Arriero, E. (2005). Haeomatological variables are good predictors of recruitment in nestling pied flycatchers (*Ficedula hypoleuca*). Ecoscience. 12, 27-34.

Love, O. P., McGowan, P. O., Sheriff, M. J. (2013). Maternal adversity and ecological stressors in natural populations: the role of stress axis programming in individuals, with implications for populations and communities. Funct. Ecol. 27, 81-92.

MacDougall-Shackleton, S. A., Dindia, L., Newman, A. E. M., Potvin, D. A., Stewart, K. A., MacDougall-Shackleton, E. A. (2009). Stress, song and survival in sparrows. Biol. Lett. 5, 746-748.

Madliger, C. L., Love, O. P. (2014). The need for a predictive, context-dependent approach to the application of stress hormones in conservation. Conserv. Biol. 28, 283-287.

Mallo, G. V., Fiedler, F., Calvo, E. L., Ortiz, E. M., Vasseur, S., Keim, V. (1997). Cloning and expression of the rat p8 cDNA, a new gene activated in pancreas during the acute phase of pancreatitis, pancreatic development, and regeneration, and which promotes cellular growth. J. Biol. Chem. 272, 32360-32369.

Marçalo, A., Mateus, L., Correia, J. H. D., Serra, P., Fryer, R., Stratoudakis, Y. (2006). Sardine (*Sardina pilchardus*) stress reactions to purse seine fishing. Mar. Biol. 149, 1509-1518.

Marcogliese, D. J., Brambilla, L. G., Gagné, F., Gendron, A. D. (2005). Joint effects of parasitism and pollution on oxidative stress biomarkers in yellow perch *Perca flavescens*. Dis. Aquat. Organ. 63, 77-84.

Marentette, J. R., Gooderham, K. L., McMaster, M. E., Ng, T., Parrott, J. L., Wilson, J. Y. (2010). Signatures of contamination in invasive round gobies (*Neogobius melanostomus*): a double strike for ecosystem health?. Ecotoxicol. Environ. Saf. 73, 1755-1764.

Martins, E. G., Hinch, S. G., Patterson, D. A., Hague, M. J., Cooke, S. J., Miller, K. M. (2012). High river temperature reduces survival of sockeye salmon (*Oncorhynchus nerka*) approaching spawning grounds and exacerbates female mortality. Can. J. Fish. Aquat. Sci. 69, 330-342.

Mazeaud, M. M., Mazeaud, F. (1981). Adrenergic responses to stress in fish. In: Pickering, A. D., *Stress and* Fish. Academic Press, London, 49-75.

Mazeaud, M. M. , Mazeaud, F. , Donaldson, E. M. (1977). Primary and secondary effects of stress in fish: some new data with a general review. Trans. Am. Fish. Soc. 106, 201-212.

McCallum, H. (2000). Population Parameters: Estimation for Ecological Models. Blackwell Publishing Ltd, Oxford.

McCallum, E. S. , Charney, R. E. , Marenette, J. R. , Young, J. A. , Koops, M. A. , Earn, D. J. (2014). Persistence of an invasive fish (*Neogobius melanostomus*) in a contaminated ecosystem. Biol. Invas. 16, 2449-2461.

McDonald, G. , Milligan, L. (1997). Ionic, osmotic and acid-base regulation in stress. In: Iwama, G. K. , Pickering, A. D. , Sumpter, J. P. , Schreck, C. B. , Fish Stress and Health in Aquaculture. Cambridge University Press, Cambridge, 119-145.

McEwen, B. S. , Wingfield, J. C. (2003). The concept of allostasis in biology and biomedicine. Horm. Behav. 43, 2-15.

Milano, D. , Lozada, M. , Zagarese, H. E. (2010). Predator-induced reaction patterns of landlocked *Galaxias maculatus* to visual and chemical cues. Aquat. Ecol. 44, 741-748.

Miller, K. M. , Schulze, A. D. , Ginther, N. , Li, S. , Patterson, D. A. , Farrell, A. P. (2009). Salmon spawning migration: metabolic shifts and environmental triggers. Comp. Biochem. Physiol. D. 4, 75-89.

Miller, K. M. , Li, S. , Kaukinen, K. H. , Ginther, N. , Hammill, E. , Curtis, J. M. (2011). Genomic signatures predict migration and spawning failure in wild Canadian salmon. Science. 331, 214-217.

Mommer, B. C. , Bell, A. M. (2014). Maternal experience with predation risk influences genome-wide embryonic gene expression in threespined sticklebacks (*Gasterosteus aculeatus*). PloS One. 9, e98564.

Mommsen, T. P. , Vijayan, M. M. , Moon, T. W. (1999). Cortisol in teleosts: dynamics, mechanisms of action, and metabolic regulation. Rev. Fish Biol. Fish. 9, 211-268.

Momoda, T. S. , Schwindt, A. R. , Feist, G. W. , Gerwick, L. , Bayne, C. J. , Schreck, C. B. (2007). Gene expression in the liver of rainbow trout, *Oncorhynchus mykiss*, during the stress response. Comp. Biochem. Physiol. D. 2, 303-315.

Monaghan, P. , Metcalfe, N. B. , Torres, R. (2009). Oxidative stress as a mediator of life history trade-offs: mechanisms, measurements and interpretation. Ecol. Lett.

12, 75-92.

Monaghan, P., Haussmann, M. F. (2006). Do telomere dynamics link lifestyle and lifespan?. Trends Ecol. Evol. 21, 47-53.

Moreno, J., Merino, S., Martinez, J., Sanz, J. J., Arriero, E. (2002). Heterophil/lymphocyte ratios and heat-shock protein levels are related to growth in nestling birds. Ecoscience. 9, 434-439.

Morrissey, M. B., Suski, C. D., Esseltine, K. R., Tufts, B. L. (2005). Incidence and physiological consequences of decompression in smallmouth bass after live-release angling tournaments. Trans. Am. Fish. Soc. 134, 1038-1047.

Moseley, P. L. (1997). Heat shock proteins and heat adaptation of the whole organism. J. Appl. Physiol. 83, 1413-1417.

Nelson, J. A., Chabot, D. (2011). General energy metabolism. In: Farrell, A. P., Academic Press, San Diego, 1566-1572.

Nelson, J. A., Gotwalt, P. S., Reidy, S. P., Webber, D. M. (2002). Beyond U_{crit}: matching swimming performance tests to the physiological ecology of the animal, including a new fish 'drag strip'. Comp. Biochem. Physiol. A. 133, 289-302.

Noga, E. J., Kerby, J. H., King, W., Aucoin, D. P., Giesbrecht, F. (1994). Quantitative comparison of the stress response of striped bass (*Morone saxatilis*) and hybrid striped bass (*Morone saxatilis* x *Morone chrysops* and *Morone saxatilis* x *Morone americana*). Am. J. Vet. Res. 55, 405-409.

O'Connor, C. M., Cooke, S. J. (2015). Ecological carryover effects complicate conservation. Ambio. 44 (6), 582-591.

O'Connor, C. M., Gilmour, K. M., Arlinghaus, R., Hasler, C. T., Philipp, D. P., Cooke, S. J. (2010). Seasonal carryover effects following the administration of cortisol to a wild teleost fish. Physiol. Biochem. Zool. 83, 950-957.

O'Connor, C. M., Gilmour, K. M., Arlinghaus, R., Matsumura, S., Suski, C. D., Philipp, D. P. (2011). The consequences of short-term cortisol elevation on individual physiology and growth rate in wild largemouth bass. Can. J. Fish. Aquat. Sci. 68, 693-705.

O'Connor, C. M., Norris, D. R., Crossin, G. T., Cooke, S. J. (2014). Biological carryover effects: linking common concepts and mechanisms in ecology and evolution. Ecosphere. 5, art28.

Odum, E. P. (1985). Trends expected in stressed ecosystems. Bioscience. 35, 419-422.

Oswald, R. L. (1978). The use of telemetry to study light synchronization with feeding and gill ventilation rates in *Salmo trutta*. J. Fish Biol. 13, 729-739.

Otto, D. M., Moon, T. W. (1996). Endogenous antioxidant systems of two teleost fish, the rainbow trout and the black bullhead, and the effect of age. Fish. Physiol. Biochem. 15, 349-358.

Øverli, Ø., Sørensen, C., Pulman, K. G., Pottinger, T. G., Korzan, W., Summers, C. H. (2007). Evolutionary background for stress-coping styles: relationships between physiological, behavioral, and cognitive traits in non-mammalian vertebrates. Neurosci. Biobehav. Rev. 31, 396-412.

Palmisano, A. N., Winton, J. R., Dickhoff, W. W. (2000). Tissue-specific induction of Hsp90 mRNA and plasma cortisol response in chinook salmon following heat shock, seawater challenge, and handling challenge. Mar. Biotechnol. 2, 329-338.

Pankhurst, N. W., Van Der Kraak, G. (1997). Effects of stress on reproduction and growth of fish. In: Iwama, G. K., Pickering, A. D., Sumpter, J. P., Schreck, C. B., . Cambridge University Press, Cambridge, 73-93.

Pickering, A. D., Pottinger, T. G., Christie, P. (1982). Recovery of the brown trout, *Salmo trutta* L., from acute handling stress: a time-course study. J. Fish. Biol. 20, 229-244.

Pine, W. E., Pollock, K. H., Hightower, J. E., Kwak, T. J., Rice, J. A. (2003). A review of tagging methods for estimating fish population size and components of mortality. Fisheries. 28, 10-23.

Plaut, I. (2001). Critical swimming speed: its ecological relevance. Comp. Biochem. Physiol. A. 131, 41-50.

Portz, D. E. (2007). Fish-Holding-Associated Stress in Sacramento River Chinook Salmon (*Oncorhynchus tshawytscha*) at South Delta Fish Salvage Operations: Effects on Plasma Constituents, Swimming Performance, and Predator Avoidance, PhD Dissertation. University of California, Davis.

Portz, D. E., Woodley, C. M., Cech, J. J. (2006). Stress-associated impacts of short-term holding on fishes. Rev. Fish Biol. Fish. 16, 125-170.

Pottinger, T. G. (2008). The stress response fish: mechanisms, effects and measurements. In: Branson, E. J., Fish Welfare. Blackwell Publishing Ltd, Oxford, 32-48.

Pottinger, T. G. (2010). A multivariate comparison of the stress response in three salmonid and three cyprinid species: evidence for inter-family differences. J. Fish Biol. 76, 601-621.

Pottinger, T. G., Carrick, T. R. (1999). Modification of the plasma cortisol response to stress in rainbow trout by selective breeding. Gen. Comp. Endo. 116, 122-132.

Pottinger, T. G., Carrick, T. R. (2000). Contrasting seasonal modulation of the stress response in male and female rainbow trout. J. Fish Biol. 56, 667-675.

Pottinger, T. G., Carrick, T. R. (2001). Stress responsiveness affects dominant-subordinate relationships in rainbow trout. Horm. Behav. 40, 419-427.

Power, M. (2002). Assessing fish population responses to stress. In: Adams, S. M., Biological Indicators of Aquatic Ecosystem Stress. American Fisheries Society, Maryland, 379-429.

Priede, I. G. (1983). Heart rate telemetry from fish in the natural environment. Comp. Biochem. Physiol. A. 76, 515-524.

Prunet, P., Cairns, M. T., Winberg, S., Pottinger, T. G. (2008). Functional genomics of stress responses in fish. Rev. Fish. Sci. 16, 157-166.

Rabben, H., Furevik, D. M. (1993). Application of heart rate transmitters in behaviour studies on Atlantic halibut (*Hippoglossus hippoglossus*). Aquac. Eng. 12, 129-140.

Raby, G. D., Donaldson, M. R., Hinch, S. G., Patterson, D. A., Lotto, A. G., Robichaud, D. (2012). Validation of reflex indicators for measuring vitality and predicting the delayed mortality of wild coho salmon bycatch released from fishing gears. J. App. Ecol. 49, 90-98.

Raby, G. D., Clark, T. D., Farrell, A. P., Patterson, D. A., Bett, N. N., Wilson, S. M. (2015). Facing the river gauntlet: understanding the effects of fisheries capture and water temperature on the physiology of coho salmon. PLoS One., e0124023.

Randall, D. J., Perry, S. F. (1992). In: Hoar, W. S., Randall, D. J., Farrell, A. P., Catecholamines. In Fish Physiology, vol. 12 Part B. Academic Press, New York, NY, 255-300.

Rankin, M. A., Burchsted, J. C. A. (1992). The cost of migration in insects. Annu. Rev. Entomol. 37, 533-559.

Reebs, S. G. (2002). Plasticity of diel and circadian activity rhythms in fishes. Rev. Fish Biol. Fish. 12, 349-371.

Reid, S. G., Bernier, N. J., Perry, S. F. (1998). The adrenergic stress response in fish: control of catecholamine storage and release. Comp. Biochem. Physiol. C. 120, 1-27.

Richter, T., von Zglinicki, T. (2007). A continuous correlation between oxidative stress and telomere shortening in fibroblasts. Exp. Gerontol. 42, 1039-1042.

Ricklefs, R. E. (2008). The evolution of senescence from a comparative perspective. Funct. Ecol. 22, 379-392.

Ricklefs, R. E., Wikelski, M. (2002). The physiology/life-history nexus. Trends Ecol. Evol. 17, 462-468.

Romero, L. M., Wikelski, M. (2001). Corticosterone levels predict survival probabilities of Galapagos marine iguanas during El Nino events. Proc. Natl. Acad. Sci. U. S. A. 98, 7366-7370.

Romero, L. M., Reed, J. M. (2005). Collecting baseline corticosterone samples in the field: is under 3 min good enough?. Comp. Biochem. Physiol. A. 140, 73-79.

Romero, L. M., Dickens, M. J., Cyr, N. E. (2009). The reactive scope model—a new model integrating homeostasis, allostasis, and stress. Horm. Behav. 55, 375-389.

Ros, A. F., Vullioud, P., Bruintjes, R., Vallat, A., Bshary, R. (2014). Intra- and interspecific challenges modulate cortisol but not androgen levels in a year-round territorial damselfish. J. Exp. Biol. 217, 1768-1774.

Ryer, C. H., Ottmar, M. L., Sturm, E. A. (2004). Behavioral impairment after escape from trawl codends may not be limited to fragile fish species. Fish Res. 66, 261-269.

Sadoul, B., Vijayan, M. M. (2016). Stress and Growth. In: Schreck, C. B., Tort, L., Farrell, A. P., Brauner, C. J., . Academic Press, San Diego, CA.

Sapolsky, R. M., Romero, L. M., Munck, A. U. (2000). How do glucocorticoids influence stress responses? Integrating permissive, suppressive, stimulatory, and preparative actions 1. Endocr. Rev. 21, 55-89.

Schreck, C. B. (2000). Accumulation and long-term effects of stress in fish. In: Moberg, G., Mench, J. A., The Biology of Animal Stress. CABI Publishing, Oxon, 147-158.

Schreck, C. B. (2010). Stress and fish reproduction: the roles of allostasis and hormesis. Gen. Comp. Endocrinol. 165, 549-556.

Schreck, C. B., Olla, B. L., Davis, M. W. (1997). Behavioral responses to stress. In: Iwama, G. K., Pickering, A. D., Sumpter, J. P., Schreck, C. B., . Cambridge University Press, Cambridge, 145-170.

Schreck, C. B., Contreras-Sanchez, W., Fitzpatrick, M. S. (2001). Effects of stress on fish reproduction, gamete quality, and progeny. Aquaculture. 197, 3-24.

Schreck, C. B., Tort, L. (2016). The Concept of Stress in Fish. In: Schreck, C. B.,

Tort, L. , Farrell, A. P. , Brauner, C. J. , . Academic Press, San Diego, CA.

Schulte, P. M. (2014). What is environmental stress? Insights from fish living in a variable environment. J. Exp. Biol. 217, 23-34.

Scott, G. R. , Sloman, K. A. (2004). The effects of environmental pollutants on complex fish behaviour: integrating behavioural and physiological indicators of toxicity. Aquat. Toxicol. 68, 369-392.

Semenza, G. L. (1998). Hypoxia-inducible factor 1: master regulator of O_2 homeostasis. Curr. Opin. Gen. Dev. 8, 588-594.

Sheriff, M. J. , Love, O. P. (2013). Determining the adaptive potential of maternal stress. Ecol. Lett. 16, 271-280.

Sheriff, M. J. , Dantzer, B. , Delehanty, B. , Palme, R. , Boonstra, R. (2011). Measuring stress in wildlife: techniques for quantifying glucocorticoids. Oecologia. 166, 869-887.

Shuter, B. J. (1990). Population-level indicators of stress. Am. Fish. Soc. Symp. 8, 145-166.

Silvestre, F. , Gillardin, V. , Dorts, J. (2012). Proteomics to assess the role of phenotypic plasticity in aquatic organisms exposed to pollution and global warming. Integr. Comp. Biol. 52, 681-694.

Sloman, K. A. , McNeil, P. L. (2012). Using physiology and behaviour to understand the responses of fish early life stages to toxicants. J. Fish Biol. 81, 2175-2198.

Sloman, K. A. , Motherwell, G. , O'Connor, K. I. , Taylor, A. C. (2000). The effect of social stress on the standard metabolic rate (SMR) of brown trout, *Salmo trutta*. Fish. Physiol. Biochem. 23, 49-53.

Somero, G. N. , Hofmann, G. E. (1996). Temperature thresholds for protein adaptation: when does temperature change start to 'hurt'?. In: Wood, C. M. , McDonald, D. G. , Cambridge University Press, Cambridge, 1-24.

Sopinka, N. M. , Marentette, J. R. , Balshine, S. (2010). Impact of contaminant exposure on resource contests in an invasive fish. Behav. Ecol. Sociobiol. 64, 1947-1958.

Sopinka, N. M. , Hinch, S. G. , Middleton, C. T. , Hills, J. A. , Patterson, D. A. (2014). Mother knows best, even when stressed? Effects of maternal exposure to a stressor on offspring performance at different life stages in a wild semelparous fish. Oecologia. 175, 493-500.

Stoot, L. J. , Cairns, N. A. , Cull, F. , Taylor, J. J. , Jeffrey, J. D. , Morin, F. (2014). Use of portable blood physiology point-of-care devices for basic and applied research on vertebrates: a review. Conserv. Physiol. 2, cou011.

Stratholt, M. L., Donaldson, E. M., Liley, N. R. (1997). Stress induced elevation of plasma cortisol in adult female coho salmon (*Oncorhynchus kisutch*), is reflected in egg cortisol content, but does not appear to affect early development. Aquaculture. 158, 141-153.

Struthers, D. P., Danylchuk, A. J., Wilson, A. D. M., Cooke, S. J. (2015). Action cameras: bringing aquatic and fisheries research into view. Fisheries. 40, 502-512.

Suski, C. D., Philipp, D. P. (2004). Factors affecting the vulnerability to angling of nesting male largemouth and smallmouth bass. Trans. Am. Fish. Soc. 133, 1100-1106.

Tomanek, L., Somero, G. N. (1999). Evolutionary and acclimation-induced variation in the heat-shock responses of congeneric marine snails (genus *Tegula*) from different thermal habitats: implications for limits of thermotolerance and biogeography. J. Exp. Biol. 202, 2925-2936.

Tort, L., Montero, D., Robaina, L., Fernández-Palacios, H., Izquierdo, M. S. (2001). Consistency of stress response to repeated handling in the gilthead sea bream *Sparus aurata* Linnaeus, 1758. Aquac. Res. 32, 593-598.

Torres, R., Velando, A. (2007). Male reproductive senescence: the price of immune-induced oxidative damage on sexual attractiveness in the blue-footed booby. J. Anim. Ecol. 76, 1161-1168.

Trushenski, J. T., Bowker, J. D., Gause, B. R., Mulligan, B. L. (2012). Chemical and electrical approaches to sedation of hybrid striped bass: induction, recovery, and physiological responses to sedation. Trans. Am. Fish. Soc. 141, 455-467.

Valavandis, A., Vlahogianni, T., Dassenakis, M., Scoullos, M. (2006). Molecular biomarkers of oxidative stress in aquatic organisms in relation to toxic environmental pollutants. Ecotoxicol. Environ. Saf. 64, 178-189.

Vamvakopoulos, N. C., Chrousos, G. P. (1994). Hormonal regulation of human corticotropin-releasing hormone gene expression: implications for the stress response and immune/inflammatory reaction. Endocr. Rev. 15, 409-420.

Venturelli, P. A., Murphy, C. A., Shuter, B. J., Johnston, T. A., van Coeverden de Groot, P. J., Boag, P. T. (2010). Maternal influences on population dynamics: evidence from an exploited freshwater fish. Ecology. 91, 2003-2012.

Vijayan, M. M., Raptis, S., Sathiyaa, R. (2003). Cortisol treatment affects glucocorticoid receptor and glucocorticoid-responsive genes in the liver of rainbow trout. Gen. Comp. Endocrinol. 132, 256-263.

Wagner, T., Congleton, J. L. (2004). Blood chemistry correlates of nutritional condition, tissue damage, and stress in migrating juvenile chinook salmon (*Oncorhynchus tshawytscha*). Can. J. Fish. Aquat. Sci. 61, 1066-1074.

Wang, C., King, W., Woods, L. C. (2004). Physiological indicators of divergent stress responsiveness in male striped bass broodstock. Aquaculture. 232, 665-678.

Webber, D. M., Boutilier, R. G., Kerr, S. R. (1998). Cardiac output as a predictor of metabolic rate in cod *Gadus morhua*. J. Exp. Biol. 201, 2779-2789.

Wedemeyer, G. A., Barton, B. A., Mcleay, D. J. (1990). Stress and acclimation. In: Schreck, C. B., Moyle, P. B., *Methods for* Fish *Biology*. American Fisheries Society, Maryland, 451-489.

Wells, R. M. G., Pankhurst, N. W. (1999). Evaluation of simple instruments for the measurement of blood glucose and lactate, and plasma protein as stress indicators in fish. J. World Aquacul. Soc. 30, 276-284.

Wendelaar-Bonga, S. (1997). The stress response in fish. Physiol. Rev. 77, 591-625.

White, A. J., Schreer, J. F., Cooke, S. J. (2008). Behavioral and physiological responses of the congeneric largemouth (*Micropterus salmoides*) and smallmouth bass (*M. dolomieu*) to various exercise and air exposure durations. Fish Res. 89, 9-16.

Wikelski, M., Cooke, S. J. (2006). Conservation physiology. Trends Ecol. Evol. 21, 38-46.

Williams, T. D., Gensberg, K., Minchin, S. D., Chipman, J. K., 2003. A DNA expression array to detect toxic stress response in European flounder (*Platichthys flesus*). Aquat. Toxicol. 65, 141-157.

Wingfield, J. C. (2013). Ecological processes and the ecology of stress: the impacts of abiotic environmental factors. Funct. Ecol. 27, 37-44.

Wingfield, J. C., Maney, D. L., Breuner, C. W., Jacobs, J. D., Lynn, S., Ramenofsky, M. (1998). Ecological bases of hormone-behaviour internactions: the ' emergency life history stage'. Am. Zool. 38, 191-206.

Wingfield, J. C., Kelley, J. P., Angelier, F. (2011). What are extreme environmental conditions and how do organisms cope with them. Curr. Zool. 57, 363-374.

Wiseman, S., Osachoff, H., Bassett, E., Malhotra, J., Bruno, J., VanAggelen, G. (2007). Gene expression pattern in the liver during recovery from an acute stressor in rainbow trout. Comp. Biochem. Physiol. D. 2, 234-244.

Wood, C. M., Turner, J. D., Graham, M. S. (1983). Why do fish die after severe exercise?. J. Fish Biol. 22, 189-201.

Woodley, C. M., Peterson, M. S. (2003). Measuring responses to simulated predation threat using behavioral and physiological metrics: the role of aquatic vegetation. Oecologia. 136, 155-160.

Woodward, J. J. (1982). Plasma catecholamines in resting rainbow trout, *Salmo gairdneri* Richardson, by high pressure liquid chromatography. J. Fish Biol. 21, 429-432.

Woodward, C. C., Strange, R. J. (1987). Physiological stress responses in wild and hatchery-reared rainbow trout. Trans. Am. Fish. Soc. 116, 574-579.

Wu, C. (1995). Heat shock transcription factors: structure and regulation. Annu. Rev. Cell Dev. Biol. 11, 441-469.

Yeom, D. H., Adams, S. M. (2007). Assessing effects of stress across levels of biological organization using an aquatic ecosystem health index. Ecotoxicol. Environ. Saf. 67, 286-295.

Zubair, S. N., Peake, S. J., Hare, J. F., Anderson, W. G. (2012). The effect of temperature and substrate on the development of the cortisol stress response in the lake sturgeon, *Acipenser fulvescens*, Rafinesque (1817). Environ. Biol. Fish. 93, 577-587.

第12章　应激管理和福利

在一系列环境下，应激对于鱼类健康和福利是一个重要的挑战。对于水产养殖、渔场管理实践、大型渔场、休闲渔业、科研和观赏鱼渔业来说，维持鱼类健康有着显而易见的益处。品质健康的鱼类可以提供更高的经济回报，有助于提高产量规模，提供完好的实验数据，这些优点极具吸引力且为大众乐于看到，不会对公共健康造成危害。然而，在鱼类所处的这些地域中，很多的生产实践通常会造成鱼类应激且其本身会影响鱼类的福利情况。对鱼类在常规过程中受到的影响进行讨论，可以更好地了解如何在人工蓄养中处理应激。如果想要减少鱼类应激和提升鱼类健康，可以通过减少鱼类在习性、发育、成长、繁殖和免疫功能上的应激及其有害的影响作为实践中的管理措施。清楚控制应激的时机或者能够对应激进行预测和准备，可以提高鱼类对于环境中所有的应激源的应对能力。然而，无法避免的、不可预测的或慢性的应激会引起鱼类失控和异型稳态负荷过大，这会导致鱼类行为异常而引发逼迫性攻击和一些定型化行为。因此，丰富的环境可以提供鱼类一些如隐藏或者改变游动方向的其他行为选择。处于这些环境中的鱼类从应激发生开始就可以根据环境信号进行应激的预测和准备，这些对于鱼类是大有裨益的。在可操作的环境中，鱼能启动自动投饵器，控制它们自己的觅食行为，这同样对鱼类福利有着积极的影响。在环境和认知正确促使的最适环境里，合适的饲养方法和社会联系对于减少人工蓄养鱼类的应激在逻辑上是重要的。在一些应激程度提高的渔业中，这些实践可以认为是一种对策。如果减少应激是重要的话，未来的研究应专注于，在不同的环境中以及不同的物种，这些因素对应激的影响。对于应激的评估和减缓，借助基于鱼类行为或环境指标而制备的稳定应激指示器和自动化警戒系统的发展，去检测鱼类的健康，将会显得十分重要。

12.1　导　言

从上一章节我们清楚知道，应激对于鱼类生物学特征有着重要影响，并且会影响到鱼类的总体健康状况。短期的急性应激可以认为是鱼类的一种适应性反应，它使鱼类个体可以对付应激状态并进一步恢复到动态平衡（Schreck，Tort，2016；见本书第1章）。所以，急性应激状态应尽量避免发生，虽然它可能并不会对鱼类的福利产生重要性损害。然而，长期慢性应激对于鱼类的生长（Sadoul，Vijayan，2016；见本书第5章）、免疫系统（Yada，Tort，2016；见本书第10章）、繁殖产量（Pankhurst，2016；见本书第8章）有着重要的影响，它会对水产养殖、渔场、休闲渔业和以鱼作为实验模型的科研工作造成消极的后果（Spagnoli 等，2016；见本书第13章）

(Davis，2006)。为了鱼类管理者和使用者的利益，我们要尽量维持鱼类的健康并使其远离应激。这样便可以使鱼维持良好的生物性特征，生长到达市场规格，繁殖效率高，并使水产养殖和渔场可以获利，亦可以把鱼类释放到野外，它们同样表现良好和外形美观，在休闲渔业中有着正常的行为。除此之外，非应激鱼类还可以提供稳定真实的实验数据以提高实验的科学有效性。本章将论述应激对于鱼类的健康和福利的影响，如何在饲养群体和适应性策略上减缓应激，提高鱼类福利，最后讨论我们知识体系的不足之处，用以论证未来科学研究的主要领域。

12.1.1　福利的定义

根据 Volpato 等（2007）的观点，福利是动物感觉良好时的一种状态，这些作者还质疑了如低福利等词组的正确性，因为这些词组存在表面矛盾的关系。除了语义外，大量的福利定义被提出，尽管该课题高度复杂，且对于动物福利应该如何去定义或测量，外界并没有清楚一致的看法。与陆地动物相比，鱼类的动物习性和生理学比较复杂，其福利测量也表现出较高的难度。不过，无论福利的辨认和测量如何困难（表 12.1；Martins 等，2012；Sneddon，2011），鱼类福利仍处于可以科学调查的范围内。某些福利计量的方法已经被提出（Stamp Dawkins，2012），尤其是基于感觉、基于功能以及与自然生活的野生动物比较法这三种方法论，已经广泛发展起来。从哲学观点到与经验性计量生物学因素相关的方面，这些方法的中心重点变化很大，有些作者则综合了多种方法。但在广义上，福利可以定义为动物是否感觉良好或者动物是否拥有它所需要的一切（Stamp Dawkins，2012）。

表 12.1　鱼类应对程序或应激源时产生的行为改变

（此表可用作养殖鱼类的物种特定福利指标，显示其积极和消极的福利后果）

福利状态	措施	程序/应激源	物种	福利指标	是否影响野生鱼类
积极的	探索性活动	饲养密度	虹鳟（*Oncorhynchus mykiss*）	饲养密度低：使用自动投喂机↑	不太可能影响自然种群。如果由于竞争、杂交和引进鱼类的捕食而增加大量养殖鱼类用于补充库存，可能会影响本地物种
	食物预期活动	投喂方法	金头鲷（*Sparus aurata*）	按时投喂（vs 随机投喂）：食物预期活动↑	除非本地食物供应受到人为因素的影响，否则不太可能，例如污染或过度捕捞造成的死亡

续表 12.1

福利状态	措施	程序/应激源	物种	福利指标	是否影响野生鱼类
消极的	攻击	投喂方法	非洲鲶鱼（*Clarias gariepinus*）	手动投喂（vs 自动投喂）：被咬鱼类百分比↑	不太可能
			大西洋鲑鱼（*Salmo salar*）	定量投喂（vs 按需投喂）：攻击性行为↑	
			大西洋鲑鱼（*S. salar*）	投喂不足时（vs 饱食状态）：攻击性行为↑	
		光周期和光强度	非洲鲶鱼（*C. gariepinus*）	光周期较长以及光强度较高：皮肤损伤↑	可能受到城市地区人为照明的影响
		饲养密度	非洲鲶鱼（*C. gariepinus*）	低放养密度：皮肤损伤↑	不太可能影响自然种群。如果由于竞争、杂交和引进鱼类的捕食而增加大量养殖鱼类用于补充库存，可能会影响本地物种
			虹鳟（*O. mykiss*）	高放养密度：攻击性行为↑	
		分级标粗	异带重牙鲷（*Diplodus sargus*）	高饲养密度：社交行为↑	不适用
			非洲鲶鱼（*C. gariepinus*）	均质的小鱼群：被咬鱼类↑	
	觅食行为	投喂方法	庸鲽（*Hippoglossus hippoglossus*）	浮性饲料（vs 沉料）：食用的颗粒饲料数量↑	除非本地食物供应受到人为因素的影响，否则不太可能；例如污染或过度捕捞造成的死亡
		RAS 中的水交换率	尼罗罗非鱼（*Oreochromis niloticus*）	水交换率高：小规格投喂时潜伏↓	不适用
		分级标粗	非洲鲶鱼（*C. gariepinus*）	均质的小鱼群：总投喂时间及摄食率↓	不适用
		禁食期	舌齿鲈（*Dicentrarchus labrax*）	停止喂食后的摄食率和日喂食次数↑	不适用

续表 12.1

福利状态	措施	程序/应激源	物种	福利指标	是否影响野生鱼类
		清洁程序	舌齿鲈（*Dicentrarchus labrax*）	自动进食↓	不适用
	游泳活动	低氧	大西洋鳕鱼（*Gadus morhua*）	游泳速度↓	水体温度升高，溶解氧减少，导致藻类大量繁殖，造成污染
			高首鲟（*Acipenser transmontanus*）	游泳速度↓	
			美洲红点鲑（*Salvelinus fontinalis*）	游泳速度↑	
			红大麻哈鱼（*Oncorhynchus nerka*）	恢复测试↓	
			尼罗罗非鱼（*O. niloticus*）	成群游泳活动↑和↓	
		高氧/过饱和	大西洋鲑鱼（*S. salar*）	游泳速度↓	可以自然发生，本地种群已经采取适应策略来应对这种情况。在突发事件中，会给野生鱼类带来风险
		污染物	红大麻哈鱼（*Oncorhynchus nerka*）	脱氢枞酸：恢复测试↓	由于人工排放水体、工业/农业用地的淋溶和水体流出，导致风险增加
			鳟鱼（*Salmo trutta*）	铜元素：恢复测试↓	
			幼年吸口鱼（*Erimyzon sucetta*）	灰分：U_{crit}↓	
		感染	越洋公鱼（*Hypomezus transpacificus*）	U_{crit}↓	自然发生的感染会因环境和生物因素干扰而恶化，这会造成应激
		寄生虫	大西洋鲑鱼（*S. salar*）	U_{crit}↓	自然发生的感染会因环境和生物因素干扰而恶化，这会造成应激

续表 12.1

福利状态	措施	程序/应激源	物种	福利指标	是否影响野生鱼类
		温度	大口黑鲈（*Micropterns salmoides*）	低温时：U_{crit} ↓	水体温度升高，溶解氧减少。低于冰点的严寒温度下，温度降低意味着冰下水体会缺氧
		运输	虹鳟（*O. mykiss*）	EMG ↑	如果用于渔业管理或科学研究，只会对野生捕捞的鱼类构成风险；如运输到实验室的鱼类
		待喂食状态	大西洋鳕鱼（*G. morhua*）	游泳速度 ↑	除非本地食物供应受到人为因素的影响，否则不太可能；例如污染造成死亡或对捕获物过度捕捞
			虹鳟（*O. mykiss*）	EMG ↑	
			金头鲷（*S. aurata*）	游泳速度 ↑ 学习操纵复杂性 ↑	
			大菱鲆（*Scophthalmus maximus*）	游泳速度 ↑	
		网箱短期和长期浸于水中	大西洋鲑鱼（*S. salar*）	游泳速度 ↑	不适用
		光周期和光强度	非洲鲶鱼（*C. gariepinus*）	光周期较长以及光强度较高：游泳活动 ↑	可能受到城市地区人为照明的影响
		光周期	非洲鲶鱼（*C. gariepinus*）	光周期较长：游泳活动 ↑	可能受到城市地区人为照明的影响
		饲养密度	非洲鲶鱼（*C. gariepinus*）	高饲养密度：游泳活动 ↑	不太可能影响自然种群。如果由于竞争、杂交和引进鱼类的捕食而增加大量养殖鱼类用于补充库存，可能会影响本地物种

续表 12.1

福利状态	措施	程序/应激源	物种	福利指标	是否影响野生鱼类
			舌齿鲈 (*D. labrax*)	高饲养密度：游泳活动↓	
			舌齿鲈 (*D. labrax*)	高饲养密度：通过肌电图（EMG）测量的肌肉活动↑	
			庸鲽 (*H. hippoglossus*)	高饲养密度：游泳活动↑	
	环境梯度		大西洋鲑鱼 (*S. salar*)	空间利用及其不调和的环境发生改变	野生鱼类对梯度的自然适应
	投喂方法		非洲鲶鱼 (*C. gariepinus*)	手工投喂（vs 自动投喂）：游泳活动↑	不适用
			大西洋鲑鱼 (*S. salar*)	定量投喂（vs 按需投喂）：游泳速度和转向角度↑	
	通气活动	饲养密度	非洲鲶鱼 (*C. gariepinus*)	高饲养密度：空气呼吸↑	不太可能影响自然种群。如果由于竞争、杂交和引进鱼类的捕食而增加大量养殖鱼类用于补充库存，可能会影响本地物种
		低氧	尼罗罗非鱼 (*O. niloticus*)	通气频率↑	可以自然发生，本地种群已经采取适应策略来应对这种情况。在突发事件中，这会给野生鱼类带来风险
		监禁	尼罗罗非鱼 (*O. niloticus*)	通气频率↑	发生在使用渔网进行捕捞后释放的钓鱼期间或渔业管理实践/科学研究期间。也可能是人为建造的结构如水坝建成时

续表 12.1

福利状态	措施	程序/应激源	物种	福利指标	是否影响野生鱼类
		手工操作	虹鳟（*O. mykiss*）	通气频率↑	发生在使用渔网进行捕捞后释放的钓鱼期间或渔业管理实践/科学研究期间
	定型化行为及异常行为	饲养密度	非洲鲶鱼（*C. gariepinus*）	高饲养密度：企图逃跑↑	不太可能影响自然种群。如果由于竞争、杂交和引进鱼类的捕食而增加大量养殖鱼类用于补充库存，可能会影响本地物种
			庸鲽（*H. hippoglossus*）	高饲养密度：垂直方向循环游泳↑；游于水体表面↑	
		饲料特性	庸鲽（*H. hippoglossus*）	浮性颗粒饲料对（沉性料）：垂直方向循环游泳↑	除非本地食物供应受到人为因素的影响，否则不太可能；例如污染造成的死亡或对捕获物过度捕捞

资料来源：改编自 Martins C I M，Galhardo L，Noble C，Damsgard B，Spedicato M T，Zupa W，等(2012). Behavioural indicators of welfare in fanned fish. Fish Physiol Biochem，38，17-41，已获 Springer 许可；（↑表示测量参数的增加，↓表示测量参数的减少；U_{crit}表示临界游泳速度）。如 Sneddon、Thomson 和 Wolfenden 在最后一栏中所示，这些程序/应激因素也可能影响野生鱼类，但这些措施在野生鱼类中还有待检验。

使用动物主观经历的定义通常都称为感觉基础定义（feelings-based），感觉基础定义的前提是动物应该感觉良好，以便应用良好的福利。但是，这也同样意味着这些个体可以经历如痛苦、恐惧以及应激等不利的福利状态（Fraser，2008）。很显然，评估任何动物内在的主观状态是很困难的，但是这些状态和选择实验有着密不可分的关系，在选择实验中，动物会自行挑选一种有价值的或者可以改善它们状态的资源（Flecknell 等，1999），或者动物必须付出代价以进入一个有价值的资源中（Mason 等，2001）。在这时，行为可以通过评估以作为动物情感状态的指标，比如，在动物遇到危险的时候，动物会表现出恐惧（Fendt，Fenselow，1999）。对鱼类的研究已证实，当面对恐吓刺激时，鱼类会出现一致的反应，如躲避新奇事物，表现冷淡而不突出；产生逃离行为；会游到远离开放或中心区域的池边；进行下沉；快速游动并且警惕被捕食（Maximino 等，2010）。对于斑马鱼，其支配恐惧反应的大脑结构和哺乳动物相似（如缰区）（Agetsuma 等，2010）。除此之外，抗焦虑药物也可以减少斑马鱼的恐惧反应（Grossman 等，2011）。虽然这些方法会对动物如何感觉做出很好的解释，但感觉基础还是有很多问题。因为其高度主观性以及观察者偏见或者拟人为的解释而备受质疑，虽然实验结果可能会显示福利缺失（Dawkins，1980），但我们经常很难去解释由感觉产生的行为问题。感觉基础定义的福利价值在外界存在很多意见。比如，一些科学家认为要知道动物如何感觉根本上是不可能的（如 Mc Glone，1993），而对于动物福利的评估，感觉是没有作用的。这个观点认为，评估动物福利状态应该完全依靠如生长率和繁殖情况的生理学标准。不过，也有的学者认为基于感觉基础判断的动物福利观点在任何讨论中都处于主流位置（如 Duncan，1993，1996）。

对动物福利评估的功能基础（function-based）和动物应对环境的能力相关。它在很多方面可以测量，比如成长率、发病率、繁殖效率以及行为性指标，还有一些如应激激素的生理学因素。这些参数可以提供稳定的福利实验测量结果，为动物的状态提供确实的证据。前提是如果动物健康，该动物应成长良好，有着健康的免疫系统，并且没有疾病，繁殖活跃，具有少的应激指标。如此，福利的生物学功能定义就可以提供福利评估的实际性测量结果。比如，Sneddon 等（2003）的一项研究调查了给虹鳟的嘴唇注射乙酸和蜂毒后的反应，该虹鳟经历了这种潜在的痛苦刺激后，出现了一系列异常行为，包括胸鳍不断摇晃，在水族箱底部不断摩擦感染区域，呼吸速率加快（Sneddon 等，2003），以及皮质醇含量增加（Ashley 等，2009）。当注射吗啡和止痛药之后，异常行为开始减少（Sneddon，2003a）。然而，尽管基于功能的福利测量方法可以提供客观的标准，但也有人反对，比如，Dawkins（1980）认为它可能有着一定的局限性。举个例子，一尾单独隔离的浅水鱼可能十分健康并且也符合所有功能基础下的福利标准，但是如果其同物种并不是这样以及它并不在浅水中，那该鱼的福利在实际上便有所减少。应激指标的改变通常都是一些稳态的挑战，但一些急性应激反应可以帮助鱼类应对一种新的威胁。因此，在上一个案例中，皮质醇含量升高只是一种自然的应对机制，而不一定意味着较低的福利状态。相反，慢性应激确实会损害免疫功能，导致产量下降，减少生长率，增加死亡率。这样，在评估福利时，功能基础确实需要谨慎地使用（Keeling，Jensen，2009）。

定量评估福利的自然生命法是一种将养殖动物与其对应的同种野生动物的行为进行直接对比的方法，由于是和养殖动物相比，野生动物生存环境多样（如食物获取能力、捕食者以及环境压力），这会增加比较的难度。福利的自然生命法的前提是当动物处于较好福利状态时，该动物的行为会和自然界中的行为相似，我们也应当提供动物所需的一切自然资源或人工制造的相类似资源，使得动物可以展现它们所有的行为天性。这样，动物便可以自然成长，也可以展现出它们的野生同伴或祖先所表现的自然行为（Rollin，1993）。然而，对于蓄养动物来说，由于有着食物的提供，其正常的进食行为如觅食的欲望会减少，所以将蓄养动物行为与野生动物行为进行比较时需谨慎对待。生产系统的条件也会影响一些行为发生的欲望。比如，成年的大西洋鲑鱼（*Salmo salar*）为了寻找食物会在海洋里迁徙数千里，而饲养的大西洋鲑鱼却能在相对小的海水网箱中享受充足的食物。这是否消除了这些鱼类迁徙的欲望（或减少了其迁徙的机会），或者，这是否解释了为什么蓄养动物要不断在海水网箱里循环游动？

无论福利如何定义，从道德、伦理和法律方面来看，使动物得到好的福利都是很重要的，同时好的福利还会有利于渔业、科研，维护生物多样性，保护野生动物数量。许多学者建议应该简化诸如动物是否健康或感觉良好以及它是否拥有所需要的一切这样的定义（Curtis，1985；Stamp Dawkins，2012），动物的需要可以根据其生理学需要、安全需要、行为需要进行全面考察（Curtis，1985）。生理学需要包括提供合适的营养、环境条件以及保证动物健康；而安全需要是包括我们要使动物远离不利的天气环境、捕食者以及装备和设施发生的事故；行为需要包括使动物远离虐待（如滥用、无视、剥夺其权利），并且为动物提供所有外部刺激和资源，使它们能展现其自然行为天性。比如，在生殖季节中，雄性三刺鱼（*Gasterosteus aculeatus*）会变得极具领土意识，并开始筑巢（Mayer，Pall，2007）。因此，对于饲养的鱼类，我们需要考虑为雄鱼（雄性标志为具有红色的咽部）提供繁殖条件，使它们拥有足够的筑巢空间和筑巢材料去适应其行为天性，从而减少应激产生。然而，这样的适应是种族特异性的。例如水产养殖中和实验条件下的鲑鱼类具有攻击性，并形成优势社会等级（如 Sneddon 等，2011）。好斗是鲑鱼类天性的一部分，这会导致从属地位种群受伤和产生长期的应激，而优势地位种群生长旺盛。这里我们不禁产生疑问，我们是否应该允许鲑鱼具有攻击性，从而满足它们的天性；或者，我们是否应该试图减少这种攻击行为，因为这损害了从属地位种群的福利。这个问题很难回答，水产养殖中的做法一般是定期将鱼类分级，分开个体较大的鱼类。因此，对于鱼类利益以及食用鱼类的人们来说，最低限度的减少或减缓应激，确保动物福利是百利而无一害的。如果我们想要动物拥有一个理想的生活条件的话，为它们提供合适的栖息地点和环境条件是特别有利的。

12.2 鱼类的应激管理

对于从野外引入蓄养环境的鱼类或那些被蓄养并准备释放到自然环境中的鱼类，它们所面临的条件差异可能会导致生存和应激管理方面的问题。一般来说，蓄养环境可能会为鱼类提供大量的益处，比如食物经常是充足的（Garner 等，2010；尽管在某些时候，比如观赏水族馆的主人会忘记饲养鱼类，导致食物供应效率低，Gronquist，Berges，2013）。另外，在蓄养环境中，鱼类没有天敌，有遮蔽处，同伴的存在使得繁殖成功率得到保障。与野生环境相反，鱼类或饲养在高密度环境下（如集约化养殖），或饲养在低密度环境下（如鱼类个体被训练，远离其他鱼类），在空间上经常严格限制（蓄养环境结构十分简单，很少甚至没有其他多余的东西），食物质量均匀，通常为人工制造的饲料。当食物投喂方法不当时，会引起个体中的争夺，并且鱼类会经常被干扰以及手工操作处理（Benhaïm 等，2013；Garner 等，2010；

Huntingford, 2004；Kulczykowska，Vázquez，2010；Newman 等，2015）。以上这些挑战会引起生理的应激，尤其是当 HPI 轴（下丘脑-垂体-肾间组织轴）活动增加时。

应激是动物生理学上的一种基础的、适应性反应，它可以引起一系列的生理性和行为性反应，这些反应可以改善恶劣环境对于鱼类的影响。应激的初始反应是警觉性增加，随之而来的是冷冻的应激反应，或者是应急反应（Galhardo，Oliveira，2009），而这些反应都伴随着长期的 HPI 活动，从而帮助机体恢复到稳态（Pankhurst，2011；Schreck，Tort，2016；见本书第 1 章）。应激的量度可以通过测量 HPI 轴活动的一些指标（如血浆皮质醇）进行判断。然而，由于除了极少数的商业性重要物种或实验重要物种外，大部分物种的血浆皮质醇标准浓度和应激后的浓度鲜为人知，加上种间和种内不同个体的浓度可能变化较大（Pankhurst，2011），因此，应激源对于个体的生理学相关影响和不同物种间 HPI 活动的变化很难进行确定，而且不同物种间对于相同的应激源反应后的皮质醇浓度会变化达百倍之多（Barton，2002；Clearwater，Pankhurst，1997）。应激反应也会随取样方法而发生变化，特别是因为鱼类对各种取样过程产生的各种干扰比较敏感（Clearwater，Pankhurst，1997；Ellis 等，2012；Pankhurst，2011）。而且，在蓄养环境中，应激源往往是慢性作用或反复起作用（Huntingford 等，2006）。慢性应激对鱼类的 HPI 反应性有不良影响，会导致 HPI 轴对接下来的应激源反应迟缓以及完全脱敏（如 Huntingford 等，2006；Pankhurst，2011；Zuberi 等，2014）。慢性应激的行为标志包括游泳模式的改变、对捕食者的异常反应、进食模式异常以及社会约束现象（Schreck 等，1997）。

许多鱼类刚进入蓄养环境时，往往由于不能适应新环境而死亡（参见本章第 12.3.4 节），所以蓄养环境具有挑选能够在这些环境中存活的鱼类的功能（Huntingford，2004；Pankhurst，1998；Zuberi 等，2014）。比如，与野生的虹银汉鱼（*Melanoteania duboulayi*）相比，饲养的虹银汉鱼对被监禁的应激以及和人工模仿的捕食者相互追逐的生理反应是减弱的（Zuberi 等，2011，2014）。引起上述两种情况生理学变化的起因是饲养鱼类一直生活在一个重复应激源的环境中，并且鱼类对于所有这样的应激的生理反应都是不良适应的。饲养鱼类与野生鱼类这样的差异被认为是环境挑战与其长期的挑选压力共同导致鱼类个体发生脱敏反应所致（Zuberi 等，2011）。因此，对于那些被挑选的野生鱼类，饲养环境可以促进它们产生不同的应对机制（Huntingford，2004；LePage 等，2000；Pampoulie 等，2006）。从野生环境进入饲养环境的鱼类发生的行为变化和生理学变化往往也仅在第一代表现（Huntingford，2004；Zuberi 等，2011），这样的适应性对鱼类存活于饲养环境是有益的，并且帮助鱼类面对一系列新的挑战时能产生合适的生理反应。

通过测量 HPI 活性的应激生理学，会表现出一种可遗传的特性（虽然可遗传性在不同物种间变化广泛）（Øverli 等，2005）。同样的，一些鱼类种群对于蓄养环境的适应能力会表现为增强的应激耐受力（Pankhurst，1998；Douxfils 等，2011），尤其是面对一些普遍的应激源如手工操作和监禁时（Cleary 等，2000；Douxfils 等，2011）。不仅如此，即使慢性应激与鱼类免疫抑制相关（Ashley，2007；Huntingford 等，

2006)，但与上一个世代相比，一些蓄养鱼类种群的选育使得鱼类后代提高了它们的免疫功能（先天免疫和后天免疫都有）(Douxfils 等，2011)。虽然应激相关的免疫抑制和皮质醇息息相关，但它也有可能被儿茶酚胺、内源性阿片类、垂体激素以及羟色胺等一些其他的激素所调节（Douxfils 等，2011，以及其中的参考文献）(Gorissen，Flik，2016；见本书第 3 章）(Yada，Tort，2016；见本书第 10 章)。

繁殖方式的变化同样也会影响蓄养鱼类和野生鱼类的遗传性健康情况。在野生环境中，鱼类多产是不可确定以及缺乏管理的，而且其雄雌配对也并非随机的（Garner 等，2010)。相反地，蓄养鱼类往往都由数量很少的亲鱼所产生，这会产生显著的建立者效应（founder effect)，限制鱼类基因的多样性，也会出现更高的近亲繁殖。这一系列做法会导致遗传多样性和基因丰富度的降低，出现遗传漂变，并且与野生鱼类相比，蓄养鱼类的杂合度降低（Douxfils 等，2011；Doyle 等，2001；Garner 等，2010)。这种遗传性变异的丧失在理论上可以导致种群适应自然环境的能力减退，尤其当环境并不稳定的时候（Pampoulie 等，2006；Douxfils 等，2011)。

繁殖期间的应激同样对后代健康有着深远的影响，虽然从表面上看，和野外捕捉和网捕的鱼类相比，这种负面影响仅会减少孵化培育的鱼（Pankhurst，2016；见本书第 8 章)。应激可通过减少鱼类血浆中的性类固醇激素来影响产卵雌鱼的繁殖数量（如 17β-雌二醇和睾酮）(Cleary 等，2000；Clearwater，Pankhurst，1997)，还会引起畸形率增加、卵大小减小、延迟排卵、减少胚胎存活率和生长率等诸多影响（Campbell 等，1992)。另外，受到应激的母本亲鱼的胚胎中含有较高的从卵腔液中转移过来的皮质醇，这也会导致幼鱼生长率和存活率下降、生长不健康的问题，还会影响到后代成年后的行为特征（Barton，2002；Sloman，2010)。然而，一些负面影响对于野生鱼类和蓄养鱼类具有显著差异，比如对应激所产生的遗传性和经验性耐受力，两种鱼类是不同的（Cleary 等，2000；Pankhurst，2011)。

不仅如此，饲养环境可能会直接影响鱼类身体成长的变化，比如大脑的大小和结构。鱼类早期发育阶段的生长环境会起动发育的“轨道”，从而使其成年后在体质上、生理上以及行为上产生差异（Kihslinger，Nevitt，2006)。这或许归因于仔鱼或幼鱼阶段的栖息地利用，更加复杂的生长环境会引起一些复杂的行为，这些行为会促使一些特殊的和这些行为相关的脑部区域如端脑、视顶盖的发育（Marchetti，Nevitt，2003)。有一些研究一直专注于端脑的神经元增殖问题，并且观察到饲养在简单环境与复杂环境中鱼类脑部的差异（Ebbesson，Braithwaite，2012)。还有一些研究发现，相比较于一个有待完善的环境，在适宜的生长环境中，神经的可塑性能够提高三刺鱼（*Gasterosteus aculeatus*）的导航能力（Salvanes 等，2013)。因此，复杂的生境可能会对脑部发育产生深刻的影响，也会对其相关的行为影响甚大。

12.2.1 野生鱼类在蓄养中照料的注意事项

将野生鱼类进行蓄养或将野生种群和蓄养种群一起蓄养时，我们都需要保证在不同环境中，鱼类福利应该到达最高标准。无论是从野生到蓄养还是相反，环境的改变势必会带来应激的问题，比如，和野生的鱼类相比，孵化培育的鱼类放养到天然生境中，其存活率十分低（McNeil，1991），这和他们缺乏捕食和避敌的经验有关（Brown 等，2003）。同样的，野生鱼类进入蓄养环境时也会呈现高死亡率。因此，对于鱼类，必须采取措施使它们能够适应新的环境。这既包括了畜牧业的一般做法，也必须注意到两种环境中潜在应激源的差异，以及以下两点：①如果鱼类将要再放养而面对自然界中一些挑战时，如何能够维持合适的生理的和行为反应；②当鱼类要进行蓄养时，如何能够适应高度管理化和应激的环境。

野生鱼类的进食可能包括连续进食和被强制禁食期分隔的间歇进食。然而，进食的周期和次数主要会随着鱼类的感觉器官不同而变化（Kulczykowska，Vázquez，2010）。鱼类进食可以分为夜间性和白天性，有些鱼类还能够在这两者中转变，如舌齿鲈（*Dicentrarcus labrax*）可以快速地改变它的进食时间以适应环境光周期（Sánchez-Vázquez 等，1995）。另外，当日间进食温度高于 10℃时，鲑鱼也可以根据环境温度改变它的进食周期（Fraser 等，1993）。但在蓄养环境中，鱼类进食被控制在一天内定时投喂。鱼类适应这种管理规则的能力也是因种族而异。一些鱼类不能调整其天生的节律（如丁鲥 *Tinca tinca*）仅在夜间进食（Herrero 等，2005），尖吻重牙鲷（*Diplodus puntazzo*）仅在白天进食（Vera 等，2006），并且其节律的破坏会导致鱼类的应激感。另外，如同在自然环境中食物来源无法确定的随机投喂，也会引起鱼类应激感。曾在以随机投喂的金头鲷（*Sparus aurata*）中检测到较高含量的皮质醇和葡萄糖，并且其反应剧烈、极具警惕性攻击。这些行为最终会导致鱼类生长缓慢。定时投喂会导致鱼类越来越警惕，而且在喂食时间来临时具有侵略性。相比于随机投喂，有些鱼类经过定时投喂后会减少应激（Kulczykowska，Vázquez，2010），但是，对于食物的争抢会给鱼类带来很大压力，并在争斗中受伤，而食物大部分都分给了体型较大、具有优势地位的鱼类。

假如把鱼类放养至野生环境中，可以预期的是会限制鱼类搜寻、觅食、捕获食物的能力，潜在性降低生长速率和肥满度（Brown 等，2003；Garner 等，2010；尽管水产养殖品种经常被选育出来提高其生长速率，但可能会导致相反效果，如 Zuberi 等，2011）。这些反应可以通过快速的后天学习而再次获得，以及在饲料中补充天然活饵料可能会唤起它们的觅食天性（Brown 等，2003；Sneddon，2003b）。一些孵化培育的鱼会比较冒失（面对新事物时警惕感降低）以及缺乏经验，在野外环境时一旦新奇的食物出现，会快速接近（Huntingford，2004），即使它们无法判断食物是否适合自己（Brown 等，2003）。对于这些鱼类，食料物质丰富是其最佳的生活环境，这有助于它们进行合理觅食以及增长区分食物的能力。孵化培育的鱼类如此冒失的另一个潜在原因可能是它们具有觅食时不顾捕食者危险的天性（详见后面的章节）。

清楚了解鱼类个体饲养时的需求对减少应激是很重要的，解决上述诸如定时投喂和饲养体系差异问题的办法之一，是制造一个控制自动投喂饲料机器的开关，鱼类可以根据自我需求打开投喂开关，这种方法可避免鱼类争斗，弱小鱼类也可以正常获取食物，或在进食时减少竞争（Huntingford，Kadri，2014）。

鱼类也会随它们个体需求不同而改变其饮食习惯，这种改变或许只是暂时的，并且具有种族特异性（就食物中的营养含量或者食物的形状和数量而言）（Huntingford，2004），其取食习惯由环境变化以及鱼的生长阶段决定（Kulczykowska，Vázquez，2010）。然而对于一些特殊品种鱼类的食物要求，它们如何利用食物中特殊的营养因子，以及这些营养因子之间如何相互影响，这些问题还是未知的（Oliva-Teles，2012）。对蓄养鱼类往往投喂先前按照配方制造的、统一的单一饮食，因此具有无法满足鱼类个体饮食需求的风险（Oliva-Teles，2012），并且这种饲料一般都因味道较差而不受鱼类喜爱（Huntingford，Kadri，2014）。对于鱼类的不同品种，这是必须要考虑到的问题，为了满足野生鱼类不同品种间的营养需求，还需做进一步的研究。

蓄养鱼类被认为是缺乏经验的，尤其是在对抗捕食者这一方面，它们反应缓慢或者不能察觉危险。与生长在野外环境的鱼类相比，蓄养鱼类被放养到野外环境，在觅食时更具风险，并且往往会被捕食（如 Benhaïm 等，2012；Garner 等，2010；Zuberi 等，2011）。在某种程度上是因为这些鱼类无法辨认自然界中的捕食者（Zuberi 等，2011），尽管一些人工驯化的鱼类依然还保留通过嗅觉辨认其天敌的能力（Huntingford，2004）。另外，自然界中的食物特性往往要求鱼类浮出水面进行捕获，这也增加了被鸟类吃掉的风险（Maynard 等，2001）。

对于那些孵化培育长大的鱼类，其被捕食的问题可通过在环境中提供嗅觉刺激来进行改善（如捕食者气味结合同种的警觉信号），继而增加鱼类与捕食者近距离接触的经验，或者饲养一些能对捕食者进行合适反应的鱼类（Huntingford，2004）。此外，一些人工驯养的鲑鱼类还保留着应对捕食者的反应，这对于再放养鱼类或将要放养的配养鱼类，是值得特别注意的（Benhaïm 等，2012）。

蓄养的环境，尤其在养殖水池中以及工厂化养殖中，往往以高密度养殖为主，这会引起鱼类应激并且对动物福利有着负面影响（Ashley，2007）。然而，鱼类是否需要社交取决于它们的品种：一些鱼类喜欢单独生活，而另一些偏爱群体生活，孵育环境或蓄养环境的单一性可能会抑制其社交需求（Garner 等，2010）。但即使对于群体生活的鱼类，水族箱式的生活环境也会给予它们一定的压力。在一些特定的训练中，单独生活会作为饲养实践的一部分，这往往也会造成应激，比如虹银汉鱼（*melannoteania duboulayi*），还未习惯这种生活状态的野生虹银汉鱼会表现出反应过激（Zuberi 等，2014）。在这种环境中，攻击也变得习以为常，尤其是在一些高密度以及低空间的围栏养殖中。一些品种会发展成优势-从属的关系或者等级制度［如大西洋鲑鱼（*Salmo salar*）、虹鳟（*Oncorhynchus mykiss*）］，在这种环境下，从属等级的鱼类会表现出 HPI 轴和羟色胺慢性增加，而下丘脑-脑垂体-性腺（HPG）轴活性下调（Cubitt 等，2008），而这些活动与动物福利息息相关。再加上鱼类饲料投喂的空间和时间都可预测，这些关系会导致竞争性行为、受伤，而鱼类之间不同的饲料摄取量，增加了不同个体间生长速率和大小的差异（如 Cubitt 等，2008；Montero 等，2009）。

蓄养鱼类和野生鱼类各自的攻击性是不同的，对于那些依赖于饲养环境以及饲料投喂方法的蓄养鱼类，可以通过增加或降低鱼类间的攻击性来挑选鱼类

（Huntingford，2004）。攻击性增加可以通过增加食物竞争以及增加冒失行为（与攻击联系十分紧密）来引起，而鱼类个体却无法表达退出竞争的决定（Huntingford，2004）。有些背景会影响斗争的结果，尤其是野生鱼类可利用其经验如领土防御等去赢过没有这些经验的孵化培育鱼类（Huntingford，2004）。

解决上述问题可能会很难，尤其是那些只关心鱼类高产的养殖产业。养殖密度减少以及养殖空间增大可能会起到改善作用，但减少攻击性的推荐密度在不同的研究中并不一致（Gronquist，Berges，2013），这表明影响攻击性的其他因素也很重要。为动物提供遮蔽处也可以减少其互相的攻击，同时对于围栏养殖来说，其空间已被充分利用（Huntingford，Kadri，2014）。

生境复杂性对许多蓄养环境来说都是一个问题，尤其在实验室环境中，减少外源性刺激是必要的。家庭水族箱以及公共水族箱通常会用一些装饰物来丰富养殖环境，但养殖工厂和孵化场往往缺乏这些。鱼类栖息地复杂度不够对生活在这些环境中的鱼类具有深远的影响，这会限制鱼类对于刺激而产生的生理的和行为的正常反应（Brydges，Braithwaite，2009）。

环境丰富度为鱼类提供了社交的和物理的刺激，而蓄养环境往往没有这些，它至少降低了环境上的单调，从而减少了应激的产生。环境丰富度对于鱼类的性格行为以及认知能力可产生较大的正面作用，这会导致动物的行为和野生动物更加相似（Brown 等，2003；Huntingford，2004；Brydges，Braithwaite，2009）。正如先前所述，环境结构复杂性可通过鱼类行为之间的交流而提高幼鱼的发育程度（Kihslinger，Nevitt，2006；Marchetti，Nevitt，2003）。甚至于增加简单的物品如用灯光照明以突出空间可以减少鱼类的拥挤感觉以及因密度过高引起的应激（Huntingford，Kadri，2014）。环境丰富度的有利之处看似已经清楚，但在构建环境结构和首选底质时我们也必须考虑到种族特异性的需求问题。

尽管鱼类运输对野生鱼类和蓄养鱼类的表型差异并不明确相关，但运输也可以使鱼类产生应激反应。鱼类运输过程包含了较多具侵略性的步骤，如手工捞取、卸货以及运输本身，并且这样的动作引起的反应会持续一段时间（Ashley，2007；Olla 等，1998）。如果鱼类不能迅速恢复过来，它们将无法适应即将到来的新环境（如野生鱼类会增加被捕食的风险）（Olla 等，1998）。在运输相关的应激达到临界点时，鱼类可能会无法再恢复到稳态，从而直接导致死亡（Olla 等，1998）。研究人员曾在两组虹鳟不同应激反应实验中观察到一个有趣的运输应激结果：尽管应激后血浆皮质醇浓度在运输过程中发生了变化，但在不同鱼中，因鱼类应激造成的行为以及外表体重损失出现了差异（Ruiz-Gomez 等，2008）。这可能表明，鱼类在应对这种特殊应激时，不同种类以及个体的应对能力是不同的。

无论鱼类是否产生应激，我们都需要采取一定的措施来降低其影响，而当把鱼类引入一个新环境时，运输是第一个减轻其负面效应的机会（Ashley，2007）。高质量的运输条件可以导致鱼类运输后恢复速度更为快捷（Tang 等，2009）；其实，装货过程是引起运输应激的主要源头，因此，在未来的动物运输福利研究中，这应该可以成为一个研究热点（Iversen 等，2005；Tang 等，2009）。鱼类可以通过水环境的逐渐改变来适应其新环境（Walster 等，2014），包括水环境中的水温以及化学/渗透压成分，但也须注意其他环境因素，如光照和噪音大小。运输过后的恢复期会使鱼重建体内的稳态，这会帮助鱼类更好地对付大量应激源（如捕食者、食物获取）（Ashley，2007）。此外，鱼类运输过后的即时照料，可以降低由于鱼类在新环境中排泄废物增加而引起一些综合征的风险（Walster 等，2014），从而进一步减少应激和降低死亡率。最后，还可以推荐对鱼使用浅麻醉来达到镇静效果，但不同物种合适的麻醉程度不同（Ashley，2007）。除此之外，鱼类食物添加剂以及对急性与轻微应激源的处理也可以减缓运输后的应激（Ashley，2007），这样的工作在运输或手工操作之前便应好好进行。鱼类运输前的禁食可以帮助鱼类舒缓肠道，减少运输过程中废物产生、排泄（Handy，Poxton，1993）以及对新陈代谢率的影响，从而减少身体上的应激且提高福利。然而，禁食的推荐持续时间在不同品种间变化不一，比如，对于一些小型品种仅要求一天（如 Froese，1988），而农场动物福利委员会（Farmed Animal Welfare Council，FAWC）对于鲑鱼的指导建议是最大限度为 72 小时（FAWC，1996）。

自然环境有着复杂的颜色以及不同波长的光照，并且该波长是连续变化的。蓄养环境则不同，其光照的时间和质量往往都是人类所需要的，尤其是在实验室环境下，强光照和单一颜色组成的高对比度背景是常见的。所以，在鱼类养殖时，我们必须考虑到光谱中的那些特殊部分，因为光照的不同波长可以依照动物的生态学特性而在不同程度上利用。例如，许多鱼类可以利用紫外波长（UV）（Losey 等，1999），但研究表明在蓄养环境下，超过紫外波长的光照无法确定是否会引起应激（Gronquist，Berges，2013，以及其中的参考文献）。另外，因外红色波长光在水中减弱的速度很快，一些鱼类无法察觉到这种颜色，因此，在该波长光下的鱼类和在黑暗环境下生长的鱼类相比，它们的行为和生理特征都相似（如 Owen 等，2010），而强白光以及水产养殖中常见的背景色都是已知的应激源（Owen 等，2010）。光质量对于卵的发育状态以及幼鱼的发育程度都有着至关重要的作用，例如，相比于那些在红光或者无光周期环境下培育的鱼卵，在与自然环境最为相似的模拟环境中（浅色或蓝色波长的光周期）培育的鱼卵发育更快（Blanco-Vives 等，2011）。此外，偏爱或来源于浑水环境下的鱼类在干净的水中会受到应激（Owen 等，2010），并且这些鱼类的福利在饲养环境下也未能到达预期效果。

用来相互交流的声音对水产动物是很重要的，并且环境中高分贝的噪音会使许多动物倍感压力（Kight，Swaddle，2011）。如水产养殖、实验室或是观赏水族馆等的蓄养环境往往都是含有噪音的，会产生吵闹的且在鱼类听力范围之内的声音，而且该声音一般比没有人为声音的自然环境要大得多（Bart 等，2001；Kight，Swaddle，2011）。噪音来源有很多，如增氧机、气泵，这些都会产生高频率的声音（大约 500 Hz），另外水循环系统、发动机以及任何用于自动化的机器如自动投喂机都会产生较低频的声音（400 Hz 以下；Bart 等，2001）。相比较于高频 110～115 dB 范围内的声音，低频变化的处于 90～130 dB 声音的声压级（Sound pressure levels，SPLs）会变得更高（Bart 等，2001）。然而，在养殖场和实验室环境下，处于 153 dB（Bart 等，2001）或 70～160 dB（Clark 等，1996；Bart 等，2001）的声音的声压级已被研究证实。与自然环境相比，蓄养环境声压级变化的可能性更大（比如，在河流以及类似的高能量环境中，声压级大约为 110 dB re 1 μpa；在湖泊以及类似的低能量环境中，其声压级大约为 100 dB re 1 μpa；而在 500 Hz 以下的主要是其频率上的变化）（Wysocki 等，2007）。鱼类听觉能力有强有弱，且随着品种以及分类群而变化不一，往往根据其解剖构造的有无及利用能力进行划分：听力最强的为听力高手，其对声音高度敏感（尤其对 100～1000 Hz 范围内的声音），可以利用与内耳连接的充气器官（如鱼鳔）辨认声音，而听力一般的鱼则缺少了这些结构（Popper 等，2003）。处于两者听力能力之间的鱼具充气结构，但一般不与内耳连接。噪音对于鱼类的影响不一致，主要和这些因素间复杂的关系有关，如今还没有完美的解决办法。但是，这方面的影响可以广义总结为，不仅会引起应激，还会造成和脱敏有关的暂时性或者永久性的身体损伤。

对于听力高手，如金鱼（*Carassius auratus*），在水族馆要保持低声压级的声音[金鱼可听见大于 110 dB re 1 μpa（Gutscher 等，2011）或大于 130 dB re 1 μpa（Smith 等，2004）的声音]，这可以防止金鱼过度利用它们的听觉器官，并且在交流中掩饰的声音很重要（Gutscher 等，2011）。声音一旦出现，听力高手类的鱼可能会表现出惊吓反应（Smith 等，2004），而听力一般的鱼则不会有此反应（Davidson 等，2009）；虽然反应会在几分钟内逐渐减弱，但当鱼处于急剧的高音环境中时，这可能就显得十分重要。听力一般的鱼，如虹鳟（*Oncorhynchus mykiss*），当噪音等级达到 130 dB re 1 μpa 时可造成应激（Kight，Swaddle，2011）。噪音引起鱼类产生应激的后果较多，会导致血浆葡萄糖和皮质醇含量增加，对 HPI 轴相关的许多下游结构造成物理性损伤（Kight，Swaddle，2011），还可以增加发育中的卵和胚胎的死亡率，对鱼苗造成长期应激影响（Kight，Swaddle，2011）。另外，这些影响若发生在鱼类的早期阶段或正在发育的幼鱼身上，其伤害是不可逆的（Kight，Swaddle，2011）。此外，异常高的噪音可以造成鱼类暂时性失聪（暂时性听阀位移，由 140 dB re 1 μpa 以上的噪音引起），以及损伤鱼类的内耳或气鳔（Kight，Swaddle，2011）。

声音特定频率的存在，而并非该声音的强弱，会给鱼带来应激感，或者使暴露在该环境的鱼产生不良的反应行为。尤其是低频率的声音，它与捕食者的声音相似，会导致鱼类产生惊吓或逃离反应（Knudsen 等，1992；Bui 等，2013）。因为在蓄养环境中次声波的刺激往往并不与捕食行为一起出现，鱼类可能会因为习惯这种声波而减少某些行为反应（如逃离动作）的发生；另外，长期待在这种环境中也会导致福利问题（Bui 等，2013）。在供观赏的水族箱中，因游客敲击玻璃产生的声音也会使鱼感到紧张，会导致其觅食减少以及血浆皮质醇含量上升（Bart 等，2001；Gronquist，Berges，2013）。由于鱼类对次声波（<20 Hz）较敏感，即使是一些低于 1 Hz 的次声波设备，也会在一些环境中用来驱赶鱼群，以防止这些种群生物毁坏一些人造的建筑如水坝（Sand 等，2001）。次声波可以由船舶的发动机、声呐以及建筑产生

（Slabbekoorn 等，2010）。幼年的大西洋鳕鱼（*Gadus morhua*）在发育过程中处于水下船舶发动机声音回放的环境中，导致其具有较低的身体宽度：长度比率，并且在躲避捕食者的实验中更容易被捕获（Nedelec 等，2015）。

声音对鱼类的影响取决于于声音强度和声音频率（无论是急性的还是慢性的）以及鱼的听觉能力和解剖结构等因素的极其复杂的相互作用。虽然本节未能充分概括这些内容，但很明显，人为噪音能够扰乱或损伤鱼类，因此，应注意确保在蓄养环境中将这类噪音降至最低。在蓄养鱼类栖息的环境中，确保噪音最小化是极其困难的，但在减少福利和应激方面可能是一个重要问题。

对于鱼类，不仅仅人为的光照和声音对它们是一个问题，鱼类其他的感觉如磁和电感受效应也会有影响，这些可能是由自然环境中一些金属结构物质以及放置于水族箱下的金属架而产生的。延绳钓渔业中添加带有磁铁的钓钩可以吸引大青鲨（*Prionace glauca*），以此来增加这种濒危保护鱼类的捕获量（Porsmoguer 等，2015）。即使磁铁移除后，这些钓钩依然具有磁化能力。实际上，科学家正在研究将永久性磁铁和正电性金属合金结合起来作为鲨鱼驱散器，已达到减少其被捕获数量的目的（O'Connell 等，2014）。然而，一些金属结构如风力涡轮机和水族箱中的货架成为一个潜在应激源的影响如今还未被研究。

12.2.2 心理应激的影响

虽然一些学者已经详细讨论了应激及其精神状态，但直到最近才在鱼类中考虑心理应激的概念（Galhardo，Oliviera，2009）。应激的认知激活理论正式界定了一些和应激体验相关的概念，如积极的（例如，应对）和消极的价值理念（例如，绝望、无助）（Ursin，Eriksen，2004）。应激反应包括神经生理活化和唤醒，如果该反应是急性的，则可以认为它是正常的。相反，正如前所述，若是长期激发的应激则该反应是有害的。面对急性应激时，初始的警觉响应对鱼类的行为反应至关重要（Ursin，Eriksen，2004）。应激激发的反应调动动物的感觉输入、感知处理和认知机制，以引起动物对应激源的全部注意力（Steckler，2005）。当遇到挑战时，动物会停止当前活动，并专注于应激源。这时往往可以观察到动物的行为性抑制：对于鱼类，其最特殊的反应是冻结行为，个体完全保持静止状态（Vilhunen，Hirvonen，2003）。冻结行为也会发生在当鱼类感到害怕时（Yue 等，2004）或是作为它的一种反捕食策略（Vilhunen，Hirvonen，2003）。另外，鱼类的搏斗或逃跑反应（即应急反应）也具有通过攻击或撤退来消除威胁的功能（Schreck 等，1997；Steckler，2005）。许多种行为模式都和应急反应相关（Ashley，Sneddon，2008；Sneddon，2013），例如逃跑、躲藏以及寻找遮蔽处或增加鱼群的凝聚力。而搏斗性行为包括追逐和攻击（Oliveira，Almada，1998）。仅在鱼类感到受伤或疼痛期间才能看到其摇摆全身和摩擦表面（Sneddon 等，2003；Sneddon，2015）。此外，鱼类的游泳或活动模式对许多应激源也很敏感（Huntingford 等，2006；Martins 等，2012）。

如果应激源是慢性的或不可避免的，鱼类的行为反应会因它而改变；这些反应使鱼类无法恢复到稳态以及无法摆脱周围的危险。面对慢性应激时，鱼类有着各种各样的行为特征，包括游泳模式的明显改变、反捕食者行为的消失、厌食、避难行为增加、社交行为发生改变以及学习能力受到损伤（Schreck 等，1997）。我们可以从这些增强的反应推断出，当鱼类面对慢性应激时其心理状态是消极的。如果保持鱼类良好的健康和福利是我们的目标，我们就应该采取办法去防止鱼类生活在任何应激环境中，但长远看来，我们应该怎样实施一些策略去帮助鱼类面对鱼类管理过程中必然产生的一些应激源呢?

鱼类是否具有内在主观体验（例如，疼痛经历）受到不同程度的争论（Rose，2002；Sneddon 等，2014；Sneddon，2015）。然而，越来越多的科学证据表明，鱼类确实存在几种和福利低下相关的负面情感状态。一些研究意图专注于几种低福利状态，如疼痛、恐惧和压力（表 12.1；Martins 等，2012）；然而，有一些有趣的研究证实，鱼类也会经历积极的状态，此时它们没有应激特征，并且正在进行一些有益行为。作为“清洁工”的裂唇鱼（*Labroides dimidiatus*）会参与种族之间的交流活动，帮助它们的“来访客户”岩礁鱼类清除其身上的寄生虫。但是，还没有确凿的证据显示裂唇鱼改善了岩礁鱼类的健康状况。从野外捕获的双色光鳃鱼（*Chromis dimidiata*）和丝鳍拟花鮨（*Pseudanthias squamipinnis*）被裂唇鱼按摩其腹鳍后，它们的应激反应会变得较低（Bshary 等，2007）。实际上，当作为“客户”的栉齿刺尾鱼（*Ctenothaetus striatus*）出现在正在游动或者静止的“清洁工”裂唇鱼附近时，栉齿刺尾鱼会主动接触裂唇鱼，并以此保持体内较低的血浆皮质醇浓度（Soares 等，2011）。这表明裂唇鱼的清洁行为或触摸行为以及其鱼类本身具有使同类平静下来的作用，从而导致其他鱼类都会寻找裂唇鱼。一些学者认为这也是鱼类具有乐趣的一个案例。鱼类具有与哺乳类等同的智力（Brown，2015），以及能够做出高度复杂的行为，这些行为正在挑战着其大脑进化及体积大小的理论教条（Abbott，2015）。诸如合作之类的复杂行为在种内和种间发生着。例如，海鳗经常与石斑鱼共同觅食，并彼此传递信号。另外，研究已经表明鱼类可分辨出与自己相关的个体和其他个体，可以采用不正当的行为操纵“客户”与裂唇鱼的关系，具有大胆和害羞的个性，大胆、具攻击性、外向的鱼类与害羞、胆小、谨慎的鱼类形成鲜明对比。鱼类还有恐惧反应，可以使用工具，学习复杂的导航能力，根据它们的经验制定未来的行为决策，学习他人的技能，有长期记忆，并可以通过嗅觉识别自己。这些研究表明，鱼具有感知能力和表现意识（Sneddon，2011）。因此，处于应激状态下的鱼类具有消极心理状态的假设是合理的，但必须注意，不要将鱼类拟人化，并且把和人类应激相关的感觉应用于鱼类当中，它们的经历可能更具原始性，尽管从健康和福祉的角度来看这显得不那么重要。

动物感觉到的应激数量取决于应激源的强度和持续时间，并且在任何特定时刻都会受到其个体心理和生理状态的影响（Curtis，1985）。考虑到鱼类性别、年龄、繁殖条件、应激应对方式的个体差异（图 12.1）（Winberg 等，2016；见本书第 2 章），还有昼夜周期，促进积极幸福的环境资源，以及环境是否可预测和可控制，而不是更加紧张而不可预测的环境，这些都应该被仔细考虑到（Curtis，1985；Sneddon，2011）。对于鱼类应激一定具有生物学相关性以及物种间的推断这类研究结果的解释应该谨慎对待。当区域性物种如虹鳟单独饲养时生长良好，而群居性鲤鱼（*Cyprinus carpio*）在单独饲养时则会生病。驯化物种或浅滩物种，例如斑马鱼（*Danio rerio*）在大型鱼群中表现出很小的攻击性，但当它们饲养在四尾鱼的鱼群甚至更小的鱼群中时，表现出高度攻击性，并导致从属地位的个体产生应激。因此，经过对比后，群居性物种和区域性物种所需求的社交环境显得尤为重要（Volpato 等，2007）。

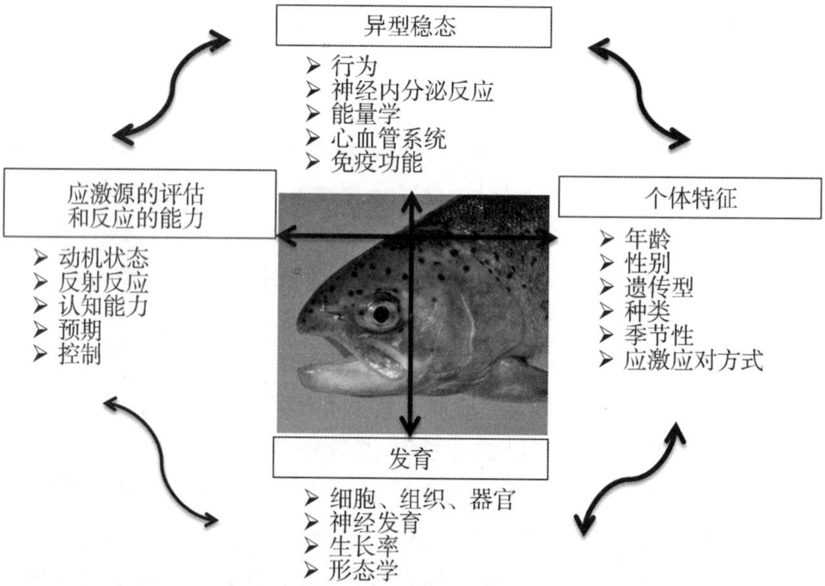

图 12.1　影响鱼类稳态的各种因素及对环境应激产生的反应

注：评估外部事件的积极价值或消极价值以及鱼类反应能力要取决于鱼类个体可利用的适应性负荷程度，所有这些都将由其个体在特定时间的状态以及之前的经验所形成。曲线和直线箭头表示四个主要的促进因素-异型稳态、应激源评估和反应能力、鱼类发育以及个体特性，这些在形成应激时都相互影响。

12.2.3　应激的控制和准备

传统动态平衡的概念已备受许多学者的质疑。Sapolsky（2004）将这种现象重新定义为异型稳态，认为动物是通过持续的变化而实现平衡，并不是依靠固定在某种状态下到达平衡，而稳态便是基于内部平衡更加灵活这一观点衍生而来（McEwen，1998）。在这种情况下，动物内部平衡会根据其特定的要求而相应形成，这个定义假定动物可以通过生理机制（如季节性变化）预测预期的应激源，以及通过其他的生理机制（如战斗或逃跑）面对不可预测的变化产生反应。根据动物面对的应激物类型，在适当的情况下合理使用这两种机制，并且这些机制具有不同的代价（异型稳态负荷）（McEwen，Wingfield，2003）。如果动物能够克服一种挑战的状态，那么它的异型稳态负荷是可控的（图 12.2）。然而，当它无法应对或正遭受慢性应激时，这会导致异型稳态负荷过载，从而导致病理变化（图 12.2）。在这种情况下，应激是不良适应的，并且不利于动物福利（Moberg，1985；McEwen，1998；Broom，2008）。当考虑到异型稳态时，两个关键问题随之要解决：①该应激是否可以预测；②动物是否可以控制其处境。这可以帮助动物在没有异型稳态负荷过载的情况下重获平衡。考虑到鱼类时，假定稳态可以预测以及了解突发事件何时发生，或者可以做出动作以便控制或逃跑。如果福利是基于异型稳态的，那么福利与应激呈倒 U 型关系；相对较少的应激（如一些小刺激）或者慢性应激（异型稳态负荷）会导致不利的福利（Korte 等，2007）。这个概念可以区分鱼类正常的急性应激反应和鱼类持续经历多少天、多少周、多少月甚至更久的慢性应激。如果慢性应激导致免疫和繁殖功能受损、生长速度降低或负增长，并可能导致死亡，则将其定义为异型稳态超负荷。反之，积极福利是指动物拥有可变化且灵活的生理预测能力、认知能力和行为能力，以预测和应对非生物的和生物的挑战，而同时满足对环境的要求（Schreck，Tort，2016；本书第 1 章已讨论过异型稳态负荷问题）。

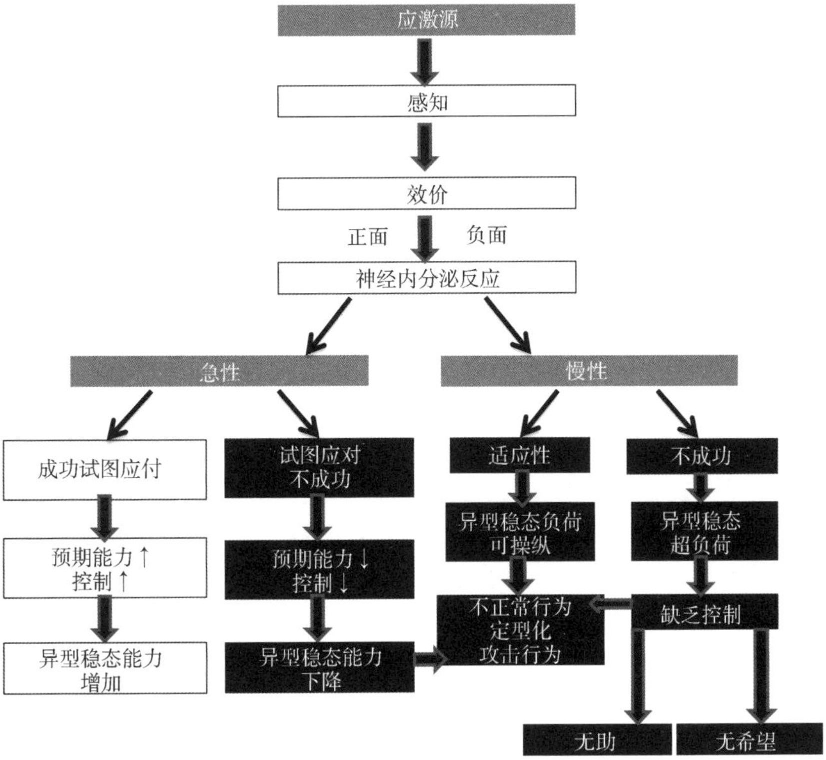

图 12.2　不同场景的图示

注：当鱼类面对应激源时，图片展示了应激源的价值、类型（急性或慢性）以及如何导致一系列的场景，包括鱼类异型稳态能力增强或减少，或异型稳态超负荷。超负荷可引发无望或无助以及潜在的死亡率。

控制可以定义为避免或减少可能影响结果的能力，因此，可以减缓应激反应

（Galhardo，Oliveira，2009）。那么，是否能够使鱼类在一定程度上获得对应激的控制呢？这样的一个案例便是为鱼类提供避难所，以便使鱼类在水族箱清洁和维护等干扰期间能主动寻求掩护或隐蔽。另一个途径是给予动物先前的应激体验以让它们能够预测将来如何应对（Ursin，Eriksen，2004）。个体先前的反应如果在减少应激上是行之有效的，可以通过控制感知使应激激发效率较低，以便于应对随后的挑战（Eriksen等，2005）。这也是为什么处于优势地位的鱼比处于弱势地位的鱼有着较低皮质醇浓度的原因（Earley 等，2006），也可以解释输赢效应（Eriksen 等，2005），赢家在未来的搏斗中胜出的概率会增加（Rutte 等，2006）。例如，大胆的虹鳟在输掉配对比赛后，因为这一段消极的经历会使得它之后变得害羞些（Frost 等，2007）。与获得控制地位不同，动物学习先前的应激应对策略也可能是徒劳的（完全失去控制），因而可能会产生无助感（Ursin，Eriksen，2004；Lovallo，2005）。无助反应与绝望反应不同，在绝望状态下，动物体内存在某种消极形式的控制愿望，并试图去应对可能会增加的应激（图 12.2）。无助和绝望反应都会激活皮质醇（Ursin，Eriksen，2004；Lovallo，2005），并且由于从属位的慢性应激，处于低等从属地位的鱼种往往最终都处于这种状态（Earley 等，2006）。

那么，是否有证据证明可预测性能够减少鱼类的应激呢？莫桑比克罗非鱼（*Oreochromis mossambicus*）便是一个例子，它可以评估可预测的环境以及不可预测的环境，并以此改变它们的应激反应。可预测性的影响主要取决于监禁或喂养实验环境的正向效价或负向效价。当发生监禁应激时，可预测的鱼类组别更倾向于视觉上的预期信号，会产生冻结反应以及比标准浓度较低的皮质醇浓度。食物可预测性会导致更明显的预期行为和活动。可预测的鱼类组别一般具有较低的皮质醇水平，但不可预测的鱼类组别在投喂饲料时也会出现皮质醇浓度降低的情况（图 12.3；Galhardo 等，2011）。这项饲料投喂的可预测性实验实际上产生了自相矛盾的结果，并且可预测性似乎还可能对金头鲷（*Sparus aurata*）（Sanchez 等，2009）等鱼类产生负面影响，这要求该实验必须以一种鱼到另一种鱼为基础进行探究。其他类型的应激可以通过一些传统的调节来减少其影响，亦可以利用一些活动或者奖励对鱼类进行暗示。Schreck等（1995）驯化了一批大鳞大麻哈鱼（*Oncorhynchus tshawytscha*），通过食物与手工操作及运输联系起来，结果显示，这些鱼对应激的生理反应降低，运输过程中的健康状况得到改善，并且当暴露于其他应激因素时可以更好地应对。因此，鱼类积极的强化训练可以调节应激源的有害影响，且有利于鱼类福利（Laule，Desmond，1998）。

提供控制的另一个较为成功的方法是使鱼类产生操作性的条件反射，这时动物会表现出新的行为，例如按键以获得奖励（如啮齿动物）。这项方法已被运用到鱼类自动投喂机上，这样鱼类便可以控制它们的进食过程。例如虹鳟，这促进了它们的生长和提高了饲料转化率（Alanärä，1996）。在行为选择之间进行选择的能力也有助于控制。进行选择可能会导致个体在应对能力上较为积极（Eriksen 等，2005）。比如，当鲤鱼察觉到危险时，为其提供一个避难所，鲤鱼的应激参数会更低，并且开始回避危险（Hoglund 等，2005）。环境丰富度可以帮助鱼类改变其应激行为，便于发泄挫败

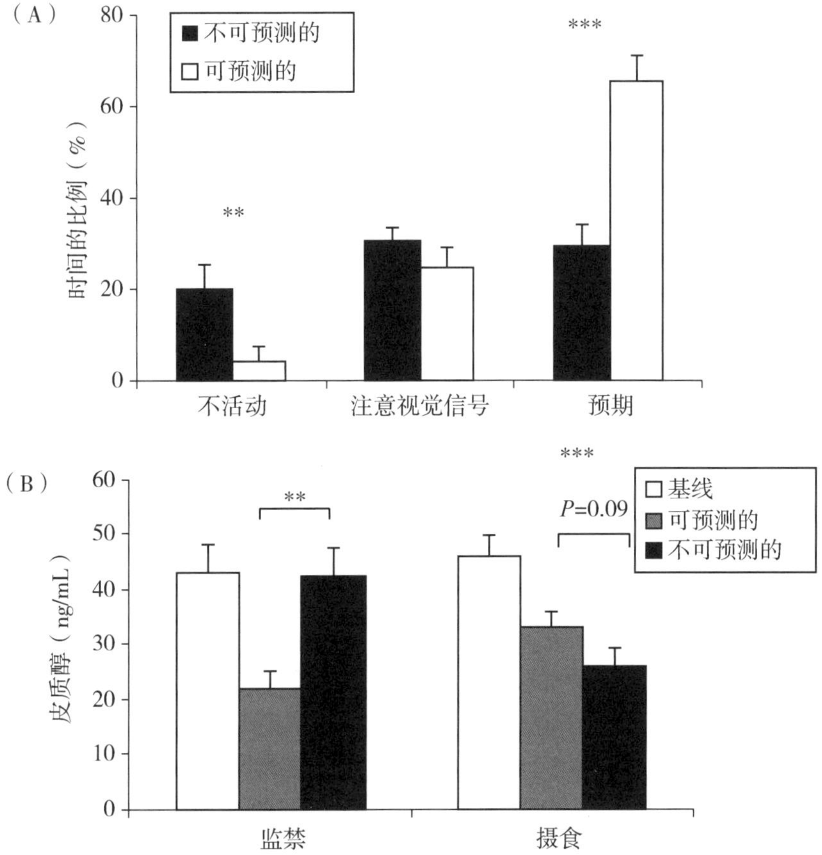

图 12.3　鱼类面对不同食物投喂方式时的行为表现及皮质醇浓度变化

(A) 为莫桑比克罗非鱼 (*Oreochromis mossambicus*) 面对可预测饲料投喂和不可预测饲料投喂时，其分别花费在静止活动、专注于视觉信号、进行预期行为的时间比例。当食物投喂可预测时，鱼花费较多时间进行预期行为。(B) 为鱼类经历监禁 (左图) 和投喂 (右图) 后，各自的血浆皮质醇标准浓度、可预测组血浆皮质醇浓度、不可预测组血浆皮质醇浓度。**表示 $P<0.01$；***表示 $P<0.001$。图片来源于 Galhardo L, Vital J and Oliveira R F. (2011). The role of predictability in the stress response of a cichlid fish. Physiol. Behav. 102, 367-372. 已获 Elsevier 许可。

感。虹鳟在抢食时表现出攻击性未果的时候，可以观察到挫败式行为，这和哺乳动物相似的脑神经化学变化有关 (Vindas 等，2014)。另外，环境底物也可以为莫桑比克罗非鱼的一些处于从属地位的个体提供挖掘的机会，避免与雄性的优势种群发生争斗，并且这种挖掘行为可以发泄因繁殖机会缺失和处于低等社会地位产生的负面情绪，从而提高动物福利 (Galhardo 等，2008)。蓄养环境具有社会稳定性特点，因此，鱼类可以根据它们的生活史以适当方式生存 (Carlstead，1996；Galhardo，Oliveira，2009)。群居物种进行隔离具有不良影响，而小的群体也可能对区域性物种有害。尽管如此，动物社交行为好像确实可以降低某些个体的恐惧感，因此，增强社交也可以提高动物福利 (Galhardo，Oliveira，2009)。如果鱼类想要应对应激并且能够预测它，那么鱼类必须接触到应激源。这使得它们能够利用以前的应对经验，或者学会预测应激情况并改善它们的适应能力（图 12.2）。

缺乏刺激的环境可能会损害鱼类的认知和生理应对机制。因此，让动物暴露于适当水平的环境刺激中是很重要的，但也应该避免过度的刺激，以防引起动物疲惫感和异型稳态负荷过载，这是不利于正面福利的（Meehan，Mench，2007；McEwen，Wingfield，2003；Galhardo，Oliveira，2009）。应激量是否过度需要根据具体情况确定。这时在其他动物的神经系统中可观察到病理性变化，表现出挫败和厌倦的异常反应（包括自我导向行为和替代性攻击）或者刻板行为（Sapolsky，2004）。对于哺乳动物，暴力的受害者经常将攻击性行为再次施加给未参与争斗的其他个体或者物体。这种心理状态称为替代性攻击，这似乎是一种进化保守的机制，也可以在鱼类中得到证实。虹鳟在受到凶猛的大鱼欺负后，会通过替代性攻击向较小个体发出攻击（Øverli 等，2004）。在这些鱼的前脑中会表现出以羟色胺周转和血浆皮质醇为特征的侵袭模式，这表明了通过向处于社会性从属地位的鱼类施加进攻可以抑制神经内分泌应激调节。这些发现证实，从属社会地位的个体可以减少优势种群的应激，并且对于从属地位种群的替代性攻击可能是一种应对应激的策略。同样地，当面对竞争对手，如鱼类摄像机时，具有领土意识的伯氏朴丽鱼（*Astatotilapia burtoni*）雄性向摄像设备发起了攻击，而无领土意识的雄鱼则向水箱中其他同类发起攻击以便发泄环境带来的应激和挫败感（Clement 等，2005）。刻板行为是指一些无用的重复性行为，通常形成于鱼类营养贫乏或慢性应激的环境中（Mason，Latham，2004）。这些行为可以作为动物精神状态受到侵害时的指标，但也有学者主张重复性动作是动物应对慢性应激的一种机制（Mason，Latham，2004）。刻板行为并未引起过多关注，但鱼类的重复性游泳活动是众所周知的，例如，大西洋鲑鱼和大西洋庸鲽（*Hippoglossus hippoglossus*）（Ashley，2007），这和哺乳动物的踱步相似。

12.3 应激对鱼类福利的影响

当考虑到一些经验性证据时（如应激中一系列生物学的和行为的措施对于鱼类的不利影响，鱼类的反应能力及尝试避免和预测负面事件发生的能力，对于自己偏爱的自然资源进行抉择的能力，以及鱼类可以感受到痛苦和恐惧），这些复杂的认知能力表明鱼类可以经历不良的福利状态（Sneddon，2011）。即使我们不认同鱼类可以经历低福利这样的概念，所有科学家也会同意这样的观点，即我们必须保持鱼类健康，以提高经济回报，保持渔业的可持续性，注重鱼类保护，提高科研有效性，并防止人畜共患疾病从不健康、应激的鱼传播给大众。在水产养殖和渔业中，大范围捕捞以及将捕鱼作为休闲娱乐或是运动项目这样的活动会造成鱼类应激。同样的，在实验室的科研活动中，鱼类也会受到应激，即使绝大部分国家对实验性鱼类都要求以人道准则去对待。最后，观赏鱼是一种流行的陪伴性动物，其很容易被一些没有养殖经验的大众所购买，这也会造成鱼类应激。关于应激和福利这些问题稍后再进行讨论，如果利用鱼类的目的是维持鱼类的身体健康，那么如今的现状迫切需要改善。

12.3.1 渔业应激

利用野生鱼类捕捞技术的渔业会严重损害鱼类福利。这已经被 Sneddon 和 Wolfenden（2012）所讨论过了，并且其中包括捕获方法所涉及的许多因素。兼捕渔获物，一般指捕鱼时不想要的鱼类，这些鱼类扔回水中时往往会造成损伤或死亡，并且鱼类捕捞后一般要经过处理、买卖以及屠杀（Sneddon，Wolfenden，2012）。据估计，1999—2007 年每年都有 2.7 万亿尾的鱼被捕捞（Fishcount. org，2015），而每年仅需 30 亿头哺乳动物（牛、绵羊、猪、山羊）以及 570 亿只家禽（家鸡、火鸡、鸭）。捕获的鱼可以作为人类的食物，但还有较大一部分被加工为饲养家畜的鱼粉和鱼油，以及鱼油营养补充剂。大规模商业化渔业捕鱼时会涉及许多媒体关注的具侵略性和破坏性的技术。2015 年的欧盟 Fish Fight 运动已有 87 万名支持者，他们希望政府颁布兼捕渔获物禁令，并且鼓励扩大公众的鱼类食用种类，而不是仅仅只关注几种已过度捕捞的品种。这项运动使得欧盟政府最终禁止了兼捕渔获物行为（Fish Fight，2015）。捕鱼有很多种方法，包括长时间拖网捕鱼法，这会迫使鱼最后精疲力竭而被捕，由于捞上船时会被其他鱼所挤压，所以这些鱼往往脊柱断裂或有其他伤病，此外，还会因为高度急速变化而带来气压性损伤。损伤也会出现在其他捕鱼方法中（如刺网捕捞和围网捕捞）。延绳钓捕捞方法往往要用鱼钩钓住鱼类几小时乃至几天，并且非目标品种如鲨鱼、海龟、海豚和海鸟也会被抓住以导致它们死亡。这种方法还会实施活体诱饵，其中较小的物种被钩在渔具上用来捕获较大的物种。捕鱼方法的不同会导致不同程度的应激（Sneddon，Wolfenden，2012）。这包括延绳钓捕捞中鱼钩和鱼线的使用、拖捕行为、竿钓行为，以及拖网、围网、刺网等的渔网技术（Sneddon，Wolfenden，2012）。通过改善渔具、捕捞方法，以及改进捕捞和屠宰过程可以减缓鱼类的捕捞应激，提高其后勤处理时的福利（Metcalfe，2009）。

兼捕时将非目标品种丢弃是一种具有争议的做法，由于配额限定、监管和经济方面的缘故（丢弃较低价值的鱼类以便它们不占用空间），这种做法在许多国家已成为惯例。兼捕的鱼类在登船时可能会窒息，而且/或者因捕获而受到前面描述的伤害（Metcalfe，2009）。此外，捕捞后水温的升高会增加丢弃鱼类释放后的死亡率，而且其对应激生理有着显著的影响（Cooke，Suski，2005；Gale 等，2013；Raby 等，2015）。鱼类被捕捞后，会出现过度运动、缺氧、重现空气、捕捞时受伤和高度快速变化带来的气压性损伤，以及神经内分泌应激反应（Davis，Olla，2001，2002；Davis 等，2001；Farrell 等，2001；Raby 等，2012）。如果将完全可销售的鱼类丢到船外，那这种兼捕时的丢弃做法显得非常浪费；仅福利而言，禁止兼捕时鱼类丢弃会影响到福利的改善问题。然而，现已证明，兼捕时鱼类丢弃禁令对于鱼类种群的总体健康有着积极的影响。丢弃禁令的试验（例如，挪威禁止在北极东北部水域丢弃渔获物）促使鱼类种群恢复，尽管经济回报率起初会下降，但市场很快适应捕获的更多样化的品种（Diamond，Beukers-Stewart，2011）。渔业中还有其他可以减少应激的策略，如减缓拖网的速度，允许非目标物种逃离，减少伤害和疲惫的发生；捕鱼时从深处缓慢上升以避免气压性损伤和温度的迅速变化；在高温度以外的水域捕鱼以减轻应激（Metcalfe，2009）。许多科学家如今正在与渔业公司共同致力于减少野生动物兼捕行为（野生动物兼捕减少协会，2016）。例如，在海洋的拖网捕捞中使用 LED 灯减少了鱼类的兼捕，使兼捕丢弃的鱼类减少了 56%～91%（Hannah 等，2015）。以及，在拖网中添加一个分隔物以形成较低的和较高的囊网，这可以分开挪威龙虾和鱼，使得鱼处于更好的状态且不容易受伤（Karlsen 等，2015）。

还有另一个问题值得关注，在有些情况下，鱼类捕获会引起应激，这可能导致死亡延缓，因此，即使鱼在释放后正常游走，它们也可能在生理上已受到损害，从而在一段时间之后再死亡。采用大型围网方法捕获银大麻哈鱼，取其血浆进行测量，结果显示其生理应激曲线和过度运动时相似，并且在缺氧期间，由于渔具缠绕时间长，导致其皮质醇和乳酸浓度显著提高（Raby 等，2012）。为了测量延缓死亡率，在实验室中进行拖网来模拟真实的拖网捕鱼，一直持续 60 天，会发现延缓死亡率发生在第 20 天后（Davis，2007）。此外，捕鱼方法也会影响延缓死亡率，如围网捕捞（9%）动物表现好于延绳钓捕捞动物（39%）（Humborstad 等，2016）。对裸盖鱼（*Anoplopoma fimbria*）的研究表明，拖网模拟实验和实验室鱼钩模拟实验会增加其应激反应，继而损伤其免疫功能并导致其出现延缓死亡（Lupes 等，2006）。在高水温环境下被捕获或者长期暴露在空气中会提高鱼类延缓死亡率（如裸盖鱼）（Davis，Parker，2004）。因此，判断鱼类是否具有延缓死亡倾向对于决定鱼类是否要丢弃，以及当鱼类快要死亡时是否要进行安乐死是有帮助的。反射动作死亡预测技术（RAMP）可以通过反射动作的有无（如轻轻掐鱼尾使之运动）对抓捕后的鱼类进行测试，得出 RAMP 分值，继而决定这条鱼的命运。这种方法已广泛用于实验室以及实地研究中，以探讨鱼类抓捕后的应激程度和预测延缓死亡率（Davis，Ottmar，2006；Lupes 等，2006；Davis，2007；Raby 等，2012；Humborstad 等，2009，2016）。如果释放鱼类的目的是要确保它们存活，那么使用 RAMP 可以做出明智的决定，确定任何能使鱼类在释放后存活的可能性。

在渔业中，对抓捕后鱼类采取的致死技术也是多种多样，可以简单地将鱼放在甲板上使其窒息，据报道这要花费 250 分钟（fishcount. org. uk，2013），或者放于冰上，或切断鱼类鳃部使其流血而死，或直接活体取其内脏。使用丹麦围网捕获的鳕鱼中，75% 鳕鱼暴露于空气 10 分钟后死亡，还有一些需要长达 20 分钟（Humborstad 等，2009）。Robb 等发现，持续流血 7 分钟后会导致脑死亡，这与脑震荡引起的脑毁坏形成鲜明对比，它会导致鱼类立即死亡。然而，对围网捕捞得来的大量鱼类都进行脑震荡在后勤上是一个巨大的挑战，因此，不同规模的渔业会使用不同的方法。任何一种方法都会对我们所关注的鱼类福利产生影响，即使目前野生鱼类屠宰期间的福利状况与养殖鱼类相比仍缺乏一定的研究。鱼类窒息死亡时间一般要 55～250 分钟，并且活体剖去内脏的鱼类在 65 分钟后依然具有反应（Fishcount. org，2015）。大菱鲆（*Scophthalmus maximus*）浸泡冰水 75 分钟后依然具有反应（Lambooij 等，2015）。人道屠宰协会（Humane Slaughter Association，HSA，2014）已对养殖鱼类提出建议，要求在使用更具侵袭性的方法加工之前，应确保已破坏大脑，使鱼变得不敏感或无意识。因此，有人提出渔业可以采用敲击式的击晕设备。然而，该方法并未被广泛应用，因为捕获的鱼类数量很多，击晕每一条鱼显得不切实际。未来的研究应该探讨更加人道的捕鱼船屠宰方法，以至于在满足后勤要求的情况下减少应激和提高动物福利。

显然，我们可以通过改进目前的捕鱼方法来减少鱼类的应激，尤其是可以减少鱼类捕捞时间使得鱼类更加快速回到船上，也可以通过改进设备来减少对鱼类造成的伤害，或者在船上使用快速而高效的人道屠宰技术，也可以减少兼捕（Metcalfe，2009）。这些可以通过以下几方面更为方便地进行考虑，一些国家可能在法律上要求（做到）这些：

（1）减少兼捕时非目标物种的初始数量。

（2）增加兼捕丢弃鱼类的生存机会。

（3）减少捕获的持续时间。

（4）减轻目标物种屠宰的应激。

（5）调整捕捞方法，以排除使用活鱼作为诱饵。

12.3.2　水产养殖中的应激

水产养殖过程中存在一系列的福利问题，其日常饲养过程和管理措施都会造成应激（表 12.2；Ashley，2007；Conte，2004）。鱼类饲养在非正常放养密度或社会环境下，由于集中饲养缘故，可能会给鱼类带来挑战。许多研究表明，放养密度会对鱼类的福利产生负面影响，而且正如 Turnbull 等讨论的那样，水体质量与社交影响和养殖鱼类的福利密切相关。然而，具体情况较为复杂，需要进一步研究以确定放养密度对鱼类福利的真实影响。例如，Ellis 等（2005）发现无论放养密度的大小如何，可能都会损害动物福利。它表明高放养密度及由此产生的次优水质可能会导致生长速率降低和身体状况不适以及出现如鳍条侵蚀等的患病症状。鳍条侵蚀，通常在水产养殖环境中称为鳍条腐烂，是指鱼鳍组织的损伤或损坏，是影响鲑鱼养殖的常见问题（Ellis 等，2008）。次优水质是其主要的应激源之一。由于鲑鱼的经济价值很高，因此水产养殖中对该鱼类可能产生有害影响的水质参数范围进行了广泛的评估（MacIntyre 等，2008）。这在循环水系统中（RAS）尤为重要，该系统每天仅补充 10% 的新水（Martins 等，2011）。如果没有监测水质情况或者没有采取足够快的补救措施，这可能会对鱼类造成严重威胁。在理想情况下，水质中 pH、氨、亚硝酸盐、硝酸盐、盐度、氧合作用、二氧化碳和温度等因素应保持在有关物种的最佳范围内。自动监测系统含有一个警报系统，其可以告知管理人员任何水质问题。有些物种在降河溯游期间会经历表型上的快速变化，比如，大西洋鲑鱼在此期间会从淡水表型变化为海水鱼表型，水产养殖管理时须加以注意。生产参数的综合分析表明，高放养密度下将大西洋鲑鱼养殖于比淡水温度更高的海水中时，鱼类的生长和存活情况会更好（Kristensen

等，2012）。这些物种的管理是复杂的，对其所有生命阶段还需要进行更多的研究，以减少关键变态期间应激源的影响。

表 12.2 鱼类在水产养殖中面临的主要福利问题，它可以导致急性应激和慢性应激，以及一些改善的建议

福利关注领域	相关福利问题	改善建议
冬季病害		
低温引发的几种疾病	虽然许多疾病与特定的细菌病原体明显相关，但免疫抑制在冬季可能会发挥很大作用	免疫接种 适应性饮食可提供需补充剂量的维生素和微量矿物质，以协助免疫系统，并且改变喂养方式以控制对病原体可用的营养素水平
鳍条腐烂		
对环境和/或侵袭性相互作用的磨损会导致鳍条损坏，这可能会继发感染	疫苗注射已取代抗生素，但是疫苗和佐剂与炎症和肉芽肿相关，以及麻醉和注射也会引起应激	疫苗可以提高功效，减少副作用，还可以口服使用
海水鱼虱		
寄生性的桡足类动物可能会造成严重的组织损伤	鱼虱对传统的化学疗法产生了抵抗力	潜在的替代性控制如疫苗接种和选择性育种，以提高对鱼虱的抵抗力 生物学控制如清洁工斑盔鱼，但还应该考虑到斑盔鱼的福利
病毒性疾病		
例如：感染性胰腺坏死病、感染性造血坏死病、病毒性出血性败血症、传染性鲑鱼贫血症、睡眠病	在过去 20 年中开发的传统疫苗仅取得了一定的成功，还有相对较少的商业疫苗和具有足够功效的特定治疗剂	开发抗病毒治疗的替代方法，如 DNA 疫苗和病毒耐受力选择育种
非传染性相关的畸形		
心脏、气鳔和脊柱畸形	由于心血管功能受损、心力衰竭或心脏破裂，心脏畸形的鱼在应激期间表现出高死亡率。遗传和环境因素都可能导致脊柱畸形	应避免在鲑鱼孵化期间出现高温。通过引入海水、疫苗接种、减少盐度和温度变化来增加鲑鱼体重可以减少脊柱畸形。来自同一家系的鱼类显示高发畸形，这些鱼不要用于繁殖

续表 12.2

福利关注领域	相关福利问题	改善建议
标粗、手工操作、拥挤环境		
固有的应激性	如标粗之类的许多程序都旨在提高福利。不同物种对于程序的应激程度存在很大差异，手工操作应激源也会影响随后的应激反应	饲料中适当添加维生素 C、维生素 E 以及葡聚糖，可避免慢性应激的不良影响。也可以合理使用良好的拥挤管理技术、合适的渔网，小心手工操作，注意恢复期，合理使用鱼泵和输送管，而且合理的饲养密度可以避免频繁标粗
运输		
固有应激性，可能涉及捕获、装货、运输、卸货及库存	运输的应激源可以影响鱼类很长时间	适当的驯化和恢复期以及适当使用麻醉和稀盐溶液可以减少不良反应
禁食		
在屠宰、运输和其他管理实践前使鱼产生饥饿感	通过减少新陈代谢，需氧量和废物产生，使其福利受益。大西洋鲑鱼和虹鳟在野外表现出长时间的厌食症状，因此水产养殖中食物匮乏的福利效应尚不清楚。在适当的条件下进行短期食物剥夺可能不会降低福利	大西洋鲑鱼的饥饿时间长达 72 小时，虹鳟的饥饿时间长达 48 小时，这应该有益于其福利，但还需要对饥饿对应激生理或行为的影响进行实证研究
屠宰		
屠宰应尽可能人道地进行，鱼类在屠宰前应该被击晕，导致其迅速失去意识，直至死亡	虹鳟和金头鲷脱水后进行冰上窒息、大西洋鲑鱼和虹鳟浸泡在饱和的 CO_2 水后进行鳃部切割或单独进行鳃部切割以及鳗鱼脱离泥土后除去内脏，这些都不符合人道屠宰的标准	通过敲击式击晕法或电击击晕法可以实现大西洋鲑鱼、金头鲷、大菱鲆、虹鳟的人道屠宰淡水中电击击晕鱼类舌头或批量电击鳗鱼时，结合氮气冲刷，可以使鱼类立刻失去意识
饲养密度		
以多种方式影响福利的关键因素（例如，通过攻击、水质和活动/喂养模式）	饲养密度的影响由众多交互作用和特定情况因素构成。海鲈高密度养殖时表现出高水平应激状态。北极红点鲑极高密度养殖时，表现出生长速率低、摄食率低的特性。	可以改变喂食方式和占地面积，以改善密度对庸鲽福利的影响；也可以看到对侵袭性的影响，如下所述，鲑鱼游泳深度和鱼群密度可受人造光源水平影响，

续表 12.2

福利关注领域	相关福利问题	改善建议
	庸鲽对高密度养殖的耐受力似乎呈现和生长阶段相关。虹鳟高密度饲养时福利降低，其中水质是一个重要的因素。大西洋鲑鱼海水网箱养殖时，高密度养殖超过临界值时会减少福利，特定位置因素对福利也会具有影响	喂食方式可以改变包括大西洋鲑鱼在内的一些物种的攻击相互作用
攻击性		
社会等级制度的形成可能导致受伤、慢性社交应激和体型大小的异质性	社会生物学、放养密度和喂养技术对社交水平有很大影响	喂养技术应适合该物种，以避免过度的竞争和攻击行为 可以考虑自动投喂机以减少攻击行为 在小型鱼群中添加少部分的大型鱼类可以减少攻击行为 在饲料中添加 L-色氨酸表明可以抑制攻击行为 对于一些物种，生长环境的底质或背景颜色可以影响攻击行为
异常行为以及表达正常行为的自由		
异常行为包括重复性行为和异常的游泳行为/模式	清楚异常行为的功能性起源是很重要的，需要进行实证研究，以确定异常行为是否代表福利减少或适应性反应，而不影响福利	丰富的饲养环境可能会改善释放后的福利，以增加野生种群 没有实证研究，对特定物种的特定行为模式的重要性尚不清楚。需要研究控制机制和/或拒绝表达关键行为的行为和生理后果 选择研究表明鱼类可以评估和特定行为或资源相关的价值，从而做出行为选择（例如，在应激事件期间避难或自我调节），所以鱼类可以预测应激源的发作

来源：数据改编自 Ashley P J（2007）. Fish welfare：current issues in aquaculture. Appl. Anim. Behav. Sci. 104，199-235. 已获得 Elsevier 许可。

水产养殖过程中鱼类可能会受到手工操作时的应激，而且这种应激对于每个个体可重复多次，如标粗、接种疫苗、屠宰等。许多鱼类研究已经证明手工操作可以引起应激，如真鲷（*Pagrus major*）（Biswas 等，2006）、大西洋鲑（McCormick 等，1998），以及鲤鱼（Saeji 等，2003）。手工操作的最少化和完善限制应激的处理技术有助于提升动物福利。例如，手工操作时鱼类所遭受的光照类型可以减少应激反应。对于舌齿鲈（*Dicentrarchus labrax*），蓝色光对其益处最大，而白色光照可以引起应激（Karakatsouli 等，2012）。而黄鲈（*Perca flavescens*）正相反，相比较于红光和全光谱光照，蓝光会使其生长更慢（Head，Malison，2000）。光照的突然变化也应在养殖过程中避免，这对鲑鱼会引起应激，因此，应逐渐改变光照强度以模拟黎明和黄昏（Mork，Gulbrandsen，1994）。网箱养殖的大西洋鲑鱼应避免高强度光照，并且它们白天一般在深处游动（Johansson 等，2006），因此，这些鱼类适宜在低光照强度下进行手工操作，以防止因为光照而产生应激。

影响硬骨鱼的疾病多种多样，它们可以由于较差水质、过度手工操作和空间拥挤这样的应激源而突然爆发以及加剧。适宜管理如水质这样的因子可以控制疾病，继而提高福利，尽管在疫情暴发时实施较多的是对许多病原体的化学控制。另外，如今疫苗接种也广泛用于应对疾病，对现代鲑鱼养殖起着至关重要的作用（Sommerset 等，2005）。生物控制也成功应用于一些寄生虫病中，尤其是通过使用“清洁工”隆头鱼中的梳隆头鱼（*Ctenolabrus rupestm*）来控制鲑鱼身上的海虱（Costello，1993）。疫苗接种如今也迎来问题：作为疫苗载体的佐剂会引发大西洋鲑鱼的腹腔性腹膜炎和应激（Bjørge 等，2011），因此还有待开发其他载体。

许多国家关于动物运输的法律已经较为完善，这也可以应用到鱼类和哺乳动物运输。鱼类对于运输的应激反应根据其运输技术而变化。由于公路运输会因空间限制和水质恶化而引起可以测量到的应激，Erikson 等（1997）建议，可以使用现代化井船来运输大西洋鲑鱼，这是挪威鲑鱼渔业的既定做法，它使许多用于应激测量的典型生理指标（如肌肉组织 pH 和腺苷酸能荷）并不会和非应激鱼类变化太大。这是由于井船可以保持适宜水质以及和屠宰场快速接手。因此，通过维持水温、水质和减少运输可以减少需要移动鱼类的应激。其他改进措施还包括使用低剂量的麻醉剂以保证鱼在旅途中镇静，并使用 5 ～ 10 ppm 盐水来预防细菌和真菌疾病的爆发（Ashley，2007）。另外，还使用 Slime Coat“黏液衣”或 Stress Coat“应激衣”等商业化产品来减轻应激，它们可以弥补鱼类丢失的所有黏液层；然而，这些做法尚未经过严格测试。水温和氧气管理是必要的，它可以通过各种因素导致应激，如温度波动大、低氧或过饱和，以及二氧化碳的积累（Tang 等，2009）。在装货和卸货过程中，鱼在拥挤状态，在渔网中和泵中也会受到应激（Nomura 等，2009）。

鱼类的攻击性是水产养殖中长期存在的问题，并且如前所述，由于低下状态的慢性应激，可能导致从属社会地位物种的异型变态超负荷。幼年鲑鱼天然具有攻击性，会防卫自己的摄食区域，这可能会导致禁闭的水产养殖环境中的问题，特别是在饲料投喂限制的期间（Brännäs，Alanärä，1994）。攻击性增加后可导致鳍条受伤更加严重

（如 Cañon Jones 等，2010），尤其是背鳍和尾鳍（Turnbull 等，2008；Persson，Alanara，2014；图 12.4）。鱼类放养后因背鳍受伤和应激会增加其死亡率（Petersson 等，2013）。为了解决这个问题，科学家在贫瘠、单调的水槽和水道中增加了环境的异质性以及美化环境。比如，在水道中使用鹅卵石作为基质可以减少虹鳟和克氏大麻哈鱼（*Oncorhynchus clarkii*）的鳍条腐烂现象（Bosakowski，Wagner，1995）。在较为异质性的养育环境中，由于为鱼类提供了避难所以及减少了攻击性的交互作用，虹鳟和大西洋鲑鱼的鳍条质量得到显著改善（Näslund 等，2013）。已有几种增加结构复杂性的方法应用到饲养环境中（Brockmark 等，2007），但结构的复杂性是否可以减少攻击和应激并继而改善其长期健康，如今对于许多物种来说还不清楚。

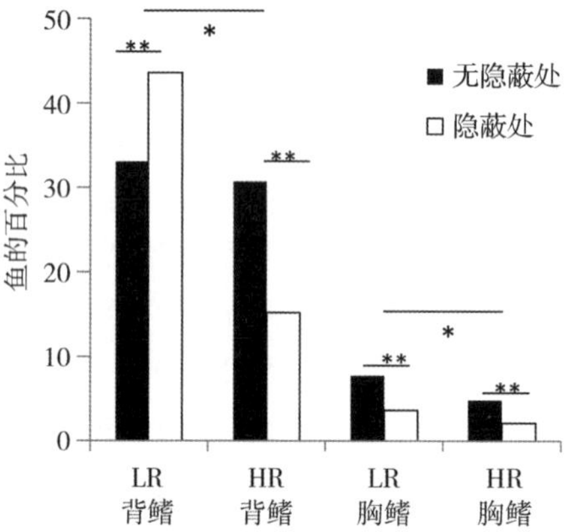

图 12.4 大西洋鲑鱼（*Salmon salar*）低频率投喂（LR）和高频率投喂（HR）时以及遮蔽物存在与否时，胸鳍和背鳍受伤鱼类所占百分比

注：结果显示，在低频率投喂、无遮蔽物时，鳍条受伤最为严重，这表示攻击性的增加。* 表示 $p<0.05$；* * 表示 $p<0.005$。数据来源于 Persson L，Alanara A（2014）. The effect of shelter on welfare of juvenile Atlantic salmon *Salmo salar* reared under a feed restriclion regimen. J. Fish. Biol. 85，645-656. 已取得 John Wiley 和 Sons 许可。

寻求食物时的觅食行为在水产养殖中是另一种应激源。当食物有限时，动物会通过争斗来获取食物，因此，在适宜时间里提供足够的、营养丰富的饲料以保证每一条鱼都可以满足它们的饥饿感是至关重要的（Martins 等，2012）。鱼类的觅食策略多种多样，并且受水产养殖系统、年龄、性别、遗传、季节性和其他因素影响。如前所述，鱼类会采用不同的摄食策略（参见本章第 12.1.1 节：底层摄食者、表层摄食者以及中层摄食者；如夜间型 vs 日间型 vs 黄昏型；定时投喂 vs 随机投喂；捕食者、腐食者或更多被动摄食者；肉食性、植食性或杂食性）。因此，用正确的方式为鱼类提供适当数量和适宜的食物是很重要的（参见本章第 12.2.1 节）。否则，便会产生攻击性和应激，并且应激可以引发厌食症从而导致鱼类生长缓慢（Wendelaar Bonga，1997）。给栖息于底层的比目鱼投喂沉料比投喂浮料更能提升福利（Kristiansen，Fernö，2007）。改变海鲷的饲料投喂时间（定时投喂 vs 随机投喂），它会因为预期食物无法提供而出现行为的和应激心理学的一系列消极变化（Sánchez 等，2009）。这可能表明，定时投喂可使鱼获得控制感，以至于它们能为下一次觅食做好准备。与随机投喂相比，定时投喂可以增加海鲷喂食前的游泳活动（如预测食物）（Sanchez 等，2009）。然而，与按需投喂相比，定时投喂会增加鳍条损伤，所以应谨慎使用（Noble 等，2007，2008）；和随机投喂相比，当预期食物未出现且个体生长速率差异较大时，大西洋鲑鱼会表现出攻击性和挫败感（Vindas 等，2012）。因此，水产管理者和护理人员应该仔细考虑何时使用何种饲养方法，这具有种族特异性，并且取决于设施的类型和管理策略。鱼类自动投喂机可以提升虹鳟的饲料转化率和生长率（Alanärä，Brännäs，1996），表明其应激性较低（Suzuki 等，2008）。这种策略使鱼类能自主控制食物，因此也有可能对福利有益。

许多日常饲养程序会导致应激和干扰觅食行为，从而减少福利。水质改变、对水池造成干扰的清洁方案、疫苗接种、攻击性和标粗程序都会影响进食行为。所有这些应激源都可以借助无害的提示让鱼类可以预测它们，并为下一次应激做好准备。另一种策略是提供丰富环境或避难所，以便鱼类在遇到干扰时可以有改变行为的选择。这些管理实践中多数对于鱼类的健康是必要且有益的，因此，在未来的研究中应该考虑如何减少这些应激源的次数、强度和持续时间。

在收获和屠宰期间，鱼类也存在福利问题。一些国家的规章规定鱼类应该快速、人道屠宰，以减少它们的应激（例如，HSA，2014）。如前所述，为此已经开发了具人道主义的杀鱼方法，例如敲击式击晕法和电击击晕法，以至于使用确保脑损伤的辅助方法时鱼是无意识的（Line，Spence，2014）。鱼类收获时采用拥挤的办法和物理办法去转移鱼类，在这之前鱼类经历禁食期以减少代谢废物以及防止食物积压在肠道从而影响肉质（Line，Spence，2014）。饲料投喂减少和收获活动都会引起鱼类应激，所以这必须快速进行。近来对虹鳟的研究表明虹鳟清理肠道需要 24 小时，但通常禁食期长达 72 小时，期间应激指标有所提升（Lopez-Luna 等，2013）。因此，更短的禁食期有待探索。鱼类收获包含将鱼捞出水体后放在一起或者用网捞鱼（Lines，Spence，2014）。氧气的减少、空间限制以及鱼类湍流对游都是应激源（Ortuno 等，2001；Brown 等，2010）。用网捞鱼时，网可以使鱼类从水中分离——湿网捞取时鱼类还保持在水体中以防止空气流动的危害，而干网捞取时鱼类可能和其他鱼发生身体上的损伤和磨损，并且该网是被放置在坚硬的地面上（Lines，Spence，2014）。因此，应该尽量缩短捞鱼时间以减少应激。另外，这些运输方法无疑会导致严重的应激，因此，应优先采用更好的运输方法，如使鱼类生活在适宜水质的井船中的运输方法。

击晕和杀死鱼类的方法不同于空气中或冰上使鱼窒息、活体冷冻、切断鳃放血、直接砍头、活体解剖内脏和使用二氧化碳饱和水这些方法，而这些都明显损害福利并且不是人道的方法（Lines，Spence，2014）。更多对福利友善的方法如敲击式击晕法和电击击晕法已被开发。敲击式击晕法主要用于鲑鱼养殖业，在其头部上方快速击打使鱼类迅速、长期或永久性无知觉。如果该打击足够强且打击位点准确，这种方法的效率是很高的。人工的敲击式击打会受到操作工人的能力、熟练度以及疲惫感的影响。这种方法主要用于鲑鱼等圆形物种，而对比目鱼、鲤鱼、鲶鱼和罗非鱼等头骨中具有坚硬骨板的物种是无效的（Lambooij 等，2007）。作为替代方法，电击击晕法的使用效率要优于物理杀鱼方法（放血）。湿电击击晕法是指对整个水体施加电流刺激，其优点是可同时击晕大量鱼；鱼无知觉之前没有必要暴露于空气中，处于拥挤环境或进行处理。这显然提升了其福利。然而，鱼并不会永久性击晕，所以工作人员必须迅速采用第二种方法，以确保鱼在击晕后不会苏醒（Line，Spence，2014）。最近的一项研究表明，施加电刺激会使莫桑比克罗非鱼（*Oreochromis mossambicus*）的应激大于单独手工操作（Roques 等，2012），因此，需仔细考虑屠宰方法。还有另一种更人道的方法，在澳大利亚、智利、新西兰、韩国、哥斯达黎加和洪都拉斯等一些国家提议使用麻醉剂。这种方法是使用异丁香酚（一种丁香油衍生物）将鱼在水池中大量麻醉，但该物质在许多其他国家还尚未批准使用。假设麻醉剂避免了应激反应（Zahl 等，2010），鱼类不反感该药物（Readman 等，2013），且对人类食用无害，麻醉将是一种更好的杀鱼方法（Kiessling 等，2009）。而且，用于人道杀伤的麻醉剂量应高于用于确保死亡的麻醉剂量（Sneddon，2011）。

12.3.3 休闲渔业中的应激

鱼类也会被休闲钓鱼者和运动性的渔民捕获。在渔业中描述的许多问题（见本章第 12.3.1 节）在这里也是相关的，只是在许多情况下这是一个人捕鱼而不是大规模捕捞。因此，我们不拟详尽的论述，只是概述和钓鱼相关的一些具体问题以及如何改进最小化应激的方法。对于一些钓鱼个体者/运动性渔民，他们认为其捕鱼乐趣高于一切成本（应激生理、行为、伤害等），并且他们的做法符合了当地准则和许可，但这还是存在道德的问题。捕获的鱼类往往被杀死，如果其捕获方法和屠宰是人道的，因为鱼类作为食物的价值大于一切成本，所以它对人类是有益的。有些鱼类也会在被捕获后释放，因为鱼类大小不合适或不适宜食用，或由于当地捕获规定或当地其他的要求（Cooke，Sneddon，2007）。如果钓鱼时捕获后释放鱼类有利于鱼群的保护和可持续性，那必须保证多数鱼类释放后可以存活下来。然而，捕获的鱼类一般会被鱼钩致伤、暴露于空气中，以及运动期间如果放在渔网中，鱼会经历缺氧（Cooke，Sneddon，2007）。大量研究为鱼类捕获后释放是否受到身体损伤，行为、生理健康上受到亚致死的变化以及导致死亡提供了分析结果。研究表明，所有的休闲渔业活动都会导致鱼类个体产生某种形式的伤害和应激（Cooke 等，2013）。然而，鱼类受伤的严重程度、应激程度以及死亡率会随着一系列环境因子以及钓鱼者的行为而有着很大差异。不过，借助该信息，我们可以通过渔具的改变（如钩子、鱼饵或网的类型）或是钓鱼行为的改变（如钓鱼或娱乐的时间、避免鱼类暴露于空气中、在极端环境下禁止钓鱼、避免繁殖期间钓鱼；表 12.3）来达到减少应激的目的。鱼类在被捕获后释放难免会引起应激和损伤（如用鱼钩钓住鱼类），但或许采用较小损伤与应激的方法会有益于鱼类健康，能提高鱼类释放后的生存机会以及提高其整体福利（表 12.3）。这将对鱼类种群和渔业有明显的好处，有利于保护和管理渔业。

表 12.3　由 Cooke and Sneddon（2007）提出的关于捕获后释放钓鱼法的改进建议，并且对该方法的益处进行了补充说明（所有方法都可以减少应激和提高鱼类福利）

改进建议	益处
使用无倒钩的鱼钩、圆钩	减少鱼类损伤
如果鱼钩在鱼体深处，取鱼时直接剪断鱼线	由于不太可能取回鱼钩，所以减少了空气对鱼类的损伤 可以考虑安乐死
避免接触到鱼鳃或鱼眼	减少鱼类损伤
与鱼类嬉戏保持最低限度	减少应激和疲惫感；为之后的行为保存体力
保证鱼类生活在水中	减少空气接触的应激
当合法时对受伤鱼类进行安乐死	提高那些释放后不能存活的鱼类的福利
使用没有结节的网	减少渔网对鱼类的磨损
从深处捞鱼时缓慢上升	减少气压性损伤
避免高温捕鱼	减少应激和死亡率

来源：数据改编自 Cooke S J, Sneddon L U（2007）. Animal welfare perspectives on recreational angling. *Appl. Anim. Behav.* Sci. 104, 176-198. 已取得 Elsevier 许可。

12.3.4　观赏鱼中的应激

观赏鱼因极具吸引力而被饲养。观赏鱼饲养开始于 1000 年前中国的驯化金鱼，到如今许多鱼类品种因其特有的标记而被高度重视（如锦鲤）。据估计，在英国有 1/10 的家庭拥有宠物鱼（Balon，2004；The Telegraph，2012），美国也是如此（Davenport，1996），并且在英国，有 2 千万～2.5 千万尾鱼饲养在水族箱中，2 千万尾鱼饲养在池塘（PFMA，2014）。大约有 4000 种淡水和海水鱼品种作为宠物在公共场所进行展出。另外，全球观赏鱼贸易量庞大，估计价值为 30～40 亿英镑（OATA，2014）。观赏鱼个体的商业价值（零售）从 1 英镑（如金鱼）到超过 10000 英镑（如高品质锦鲤）不等。

宠物鱼品种繁多且来自不同的自然栖息地，因此，之前提及的野生鱼类蓄养（参见本章第 12.2 节）和水产养殖中（参见本章第 12.3.2 节）的所有福利问题也适用于观赏鱼。许多观赏鱼品种都是野生抓捕的，然后从亚马逊和珊瑚礁等地区长途运输多达 3 天（Walster，2008）。用船运输野生捕获的鱼类，其死亡率一般为 1%～30%（Ploeg，2005），因为当鱼类经历活鱼产业链的诸多环节时，环境恶劣以及运输和驯化都可以引发应激。死亡率在交货后也会变得较高（Ploeg，2005）。一般来说，90% 淡水观赏鱼都是养殖的，只有 10% 是在野外捕获的。养殖的观赏鱼往往来自以色列和远东地区的大型工厂，并且就疾病和死亡率而言，这些鱼或许存在着最大的福利隐患。例如，这些密集化工厂中的某些部分经常报告可能具有破坏性的病毒性疾病暴发（Chua，1996；Paperna 等，2001）。而对于海水性观赏鱼，95%的鱼类是野外抓捕的，只有5%是人工养殖的（Oliver，2001，2003；Sea Shepherd，2014）。因此，热带海水鱼类贸易可能会影响自然种群并对其保护和生物多样性产生负面影响。许多极端形态的观赏鱼已被观赏鱼饲养者培育出来，如水泡眼金鱼（UFAW，2013）和球花鳉（身体形态矮小且腹部胀大的玛丽鱼品种）（UFAW，2013）。由于应激和影响健康的缘故，这些都损害了其福利，加上由于观赏价值较高，它们的许多杂交种已培育出来（例如鹦鹉丽鱼），而这些品种的福利后果是未知的，因此还需要更多的证据。这和注重某些猫狗类的血统，想要选育出一些极端特征的情况相似，会导致动物健康问题（如有些狗类品种的鼻部缩短导致其患上短头阻塞性气道综合征）（Packer 等，2015）。

鱼类作为宠物的价格相对便宜，并且一般来说，任何成年人都能够买到一尾水族馆中或池塘中的鱼，即使他们除了对水质和食物要求之外，对养鱼一无所知（这与饲养哺乳类宠物相反，因为人类也是哺乳类，所以对哺乳类宠物的饲养和照顾有着更好的理解）。信誉良好的宠物商店往往会询问潜在客户的饲养水平，以便对鱼缸尺寸、水环境以及其他和特定种类相关的因素提供咨询。然而，并不是所有的店主都会这样做，以至于一些没有经验的饲养者在操作过程中会引起应激和较低的福利。

通过立法禁止残害宠物鱼的商业贸易（如染料文身、染料注射以及“整容”手术）是比较可取的，并且如果考虑到鱼类福利的重要性，也可以对事关福利的人造品系进行禁止或限制（如望天眼金鱼、脊柱侧突的鹦鹉丽鱼杂交种，以及由于不正常的身体形状而出现游泳问题的球花鳉）。另外，还可以对体积太大，不能养在家庭水族箱中的品种进行进口限制，如双齿巨脂鲤和一些大型鲶鱼品种（如鲑鲶、红尾鲶和铲鼻鲶），这可以防止它们生活在具紧张感的环境，否则它们会过早死亡或重新回到公共水族馆中。

野生抓捕的观赏鱼死亡率可能会比较高（Townsend，2011），可高达 30%（Ploeg，2005），这是由于船运的应激环境和动物的适应性应激造成的。对于常见的小丑鱼，运输时间超过 24 小时会导致应激，尽管其运输指南允许运输长达 3 天（Wolfenden 等，未发表数据）。除了从野外捕获鱼类之外，在进口国家研究可持续的育种计划，可以减少这些问题以及使用更短的运输时间，从而改善鱼类福利。

12.3.5 实验室环境中的科研应激

就鱼类在生物学和医学研究中的作用而言，其作为模式动物的使用机会正在增加。斑马鱼是一种重要的模式动物（Clark，Ekker，2015；Spagnoli 等，2016；见本书第 13 章）。就英国监管动物使用数量而言，鱼类仅次于小鼠的使用数量（Home Office，2014），并且鱼类在欧洲的使用情况正在增加（2008—2011 年期间，增长了 28%，如今约 130 万尾，欧洲，2015）。鱼类品种使用情况的增加可能反映了全球实验性动物的使用情况（如 USDA，2013）。Spagnoli 及其同事已介绍了应激的确切性质及其对鱼类福利的影响（2016 年；见本书第 13 章），但上述所有关于饲养、环境、营养和社会的应激源同样适用于实验性鱼类，因为实验性动物一般也处于蓄养环境下（参见本章第 12.2.1 节和第 12.3.2 节）（Reinhardt，2004）。例如，实验室水族箱可能会缺乏任何用于丰富环境的物体，尤其在斑马鱼的养殖设施中，即使斑马鱼饲养在贫瘠环境中 7 个月后，对于贫瘠环境和丰富环境的选择，斑马鱼还是选择底质和植物丰富的环境（Schroeder 等，2014）。欧洲法律严格规定了鱼类的健康和福利，并制定了明确的指导方针，以改善蓄养实验鱼的健康，因为在对这些鱼的日常处理中可能会导致应激、疼痛或持久伤害（European Directive，2010/63/EU）。其他的许多国家同样也有明确的准则用来提高鱼类福利，因为这有益于科学数据的收集。比如，美国渔业协会发布了鱼类在研究中的应用指南（AFS，2014），这对于北美科学家及护理人员（技术工人）是有帮助的。尽管实验室鱼类福利远远落后于哺乳动物，特别是啮齿类动物，但通过回顾有关啮齿动物居住和饲养的文献可以从中学到很多东西（Singhal 等，2014）。适度社交、感官刺激、鱼缸设计、营养、物理丰富度以及提供精神刺激的新物体是改善其幸福感的关键因素。然而，我们需要进行更多的研究才能真正了解实验室鱼类的需求，并从物种特异性的角度了解它们的生活史、环境耐受力和行为。这将确保所收集数据的完整性和对科学研究成果的相关性（Spence 等，2008）。

12.3.6　野生鱼类的应激和福利

12.3.6.1　鱼类调查

生物研究和渔业管理调查通常需要从野生种群中捕获和/或收集动物，任何这类的干预都可能对动物造成应激。最小化的应激或其效应应当不只是出于人道原因，还可能最大化地利用标记再捕获技术研究的成功率，可以确保鱼类在重新释放时或转移到别处时（如实验室、水族馆）保持最大的存活率。此外，当鱼类遭受不适宜的或落后的技术处理时，其行为的和生理的测量结果会受到低质量数据的影响，因此，我们应该关爱鱼类，以确保野外调查时鱼类处于最小的应激状态。

鱼类诱捕过程中包括许多应激源，这可能涉及中断鱼类自然行为的初始阶段（如防止鱼类沿其路线进行迁移）、实际诱捕阶段和最后处理阶段（Clements 等，2002）。捕获方法本身对鱼类健康和存活率有着直接的影响，尽管对如今各种捕获技术的弊端可能还没有全面了解（Clements 等，2002）。本节并非想要详尽列出捕获类型，但总结了最常见的捕获方法中的一些主要应激和福利问题来源。捕获一般使用渔网或者金属丝渔笼，无论是被动捕鱼（如放置好设备后等待鱼类游入）还是主动捕鱼（有目标的进行拖网捕鱼），这都是常见的方法，尤其在商业性捕鱼中。鱼类甚至在捕获前便会产生应激；另外，阻止鱼类达到其预期目标（如在其迁移路线上设置移动的障碍）会导致爆发性活动增加以及激活应激反应（Clements 等，2002）。鱼类在被捕获后同样面临着应激，可能由于被监禁而导致逃生无望的缘故（Clements 等，2002）。

电渔法是一种流行的替代方法，其通过对水域施加强电流使得鱼类迷失方向、昏迷，继而快速收集到大量鱼类样本，具有容易捕获和操作简单的优点。而一旦电场消失，鱼类会重新定向，并且恢复过来，这种方法往往对鱼不会造成外部性损伤，表明该方法对鱼类生理的影响有限。然而，对使用电渔法之后的鱼类进行 X 光检查，发现44%～67%的鱼类脊椎发生偏移和骨折损伤（Sharber，Carothers，1988）。电渔过程中非致死性损伤的程度和致死率随各种知之甚少的因素而变化，包括电场的性质和鱼类生物学特征，如个体大小和物种（Snyder，2003）。由于鱼类全身突然出现电压差而导致其惊厥，鱼类脊柱损伤和肌肉出血一般认为是在此期间发生的，而电压差可能发生在电渔设备开启或关闭时，因为电场具有脉冲特性，或者是当鱼类离开又重新进入水体时（如鱼类逃离反应时跳出水面或移除电场用网捕获时），由于这些证据的性质和质量问题，从而很难直接改善鱼类潜在的损伤和应激。电渔时建议使用连续电流（例如直流电）和低功率电流，以及改进鱼类捕获后的护理方法，包括减少处理时间并确保高质量的养育水池（Snyder，2003）。

12.3.6.2 维持环境

鱼类捕获后的维持环境应优先考虑鱼类福利，因此，了解如何改善维持环境对于相关的从业者是很重要的。尤其对于一些重视鱼类存活率和健康的行为（如鱼群的供应、鱼类调查、抓捕后释放行为，等等）。这里主要关注的是确保捕获和维持期间的应激最小化。相比之下，在捕捞后很快被屠宰的商业渔业中，维持鱼类福利可能被认为不那么重要，而且其船上也不适合装有确保良好福利标准的设施。然而，即使在这类渔业中，相关从业者也应注意鱼类死亡前对于其肉质的影响（如 Poli 等，2005），因此，无论从道德角度还是商业角度，我们都应该尽可能维持较高的福利标准。不过，本节将会着重关注于急需的持续性福利情况，当然，这些方法也适用于所有需要改善福利的方面。

陷阱捕获时以及之后的养育池或养育室中都可能会引发应激，在其过程中，鱼类血浆皮质醇浓度上升，而且可能由鱼类监禁所导致（Clements 等，2002；Portz 等，2006；Sharpe 等，1998）。实际上，监禁被认为是用于测量鱼类应激后皮质醇反应的一种标准化应激源（Pottinger 等，1992；Sharpe 等，1998），并且其根本原因被认为是过度拥挤以及水质影响（Portz 等，2006）。监禁造成鱼类不良影响的程度与监禁的持续时间和强度紧密相关，其高的维持密度（如鱼类在水族箱中即将被处理时）可以导致明显偏高的应激反应，而且该反应与慢性应激反应强度相当（Portz 等，2006）。因此，鱼类在维持养殖时应尽可能保持低密度养殖，合理养殖特定物种以及尽可能缩短养殖时间。

水溶氧量、温度、光照强度、放养密度（关于放养密度的更多细节参见本章第 12.2.1 节）和捕获的持续时间都被认为是所有与应激相关的维持环境中最重要的特征之一（Oldenburg 等，2011），并且有些因素取决于其维持的物种。水中溶氧量接近饱和（还未到引发气泡病的程度）是很重要的，尤其因为鱼类应激时其耗氧量会增加（如 Barton，Schreck，1987）。鱼类也应该养在对其物种最适宜的温度下。当热应激减少时，那些经历了外部损伤的鱼类（如用于标记，详见本章第 12. 3. 6. 3 节）可以提高愈合能力，在适当温度下可以降低免疫抑制的可能性而使鱼受益（Oldenburg 等，2011）。此外，温度的快速变化对鱼类健康也是不利的。因此，养育水族箱或养育栏应保持温度的稳定，并且和鱼类捕获时的当地温度相似（Oldenburg 等，2011）。

最后，尽管捕获时间应该最短化，但维持养殖时的人工条件难免会引起应激（Portz 等，2006），所以，在任何侵袭性操作之前，我们应为鱼类提供一段恢复期（Oldenburg 等，2011）。恢复期应当足以使鱼恢复体内稳态，因此取决于应激反应的程度和持续时间，而应激本身由操作过程决定。然而，我们明确建议要确保鱼类能尽快返回野外，因此要迅速进行操作，并在适当的环境中恢复，而且如果我们希望减少应激，则应在鱼类恢复期后立即将它们释放（Clements 等，2002；Oldenburg 等，2011）。

12. 3. 6. 3 标记及入侵性处理

鱼类标记已成为监视鱼类种群和/或其个体的重要方面。然而，除了捕获和维持养育外，标记的植入也会引起应激。根据处理难度可将鱼类标记分为多种，还可以分为可快速恢复的短期效应类型，以及伴随鱼类终生直至标记脱落的长期效应类型，不同的鱼类标记有着不同的效应。标记可以是外部类型的，如弹出式卫星档案标记以及锚/T 形标记（尽管属于外部标记，但还需要用枪注射进组织中），或者是一些简单的处理，如用夹子夹住鳍条；标记还可以是皮下注射类型的，如荧光色素可见标记；还可以是通过插入胃部或者通过手术植入体腔的内部标记，如被动整合雷达标记（Furtbauer 等，2015；Hoolihan 等，2011；Newby 等，2007）。研究人员很早就知道标记可以造成短期行为改变（通过捕获和处理时鱼类的一些生理反应）以及长期行为改变（通过受伤、对标签的适应或者带上有害的标签），然而应激和行为改变的程度往往并不知道或者不够清楚（Hoolihan 等，2011）。

外部标签操作迅速且易于应用，并且手术相关部分的侵入性较低。然而，它们依然能引起鱼类明显的应激和/或疼痛。当尼罗罗非鱼（*Oreochromis niloticus*）和鲤鱼（*Cyprinus carpio*）尾鳍夹上标记后，其皮质醇反应并不会强于施加相关处理程序的鱼类；然而，该鳍条夹片确实造成了鱼类鳃部生理和疼痛的变化，以至于鱼类恢复期至少要 6 小时（Roques 等，2010）。和对照组相比，实验组鱼类产生了行为上的改变，包括游泳活动增加以及不会待在对照组个体经常出没的水族箱区域，而在 6 小时后这些差异便消失了。此外，鱼类 C 类神经纤维和 A-δ 类神经纤维也表明在鱼类行为恢复期间，其伤害感受能力和所观察到的哺乳动物隐痛状况相似（Roques 等，2010）。T 型标记可使用或者不使用麻醉，这取决于具体情况。被标记鱼类的死亡率可能会更高（Jolivet 等，2009）。在这项研究中，标记对生长的任何影响都被标记鱼类的高死亡率所掩盖。然而，其他研究表明外部标记对鱼类摄食或游泳行为是没有影响的（Berumen，Almany，2009）。因此，外部标签对鱼类的影响可能是广泛范围的，并且鱼类标记后生存能力的决定因素尚不清楚。

VIE 标记由一种彩色的流动弹性体组成，这需要通过皮下注射。为此，鱼类往往需要先麻醉，但也并不是所有的 VIE 标记都需要麻醉。三刺鱼（*Gasterosteus aculeatus*）标记时会呈现应激状态，其反应的生理变化类似暴露于捕食者后的生理变化（Fürtbauer 等，2015）。然而，这不一定意味着标记处理是一种尝试性的捕食活动，并且其标记后的恢复率也不同于经历捕食后依旧存活的状况。

内标法必定是具侵袭性的，并且对于鱼类应激和健康的影响呈现多样化。胃部标记不需要借助手术；相反地，标记是通过口/食道引入而摄取的，使鱼强行或者自愿咽下标记（尽管这仍然需要哄骗并将标签引入口中）。但无论是强制还是自愿行为，都会产生应激，与自愿吞咽标记的鱼类相比，强制吞喂的反应可能会更加强烈，导致更快速的胃部反刍（约 30 天而不是 60 天），还会产生厌食以及降低鱼类的存活率（Winger，Walsh，2001）。将鱼类仔细处理，尽量减少接触，处理人员间良好配合，可以使胃部标记植入时鱼类的应激最小化，且相对于未标记的同种鱼类，其行为的改变也可以减少（Smith 等，2009）。相比之下，腹腔内/体腔内的标记必须通过手术植入，而且往往要使用到麻醉（但是，尤其对于食用性鱼类，其化学麻醉剂的使用受到严格管制，通常进行手术后需要长时间停药）（Trushenski 等，2013）。与经过处理的对照组鱼类或麻醉状态下处理的鱼类相比，经历标记植入过程的鱼类的血浆皮质醇浓度较高，显示其产生了应激反应（Lower 等，2005）。从长远来看，与手术进行部位相关的身体组织的健康状况存在明显问题（Caputo 等，2009）。尽管关于生理健康的报告结果各不相同，如一些研究表明，标记植入导致的长期生理健康状况不佳（Montoya 等，2012），而也有一些报告显示，该过程对于鱼类并无长期的生理影响（Caputo 等，2009；Loher，Rensmeyer，2011）。

因此，标记的使用需要仔细考虑最合适的标记类型，如果我们重视该问题，不仅需要考虑到实验因素，如需要收集的数据类型和不同类型标记对于鱼类的恢复速率，还需要考虑到标记过程对于鱼类及其健康的直接影响和长期影响。

疫苗接种也是一个重要的操作程序，特别是在水产养殖中，它可以通过提供抗生素来避免疾病的感染并最大限度地减少对当地生态系统可能产生的影响（参见本章第 12. 3. 2 节）。然而，疫苗接种需要对鱼类暂时进行手工操作和监禁，并给予注射，这一系列行为会对鱼类福利和应激产生影响。而且，腹腔注射疫苗会使注射部位受伤，这也会导致应激和引发福利问题（Poli，2009）。这会导致鱼类行为异常、食欲降低，从而影响其生长。显然，疫苗接种的好处（提高免疫能力和减少对抗生素的依赖）必须根据具体情况与应激成本进行权衡。重要的是，应仔细考虑整个过程中的护理水平，包括从诱捕到后续的释放，尽量减少对鱼类的影响。

鱼类的处理包含一系列动作，如测量、夹上鳍条、检查标记/夹片，这是所有鱼类调查的一部分，并且可以产生显著的应激反应，而这些反应和鱼类在诱捕陷阱或者水池中被监禁时的情形是不一样的（Clements 等，2002）。除了手工操作（参见本章第 12. 3. 2 节），鱼类露出水面被认为是一种急性应激源，它可以产生非常高的血浆皮质醇浓度（如 Pickering，Pottinger，1989）。此外，企图捕获鱼类（如用来自鱼缸的渔网捕鱼）可以模拟捕食者追逐运动，这会引起鱼类的生理反应（Brown 等，2007）。

还有，任何手工操作都可能破坏鱼类的黏液上皮层，导致伤害和感染。因此，这时的手工操作必须快速进行，减少接触，并且必须操作最短的时间从而最大限度地减少任何负面生理效应。处理人员之间的有效组织和协调可以减少因手工操作而产生的任何应激（Smith 等，2009）。使用一些特殊的网具（可以是某些品种专用的网）或用湿毛巾/进行手工操作也可以最大限度地减少对鱼类组织或黏膜层的物理损害（Ashley，2007）。

12.3.6.4　水坝和其他障碍物

鱼类可能经常接触一些阻碍它们沿着预定路径移动的障碍物。这些可能是人为的，如水坝和水力涡轮机，也可能是天然屏障，如瀑布、高陡坡或者一些废弃物（如河狸坝）。尤其对于迁移物种，这些都是它们前进路上影响重大的障碍物，阻止它们往上或往下游动。然而，在多数情况下，障碍物会导致鱼类产生应激和损伤，其效果会根据替代路线的可用性而变化，并且通过大坝时根据其所使用的涡轮机类型，效应也会发生改变（Brown 等，2012）。鱼类群落可能围绕天然屏障进行演化，使得特定物种/种群在地理上被隔离，而其他的群体能够穿越这些障碍（Torrente-Vilara 等，2011）。然而，在关于人为障碍物影响的研究文献中，基本上忽视了天然屏障对于鱼类迁徙的影响（Thorstad 等，2008）。天然屏障可能导致溪流栖息地的显著分裂，尽管我们对这种分裂后果的了解目前还是有限的（Bourne 等，2011）。此外，水坝对鱼类的影响将取决于鱼类品种及其生活史（例如迁徙与非迁徙，以及不同发育阶段的迁移方向）（Coutant，Whitney，2000）；无论天然屏障还是人为障碍物，都会出现相似的变化。本节主要侧重于改善人为障碍物的影响后果。

水坝建立时往往带有通道（如溢洪道以及排污架），和鱼类试图强行从水力涡轮机冲出去相比（死亡率为 5%～15%）（Muir 等，2001；Whitney 等，1997），在通道中几乎可以完全存活下来（死亡率 <2%）（Muir 等，2001；Whitney 等，1997）。由于鱼类在水柱中无法定向以及维持自己的位置，它们会和一些坚固结构的物质（如水坝的墙体）或者是涡轮机本身发生摩擦，导致死亡率增高和受伤（Coutant，Whitney，2000；Brown 等，2012），并且会使其气鳔出血、破裂（Brown 等，2012）。鱼类通过涡轮机的能力在很大程度上取决于它们的生物学特征：鱼类在水柱中的垂直位置受其所受浮力影响，而这会影响到它们的通过路径，因此，浮力方法可以直接决定鱼类是否能安全通过涡轮机。带有气鳔的鱼类无法迅速调整浮力，从而不能解决通过涡轮机时向下的拉力（Alexander，1993），导致其迷失方向。这些鱼往往会通过往上游动以获得浮力平衡，使得它们相对于涡轮机而重新定向（Coutant，Whitney，2000）。不同的是，善于利用动态升力的鱼类通过涡轮机时会下沉，因为它们穿过湍急水流时无法保持向前的推力（Coutant，Whitney，2000）。另外，任何植入了体腔遥感标签的鱼类都可能会面临更高的损伤或死亡风险，标签相对于鱼类的体积越大，风险就越高（Brown 等，2012）。通过平衡浮力，外部连接的声波辐射器，可以在很大程度上消除这种危险（Brown 等，2012）。尽管这些标记会造成组织损伤，并且和游泳能力降低、生长减缓和存活率降低有关，但使用适合于该物种且精心挑选的标记方

法可以减少这些负面影响（Jepsen 等，2015）。此外，专门为通过水坝而开发的标记已初步具有可观的结果，但还需要进一步开发（Brown 等，2013）。

因此，最安全的水坝通道可能是那些绕过涡轮机的路线，但这些路线的作用同样取决于生物学特征，尤其是鱼类的游泳模式。例如，向下游迁移的幼年鲑鱼经常会利用水柱上层，因此，它们比成年鲑鱼更有可能利用溢洪道（Giorgi，Stevenson，1995；Coutant，Whitney，2000）。如果没有这样的路线，这些鱼通常会下降到更深的水域寻找通道（Coutant，Whitney，2000）。溢洪道的功效同样取决于其结构。增加溢洪道的排出量（通过溢洪道的水量而不是通过涡轮机的水量）会增加间接利用溢洪道的鱼类数量（Coutant，Whitney，2000）。溢洪道的水流速度更快时也可以增加鱼类成功通过的机会，这可能与季节因素相关。许多迁徙的鱼类会在水流最快的水层中排成一列，这些地方一般在河流中央上层，和通道的摩擦力是最小的（Coutant，Whitney，2000）。最后，与河流流动相一致的溢洪道也可以为数量最多的鱼类所利用（Willis，1982；Coutant，Whitney，2000）。因此，根据鱼类如何使用溢洪道，合理放置其位置可以最大限度地减少强行通过涡轮机的鱼类数量，提高通过水坝的鱼类的存活率。此外，如进口拦网、施加电场以及水下声音和光照之类的一些引导手段也可以引导鱼类沿特定路线游动。

成年鲑鱼返回其出生的河流进行产卵时会出现一系列不同的问题。它们通过涡轮机以及水坝的通道一般是借助鱼道或鱼梯（Pon 等，2009）。通过这种方式穿越水坝仍然是一个重大困难。因为血浆皮质醇反应比较短暂，或者相对于迁徙的整体应激其效应较低，穿过水坝引起的应激差异就可能很难测量出来（Pon 等，2009；Roscoe 等，2011）。然而，和上游迁徙过程中没有水坝的鱼类相比，经历水坝穿越的产卵鲑鱼的死亡率显著升高（Pon 等，2009；Roscoe 等，2011）。尽管如此，至今为止还是没有明确的生理相关性或行为相关性证据显示鱼类能否成功穿越水坝（Pon 等，2009）。鱼类性别可能是一个因素。一项研究发现，穿越水坝后的雌性红大麻哈鱼（*Oncorhynchus nerka*）要少于雄性红大麻哈鱼（Roscoe 等，2011）。鱼类无法通过水坝时应激可能会增加。而如果河流被网状物阻塞时，鱼类无法继续前进，便可能会增加其游泳行为，血浆皮质醇也随之增加（Clements 等，2002）。因此，过激的无氧游泳运动可能会抑制接下来的行为。然而，对于那些在水坝前游动却又无法找到通道的成年鲑鱼，并没有发现任何剧烈游泳或代谢状态受损的证据，除了血浆 Na^+ 浓度有所降低外，其他仍然在可接受的标准范围内（Pon 等，2009）。水坝、江堰和发电站会耽误鱼类迁徙数天至数周；企图移动的大西洋鲑鱼（*Salmo salar*）往往会忽略绕行路线，这可能是因为沿着水坝/发电站游动的模式已形成固定的印记，或者是因为相较于绕行路线的低排水量，水坝高排水量显得更具吸引力（Thorstad 等，2008）。此外，水坝对于那些不进行迁徙但又偶尔出现在其中的鱼类是有风险的，特别是游泳能力较差或较弱的幼鱼，它们可能会被意外地吸入涡轮机里（Coutant，Whitney，2000）。为减少鱼类这种风险而提供办法将会非常困难。

数据表明，水坝为上下游的鱼类提供了重要屏障，但也对栖息在附近地区的鱼类构成了威胁。水坝安全通道的完善似乎可以减轻对鱼类造成的大部分损害，并且实验观察到确实降低了死亡率。但是，尤其对于向水坝上游迁移的鱼类，这仍然需要改进。

12.3.6.5　水产养殖以及对野生鱼类的影响

饲养鱼类与其所在的当地自然种群往往会发生相互作用（Skaala 等，1990），这对自然种群的健康有着重要的影响。这样的相互作用可能是由于养殖场附近存在被流出网箱的未食用饲料所吸引的野生鱼类（Dempster 等，2009），或者是由于饲养鱼类因拦网毁坏或人员失误而从养殖场逃逸出来（Grigorakis，Rigos，2011；Johanesen 等，2011）。相互作用的性质相当多样化，包括繁殖以及因此造成的两方面种群基因组变化。饲养鱼类通过近亲繁殖会导致基因多样性的丢失，并且如果这些基因引入野生鱼类，一些不良的遗传特征（如病害耐受力降低、生长速率降低或过快、繁殖力增加或减少、配子/后代质量较差）对野生种群的遗传是有害的。这可能发生在饲养鱼类放生或鱼类逃出养殖场时，但对于促使或限制饲养鱼类与野生鱼类基因混合的因素，目前仍不清楚（Skaala 等，1990）。然而，其他的一些相互作用如竞争、捕食和病原体引入（Skaala 等，1990），对野生种群有着更直接的影响。例如，养殖场的水排放是通过开放的拦网，这对于病原体的通过是根本无法限制的（Johanesen 等，2011），而由于野生鱼类分布较为分散，病原体可能很容易传播。

水中废弃物、未食用的饲料与未被消化的饲料的混合物以及鱼类的排泄物可通过有机负荷方式导致水质恶化（Grigorakis，Rigos，2011），包括增加养殖场自然环境中的可溶性氮和磷化合物含量。这对当地生态环境的影响最为强烈，会造成微生物负荷和污染环境的有机体增加，从而使底栖息生物群落发生变化。养殖场持续使用抗生素也是需要考虑的重要方面，这和饲养鱼类对于病原体的耐药性增加有关；不过，药物的毒性似乎对当地生态系统的影响有限（Grigorakis，Rigos，2011）。但是，这些作用会直接影响到栅栏内的饲养动物，尤其是关系到水质，会造成动物福利问题。

在水产养殖和养鱼场内，必须制定明确的管理措施，以确保干净水源的提供并使其流通整个养殖系统，清除废物并补充氧气。疾病和病原体也必须妥善处理，但在所有情况下都必须确保不对当地的环境健康造成损害。

12.3.7　手术和麻醉

麻醉（Sneddon，2011）和手术会引起鱼的应激反应。在野生鱼类中，麻醉和手术通常用于植入侵入性标记（参见本章第 12.3.6.3 节），从中获得了大量关于渔业管理和鱼类生物研究的重要生物学知识（Loher，Rensmeyer，2011）。对比目鱼进行内部标记时，会导致其发生腹腔粘连，以此可能造成疼痛感，但和对照组相比，这些鱼的生长速率并无差异，所以不太可能造成应激（Loher，Rensmeyer，2011）。体腔内标记程序应包括良好的外科手术，包括术前和术后护理、灭菌器械、麻醉、使用吸收性缝合线进行伤口闭合，以及使用抗生素（尽管对其效用存在争议）（Brown 等，2011）。实验室环境中，鱼可能会受到诸多的侵入性手术，如标记植入、鳍片夹带标记、插管、脊柱神经和视神经的挤压、脑部损伤、肠道切除以及心脏手术（如 Itou 等，2014；SaeraVila 等，2015；Schall 等，2015）。此外，良好的外科手术对于促进伤口愈合和减少继发性感染至关重要。鱼类手术过程中并不经常使用镇痛程序，但最近的研究表明，镇痛可能会在鱼类潜在的痛苦时期改善其相应的行为和生理变化（Sneddon，2011，2015）。

由于许多药剂对鱼类应激会产生影响，我们应当谨慎考虑麻醉（Sneddon，2011）。多数麻醉药物通过浸泡作用于鱼类。正确的给药可以导致急性手术时有效麻醉以及鱼类失去意识。然而，需要注意的是，不同鱼类品种间，麻醉药物种类和剂量是不同的，并且麻醉药物的效用可能受到各种生理因素（例如体重、应激状态）以及环境条件（例如温度）的影响。目前，在实验室、兽医和水产养殖环境中已使用的药物偏多。最常见的是缓冲的 MS-222（三卡因）、苯佐卡因、异丁香酚、美托咪酯、2-苯氧基乙醇和喹哪啶（Ackerman 等，2005；Ross，Ross，2008；Neiffer，Stamper，2009）。目前对这些药物的诱导作用、回收率和药物动力学以及不良作用或副作用已探究清楚（Sneddon，2011）。然而，这些麻醉药物仅限于少数物种，所以应该慎重地对未经测试的物种使用这些物质。最近的研究表明，MS-222 和苯佐卡因会使斑马鱼产生厌恶情绪，而美托咪酯和 2,2,2-三溴乙醇（TBE）则不会引起该反应（Readman 等，2013）。但是，这些非厌恶性麻醉药物并不具备镇痛功能，而 MS-222 和苯佐卡因是具有缓解疼痛特性的局部麻醉剂，因此，问题仍然是在侵入性操作中使用这些药物是否会有利于鱼类。其他的一些麻醉方法如二氧化碳法和低温冷却法都备受争议，麻醉时应当避免二氧化碳法的使用，因为这会导致鱼类外周组织酸中毒，在虹鳟中会激发其伤害感受能力，如今该方法已被欧盟禁止（Mettam 等，2012）。低温冷却法或许仅能产生麻痹功效，但已有研究证实，对于如斑马鱼这样的小型热带物种，该法比 MS-222 麻醉导致的应激更小（Wilson 等，2009）。低温冷却法必须使水温保持在 4℃，并且鱼类不可以接触到冰以防体内冰晶的形成。为了确保鱼类的良好福利，选择麻醉方法前应充分了解该方法的使用对象及它们对特定方法或药物的反应。

12.4 结论和未来方向

显然，鱼类所处的蓄养环境和野外环境会产生多种生理和行为差异。蓄养鱼类通过发育、经验和遗传因素来缓解其应激反应和行为特征，这可能会减少将它们释放到天然环境时的存活率或成功率。同样的，当鱼类从野外环境转到实验室环境、观赏鱼环境或养殖场时，它们要面临一种不同的社会和物理环境，这种环境更有利于一些不同的表型生存，而不是野外环境成功存活的鱼类表型。鱼类在不同环境下的应激管理极大地方便了其相关研究人员和护理人员。鱼类的异型稳态反应与应激和鱼类幸福健康的心理学影响相关，我们对于这些研究已有进展，这意味着我们可以探索不同的策略来改善饲养鱼类的福利。显然，减少鱼类异型稳态的负荷是至关重要的，尤其是当应激即将发生时，这可以避免应激叠加性效应的产生。因此，相关人员要谨慎考虑活动程序的时序安排。在大西洋鲑鱼疫苗接种前后，其会产生应激的累加效应（图 12.5），这时可以观察到鲑鱼的 HPI 轴活动和死亡率都会升高，最终导致异型稳态超负荷（Iversen，Eliassen，2014）。如果把最大限度地减少应激作为目标，那么确保鱼类有足够的时间从应激中恢复过来是必要的。进一步使鱼类能够控制或预测应激事件则应该是使用鱼类的科学家的目标。蓄养环境中的丰富度可以为鱼类的挫败感及其一些替代行为提供发泄口，或者说这是一种可以使鱼类个体逃离或者躲避应激的方法。一些简单的传统调节方法（如光照信号）可以预示应激源的发生，这在蓄养环境中可以使鱼类预测或准备好应激的到来。此外，操作性条件反射也是一种给予鱼类控制感的方法，这是可以让鱼类自主控制的自动投喂机，有益于鱼类福利。这些认知性策略确实还需要在不同的环境中以及对不同的物种进行研究。由于受捕获环境调节的鱼类行为、生态及其生活史具有多样性，鱼类放养密度及其对觅食行为的影响也是如此。了解清楚异型稳态负荷可接受范围将是向前迈出的重要一步，因为过少的应激可能也会如同过多的应激一般导致负面的福利。

许多其他的因素也会影响鱼类个体应对应激的能力，如基因型、应激应对方式、既往经验，以及从亲本到子代的应激历史性转移，这会改变鱼类直至成年的发育和行为（Andersson 等，2011）。我们需要进行研究以了解这些因素在表型形成中的作用。耐受应激鱼类的产生（例如，高应激反应虹鳟品系和低应激反应虹鳟品系的繁殖）（Pottinger，Carrick，1999）可能是另一种选择。否则，如果想要避免应激以及可能的死亡情况，就需要在养殖管理方面进行护理，特别是在运输过程中以及在野生栖息环境与蓄养栖息环境之间一些不同的地方，这可以最大限度地提高鱼类存活率以及提供最合适的福利。这在确保捕获的实验室鱼类可以展现野外类型的反应时尤为重要，它关系到科学数据能否具有从实验室环境到野外环境的广泛性，以防止实验室鱼类受到驯化后会影响到它们的自然反应，从而无法推断出它们处于野外环境时的反应（Christie 等，2012）。

最后，我们还需要一种识别早期应激的方法，以便采取补救措施。例如，在运输大西洋鲑鱼时，可以利用耗氧量和鱼类行为来监视其健康状况（Farrell，2006）。当行为偏离范围时，可以提供警报的自动监测系统将是一种有价值的工具。这需要在养育水池的侧面、上方甚至其内部安装摄像机来记录其行为，但还需要清楚每个物种的正常行为。此外，还可以对鱼类进行生理测量。这有助于了解鱼类何时从应激状态中恢复过来，以便下一步程序可以有效按时进行，以减少对鱼类的影响。行为监视系统可以用于监视途径以及整个 RAS 系统中虹鳟的福利情况（Colson 等，2015），尽管每个水族箱鱼的数量很少，仅有 50 条（放养密度：30 kg/m^3）。开发此类系统对水产养殖、实验室水族箱和观赏鱼产业都十分有益，将对减轻鱼类应激和改善其福利产生广泛影响。除此之外，为鱼类提供合适的环境、水质、社交、正确的喂食机会以及饲养时使用一些干扰性行为使得鱼类能获得预期、准备和控制能力，这都可以减少鱼类的应激以及提高其福利。而且，当鱼类受到水产养殖、渔

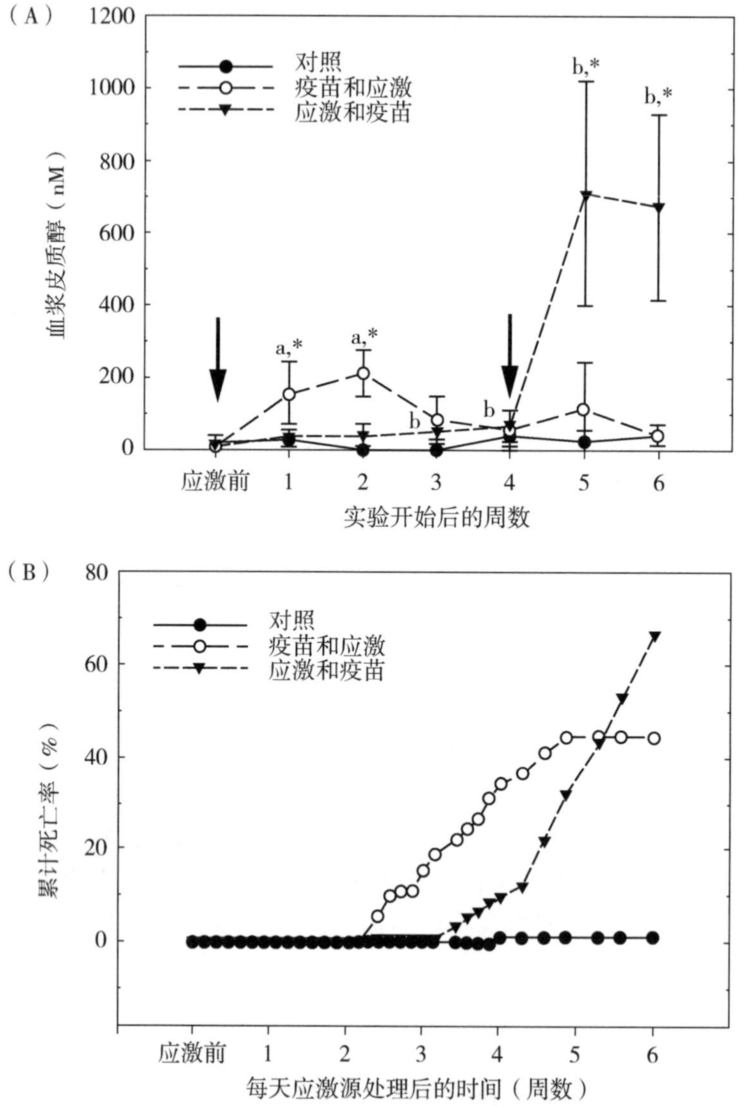

图 12.5 大西洋鲑鱼接种疫苗前后产生应激的累加效应

注：（A）在 4 周内，每天施加应激源对于先前二次疫苗接种的大西洋鲑鱼（*Salmo salar*）（疫苗应激组）和之后二次疫苗接种的大西洋鲑鱼（*Salmo salar*）（应激疫苗组）分别的影响。对照组不施加二次疫苗接种及日常应激源。数值以 mean ± SD（n = 6）表示。应激的显著性变化（$p < 0.05$）以字母表示，组别间的显著性变化（$p < 0.05$）以星号表示。黑色箭头表示疫苗应激组二次接种的时间，灰色箭头表示应激疫苗组二次接种的时间。（B）4 周内日常施加应激源对于疫苗应激组和应激疫苗组分别的影响（总的累加死亡率%），对照组不施加二次疫苗接种及日常应激源。数据来源于 Iversen M H and Eliassen R A（2014）. The effect of allostatic load on hypothalamic-pituitary-interrenal（HPI）axis before and after secondary vaccination in Atlantic salmon postsmolts（Salmo salar L.）. *Fish Physiol. Biochem.* 40，527-538. 已获 *Springer* 许可。

业、捕捞和释放、实验以及作为观赏鱼的应激时，继续使用最人道的方法或者减少应激后的操作，可以保证我们所使用鱼类的健康幸福。对于任何蓄养的鱼类种群，开发有效的应激指标对于管理鱼类健康是重要的，这可以使鱼类面对应激挑战时保持合适的行为性和生理性反应。这些指标可用于提醒调查人员或相关养殖管理人员何时会出现与预期标准不同的偏差，从而进行干预和调整。因此，如果我们的目标是尽量减少应激以确保鱼类健康，未来的研究应着重于开发有意义的、实用的鱼类福利指标。

L. U. 斯纳顿　D. C. C. 沃尔芬登　J. S. 汤姆森　著
李水生　译
林浩然　校

参考文献

Abbott, A. (2015). Clever fish. Nature. 521, 412-414.

Ackerman P. A., Morgan J. D., Iwama G. K. (2005) Anesthetics. CCAC guidelines on: The care and use of fish in research, teaching and testing, Canadian Council on Animal Care, Ottawa Canada. Available at: http://www.ccac.ca/Documents/Standards/Guidelines/Fish.pdf.

AFS. (2014). http://fisheries.org/guide-for-the-use-of-fishes-in-research.

Agetsuma, M., Aizawa, H., Aoki, T., Nakayama, R., Takahoko, M., Goto, M. (2010). The habenula is crucial for experience-dependent modification of fear responses in zebrafish. Nat. Neurosci. 13, 1354-1356.

Alanärä, A. (1996). The use of self-feeders in rainbow trout (*Oncorhynchus mykiss*) production. Aquaculture. 145, 1-20.

Alanärä, A., Brännäs, E. (1996). Dominance in demand-feeding behaviour in Arctic charr and rainbow trout: the effect of stocking density. J. Fish. Biol. 48, 242-254.

Alexander, R. M. (1993). Buoyancy. In: Evans, D. H., The Physiology of Fishes. CRC Press, Boca Raton, FL, 75-97.

Andersson, M. A., Silva, P. I. M., Steffensen, J. F., Hoglund, E. (2011). Effects of maternal stress coping style on offspring characteristics in rainbow trout (*Oncorhynchus mykiss*). Horm. Behav. 60, 699-705.

Ashley, P. J. (2007). Fish welfare: current issues in aquaculture. Appl. Anim. Behav. Sci. 104, 199-235.

Ashley, P. J., Sneddon, L. U. (2008). Pain and fear in fish. In: Branson, E. J., Fish Welfare. Blackwell Publishing, Oxford, 49-77.

Ashley, P. J., Ringrose, S., Edwards, K. L., McCrohan, C. R., Sneddon, L. U. (2009). Effect of noxious stimulation upon antipredator responses and dominance status in rainbow trout. Anim. Behav. 77, 403-410.

Balon, E. K. (2004). About the oldest domesticates among fishes. J. Fish. Biol. 65, 1-27.

Bart, A. N., Clark, J., Young, J., Zohar, Y. (2001). Underwater ambient noise measurements in aquaculture systems: a survey. Aquacul. Eng. 25, 99-110.

Barton, B. A. (2002). Stress in fishes: a diversity of responses with particular reference to changes in circulating corticosteroids. Integr. Comp. Biol. 42, 517-525.

Barton, B. A., Schreck, C. N. (1987). Metabolic cost of acute physical stress in juvenile steelhead. Trans. Am. Fish. Soc. 116, 257-263.

Benhaïm, D., Péan, S., Lucas, G., Blanc, N., Chatain, B., Bégout, M.-L. (2012). Early life behavioural differences in wild caught and domesticated sea bass (*Dicentrarchus labrax*). Appl. Anim. Behav. Sci. 141, 79-90.

Benhaïm, D., Guyomard, R., Chatain, B., Quillet, E., Bégout, M.-L. (2013). Genetic differences for behaviour in juveniles from two strains of brown trout suggest an effect of domestication history. Appl. Anim. Behav. Sci. 147, 235-242.

Berumen, M. L., Almany, G. R. (2009). External tagging does not affect the feeding behavior of a coral reef fish, *Chaetodon vagabundus* (Pisces: Chaetodontidae). Environ. Biol. Fish. 86, 447-450.

Biswas, A. K., Seoka, M., Takii, K., Maita, M., Kumai, H. (2006). Stress response of red sea bream *Pagrus major* to acute handling and chronic photoperiod manipulation. Aquaculture. 252 (2-4), 566-572.

Bjørge, M. H., Nordgreen, J., Janczak, A. M., Poppe, T., Ranheim, B., Horsberg, T. E. (2011). Behavioural changes following intraperitoneal vaccination in Atlantic salmon (*Salmo salar*). Appl. Anim. Behav. Sci. 133, 127-135.

Blanco-Vives, B., Aliaga-Guerrero, M., Cañavate, J. P., Muñoz-Cueto, J. A., Sánchez-Vázquez, F. J. (2011). Does lighting manipulation during incubation affect hatching rhythms and early development of sole?. Chronobiol. Int. 28 (4), 300-306.

Bosakowski, T., Wagner, E. J. (1995). Experimental use of cobble substrates in concrete raceways for improving fin condition of cutthroat (*Oncorhynchus clarkii*) and rainbow trout (*O. mykiss*). Aquaculture. 130, 159-165.

Bourne, C. M., Kehler, D. G., Wiersma, Y. F., Cote, D. (2011). Barriers to fish passage and barriers to fish passage assessments: the impact of assessment methods and assumptions on barrier identification and quantification of watershed connectivity. Aquat. Ecol. 45, 389-403.

Brännäs, E., Alanärä, A. (1994). Effect of reward level on individual variability in demand feeding activity and growth rate in Arctic charr and rainbow trout. J. Fish Biol. 45, 423-434.

Brockmark, S. , Neregard, L. , Bohlin, T. , Bjornsson, B. T. , Johnsson, J. I. (2007). Effects of rearing density and structural complexity on the pre-and postrelease performance of Atlantic salmon. Trans. Am. Fish. Soc. 136, 1453-1462.

Broom, D. M. (2008). Welfare assessment and relevant ethical decisions: key concepts. Annu. Rev. Biomed. Sci. 10, T79-T90.

Brown, C. (2015). Fish intelligence, sentience and ethics. Anim. Cogn. 18, 1-17.

Brown, C. , Davidson, T. , Laland, K. (2003). Environmental enrichment and prior experience improve foraging behaviour in hatchery-reared Atlantic salmon. J. Fish. Biol. 63 (S1), 187-196.

Brown, C. , Burgess, F. , Braithwaite, V. A. (2007). Heritable and experiential effects on boldness in a tropical poeciliid. Behav. Ecol. Sociobiol. 62, 237-243.

Brown, J. A. , Watson, J. , Bourhill, A. , Wall, A. (2010). Physiological welfare of commercially reared cod and effects of crowding for harvesting. Aquaculture. 298, 315-324.

Brown, R. S. , Eppard, M. B. , Murchie, K. J. , Nielsen, J. L. , Cooke, S. J. (2011). An introduction to the practical and ethical perspectives on the need to advance and standardize the intracoelomic surgical implantation of electronic tags in fish. Rev. Fish Biol. Fisher. 21, 1-9.

Brown, R. S. , Pflugrath, B. D. , Carlson, T. J. , Deng, Z. D. (2012). The effect of an externally neutrally buoyant transmitter on mortal injury during simulated hydroturbine passage. J. Renew. Sustain. Energy. 4 (1), 013107.

Brown, R. S. , Deng, Z. D. , Cook, K. V. , Pflugrath, B. D. , Li, X. , Fu, T. (2013). A field evaluation of an external and neutrally buoyant acoustic transmitter for juvenile salmon: implications for estimating hydroturbine passage survival. PLoS One. 8 (10), e77744.

Brydges, N. M. , Braithwaite, V. A. (2009). Does environmental enrichment affect the behaviour of fish commonly used in laboratory work?. Appl. Anim. Behav. Sci. 118, 137-143.

Bshary, R. , Oliveira, R. F. , Oliveira, T. S. F. , Canário, A. V. M. (2007). Do cleaning organisms reduce the stress response of client reef fish?. Front. Zool. 4, 21.

Bui, S. , Oppedal, F. , Korsøen, Ø. J. , Sonny, D. , Dempster, T. (2013). Group behavioural responses of Atlantic salmon (*Salmo salar* L.) to light, infrasound and sound stimuli. PLoS One. 8 (5), e63696.

Campbell, P. M. , Pottinger, T. G. , Sumpter, J. P. (1992). Stress reduces the quality of gametes produced by rainbow trout. Biol. Reprod. 47, 1140-1150.

Cañon Jones, H. A. , Hansen, L. , Noble, C. , Damsgård, B. , Broom, D. M. , Pearce, G. P. (2010). Social network analysis of behavioural interactions influencing fin damage development in Atlantic salmon (*Salmo salar*) during feed-restriction. Appl. Anim. Behav. Sci. 127, 139-151.

Caputo, M., O'Connor, C. M., Hasler, C. T., Hanson, K. C., Cooke, S. J. (2009). Long-term effects of surgically implanted telemetry tags on the nutritional physiology and condition of wild freshwater fish. Dis. Aquat. Organ. 84 (1), 35-41.

Carlstead, K. (1996). Effects of captivity on the behavior of wild mammals. In: Kleiman, D. G., Allen, M. E., Thompson, K. V., Lumpkin, S., Wild Mammals in Captivity: Principles and Techniques. The University of Chicago Press, Chicago.

Christie, M. R., Marine, M. L., French, R. A., Blouin, M. S. (2012). Genetic adaptation to captivity can occur in a single generation. PNAS. 109, 238-242.

Chua, F. H. C. (1996). Aquaculture health management in Singapore: current status and future directions. In: Subasinghe, R. P., Arthur, J. R., Shariff, M., Health Management in Asian Aquaculture. Proceedings of the Regional Expert Consultation on Aquaculture Health Management in Asia and the Pacific. FAO, Fisheries Technical paper No. 360, Rome, 115-126.

Clark, K. J., Ekker, S. C. (2015). How zebrafish genetics informs human biology. Nat. Educ. 8 (4), 3.

Clark, J., Young, J., Bart, A. N., Zohar, Y. (1996). Underwater ambient noise measurements. In: 30th Proceedings of the Acoustical Society of America, St. Louis, MO.

Clearwater, S. J., Pankhurst, N. W. (1997). The response to capture and confinement stress of plasma cortisol, plasma sex steroids and vitellogenic oocytes in the marine teleost, red gurnard. J. Fish. Biol. 50, 429-441.

Cleary, J. J., Pankhurst, N. W., Battaglene, S. C. (2000). The effect of capture and handling stress on plasma steroid levels and gonadal condition in wild and farmed snapper *Pagrus auratus*. J. World Aquacul. Soc. 31, 558-569.

Clement, T. S., Parikh, V., Schrumpf, M., Fernald, R. D. (2005). Behavioral coping strategies in a cichlid fish: the role of social status and acute stress response in direct and displaced aggression. Horm. Behav. 47, 336-342.

Clements, S. P., Hicks, B. J., Carragher, J. F., Dedual, M. (2002). The effect of a trapping procedure on the stress response of wild rainbow trout. N. Am. J. Fish. Manage. 22 (3), 907-916.

Colson, V., Sadoul, B., Valotaire, C., Prunet, P., Gaume, M., Labbe, L. (2015). Welfare assessment of rainbow trout reared in a recirculating aquaculture system: comparison with a flow-through system. Aquaculture. 436, 151-159.

Consortium for Wildlife Bycatch Reduction. (2016). http://www.bycatch.org/publications.

Conte, F. S. (2004). Stress and the welfare of cultured fish. Appl. Anim. Behav. Sci. 86, 205-223.

Cooke, S. J., Sneddon, L. U. (2007). Animal welfare perspectives on recreational angling. Appl. Anim. Behav. Sci. 104, 176-198.

Cooke, S. J., Suski, C. D. (2005). Do we need species-specific guidelines for catch-and release recreational angling to effectively conserve diverse fishery resources?. Biodiv. Cons. 14, 1195-1209.

Cooke, S. J., Raby, G. D., Donaldson, M. R., Hinch, S. G., O'Connor, C. M., Arlinghaus, R. (2013). The physiological consequences of catch-and-release angling: perspectives on experimental design, interpretation, extrapolation and relevance to stakeholders. Fish. Manage. Ecol. 20, 268-287.

Costello, M. J. (1993). Review of methods to control sea lice (Caligidae: Crustacea) infestations on salmon (*Salmo salar*) farms. In: Boxshall, G. A., Defaye, D., Pathogens of Wild and Farmed Fish: Sea Lice. Ellis Horwood, Chichester, 219-254.

Coutant, C. C., Whitney, R. R. (2000). Fish behaviour in relation to passage through hydropower turbines: a review. Trans. Am. Fish. Soc. 192, 351-380.

Cubitt, K. F., Winberg, S., Huntingford, F. A., Kadri, S., Crampton, V. O., Øverli, Ø. (2008). Social hierarchies, growth and brain serotonin metabolism in Atlantic salmon (*Salmo salar*) kept under commercial rearing conditions. Physiol. Behav. 94, 529-535.

Curtis, S. E. (1985). What constitutes animal well-being?. In: Moberg, G. P., Animal Stress. American Physiological Society, Bethesda, 1-14.

Davenport, K. E. (1996). Characteristics of the current international trade in ornamental fish, with special reference to the European Union. Rev. Sci. Tech. de l'Office International des Epizooties. 15, 435-443.

Davidson, J., Bebak, J., Mazik, P. (2009). The effects of aquaculture production noise on the growth, condition factor, feed conversion and survival of rainbow trout, *Oncorhynchus mykiss*. Aquaculture. 288, 337-343.

Davis, K. B. (2006). Management of physiological stress in finfish aquaculture. N. Am. J. Aquacul. 68, 116-121.

Davis, M. W. (2007). Simulated fishing experiments for predicting delayed mortality rates using reflex impairment in restrained fish. ICES J. Mar. Sci. 64, 1535-1542.

Davis, M. W., Olla, B. L. (2001). Stress and delayed mortality induced in Pacific halibut by exposure to hooking, net towing, elevated seawater temperature and air: Implications for management of bycatch. North Am. J. Fisheries Manag. 21, 725-732.

Davis, M. W., Olla, B. L. (2002). Mortality of lingcod towed in a net as related to fish length, seawater temperature, and air exposure: A laboratory bycatch study.

North Am. J. Fisheries Manag. 22, 1095-1104.

Davis, M. W., Olla, B. L., Schreck, C. B. (2001). Stress induced by hooking, net towing, elevated sea water temperature and air in sablefish: lack of concordance between mortality and physiological measures of stress. J. Fish Biol. 58, 1-15.

Davis, M. W., Parker, S. J. (2004). Fish size and exposure to air: potential effects on behavioral impairment and mortality rates in discarded sablefish. N. Am. J. Fish. Manage. 24, 518-524.

Davis, M. W., Ottmar, M. L. (2006). Wounding and reflex impairment may be predictors for mortality in discarded or escaped fish. Fish. Res. 82, 1-6.

Dawkins, M. S. (1980). Animal Suffering: The Science of Animal Welfare. Springer, Netherlands.

Dempster, T., Uglem, I., Sanchez-Jerez, P., Fernandez-Jover, D., Bayle-Sempere, J., Nilsen, R. (2009). Coastal salmon farms attract large and persistent aggregations of wild fish: an ecosystem approach. Mar. Ecol. Prog. Ser. 385, 1-14.

Diamond, B., Beukers-Stewart, B. D. (2011). Fisheries discards in the north sea: waste of resources or necessary evil?. Rev. Fish. Sci. 19, 231-245.

Douxfils, J., Mathieu, C., Mandiki, S. N. M., Milla, S., Henrotte, E., Wang, N. (2011). Physiological and proteomic evidences that domestication process differentially modulates the immune status of juvenile Eurasian perch (*Perca fluviatilis*) under chronic confinement stress. Fish. Shellfish Immunol. 31, 1113-1121.

Doyle, R. W., Perez-Enriquez, R., Takagi, M., Taniguchi, M. (2001). Selective recovery of founder genetic diversity in aquacultural broodstocks and captive, endangered fish population. Genetica. 111, 291-304.

Duncan, J. H. (1993). Welfare is to do with what animals feel. J. Agric. Environ. Ethics. 6 (Suppl. 2), 8-14.

Duncan, J. H. (1996). Animal welfare is defined in terms of feelings. *Acta Agricul. Scand.* A Anim. Sci. 27 (Suppl.), 29-35.

Earley, R. L., Edwards, J. T., Aseem, O., Felton, K., Blumer, L. S., Karom, M. (2006). Social interactions tune aggression and stress responsiveness in a territorial cichlid fish (*Archocentrus nigrofasciatus*). Physiol. Behav. 88, 353-363.

Ebbesson, L. O. E., Braithwaite, V. A. (2012). Environmental effects on fish neural plasticity and cognition. J. Fish. Biol. 81 (7), 2151-2174.

Ellis, T., James, J. D., Scott, A. P. (2005). Branchial release of free cortisol and melatonin by rainbow trout. J. Fish. Biol. 67, 535-540.

Ellis, T., Oidtmann, B., St-Hilaire, S., Turnbull, J. F., North, B. P., MacIntyre, C. M. (2008). Fin erosion in farmed fish. In: Branson, E. J., Fish Welfare.

Blackwell, Oxford, 121-149.

Ellis, T. , Yildiz, H. Y. , López-Olmeda, J. , Spedicato, M. T. , Tort, L. , Øverli, Ø. (2012). Cortisol and finfish welfare. Fish Physiol. Biochem. 38, 163-188.

Eriksen, H. R. , Murison, R. , Pensgaard, A. M. , Ursin, H. (2005). Cognitive activation theory of stress (CATS): from fish brains to the Olympics. Psychoneuroendocrinology. 30, 933-938.

Erikson, U. , Sigholt, T. , Seland, A. (1997). Handling stress and water quality during live transportation and slaughter of Atlantic salmon (*Salmo salar*). Aquaculture. 149 (3), 243-252.

Europa. (2015). http://eur-lex. europa. eu/legal-content/EN/TXT/? uri = CELEX: 52013SC0497.

European Directive. (2010/63/EU). http://eur-lex. europa. eu/legal-content/EN/TXT/? uri = CELEX: 32010L0063.

Farrell, A. P. (2006). Bulk oxygen uptake measured with over 60, 000 kg of adult salmon during live-haul transportation at sea. Aquaculture. 254, 646-652.

Farrell, A. P. , Gallaugher, P. E. , Routledge, R. (2001). Rapid recovery of exhausted adult coho salmon after commercial capture by troll fishing. Can. J. Fish. Aquat. Sci. 58, 2319-2324.

FAWC. (1996). FAWC (Farmed Animal Welfare Council), 1996. Report on the Welfare of Farmed Fish. Surbiton, Surrey.

Fendt, M. , Fenselow, M. S. (1999). The neuroanatomical and neurochemical basis of conditioned fear. Neurosci. Biobehav. Rev. 23, 743-760.

Fishcount. org. (2015). http://fishcount. org. uk.

Fish Fight. (2015). http://www. fishfight. net/story. html.

Flecknell, P. A. , Roughan, J. V. , Stewart, R. (1999). Use of oral buprenorphine ('buprenorphine jello') for postoperative analgesia in rats: a clinical trial. Lab. Anim. 33, 169-174.

Fraser, D. (2008). Understanding Animal Welfare, The Science in its Cultural Context. Wiley Blackwell, Chichester.

Fraser, N. H. C. , Metcalfe, N. B. , Thorpe, J. E. (1993). Temperature-dependent switch between diurnal and nocturnal foraging in salmon. Proc. R. Soc. Lond. B Biol. 252 (1334), 135-139.

Froese, R. (1988). Relationship between body weight and loading densities in fish transport using the plastic bag method. Aquacul. Fish. Manage. 19, 275-281.

Frost, A. J. , Winrow-Giffen, A. , Ashley, P. J. , Sneddon, L. U. (2007). Plasticity in animal personality traits: does prior experience alter the degree of boldness?. Proc. R. Soc. Lond. B Biol. Sci. 274, 333-339.

Fürtbauer, I., King, A. J., Heistermann, M. (2015). Visible implant elastomer (VIE) tagging and simulated predation risk elicit similar physiological stress responses in three-spined stickleback *Gasterosteus aculeatus*. J. Fish. Biol. 86, 1644-1649.

Gale, M. K., Hinch, S. G., Donaldson, M. R. (2013). The role of temperature in the capture and release of fish. Fish Fish. 14, 1-33.

Galhardo, L., Oliveira, R. F. (2009). Psychological stress and welfare in fish. ARBS Ann. Rev. Biomed. Sci. 11, 1-20.

Galhardo, L., Correia, J., Oliveira, R. F. (2008). The effect of substrate availability on behavioural and physiological indicators of welfare in the African cichlid (*Oreochromis mossambicus*). Anim. Welfare. 17, 239-254.

Galhardo, L., Vital, J., Oliveira, R. F. (2011). The role of predictability in the stress response of a cichlid fish. Physiol. Behav. 102, 367-372.

Garner, S. R., Madison, B. N., Bernier, N. J., Neff, B. D. (2010). Behavioural interactions and hormones in naturally and hatchery-spawned Chinook salmon. Ethology. 117, 37-48.

Giorgi, A. E., Stevenson, J. R. (1995). A Review of Biological Investigations Describing Smolt Passage Behavior at Portland District Corps of Engineers Projects: Implications to Surface Collection Systems. Don Chapman Consultants, Boise, Idaho.

Gorissen, M., Flik, G. (2016). Endocrinology of the Stress Response in Fish. In: Schreck, C. B., Tort, L., Farrell, A. P., Brauner, C. J., . Academic Press, San Diego, CA.

Grigorakis, K., Rigos, G. (2011). Aquaculture effects on environmental and public welfare-the case of Mediterranean mariculture. Chemosphere. 855, 899-919.

Gronquist, D., Berges, J. A. (2013). Effects of aquarium-related stressors on the zebrafish: a comparison of behavioural, physiological, and biochemical indicators. J. Aquatic Health. 25, 53-65.

Grossman, L., Stewart, A., Gaikwad, S., Utterback, E., Wu, N., DiLeo, J. (2011). Effects of piracetam on behavior and memory in adult zebrafish. Brain Res. Bull. 85, 58-63.

Gutscher, M., Wysocki, L. E., Ladich, F. (2011). Effects of aquarium and pond noise on hearing sensitivity in an otophysine fish. Bioacoustics. 20, 117-136.

Handy, R. D., Poxton, M. G. (1993). Nitrogen pollution in mariculture: toxicity and excretion of nitrogenous compounds by marine fish. Rev. Fish. Biol. Fisher. 3, 205-241.

Hannah, R. W., Lomeli, M. J. M., Jones, S. A. (2015). Tests of artificial light for bycatch reduction in an ocean shrimp (*Pandalus jordani*) trawl: strong but opposite effects at the footrope and near the bycatch reduction device. Fish. Res. 170, 60-67.

Head, A. B., Malison, J. A. (2000). Effects of lighting spectrum and disturbance level on the growth and stress responses of yellow perch Perca flavescens. J. World Aquacult. Soc. 31, 73-80.

Herrero, M. J., Madrid, J. A., Sánchez-Vázquez, F. J. (2005). Demand-feeding rhythms and feeding-entrainment of locomotor rhythms in tench (*Tinca tinca*). Physiol. Behav. 84, 595-605.

Hoglund, E., Weltzien, F. A., Schjolden, J., Winberg, S., Ursin, H., Doving, K. B. (2005). Avoidance behavior and brain monoamines in fish. Brain Res. 1032, 104-110.

Home Office. (2014). https://www.gov.uk/government/statistics/statistics-of-scientific-procedures-on-living-animals-great-britain-2013.

Hoolihan, J. P., Luo, J., Abascal, F. J., Campana, S. E., De Metrio, G., Dewar, H. (2011). Evaluating post-release behavior modification in large pelagic fish deployed with pop-up satellite archival tags. ICES J. Mar. Sci. 68 (5), 880-889.

HSA. (2014). Humane Harvesting of Fish, http://www.hsa.org.uk/downloads/related-items/harvesting-of-fish.pdf.

Humborstad, O. B., Davis, M. W., Lokkeborg, S. (2009). Reflex impairment as a measure of vitality and survival potential of Atlantic cod (*Gadus morhua*). Fish. Bull. 107, 395-402.

Humborstad, O. B., Breen, M., Davis, M. W., Lokkeborg, S., Mangor-Jensen, A., Midling, K. O. (2016). Survival and recovery of longline-and pot-caught cod (*Gadus morhua*) for use in capture-based aquaculture (CBA). Fish. Res. 174, 103-108.

Huntingford, F. A. (2004). Implications of domestication and rearing conditions for the behaviour of cultivated fishes. J. Fish. Biol. 65 (Suppl. A), 122-142.

Huntingford, F. A., Kadri, S. (2014). Defining, assessing and promoting the welfare of farmed fish. Rev. Sci. Tech. de L'Office International des Epizooities. 33, 233-244.

Huntingford, F. A., Adams, C., Braithwaite, V. A., Kadri, S., Pottinger, T. G., Sandoe, P. (2006). Current issues in fish welfare. J. Fish. Biol. 68, 332-372.

Itou, J., Akiyama, R., Pehoski, S., Yu, X., Kawakami, H., Kawakami, Y. (2014). Regenerative responses after mild heart injuries for cardiomyocyte proliferation in zebrafish. Developmental Dynamics. 243, 1477-1486.

Iversen, M. H., Eliassen, R. A. (2014). The effect of allostatic load on hypothalamic-pituitary-interrenal (HPI) axis before and after secondary vaccination in Atlantic salmon postsmolts (*Salmo salar* L.). Fish Physiol. Biochem. 40, 527-538.

Iversen, M., Finstad, B., Nilssen, K. J. (2005). Recovery from loading and transport stress in Atlantic salmon (*Salmo salar* L.) smolts. Aquaculture. 168, 387-394.

Jepsen, N., Thorstad, E. B., Havn, T., Lucas, M. C. (2015). The use of external electronic tags on fish: an evaluation of tag retention and tagging effects. Anim. Biotelem. 3, 49.

Johansson, D., Ruohonen, K., Kiessling, A., Oppedal, F., Stiansen, J. E., Kelly, M. (2006). Effect of environmental factors on swimming depth preferences of Atlantic salmon (Salmo salar L.) and temporal and spatial variations in oxygen levels in sea cages at a fjord site. Aquacult. 254, 594-605.

Johanesen, L.-H., Jensen, I., Mikkelsen, H., Bjørn, P. A., Jansen, P. A., Bergh, Ø. (2011). Disease interaction and pathogens exchange between wild and farmed fish populations with special reference to Norway. Aquaculture. 315, 167-186.

Jolivet, A., de Pontual, H., Garren, F., Bégout, M. L. (2009). Effects of T-bar and DST tagging on survival and growth of European hake. In: Nielsen, J. L., Arrizabalaga, H., Fragosos, N., Hobday, A., Lutcavage, M., Sibert, L., Tagging and Tracking of Marine Animals with Electronic Devices. Springer, Dordrecht, Netherlands, 181-193.

Karakatsouli, N., Katsakoulis, P., Leondaritis, G., Kalogiannis, D., Papoutsoglou, S. E., Chadio, S. (2012). Acute stress response of European sea bass Dicentrarchus labrax under blue and white light. Aquacult. 364, 48-52.

Karlsen, J. D., Krag, L. A., Albertsen, C. M., Frandsen, R. P. (2015). From fishing to fish processing: separation of fish from crustaceans in the Norway lobster-directed multispecies trawl fishery improves seafood quality. PLoS One. 10, e0140864.

Keeling, L., Jensen, P. (2009). Behavioural disturbances, stress and welfare. In: Jensen, P., The Ethology of Domestic Animals: an Introductory Text. CABI Publishing, Wallingford, Oxfordshire, 85-101.

Kiessling, A., Johansson, D., Zahl, L. H., Samuelsen, O. B. (2009). Pharmacokinetics, plasma cortisol and effectiveness of benzocaine, MS-222 and isoeugenol measured in individual dorsal aorta-cannulated Atlantic salmon (*Salmo salar*) following bath administration. Aquaculture. 286, 301-308.

Kight, C. R., Swaddle, J. P. (2011). How and why environmental noise impacts animals: an integrative, mechanistic review. Ecol. Lett. 14, 1052-1061.

Kihslinger, R. L., Nevitt, G. A. (2006). Early rearing environment impacts cerebellar growth in juvenile salmon. J. Exp. Biol. 209, 504-509.

Knudsen, F. R., Enger, P. S., Sand, O. (1992). Awareness reactions and avoidance responses to sound in juvenile Atlantic salmon, *Salmo salar* L. J. Fish. Biol. 40, 523-534.

Korte, S. M., Olivier, B., Koolhaas, J. M. (2007). A new animal welfare concept based on allostasis. Physiol. Behav. 92 (3), 422-428.

Kristensen, T., Haugen, T. O., Rosten, T., Fjellheim, A., Atland, A., Rosseland, B. O. (2012). Effects of production intensity and production strategies in commercial Atlantic salmon smolt (*Salmo salar* L.) production on subsequent performance in the early sea stage. Fish Physiol. Biochem. 38, 273-282.

Kristiansen, T. S., Fernö, A. (2007). Individual behaviour and growth of halibut (*Hippoglossus hippoglossus* L.) fed sinking and floating feed: evidence of different coping styles. Appl. Anim. Behav. Sci. 104, 236-250.

Kulczykowska, E., Vázquez, F. J. S. (2010). Neurohormonal regulation of feed intake and response to nutrients in fish: aspects of feeding rhythm and stress. Aquacul. Res. 41, 654-667.

Lambooij, E., Eilarczyk, M., Bialowas, H., van den Boogaart, J. G. M., van de Vis, J. W. (2007). Electrical and percussive stunning of the common carp (*Cyprinus carpio* L.): neurological and behavioural assessment. Aquacul. Eng. 37, 171-179.

Lambooij, B., Bracke, M., Reimert, H., Foss, A., Imsland, A., vande Vis, H. (2015). Electrophysiological and behavioural responses of turbot (*Scophthalmus maximus*) cooled in ice water. Physiol. Behav. 149, 23-28.

Laule, G., Desmond, T. (1998). Positive reinforcement training as an enrichment strategy. In: Shepherdson, D. J., Mellen, J. D., Hutchins, M., Second Nature: Environmental Enrichment for Captive Animals. Smithsonian Institution Press, Washington.

LePage, O., Øverli, Ø., Petersson, E., Järvi, T., Winberg, S. (2000). Differential stress coping in wild and domesticated sea trout. Brain Behav. Evol. 56, 259-268.

Lines, J. A., Spence, J. (2014). Humane harvesting and slaughter of farmed fish. Rev. Sci. Tech. Off. Int. des Epizooties. 33, 255-264.

Loher, T., Rensmeyer, R. (2011). Physiological responses of Pacific halibut, *Hippoglossus stenolepis*, to intracoelomic implantation of electronic archival tags, with a review of tag implantation techniques employed in flatfish. Rev. Fish Biol. Fisher. 21 (1), 97-115.

Lopez-Luna, J., Yásquez, L., Torrent, E., Villarroel, M. (2013). Short-term fasting and welfare prior to slaughter in rainbow trout, *Oncorhynchus mykiss*.

Aquaculture. 400-401, 142-147.

Losey, G. S., Cronin, T. W., Goldsmith, T. H., Hyde, D., Marshall, N. J., McFarland, W. N. (1999). The UV visual world of fishes: a review. J. Fish. Biol. 54, 921-943.

Lovallo, W. R. (2005). Stress & Health: Biological and Psychological Interactions. Sage Publications, Thousand Oaks.

Lower, N., Moore, A., Scott, A. O., Ellis, T., James, J. D., Russell, I. C. (2005). A non-invasive method to assess the impact of electronic tag insertion on stress levels in fishes. J. Fish. Biol. 67 (5), 1202-1212.

Lupes, S. C., Davis, M. W., Olla, B. L., Schreck, C. B. (2006). Capture-related stressors impair immune system function in sablefish. Trans. Am. Fish. Soc. 135, 129-138.

MacIntyre, C. M., Ellis, T., North, B. P., Turnbull, J. F. (2008). The influences of water quality on the welfare of farmed rainbow trout: a review. In: Branson, E. J., Fish Welfare. Blackwell, Oxford, 150-184.

Marchetti, M. P., Nevitt, G. A. (2003). Effects of hatchery rearing practices on brain structures of rainbow trout (*Oncorhynchus mykiss*). Environ. Biol. Fishes. 66, 9-14.

Martins, C. I. M., Eding, E. H., Verreth, J. A. J. (2011). Stressing fish in recirculating aquaculture systems (RAS): does stress induced in one group of fish affect the feeding motivation of other fish sharing the same RAS?. Aquacul. Res. 42, 1378-1384.

Martins, C. I. M., Galhardo, L., Noble, C., Damsgård, B., Spedicato, M. T., Zupa, W. (2012). Behavioural indicators of welfare in farmed fish. Fish Physiol. Biochem. 38, 17-41.

Mason, G. J., Latham, N. R. (2004). Can't stop, won't stop: is stereotypy a reliable animal welfare indicator?. Anim. Welfare. 13, S57-S69.

Mason, G. J., Cooper, J. J., Clareborough, C. (2001). Frustrations of fur-farmed mink. Nature. 410, 35-36.

Maximino, C., Marques de Brito, T., Waneza da Silva Batista, A., Herculano, A. M., Morato, S., Gouveia, A. (2010). Measuring anxiety in zebrafish: a critical review. Behav. Brain Res. 215, 157-171.

Mayer, I., Pall, M. (2007). Hormonal control of reproductive behaviour in the stickleback. In: Ostlund-Nilsson, S., Mayer, I., Huntingford, F. A., Biology of the Three-Spined Stickleback. [Marine Biology Series.]. CRC Press, Boca Raton, FL, 249-269.

Maynard, D. J., Berejikian, B. A., Flagg, T. A., and Mahnken, C. V. W. (2001). Development of a natural rearing system to improve supplemental fish quality, 1996-

1998 Progress Report, Project No. 199105500, 174 electronic pages (BPA Report DOE/BP-00004768-1). http: //www. nwfsc. noaa. gov/assets/26/5432_ 04302014 _ 123559_ Maynard. et. al. 2001-NATURES-1996-1998. pdf.

McCormick, S. D., Shrimpton, J. M., Carey, J. B., O'Dea, M. F., Sloan, K. E., Moriyama, S. (1998). Repeated acute stress reduces growth rate of Atlantic salmon parr and alters plasma levels of growth hormone, insulin-like growth factor I and cortisol. Aquaculture. 168, 221-235.

McEwen, B. S. (1998). Stress, adaptation, and disease-allostasis and allostatic load. Neuroimmunomodulation. 840, 33-44.

McEwen, B. S., Wingfield, J. C. (2003). The concept of allostasis in biology and biomedicine. Horm. Behav. 43, 2-15.

McGlone, J. J. (1993). What is animal welfare?. J. Agric. Environ. Ethics. 6 (Suppl. 2), 22-36.

McNeil, W. J., 1991. Expansion of cultured Pacific salmon into marine ecosystems. Aquaculture. 98, 173-183.

Meehan, C. L., Mench, J. A. (2007). The challenge of challenge: can problem solving opportunities enhance animal welfare?. Appl. Anim. Behav. Sci. 102, 246-261.

Metcalfe, J. D. (2009). Welfare in wild-capture marine fisheries. J. Fish. Biol. 75, 2855-2861.

Mettam, J. J., McCrohan, C. R., Sneddon, L. U. (2012). Characterisation of chemosensory trigeminal receptors in the rainbow trout (Oncorhynchus mykiss): responses to irritants and carbon dioxide. J. Exp. Biol. 215, 685-693.

Moberg, G. P. (1985). Biological response to stress: key to assessment of animal well-being?. In: Moberg, G. P., Animals Stress. American Physiological Society, Bethesda, 27-49.

Montero, D., Lalumera, G., Izquierdo, M. S., Caballero, M. J., Saroglia, M., Tort, L. (2009). Establishment of dominance relationships in gilthead sea bream *Sparus aurata* juveniles during feeding: effects on feeding behavior, feed utilization and fish health. J. Fish. Biol. 74, 790-805.

Montoya, A., López-Olmeda, J. F., Lopez-Capel, A., Sánchez-Vázquez, F. J., Pérez-Ruzafa, A. (2012). Impact of a telemetry-transmitter implant on daily behavioural rhythms and physiological stress indicators in gilthead seabream (*Sparus aurata*). Mar. Environ. Res. 79, 48-54.

Mork, O. I., Gulbrandsen, J. (1994). Vertical activity of 4 salmonid species in response to changes between darkness and 2 intensities of light. Aquacult. 127, 317-328.

Muir, W. D. , Smith, S. G. , Williams, J. G. , Sandford, B. P. (2001). Survival of juvenile salmonids passing through bypass systems, turbines, and spillways with and without flow detectors at Snake River dams. N. Am. J. Fish. Manage. 21 (1), 135-146.

Näslund, J. , Rosengren, M. , Del Villar, D. , Gansel, L. , Norrgård, J. R. , Persson, L. (2013). Hatchery tank enrichment affects cortisol levels and shelter-seeking in Atlantic salmon (*Salmo salar*). Can. J. Fish. Aquat. Sci. 70, 585-590.

Nedelec, S. L. , Simpson, S. D. , Morley, E. L. , Nedelec, B. , Radford, A. N. (2015). Impacts of regular and random noise on the behavior, growth and development of larval Atlantic cod (*Gadus morhua*). Proc. R. Soc. B Biol. 282 (1817), 20151943.

Neiffer, D. L. , Stamper, M. A. (2009). Fish sedation, anesthesia, analgesia, and euthanasia: considerations, methods, and types of drugs. ILAR J. 50, 343-360.

Newby, N. , Binder, T. R. , Stevens, D. (2007). Passive integrated transponder (PIT) tagging did not negatively affect the short-term feeding behavior or swimming performance of juvenile rainbow trout. Trans. Am. Fish. Soc. 136 (2), 341-345.

Newman, A. E. M. , Edmunds, N. B. , Ferraro, S. , Heffell, Q. , Merritt, G. M. , Pakkala, J. J. (2015). Using ecology to inform physiology studies: implications of high population density in the laboratory. Am. J. Physiol. 308, R449-R454.

Noble, C. , Kadri, S. , Mitchell, D. F. , Huntingford, F. A. (2007). Influence of feeding regime on intraspecific competition, fin damage and growth in 1 + Atlantic salmon parr (*Salmo salar* L.) held in freshwater production cages. Aquacul. Res. 38, 1137-1143.

Noble, C. , Kadri, S. , Mitchell, D. F. , Huntingford, F. A. (2008). Growth, production and fin damage in cage-held 0 + Atlantic salmon pre-smolts (*Salmo salar* L.) fed either (a) on-demand, or (b) to a fixed satiation-restriction regime: data from a commercial farm. Aquaculture. 275, 163-168.

Nomura, M. , Sloman, K. A. , von Keyserlingk, M. A. G. and Farrell, A. P. (2009). Physiology and behaviour of Atlantic salmon (Salmo salar) smolts during commercial land and sea transport. Physiol. Behav. 96, 233-243.

OATA. (2014). Available from: http://www. ornamentalfish. org/useful-links (retrieved 09.02.14.).

O'Connell, C. P. , Gruber, S. H. , O'Connell, T. J. , Johnson, G. , Grudecki, K. , He, P. (2014). The Use of Permanent Magnets to Reduce Elasmobranch Encounter with a Simulated Beach Net. 1. The Bull Shark (Carcharhinus leucas). Ocean & Coastal Management. 97, 12-19.

Oldenburg, E. W. , Colotelo, A. H. , Brown, R. S. , Eppard, M. B. (2011). Holding of juvenile salmonids for surgical implantation of electronic tags: a review and recommendations. Rev. Fish Biol. Fisher. 21, 35-42.

Oliva-Teles, A. (2012). Nutrition and health of aquaculture fish. J. Fish. Dis. 35, 83-108.

Oliveira, R. F., Almada, V. C. (1998). Dynamics of social interactions during group formation in males of the cichlid fish *Oreochromis mossambicus*. Acta Ethol. 1, 57-70.

Oliver, K. (2001). The ornamental fish market. FAO, Rome.

Oliver, K. (2003). World trade in ornamental species. In: Cato, J. C., Brown, C. L., Collection, Culture and Conservation. Blackwell Publishing, Iowa, 49-64.

Olla, B. L., Davis, M. W., Ryer, C. H. (1998). Understanding how the hatchery environment represses or promotes the development of behavioral survival skills. Bull. Mar. Sci. 62, 531-550.

Ortuño, J., Esteban, M., Meseguer, J. (2001). Effects of short-term crowding stress on the gilthead seabream (*Sparus aurata* L.) innate immune response. Fish Shellfish Immun. 11 (2), 187-197.

Øverli, Ø., Korzan, W. J., Larson, E. T., Winberg, S., Lepage, O., Pottinger, T. G. (2004). Behavioral and neuroendocrine correlates of displaced aggression in trout. Horm. Behav. 45 (5), 324-329.

Øverli, Ø., Winberg, S., Pottinger, T. G. (2005). Behavioral and neuroendocrine correlates of selection for stress responsiveness in rainbow trout-a review. Integrat. Comp. Biol. 45, 463-474.

Owen, M. A. G., Davies, S. J., Sloman, K. A. (2010). Light colour influences the behaviour and stress physiology of captive tench (*Tinca tinca*). Rev. Fish Biol. Fisheries. 20, 375-380.

Packer, R. M. A., Hendricks, A., Burn, C. C. (2015). Impact of facial conformation on canine health: brachycephalic obstructive airway syndrome. PLoS One. 10, e0137496.

Pampoulie, C., Jörunsdóttir, T. D., Steinarsson, A., Pétursdóttir, G., Stefánsson, M. O., Daníelsdóttir, A. K. (2006). Genetic comparison of experimental farmed strains and wild Icelandic populations of Atlantic cod (*Gadus morhua* L.). Aquaculture. 261, 556-564.

Pankhurst, N. W. (1998). Reproduction. In: Black, K., Pickering, A. D., Biology of Farmed Fish. Sheffield Academic Press, Sheffield, 1-26.

Pankhurst, N. W. (2011). The endocrinology of stress in fish: an environmental perspective. Gen. Comp. Endocrinol. 170, 265-275.

Pankhurst, N. W. (2016). Reproduction and Development. In: Schreck, C. B., Tort,

L. , Farrell, A. P. , Brauner, C. J. , . Academic Press, San Diego, CA.

Paperna, I. , Vilenkin, M. , Alves de Matos, P. (2001). Iridovirus infections in farm-reared tropical ornamental fish. Dis. Aquatic Orgs. 48, 17-25.

Persson, L. , Alanara, A. (2014). The effect of shelter on welfare of juvenile Atlantic salmon *Salmo salar* reared under a feed restriction regimen. J. Fish. Biol. 85, 645-656.

Petersson, E. , Karlsson, L. , Ragnarsson, B. , Bryntesson, M. , Berglund, A. , Stridsman, S. (2013). Fin erosion and injuries in relation to adult recapture rates in cultured smolts of Atlantic salmon and brown trout. Can. J. Fish. Aquat. Sci. 70, 915-921.

PFMA. (2014). http://www.pfma.org.uk/pet-population-2014.

Pickering, A. D. , Pottinger, T. G. (1989). Stress responses and disease resistance in salmonid fish: effects of chronic elevation of plasma cortisol. Fish Physiol. Biochem. 7, 253-258.

Ploeg, A. (2005). Facts on mortality with shipments of ornamental fish. In: Ornamental Fish International. <http://www.ornamental-fish-int.org/files/files/mortality.pdf/>.

Poli, B. M. (2009). Farmed fish welfare-suffering assessment and impact on product quality. Ital. J. Anim. Sci. 8 (Suppl. 1), 139-160.

Poli, B. M. , Parisi, G. , Scappini, F. , Zampacavallo, G. (2005). Fish welfare and quality as affected by pre-slaughter and slaughter management. Aquacul. Int. 13, 29-49.

Pon, L. B. , Hinch, S. G. , Cooke, S. J. , Patterson, D. A. , Farrel, A. P. (2009). Physiological, energetic and behavioural correlates of successful fishway passage of adult sockeye salmon *Oncorhynchus nerka* in the Seton River, British Columbia. J. Fish. Biol. 74, 1323-1336.

Popper, A. N. , Fay, R. R. , Platt, C. , Sand, O. (2003). Sound detection mechanisms and capabilities of teleost fishes. In: Collin, S. P. , Marshall, N. J. , Sensory Processing in Aquatic Environments. Springer-Verlag, New York, NY, 3-38.

Porsmoguer, S. B. , Banaru, D. , Boudouresque, C. F. , Dekeyser, I. , Almarcha, C. (2015). Hooks equipped with magnets can increase catches of blue shark (Prionace glauca) by longline fishery. Fish. Res. 172, 345-351.

Portz, D. E. , Woodley, C. M. , Cech, J. J. (2006). Stress-associated impacts of short-term holding on fishes. Rev. Fish. Biol. Fisher. 16, 125-170.

Pottinger, T. G. , Carrick, T. R. (1999). Modification of the plasma cortisol response to stress in rainbow trout by selective breeding. Gen. Comp. Endocrinol. 116,

122-132.

Pottinger, T. G. , Pickering, A. D. , Hurley, M. A. (1992). Consistency in the stress response of individuals of two strains of rainbow trout, *Oncorhynchus mykiss*. Aquaculture. 103, 275-289.

Raby, G. D. , Donaldson, M. R. , Hinch, S. G. , Patterson, D. A. , Lotto, A. G. , Robichaud, D. (2012). Validation of reflex indicators for measuring vitality and predicting the delayed mortality of wild coho salmon bycatch released from fishing gears. J. Appl. Ecol. 49, 90-98.

Raby, G. D. , Clark, T. D. , Farrell, A. P. , Patterson, D. A. , Bett, N. N. , Wilson, S. M. (2015). Facing the river gauntlet: understanding the effects of fisheries capture and water temperature on the physiology of coho salmon. PLoS ONE. 10, e0124023.

Readman, G. D. , Owen, S. E. , Murrell, J. G. , Knowles, T. G. (2013). Do fish perceive anaesthetics as aversive?. PLoS One. 8 (9), e73773.

Reinhardt, V. (2004). Common husbandry-related variables in biomedical research with animals. Lab. Anim. 38, 213-235.

Robb, D. H. F. , Wotton, S. B. , McKinstry, J. L. , Sorensen, N. K. , Kestin, S. C. (2000). Commercial slaughter methods used on Atlantis salmon: determination of the onset of brain failure by electroencephalography. Vet. Rec. 147, 298-303.

Rollin, B. E. (1993). Animal welfare, science and value. J. Agricult. Environ. Ethics. 6 (Suppl. 2), 44-50.

Roques, J. A. C. , Abbink, W. , Geurds, F. , van de Vis, H. , Flik, G. (2010). Tailfin clipping, a painful procedure: studies on Nile tilapia and common carp. Physiol. Behav. 101, 533-540.

Roques, J. A. C. , Abbink, W. , Chereau, G. , Fourneyron, A. , Spanings, T. , Burggraaf, D. (2012). Physiological and behavioral responses to an electrical stimulus in Mozambique tilapia (*Oreochromis mossambicus*). Fish Physiol. Biochem. 38, 1019-1028.

Roscoe, D. W. , Hinch, S. G. , Cooke, S. J. , Patterson, D. A. (2011). Fishway passage and post-passage mortality of up-river migrating sockeye salmon in the Seton River, British Columbia. River Res. Appl. 27, 693-705.

Rose, J. D. (2002). The neurobehavioral nature of fishes and the question of awareness and pain. Rev. Fish. Sci. 10, 1-38.

Ross, L. G. , Ross, B. (2008). Anaesthetic and Sedative Techniques for Aquatic Animals. Ross and Ross, 2008 Ross, L. G. , Ross, B. , Anaesthetic and Sedative Techniques for Aquatic Animals3rd Edition2008Blackwells, Oxford280, Blackwells, Oxford, 280.

Ruiz-Gomez, M. deL. , Kittilsen, S. , Höglund, E. , Huntingford, F. A. , Sørensen, C. , Pottinger, T. G. (2008). Behavioral plasticity in rainbow trout (*Oncorhynchus mykiss*) with divergent coping styles: when doves become hawks. Horm. Behav. 54, 534-538.

Rutte, C. , Taborsky, M. , Brinkhof, M. W. (2006). What sets the odds of winning and losing?. Trends. Ecol. Evol. 21, 16-21.

Saeji, J. P. J. , Vervurg-van-Kemenade, L. B. M. , Van Muiswinkel, W. B. , Wiegertjes, G. F. (2003). Daily handling stress reduces resistance of carp to *Trypanoplasma borreli*: in vitro modulatory effects of cortisol on leukocyte function and apoptosis. Dev. Comp. Immunol. 27 (3), 233-245.

Saera-Vila, A. , Kasprick, D. S. , Junttila, T. L. , Grzegorski, S. J. , Louie, K. W. , Chiari, E. F. (2015). Myocyte dedifferentiation drives extraocular muscle regeneration in adult zebrafish. Investigative Ophthalm. Visual Sci. 56, 4977-4993.

Sadoul, B. , Vijayan, M. M. (2016). Stress and Growth. In: Schreck, C. B. , Tort, L. , Farrell, A. P. , Brauner, C. J. , . Academic Press, San Diego, CA.

Salvanes, A. G. V. , Moberg, O. , Ebbesson, L. O. E. , Nilsen, T. O. , Jensen, K. H. , Braithwaite, V. A. (2013). Environmental enrichment promotes neural plasticity and cognitive ability in fish. Proc. R. Soc. B. Biol. 280, 20131331.

Sánchez, J. A. , López-Olmeda, J. F. , Blanco-Vives, B. , Sánchez-Vázquez, F. J. (2009). Effects of feeding schedule on locomotor activity rhythms and stress response in sea bream. *Physiol. Behav.* 98, 125-129.

Sánchez-Vázquez, F. J. , Madrid, J. A. , Zamora, S. (1995). Circadian rhythms of feeding activity in sea bass, *Dicentrarchus labrax* L. : dual phasing capacity of diel demand-feeding pattern. J. Biol. Rhythm. 10, 256-266.

Sand, O. , Enger, P. S. , Karlsen, H. E. , Knudsen, F. R. (2001). Detection of infrasound in fish and behavioral responses to intense infrasound in juvenile salmonids and European silver eels: a minireview. Am. Fish. Soc. Symp. 26, 183-193.

Sapolsky, R. M. (2004). Why Zebras Don't Get Ulcers. Henry Holt and Company, New York, NY.

Schall, K. A. , Holoyda, K. A. , Grant, C. N. , Levin, D. E. , Torres, E. R. , Maxwell, A. (2015). Adult zebrafish intestine resection: a novel model of short bowel syndrome, adaptation, and intestinal stem cell regeneration. Am. J. Physiol. -Gastrointestinal Liver Physiol. 309, G135-G145.

Schreck, C. B. , Jonsson, L. , Feist, G. , Reno, P. (1995). Conditioning improves performance of juvenile Chinook salmon, *Oncorhynchus tshawytscha*, to transportation stress. Aquaculture. 135, 99-110.

Schreck, C. B. , Olla, B. L. , Davis, M. W. (1997). Behavioral responses to stress. In: Iwama, G. K. , Pickering, A. D. , Sumpter, J. P. , Schreck, C. B. , Fish Stress and Health in Aquaculture. Cambridge University Press, Cambridge, 145-170.

Schreck, C. B. , Tort, L. (2016). The Concept of Stress in Fish. In: Schreck, C. B. , Tort, L. , Farrell, A. P. , Brauner, C. J. , . Academic Press, San Diego, CA.

Schroeder, P. , Jones, S. , Young, I. Y. , Sneddon, L. U. (2014). What do zebrafish want? Impact of social grouping, dominance and gender on preference for enrichment. Lab. Anim. 48, 328-337.

Sea Shepherd. (2014). http: //www. seashepherd. org/reef-defense/aquarium-trade. html.

Sharber, N. G. , Carothers, S. W. (1988). Influence of electrofishing pulse shape on spinal injuries in adult rainbow trout. N. Am. J. Fish. Manage. 8, 117-122.

Sharpe, C. S. , Thompson, D. A. , Blankenship, H. L. , Schreck, C. B. (1998). Effects of routine handling and tagging procedures on physiological stress responses in juvenile Chinook salmon. Prog. Fish Cult. 60 (2), 81-87.

Singhal, G. , Jaehne, E. J. , Corrigan, F. , Baune, B. T. (2014). Cellular and molecular mechanisms of immunomodulation in the brain through environmental enrichment. Front. Cell. Neurosci.

Skaala, Ø. , Dahle, G. , Jørstad, K. E. , Nævdal, G. (1990). Interactions between natural and farmed fish populations: information from genetic markers. J. Fish. Biol. 36, 449-460.

Slabbekoorn, H. , Bouton, N. , van Opzeeland, I. , Coers, A. , ten Cate, C. , Popper, A. N. (2010). A noisy spring: the impact of globally rising underwater sound levels on fish. Trends. Ecol. Evol. 25 (7), 419-427.

Sloman, K. A. (2010). Exposure of ova to cortisol pre-fertilisation affects subsequent behaviour and physiology of brown trout. Horm. Behav. 58, 433-439.

Smith, M. E. , Kane, A. S. , Popper, A. N. (2004). Acoustical stress and hearing sensitivity in fishes: does the linear threshold shift hypothesis hold water?. J. Exp. Biol. 207, 3591-3602.

Smith, J. M. , Mather, M. E. , Frank, H. J. , Muth, R. M. , Finn, J. T. , McCormick, S. D. (2009). Evaluation of a gastric radio tag insertion technique for anadromous river herring. N. Am. J. Fish. Manage. 29 (2), 367-377.

Sneddon, L. U. (2003). The evidence for pain in fish: the use of morphine as an analgesic. Appl. Anim. Behav. Sci. 83 (2), 153-162.

Sneddon, L. U. (2003). The bold and the shy: individual differences in rainbow trout. J. Fish. Biol. 62, 971-975.

Sneddon, L. U. (2011). Cognition and welfare. In: Brown, C., Laland, K., Krause, J., Fish Cognition and Behavior. Wiley-Blackwell, Oxford, 405-434.

Sneddon, L. U. (2013). Do painful sensations and fear exist in fish?. In: van der Kemp, T. A., Lachance, M., Animal Suffering: From Science to Law, International Symposium. Carswell, Toronto, 93-112.

Sneddon, L. U. (2015). Pain in aquatic animals. J. Exp. Biol. 218, 967-976.

Sneddon, L. U., Wolfenden, D. (2012). How are fish affected by large scale fisheries: pain perception in fish?. In: Soeters, K., See the Truth. Nicolaas G. Pierson Foundation, Amsterdam.

Sneddon, L. U., Braithwaite, V. A., Gentle, M. J. (2003). Do fish have nociceptors? Evidence for the evolution of a vertebrate sensory system. Proc. R. Soc. Lond. B. 270, 1115-1121.

Sneddon, L. U., Schmidt, R., Fang, Y., Cossins, A. R. (2011). Molecular correlates of social dominance: a novel role for ependymin in aggression. PLoS One. 6 (4), e18181.

Sneddon, L. U., Elwood, R. W., Adamo, S. A., Leach, M. C. (2014). Defining and assessing pain in animals. Anim. Behav. 97, 201-212.

Snyder, D. (2003). Inited overview: conclusions from a review of electrofishing and its harmful effects on fish. Rev. Fish Biol. Fish. 13, 445-453.

Soares, M. C., Oliveira, R. F., Ros, A. F. H., Grutter, A. S., Bshary, R. (2011). Tactile stimulation lowers stress in fish. Nat. Commun. 2, 534.

Sommerset, I., Krossøy, B., Biering, E., Frost, P. (2005). Vaccines for fish in aquaculture. Expert. Rev. Vaccines. 4 (1), 89-101.

Spagnoli, S., Lawrence, C., Kent, M. L. (2016). Stress in Fish as Model Organisms. In: Schreck, C. B., Tort, L., Farrell, A. P., Brauner, C. J., . Academic Press, San Diego, CA.

Spence, R., Gerlach, G., Lawrence, C., Smith, C. (2008). The behaviour and ecology of the zebrafish. Danio Rerio. Biol. Rev. Cam. Philos. Soc. 83, 13-34.

Stamp Dawkins, M. (2012). Why animals matter: animal consciousness, animal welfare, and human well-being. Oxford University Press, Oxford.

Steckler, T. (2005). The neuropsychology of stress. In: Steckler, T., Kalin, N. H., Reul, J. M. H. N., Handbook of Stress and the Brain. Part 1: The Neurobiology of Stress. Elsevier, Amsterdam.

Suzuki, K., Mizusawa, K., Noble, C., Tabata, M. (2008). The growth, feed conversion ratio and fin damage of rainbow trout *Oncorhynchus mykiss* under self-feeding and hand-feeding regimes. Fish. Sci. 74, 941-943.

Tang, S., Brauner, C. J., Farrell, A. P. (2009). Using bulk oxygen uptake to assess the welfare of adult Atlantic salmon, *Salmo salar*, during commercial live-haul transport. Aquaculture. 286, 318-323.

The Telegraph. (2012). Available from: http://www.telegraph.co.uk/lifestyle/pets/9217643/One-in-ten-Britons-now-have-pet-fish.html (retrieved 09.02.14.).

Thorstad, E. B., Økland, F., Aarestrup, K., Heggberget, T. G. (2008). Factors affecting the within-river spawning migration of Atlantic salmon, with emphasis on human impacts. Rev. Fish Biol. Fisher. 18, 345-371.

Torrente-Vilara, G., Zuanon, J., Leprieur, F., Oberdorff, T., Tedesco, P. A. (2011). Effects of natural rapids and waterfalls on fish assemblage structure in the Madeira River (Amazon Basin). Ecol. Freshw. Fish. 20, 588-597.

Townsend, D. (2011). Sustainability, equity and welfare: a review of the tropical marine ornamental fish trade. SPC Live Reef Fish Inform. Bull. 20, 2-12.

Trushenski, J. T., Bowker, J. D., Cooke, S. J., Erdahl, D., Bell, T., MacMillan, J. R. (2013). Issues regarding the use of sedatives in fisheries and the need for immediate-release options. Trans. Am. Fish. Soc. 142 (1), 156-170.

Turnbull, J. F., North, B. P., Ellis, T., Adams, C. E., Bron, J., MacIntyre, C. M. (2008). Stocking density and the welfare of farmed salmonids. In: Branson, E. J., Fish Welfare. Blackwell, Oxford, 111-120.

UFAW. (2013). Available from: http://www.ufaw.org.uk/fish/fish (retrieved 31.08.14.).

Ursin, H., Eriksen, H. R., 2004. The cognitive activation theory of stress. Psychoneuroendocrinology. 29, 567-592.

USDA. (2013). https://speakingofresearch.files.wordpress.com/2008/03/usda-animal-research-use-2011-13.pdf.

Vera, L. M., Madrid, J. A., Sánchez-Vázquez, F. J. (2006). Locomotor, feeding and melatonin daily rhythms in sharpsnout seabream (*Diplodus puntazzo*). Physiol. Behav. 88 (1-2), 167-172.

Vilhunen, S., Hirvonen, H. (2003). Innate antipredator responses of Arctic charr (*Salvelinus alpinus*) depend on predator species and their diet. Behav. Ecol. Sociobiol. 55, 1-10.

Vindas, M. A., Folkedal, O., Kristiansen, T. S., Stien, L. H., Braastad, B. O., Mayer, I. (2012). Omission of expected reward agitates Atlantic salmon (*Salmo salar*). Anim. Cogn. 15, 903-911.

Vindas, M. A., Johansen, I. B., Vela-Avitua, S., Nørstrud, K. S., Aalgaard, M., Braastad, B. O. (2014). Frustrative reward omission increases aggressive behaviour of inferior fighters. Proc. R. Soc. B. 281, 20140300.

Volpato, G. L., Gonçalves-de-Freitas, E., Fernandes-de-Castilho, M. (2007). Insights into the concept of fish welfare. Dis. Aquat. Orgs. 75, 165-171.

Walster, C. (2008). The welfare of ornamental fish. In: Branson, E. J., Fish Welfare. Blackwell, Oxford, 271-290.

Walster, C., Rasidi, E., Saint-Erne, N., Loh, R. (2014). The welfare of ornamental fish in the home aquarium. Companion Anim. 20, 302-306.

Wendelaar Bonga, S. E. (1997). The stress response in fish. Physiol. Rev. 77, 591-625.

Wilson, J. M., Bunte, R. M. and Carty, A. J. Evaluation of rapid cooling and tricaine methanesulfonate (MS222) as methods of euthanasia in zebrafish (Danio rerio). JAALAS 48, 785-789.

Whitney, R. R., Calvin, L. D., Erho Jr., M. W., and Coutant, C. C. (1997). Downstream passage for salmon at hydroelectric projects in the Columbia River basin: development, installation, evaluation. North-West Power Planning Council, NPCC Resport 97-15, Portland, Oregon.

Willis, C. F. (1982). Indexing of juvenile salmonids migrating past The Dalles Dam, 1982. Report of Oregon Department of Fish and Wildlife (Contract DACW57-78-C-0056) to U. S. Army Coprs of Engineers, Portland, Oregon.

Winger, P. D., Walsh, S. J. (2001). Tagging of Atlantic cod (*Gadus morhua*) with intragastric transmitters: effects of forced insertion and voluntary ingestion on retention, food consumption and survival. J. Appl. Ichthyol. 17 (5), 234-239.

Winberg, S., Höglund, E., Øverli, Ø. (2016). Variation in the Neuroendocrine Stress Response. In: Schreck, C. B., Tort, L., Farrell, A. P., Brauner, C. J., . Academic Press, San Diego, CA.

Wysocki, L. E., Amoser, S., Ladich, F. (2007). Diversity in ambient noise in European freshwater habitats: noise levels, spectral profiles, and impact on fishes. J. Acoust. Soc. Am. 121 (5), 2559-2566.

Yada, T., Tort, L. (2016). Stress and Disease Resistance: Immune System and Immunoendocrine Interactions. In: Schreck, C. B., Tort, L., Farrell, A. P., Brauner, C. J., . Academic Press, San Diego, CA.

Yue, S., Moccia, R. D., Duncan, I. J. H. (2004). Investigating fear in domestic rainbow trout, Oncorhynchus mykiss, using an avoidance learning task. Appl. Anim. Behav. Sci. 87, 343-354.

Zahl, H. L., Kiessling, A., Samuelsen, O. B., Olsen, R. E. (2010). Anaesthesia induces stress in Atlantic salmon (*Salmo salar*), Atlantic cod (*Gadus morhua*) and

Atlantic halibut (*Hippoglossus hippoglossus*). Fish Physiol. Biochem. 36, 719-730.

Zuberi, A., Ali, S., Brown, C. (2011). A non-invasive assay for monitoring stress responses: a comparison between wild and captive-reared rainbowfish (*Melanoteania duboulayi*). Aquaculture. 321, 267-272.

Zuberi, A., Brown, C., Ali, S. (2014). Effect of confinement on water-borne and whole body cortisol in wild and captive-reared rainbowfish (*Melanoteania duboulayi*). Int. J. Agricul. Biol. 16, 183-188.

第 13 章　鱼类作为模式生物的应激

随着斑马鱼、三刺鱼和青鳉成为常用的实验动物，鱼类作为模式生物可用来进行越来越多的实验研究，而与此同时，这也给研究者和水产养殖人员提出了新的挑战。对于这些鱼类的应激，我们必须转变考虑的重点，由关注繁殖力和生长性状这些传统的生产定义转为以实验的一致性、动物福利以及构建模型的稳定性为目标。为了将鱼类发展为良好的模式动物，水产养殖人员和研究者必须遵循这些生物的自然史以建立适宜的饲养操作过程。本章中我们总结以鱼类进行应激实验的数据来为建立更好的饲养方式打下基础，也为以后能更好地将鱼类改善为应激研究的模式生物。

13.1　导　言

在过去的几十年里，将水生动物模型作为综合途径的一部分应用于生物医学研究得到迅速发展。斑马鱼作为一种普遍的实验动物，具有众多优点来进行高通量的研究：基因组已知；生命周期短；比小鼠更少的饲养成本；幼体透明，偶尔也出现透明的成体。此外，它们既有所谓低等生物体的作为实验生物的实用性和高效率，同时也保持了和人类这种所谓高等生物体的生理相似性（Lieschke，Currie，2007；Allen，Neely，2010；Harland，Grainger，2011；Phillips，Westerfield，2014）。斑马鱼（*Danio rerio*）作为模式生物被广泛应用最显著的例子就是斑马鱼模式生物数据库网站（http：//zfin. org）列出了大约 1000 家使用斑马鱼的实验室，以及 2014 年在美国国立卫生研究院（National Institutes of Health）网站上进行的调查发现，对含有“斑马鱼”一词的研究项目的拨款资助达到了 735 笔。事实上，在 2011 年，斑马鱼在 PubMed 上的索引条目已经超过果蝇（2279 vs 2265）。

对研究者和水产养殖人员来说，必须把他们两者的主要目标灵活地统一在一起，不同的目标就需要在不同的侧重点间进行转换，包括生产上鱼的最终产量、资源保育及实验室条件下不同的实验目的。本书大部分的内容都在讨论应激，因为这与鱼类的行为表现密切相关。在生产方面，鱼类行为表现的评价指标可归结于其繁殖力、生长性状和存活率。在资源保育方面，行为表现可用群体的稳定性、恢复能力、回归率和偏离率来进行评价。而在实验室方面，行为表现的含义是不固定的，要根据不同的实验目的进行不同的评价。当然，也有一些适用于所有实验生物的基础目标：除了实验操作之外，还有实验项目的相似性、动物福利以及研究中模型的稳定性——这既是所有实验普遍共同的基础目标，也关系到特定研究的最终目标。

很久之前，研究人员就用成年的鳟鱼（Bailey 等，1996；Williams 等，2003）、青鳉（Takeda，Shimada，2010）以及剑尾鱼的杂交后代（Patton 等，2010）作为模型来进行基础生物医学研究。用斑马鱼做实验最初是由发育遗传学家引领的，他们的主要兴趣是胚胎或幼鱼的遗传发育性状。随着逐渐发掘斑马鱼作为研究衰老、慢性疾病和复杂神经行为综合征的模型的实用性，人们越来越多地关注到幼年和成年期的斑马鱼（Dooley，Zon，2000）。利用胚胎和新孵出的仔鱼进行实验和用幼鱼到成鱼阶段的鱼进行实验具有同样的难处，因为它们都需要适宜的饲养条件和维持群体环境。然而，后者还需要做更多的努力以避免在体实验中非实验因素导致的差异。在应对饲养和群体维持过程中的紧急状态时，相比于胚胎和仔鱼，饲养幼鱼或成鱼的实验更需要接受兽医专家关于比较病理学和实验动物医学的相关训练指导。

利用斑马鱼幼鱼和成鱼进行的具体研究领域十分广泛，包括传染病的易感性与免疫系统功能（Novoa，Figueras，2012）、衰老（Gerhard，2003；Kishi 等，2009）、毒理学（Truong 等，2011）、肿瘤学（Liu，Leach，2011；Ceol 等，2011）以及行为学（Sisson 等，2006；Wong 等，2010；Stewart 等，2012）等方面的研究。在对传染病的研究上，利用斑马鱼模型已对大约 30 种不同的细菌病原体进行研究（Meijer，Spaink，2011）。有许多研究使用分枝杆菌类病菌去感染斑马鱼进行实验（Prouty 等，2003；Meijer 等，2005；Swaim 等，2006），然而鱼在饲养过程中可能就已被感染这类病菌，导致实验结果不准确。此外，还利用鱼类进行肿瘤方面的研究，包括将人类肿瘤移植到免疫缺陷的斑马鱼以建立异源移植模型等（Patton 等，2011；White 等，2008；Patton，Zon，2005；Taylor，Zon，2009）。

其他常用作模式生物的水生物种包括三刺鱼（*Gasterosteus aculeatus*）、日本青鳉（*Oryzias latipes*）、胖头鲅（*Pimephales promelas*）以及一些剑尾鱼。三刺鱼被认为是研究进化和生态学的良好模型，特别是在表型和基因组变异、物种形成和自然选择方面（Hendry 等，2013）。青鳉是研究基因组学和毒理学不可缺少的物种，此外，它还是第一个成功转基因的鱼类（Shima，Mitani，2004）。胖头鲅则作为生物指标被用于大量毒理学的研究中（Ankley，Villeneuve，2006）。剑尾鱼是研究黑色素瘤和性发育障碍的极佳模型（Schartl，2014）。其他用鱼类模型进行的研究包括，利用鲤鱼（*Cyprinus carpio*）研究环境毒理学，利用斑点叉尾鮰进行进化基因组学研究（Liu，2003），以及利用莫桑比克罗非鱼（*Oreochormis mossambicus*）进行渗透压调节相关研究（Gardell 等，2013）。

尽管这些生物具有广泛的用途，但研究人员直到现在才意识到利用仔鱼进行中长期实验研究的过程中所出现的潜在应激和慢性疾病可能对实验结果产生影响。随着越来越多的神经行为学实验将鱼类作为模型进行实验（Rihel，Schier，2012；Spence，2011；Bailey，2013），大量实验结果表明慢性应激可能对行为学或心理学实验结果产生不好的影响。Kent 等（2012）曾就慢性疾病在斑马鱼实验中造成非实验因素误差的影响做过综述。本章将展开这一主题，以深入阐述应激对模型实验鱼的潜在的和已证明的影响，并对这些动物的应激因素进行概述。

13.2 实验室鱼类应激的指标

研究人员和水产养殖者之间存在一个相互矛盾的关系：研究人员将适应于自然环境的动物饲养于完全人工制造的环境中，以消除潜在的混淆不清的变动，包括能对动物产生应激的环境嗅觉信号、当地的环境微生物群和野外环境的营养状况（Wedemeyer，1997）。水产养殖者则会从健康、福利和环境适应方面来进行考虑，认为对鱼类有利的应该是引入一些混杂的变动，如加入多种细菌群、改变鱼的行为症状以及一些环境变动因子，以便使养殖环境更接近自然环境。当我们在实验室中进行鱼类研究实验时，我们必须将它们的生活自然史知识和它们在实验室环境世代的潜在适应能力结合起来，以满足现行实验的需求，并在不牺牲动物福利的情况下产生可靠的科学模型。的确，密切关注动物福利可以减少或消除这一不可控的变动，而在实验室中减少鱼类应激的第一步是测定和监测应激反应的能力（Sneddon 等，2016；见本书第 12 章）。

鱼类中最常用的应激指标是皮质醇，它参与主要的神经内分泌应激反应且易于检测（Ellis 等，2012；Sopinka 等，2016；见本书第 11 章）。尽管有许多其他激素和生理变化与应激有关，但血清中皮质醇升高是鱼类应激的一个很好的指标（Martinez-Porchas 等，2009；Ellis 等，2012；Schreck，Tort，2016；见本书第 1 章）。对于小的鱼来说，其体内血液较少，不足以支持进行血清皮质醇的检测，因此，我们会对其全身的皮质醇浓度进行检测（Barry 等，1995；Feist，Schreck，2001）。这样的检测方法也已推广到斑马鱼的研究中（Canavello 等，2011；Ramsay 等，2006，2009a，b，c；Egan 等，2009）。Pavlidis 等（2011，2013）曾用类似的方法对斑马鱼进行研究，但他们检测的区域限定在身体的躯干部分。作为全身或部分身体匀浆的替代方法，Gesto 等（2015）曾提出用鳃组织进行检测，其优点是无须对鱼进行安乐死即可获得鳃组织检测，而且取样所引起的应激反应远小于抽血。实验常用鱼类如斑马鱼，通常安放在相对较小的容器中饲养，而现在已经开发了测试水中皮质醇水平的方法（Ellis 等，2004；Gronquist，Berges，2013），这和在血清中测得的皮质醇水平有很好的相关性（Felix 等，2013）。这种非致死性的方法还可以只需进行一次检测就可以对整个缸中的应激情况进行全面评估。这些评估皮质醇/应激相关性的研究表明，研究设施中鱼群的拥挤、短暂对鱼的手工操作处理（网捕）或潜在的两种常见的斑马鱼病原体的慢性感染（*Pseudoloma neurophilia* 或 *Mycobacterium marinum*）都和皮质醇水平升高有关。

最近，Aerts 等证明了鲤鱼（*Cyprinus carpio*）个体发育中和再生鳞片中的皮质醇含量可以作为该物种慢性应激的可靠和可定量的生物学标志（Aerts 等，2015）。传统检测皮质醇水平的方法是从血清、粪便和水中进行检测，这适用于特定的实验条件下动物短期的反应情况，而这种新方法则可在更长的时间范围内检测鱼的应激水平。

皮质醇虽然是最容易量化的应激反应指标，但它并不能说明应激下的全部情况。

因为应激从被感知到机体做出反应，涉及许多神经回路和内分泌回路的调控（Maximino 等，2010）。这些网络回路中的每一条通路都正在研究探索中，而行为作为应激反应最终输出的形式之一，可以作为一个应激的定量和定性的指标。行为是应激反应各方面最终的综合表现，由于斑马鱼对某些刺激可做出相对明显且可重复的反应，因此，其行为可以作为应激反应的软指标。行为反应虽然不像皮质醇那样是一种具体物质，但它可能是一种更好的表现鱼类在体的偏爱方法，可用于改善鱼的福利和提高实验的一致性。

大多数与应激有关的行为在脊椎动物中是保守的，由种类特异性的自然历史指导其个体的反应。以斑马鱼为例，它是在印度、孟加拉国和尼泊尔的主要河流流域进化而来的一种被捕食型的物种，因此，不难推测这种动物在应激的环境中更喜欢在黑暗和深水中生活——这样可以更好地躲避捕食者（Engeszer 等，2007）。Egan（2009）等发现诸如减少在缸中的游动、趋暗（避光性）、贴壁游动（贴壁性）、无规则的游动、下潜和静止（突然、短暂的不动）等行为都与皮质醇水平升高有关，这些都是应激的指标。作者们还发现，不同品系的斑马鱼表现出不同的内在焦虑水平，这对经过基因修饰的动物来说尤为重要，因为这可以更好地解释不同遗传背景的鱼应激水平的差别。此外，还有一些更为复杂的情况，不同个体之间的行为表现有很大的差异，而这可归结于应激反应的高度差异（Dugatkin 等，2004；Conrad 等，2011）。

虽然在实验室环境下进行应激的诊断测试显得不太切合实际，但是使用和应激与焦虑相关的药理学物质进行处理的实验可以让研究人员更好地鉴别皮质醇诱导下的行为偏爱，并将这些行为和应激相关的皮质醇水平联系起来。例如，在一个偏爱的实验中，斑马鱼必须在一个白的和黑的隔室中做出选择。当用抗焦虑治疗药物（如选择性羟色胺再摄取抑制剂）进行处理时，斑马鱼会在白色的隔室停留更多的时间。而当用产生焦虑的化合物（如咖啡因和安非他命替代物）进行处理时，鱼会花更多的时间在黑色的隔室停留（Araujo 等，2012）。Tudorache 等（2015）用斑马鱼幼鱼进行行为和皮质醇水平之间的关系研究，他们发现在 8 日龄的幼鱼中，早孵化的鱼比晚孵化的鱼表现出更高的皮质醇峰值，但恢复时间亦较快。在使用斑马鱼作为模型生物进行药物研究，特别是进行药物筛选时，对实验动物应激相关行为的密切评估尤其重要。如果确认某些行为与应激有关，那么药物处理后的行为反应就可以用来描述药物的神经行为特征（Stewart 等，2015）。应激的行为指标还有一个优点，即在进行检测时，不会干扰动物个体和整个动物生活的群体环境。

13.3 对鱼类进行实验室操作过程中影响应激的因素

人们对鱼进行手工操作时通常需要将鱼拿出水体，而这可能是大部分实验鱼类面临的最大的应激源（Ghisleni 等，2012）。因此，对鱼的手工操作应该保持对鱼产生尽量小的影响。因为不同的实验通常需要在实验鱼生命的不同时期进行多次处理，所以，了解实验鱼类对实验处理的反应有助于减少鱼类应激并提高处理后鱼的存活率。

幸运的是，由于鱼类对应激的感知和预期构成了应激反应的一个很大的组成部分，对这两者进行实验处理可以大大降低某些物种的整体应激水平。例如，在用网捞起一龄的大鳞大麻哈鱼之前，对其进行麻醉可以降低皮质醇水平，提高捞起处理后鱼的存活率。而对于实验中用到的非鲑科鱼类，比如金鱼，对其进行快速捕捞、冰冻麻醉、3-氨基苯甲酸乙酯甲基磺酸盐（MS-222）麻醉或电击麻醉等操作时，并不会影响其血清皮质醇水平（Singley 和 Chavin，1975）。

相比之下，斑马鱼在急性网捕应激中，皮质醇水平会迅速升高并随后逐渐恢复（Ramsay 等，2009a，b，c；Tran 等，2014）。Brydges 和 Braithwait（2009）发现，在水中将鱼舀出，而不是用网捞鱼，会使皮质醇反应减弱。此外，还有一些报道显示一些实验操作会导致繁殖力的下降，这将在稍后进行讨论。

在实验室环境中处理鱼类通常涉及比使用渔网捞鱼更复杂和更易产生应激的操作，比如实验室鱼类经常通过浸泡、灌喂或各种注射途径接触病原体或化学制剂。因此，了解这些过程所产生的应激是很重要的。然而这类数据非常有限，只能从鱼类整体处理后的恢复状况和生存率来进行推断。在为数不多的此类研究中，Kahl 等（2001）发现，MS-222 麻醉、玉米油和 10% 乙醇腹腔注射和行为、性逆转、生存率降低或繁殖力降低无关。

13.4 饲养环境

用于研究的鱼通常被饲养在流式或循环式养殖系统（RAS）中（Lawrence，Mason，2012）。虽然关于不同类型养殖系统对应激的不同影响的文献很少，但已经有人提出，在循环水系统中饲养的动物可能会因为代谢产物、皮质醇和警戒物质在密集系统中的积累而引起应激反应的升高（Martins 等，2011）。这一假设在尼罗罗非鱼中进行了试验，但结果并不能得出确切的结论（Martins 等，2009）。由于现有的研究数据有限，因而值得做进一步研究，特别是像斑马鱼这样的实验动物几乎都饲养在循环水系统中。

Parker 等（2012）进行了新的饲养环境探索实验，检测不同饲养环境条件对应激产生的影响。他们发现群体大小、水体更换尤其是视觉接触的能力会对斑马鱼在新环境的探索产生影响。他们还发现，单独饲养的鱼基础皮质醇水平低于群体饲养的鱼。这些数据表明饲养环境能介导该物种的应激反应，因此，其饲养环境需要进行标准化以改进研究设计方案。

研究人员还关注一些其他的饲养环境条件，研究它们是否也对斑马鱼应激和焦虑产生影响。Blaser 和 Rosemberg（2012）关注到鱼缸深度和缸壁的颜色，因而设计实验以观测鱼对水体深度及趋暗的偏爱。此外，在鱼的饲养过程中经常被忽略的一个问题是气泵和水泵的噪音。已有一些关于噪音对鲑科鱼类影响的实验表明，特定频率的噪音会使大西洋鲑（*Salmo salar L.*）幼鱼产生躲避行为（Knudsen 等，1992）。虽然关于饲养过程中噪声对斑马鱼行为影响的具体研究还处于初步阶段，但 Smith 等

（2003）证明金鱼（*Carassius auratus*），一种和斑马鱼相似的鲤科鱼类，在面对长期的噪声时可能会产生应激反应并造成听力损失。最近，Neo 等（2015）发现在实验用斑马鱼中，中等强度噪音处理会改变群体聚集程度、游泳速度和鱼的深度偏爱。有趣的是，这项研究还发现，鱼群对慢性和高强度的噪音处理并没有产生躲避行为。这些研究结果提供了进一步的证据证明饲养环境的条件配置是斑马鱼应激和行为反应的一个重要因素。

除斑马鱼外，外界声音对鱼类行为的影响已经在许多模型和非模型生物中进行过研究，总体结果表明声音对鱼类的生存和生态有着微妙而潜在的影响。Purser 和 Radford（2011）研究发现，监禁的三刺鱼处在短暂和长时间的噪音中时，它们的总食量没有减少，但对获取食物时的出错率却增加了，它们难以分辨食物和非食物之间的区别，这表明噪音分散了它们的注意力。类似的，欧洲鳗鲡（*Anguilla anguilla*）在受到人为噪音影响时，会减少其抗捕食者行为（Simpson 等，2014）。此外，有充分的证据表明，人为噪音干扰了鱼类之间的听觉交流，而这种交流也是一种和社会行为相关的鱼类的重要生存策略（Radford 等，2014）。

13.4.1　密度

大多数在实验室中利用鱼类进行的研究，除了营养或饲养研究外，基本不涉及对其繁殖力或食物效率的考虑。然而，单位体积内的动物数量以及水的流动交换率与鱼类的应激水平密切相关，并可能导致研究中的检测结果（包括免疫、代谢和行为反应）出现非实验因素引起的误差。

单位体积的鱼的数量、水流和应激之间的关系可能既不直接也不直观。一般来说，对于饲养密度高的斑马鱼，在一定程度上其数量的减少和皮质醇水平的降低有关（Ramsay 等，2006；Pavlidis 等，2013）。这之间存在一个较低的密度界限，当超过这个界限后，皮质醇水平开始随着鱼类数量的减少而增加（图 13.1）。这可能是由于在小的鱼群群体间建立了优势等级制度，这些鱼群缺乏将追逐和攻击行为扩展到大的个

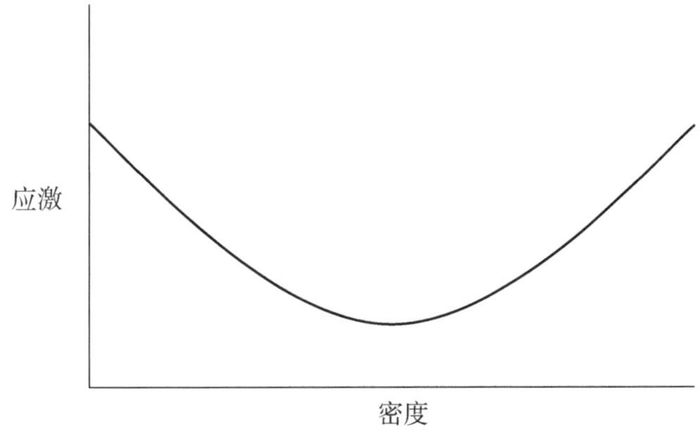

图 13.1　斑马鱼的皮质醇水平和其种群密度有关

来源：Harper，Lawernce（2010）. *The Laboratory Zebrafish. CRC Press*，*Boca Raton*，*FL.*

体群体的能力（Filby 等，2010；Pavlidis 等，2013）。值得注意的是，尽管斑马鱼是一种群居鱼类，Parker 等（2012）和 Giacomini 等（2015）研究表明，对斑马鱼进行个体隔离实际上会降低皮质醇水平，这可能由于减少了群体内争斗行为而导致的。后者还指出，这些应激反应的减少可能和同一个群体内的其他鱼缺少行为的和化学的信号有关。相应地，当斑马鱼接触到群体内死鱼释放的化学信号时，其体内皮质醇水平也会升高（Oliveira 等，2014）。这一点在斑马鱼养殖中尤其值得注意，因为绝大多数实验室都使用循环水系统进行养殖，一些对鱼类有预警作用的信息素会持续存在并通过这些系统传播，从而导致整个系统应激水平升高。

一些文献表明，养殖密度与应激之间的关系是极为复杂多变的。例如，和刚刚提到的所有观察结果相反，Pangnussat 等（2013）发现，隔离饲养的斑马鱼比在小群体（3 条鱼）中饲养的斑马鱼皮质醇水平更高。类似的，一些研究表明养殖密度与鱼缸内皮质醇水平没有相关性（Gronquist，Berges，2013）。Ramsay 等（2006）发现，虽然养殖密度相同（4 条鱼/升），但在 76 L 的鱼缸里的鱼比在 4 L 的鱼缸里的鱼会表现出更高的皮质醇水平。这些研究表明，不同的环境条件、饲养方法和实验设计都可能会导致非常不同的应激反应，尤其是当以皮质醇为主要检测指标时。

正如前面所讨论的，皮质醇水平可能并不能完全说明应激和社会行为的关系。已证实斑马鱼具有群体行为偏爱，当让它们在空的鱼缸和满是同类的鱼缸之间做出选择时，它们倾向于在鱼群附近停留更长的时间（Stewart 等，2012）。此外，许多动物护理和实验规程认为，尽量减少群体中动物的数量能改善饲养环境条件。尽管有研究表明单独一条鱼的皮质醇水平可能较低，但斑马鱼对鱼群有一种行为偏爱，这可能表明从鱼的角度出发，在某种程度上，社会性应激可能是一种积极的，而不是消极的应激源。

有趣的是，有研究表明斑马鱼早在受精卵时期，种群密度就可能对斑马鱼的行为产生影响。Steenbergen 等（2011）进行鱼对黑暗脉冲测定实验，发现单独培养长大的胚胎与共同培养的一群胚胎（25 个）相比，会做出不同的反应。这种现象的出现是因为胚胎间的相互接触导致的，当在胚胎孵化的前两天对单独培养的胚胎进行人为的接触刺激，可以消除单独培养导致的对黑暗反应的差异。因此，胚胎培养的密度可能对幼鱼甚至成鱼的发育以及行为表型有很重要的影响。

13.4.2 环境优化改善

本章相当大一部分已经讨论关于实验鱼类生产实践的优化。有必要记住的是，“快乐”的鱼通常是多产的鱼，动物福利是提高产量的基石，尤其是在实验室环境中。优化饲养不仅仅是简单地满足动物的生理需求，如饲养密度、营养和水的质量。在这方面，动物保护委员会和研究机构的管理监督机构要求对实验哺乳动物的生活环境进行各种形式的改善（Baumans，Van Loo，2013；Hutchinson 等，2005；Institute of Laboratory Animal Resources Committee，2011）。对所有动物来说，空间结构的环境优化改善所带来的好处具有物种的和环境的密切相关性。已有许多研究表明，改善优化环境会影响鱼类特定的行为性状（Brydges，Braithwait，2009）。比如，Kistler 等（2011）研究发现斑马鱼会被环境中的物件（人造植物）所吸引，当在环境中放入稀鳞无须鲃（*Puntius oligolepis*）时，对斑马鱼的吸引作用更为强烈。对斑马鱼来说，更复杂的空间会减少鱼之间的攻击行为，而攻击性的降低和繁殖力的增加有关（Carfagnini 等，2009）。然而，环境的优化改善与应激减少之间的直接相关性尚未得到证实。比如，Wilkes 等（2012）研究发现，用玻璃棒模拟植物茎部加入养殖环境，并不能降低斑马鱼的皮质醇水平。在另一项研究中，Keck 等（2015）发现虽然加入塑料植物并没有降低成对饲养的斑马鱼体内的皮质醇水平，但确实降低了在这种环境中与攻击性有关的死亡率。相比之下，Pavlidis 等（2013）研究表明，在较黑暗的环境中的斑马鱼和在光亮环境中的斑马鱼相比，具有更低的皮质醇水平。这样一些发现以及类似的研究突出地显示了鱼类与饲养环境有关的行为复杂性。因此，还需要进行更多的研究，探讨实验室鱼类的具体环境优化需求，以便改善它们的动物福利和作为模型生物的有效性。

13.4.3　光/暗周期

鱼和所有动物一样，都深受光周期的影响（Pavlidis 等，1999；Valenzuela 等，2012；Boeuf 和 Le Bail，1999），因此，了解和调整实验鱼类生活环境中的光周期至关重要。鉴于光周期对鱼的生理有如此深刻的影响，因此，除了动物进化中产生的光偏爱之外，在“一天”的长度范围内，光的变化可能成为鱼类痛苦（亦即恶性应激）的来源。比如，当非洲胡鲶（*Clarias gariepinus*）处于相比自然状态更长时间的光照时，会表现出体内皮质醇水平上升的情况（Almazan-Rueda 等，2005）。此外，Mustapha 等（2014）研究发现，尼罗罗非鱼和非洲胡鲶在不间断的持续光照下（没有黑暗处理时期），其死亡率会增加（Mustapha 等，2014）。和长期处于黑暗中的鱼类相比，处在较长的光照周期中的雌性大头丽脂鲤（*Astyanax bimaculatus*）表现出皮质醇水平升高、攻击性增强及存活率下降。

虽然斑马鱼已被证实是一种可依据光周期进行繁殖的物种（Lawrence，2007；Westerfield，2007），但几乎没有相关实验进行白天长短对应激影响的研究。现有的对光周期的研究并没有特别关注到应激的方面，但它们的研究结果和之前提到的是相对一致的：幼鱼在黑暗中游动的频率比在光照下要高（De Esch 等，2012），此外幼鱼和成鱼在关灯的瞬间都会增加活动强度（Vignet 等，2013）。从这些研究可以推断，光亮和黑暗对斑马鱼的神经行为有影响，但是还需要更多的研究来确定光周期和应激之间的潜在相关性。

13.5 喂食和应激

许多动物保护委员会认为每天喂鱼是人道待遇的一项要求。对于小型水缸，尤其是静水养殖，每天投喂食物可能会导致水质问题，并增加高密度养殖的不良影响。当突然停止投喂时，鲑鳟鱼类可能表现出皮质醇水平上升，这可能是由于彼此间攻击性增加所致（Brannass 等，2003）。相比之下，Gronquist 和 Berges（2013）研究发现，在连续 9 天的停止投喂实验中，斑马鱼的皮质醇水平明显下降，这可能是由于水质改善的影响超过了相对静止少动的斑马鱼代谢需求所致。对此的另一种可能是，这些蓄养的鱼没有自然活动模式而被长期过量投喂进食，而这也可能是干扰实验结果的因素。相反，在许多物种中，应激会抑制食物的摄入（Steenbergen 等，2011）。有趣的是，习惯于特定时间间隔喂食的一群丽鱼，改变其进食时间会导致皮质醇水平上升（Galhardo 等，2011），这表明在实验室条件下饲养的鱼，进食习惯的改变可能是应激的来源之一。

13.6 性别与社会等级

在水族箱中养殖的斑马鱼以及包括鲑鱼在内的许多其他物种，群体中建立了优势社会等级，这在雄鱼中尤为明显（Spence 等，2008；Paull，2010；Ejike，Schreck，1980）。Filby 等（2010）研究发现，相比于处于从属社会等级地位的雌鱼，处于从属社会等级地位的雄鱼皮质醇水平上升得较多。这说明雄鱼比较普遍地处于优势社会等级地位。

在实验室养殖的斑马鱼中，和优势等级有关的生理因素包括鱼的大小和争斗行为（Paull 等，2010）。在整个种群中，处于从属社会等级地位会比处于优势社会等级地位的鱼承受更多的应激。比如，对伯氏朴丽鱼（*Haplochromis burtoni*）来说，相比于拥有势力范围的雄鱼，没有自己势力范围的雄鱼表现出较高的皮质醇水平（Fox 等，1997）。虹鳟的幼鱼也是如此，其顺从的行为表型被认为是慢性的应激源。有趣的是，皮质醇的升高和顺从行为之间可能存在一种正强化的相互作用关系，顺从屈服行为会导致应激，而由于健康状况不佳、体型较小等，皮质醇的升高又会导致顺从行为的增加（Gilmour 等，2005）。对于是什么因素使鱼更有可能成为群体中优势者或顺从者的问题，这可能和外界刺激引起皮质醇反应增高的遗传易感性（genetic predisposition）有关，尽管这一假说仍在被探讨与证实之中（Ramsay 等，2009a，b，c）。

为了改善鱼群中个体行为表现的一致性以及减少和社会等级有关的应激，充足的喂养被认为是潜在的一种解决方案。因为它可以使鱼缸内个体大小的分布一致，同时减少因争夺食物而产生的攻击行为。当然，拥挤与资源分配和等级形成是密切相关的，因此，适当的饲养密度可能有助于减少等级引起的应激（Ramsay 等，2009a，b，c）。

13.7　性别决定与性逆转

鱼类常用于进行性腺表型为终点的研究，如内分泌干扰物对性腺表型变化的评定。日本青鳉（*Oryzias latipes*）是毒理学研究的重要物种。Hayashi 等（2010）研究表明，皮质醇可以导致雌鱼向雄鱼的性逆转，而甲吡酮（一种皮质醇合成抑制剂）可抑制雌性青鳉由 HT-诱导的雄性化。因为皮质醇也被认为可以影响银汉鱼（*Odontesthes bonariensis*）（Hattori 等，2009）和日本牙鲆（*Paralichthys olivaceus*）等其他物种的性腺表型，所以，对于像斑马鱼这样的实验室鱼类来说，皮质醇可能也具有类似的作用。

13.8　应激、皮质醇和生殖

大量利用鱼类进行的生物医学研究主要都集中于其早期发育阶段，尤其是胚胎期。与传统的哺乳动物模型相比，斑马鱼具有繁殖力高、可体外受精、发育快速、幼鱼全身透明等优点，非常适合进行高通量的实验（Murphey，Zon，2006）。所以，在许多实验室中，一些成鱼专门被用于进行胚胎生产，在这种情况下，应激对繁殖力和胚胎发育的影响尤其引人关注。

由于繁殖力在养殖上是一个最终的生产指标，大量关于应激和繁殖力之间关系的研究已在虹鳟和其他鲑鳟鱼类中开展。这些研究已被 Schreck 等（2001）进行了很好的综述。类似的研究在实验室鱼类中还十分缺乏。Cloud（1981）研究发现，用脱氧皮质酮孵育青鳉受精卵可以加快其孵化率。

关于应激和繁殖力的讨论还必须关注到母体应激对后代的影响，这些影响大多数是由表观遗传的。Schreck 和 Tort（2016；见本书第 1 章）已经讨论了关于母体应激对鲑鳟鱼类后代的影响，但是母体应激在实验鱼类中影响的研究数据非常少。Nesan 和 Vijayan（2012）研究发现，在孵化前用皮质醇处理斑马鱼胚胎会导致心脏功能障碍和其他发育缺陷。在这项研究中，因为斑马鱼在孵化后才会产生自己独立的皮质醇，所以研究者将皮质醇注入单细胞期的胚胎中来模拟母体皮质醇传递到受精卵中的过程。在三刺鱼中，当母体处于被捕食的状态下，会增加胚胎中关于代谢、表观遗传和神经发育的基因转录。上述这种雌鱼产下的胚胎会比对照组雌鱼产下的胚胎个体更大（Mommer，Bell，2014）。

幼鱼和胚胎期受到应激的后果可以延伸到成年期，例如，虹鳟在胚胎早期受到应激后（包括母体应激和内源性皮质醇），其在成年后的应激反应会降低（Steenbergen 等，2011）。这些结果着重说明，对母体的来源管理和饲养的稳定优化，以及对胚胎和幼鱼培育稳定优化，对于减少此类因素产生的应激影响是非常必要的，这也会减少此类非实验因素引起的实验误差。

13.9 麻 醉

实验用鱼通常会被麻醉以便进行各种各样的实验操作，包括注射、标记和埋植。Palic 等（2006）在胖头鲹中曾经评估 MS-222、美托咪酯以及丁香酚作为麻醉剂对体内皮质醇水平的影响。和对照组相比，用美托咪酯或丁香酚（MS-222 除外）麻醉的鱼在之后的操作过程以及后续置于拥挤环境中时，表现出较低的皮质醇水平。最近，两项不同的实验证明在斑马鱼实验中常用的麻醉剂 MS-222 会使斑马鱼感到厌恶。这里所说的厌恶意味着鱼类试图避开研究中使用的 MS-222——如果有选择的话，它们更喜欢在含低浓度或完全不含 MS-222 的水中。利用可以量化游泳行为空间偏好的分析软件，Readman 等（2013）研究发现在 9 种常用的麻醉剂中，斑马鱼对其中 7 种都表现出了厌恶行为，这其中就包括 MS-222 和苯佐卡因（Readman 等，2013）。他们还建议用依托咪酯作为代替品对斑马鱼进行麻醉。在另一个实验中，Wong 等（2014）观察在不同条件下斑马鱼的空间位置来证明其对 MS-222 的厌恶行为（Wong 等，2014）。虽然这些数据只适用于斑马鱼，但它们清楚地表明，麻醉剂的选择应根据物种和实验实际情况进行慎重考虑。同时，对鱼来说，这种麻醉带来的短暂厌恶感和不麻醉进行实验操作或手术带来的应激和痛感也值得我们去认真考虑取得平衡。当然，我们不建议在实验操作之前不使用麻醉——事实上，在许多情况下麻醉对鱼的生存是必要的——我们仅仅是建议应该探讨较为广泛的选择，以期得到不会使鱼产生那么明显的厌恶感或应激的麻醉方法。

13.10 潜在的疾病

Kent 等（2012）曾综述在斑马鱼研究中潜在的病变和感染对实验产生的影响。回归到本书的主题，应激与潜在疾病具有相互促进的关系，疾病有可能提高初期应激水平，应激反过来又加剧病原体的致病性。例如，*Pseudoloma neurophilia* 是斑马鱼养殖系统中最常见的微孢子虫（Sanders 等，2012），我们最近发现这种微孢子虫可以感染诸如青鳉和胖头鲹等许多实验鱼类的中枢神经系统。大部分这类感染的病征都不明显（Spagnoli 等，2015a，b），但当鱼类受到拥挤带来的应激时会容易受到更严重的感染（Ramsay 等，2009a，b，c）。研究者已把 *Pseudoloma neurophilia* 感染与评估鱼类对应激和焦虑反应的变化联系起来，也就是说，被感染的鱼会出现高度应激反应行为（Spagnoli 等，2015a，b）。相比于没有应激的对照组，被 *Mycobacterium marinum* 感染的斑马鱼在拥挤的应激中可能表现出较高的感染死亡率。这种情况不是由病原体在鱼之间传播引起的，因为这种病原体需要其他的鱼将感染的组织或无脊椎动物摄入体内才能被传播感染（Peterson 等，2013）。更确切地说，死亡率的增加是由于鱼类因应激而加剧了先前存在的感染状况。关于应激、拥挤和皮质醇升高会促进鱼体内病原体的增殖和毒性的例子举不胜举，众所周知，因为皮质醇具有免疫抑制作用（Cortes

等，2013）。相反地，潜在的寄生虫感染本身就是一种应激源，它可导致皮质醇水平升高（Hoole，Williams，2004）。皮质醇作为一种免疫抑制剂，在有关鱼类应激的炎症和免疫反应的研究中尤其重要（Yada 和 Tort，2016；见本书第 10 章）。

13.11　一致性

虽然前面讨论的所有特异的方面单独来看都是很重要的，但是有一个因素应该把它们统一结合起来，用以指导人们的决策过程：鱼类照料和饲养过程的一致性。鱼类从幼年到成年的过程中，不同阶段饲养条件变化显然是必要的，但是一旦确定了特定生命阶段的具体饲养方法，就应尽量减少摄食、温度、环境噪声和水质等因素的剧烈变化。Piato 等（2011）进行的一项有趣的研究发现，斑马鱼有潜力成为研究人类中不可预知的慢性应激的优秀模型。当斑马鱼在不规律的时间间隔和不确定的时间点中处在不可预知的多重应激物影响下时，实验 7 天后，斑马鱼表现出焦虑增加、认知功能下降以及皮质醇增加。尽管这对于斑马鱼作为人类神经行为疾病的模型来说是一个极好的消息，但这可能意味着饲养过程中连续的、随机的变化既可能产生过度的应激，也可能改变研究中的神经行为反应。因此，应尽可能避免饲养过程中的变化或频繁产生应激的状况，如设备的维修或更换，以及警报噪声或照明的变化等。

13.12　结束语和关键的未知因素

未来的研究应该探讨特定的养殖操作对应激的实际效应，以及它们对鱼类作为模式生物的影响。随着对这一问题收集的数据越来越多，科学界很可能会开始发掘一些迄今未知的复杂因素。例如，鲑鱼的驯化选育是一个建立得很好的技术（Araki 等，2008），其显著的遗传和行为变化可以快速发生在一个世代的时间内。因此，在完全人工的实验室环境中，驯化选育可能对模型鱼类的行为性状产生强烈影响。为了支持这个说法，研究者已经通过实验观察到 AB 和 TU/TL 斑马鱼品系之间的行为差异（Varga，2011）。这种人工选育实际上可能是有益的，因为它可以减少在实验室中鱼的遗传变异。然而，在不同实验室和培养者之间，当无意中选育了不同行为表现型的鱼时，可能产生不同的应激反应。例如，Spagnoli 等（2015a，b）观察到的和 *Pseudoloma neurophilia* 感染相关的行为转变，可能是由于在选育被寄生虫感染而产生特殊行为表型的鱼时其产生的选育压力所致。如果是这样的话，那么被这种非常普遍的寄生虫污染的实验室就有可能无意中选育出具有高度应激反应而无关感染死亡的行为表型的斑马鱼。

本章对实验鱼类应激的来源、检测和影响进行了详细的讨论。这里有必要重申用饲养的动物进行研究时的三个基本目标：研究对象的一致性、研究模型的稳定性和动物福利。应激是所有脊椎动物连接大脑和身体的一种基本神经生理状态。因此，大量利用实验室鱼类实验的最终结果，无论它们是神经行为领域的、病理领域的、分子领域的还是代谢领域的，都可能受到应激的影响。全面了解实验室鱼类应激产生的原因和后果，并结合动物福利以及实验的一致性，将最终引领鱼类饲养技术的进一步发展。

S. 斯帕诺里　C. 劳伦斯　M. L. 肯特　著
李水生　译
林浩然　校

参考文献

Aerts, J., Metz, J. R., Ampe, B., Decostere, A., Flik, G., De Saeger, S. (2015). Scales tell a story on the stress history of fish. PLoS One. 10, e0123411.

Allen, J. P., Neely, M. N. (2010). Trolling for the ideal model host: zebrafish take the bait. Future Microbiol. 5, 563-569.

Almazán-Rueda, P., Van Helmond, A. T. M., Verreth, J. A. J., Schrama, J. W. (2005). Photoperiod affects growth, behaviour and stress variables in *Clarias gariepinus*. J. Fish. Biol. 67, 1029-1039.

Ankley, G. T., Villeneuve, D. L. (2006). The fathead minnow in aquatic toxicology: past, present and future. Aquat. Toxicol. 78, 91-102.

Araki, H., Berejikian, B. A., Ford, M. J., Blouin, M. S. (2008). Fitness of hatchery-reared salmonids in the wild. Evol. Appl. 1, 342-355.

Araujo, J., Maximino, C., de Brito, T. M., da Salva, A. W. B., Oliveira, K. R. M., Batista, E. J. O. (2012). Behavioral and pharmacological aspects of anxiety in the light/dark preference test. In: Kalueff, A. V., Stewart, A. M., Zebrafish Protocols for Neurobehavioral Research. Springer, New York, NY, 191-203.

Bailey, G. S., Williams, D. E., Hendricks, J. D. (1996). Fish models for environmental carcinogenesis: the rainbow trout. Environ. Health Perspect. 104, 5-21.

Bailey, J., Oliveri, A., Levin, E. D. (2013). Zebrafish model systems for developmental neurobehavioral toxicology. Birth Defects Res. C Embryo Today. 99, 14-23.

Barry, T. P., Malison, J. A., Held, J. A., Parrish, J. J. (1995). Ontogeny of the cortisol stress response in larval rainbow trout. Gen. Comp. Endocrinol. 97, 57-65.

Baumans, V., Van Loo, P. L. P. (2013). How to improve housing conditions of laboratory animals: the possibilities of environmental refinement. Vet. J. 195, 24-32.

Blaser, R. E., Rosemberg, D. B. (2012). Measures of anxiety in zebrafish (*Danio rerio*): dissociation of black/white preference and novel tank test. PLoS One. 7, e36931.

Boeuf, G., Le Bail, P. Y. (1999). Does light have an influence on fish growth?.

Aquaculture. 177, 129-152.

Brännäss, E., Jonsson, S., Lundqvist, H. (2003). Influence of food abundance in individual behaviour strategy and growth rate in juvenile Brown Trout (*Salmo trutta*). Can. J. Zool. 81, 684-692.

Brydges, N. M., Braithwait, V. A. (2009). Does environmental enrichment affect the behaviour of fish commonly used in laboratory work?. Appl. Anim. Behav. Sci. 118, 137-143.

Canavello, P. R., Cachat, J. M., Beeson, E. C., Laffoon, A. L., Grimes, C., Haymore, W. A. M. (2011). Measuring endocrine (cortisol) responses of zebrafish to stress. In: Kalueff, A. V., Stewart, A. M., Zebrafish Neurobehavioral Protocols. Springer, New York, NY, 135-142.

Carfagnini, A. G., Rodd, F. H., Jeffers, K. B., Bruce, A. E. E. (2009). The effects of habitat complexity on aggression and fecundity in zebrafish (*Danio rerio*). Environ. Biol. Fish. 86, 403-409.

Ceol, C. J., Houvras, Y., Jane-Valbuena, J., Bilodeau, S., Orlando, D. A., Battisti, V. (2011). The histone methyltransferase SETDB1 is recurrently amplified in melanoma and accelerates its onset. Nature. 471, 513-517.

Cloud, J. G. (1981). Deoxycorticosterone-induced precocious hatching of teleost embryos. J. Exp. Zool. 216, 197-199.

Conrad, J. L., Weinersmith, K. L., Brodin, T., Saltz, J. B., Sih, A. (2011). Behavioural syndromes in fishes: a review with implications for ecology and fisheries management. J. Fish. Biol. 78, 395-435.

Cortés, R., Teles, M., Tridico, R., Acerete, L., Tort, L. (2013). Effects of cortisol administered through slow release implants on innate immune responses in rainbow trout (*Onchorhynchus mykiss*). Int. J. Genomics. 2013, 619714.

De Esch, C., van der Linde, H., Slieker, R., Willemsen, R., Wolterbeek, A., Woutersen, R. (2012). Locomotor activity assay in zebrafish larvae: influence of age, strain and ethanol. Neurotoxicol. Teratol. 34, 425-433.

Dooley, K., Zon, L. I. (2000). Zebrafish: a model for the study of human disease. Curr. Opin. Genet. Dev. 10, 252-256.

Dugatkin, L., McCall, M. A., Gregg, R. G., Cavanaugh, A., Christensen, C., Unseld, M. (2004). Zebrafish (*Danio rerio*) exhibit individual differences in risk-taking behavior during predator inspection. Ethol. Ecol. Evol. 17, 77-81.

Egan, R. J., Bergner, C. L., Hart, P. C., Cachat, J. M., Canavello, P. R., Elegante, M. F. (2009). Understanding behavioral and physiological phenotypes of stress and anxiety in zebrafish. Behav. Brain Res. 205, 38-44.

Ejike, C., Schreck, C. B. (1980). Stress and social hierarchy rank in coho salmon.

Trans. Am. Fish. Soc. 109, 423-426.

Ellis, T. , James, J. D. , Stewart, C. , Scott, A. P. (2004). A non-invasive stress assay based upon measurement of free cortisol released into the water by rainbow trout. J. Fish. Biol. 65 (5), 1233-1252.

Ellis, T. , Yildiz, H. Y. , Lopez-Olmeda, J. , Spedicato, M. T. , Tort, L. , Overli, O. (2012). Cortisol and finfish welfare. Fish. Physiol. Biochem. 38, 163-188.

Engeszer, R. E. , Patterson, L. B. , Rao, A. A. , Parichy, D. M. (2007). Zebrafish in the wild: a review of natural history and new notes from the field. Zebrafish. 4, 21-40.

Feist, G. , Schreck, C. B. (2001). Ontogeny of the stress response in chinook salmon, Oncorhynchus tshawytscha. Fish. Physiol. Biochem. 25, 31-40.

Félix, A. S. I. , Faustino, A. I. , Cabral, E. M. , Oliveira, R. F. (2013). Noninvasive measurement of steroid hormones in zebrafish holding-water. Zebrafish. 10, 110-115.

Filby, A. L. , Paull, G. C. , Bartlett, E. J. , Van Look, K. J. W. , Tyler, C. R. (2010). Physiological and health consequences of social status in zebrafish (*Danio rerio*). Physiol. Behav. 101, 576-587.

Fox, H. E. , White, S. A. , Kao, M. H. F. , Fernald, R. D. (1997). Stress and dominance in a social fish. J. Neurosci. 17, 663-6469.

Galhardo, L. , Vital, J. , Oliveira, R. F. (2011). The role of predictability in the stress response of a cichlid fish. Physiol. Behav. 102, 367-372.

Gardell, A. M. , Yang, J. , Sacchi, R. , Fangue, N. A. , Hammock, B. D. , Kultz, D. (2013). Tilapia (*Oreochromis mossambicus*) brain cells respond to hyperosmotic challenge by inducing myo-inositol biosynthesis. J. Exp. Biol. 216, 4615-4625.

Gerhard, G. S. (2003). Comparative aspects of zebrafish (*Danio rerio*) as a model for aging research. Exp. Gerontol. 38, 1333-1341.

Gesto, M. , Hernandez, J. , Lopez-Patino, M. A. , Soengas, J. L. , Miguez, J. M. (2015). Is gill cortisol concentration a good acute stress indicator in fish? A study in rainbow trout and zebrafish. Comp. Biochem. Physiol. A Mol. Integr. Physiol. 188, 65-69.

Ghisleni, G. , Capiotti, K. M. , Da Silva, R. S. , Oses, J. P. , Piato, A. L. , Soares, V. (2012). The role of CRH in behavioral responses to acute restraint stress in zebrafish. Prog. Neuro-Psychopharmacol. Biol. Psychiatry. 36, 176-182.

Giacomini, A. C. , de Abreu, M. S. , Koakoski, G. , Idalencio, R. , Kalichak, F. , Oliveira, T. A. (2015). My stress, our stress: blunted cortisol response to stress in isolated zebrafish. Physiol. Behav. 139, 182-187.

Gilmour, K. M. , DiBattista, J. D. , Thomas, J. B. (2005). Physiological causes and consequences of social status in salmonid fish. Integ. Comp. Biol. 45, 263-273.

Gronquist, D., Berges, J. A. (2013). Effects of aquarium-related stressors on the zebrafish: a comparison of behavioral, physiological, and biochemical indicators. J. Aquat. Anim. Health. 25, 53-65.

Harland, R. M., Grainger, R. N. (2011). *Xenopus* research: metamorphosed by genetics and genomics. Trends Gen. 27, 507-515.

Hattori, R. S., Fernandino, J. I., Kishii, A., Kimura, H., Kinno, T., Oura, M. (2009). Cortisol-induced masculinization: does thermal stress affect gonadal fate in pejerrey, a teleost fish with temperature-dependent sex determination?. PLoS One. 4 (8), e6548.

Hayashi, Y., Kobira, J., Yamaguchi, T., Shiraishi, E., Yazawa, T., Hirai, T. (2010). High temperature causes masculinization of genetically female medaka by elevation of cortisol. Mol. Rep. Dev. 77, 679-686.

Hendry, A. P., Peichel, C. L., Matthews, B., Boughman, J. W., Nosil, P. (2013). Stickleback research: the now and the next. Evol. Ecol. Res. 15, 111-141.

Hoole, D., Williams, G. T. (2004). The role of apoptosis in non-mammalian host-parasite relationships. In: Fli, G., Wiegertjes, G., Host-Parasite Interactions. BIOS Scientific Publishers, , 13-45.

Hutchinson, E., Avery, A., VandeWoude, S. (2005). Environmental enrichment for laboratory rodents. ILAR J. 46, 148-161.

Institute of Laboratory Animal Resources Committee. (2011). Guide for the Care and Use of Laboratory Animals.

Kahl, M. D., Jensen, K. M., Korte, J. J., Ankley, G. T. (2001). Effects of handling on endocrinology and reproductive performance of the fathead minnow. J. Fish. Biol. 59, 515-523.

Keck, V. A., Edgerton, D. S., Hajizadeh, S., Swift, L. L., Dupont, W. D., Lawrence, C., Boyd, K. L. (2015). Effects of habitat complexity on pair-housed zebrafish. J. Am. Assoc. Lab. Anim. Sci. 54, 378-383.

Kent, M. L., Harper, C., Wolf, J. C. (2012). Documented and potential research impacts of subclinical diseases in zebrafish. ILAR J. 53, 126-134.

Kishi, S., Slack, B. E., Uchiyama, J., Zhdanova, I. V. (2009). Zebrafish as a genetic model in biological and behavioral gerontology: where development meets aging in verterates-a mini-review. Gerontology. 55, 430-441.

Kistler, C., Hegglin, D., Würbel, H., König, B. (2011). Preference for structured environment in zebrafish (*Danio rerio*) and checker barbs (*Puntius oligolepis*). Appl. Anim. Behav. Sci. 135, 318-327.

Knudsen, F. R., Enger, P. S., Sand, O. (1992). Awareness reactions and avoidance responses to sound in juvenile Atlantic salmon, Salmo salar L. J. Fish. Biol. 40, 523-534.

Lawrence, C. (2007). The husbandry of zebrafish (*Danio rerio*): a review. Aquaculture. 269, 1-4.

Lawrence, C., Mason, T. (2012). Zebrafish housing systems: a review of basic operating principles and considerations for design and functionality. ILAR J. 53, 179-191.

Lieschke, G. J., Currie, P. D. (2007). Animal models of human disease: zebrafish swim into view. Nat. Rev. Genet. 8, 353-367.

Liu, Z. (2003). A review of catfish genomics: progress and perspectives. Comp. Funct. Genom. 4, 259-265.

Liu, S., Leach, S. D. (2011). Zebrafish models for cancer. Ann. Rev. Pathol. Mech. Dis. 6, 71-93.

Martinez-Porchas, M., Martinez-Cordova, L. T., Ramos-Enriquez, R. (2009). Cortisol and glucose: reliable indicators of fish stress?. J. Aquat. Sci. 4, 158-178.

Martins, C. I. M., Ochola, D., Ende, S. S. W., Eding, E. H., Verreth, J. A. J. (2009). Is growth retardation present in Nile tilapia *Oreochromis niloticus* cultured in low water exchange recirculating aquaculture systems?. Aquaculture. 298, 1-2.

Martins, C. I. M., Eding, E. H., Verreth, J. A. J. (2011). Stressing fish in recirculating aquaculture systems (RAS): does stress induced in one group of fish affect the feeding motivation of other fish sharing the same RAS?. Aquac. Res. 42, 1378-1384.

Maximino, C., deBrito, T. M., da Silva Batista, A. W., Herculano, A. M., Morato, S., Gouveia, A. (2010). Measuring anxiety in zebrafish: a critical review. Behav. Brain Res. 214, 157-171.

Meijer, A. H., Spaink, H. P. (2011). Host-pathogen interactions made transparent with the zebrafish model. Curr. Drug. Targets. 12, 1000-1017.

Meijer, A. H., Verbeek, F. J., Salas-Vidal, E., Corredor-Adamez, M., Bussman, J., van der Sar, A. M. (2005). Transcriptome profiling of adult zebrafish at the late stage of chronic tuberculosis due to *Mycobacterium marinum* infection. Mol. Immunol. 42, 1185-1203.

Mommer, B. C., Bell, A. M. (2014). Maternal experience with predation risk influences genome-wide embryonic gene expression in threespined sticklebacks (*Gasterosteus aculeatus*). PLoS One. 9.

Murphey, R. D., Zon, L. I. (2006). Small molecule screening in the zebrafish. Methods. 39, 255-261.

Mustapha, M. K., Oladokun, O. T., Salman, M. M., Adeniyi, I. A., Ojo, D.

(2014). Does light duration (photoperiod) have an effect on the mortality and welfare of cultured *Oreochromis niloticus* and *Clarias gariepinus*?. Turk. J. Zool. 38, 466-470.

Neo, Y. Y., Parie, L., Bakker, F., Snelderwaard, P., Tudorache, C., Schaaf, M. (2015). Behavioral changes in response to sound exposure and not special avoidance of noisy conditions in captive zebrafish. Front. Behav. Neurosci. 9, 28.

Nesan, D., Vijayan, M. M. (2012). Embryo exposure to elevated cortisol level leads to cardiac performance dysfunction in zebrafish. Mol. Cell. Endocrinol. 363, 85-91.

Novoa, B., Figueras, A. (2012). Zebrafish: model for the study of inflammation and the innate immune response to infectious diseases. Adv. Exp. Med. Biol. 946, 253-275.

Oliveira, T. A., Koakoski, G., da Motta, A. C., Piato, A. L., Barreto, R. E., Volpato, G. L. (2014). Death-associated odors induce stress in zebrafish. Horm. Behav. 65, 340-344.

Palić, D., Herolt, D. M., Andreasen, C. B., Menzel, B. W., Roth, J. A. (2006). Anesthetic efficacy of tricaine methanesulfonate, metomidate and eugenol: effects on plasma cortisol concentration and neutrophil function in fathead minnows (*Pimephales promelas* Rafinesque, 1820). Aquaculture. 254, 675-685.

Pangnussat, N., Piato, A. L., Schaefer, I. C., Blank, M., Tamborski, A. R., Guerim, L. D. (2013). One for all and all for one: the importance of shoaling on behavioral and stress responses in zebrafish. Zebrafish. 10, 338-342.

Parker, M. O. I., Millington, M. E., Combe, F. J., Brennan, C. H. (2012). Housing conditions differentially affect physiological and behavioural stress responses of zebrafish, as well as the response to anxiolytics. PLoS One. 7 (4), e34992.

Patton, E. E., Zon, L. I. (2005). Taking human cancer genes to the fish: a transgenic model of melanoma in zebrafish. Zebrafish. 1, 363-368.

Patton, E. E., Mitchell, D. L., Nairn, R. S. (2010). Genetic and environmental melanoma models in fish. Pigment. Cell. Melanoma. Res. 23, 314-337.

Patton, E. E., Mathers, M. E., Schartl, M. (2011). Generating and analyzing fish models of melanoma. Methods Cell. Biol. 105, 339-366.

Paull, G. C., Filby, A. L., Giddins, H. G., Coe, T. S., Hamilton, P. B., Tyler, C. R. (2010). Dominance hierarchies in zebrafish (*Danio rerio*) and their relationship with reproductive success. Zebrafish. 7, 109-117.

Pavlidis, M. (1999). The effect of photoperiod on diel rhythms in serum melatonin, cortisol, glucose, and electrolytes in the common dentex, *Dentex dentex*. Gen. Comp. Endocrinol. 113, 240-250.

Pavlidis, M., Digka, N., Theodoridi, A., Campo, A., Brasakis, K., Skouradakis,

G. (2013). Husbandry of zebrafish, *Danio rerio*, and the cortisol stress response. Zebrafish. 10, 524-531.

Peterson, T. S., Ferguson, J. A., Watral, V. G., Mutoji, K. N., Ennis, D. G., Kent, M. L. (2013). *Paramecium caudatum* enhances transmission and infectivity of *Mycobacterium marinum* and *Mycobacterium chelonae* in zebrafish (*Danio rerio*). Dis. Aquat. Org. 106, 229-239.

Phillips, J. B., Westerfield, M. (2014). Zebrafish models in translational research: tipping the scales towards advancements in human health. Dis. Model Mech. 7, 739-743.

Piato, A. L., Capiotti, K. M., Tamborski, A. R., Oses, J. P., Barcellos, L. J. G., Bogo, M. R. (2011). Unpredictable chronic stress model in zebrafish (*Danio rerio*): behavioral and physiological responses. Prog. Neuro-Psychopharmacol. Biol. Psychiatry. 35, 561-567.

Prouty, M. G., Correa, N. E., Barker, L. P., Jagadeeswaran, P., Klose, K. E. (2003). Zebrafish-*Mycobacterium marinum* model for mycobacterial pathogenesis. FEMS Microbiol. Lett. 225, 177-182.

Purser, J., Radford, A. N. (2011). Acoustic noise induces attention shifts and reduces foraging performance in three-spined sticklebacks (*Gasterosteus aculeatus*). PLoS One.

Radford, A. N., Kerridge, E., Simpson, S. D. (2014). Acoustic communication in a noisy world: can fish compete with anthropogenic noise?. Behav. Ecol. 25 (5), 1022-1030.

Ramsay, J. M., Feist, G. W., Varga, Z. M., Westerfield, M., Kent, M. L., Schreck, C. B. (2006). Whole-body cortisol is an indicator of crowding stress in adult zebrafish, *Danio rerio*. Aquaculture. 258, 565-574.

Ramsay, J. M., Feist, G. W., Varga, Z. M., Westerfield, M., Kent, M. L., Schreck, C. B. (2009). Rapid cortisol response of zebrafish to acute net handling stress. Aquaculture. 297, 157-162.

Ramsay, J. M., Watral, V., Schreck, C. B., Kent, M. L. (2009). Husbandry stress exacerbates mycobacterial infections in adult zebrafish, *Danio rerio* (Hamilton). J. Fish. Dis. 32, 931-941.

Ramsay, J. M., Watral, V., Schreck, C. B., Kent, M. L. (2009). *Pseudoloma neurophilia* (Microsporidia) infections in zebrafish (*Danio rerio*): effects of stress on survival, growth and reproduction. Dis. Aquat. Org. 88, 69-84.

Readman, G. D., Owen, S. F., Murrell, J. C., Knowles, T. G. (2013). Do fish perceive anaesthetics as aversive?. PLoS One. 8 (9), e73773.

Rihel, J., Schier, A. F. (2012). Behavioral screening for neuroactive drugs in zebrafish. Dev. Neurobiol. 72, 373-385.

Sanders, J. L., Watral, V., Kent, M. L. (2012). Microsporidiosis in zebrafish research facilities. ILAR J. 53, 106-113.

Schartl, M. (2014). Beyond the zebrafish: diverse fish species for modeling human disease. Dis. Model Mech. 7, 181-192.

Schreck, C. B., Tort, L. (2016). The Concept of Stress in Fish. In: Schreck, C. B., Tort, L., Farrell, A. P., Brauner, C. J., . Academic Press, San Diego, CA.

Schreck, C. B., Contreas-Sanchez, W., Fitzbatrick, M. S. (2001). Effects of stress on fish reproduction, gamete quality, and progeny. Aquaculture. 197, 3-24.

Shima, A., Mitani, H. (2004). Medaka as a research organism: past, present and future. Mech. Dev. 121, 599-604.

Simpson, S. D., Purser, J., Radford, A. N. (2014). Anthropogenic noise compromises antipredator behavior in European eels. Glob. Change Biol. 21 (2), 586-593.

Singley, J. S., Chavin, W. (1975). Serum cortisol in normal goldfish (*Carassius auratus L.*). Comp. Biochem. Physiol. A Physiol. 50, 77-82.

Sisson, M., Cawker, J., Buske, C., Gerlai, R. (2006). Fishing for genes influencing vertebrate behavior: zebrafish making headway. Lab Anim (NY). 35 (5), 33-39.

Smith, M. E., Kane, A. S., Popper, A. N. (2003). Noise-induced stress response and hearing loss in goldfish (*Carassius auratus*). J. Exp. Biol. 207, 427-435.

Sneddon, L. U., Wolfenden, D. C. C., Thomson, J. S. (2016). Stress Management and Welfare. In: Schreck, C. B., Tort, L., Farrell, A. P., Brauner, C. J., . Academic Press, San Diego, CA.

Spagnoli, S., Xue, L., Kent, M. L. (2015). The common neural parasite *Pseudoloma neurophilia* is associated with altered startle response habituation in adult zebrafish (*Danio rerio*): implications for the zebrafish as a model organism. Behav. Brain Res. 291:, 351-360.

Spagnoli, S., Xue, L., Murray, K. M., Chow, F., Kent, M. L. (2015). *Pseudoloma neurophilia*: a retrospective and descriptive study with new implications for pathogenesis and behavioral phenotypes. Zebrafish. 12, 189-201.

Spence, R. (2011). Zebrafish models in neurobehavioral research. Neuromethods. 52, 211-222.

Spence, R., Gerlach, G., Lawrence, C., Smith, C. (2008). The behaviour and ecology of the zebrafish, *Danio rerio*. Biol. Rev. Camb. Philos. Soc. 83, 13-34.

Steenbergen, P. J., Richardson, M. K., Champagne, D. L. (2011). The use of the zebrafish model in stress research. Prog. Neuro-Psychopharmacol. Biol. Psychiatry.

35, 1432-1451.

Stewart, A., Gaikwad, S., Kyzar, E., Green, J., Roth, A., Kalueff, A. V. (2012). Modeling anxiety using adult zebrafish: a conceptual review. Neuropharmacology. 62, 135-143.

Stewart, A. M., Gerlai, R., Kalueff, A. V. (2015). Developing higher throughput zebrafish screens for in-vivo CNS drug discovery. Front Behav. Neurosci. 9, 1-8.

Swaim, L. E., Connoly, L. E., Volkman, H. E., Humbert, O., Born, D. E., Ramakrishnan, L. (2006). *Mycobacterium marinum* infection of adult zebrafish causes caseating granulomatous tuberculosis and is moderated by adaptive immunity. Infect. Immun. 74, 6108-6117.

Takeda, H., Shimada, A. (2010). The art of medaka genetics and genomics: what makes them so unique?. Annu. Rev. Genet. 44, 217-241.

Taylor, A. M., Zon, L. I. (2009). Zebrafish tumor assays: the state of transplantation. Zebrafish. 6, 339-346.

Tran, S., Chatterjee, D., Gerali, R. (2014). Acute net stressor increases whole-body cortisol levels without altering whole-brain monoamines in zebrafish. Behav. Neurosci. 128, 621-624.

Truong, L., Harper, S. L., Tanguay, R. L. (2011). Evaluation of embryo toxicity using the zebrafish model. Methods Mol. Biol. 691, 271-279.

Tudorache, C., Ter Braake, A., Tromp, M., Slabbekoom, H., Schaaf, M. J. (2015). Behavioral and physiological indicators of stress coping styles in larval zebrafish. Stress. 18, 121-128.

Valenzuela, A., Campos, V., Yanez, F., Alveal, K., Gutierrez, P., Rivas, M. (2012). Application of artificial photoperiod in fish: a factor that increases susceptibility to infectious diseases?. Fish Physiol. Biochem. 38, 943-950.

Varga, Z. M. (2011). Aquaculture and husbandry at the zebrafish international resource center. Methods Cell. Biol. 104, 453-478.

Vignet, C., Begout, M. L., Pean, S., Lyphout, L., Leguay, D., Cousin, X. (2013). Systematic screening of behavioral responses in two zebrafish strains. Zebrafish. 10, 365-375.

Yada, T., Tort, L. (2016). Stress and Disease Resistance: Immune System and Immunoendocrine Interactions. In: Schreck, C. B., Tort, L., Farrell, A. P., Brauner, C. J., . Academic Press, San Diego, CA.

Wedemeyer, G. A. (1997). Rearing conditions: effects on fish in intensive culture. In: Iwama, G. K., Pickering, A. D., Sumpter, J. P., Schreck, C. B., Fish Stress and Health in Aquaculture. Cambridge University Press, New York, NY, 35-73.

Westerfield, M. (2007). The Zebrafish Book. A Guide for the Laboratory Use of

Zebrafish (*Danio rerio*). Westerfield, 2007 Westerfield, M., The Zebrafish Book. A Guide for the Laboratory Use of Zebrafish (*Danio rerio*) fifth ed2007University of Oregon Press, Eugene, University of Oregon Press, Eugene.

White, R. M., Sessa, A., Burke, C., Bowman, T., LeBlanc, J., Ceol, C. (2008). Transparent adult zebrafish as a tool for in vivo transplantation analysis. Cell. Stem. Cell. 2, 183-189.

Wilke, L., Owen, S. F., Readman, G. D., Sloman, K. A., Wislon, R. W. (2012). Does structural enrichment for toxicology studies improve zebrafish welfare?. Appl. Anim. Behav. Sci. 129, 143-150.

Williams, D. E., Bailey, G. S., Reddy, A., Hendricks, J. D., Oganesian, A., Orner, G. A. (2003). The rainbow trout (*Oncorhynchus mykiss*) tumor model: recent applications in low-dose exposures to tumor initiators and promoters. Toxicol. Pathol. 31, 58-61.

Wong, K., Eegante, M., Bartels, B., Elkhayat, S., Tien, D., Roy, S. (2010). Analyzing habituation responses to novelty in zebrafish (*Danio rerio*). Brain. Res. 208, 450-457.

Wong, D., von Keyserlingk, M. A., Richards, J. G., Weary, D. M. (2014). Conditioned place avoidance of zebrafish (*Danio rerio*) to three chemicals used for euthanasia and anaesthesia. PLoS One. 9 (2), e88030.

索　引

注：页数后带有“*f*”或“*t*”的分别表示“图”或“表格”。

A

Abnormal behavior　异常行为　460t
Acidic stress　酸性应激　211
Actinopterygian teleostean fishes　辐鳍硬骨鱼类　77
Activation protein-1(AP-1)　活化蛋白质　395
Acute stressors　急性应激物(急性应激源)　14 – 18
　temporal patterns　时间格局　15f
Adaptation　适应　19 – 20
Adenosine diphosphate(ADP)　腺苷二磷酸　244, 246
Adenosine triphosphate(ATP)　腺苷三磷酸　75 – 76, 244, 246
Adrenaline(AD)　肾上腺素　85 – 86, 88 – 89, 293
Adrenergic fibers　肾上腺素能纤维　249
Adrenergic innervations　肾上腺素能神经分布　243, 249
Adrenocorticotropic hormone(ACTH)　促肾上腺皮质激素　13, 33 – 34, 77, 86, 109, 170, 283, 348 – 349
α-Adrenoreceptors　α-肾上腺素受体　249
β-Adrenoreceptors　β-肾上腺素受体　249
Adult salmonids　成年鲑鳟鱼类　474
Aerobic metabolism　有氧代谢　246
Aerobic phenotype　需氧表现型　256 – 257
　acquisition in skeletal muscle　骨骼肌获得　256 – 257
　association with performance　和行为表现联系　257
African catfish(*Clarias gariepinus*)　非洲胡鲶　509
African cichlid(*Haplochromis burtoni*)　伯氏朴丽鱼　510
Aggression　攻击　46, 460t, 463 – 464
Agonistic interactions　对抗相互作用　44
　aggression　攻击　46
　behavioral and physiological characteristics　行为和生理学特征　44
　social interactions　社会相互作用　45

subordinate individuals 从属个体 45
AKT/mTOR pathway AKT/mTOR 通道 169
Alarm reaction 预警反应 320
Allostasis 异型稳态 5, 7, 90, 452 - 453
allostatic load 异型稳态负荷 452 - 453
allostatic state 异型稳态的状态 90
Alpha-melanocyte-stimulating hormone(α-MSH) α-黑色素细胞刺激激素 33, 87, 109
Amino acid(AA) 氨基酸 46 - 48, 169
AMPK 腺苷酸依赖蛋白激酶 254 - 255
Anaerobic glycolysis 无氧糖酵解 246
Anesthesia in fish 鱼麻醉 475 - 476
Anesthetics in fish 鱼麻醉剂 512
Angiogenic response 血管生成反应 252 - 253
Angiotensin II(Ang II) 血管紧张素 II 209
Animal personalities 动物品格 37
conserved physiology of contrasting 对比的保守生理学 38
HR-LR rainbow trout model HR-LR 虹鳟模式 39 - 40
HR-LR strain selection HR-LR 品系选育 39 - 40
membrane receptors 膜受体 40
physiological stress responses and behavioral patterns 生理的应激反应和行为型式 38t
genetic basis for individuality 个性的遗传基础 42 - 43
hypothetical pathways 假设的通道 43f
stress, neuroplasticity, and coping style 应激、神经塑性和应对方式 41 - 42
stress coping and life history 应激应对和生活史 43 - 44
Animal's stress response 动物的应激反应 32
Anion exchanger(AE) 阴离子交换剂 205 - 206
Anterior pituitary 前脑垂体 111 - 112
Anxiolytic compounds 抗焦虑化合物 505
Aquaculture 水产养殖 459
and impact on wild fish 对野生鱼类的影响 475
stress in 应激 459 - 466
Aquaporin 2(AQP2) water channel 水通道蛋白 2 的水通道 216
Arachidonic acid(ARA) 花生四烯酸 48
Arctic charr(*Salvelinus alpinus*) 北极红点鲑 35, 282
Arginine vasopressin(AVP) 精氨酸加压素 77
Arginine vasotocin(AVT) 精氨酸催产素 35 - 36

Assimilation 同化作用 164 - 165
Astatotilapia burtoni 伯氏妊丽鱼 456
Asynchronous gamete development 不同步配子发育 279
Atlantic cod(*Gadus morhua*) 大西洋鳕鱼 47, 283, 450
Atlantic croaker(*Micropogonias undulatus*) 大西洋绒须石首鱼 288
Atlantic halibut(*Hippoglossus hippoglossus*) 庸鲽 456
Atlantic salmon(*Salmo salar*) 大西洋鲑 35, 171, 252, 283, 441, 446, 464f, 474, 506
Atrial natriuretic peptide 心钠素 209
Australasian snapper(*Pagrus auratus*) 金赤鲷 282
Autophagic-lysosomal system(ALS) 自体吞噬性溶酶体系统 166
Autophagy-related proteins(ATGs) 自体吞噬相关蛋白 167
Aversive 恶劣 313

B

Barfin flounder(*Verasper moseri*) 条纹星鲽 87
Behavior 行为 257 - 260
  behavioral needs 行为需要 442
  cognition, and learning relationships in fish 鱼类认知和学习的关系 310
  critical knowledge gaps 关键的知识空白 323 - 324
  hypothetical tolerance 假设的耐性 311f
  learning in relation to stress in fishes 鱼类学习和应激的关系 317 - 322
    optimality, preferences, and decision-making 最优化、偏爱和做出决定 311 - 316
    salmon as model species 鲑鱼作为模式种类 316 - 317
Behavioral syndromes 行为综合征 37
  conserved physiology of contrasting 对比的保守生理学 38
      HR-LR rainbow trout model HR-LR 虹鳟模式 39 - 40
      HR-LR strain selection HR-LR 品系选育 39 - 40
      membrane receptors 膜受体 40
      physiological stress responses and behavioral patterns 生理的应激反应和行为型式 38t
  genetic basis for individuality 个性的遗传基础 42 - 43
  hypothetical pathways 假设的通道 43f
  stress, neuroplasticity, and coping style 应激、神经塑型和应对方式 41 - 42
  stress coping and life history 应激应对和生活史 43 - 44
Benzocaine 苯佐卡因,氨基苯甲酸乙酯 512

Bicarbonate buffers 碳酸氢盐缓冲液 205
Bicolor damselfish(*Stegastes partitus*) 双色雀鲷 294
Bioenergetics 生物能量学 397-398
Biological functioning definitions 生物学机能定义 441
Black bream(*Acanthopagrus butcheri*) 布氏棘鲷 282
Blue banded goby(*Lythrypnus dalli*) 蓝带虾虎鱼 286
Bluegill sunfish(*Lepomis macrochirus*) 蓝鳃太阳鱼 294
Blue shark(*Prionace glauca*) 大青鲨 450
Body fluid balance 体液平衡 203
Bone morphogenetic proteins(BMPs) 骨骼形态发生蛋白 120-121
Brain, stress and 脑,应激 76
  actinopterygian teleostean fishes 辐鳍硬骨鱼类 77
  control over pituitary gland 对脑垂体的调控 81-85
  CRF system CRF 系统 79-80
    and behavior 和行为 85
    CRF-related peptides CRF 相关肽 77
    ontogeny 个体发育 80
  fundamental axes interact 基本轴相互作用 78-79
  hypothalamic portal system 下丘脑门脉系统 76
Brain aromatase genes 脑芳化酶基因 286
Brain serotonin system 脑羟色胺系统 40
Brain-derived neurotrophic factor(BDNF) 脑衍生的神经营养因子 47, 261
8-BromocAMP 8-溴环 AMP 292
Brook trout(*Salmo fontinalis*) 美洲红点鲑 36, 293
Brown trout(*Salmo trutta*) 褐鳟 129, 257, 282, 357
Burst-and-glide swimming behaviors 突发和爆发式的游动行为 251-252
Bycatch fish 副渔获物,兼捕鱼类 457-458

C

Ca^{2+}-sensing receptor(CaSR) Ca^{2+}感觉受体 214, 215
Calcitonin(CT) 降钙素 213
Callorhynchus milii 米氏叶吻银鲛 114
cAMP response binding protein(CREB) cAMP 反应结合蛋白 107-109
CaO_2-CvO_2 CaO_2-CvO_2 248-249
Captive environments 蓄养环境 446
  habitat complexity 生境复杂性 447

Captive fish 蓄养鱼类 446

Captive jack mackerel(*Trachurus japonicus*) 蓄养的日本竹筴鱼 285

Captive jundia(*Rhamdia quelen*) 蓄养的克林雷氏鲇 282, 405

Captive pike(*Esox lucius*) 蓄养的白斑狗鱼 281

Captivity 蓄养 443

care of wild fish in 野生鱼类的爱护 444 – 450

Carbonic anhydrase 碳酸酐酶 206, 212f

Cardiac growth 心肌生长 252 – 253

Cardiac pumping capacity 心脏泵动能力 250

Cardiovascular adjustments 心血管调节 249 – 250

to stress and exercise 应激和运动 248 – 251

Cardiovascular system, swim training on 心血管系统,游泳训练 252 – 253

Care of wild fish in captivity 野生鱼类在蓄养中的照料 444 – 445

aggression 攻击 446 – 447

captive environments 蓄养环境 446

captive fish 蓄养鱼类 445 – 446

enrichment 富集 447

feeding in wild fish 野生鱼类摄食 444

habitat complexity 生境复杂性 447

hatchery-reared fish 孵化-养育鱼类 446

hearing specialists 听觉高手 449

individual needs 个体需要 443

magneto-and electroreception effect 磁和电感受效应 450

maladaptive behaviors in fish 鱼类的不良适应行为 449 – 450

natural environment 自然环境 448

predictability 预期能力 445

sound effects on fish 鱼类的声音效应 450

sound use for communication 用于沟通的声音 448 – 449

transportation of fish 鱼类的运输 447

Carnitine palmitoyl transferase 肉碱棕榈酰转移酶 220

Catch-and-release angling 捕获后释放 466, 467t

Catecholamine-producing cells 儿茶酚胺生产细胞 88

Catecholamines 儿茶酚胺 1, 3, 5, 13, 32, 37, 46, 48, 73, 88, 107, 202, 213, 240, 242 – 243, 243f, 244, 246 – 251, 293, 295, 344, 347, 349, 351 – 353, 385t, 397, 405, 444

Caudal neurosecretory system(CNSS) 尾神经内分泌系统 77, 110

Cellular/cell 细胞的/细胞 206

energy status 能量状态 246
and humoral immune response 体液免疫反应 347 – 348
mechanisms 作用机理 208
protein homeostasis 蛋白质稳态 175
shrink 收缩 206
stress indicators 应激指标 383t, 385 – 396
volume regulation 体积调节 206 – 207
Centropomus parallelus 小锯盖鱼 219
Centropomus undecimalis 锯盖鱼 219
Cerebrospinal fluid(CSF) 脑脊液 214
Channel catfish(*Ictalurus punctatus*) 斑点鮰 172, 503
Checkered barb(*Puntius oligolepis*) 稀鳞无须鲃 509
Chemoreception 化学感受 245
Chemoreceptors 化学感受器 215
Chinook salmon(*Oncorhynchus tshawytscha*) 大鳞大麻哈鱼 34, 40, 80, 405, 454, 506
Chocolate dip chromis(*Chromis dimidiata*) 双色光鳃鱼 451
Cholecystokinin 缩胆囊素 78, 177 – 178
Cholinergic fibers 胆碱能纤维 249
Chromaffin cells 嗜铬细胞 87 – 89, 202, 213, 351
Chronic stress 慢性应激 3, 6, 9, 17, 441, 443, 450 – 451
Chronic stressors 慢性应激物(慢性应激源) 6, 14 – 18
exposure 暴露,接触 402
temporal pattern 时间格局 15f, 16f
Chub(*Leuciscus cephalus*) 圆鳍雅罗鱼 353
Chum salmon(*Oncorhynchus keta*) 大麻哈鱼 218, 353
Cichlid(*Neolamprologus pulcher*) 丽鱼科的一种 283
Citric acid cycle 三羧酸循环 247
Classic winner-loser effects 传统的胜者-败者效应 49
Classical conditioning 经典性条件反射 318
Cleaner wrasse(*Labroides dimidiatus*) 裂唇鱼 451
Clustered regularly interspaced short palindromic repeat(CRISPR) 簇状正规间隙短回文重复 132 – 133
Cognition 认知 447
critical knowledge gaps 关键的知识空白 323 – 324
hypothetical tolerance 假设的耐性 311f
learning, and behavior relationships in fish 鱼类学习和行为的关系 310
learning in relation to stress in fishes 鱼类学习和应激的关系 317 – 322

optimality, preferences, and decision-making 最优化、偏爱和做出决定 311 - 316

salmon as model species 鲑鱼作为模式种类 316 - 317

Cognitive Activation Theory of Stress 应激的认知激活理论 450

Coho salmon(*Oncorhynchus kisutch*) 银大麻哈鱼 129, 344, 352, 353, 458

Common carp(*Cyprinus carpio*) 鲤鱼 34, 81, 83f, 116, 288, 452, 471, 503, 504

system 系统 88 - 89

Compensation 补偿 5 - 6

Context-specific differences 前后(序列)效应-特异性差别 401 - 402

Coping style 应对方式 41 - 42

Coregonus lavaretus 白鲑 218, 257

Corticosteroids 皮质类固醇 3, 11, 33, 38, 40, 107, 113 - 119, 278, 286, 289, 294 - 295, 352

corticosteroid receptors 皮质类固醇受体 354

corticosteroid-regulated mechanisms 皮质类固醇激素调节机制 287

immune modulation by 免疫调节 352 - 353

Corticotropin releasing factor(CRF) 促肾上腺皮质激素释放因子 13, 35, 36, 77, 79, 109, 170

and behavior 行为 85

ontogeny 个体发育 80

Corticotropin releasing factor receptor 1(CRFR1) 促肾上腺皮质激素释放因子受体 1 285

Corticotropin releasing hormone(CRH) 促肾上腺皮质激素释放激素 13, 33, 344, 350

Corticotropin-like intermediate peptide(CLIP) 中间叶促皮质样肽 85

Cortisol 皮质醇 85, 88, 89, 209, 243 - 244, 247 - 248, 348, 353, 395, 397, 504, 505

action 作用 109

effect 效应 178f

injection into single-celled eggs 注入单细胞期的胚胎中 511

levels in zebrafish 斑马鱼水平 507 - 508, 507f

receptor-mediating action in fish immunity 在鱼类免疫中受体介导的作用 353

treatment 处理 129

in vitro protocols 离体方案 291 - 292

in vivo protocols 在体方案 290 - 291

Cortisol genomic effects 皮质醇基因组效应 120

cellular adjustments 细胞调节 130 - 132

molecular adjustments during stress 在应激期间的分子调节 122 - 130

changes in growth and immune modulation 生长和免疫调节的变化 129 - 130

cortisol-responsive gene 皮质醇反应基因 128t
enzyme activity 酶活性 123 – 124
intracellular cross-talk 细胞内的互相作用 130f
liver 肝脏 122
effect of stress/cortisol on gene changes 应激/皮质醇对基因变化的作用 124t
target tissue molecular responses 靶组织分子反应 127
temporal gene transcript changes 时序基因转录变化 123f
tissues 组织 124
upregulation of transcript abundance 转录丰度的上调 123 – 124
work in rainbow trout hepatocytes 虹鳟肝细胞作用 129
stress axis development 应激轴发育 120 – 122
Cortisol-glucocorticoid receptor（Cortisol-GR） 皮质醇-糖皮质激素受体 213
CPT1 线粒体脂肪酸转运 253
Creatine phosphate（CrP） 磷酸肌酸 246
CRF-binding protein（CRFBP） CRF-结合蛋白 77, 80
CRF-receptors（CRF-Rs） CRF-受体 111 – 112
CRISPR-Cas9 簇状正规间隙短回文重复（CRISPR）-Cas9-基因编辑工具 133 – 134
Critical swimming speed（U_{crit}） 临界游泳速度 249
Crowding 拥挤 460t, 510
Cutthroat trout（*Oncorhynchus clarkia*） 克氏鳟 131
Cyclic adenosine monophosphate（cAMP） 环腺苷酸 108, 247, 281
Cyclooxygenases（COX） 环加氧酶 48
Cyprinid（*Chalcalburnus tarichi*） 塔氏油白鱼 289
Cystic fibrosis transport regulator（CFTR） 囊性纤维化转运调节蛋白 204, 208, 218
Cytokines 细胞因子 348 – 349

D

Dams 水坝,水闸 473 – 475
Damselfish（*Pomacentrus amboinensis*） 雀鲷属鱼类中的一种 283
Decision-making 决策 311 – 312, 323
adaptive responses of ancestors 祖先的适应性反应 315
aversive 恶劣 313
eustress 真正应激 316
fight-or-flight situation 搏斗或逃逸状态（应急状态） 315
fish behavior 鱼类行为 314
fishes behavioral response 鱼类行为的反应 315

maladaptive　不良适应　315
metabolism of fish　鱼的新陈代谢　312
physiological response　生理反应　313, 314f
physiological state of fish　鱼类的生理状态　312f
Definition of stress　应激的定义　2 – 3, 75, 311 – 312
Density　密度　507 – 508
11-Deoxycortisol　11-脱氧皮质类固醇　289
Desert pupfish(*Cyprinodon nevadensis*)　内华达鳉　281
Dietary composition　食物组成　50
Dietary FA composition　食物的脂肪酸组成　48
17, 20β-Dihydroxy-4-pregnen-3-one(17, 20βP)　17, 20β-二羟基-4-孕烯-3-酮　281
Disease resistance　疾病抵抗力　342
"Disrupted negative feedback"　被打乱的负反馈作用　402
Distress　痛苦　316
eustress *vs.*　相对于真正应激　18
primary, secondary, and tertiary responses of fish　鱼类的初级、次级和三级反应　4f
Diuretic hormone(DH)　利尿激素　80
Divalent ions as stressors　二价离子的应激源　208 – 209
DMY gene　DMY 基因　288
DNA methylation　DNA 甲基化　134 – 135, 283
DNA repair　DNA 修复　218
Docosahexaenoic acid(DHA)　二十二碳六烯酸　48
Dominance relationship　优势关系　286
Dopamine(DA)　多巴胺　39, 78, 280
double-stranded RNAs(dsRNAs)　双链 RNAs　135
Dynamic Energy Budget(DEB) Model　动态能量收支(DEB)模型　163
Dynamics of stress response and effects on performance　应激反应的动态和对行为表现的效应　9 – 10

E

Ecosystem health, from individual indicators to　生态系统健康,个体指标　405 – 406
EDNRAa　内皮缩血管肽受体之一　213
Eicosapentaenoic acid(EPA)　十二碳五烯酸　48
Electrical stunning　电击击晕　466
Electrofishing　电鱼　470
Elevated temperature　升高温度　287

Endocrine　内分泌　3 - 4, 13, 19, 295, 342
　control of reproduction　生殖调控　279 - 281, 280f
　factors　因子　351 - 352
　response　反应　345
　stress axis　应激轴　242
　system　系统　202
Endocrinology of stress　应激的内分泌学　3
Endocrinology of stress response in fish　鱼类应激反应的内分泌学　77
　adaptation　适应　73
　allostatic landscape　异型稳态景观　91f
　fish forebrain　鱼前脑　74 - 75
　(neuro-)endocrine hypothalamus　神经内分泌下丘脑　76 - 85
　physiological stress response　生理应激反应　72
　stress　应激　75 - 76
　　and head kidney　头肾　87 - 89
　　and pituitary gland　脑垂体　85 - 87
β-Endorphins　β-内啡肽　85 - 86
Endothelin 1(EDN1)　内皮素 1　213
Energetics　能量学　7
　allostatic load　异型稳态负荷　7 - 8
　capacity of fish　鱼的能力　8 - 9
　chronic stress　慢性应激　9
　cost of stress　应激的代价　8
Energy　能量　183
　demand for maintenance　对维护的需要　174 - 175
　fluxes　流量　165f, 171
　homeostasis　稳态　244
　stress and　应激　89
　substrate absorption at gut　消化道的底物吸收　173 - 174
　supply　供应　218, 244
Energy metabolism　能量代谢　271
　effects of swim training on feeding and　游泳训练对摄食的效应　253 - 255
　　AMPK　腺苷酸依赖蛋白激酶　254 - 255
　during exercise and stress　在运动和应激期间　244 - 248
　in response of osmotic stress　对渗透性应激的反应　218 - 222, 221f
　　metabolism modifications　代谢修饰　219 - 220
　　metabolites transport　代谢物转运　221 - 222

oxygen consumption 氧消耗 218 – 219
stress and exercisc 应激和运动 244 – 248
cardiovascular and respiratory adjustments to 心血管和呼吸的调节 248 – 251
energy production and utilization 能量的产生和利用 246 – 248
exercise effects on water and ion homeostasis 运动对水和离子稳态的效应 244
limits of swimming exercise and stress 游泳运动和应激的限制 251
muscle activity and locomotion 肌肉活性和运动 244 – 245
Enhancing effects, suppressive effects *vs.* 增强效应对抑制效应 344
Enrichment 丰富 447, 508 – 509
Environmental regulation of reproduction 生殖的环境调节 279 – 280
Environmental stressors 环境应激物(环境应激源) 386
environmental salinity 环境盐度 357
seasonality 季节性 357 – 359
temperature 温度 357 – 359
Epigenetic(s) 外遗传 134
regulation of stress response 应激反应的调节 134
DNA methylation DNA 甲基化 134
ncRNAs 非编码 RNAs 135
Epinephrine 肾上腺素 3 – 4
signals 信号 107 – 108
Epithelial Ca^{2+} channel(ECaC) 上皮细胞中的钙离子通道 210
Epithelial cells 上皮细胞 203
Erythrocytes 红细胞 206
Estradiol(E2) 雌二醇 281, 284f, 287, 288, 290 – 294
Estrogen receptor(ER) 雌激素受体 281
Estuarine species 河口物种 288
Ethogram 行为谱 315
European eels(*Anguilla anguilla*) 欧洲鳗鲡 290, 507
European sea bass (*Dicentrarchus labrax*) 舌齿鲈 48, 116, 202, 293, 342, 400, 445, 463
Euryhaline fish(*Gillichthyes mirabilis*) 长颌姬虾虎鱼 217
Eustress 真正应激 75, 316
Eversion 外翻 74
Excess postexercise oxygen consumption(EPOC) 过量的运动后氧耗 248
Excitation-contraction coupling 兴奋收缩偶联 255 – 256, 257
Extant bony fishes(*Osteichthyes*) 现存的硬骨鱼类 471
Extrapituitary expression 脑垂体外表达 356

F

Farmed Animal Welfare Council(FAWC) 农场动物福利理事会 448
Fathead minnow(*Pimephales promelas*) 胖头鲹 116, 283, 503
Fatty acids(FA) 脂肪酸 46, 48
Fecundity 生育力 12, 76, 283, 286, 396
Feeding 摄食 8, 46, 75, 76, 84, 111, 164, 240, 244
  feelings-based definitions 感觉基础的定义 440
  and stress 和应激 510
  swim training on 游泳训练 253 - 255
  in wild fish 在野生鱼类 445
Fick principle states 菲克原理 248 - 249
Field measurements of stress, laboratory measurements of stress *vs.* 应激的野外测定对应激的实验室测定 403 - 404
Field-oriented mechanistic approach 野外的机械式应激检测 404
Fight-or-flight situation 应急反应状态 315
Fighting of fish 鱼类的搏斗 442
Fin rot 烂鳃病 460t
Fish 鱼类 240 - 241, 241 - 242, 243, 246, 247 - 248, 255, 466 - 467
  aggression 攻击 446 - 447
  allostasis 异型稳态 452f
  anesthetized with metomidate or eugenol 使用美托咪酯或丁香酚进行麻醉 512
  in aquaculture 在水产养殖中 463
  behavior 行为 314, 315
  diet dependency 饮食习惯 446
  forebrain 前脑 74 - 75
  gonadal phenotype 性腺表型 511
  handling in laboratory setting 实验室装置的手工操作 506
  health 健康 451, 466 - 467, 469 - 470, 471
  immune system 免疫系统 355 - 356
  immunity 免疫 386
    environmental salinity 环境盐度 357
    seasonality 季节性 357 - 358
    temperature 温度 357 - 358
  life 生命 1
  maladaptive behaviors 不良适应的行为 449 - 450

physiology 生理学 509
stress management in 应激管理 442 - 456
stunning and killing 击昏和杀死 465 - 466
surveys 调查 469 - 470
Fish welfare 鱼类福利 11, 20, 260, 381, 434 - 435, 444, 454, 456 - 459, 468 - 470, 472, 477
effects of swim training on 游泳训练的效应 260
future directions 未来的方向 476 - 479
issues in aquaculture 水产养殖的问题 460t
stress impact 应激影响 456 - 457
stress and welfare in wild fish 应激和野生鱼类的福利 469 - 475
stress in aquaculture 水产养殖的应激 459 - 466
stress in fisheries 渔业的应激 457 - 459
stress in ornamental fish 观赏鱼类的应激 467 - 468
stress in recreational fishing 休闲渔业的应激 466 - 467
stress in research within laboratory context 在实验室前后序列内研究的应激 468 - 469
Fisheries, stress in 渔业,应激 457 - 459
FK506-binding protein 51(FKBP-51) FK506-结合蛋白 217
Focal adhesion kinase 黏着斑激酶 207
Follicle stimulating hormone(FSH) 促卵泡激素 281
Follicle stimulating hormone receptor(FSHR) 促卵泡激素受体 286 - 287
Food intake 食物摄取 171 - 172
anorexigenic peptides 厌食性多肽 173
leptin 瘦素 173
loss of appetite 食欲不振 171
role of stress 应激的作用 172f
strong orexigenic hormone 强的食欲性激素 173
Food withdrawal 食物取消 460t
Foraging behavior 觅食行为 322
Fork-head box O(FOXO) 分叉头框 O 169 - 170
Forkhead transcription factor(fox12) 分叉头 12 转录因子 287
FOXI3a 控制离子细胞分化的一种转录因子 213
Free-swimming fish 自由游泳鱼类 245
Freshwater(FW) 淡水 202, 203, 210
pH regulationin pH 调节 205 - 206
Fugu(*Takifugu rubripes*) 红鳍东方鲀 86

Function-based definitions 功能基础的定义 441

G

Gamma-aminobutyric acid(GABA) γ-氨基丁酸 13
Gene expression 基因表达 382, 396, 402, 409
General adaptation syndrome(GAS) 普遍性适应综合征 1, 3, 18
Genetic basis for individuality 个性的遗传基础 42
Genetic manipulations 基因操作 109
Genomic cortisol signaling 基因组的皮质醇信号 116
  GR 糖皮质类固醇激素受体 116 - 119
  MR 盐皮质类固醇激素受体 119 - 120
Genomics tools 基因组工具 396
GH-IGF axis, stress regulation of GH-IGF 轴,应激调节 177 - 178
  activity of IGFs on muscle function IGF 对肌肉功能的活性 179
Gills-associated lymphoid tissue(GIALT) 鳃-相联系的淋巴组织 346
Gilthead sea bream(*Sparus aurata*) 金头鲷 222
Glial cell missing 2(GCM2) 神经胶质细胞丧失 2 211
Glucocorticoid receptor(GR) 糖皮质激素受体 34, 109, 116 - 119, 173 - 174, 286, 344, 353
Glucocorticoid response element(GRE) 糖皮质激素反应原件 115, 176
Glucocorticoids 糖皮质激素 32, 40, 247, 353
Glucokinase(GK) 葡萄糖激酶 124
Gluconeogenesis 糖异生 343 - 344, 246, 246 - 247
Glucose 葡萄糖 246
Glucose transporter(GLUT) 葡萄糖转运蛋白 216, 222
Glycogen phosphorylase(GP) 糖原磷酸化酶 247
Glycogen-rich cell(GLR cell) 富含糖原细胞 221 - 222
Glycogenolysis 糖原分解 220, 221 - 222, 246
Glycolysis 糖酵解 246 - 247
Goldfish(*Carassius auratus*) 金鱼 165, 291, 319, 449, 507
Goldsinny(*Ctenolabrus rupestris*) 梳隆头鱼 463
Gonadal aromatase genes 性腺芳化酶基因 286
Gonadal steroids 性腺类固醇 36, 284
Gonadosomatic index(GSI) 性腺成熟系数 288
Gonadotropin-releasing hormone(GnRH) 促性腺激素释放激素 280, 350, 356
Gonadotropin-producing cells(GTH-producing cells) 促性腺激素生产细胞 280

G-protein coupled receptors(GPCRs)　G-蛋白偶联受体　107 – 108
Grading　分级　460t
Growth　生长　162 – 163
conceptual framework　概念组成　163
DEB model for　DEB 模式　163 – 165, 163t
myocyte growth　肌细胞生长　165 – 170
protein balance model　蛋白质平衡模式　180f
stress effect on energy for　应激对能量的效应　170 – 175
Growth hormone(GH)　生长激素　73 – 74, 168, 170, 209, 349
Growth hormone receptor(GHR)　生长激素受体　170
Growth hormone-releasing hormone(GHRH)　生长激素释放激素　177 – 178
Growth-promoting effects　促进生长的效应　253
Gulf killifish(*Fundulus grandis*)　大底鳉　288
Gut, energy substrate absorption at　消化道、能量底物吸收　173 – 174
Gut-associated lymphoid tissue(GALT)　肠道相关的淋巴组织　346

H

H^+-ATPase(HA)　H^+-ATP 轴　206
H^+-ATPase-rich(HR) ionocytes　H^+-ATP 酶富含离子细胞　211
Habitat complexity　生境复杂性　447
Habituation　适应,驯化　32, 317 – 318, 381 – 382, 402 – 403
Handling　手工操作　460t, 463
Hatchery-reared fish　孵化养育鱼类　446
Head kidney　头肾　113 – 114, 349
activation of cortisol biosynthetic machinery　皮质醇生物合成作用机理的激活　115 – 116
contaminants disrupt cortisol production　污染物破坏皮质醇生成　115 – 116
cortisol production　皮质醇生成　114 – 115
MRAP1 transcript level　MRAP1 转录水平　114
StAR and P450scc genes　StAR 和 P450scc 基因　115
steroidogenic gene expression patterns　类固醇生成基因表达型式　115
stress and　应激　87
catecholamine-producing cells　儿茶酚胺产生细胞　88
communication within　内部交流沟通　88
steroid-producing cells　类固醇产生细胞　88
stress and energy　应激和能量　89

Health assessment index(HAI) 健康评价指数 398
Health in fish 鱼类健康 257 - 260
Hearing specialists 听力高手 449
Heat shock element 热休克原件 130 - 131
Heat shock factors(HSFs) 热休克因子 130 - 131
  HSF1 热休克因子1 395 - 396
Heat shock proteins(HSPs) 热休克蛋白 14, 130 - 131, 174, 355, 395 - 396
  intracellular chaperoning roles 细胞内伴侣作用 131
Hemoglobin(Hb) 血红蛋白 248 - 249
Heterostasis 异型稳态 5
Hexokinase(HK) 已糖激酶 123 - 124
High responsive(HR) cortisol 高度反应的皮质醇 36
Higher organisms 较高等的生物有机体 502
Holding 保持 470
  conditions 状况 470 - 471
Homeostasis 稳态 5
  dynamic process 动态过程 5
  homeostatic mechanisms 稳态作用机理 251
  homeostatic responses 稳态反应 313
  hormone binding 激素结合 6 - 7
  nonspecific stress response 非特异性应激反应 6 - 7
  recovery process 恢复过程 5 - 6
  stress response 应激反应 5
  stress sensing 应激感受 214
    osmosensors 渗透传感器 214 - 215
    signal transduction from sensors 来自传感器的信号转导 215 - 216
    targets of intracellular signaling cascade 细胞内信号级联放大的靶标 216 - 218
Hormonal actions 激素的作用 211 - 212
Hormonal stress response 激素的应激反应 107
Hormone 激素 4, 6
  effects on immune system 对免疫系统的效应 350
    fish immune system 鱼类免疫系统 355 - 356
    hypothalamic hormones 下丘脑的激素 350
    interrenal hormones 肾间腺激素 351 - 353
    pituitary hormones 脑垂体激素 350 - 351
    receptor-mediating action of cortisol 受体介导的皮质醇作用 353 - 355
    somatotropic axis 促生长轴 355 - 356

signaling 信号 211 - 212, 212f
Hours postfertilization(hpf) 孵化后的小时数 33, 80 - 81
Housing 家庭 506 - 509
density 密度 507 - 508
enrichment 富集 508 - 509
light/dark cycle 光亮/黑暗周期 509
HR-LR rainbow trout model HR-LR 虹鳟模式 40
90 kDa HSP(HSP90) 90 kDa 热休克蛋白 355
Human chorionic gonadotropin(hCG) 人体绒毛膜促性腺激素 282
3-Hydroxyacyl CoA dehydrogenase(HOAD) 3-羟酰辅酶 A 脱氢酶 220
5-Hydroxyindoleacetic acid(5-HIAA) 5-羟基吲哚乙酸 39 - 40
11β-Hydroxylase(11βH) 11β-羟化酶 286
17-Hydroxyprogesterone(17 P) 17-羟基黄体酮 291
11β-Hydroxysteroid dehydrogenase(11βHSD) 11β-羟类固醇脱氢酶 286
11β-HSD2 11β-羟类固醇脱氢酶 2 34, 115 - 116, 119
20β-Hydroxystcroid dehydrogenase(20βHSD) 20β-羟类固醇脱氢酶 121 - 122, 288
Hypercalcemic hormones 高血钙激素 213
Hyperosmotic stress responses 高钙应激反应 207
divalent ions as stressors 三价离子应激物(应激源) 208 - 209
involvement of hormones 激素参与 209
monovalent ions as stressors 单价离子应激物(应激源) 207 - 208
Hyperplasia 增生 166
Hyperplastic muscle growth 增生的肌肉生长 166
Hyperresponsive periods 反应过度的时期 405
Hypertrophic growth 过度的生长 256
Hypertrophic response 过度生长的反应 252 - 253
Hyperventilation 通氧过度 250
Hypoglycemic effects 低血糖效应 253
Hypoosmotic stress responses 低渗应激反应 209 - 213, 212f
involvement of hormones 激素参与 211 - 213
ionic compositions as stressors 离子组成作为应激物(应激源) 210
low pH as stressors 低 pH 作为应激物(应激源) 210 - 211
Hypostimulation 低刺激作用 452 - 453
Hypothalamic hormones 下丘脑激素 350
Hypothalamic-pituitary-adrenocortical(HPA) axis 下丘脑-脑垂体-肾上腺皮质轴 34, 109
Hypothalamic-pituitary-gonadal(HPG) axis 下丘脑-脑垂体-性腺轴 279 - 280, 350

Hypothalamic-pituitary-interrenal(HPI) axis 下丘脑-脑垂体-肾间腺轴 17, 32, 34, 35, 109, 162, 287, 344, 395, 403
activity 活性 399-400, 401, 442-443
molecular regulation of 分子调节 109-116
Hypothalamic-sympathetic-chromaffin cell axis, immune modulation by 下丘脑-交感神经-嗜铬细胞轴,免疫调节 351-352
Hypothalamus 下丘脑 110, 350
CRF transcript levels CRF 转录水平 111
nutritional status 营养状态 111
stressor-mediated upregulation of CRF CRF 的应激物介导上调 110
Hypoxia 低氧 288
effects 效应 288
Hypoxia-inducible factor 3α(HIF3A) 低氧诱导因子 3α 218

I

IGF binding protein(IGFBP) IGF 结合蛋白 178f
IGF-2 transcript level IGF-2 转录水平 129
Immune modulation 免疫调节 350-352, 354-355
by corticosteroids 由皮质类固醇 352-353
by hypothalamic-sympathetic-chromatffin cell axis 由下丘脑-交感神经-嗜铬细胞轴 351-352
Immune system 免疫系统 165, 343
activation 激活 164-165
and immunoendocrine interactions 和免疫内分泌相互作用 342
effects of stressors on immune response 应激物对免疫反应的效应 343-348
environmental stressors 环境应激物(应激源) 357-359
fish immunity 鱼类免疫 357-359
hormone effects on 激素效应 350-356
organization of immune response 免疫反应的组成 348-349
stress and disease resistance 应激和疾病抵抗力 342-343
Immune-endocrine response 免疫内分泌反应 345-346
Immunocompetence 免疫活性 260
Immunoglobulin M(IgM) 免疫球蛋白 M 347
Immunoglobulin Z(IgZ) 免疫球蛋白 Z 347
Immunoresponsiveness 免疫反应 260
Imprinting 印记 320

In vitro protocols 离体方案 291
follicles harvest 滤泡收获 291 – 292
GR-mediated suppression GR-介导的抑制作用 292
hCG-stimulated E2 production hCG-刺激 E2 产生 291
incubates ovarian follicles of rainbow trout 虹鳟卵泡的孵育 291
In vivo protocols 在体方案 290
juvenile common carp 幼鲤 290
SBP 类固醇结合蛋白 290
Individual indicators to ecosystem health 对生态系统健康的个体指标 405 – 406
Individuality, genetic basis for 个性、遗传基础 42
Indoleamine dioxygenase(IDO) 吲哚胺双加氧酶 47 – 48
Infectious pancreatic necrosis virus(IPNV) 传染性胰脏坏死病毒 259 – 260
Inosine monophosphate(IMP) 肌苷酸 247
Instrumental conditioning 工具条件反射,操作性条件反射 319
Insulin-like growth factors(IGFs) 胰岛素样生长因子 168
IGF-I 胰岛素样生长因子-1 355 – 356
Interferon-α(IFN-α) 干扰素-α 349
Interferon-γ(IFN-γ) 干扰素-γ 348
Interleukin(IL) 白细胞介素 343 – 344
Internal tags 内标记 472
Interrenal cells 肾间腺细胞 86 – 87
Interrenal hormones 肾间激素 351
immune modulation 免疫调节 352
by corticosteroids 由皮质类固醇 352 – 353
by hypothalamic-sympathetic-chromaffin cell axis 由下丘脑-交感神经-嗜铬细胞轴 351 – 352
Interrenal tissue 肾间腺组织 87 – 88, 113 – 116
Interspecific differences 种间差别 399 – 400
Intraperitoneal injections 腹腔注射 506
Intraperitoneal/intracoelomic tags 腹腔内/体腔内标记 472
Intraspecific differences 种内差别 400 – 401
Invasive procedures 入侵过程 471 – 473
Ionocytes 离子细胞 202, 204, 208, 213, 222
Isotocin(IST) 硬骨鱼催产素 35 – 36, 77

J

Japanese flounder(*Paralichthys olivaceus*) 牙鲆 286 - 287, 511
Japanese medaka(*Oryzias latipes*) 青鳉 85, 169, 218, 286 - 287, 503, 511

K

11-Ketoandrostenedione(11KA) 11-酮基雄烷二酮 286
11-Ketotestosterone(11-KT) 11-酮基睾酮 36, 281
Kidney 肾脏 203
Killifish(*Fundulus heteroclitus*) 底鳉 207, 400
Kisspeptin-kissreceptor system 亲吻激素-亲吻受体系统 279 - 280
Kynurenic acid 犬尿喹啉酸 49

L

Laboratory fishes 实验室鱼类 504
  factors impacting stress in handling 手工操作中影响应激的因素 505 - 506
  stress indicators in 应激指标 504 - 505
Laboratory measurements of stress 应激的实验室测定 403 - 404
  field measurements of stress *vs.* 对应激的野外测定 403 - 404
Laboratory organism, zebrafish as 实验室生物有机体,如斑马鱼 502
Laboratory-based indicators 以实验室为基础的指标 403
Large neutral amino acids(LNAAs) 大型中性氨基酸 46 - 47
Largemouth bass(*Micropterus salmoides*) 大口黑鲈 290 - 291
Learning 学习 310
  and cognition, behavior relationships in fish 鱼类学习和认知、行为的关系 310
  critical knowledge gaps 关键的知识空白 323 - 324
  hypothetical tolerance 假设的耐性 311f
  optimality, preferences, and decision-making 最优化、偏好性和决策 311 - 316
  in relation to stress in fishes 和鱼类应激的关系 320
    alarm reaction 警戒反应 320
    classical conditioning 经典性条件 318
    classical conditioning of heart rate 心率的经典性条件 318
    elaborate parental behavior 精细的抚幼行为 321
    feeding fish species 摄食鱼类 322

habitat selection 生境选择 322
habituation 驯化,适应 317 - 318
instrumental conditioning 工具条件反射 319
natal streams 出生的河流 320
paradoxical situation 反常的状态 319
physiological mechanisms 生理作用机理 320
plasticity, and problem solving 可塑性和问题解决 317
salmon as model species 鲑鱼作为模式种类 316 - 317
Leptin 瘦素 89, 173
Light : dark cycles 光亮 : 黑暗周期 445, 509
LIM domain-binding protein 1(LDB1) LIM-区结合蛋白 1 218
Lipid catabolism 脂类分解代谢 247
Lipogenesis 脂肪生成 220
Lipolysis 脂解作用 244
Lipopolysaccharide(LPS) 脂多糖 129 - 130, 343 - 344
β-Lipotropic hormone(LPH) β-促脂解素 85 - 86
Liver 肝脏 122 - 123
Lobe-finned fish(*Sarcopterygii*) 叶鳍鱼类 72
Local responses 局部反应 345
pattern of mucosal and systemic responses 黏膜的和身体的反应型式 346f
systemic responses *vs.* 全身的反应 345 - 347
Locomotor activity 运动活性 398 - 399
Long noncoding RNAs(lncRNAs) 长非编码 RNAs 135
Long-term chronic stress 长时间的慢性应激 434 - 435
Low pH as stressors 低 pH 作为应激物(应激源) 210 - 211
Low responsive(LR) 低反应性 36
Lower organisms 较低等生物有机体 502
Lucania goodie 蓝鳍卢氏鳉 217 - 218
Lungfish(*Ceratodontiformes*) 澳大利亚肺鱼 72
Luteinizing hormone(LH) 促黄体激素 281
Lyretail anthias(*Pseudanthias squamipinnis*) 拟花鮨 451

M

Magnocellular vasopressin neurons 大细胞加压素神经元 215
Mammalian models 哺乳类模式 47
Masculinizing effect 雄性化效应 286 - 287

Maternal stress 母体的应激 35
Matrinxa(*Brycon amazonicus*) 亚马逊石脂鲤 255 - 256
Mechanistic studies 机制研究 381 - 382
mechanistic target of rapamycin(mTOR) 雷帕酶素蛋白的机制性靶标 169
Melanin concentracting hormone(MCH) 黑色素浓集激素 35 - 36, 78, 350
Melanocortin 2 receptor(MC2R) 黑皮质素受体 2 33, 86, 109, 292
Melanocortin receptor accessory proteins(MRAP) 黑皮质素受体辅助蛋白 113 - 114
  MRAP1 黑皮质素受体辅助蛋白 1 87
Melanophore stimulating hormone(MSH) 促黑色素细胞激素 77
Metabolism 代谢 398 - 399
  modifications 修饰 219 - 220
Metabolites transport 代谢物运输 221 - 222
  carbohydrate energy supply 碳水化合物能量供给 221f
3-Methoxy-4-hydroxyphenylglvcol 3-甲氧基-4-羟苯乙醇 39 - 40
Microarrays 微点阵 396
microRNAs(miRNAs) 微小 RNAs 131 - 132
Mid-blastula transition(MBT) 中囊胚期过渡(转换) 120 - 121
Million years ago(Mya) 百万年前 72 - 73
Mineralocorticoid receptor(MR) 盐皮质素受体 34, 109, 119
Mitochondria 线粒体 247
Mitogen-activated protein kinase(MAPK) 促细胞分裂原激活蛋白激酶 207, 216
Molecular mechanisms 分子作用机理 121 - 122, 135, 208
Molecular stress indicators 分子应激指标 383t, 395 - 396
Molecular stress response 分子应激反应 133
  approaches 途径 133
  elevated plasma cortisol level 提高血浆皮质醇水平 109
  epigenetic regulation of stress response 应激反应的外遗传调节 134 - 135
  epinephrine signals 肾上腺素信号 107 - 109
  genomic cortisol signaling 基因组的皮质醇信号 116
  genomic effects of cortisol 基因组的皮质醇效应 120
    cellular adjustments 细胞的调节 130 - 132
    molecular adjustments during stress 应激期间的分子调节 122 - 130
    stress axis development 应激轴的发育 120 - 122
  hormonal stress response 激素的应激反应 107
  mechanistic studies using targeted mutagenesis 使用定向诱变剂的机理研究 133 - 134

molecular regulation of HPI axis HPI 轴的分子调控 109
head kidney 头肾 113 - 116
hypothalamus 下丘脑 110 - 111
pituitary 脑垂体 111 - 113
stress-related genes 应激相关基因 110f
protein expression 蛋白质表达 132
stress hormone regulation 应激的激素调节 108f
transcript levels of genes 基因的转录水平 132 - 133
Molecular techniques 分子技术 279
Monoamine oxidase(MAO) 单胺氧化酶 39 - 40
Monopterus albus 黄鳝 220
Monovalent ions as stressors 单价离子作为应激物(应激源) 207 - 208
Mortality 死亡率 251
Mozambique tilapia(*Oreochromis mossambicus*) 莫桑比克罗非鱼 34, 77, 81, 84, 87, 215, 290, 454, 455f, 466, 503
pituitary cells 脑垂体细胞 214 - 215
MS-222 麻醉剂的一种 476, 512
Mucosal responses pattern 黏膜反应型式 346f
Mucosal surfaces 黏膜表面 346
Mucosal-associated lymphoid tissues(MALT) 黏膜相联系的淋巴组织 346
Muscle growth 肌肉生长 166, 169
Muscle protein 肌肉蛋白 168
dynamic regulation 动态调节 168 - 170, 170f
synthesis 合成 169 - 170
Muscle proteolysis 肌肉蛋白酶解 167, 167t, 168f
Muscle RING finger protein-l(MURF1) 肌肉环指蛋白-1 166 - 167
Mycobacterium marinun infections 分枝杆菌感染 512 - 513
Myoblast differentiation and fusion 成肌细胞分化和融合 166
Myocyte growth 肌细胞生长 166
muscle proteolysis 肌肉蛋白酶解 166 - 168, 167t, 168f
myoblast differentiation and fusion 成肌细胞分化和融合 166
regulation of muscle protein dynamics 肌肉蛋白动态的调节 168 - 170, 170f
Myogenesis, stress effects on 肌发生,应激作用 176 - 177, 176t
Myogenic differentiation protein(MYOD) 生肌分化蛋白 166
Myogenic regulatory factors(MRFs) 生肌调节因子 166
myo-inositol biosynthetic pathway 肌醇生物合成通路 216
Myosin heavy chain(MHC) 肌球蛋白重链 163

Myotomal muscle 生肌节的肌肉 245

N

Na^+ efflux Na^+流出 211
Na^+ sensor Na^+传感器 215
Na^+-$C1^-$ cotransporter(NCC) Na^+-$C1^-$协同转运蛋白 204
Na^+-K^+-$2Cl^-$ cotransporter type 1(NKCC1) Na^+-K^+-$2Cl^-$共转运蛋白类型 1 204
Na^+/H^+ exchanger(NHE) Na^+/H^+交换剂 204
Na^+/K^+-ATPase(NKA) Na^+/K^+-ATP 酶 204
  inhibitor 抑制剂 218 – 219
NaCl solutions NaCl 溶液 214
Natal streams 出生溪流 320
Natural lives approach 自然生活途径 441 – 442
NCXlb 一种基底外侧转运蛋白 213
Nervous system 神经系统 348 – 349
Neuroendocrine hypothalamus 神经内分泌下丘脑 76
  actinopterygian teleostean fishes 辐鳍硬骨鱼类 77
  control over pituitary gland 对脑垂体的调控 81 – 85
  CRF system CRF 系统 79 – 80
  CRF-related peptides CRF 相关肽 77
  fundamental axes interact 基本轴的相互作用 78 – 79
  hypothalamic portal system 下丘脑门脉系统 76 – 77
Neuroendocrine aspects of stress and exercise 应激和运动的神经内分泌方面 242 – 244
Neuroendocrine stress response 神经内分泌的应激反应 32
  agonistic interactions 对抗的相互作用 44 – 46
  animal personalities 动物的个性 37 – 44
  behavioral syndromes 行为综合征 37 – 44
  divergent stress coping styles 趋异的应激应对方式 37 – 44
  neuronal substrate for stress 应激的神经元底物 35 – 36
  nutritional factors affecting stress responses 影响应激反应的营养因子 46 – 49
  ontogeny of teleost stress response 硬骨鱼类应激反应的个体发育 33
  stress response profiles 应激反应型式 32 – 33
  variation in stress responses 应激反应的变动 35 – 36
Neuroendocrine system 神经内分泌系统 32
Neurogenesis 神经发生 260
Neurohypophysis 神经脑垂体 214

Neuroimmunoendocrine 神经免疫内分泌 348 - 349
Neuronal signaling 神经元信号 12 - 13
Neuronal substrate for stress 应激的神经元底物 35 - 36
Neuronal system 神经元系统 32
Neuropeptide Y(NPY) 神经肽 Y 35 - 36, 172
Neuropeptides 神经肽 12 - 13, 348 - 349, 352
Neuroplasticity 神经塑型 41 - 42
Neurotransmitters 神经递质 352
Newts(*Taricha granulosa*) 蝾螈 40
Nile tilapia(*Oreochromis ntloticus*) 尼罗罗非鱼 173 - 174, 471
Nitric oxidc signaling cascades 一氧化氮信号级联 250
NKCC2 Na^+-K^+-$2Cl^-$共转运蛋白 2 215, 218
Nonadaptation, adaptation *vs.* 非适应性对适应性 19 - 20
Nonbicarbonate buffers 非碳酸氮盐缓冲剂 205
noncoding RNAs(ncRNAs) 非编码 RNA 135
Nonendocrine factors 非内分泌因子 351 - 352
Nongenomic mechanisms 非基因组作用机理 250
Noninfectious production-related deformities 非传染性生产相关的畸形 460t
Noradrenaline(NA) 去甲肾上腺素 293
Norepinephrine(NE) 去甲肾上腺素 3 - 4, 39 - 40
Nose-associated lymphoid tissue(NALT) 鼻腔相联系的淋巴组织 346
Nucleus lateralis tuberis (*nlt*) 外侧结节核 77
Nucleus preopticus(npo) 视前核 77
NUPR1 transcription factor 一种转录因子 396
Nutritional factors 营养因素 46
  affecting stress responses 影响应激反应 46

O

Odorant receptors 嗅觉受体 215
Olfactory imprinting 嗅觉印记 321
Omega-3 FAs ω-3 脂肪酸 49
Omega-3 rich fish oil 富含 ω-3 的鱼油 48 - 49
Ontogeny 个体发育 80
  CRF system CRF 系统 80
  teleost stress response 硬骨鱼类应激反应 33
Optimal swimming speed(U_{opt}) 最优的游泳速度 251 - 252

Optimality 最优化 311 - 312
adaptive responses of ancestors 祖先的适应性反应 315
aversive 恶劣 313
fight-or-flight situation 应急状态 315
fish behavior 鱼类行为 314
fishes behavioral response 鱼类行为反应 315
maladaptive 不良适应 315
metabolism of fish 鱼类的代谢 312
physiological response 生理反应 313, 314f
physiological state of fish 鱼类的生理状态 312f
Orange roughy(*Hoplostethus atlanticus*) 大西洋棘鲷 289
Organosomatic indices 器官身体指数 398
Organum vasculosum of lamina terminalis(OVLT) 终板血管器 214
Ornamental fish 观赏鱼类 456 - 457
stress in 应激 467 - 468
Orthopedia transcription factors(Otp's) 一种转录因子 77 - 78
Osmoregulation 渗透调节 165, 248
in fishes 在鱼类 203 - 205
Osmosensing 渗透感应 203
Osmosensors 渗透传感器 214 - 215
Osmotic stress 渗透性应激 206
cell volume regulation 细胞体积调节 206 - 207
energy metabolism in response 反应中的能量代谢 218 - 222
hormonal and nervous responses to stressors 对应激物的激素和神经反应 205f
hyperosmotic stress responses 高渗性应激反应 207 - 209
hypoosmotic stress responses 低渗性应激反应 209 - 213
osmoregulation in fishes 鱼类的渗透压调节 203 - 205
pH regulation in FW and SW 在淡水和海水中的 pH 调节 205 - 206
sequential events 顺序的过程 204f
stress sensing to homeostasis 对稳压的应激感受 214 - 218
Osmotic transcription factor-l(OSTF-1) 渗透性转录因子-1 217
Ovarian atresia 卵巢闭锁(退化) 281
Oxidative phosphorylation 氧化磷酸化 247 - 248
Oxidative stress 氧化应激 395 - 396
Oxidative stress-response kinase(OSR1) 氧化应激反应激酶 206 - 207
Oxygen consumption 氧消耗 218 - 219
measurements 测定 248

Oxygen-to-nitrogen quotient(O : N) 氧对氮系数 219 - 220
Oxytocin in zebrafish 斑马鱼的催产素 77 - 78

P

Pacific masou salmon(*Oncorhynchus masou*) 马苏大麻哈鱼 77
Pacific salmon(*Oncorhynchus* species) 太平洋鲑鱼类 314 - 315, 316 - 317, 320
Parathyroid hormone(PTH) 甲状旁腺激素 213
Parental stress 亲代应激 34
Pars distalis(PD) 远侧部 77, 83f
Pars intermedia(PI) 中间部 77, 81, 83f, 85 - 86
Pars nervosa(PN) 神经部 77, 83f, 84
Pathogen associated molecular patterns(PAMP) 病原体相联系的分子型式 348 - 349
Pejerrey(*Odontesthes bonariensis*) 银汉鱼 286 - 287, 511
Periophthalmus modestus 弹涂鱼 209
Pet fish 观赏鱼 467 - 468
pH regulation pH 调节 205
  in FW 在淡水 205 - 206
  in SW 在海水 205 - 206
Phenotypic plasticity 表型的可塑性 403
Pheromone receptors 信息素受体 215
Phosphagens 磷酸原 246
Phosphatidylinositol 3-kinase(PI3K) 磷脂酰肌醇 3-激酶 167
Phosphatidylinositol 4-kinaseα(PI4Kα) 磷脂酰肌醇 4-激酶 α 168
Phosphoenolpyruvate carboxykinase(PEPCK) 烯醇丙酮酸磷酸羟激酶 122
Phosphoenol-pyruvate carboxykinase 1(PCK1)-mediated gluconeogenesis 烯醇丙酮酸磷酸羟激酶介导的糖异生 220
Phosphoinositide 3-kinase/AKT(PI3K/AKT) 磷酸肌醇 3-激酶 169 - 170
Physiological adaptation 生理学的适应 19
Physiological adaptations and stress 生理适应和应激 251
  cardiovascular system, swim training on 心血管系统,游泳训练 252 - 253
  feeding and energy metabolism, swim training on 摄食和能量代谢,游泳训练 253 - 255
  skeletal muscle growth, swim training on 骨骼肌生长,游泳训练 255 - 257
  stress, swim training on 应激,游泳训练 257 - 260
Physiological demands of exercise and stress 运动和应激的生理需求 242
  cardiovascular adjustments 心血管的调节 248 - 251

energy metabolism 能量代谢 244 – 248
limits 极限 251
neuroendocrine aspects 神经内分泌方面 242 – 244
respiratory adjustments 呼吸的调节 248 – 251
Physiological needs 生理需要 442
Physiological performance 生理的行为表现 245
Physiological stress response 生理的应激反应 3 – 5
Physiological systems response 生理的系统反应 12 – 14
Pikeperch(*Sander lucioperca*) 梭鲈 49
Pituitary 脑垂体 111 – 112
hormones 激素 350 – 351
pharmacological and genomic techniques 药理学和基因组学的技术 113t
Pituitary adenylate cyclase activating polypeptide(PACAP) 脑垂体腺苷酸环化酶激活多肽 79
Pituitary gland 脑垂体腺 81
ACTH-releasing function of CRF CRF 的 ACTH 释放功能 84
actinopterygian/teleostean HPI axis 辐鳍鱼类/硬骨鱼类 HPI 轴 82f
control over 控制住 81
CRF-and UI-expressing neurons CRF-和 UI-表达神经元 84
immunohistochemistry of hypothalamo-pituitary complex 下丘脑-脑垂体复合物的免疫组织化学 83f
in Mozambique tilapia 在莫桑比克罗非鱼 84
Piwi-interacting RNAs(piRNAs) Piwi-相互作用 RNA 135
Plasma 血浆 397
Point-of-care devices 监护点设备 404
Polyamine signaling cascades 多胺信号级联 250
Polychlorinated biphenyl(PCBs) 多氯联苯 132 – 133
Population-and ecosystem-level stress indicators 种群和生态系统水平的应激指标 406
Positive energy balance 正的能量平衡 246
Predictability 可预测的 445
food presentation 食物的提供 454
Preoptic area(POA) 视前区 35 – 36, 110
Prerequisite 前提 260
Primary physiological stress indicators 初级的生理应激指标 385t, 397 – 398
Proactive coping styles 作用前的应对方式 37
Prolactin(PRL) 催乳素 213, 351
Proopiomelanocortin(POMC) 阿黑皮素原 33, 85 – 86, 109, 111 – 112, 170 –

171, 351
Protein 蛋白质 352
kinases 激酶 352
synthesis 合成 248
Proteolysis 蛋白质水解 169 - 170
Proteolytic systems 蛋白酶解系统 253 - 254
Proteotoxicity 蛋白质毒性 131
Protogynous sandperch(*Parapercis cylindrica*) 圆拟鲈 286
Pseudoloma neurophilia 一种微孢子虫 512 - 513
Psychological stress impact in fish 鱼类心理应激影响 450 - 452
Pyruvate kinase(PK) 丙酮酸激酶 123 - 124

Q

Qingbo(*Spinibarbus sinensis*) 中华倒刺鲃 253
Quantification of stress 应激的量化 381 - 382
Quinolinic acid 喹啉酸 49

R

RAD54-Iike 2(RAD54L2) 一种应激特异性的基因 218
Rainbow trout(*Oncorhynchus mykiss*) 虹鳟 34, 80 - 81, 111, 281 - 282, 446, 456
Rainbow fish(*Melanoteania duboulayi*) 虹银汉鱼 443, 446
Ray-finned fish(*Actinopterygii*) 辐鳍鱼类 72
Reactive coping styles 反应应对方式 37
Reactive oxygen species(ROS) 活性氧类别 395 - 396
Reactive scope model 活性范围模式 5
Receptor-mediating action 受体介导活动 353
cortisol in fish immunity 在鱼类免疫中的皮质醇 353
in vitro studies 离体实验 355
in vivo studies 在体实验 353 - 354, 354f
Recirculating aquaculture systems(RAS) 循环水产养殖系统 459 - 460, 506
Recovery process 恢复过程 5 - 6
"Recovery ratio"approach "恢复率"方法 398
Recreational fishing, stress in 休闲渔业,应激 466 - 467
Red belly tilapia(*Tilapia zillii*) 红腹罗非鱼 282
Red fibers 红纤维 245

Red gurnard(*Chelidonychthys kumu*) 绿鳍鱼 282
Red-oxidative muscle 红氧化肌肉 245
Reflex action mortality predictors(RAMP) 反射动作致死预测法 458
Reflex impairment 反射性损伤 399, 408
Reninangiotensin system 肾素血管紧张素系统 213
Reproduction 生殖 278
  and development 和发育 278 - 279
  future directions 未来方向 295 - 296
  mechanisms of stress action 应激活动的作用机理 289 - 290
    cortisol 皮质醇 290 - 291, 291 - 292
    effects of stress factors 应激因子的效应 292 - 293
  patterns 模式 279
  regulation 调节 279
    endocrine control 内分泌调控 279 - 281
    environmental regulation 环境调节 279
  stress effects on 应激的作用 281
    effects of hypoxia 低氧影响 288 - 289
    in natural environments 在自然环境中 294 - 295
    on reproductive endocrine system 对生殖内分泌系统 283 - 286
  reproductive performance 生殖的行为表现 281 - 283
  stimulatory effects of stress on 应激的刺激效应 289
  thermal stress 热应激 286 - 287
Respiratory adjustments 呼吸的调节 248 - 251
  to stress and exercise 对应激和运动 248 - 251, 250 - 251
Respiratory-osmoregulatory 呼吸-渗透压调节 244
Reverse transcription polymerase chain reaction 逆转录多聚酶链式反应 33
Riverine specie 河流鱼类 288
Roach(*Rutilus rutilus*) 拟鲤 284 - 285
Rostral *Pars distalis*(rPD) 吻远侧部 79

S

Sablefish(*Anoplopoma fimbria*) 裸盖鱼 458
Salmo species 鲑鱼类 314 - 315
Salmon as model species 鲑鱼作为模式鱼类 316 - 317
Salmonids 鲑鳟鱼类 43 - 44, 442
Scanning ion-selective electrode techno1ogy(SIET) 扫描离子选择电极技术 210, 213

Sea lice 海虱 460t
Sea trout(*Cynoscion nebulosus*) 云纹犬牙石首鱼 284-285
Seasonality 季节性 357-359
Seawater(SW) 海水 202, 203, 206, 208-209
  pH regulation in pH 调节 205-206
Secondary physiological stress indicators 次级生理应激指标 385t, 397
Selective serotonin reuptake inhibitor(SSRI) 选择性羟色胺重摄取抑制剂 47, 505
Semelparous species 终生一胎的种类 279
Sensitization 致敏 32
Seriola lalandi 黄尾鰤 255-256
Serotonin(5-HT) 5-羟色胺 34, 39-40
  biosynthesis 生物合成 46-47
Serotonin transporter(5-HTT) 羟色胺转运蛋白 40
Serum/glucocorticoid regulated kinase 1(SKG1) 血清/糖皮质激素调节激酶1 218
Several somatostatins(SSTs) 几种生长抑素 177-178
Sex 性别 511
  determination and reversal 决定和反转 511
  differences 差别 400-401
  and hierarchies 等级 510
Shark(*Squalus acanthias*) 白斑角鲨 215
short-interfering RNAs(siRNAs) 短-干扰 RNAs 135
Short-term acute stress 短期急性应激 434-435
Signaling system 信号系统 209
Skeletal muscle 骨胳肌 245, 253
  aerobic swimming and regulation of mass 需氧游泳和质量调节 256
  effects of swim training on 游泳训练的效应 255-256
    acquisition of aerobic phenotype in 获得有氧呼吸表型 256-257
    aerobic swimming and regulation of 有氧游动及调节 255-256
  energy sensing mechanisms in swimming fish 游泳鱼类能量感受的作用机理 254-255
Skeletal muscle growth, swim training on 骨骼肌生长,游泳训练 255-256
Skin-associated lymphoid tissue(SALT) 皮肤相联系的淋巴系统 346
Skinner box 皮革箱 319
Smallmouth bass(*Micropterus dolomieu*) 小口黑鲈 284-285
Sockeye salmon (*Oncorhynchus nerka*) 红大麻哈鱼 217-218, 241, 283-284, 396, 474
Somatic growth 机体生长 255

Somatolactin(SL) 生长促乳素 351
Somatotropic axis 促生长轴 355 - 356
Spawning patterns 产卵型式 279
Specific dynamic action(SDA) 特异性动态动作 76
Spiny damselfish(*Acanthochromis polyacanthus*) 棘光鳃鱼 284 - 285
Staghorn damselfish(*Amblyglyphidodon curacao*) 雀鲷科鱼类中的一种 294 - 295
Standard metabolic rate(SMR) 标准代谢率 8
Stanniocalcin(STC) 司腺钙蛋白 213
Ste20p-related proline alanine-rich kinase(SPAK) complex Ste20p-相关脯氨酸富含丙氨酸激酶复合物 206 - 207
Steroid acute regulatory(StAR) protein 类固醇急性调节蛋白 114 - 115, 281
Steroid binding protein(SBP) 类固醇结合蛋白 290
Steroid-producing cells 类固醇产生细胞 88
Steroid-suppressing effects 类固醇抑制效应 290 - 291
Sticklebacks(*Gasterosteus aculeatus*) 三刺鱼 320
Stimulus filtering 刺激过滤 318
Stocking density 养殖密度 460t
Strain 品系 2
Striped bass(*Morone saxatilis*) 条纹狼鲈 255 - 256, 284 - 285
Stress 应激 1, 2, 75 - 76, 162 - 163, 240 - 241, 258f, 311 - 312, 321, 342 - 343, 347 - 348, 443
  adaptation *vs.* nonadaptation 适应性对非适应性 19 - 20
  in aquaculture 水产养殖 459 - 466
  axis development 轴的发育 120
  chronic 慢性的 443
  classical and contemporary approaches in fish 在鱼类传统的和当代的途径 20t
  definition 定义 2 - 3, 75, 311 - 312
  dynamics of stress response and effects on performance 应激反应动态和行为表现的效应 9 - 10
    acute stressors 急性应激物(应激源) 14 - 17
    chronic stressors 慢性应激物(应激源) 14 - 17
    final state of fish 鱼的最后状态 10f
    fish's prior experience 鱼的先前经历 11
    multiple stressors 多个应激物(应激源) 14 - 17
    response of physiological systems 生理系统的反应 12 - 14
    resting and distressed fish cortisol levels 静止和痛苦鱼类的皮质醇水平 12f
    sexual dimorphisms 性别二态 10 - 11

stress 应激 12
effect on energy for growth 能量对生长的效应 170 – 171
energy demand for maintenance 对维持的能量需求 174 – 175
energy substrate absorption at gut 消化道对能量底物的吸收 173 – 174
food intake 食物摄取 171 – 173, 172f
model for stress effect 应激效应的模式 171f
effects on promoters of muscle formation 肌肉形成的启动子影响 175
on myogenesis 肌发生 176 – 177, 176t
stress regulation of GH-IGF axis GH-IGF 轴的应激调节 177 – 179
effects on reproductive 对生殖的影响 281 – 289
effects of hypoxia 低氧影响 288 – 289
endocrine system 内分泌系统 283 – 286
in natural environments 在自然环境中 294 – 295
performance 行为 281 – 283
stimulatory effects of stress on 应激的刺激作用 289
thermal stress 热应激 286 – 287
and energetics 能量学 7 – 9
in fish 在鱼类 1
in fisheries 在渔业 457 – 459
GAS 普通适应综合征 18
and homeostasis 稳态 5 – 7
in ornamental fish 观赏鱼类 467 – 468
physiology 生理学 2, 443
physiological response to stressor 对应激物的生理反应 3
physiological stress response 生理应激反应 3 – 5
in recreational fishing 休闲渔业 466 – 467
in research within laboratory context 实验室环境中的科研 468 – 469
sensing to homeostasis 对稳态的感知 214
osmosensors 渗透压传感器 214 – 215
signal transduction from sensors 从传感器的信号传导 215 – 216
targets of intracellular signaling cascade 细胞内信号串联的靶标 216 – 218
sensory systems and perception 感觉系统和感受 19 – 20
stress-associated behaviors 应激相关的行为 505
stress-induced 应激诱导的 505
behavioral preferences 行为偏爱 505
mechanisms 作用机理 286
stress-modifying factor 应激缓解因子 241

stress-related behaviors 应激-相关行为 505
swim training effects on 游泳训练效应 257
behavior and stress 行为和应激 257 - 258, 259f
disease resistance 疾病抵抗力 259 - 260
fish welfare 鱼类福利 260
variation in stress responses 应激反应的变动 35 - 36
and welfare in wild fish 野生鱼类福利 469 - 475
Stress action mechanism 应激作用机理 289 - 290
cortisol 皮质醇 290 - 291, 291 - 292
effects of stress factors 应激因子的效应 292
Stress coping 应激应对 85 - 86
conserved physiology of contrasting 对比的保守生理学 38
HR-LR rainbow trout model HR-LR 虹鳟模式 40
HR-LR strain selection HR-LR 品系发育 39 - 40
membrane receptors 膜受体 40
physiological stress responses and behavioral patterns 生理应激反应和行为模式 38t
genetic basis for individuality 个性的遗传基础 42
hypothetical pathways 假设的通道 43f
and life history 和生活史 43 - 44
stress, neuroplasticity, and coping style 应激,神经塑性和应对方式 41 - 42
styles 方式 37
Stress indicators in fish 鱼类应激指标 380
future 未来 407 - 409
from individual indicators to ecosystem health 从单个指标到生态系统健康 405 - 406
measures of fish stress 鱼类应激的测定 382 - 383
cellular indicators 细胞指标 383t, 395 - 396
molecular indicators 分子指标 383t, 395 - 396
primary physiological stress indicators 初级生理应激指标 385t, 397
secondary physiological stress indicators 次级生理应激指标 385t, 397
whole-organism stress indicators 整个生物有机体应激指标 390t, 398 - 399
measuring and interpreting stress 测量和解读应激 399
context-specific differences 邻近序列特异性差别 401 - 402
field *vs.* laboratory 野外对实验室 403 - 404
interspecific differences 种间差别 399 - 400
intraspecific differences 种内差别 400 - 401
stressor severity 应激物的激烈程度 402 - 403

temporal aspects 时间方面 404 - 405
measuring stress 测定应激 380 - 381
quantification of stress 应激的定量 381 - 382
quantifying and interpreting indicators 指标定量和说明 410f
stress indicators for use in fish 在鱼类使用的应激指标 407f
Stress management in fish 鱼类的应激管理 442
anesthetics 麻醉剂 512
aquatic species 水生种类 503
captivity 监禁 443
care of wild fish in 野生鱼类的关爱 444 - 450
chronic stress 慢性应激 443
consistency 连贯性，一致性 513
controlling and preparing for stress 对应激的控制和准备 452 - 456
factors impacting stress in laboratory fish handling 在实验室鱼类手工操作中影响应激的因子 505 - 506
feeding and stress 摄食和应激 510
housing 家居 506 - 510
density 密度 507 - 508
enrichment 优化改善 508 - 509
light/dark cycle 光/黑周期 509
HPI axis activity HPI 轴活性 443
impacts on health of progeny 对子代健康的影响 444
as model organisms 作为模式的生物有机体 502
psychological stress impact 心理应激影响 450 - 452
rearing environment effect 养育环境影响 444
sex and hierarchies 性别与社会等级 510
sex determination and reversal 性别决定和性逆转 511
stress, cortisol, and reproduction 应激、皮质醇和生殖 511
stress indicators in laboratory fish 实验室鱼类应激的指标 504 - 505
stress physiology 应激生理学 443 - 444
underlying diseases 隐在的疾病 512 - 513
variation in reproductive strategy 生殖策略的变动 444
zebrafish 斑马鱼 502, 503
Stress-activated protein kinase(SAPK)pathways 应激激活蛋白激酶通道 215 - 216
Stressor(s) 应激物(应激源) 2, 3, 203, 240 - 241, 311 - 312, 342
divalent ions as 二价离子 208 - 209
effects on immune response 对免疫反应的效应 343, 343f

cellular and humoral immune response 细胞和体液的免疫反应 347 – 348
stress 应激 347 – 348
suppressive effects *vs*. enhancing effects 抑制效应对增强效应 344 – 345
systemic responses *vs*. local responses 系统反应对局部反应 345 – 347, 345f
ionic compositions as 离子组成 210
low pH as 低 pH 210 – 211
monovalent ions as 单价离子 207 – 208
severity 激烈程度 402 – 403
stressor-mediated molecular response 应激物介导的分子反应 109
Striped trumpeter(*Latris lineata*) 条纹婢鲻 282
Stunning fish 被击昏的鱼 465 – 466
Subfornical organ(SFO) 窟窿下器 214
Suppressive effects, enhancing effects, *vs*. 抑制作用与增强作用 344 – 345
Suppressor of cytokine signaling(SOCS) 细胞因子信号的抑制剂 129 – 130, 178 – 179
Surgeonfish(*Ctenochiaetus striatus*) 栉齿刺尾鱼 451
Surgery in fish 鱼类的外科手术 473 – 475
Swim training effects on cardiovascular system 游泳训练对心血管系统的影响 252 – 253
Swimming 游泳 240 – 241
circulatory, respiratory, and metabolic responses 血液循环、呼吸和代谢反应 404f
performance 行为表现 245, 257, 398
physiological adaptations and stress 生理适应和应激 251
swim training on cardiovascular system 游泳训练对心血管系统 252 – 253
swim training on feeding and energy metabolism 游泳训练对摄食和能量代谢 253 – 255
swim training on skeletal muscle growth 游泳训练对骨骼肌生长 255 – 257
swim training on stress 游泳训练对应激 257 – 260
physiological demands of exercise and stress 运动和应激的生理需求 242
cardiovascular adjustments 心血管调节 248 – 251
energy metabolism 能量代谢 244 – 248
limits 极限 251
neuroendocrine aspects 神经内分泌方面 242 – 244
respiratory adjustments 呼吸调节 248 – 251
swimming-induced growth 游泳诱导生长 255 – 256
Sympathetic nervous system 交感神经系统 202
Sympathetic-chromaffin(SC) axis 交感-嗜铬组织轴 32
Synchronous gamete development 同步的配子发育 279

Systemic responses 全身的反应 345
local responses *vs.* 对局部的反应 345 - 347, 345f
pattern 型式 346f

T

Tagging 标记 471 - 473
Target osmoregulatory genes 渗透压调节靶基因 209
Targeted mutagenesis, mechanistic studies using 靶标突变发生、使用机械论研究 133 - 134
Temperature 温度 357 - 359
Temporal aspects 时间方面 404 - 405
Tetraiodothyronine(T4) 四碘甲腺原氨酸 78
Thermal stress 热应激 286 - 287
Thick ascending limb of Henle's loop(TAHL) 亨勒襻的厚升支 215
Thigmotaxis 趋触性 440 - 441
Thyroid hormones 甲状腺激素 356
Thyroid-stimulating hormone(TSH) 甲状腺刺激素 356
Thyrotropin-releasing hormone(TRH) 促甲状腺素释放激素 77 - 78
Time-course sampling 取样时程 405
Tolerance capacity 抵抗力 260
Transcription factor NUPR1 转录因子 NUPR1 396
Transforming growth factor-β(TGF-β) 转化生长因子-β 168, 348
Transient receptor potential vallinoid-type channel(TRPV channel) 瞬时受体潜在香草酸型通道 206 - 207, 214
Transport proteins 转运蛋白 202, 206 - 207, 216
Transportation 运输 460t
stress 应激 447
2, 2, 2 Tribromoethanol(TBE) 三溴乙醇 476
Tricarboxylic acid cycle 三羧酸循环 247
Triiodothyronine(T3) 三碘甲腺原氨酸 78
Trinidad gupp(*Poecilia reticulata*) 网纹花鳉 42, 320
Tryptophan(TRP) 色氨酸 46
Tryptophan dioxygenase(TDO) 色氨酸双加氧酶 47 - 48
Tryptophan hydroxylase(TPH) 色氨酸脱水羟化酶 46
TSC22 domain family protein 3(TSC22D3) TSC22 区家族蛋白 3 218
Tumor necrosis factor-alpha(TNF-α) α 肿瘤坏死因子 343 - 344

Turbot(*Scophthalmus maximus*) 大菱鲆 458 – 459
Tyrosine(TYR) 酪氨酸 46

U

Ubiquitin-proteasome system(UPS) 泛素蛋白酶体系统 166, 253 – 254
Ultraviolet(UV) 紫外线 448
Urocortin-1(Ucnl) 尿皮质素 79 – 80
Urotensin I(UI) 硬骨鱼类紧张肽 I 35 – 36, 77, 110 – 111, 294 – 295

V

Vaccination 接种疫苗 463, 472
vasa gene expression 血管基因表达 288
Vasopressin 血管升压素 77 – 78
Viral diseases 病毒病 460t
Visible implant elastomer(VIE) tags VIE 标记 472
Vitamin D(VitD) 维生素 D 213
VitD-receptor(VDRa) 维生素 D-受体 213
Vitellogenin(VTG) 卵黄蛋白原 281

W

Walleye(*Sander vitreus*) 玻璃梭鲈 405 – 406
Welfare 福利 257 – 260, 435
  behavioral changes 行为变化 435t
  behavioral needs 行为需求 442
  biological functioning definitions 生物功能定义 441
  feelings-based definitions 感觉基础定义 440 – 441
  function-based definitions 功能基础定义 441
  natural lives approach 自然生命途径 441 – 442
  physiological needs 生理需求 442
  status 状态 257, 261
  stress in wild fish 野生鱼类应激 469 – 475
White fibers 白纤维 245
White suckers(*Catostomus commersoni*) 白亚口鱼 284 – 285
Whole genome duplication events(WGDs) 整体基因组复制 72 – 73

Whole-body cortisol concentrations 整体皮质醇浓度 504
Whole-organism stress indicators 整个生命有机体应激指标 390t
Wild fish 野生鱼类 251, 278, 281, 294
care in captivity 在监禁中的爱护 398 – 399
stress in welfare 在福利中的应激 469 – 475
Wild-caught fish 野外捕获鱼类 468
Winter diseases 冬季疾病 460t

Y

Yellow perch (*Perca flavescens*) 黄鲈 34

Z

Zebrafish (*Danio rerio*) 斑马鱼 33, 79, 112, 252 – 253, 288 – 289, 312, 321 – 322, 323 – 324, 451 – 452, 502
behaviors 行为 505
cortisol elevation 皮质醇升高 506, 507 – 508, 507f
dominance hierarchies 优势等级 510, 511
experimentation 实验法 502 – 503
housing characteristics 饲养环境条件 506 – 507
with *Mycobacterium marinum* infections 为分枝杆菌感染 512 – 513
oxytocin in 催产素 77 – 78
stocking density effect on behavior 养殖密度对行为的影响 508
study areas 研究领域 503
Zona pellucida (ZP) 透明带 281

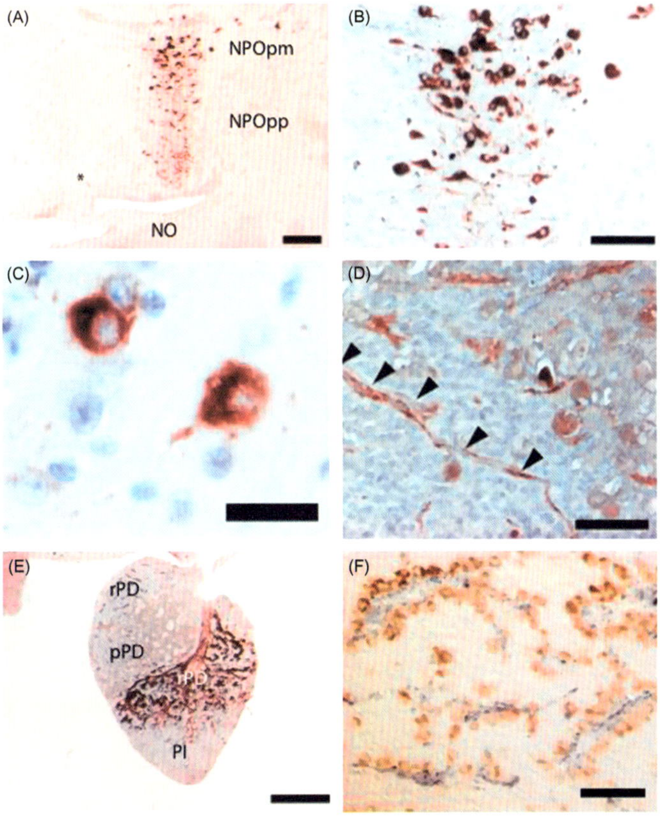
(A)
NPOpm
NPOpp
NO
(B)
(C)
(D)
(E)
rPD
pPD
PD
PI
(F)

图 3. 2

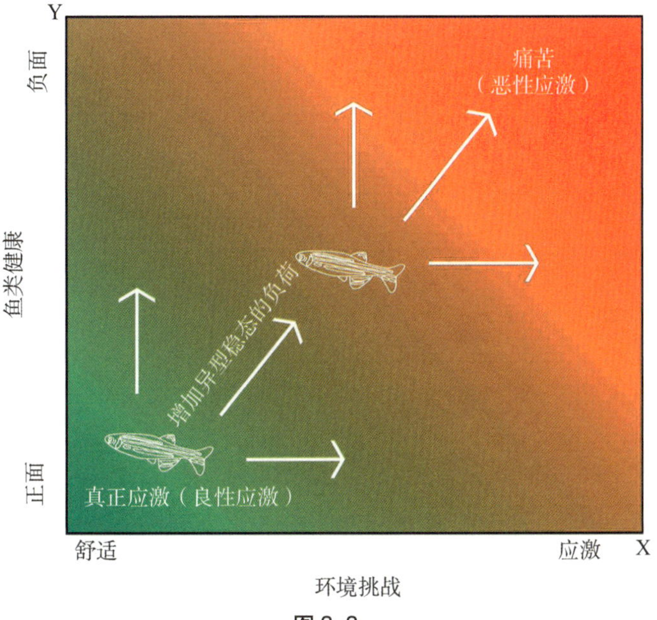
Y
负面
鱼类健康
正面
痛苦
（恶性应激）
增加异型稳态的负荷
真正应激（良性应激）
舒适
应激
X
环境挑战

图 3. 3

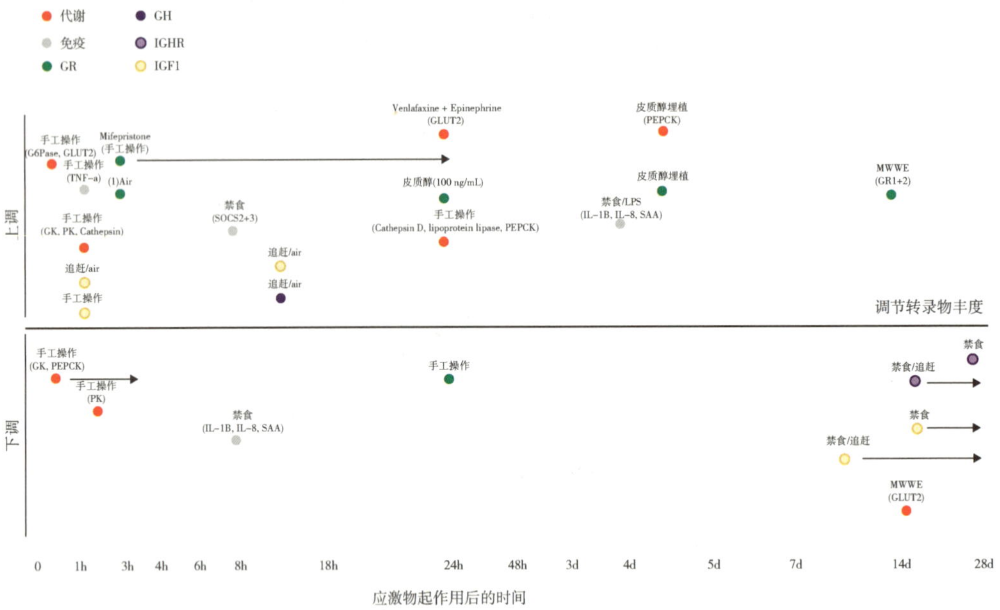
代谢
GH
免疫
IGHR
GR
IGF1
上调
手工操作
(G6Pase, GLUT2)
Mifepristone
(手工操作)
手工操作
(TNF-a)
(1)Air
手工操作
(GK, PK, Cathepsin)
追赶/air
手工操作
禁食
(SOCS2+3)
追赶/air
追赶/air
Venlafaxine + Epinephrine
(GLUT2)
皮质醇(100 ng/mL)
手工操作
(Cathepsin D, lipoprotein lipase, PEPCK)
皮质醇埋植
(PEPCK)
皮质醇埋植
禁食/LPS
(IL-1B, IL-8, SAA)
MWWE
(GR1+2)
调节转录物丰度
下调
手工操作
(GK, PEPCK)
手工操作
(PK)
禁食
(IL-1B, IL-8, SAA)
手工操作
禁食
禁食/追赶
禁食
禁食/追赶
MWWE
(GLUT2)
0
1h
3h
4h
6h
8h
18h
24h
48h
3d
4d
5d
7d
14d
28d
应激物起作用后的时间

图 4. 3

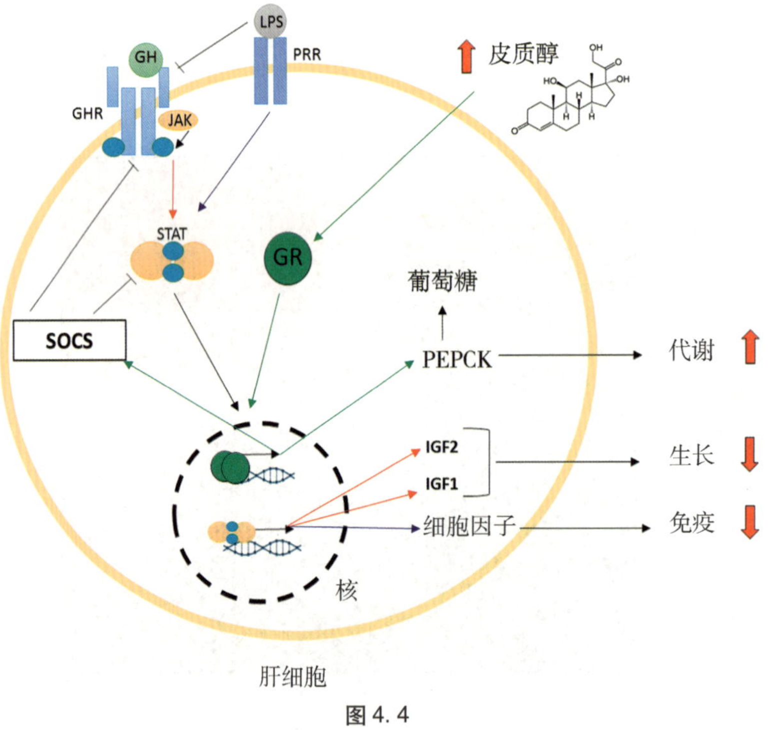
LPS
GH
PRR
GHR
JAK
皮质醇
OH
HO
OH
H
H
H
O
STAT
GR
葡萄糖
SOCS
PEPCK
代谢
IGF2
IGF1
生长
细胞因子
免疫
核
肝细胞

图 4. 4

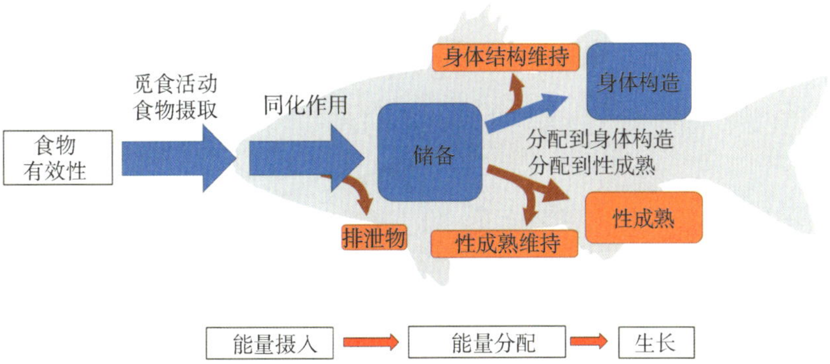
食物
有效性
觅食活动
食物摄取
同化作用
储备
排泄物
身体结构维持
身体构造
分配到身体构造
分配到性成熟
性成熟
性成熟维持
能量摄入
能量分配
生长

图 5. 1

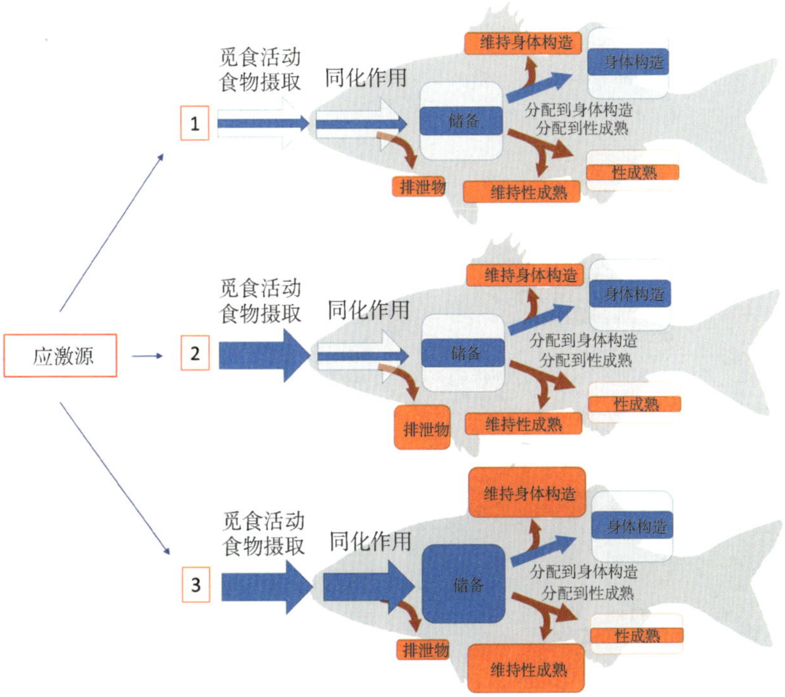
应激源
1
2
3
觅食活动
食物摄取
同化作用
储备
排泄物
维持身体构造
身体构造
分配到身体构造
分配到性成熟
性成熟
维持性成熟

图 5. 4

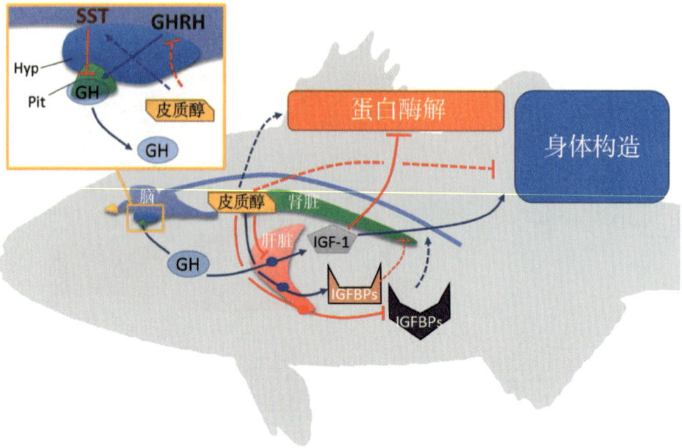
SST
GHRH
Hyp
Pit
GH
皮质醇
GH
蛋白酶解
身体构造
脑
皮质醇
肾脏
肝脏
IGF-1
GH
IGFBPs
IGFBPs

图 5. 6

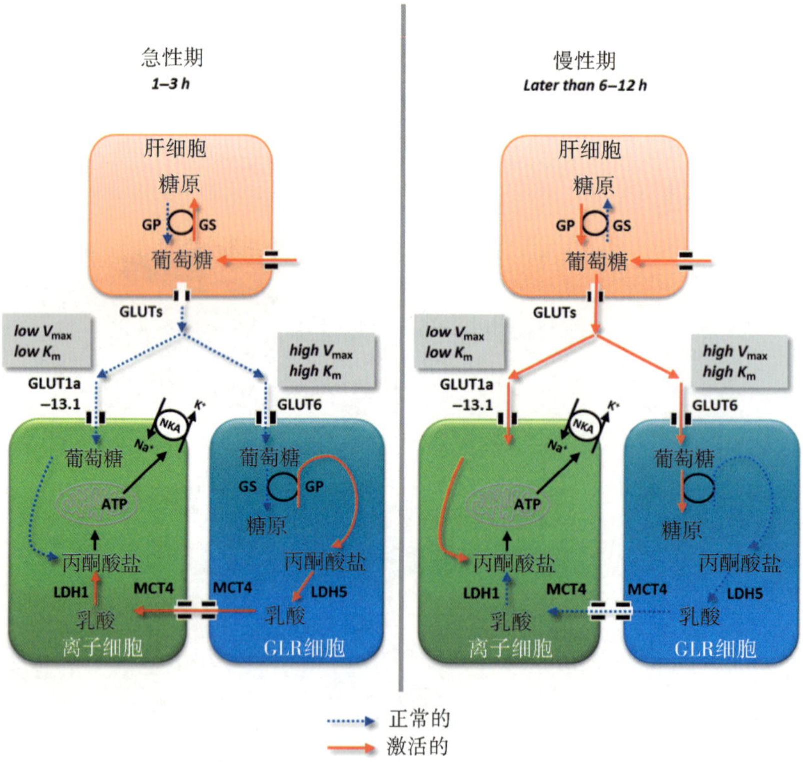
急性期
1–3 h
慢性期
Later than 6–12 h
肝细胞
糖原
GP
GS
葡萄糖
GLUTs
low Vmax
low Km
high Vmax
high Km
GLUT1a
–13.1
GLUT6
NKA
K+
Na+
葡萄糖
ATP
丙酮酸盐
LDH1
MCT4
乳酸
离子细胞
葡萄糖
GS
GP
糖原
丙酮酸盐
MCT4
LDH5
乳酸
GLR细胞
正常的
激活的

图 6. 4

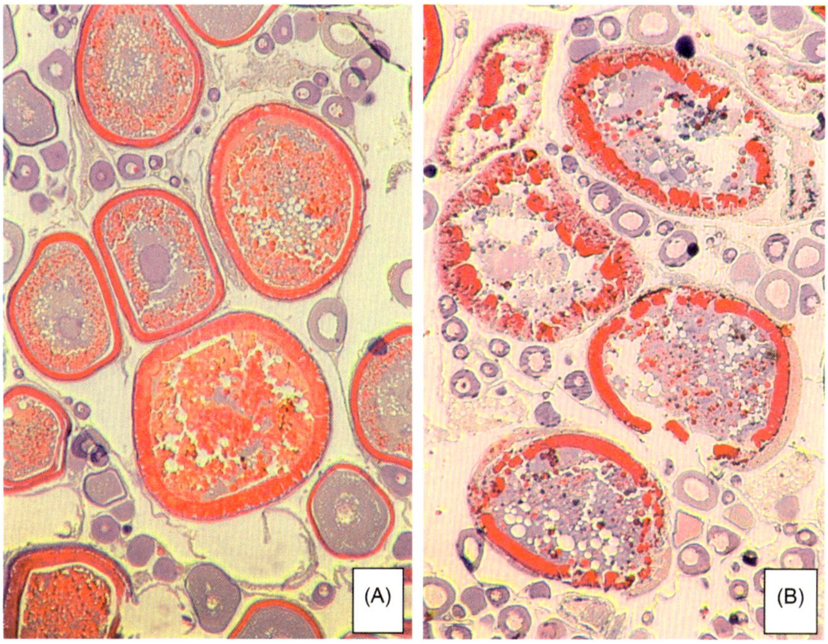

图 8. 2